Basic College Mathematics

Seventh Edition

John Tobey

North Shore Community College
Danvers, Massachusetts

Jeffrey Slater

North Shore Community College
Danvers, Massachusetts

Jamie Blair

Orange Coast College
Costa Mesa, California

Jennifer Crawford

Normandale Community College
Bloomington, Minnesota

PEARSON

Boston   Columbus   Indianapolis   New York   San Francisco   Upper Saddle River
Amsterdam   Cape Town   Dubai   London   Madrid   Milan   Munich   Paris   Montréal   Toronto
Delhi   Mexico City   São Paulo   Sydney   Hong Kong   Seoul   Singapore   Taipei   Tokyo

Editorial Director, Mathematics: *Christine Hoag*
Editor in Chief: *Paul Murphy*
Acquisitions Editor: *Dawn Giovanniello*
Executive Content Editor: *Kari Heen*
Senior Content Editor: *Lauren Morse*
Editorial Assistant: *Chelsea Pingree*
Vice President, Executive Director of Development: *Carol Trueheart*
Senior Development Editor: *Elaine Page*
Senior Managing Editor: *Karen Wernholm*
Senior Production Supervisor: *Ron Hampton*
Design Manager: *Andrea Nix*
Interior Design: *Tamara Newnam*
Senior Design Specialist: *Barbara Atkinson*
Digital Assets Manager: *Marianne Groth*
Supplements Production Project Manager: *Katherine Roz*
Content Development Manager: *Rebecca E. Williams*
Senior Content Developer: *Mary Durnwald*
Executive Manager, Course Production: *Peter Silvia*
Media Producers: *Audra Walsh and Vicki Dreyfus*
Executive Marketing Manager: *Michelle Renda*
Marketing Manager: *Rachel Ross*
Marketing Assistant: *Ashley Bryan*
Senior Author Support/Technology Specialist: *Joe Vetere*
Procurement Manager/Boston: *Evelyn M. Beaton*
Procurement Specialist: *Debbie Rossi*
Senior Media Buyer: *Ginny Michaud*
Permissions Project Supervisor: *Michael Joyce*
Production Management, Composition, and Answer Art: *Integra*
Text Art: *Scientific Illustrators*
Cover Images: *Illustration by Amy DeVoog*

Many of the designations used by manufacturers and sellers to distinguish their products are claimed as trademarks. Where those designations appear in this book, and Pearson Education was aware of a trademark claim, the designations have been printed in initial caps or all caps.

Library of Congress Cataloging-in-Publication Data

Basic college mathematics / John Tobey ... [et al.].—7th ed.
   p. cm.
 Includes index.
 ISBN 978-0-321-74759-4 (alk. paper)
1. Mathematics—Textbooks. I. Tobey, John.
2. QA39.3.T63 2012
 513'.1—dc23   2011023690

1 2 3 4 5 6 7 8 9 10—DOW—16 15 14 13 12

ISBN-10: 0-321-74759-3 (paperback)
ISBN-13: 978-0-321-74759-4 (paperback)

pearsonhighered.com

This book is dedicated to Nancy Tobey
A loving wife for forty-four years,
An outstanding mother of three children,
A joyful and thankful grandmother of seven children,
A dedicated but retired elementary teacher,
My closest friend in all the world.

Contents

Preface

TO THE INSTRUCTOR

One of the hallmark characteristics of *Basic College Mathematics* that makes the text easy to learn and teach from is the building-block organization. Each section is written to stand on its own, and every homework set is completely self-testing. Exercises are paired and graded and are of varying levels and types to ensure that all skills and concepts are covered. As a result, the text offers students an effective and proven learning program suitable for a variety of course formats—including lecture-based classes; discussion-oriented classes; modular, self-paced courses; distance learning; mathematics laboratories; and computer-supported centers.

We have visited and listened to teachers across the country and have incorporated a number of suggestions into this edition to help you with the particular learning delivery system at your school.

WHAT'S NEW IN THE SEVENTH EDITION?

- **Chapter Organizers** have been updated to include a You Try It column that provides additional opportunity for students to practice relevant chapter topics and procedures.

- A solid correlation has been made between the material on the **How Am I Doing? Chapter Test** and the examples, exercises, Chapter Review, and Cumulative Review. Each Chapter Test problem has at least 1 example, 2 Chapter Review exercises, and 2 Cumulative Review exercises that represent the same problem type. New assessment check boxes allow students to tally their answers and gauge their preparedness for the actual test.

- Following each Chapter Test, the new **Math Coach** provides students with a personal office-hour experience by walking them through some helpful hints to keep them from making common errors on test problems. For additional help, students can also watch the authors work through these problems on the accompanying Math Coach videos, available on YouTube and in MyMathLab.

- Select **Examples and Student Practice** problems, representing some of the most difficult concepts for students to master in a chapter, have been placed side by side to encourage students to work through each step of these problems to gain further understanding. These concepts are also covered on the Chapter Test and in the Math Coach.

- Enhanced emphasis on **Steps to Success boxes** (formerly Developing Your Study Skills) have been integrated throughout the text to provide students with more guided techniques for improving their study skills and succeeding in math.

- The **Use Math to Save Money** features are now assignable so that students can apply this new knowledge to their everyday lives. All of the topics have been chosen based on a student survey of over 1000 developmental math college students. These give practical, realistic examples of how students can use math to cut costs and spend less.

- Ten percent of the exercises throughout the text have been refreshed.

- All real-world application problems have been updated.

- *New* **The Lecture Series on DVD** has been completely revised to provide students with extra help for each section of the textbook. The videos include

  - **Interactive Lectures** that highlight key examples and exercises from every section of the textbook. A new interface allows for easy navigation to sections, objectives, and examples.

  - **Math Coach Videos**, featuring the text authors (John Tobey, Jeffrey Slater, Jamie Blair, and Jennifer Crawford), coach students in avoiding the most commonly made mistakes in a particular problem when students need the most help: the night before an exam.

  - **Chapter Test Prep Videos** provide step-by-step video solutions to every problem in each How Am I Doing? Chapter Test in the textbook.

Student and Instructor Resources

Worksheets with the Math Coach

Provides extra vocabulary and practice exercises for every section of the text. Each chapter also includes the Math Coach problems with ample space for students to show their work. The worksheets can be packaged with the textbook or with the MyMathLab access kit.

Student Solutions Manual

Provides worked-out solutions to all odd-numbered section exercises, even and odd exercises in the Quick Quiz, mid-chapter reviews, chapter reviews, chapter tests, Math Coach, and cumulative reviews

Lecture Series on DVD Featuring Math Coach and Chapter Test Prep Videos

Provides students with extra help for each section of the textbook. The videos include

- A complete lecture for each section of the text-book. The new interface allows easy navigation to objectives and examples.

- Math Coach videos that coach students in avoiding the most commonly made mistakes in a particular problem.

- Step-by-step video solutions to every problem in each How Am I Doing? Chapter Test.

Math Coach and Chapter Test Videos are also available in MyMathLab and on YouTube.

All Student Resources are available for purchase at www.mypearsonstore.com

Annotated Instructor's Edition

Contains all of the content found in the student edition, plus the following:

- Answers to all Student Practice problems, section exercises, mid-chapter reviews, chapter reviews, chapter tests, cumulative tests, and practice final exam

- Teaching Tips placed in the margin at key points where students historically need extra help

- Teaching Examples placed in the margins to accompany each example

Instructor's Solutions Manual

- Detailed step-by-step solutions to the even-numbered section exercises

- Solutions to every exercise (odd and even) in the Classroom Quiz, mid-chapter reviews, chapter reviews, chapter tests, cumulative tests, and practice final

(Available for download from the Instructor's Resource Center)

Instructor's Resource Manual with Tests and Mini-Lectures

- Mini-lecture for each text section

- Two short group activities per chapter

- Three forms of additional practice exercises

- Two pretests per chapter—free response and multiple choice

- Six tests per chapter—free response and multiple choice

- Two cumulative tests per even-numbered chapter—free response and multiple choice

- Two final exams—free response and multiple choice

- Answers to all items

(Available for download from the Instructor's Resource Center)

MyMathLab® Online Course (access code required)

MathXL® Online Course (access code required)

TestGen® (Available for download from the Instructor's Resource Center)

Diagnostic Pretest: Basic College Mathematics

1. _____

2. _____

3. _____

4. _____

5. _____

6. _____

7. _____

8. _____

9. _____

10. _____

11. _____

12. _____

Chapter 1

1. Add. $3846 + 527$

2. Divide. $58\overline{)1508}$

3. Subtract.
$$\begin{array}{r} 12{,}807 \\ -11{,}679 \end{array}$$

4. The highway department used 115 truckloads of sand. Each truck held 8 tons of sand. How many tons of sand were used?

Chapter 2

5. Add. $\dfrac{3}{7} + \dfrac{2}{5}$

6. Multiply and simplify. $3\dfrac{3}{4} \times 2\dfrac{1}{5}$

7. Subtract. $2\dfrac{1}{6} - 1\dfrac{1}{3}$

8. Mike's car traveled 237 miles on $7\frac{9}{10}$ gallons of gas. How many miles per gallon did he achieve?

Chapter 3

9. Multiply.
$$\begin{array}{r} 51.06 \\ \times\ 0.307 \end{array}$$

10. Divide. $0.026\overline{)0.0884}$

11. The copper pipe was 24.375 centimeters long. Paula had to shorten it by cutting off 1.75 centimeters. How long will the copper pipe be when it is shortened?

12. Russ bicycled 20.5 miles on Monday, 5.8 miles on Tuesday, and 14.9 miles on Wednesday. How many miles did he bicycle on those three days?

Chapter 4

Solve each proportion problem. Round to the nearest tenth if necessary.

13. $\dfrac{3}{7} = \dfrac{n}{24}$

14. $\dfrac{0.5}{0.8} = \dfrac{220}{n}$

15. Wally's Landscape earned $600 for mowing lawns at 25 houses last week. At that rate, how much would he earn for doing 45 houses?

16. Two cities that are actually 300 miles apart appear to be 8 inches apart on the road map. How many miles apart are two cities that appear to be 6 inches apart on the map?

Chapter 5

Round to the nearest tenth if necessary.

17. Change to a percent: $\dfrac{3}{8}$

18. 138% of 5600 is what number?

19. At Mountainview College 53% of the students are women. There are 2067 women at the college. How many students are at the college?

20. At a manufacturing plant it was discovered that 9 out of every 3000 parts made are defective. What percent of the parts are defective?

Chapter 6

21. 15 qt = _____ gal

22. 3 cm = _____ meter

23. 1.56 tons = _____ lb

24. 4900 kg = _____ milligrams

Chapter 7

Round to the nearest hundredth when necessary. Use $\pi \approx 3.14$ when necessary.

▲**25.** Find the area of a triangle with a base of 34 meters and an altitude of 23 meters.

▲**26.** Find the cost to install carpet in a circular area with a radius of 5 yards at a cost of $35 per square yard.

27. In a right triangle the longest side is 15 meters and the shortest side is 9 meters. What is the length of the other side of the triangle?

▲**28.** How many pounds of fertilizer can be placed in a cylindrical tank that is 4 feet tall and has a radius of 5 feet if one cubic foot of fertilizer weighs 70 pounds?

▲ represents geometry-related content.

13. _____

14. _____

15. _____

16. _____

17. _____

18. _____

19. _____

20. _____

21. _____

22. _____

23. _____

24. _____

25. _____

26. _____

27. _____

28. _____

Chapter 8

The following double bar graph indicates the sale of Dodge Calibers for Westover County as reported by the district sales managers. Use this graph to answer questions 29–32.

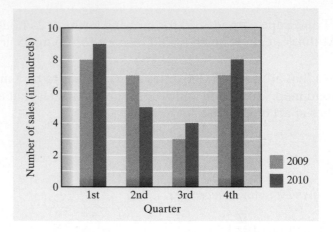

29. How many Dodge Calibers were sold in the second quarter of 2010?

30. How many more Dodge Calibers were sold in the fourth quarter of 2010 than were sold in the fourth quarter of 2009?

31. In which year were more Dodge Calibers sold, in 2009 or 2010?

32. What is the *mean* number of Dodge Calibers sold per quarter in 2009?

Chapter 9

Perform the following operations.

33. $-5 + (-2) + (-8)$

34. $-8 - (-20)$

35. $\left(-\dfrac{3}{4}\right) \div \left(\dfrac{5}{6}\right)$

36. $(-3)(2)(-1)(-3)$

Chapter 10

Simplify.

37. $9(x + y) - 3(2x - 5y)$

In exercises 38–39, solve for x.

38. $3x - 7 = 5x - 19$

39. $2(x - 3) + 4x = -2(3x + 1)$

▲ **40.** A rectangle has a perimeter of 134 meters. The length of the rectangle is 4 meters longer than double the width of the rectangle. What are the length and the width of the rectangle?

For many years there were many more drivers in the United States than there were passenger cars. Over the years, that trend has changed. Now there are more passenger cars than there are drivers in the United States. When did that change occur? How many more cars are there than drivers? The mathematics you learn in this chapter will help you to answer these kinds of questions.

Whole Numbers

1.1 Understanding Whole Numbers

Student Learning Objectives

After studying this section, you will be able to:

① Write numbers in expanded notation.

② Write whole numbers in standard notation.

③ Write a word name for a number and write a number for a word name.

④ Read numbers in tables.

① Writing Numbers in Expanded Notation

To count a number of objects or to answer the question "How many?" we use a set of numbers called **whole numbers.** These whole numbers are as follows.

0, 1, 2, 3, 4, 5, 6, 7, 8, 9, 10, 11, 12, 13, 14, 15, . . .

There is no largest whole number. The three dots . . . indicate that the set of whole numbers goes on indefinitely. Our number system is based on tens and ones and is called the **decimal system** (or the **base 10 system**). The numbers 0, 1, 2, 3, 4, 5, 6, 7, 8, 9 are called **digits.** The position, or placement, of the digits in a number tells the value of the digits. For example, in the number 521, the "5" means 5 hundreds (500). In the number 54, the "5" means 5 tens (50).

521
↑
5 means 5 hundreds or 500

54
↑
5 means 5 tens or 50

For this reason, our number system is called a **place-value system.**

Consider the number 5643. We will use a place-value chart to illustrate the value of each digit in the number 5643.

Place-value Chart

Millions			Thousands			Ones		
					5	6	4	3
Hundred millions	Ten millions	Millions	Hundred thousands	Ten thousands	Thousands	Hundreds	Tens	Ones

The value of the number is 5 thousands, 6 hundreds, 4 tens, 3 ones.

The place-value chart shows the value of each place, from ones on the right to hundred millions on the left. When we write very large numbers, we place a comma after every group of three digits, called a **period,** moving from right to left. This makes the number easier to read. It is usually agreed that a four-digit number does not have a comma, but that numbers with five or more digits do. So 32,000 would be written with a comma but 7000 would not.

To show the value of each digit in a number, we sometimes write the number in expanded notation. For example, 56,327 is 5 ten thousands, 6 thousands, 3 hundreds, 2 tens, and 7 ones. In **expanded notation,** this is

$$50{,}000 + 6000 + 300 + 20 + 7.$$

EXAMPLE 1   Write each number in expanded notation.

(a) 2378   **(b)** 538,271   **(c)** 980,340,654

Solution

(a) Sometimes it helps to say the number to yourself.

$$\underset{\text{two thousand}}{2000} + \underset{\text{three hundred}}{300} + \underset{\text{seventy}}{70} + \underset{\text{eight}}{8}$$

2378 = 2000 + 300 + 70 + 8

(b)

Expanded notation

538,271 = 500,000 + 30,000 + 8000 + 200 + 70 + 1

(c) When 0 is used as a placeholder, you do not include it in the expanded notation.

Expanded notation

980,340,654 = 900,000,000 + 80,000,000 + 300,000 + 40,000 + 600 + 50 + 4

Student Practice 1   Write each number in expanded notation.

(a) 3182   **(b)** 520,890   **(c)** 709,680,059

NOTE TO STUDENT: *Fully worked-out solutions to all of the Student Practice problems can be found at the back of the text starting at page SP-1.*

② Writing Whole Numbers in Standard Notation

The way that you usually see numbers written is called **standard notation.** 980,340,654 is the standard notation for the number nine hundred eighty million, three hundred forty thousand, six hundred fifty-four.

EXAMPLE 2   Write each number in standard notation.

(a) 500 + 30 + 8   **(b)** 300,000 + 7000 + 40 + 7

Solution

(a) 538

(b) Be careful to keep track of the place value of each digit. You may need to use 0 as a placeholder.

$$\underset{\text{7 thousand}}{\overset{\text{3 hundred thousand}}{300,000 + 7000 + 40 + 7}} = 307,047$$

We needed to use 0 in the ten thousands place and in the hundreds place.

Student Practice 2   Write each number in standard notation.

(a) 400 + 90 + 2   **(b)** 80,000 + 400 + 20 + 7

EXAMPLE 3   Last year the population of Central City was 1,509,637. In the number 1,509,637

(a) How many ten thousands are there?   **(b)** How many tens are there?

(c) What is the value of the digit 5?   **(d)** In what place is the digit 6?

Continued on next page

Solution   A place-value chart will help you identify the value of each place.

(a) Look at the digit in the ten thousands place. There are 0 ten thousands.

(b) Look at the digit in the tens place. There are 3 tens.

(c) The digit 5 is in the hundred thousands place. The value of the digit is 5 hundred thousand or 500,000.

(d) The digit 6 is in the hundreds place.

Student Practice 3   The campus library has 904,759 books.

(a) What digit tells the number of hundreds?

(b) What digit tells the number of hundred thousands?

(c) What is the value of the digit 4?

(d) What is the value of the digit 9? Why does this question have two answers?

③ Writing Word Names for Numbers and Numbers for Word Names

A number has the same *value* no matter how we write it. For example, "a million dollars" means the same as "$1,000,000." In fact, any number in our number system can be written in several ways or forms:

- Standard notation   521
- Expanded notation   500 + 20 + 1
- Word name   five hundred twenty-one

You may want to write a number in any of these ways. To write a check, you need to use both standard notation and words.

To write a word name, start from the left. Name the number in each period, followed by the name of the period, and a comma. The last period name, "ones," is not used.

EXAMPLE 4   Write a word name for 364,128,957.

Solution

Place-value Chart

Billions			Millions			Thousands			Ones		
			3	6	4	1	2	8	9	5	7
Hundreds	Tens	Ones	Hundreds	Tens	Ones	Hundreds	Tens	Ones	Hundreds	Tens	Ones

We want to write a word name for 364, 128, 957.

   three hundred sixty-four million,⌐

     one hundred twenty-eight thousand,⌐

       nine hundred fifty-seven

    The answer is three hundred sixty-four million, one hundred twenty-eight thousand, nine hundred fifty-seven.

Student Practice 4   Write a word name for 267,358,981.

EXAMPLE 5   Write the word name for each number.

(a) 1695   **(b)** 200,470   **(c)** 7,003,038

Solution   Look at the place-value chart if you need help identifying the place for each digit.

(a) To help us, we will put in the optional comma: 1,695.

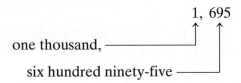

    The word name is one thousand, six hundred ninety-five.

(b)

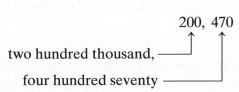

    The word name is two hundred thousand, four hundred seventy.

Continued on next page

(c)

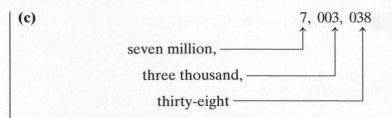

The word name is seven million, three thousand, thirty-eight.

Student Practice 5   Write the word name for each number.

(a) 2736   **(b)** 980,306   **(c)** 12,000,021

CAUTION: DO NOT USE THE WORD <u>AND</u> FOR WHOLE NUMBERS.
Many people use the word *and* when giving the word name for a whole number. For example, you might hear someone say the number 34,507 as "thirty-four thousand, five hundred *and* seven." However, this is not technically correct. In mathematics we do NOT use the word *and* when writing word names for whole numbers. In Chapter 3 we will use the word *and* to represent the decimal point. For example, 59.76 will have the word name "fifty-nine *and* seventy-six hundredths."

Very large numbers are used to measure quantities in some disciplines, such as distance in astronomy and the national debt in macroeconomics. We can extend the place-value chart to include these large numbers.

The national debt in the United States as of June 4, 2010, was $13,058,031,332,634. This number is indicated in the following place-value chart.

Place-value Chart

Trillions		Billions			Millions			Thousands			Ones		
	1	3	0	5	8	0	3	1	3	3	2	6	3	4

EXAMPLE 6   Write the number for the national debt in the United States as of June 4, 2010, in the amount of $13,058,031,332,634 using a word name.

Solution   The national debt on June 4, 2010, was thirteen trillion, fifty-eight billion, thirty-one million, three hundred thirty-two thousand, six hundred thirty-four dollars.

Student Practice 6   As of July 16, 2010, the estimated population of the world was 6,856,340,189. Write this world population using a word name.

Occasionally you may want to write a word name as a number.

EXAMPLE 7   Write each number in standard notation.

(a) twenty-six thousand, eight hundred sixty-four

(b) two billion, three hundred eighty-six million, five hundred forty-seven thousand, one hundred ninety

Solution

(a) twenty-six thousand, ⎯⎯⎯⎯⎯⎯⎯⎯⎯

⎯ eight hundred sixty-four

26, 864   Thus we have 26,864.

(b) two billion, ⎯⎯⎯⎯⎯⎯⎯⎯⎯

⎯ three hundred eighty-six million,

⎯five hundred forty-seven thousand,

⎯ one hundred ninety

2, 386, 547, 190   Thus we have 2,386,547,190.

Student Practice 7   Write in standard notation.

(a) eight hundred three

(b) thirty thousand, two hundred twenty-nine

④ Reading Numbers in Tables

Sometimes numbers found in charts and tables are abbreviated. Look at the chart below from the U.S. Bureau of the Census. Notice that the second line tells us the numbers represent thousands. To understand what these numbers mean, think "thousands." If the number 23 appears across from 1740 for New Hampshire, the 23 represents 23 thousand. 23 thousand is 23,000. Note that census figures for some colonies are not available for certain years.

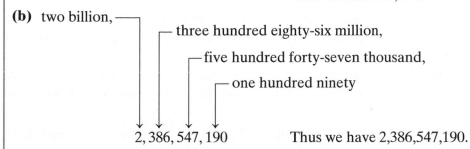

Estimated Population of the American Colonies from 1650 to 1780 (in thousands)

	Maine	New Hampshire	Vermont	Plymouth & Massachusetts	Rhode Island	Connecticut
1650	1	1	★	16	1	4
1670	★	2	★	35	2	13
1690	★	4	★	57	4	22
1700	★	5	★	56	6	26
1720	★	9	★	91	12	59
1740	★	23	★	152	25	90
1750	★	28	★	188	33	111
1770	31	62	10	235	58	184
1780	49	88	48	269	53	207

EXAMPLE 8   Refer to the chart on the previous page to answer the following questions. Write each number in standard notation.

(a) What was the estimated population of Maine in 1780?

(b) What was the estimated population of Plymouth and Massachusetts in 1720?

(c) What was the estimated population of Rhode Island in 1700?

Solution

(a) To read the chart, first look for Maine along the top. Read down to the row for 1780. The number is 49. In this chart 49 means 49 thousands.

$$49 \text{ thousands} \Rightarrow 49{,}000$$

(b) Read the column of the chart for Plymouth and Massachusetts. The number for Plymouth and Massachusetts in the row for 1720 is 91. This means 91 thousands. We will write this as 91,000.

(c) Read the column of the chart for Rhode Island. The number for Rhode Island in the row for 1700 is 6. This means 6 thousands. We will write this as 6000.

TO THINK ABOUT: Interpreting Data in a Table Why do you think Plymouth and Massachusetts had the largest population for the years shown in the table?

Student Practice 8   Refer to the chart on the previous page to answer the following questions. Write each number in standard notation.

(a) What was the estimated population of Connecticut in 1670?

(b) What was the estimated population of New Hampshire in 1780?

(c) What was the estimated population of Vermont in 1770?

STEPS TO SUCCESS Be Involved.

If you are in a traditional class:

Don't just sit on the sidelines of the class and watch. Take part in the classroom discussion. People learn mathematics best through active participation. Whenever you are not clear about something, ask a question. Usually your questions will be helpful to other students in the room. When the teacher asks for suggestions, be sure to contribute your own ideas. Sit near the front where you can see and hear well. This will help you to focus on the material being covered

Making it personal: Which of the suggestions above is the one you most need to follow? Write down what you need to do to improve in this area. ▼

If you are in an online class or a nontraditional class:

Be sure to e-mail the teacher. Talk to the tutor on duty. Ask questions. Think about concepts. Make your mind interact with the textbook. Be mentally involved. This active mental interaction is the key to your success.

Making it personal: Which of the suggestions is the one you most need to follow? Write down what you need to do to improve in this area. ▼

Write each number in expanded notation.

1. 6731

2. 9519

3. 108,276

4. 701,285

5. 23,761,345

6. 46,198,253

7. 103,260,768

8. 820,310,574

Write each number in standard notation.

9. $600 + 70 + 1$

10. $500 + 90 + 6$

11. $9000 + 800 + 60 + 3$

12. $7000 + 600 + 50 + 2$

13. $40,000 + 800 + 80 + 5$

14. $60,000 + 7000 + 200 + 4$

15. $700,000 + 6000 + 200$

16. $300,000 + 40,000 + 800$

Verbal and Writing Skills, Exercises 17–20

17. In the number 437,521
   (a) What digit tells the number of thousands?
   (b) What is the value of the digit 3?

18. In the number 805,712
   (a) What digit tells the number of ten thousands?
   (b) What is the value of the digit 8?

19. In the number 1,214,847
   (a) What digit tells the number of hundred thousands?
   (b) What is the value of the digit?

20. In the number 6,789,345
   (a) What digit tells the number of thousands?
   (b) What is the value of the digit?

Write a word name for each number.

21. 142

22. 376

23. 9304

24. 7606

25. 36,118

26. 55,742

27. 105,261

28. 370,258

29. 14,203,326

30. 68,089,213

31. 4,302,156,200

32. 7,436,210,400

Write each number in standard notation.

33. one thousand, five hundred sixty-one

34. three thousand, one hundred eighty-nine

35. thirty-three thousand, eight hundred nine

36. two hundred three thousand, three hundred seventy-four

37. one hundred million, seventy-nine thousand, eight hundred twenty-six

38. four hundred fifty million, three hundred thousand, two hundred forty-nine

Applications *When writing a check, a person must write the word name for the dollar amount of the check.*

39. *Personal Finance* Alex bought new equipment for his laboratory for $1965. What word name should he write on the check?

40. *Personal Finance* Alex later bought a new desktop computer for $1749. What word name should he write on the check?

In exercises 41–44, use the following chart prepared with data from the U.S. Bureau of the Census. Notice that the second line tells us that the numbers represent millions. These values are only approximate values representing numbers written to the nearest million. They are not exact census figures.

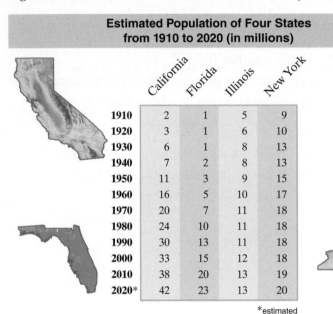

	California	Florida	Illinois	New York
Estimated Population of Four States from 1910 to 2020 (in millions)				
1910	2	1	5	9
1920	3	1	6	10
1930	6	1	8	13
1940	7	2	8	13
1950	11	3	9	15
1960	16	5	10	17
1970	20	7	11	18
1980	24	10	11	18
1990	30	13	11	18
2000	33	15	12	18
2010	38	20	13	19
2020*	42	23	13	20

*estimated

Source: U.S. Bureau of the Census

41. *Historical Analysis* What was the estimated population of New York in 1910?

42. *Historical Analysis* What was the estimated population of Florida in 1970?

43. *Historical Analysis* What is the estimated population of California in 2020?

44. *Historical Analysis* What was the estimated population of Illinois in 1940?

In exercises 45–48, use the following chart:

Number of Flights and Passengers for Selected Airlines in 2000, 2005, and 2009 (in thousands)

	2000		2005		2009	
Airline	Flights*	Passengers	Flights*	Passengers	Flights*	Passengers
American	864	86,214	818	97,950	680	85,674
Continental	458	45,297	368	42,736	344	43,933
Delta	946	105,556	700	85,824	503	67,694
Northwest	584	56,689	519	54,558	328	39,243

*Includes passenger and freight flights
Source: Bureau of Transportation Statistics

45. *Airline Travel* How many flights did Delta have in 2000?

46. *Airline Travel* How many passengers flew on American flights in 2005?

47. *Airline Travel* How many passengers flew on Northwest flights in 2005?

48. *Airline Travel* How many flights did Continental have in 2009?

49. *Physics* The speed of light is approximately 29,979,250,000 centimeters per second.

   (a) What digit tells the number of ten thousands?

   (b) What digit tells the number of ten billions?

▲ **50.** *Earth Science* The circumference of Earth at the equator is 131,480,184 feet.

   (a) What digit tells the number of ten millions?

   (b) What digit tells the number of hundred thousands?

51. *Blood Vessels* There are about 316,820,000 feet of blood vessels in an adult human body.

   (a) What digit tells the number of ten thousands?

   (b) What digit tells the number of ten millions?

52. *U.S. Currency* In 2009, the value of all $100 bills in circulation in the United States was $646,023,211,900.

   (a) Which digit tells the number of millions?

   (b) Which digit tells the number of ten billions?

53. Write in standard notation: six hundred thirteen trillion, one billion, thirty-three million, two hundred eight thousand, three.

54. Write in standard notation: nine hundred fourteen trillion, two billion, fifty-two million, four hundred nine thousand, six.

To Think About

55. Write a word name for 3,682,968,009,931,960,747. (*Hint:* The digit 1 followed by 18 zeros represents the number *1 quintillion*. 1 followed by 15 zeros represents the number *1 quadrillion*.)

56. The number 50,000,000,000,000,000,000 is represented on some scientific calculators as 5 E 19. We will cover this in more detail in a later chapter. However, for the present we can see that this is a convenient notation that allows us to record very large whole numbers. Note that this number (50 quintillion) is a 5 followed by 19 zeros. Write in standard notation the number that would be represented on a calculator as 6 E 22.

57. Think about the discussion in exercise 56. If the number 4 E 20 represented on a scientific calculator was divided by 2, what number would be the result? Write your answer in standard notation.

58. Consider all the whole numbers between 200 and 800 that contain the digit 6. How many such numbers are there?

Quick Quiz 1.1

1. Write in expanded notation.   73,952

2. Write a word name.   8,932,475

3. Write in standard notation.
   Nine hundred sixty-four thousand, two hundred fifty-seven

4. **Concept Check** Explain why the zeros are needed when writing the following number in standard notation: three hundred sixty-eight million, five hundred twenty-two.

▲ represents geometry-related content.

1.2 Adding Whole Numbers

NOTE TO STUDENT: *Fully worked-out solutions to all of the Student Practice problems can be found at the back of the text starting at page SP-1.*

① Mastering Basic Addition Facts

We see the addition process time and time again. Carpenters add to find the amount of lumber they need for a job. Auto mechanics add to make sure they have enough parts in the inventory. Bank tellers add to get cash totals.

What is addition? We do addition when we put sets of objects together.

■■■■■   ■■■■■■■   ■■■■■■■■■■■■

5 objects   +   7 objects   =   12 objects

$$5 + 7 = 12$$

Usually when we add numbers, we put one number under the other in a column. The numbers being added are called **addends.** The result is called the **sum.**

Suppose that we have four pencils in the car and we bring three more pencils from home. How many pencils do we have with us now? We add 4 and 3 to obtain a value of 7. In this case, the numbers 4 and 3 are the addends and the answer 7 is the sum.

$$\begin{array}{rl} 4 & \text{addend} \\ +\ 3 & \text{addend} \\ \hline 7 & \text{sum} \end{array}$$

Think about what we do when we add 0 to another number. We are not making a change, so whenever we add zero to another number, that number will be the sum. Because this is always true, this is called a *property.* Because the sum is identical to the number that is added to zero, this is called the **identity property of zero.**

EXAMPLE 1   Add.

(a) $8 + 5$ **(b)** $3 + 7$ **(c)** $9 + 0$

Solution

(a) $\begin{array}{r} 8 \\ +\ 5 \\ \hline 13 \end{array}$ **(b)** $\begin{array}{r} 3 \\ +\ 7 \\ \hline 10 \end{array}$ **(c)** $\begin{array}{r} 9 \\ +\ 0 \\ \hline 9 \end{array}$ ←

Note: When we add zero to any other number, that number is the sum.

Student Practice 1   Add.

(a) $\begin{array}{r} 6 \\ +\ 5 \end{array}$ **(b)** $\begin{array}{r} 9 \\ +\ 4 \end{array}$ **(c)** $\begin{array}{r} 3 \\ +\ 0 \end{array}$

The following table shows the basic addition facts. You should know these facts. If any of the answers don't come to you quickly, now is the time to learn them. To check your knowledge try Exercises 1.2, exercises 3 and 4.

Basic Addition Facts

+	0	1	2	3	4	5	6	7	8	9
0	0	1	2	3	4	5	6	7	8	9
1	1	2	3	4	5	6	7	8	9	10
2	2	3	4	5	6	7	8	9	10	11
3	3	4	5	6	7	8	9	10	11	12
4	4	5	6	7	8	9	10	11	12	13
5	5	6	7	8	9	10	11	12	13	14
6	6	7	8	9	10	11	12	13	14	15
7	7	8	9	10	11	12	13	14	15	16
8	8	9	10	11	12	13	14	15	16	17
9	9	10	11	12	13	14	15	16	17	18

To use the table to find the sum $4 + 7$, read across the top of the table to the 4 column, and then read down the left to the 7 row. The box where the 4 and 7 meet is 11, which means that $4 + 7 = 11$. Now read across the top of the table to the 7 column and down the left to the 4 row. The box where these numbers meet is also 11. We can see that the order in which we add the numbers does not change the sum. $4 + 7 = 11$, and $7 + 4 = 11$. We call this the **commutative property of addition.**

This property does not hold true for everything in our lives. When you put on your socks and then your shoes, the result is not the same as if you put on your shoes first and then your socks! Can you think of any other examples where changing the order in which you add things would change the result?

② Adding Several Single-Digit Numbers

If more than two numbers are to be added, we usually add from the first number to the next number and mentally note the sum. Then we add that sum to the next number, and so on.

EXAMPLE 2   Add. $3 + 4 + 8 + 2 + 5$

Solution   We rewrite the addition problem in a column format.

$$
\begin{array}{r}
3 \\
4 \\
8 \\
2 \\
+ 5 \\
\hline
22
\end{array}
$$

$\left.\begin{array}{l}3\\4\end{array}\right\}\ 3 + 4 = 7$ Mentally, we do these steps.

$7 + 8 = 15$

$15 + 2 = 17$

$17 + 5 = 22$

Student Practice 2   Add. $7 + 6 + 5 + 8 + 2$

Because the order in which we add numbers doesn't matter, we can choose to add from the top down, from the bottom up, or in any other way. One short-cut is to add first any numbers that will give a sum of 10, or 20, or 30, and so on.

EXAMPLE 3   Add.   3
                      4
                      8
                      2
                    + 6

Solution   We mentally group the numbers into tens.

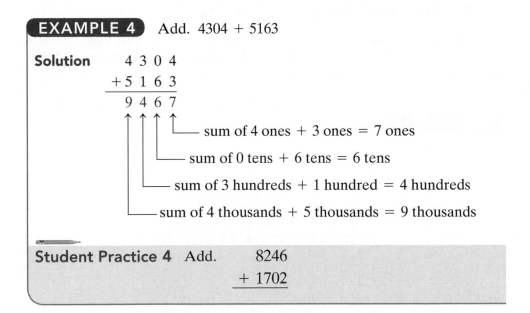

The sum is $10 + 10 + 3$ or 23.

Student Practice 3   Add. $1 + 7 + 2 + 9 + 3$

③ Adding Several-Digit Numbers When Carrying Is Not Needed

Of course, many numbers that we need to add have more than one digit. In such cases, we must be careful to first add the digits in the ones column, then the digits in the tens column, then those in the hundreds column, and so on. Notice that we move from *right to left*.

EXAMPLE 4   Add. $4304 + 5163$

Solution   4 3 0 4
                + 5 1 6 3
                  9 4 6 7
                  ↑ ↑ ↑ ↑
                          └── sum of 4 ones + 3 ones = 7 ones
                        └──── sum of 0 tens + 6 tens = 6 tens
                      └────── sum of 3 hundreds + 1 hundred = 4 hundreds
                    └──────── sum of 4 thousands + 5 thousands = 9 thousands

Student Practice 4   Add.   8246
                                 + 1702

④ Adding Several-Digit Numbers When Carrying Is Needed

When you add several whole numbers, often the sum in a column is greater than 9. However, we can only use *one* digit in any one place. What do we do with a two-digit sum? Look at the following example.

EXAMPLE 5 Add. 45 + 37

Solution

```
      1
    4 5
  + 3 7
      2
```

5 ones and 7 ones = 12.
We rename 12 in expanded notation: 1 ten + 2 ones.
We place the 2 ones in the ones column.
We carry the 1 ten over to the tens column.

Note: Placing the 1 in the next column is often called "carrying the one."

```
    1
  4 5
+ 3 7
  8 2
```

Now we can add the digits in the tens column.

Thus, 45 + 37 = 82.

Student Practice 5 Add.
```
    56
  + 36
```

Often you must use carrying several times by bringing the left digit into the next column to the left.

EXAMPLE 6 Add. 257 + 688 + 94

Solution

Thousands Column · Hundreds Column · Tens Column · Ones Column

```
    2 1
    2 5 7
    6 8 8
  +   9 4
  1 0 3 9
```

In the ones column we add 7 + 8 + 4 = 19. Because 19 is 1 ten and 9 ones, we place 9 in the ones column and carry 1 to the top of the tens column.

In the tens column we add 1 + 5 + 8 + 9 = 23. Because 23 tens is 2 hundreds and 3 tens, we place the 3 in the tens column and carry 2 to the top of the hundreds column.

In the hundreds column we add 2 + 2 + 6 = 10 hundreds. Because 10 hundreds is 1 thousand and 0 hundreds, we place the 0 in the hundreds column and place the 1 in the thousands column.

Student Practice 6 Add. 789 + 63 + 297

We can add numbers in more than one way. To add $5 + 3 + 7$ we can first add the 5 and 3. We do this by using parentheses to show the first operation to be done. This shows us that $5 + 3$ is to be grouped together.

$$5 + 3 + 7 = (5 + 3) + 7$$
$$= \quad 8 \quad + 7$$

We could add the 3 and 7 first. We use parentheses to show that we group $3 + 7$ together and that we will add these two numbers first.

$$5 + 3 + 7 = 5 + (3 + 7) = 15$$
$$= 5 + \quad 10 \quad = 15$$

The way we group numbers to be added does not change the sum. This property is called the **associative property of addition.**

⑤ Reviewing the Properties of Addition

Look again at the three properties of addition we have discussed in this section.

1. **Identity Property of Zero** When zero is added to a number, the sum is that number.	$8 + 0 = 8$ $0 + 5 = 5$
2. **Commutative Property of Addition** Two numbers can be added in either order with the same result.	$5 + 12 = 12 + 5$ $17 = 17$
3. **Associative Property of Addition** When we add three numbers, we can group them in any way.	$(8 + 2) + 6 = 8 + (2 + 6)$ $10 + 6 = 8 + 8$ $16 = 16$

Because of the commutative and associative properties of addition, we can check our addition by adding the numbers in the opposite order.

EXAMPLE 7

(a) Add the numbers. $39 + 7284 + 3132$

(b) Check by reversing the order of addition.

Solution **(a)**
$$\overset{1\,1}{39}$$
$$7284$$
$$+\ 3132$$
$$\overline{10{,}455}$$
Addition

(b)
$$\overset{1\,1}{3132}$$
$$7284$$
$$+\quad 39$$
$$\overline{10{,}455}$$
Check by reversing the order.

> The sum is the same in each case.

Student Practice 7

(a) Add.
$$127$$
$$9876$$
$$+\ \ 342$$

(b) Check by reversing the order.
$$342$$
$$9876$$
$$+\ \ 127$$

⑥ Applying Addition to Real-Life Situations

We use addition in all kinds of situations. There are several key words in word problems that imply addition. For example, it may be stated that there are 12 math books, 9 chemistry books, and 8 biology books on a book shelf. To find the *total* number of books implies that we add the numbers $12 + 9 + 8$. Other key words are *how much, how many,* and *all.*

Sometimes a problem will have more information than you will need to answer the question. If you have too much information, to solve the problem you will need to separate out the facts that are not important. The following three steps are involved in the problem-solving process.

Step 1 Understand the problem.
Step 2 Calculate and state the answer.
Step 3 Check.

We may not write all of these steps down, but they are the steps we use to solve all problems.

EXAMPLE 8   The bookkeeper for Smithville Trucking was examining the following data for the company checking account.

Monday:	$23,416 was deposited and $17,389 was debited.
Tuesday:	$44,823 was deposited and $34,089 was debited.
Wednesday:	$16,213 was deposited and $20,057 was debited.

What was the total of all deposits during this period?

Solution

Step 1 *Understand the problem.*
Total implies that we will use addition. Since we don't need to know about the debits to answer this question, we use only the *deposit* amounts.

Step 2 *Calculate and state the answer.*

Monday:	$23,416 was deposited.	$\overset{11\ 1}{23,}416$
Tuesday:	$44,823 was deposited.	$44,823$
Wednesday:	$16,213 was deposited.	$\underline{+\ 16,213}$
		$84,452$

A total of $84,452 was deposited on those three days.

Step 3 *Check.*
You may add the numbers in reverse order to check. We leave the check up to you.

Student Practice 8 North University has 23,413 men and 18,316 women. South University has 19,316 men and 24,789 women. East University has 20,078 men and 22,965 women. What is the total enrollment of women at the three universities?

▲ **EXAMPLE 9**   Mr. Ortiz has a rectangular field whose length is 400 feet and whose width is 200 feet. What is the total number of feet of fence that would be required to fence in the field?

Solution

1. ***Understand the problem.*** To help us to get a picture of what the field looks like, we will draw a diagram.

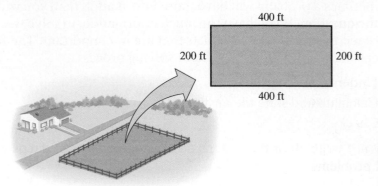

Note that ft is the abbreviation for feet. ft means feet.

2. ***Calculate and state the answer.***
Since the fence will be along each side of the field, we add the lengths all around the field.

$$\begin{array}{r} 200 \\ 400 \\ 200 \\ + 400 \\ \hline 1200 \end{array}$$

The amount of fence that would be required is 1200 feet.

3. ***Check.***
Regroup the addends and add.

$$\begin{array}{r} 200 \\ 200 \\ 400 \\ + 400 \\ \hline 1200 \;\checkmark \end{array}$$

▲ **Student Practice 9**   In Vermont, Gretchen fenced the rectangular field on which her sheep graze. The length of the field is 2000 feet and the width of the field is 1000 feet. What is the perimeter of the field? (*Hint:* The "distance around" an object [such as a field] is called the *perimeter.*)

 STEPS TO SUCCESS Review a Little Every Day.

Successful students find that review is not something you do the night before the test. Take time to review a little each day. When you are learning new material, take a little time to look over the concepts previously learned in the chapter. By this continual review you will find the pressure is reduced to prepare for a test. You need time to think about what you have learned and make sure you really understand it. This will help to tie together the different topics in the chapter. A little review of each idea and each kind of problem will enable you to

feel confident. You will think more clearly and have less tension when it comes to test time.

Making it personal: Which of these suggestions is the one you most need to follow? Write down what you need to do to improve in this area. ▼

Verbal and Writing Skills, Exercises 1 and 2

1. Explain in your own words.
   (a) the commutative property of addition
   (b) the associative property of addition

2. When zero is added to any number, it does not change that number. Why do you think this is called the identity property of zero?

Complete the addition facts for each table. Strive for total accuracy, but work quickly. Allow a maximum of five minutes for each table.

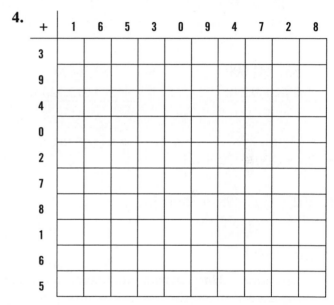

3.

+	3	5	4	8	0	6	7	2	9	1
2										
7										
5										
3										
0										
4										
1										
8										
6										
9										

4.

+	1	6	5	3	0	9	4	7	2	8
3										
9										
4										
0										
2										
7										
8										
1										
6										
5										

Add.

5.	6.	7.	8.	9.	10.
4	4	2	1	18	63
2	6	6	5	36	11
8	2	7	5	+ 3	+ 6
+ 9	+ 7	8	9		
		+ 3	+ 9		

11.	12.	13.	14.	15.	16.
63	54	3315	5773	5631	5017
24	21	726	425	2344	2984
+ 12	+ 23	+ 84	+ 67	+ 2019	+ 1328

17.	18.	19.	20.
8235	6753	62,504	83,596
+ 5626	+ 3265	+ 54,736	+ 56,384

Add from the top. Then check by adding in the reverse order.

21.
```
   36
   41
   25
    6
+ 13
```

22.
```
   24
   39
   16
   14
+  9
```

23.
```
  207
   15
    3
   57
+ 861
```

24.
```
  426
   39
    6
   52
+ 802
```

Add.

25.
```
     85
    256
     55
+ 9734
```

26.
```
    582
   1674
    336
+ 8458
```

27.
```
  1,362,214
  7,002,316
+ 3,214,896
```

28.
```
  4,002,983
  2,134,702
+ 3,592,001
```

29.
```
  837,241,000
+ 298,039,240
```

30.
```
  982,306,000
+ 583,215,320
```

31.
```
    516,208
     24,317
+ 1,763,295
```

32.
```
     32,500
    763,420
+ 2,837,667
```

33. $25 + 130 + 70 + 75$

34. $110 + 20 + 280 + 90$

35. $102 + 50 + 98 + 35 + 50$

36. $30 + 210 + 70 + 58 + 90$

Applications

37. ***Consumer Mathematics*** Stephanie took her triplets shopping for the new school year. She spent $455 on clothes, $186 on shoes, and $82 on supplies. What was the total amount of money Stephanie spent?

38. ***Consumer Mathematics*** At the beginning of fall semester, Felipe bought a graphing calculator for $95, textbooks for $245, and other supplies for $58. What is the total amount Felipe spent on school supplies?

39. ***Personal Finance*** Stella owns a studio where she teaches music classes to children. Two months ago she made a profit of $1875. Last month she made $1930 and this month she earned $1744. What is the total amount for the three months?

40. ***Personal Finance*** Paul is a soccer coach and teaches kids' classes in the summer. Last summer he earned $2025 in June, $2650 in July, and $1960 in August. What is the total amount Paul earned last summer?

▲ **41.** ***Geometry*** Nathaniel wants to put a fence around his backyard. The sketch below indicates the length of each side of the yard. What is the total number of feet of fence he needs for his backyard? (Find the perimeter of the yard.)

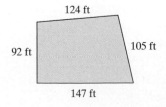

▲ **42.** ***Geometry*** Josiah has a field with the length of each side as labeled on the sketch. What is the total number of feet of fence that would be required to fence in the field? (Find the perimeter of the field.)

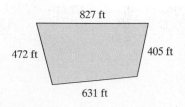

▲ **43.** *Geography* The largest island in the Atlantic Ocean is Greenland, with an area of 840,000 square miles. The second largest island, Great Britain, has an area of 88,407 square miles and the third largest is Iceland, with 39,699 square miles. What is the total area of these three islands?

▲ **44.** *Geography* The largest group of islands in the Pacific Ocean is the Japanese islands, with an area of 145,850 square miles. The second largest group of islands, the Philippine islands, has an area of 115,860 square miles. The third largest group is New Zealand, with 104,454 square miles. What is the total area of these groups of islands?

45. *Geography* The Nile River is Africa's longest river, measuring 7,272,320 yards. The second and third longest rivers in Africa are the Congo River, measuring 5,104,000 yards, and the Niger River, which measures 4,558,400 yards. What is the total length of these rivers?

▲ **46.** *Geography* The three largest of the Great Lakes are Lake Superior at 81,000 square miles, Lake Michigan at 67,900 square miles, and Lake Huron at 74,700 square miles. What is the total area of these three lakes?

In exercises 47–48, be sure you understand the problem and then choose the numbers you need in order to answer each question. Then solve the problem.

47. *Education* The admissions department of a competitive university is reviewing applications to see whether students are *eligible* or *ineligible* for student aid. On Monday, 415 were found eligible and 27 ineligible. On Tuesday, 364 were found eligible and 68 ineligible. On Wednesday, 159 were found eligible and 102 ineligible. On Thursday, 196 were found eligible and 61 ineligible.

(a) How many students were eligible for student aid over the four days?

(b) How many students were considered in all?

48. *Manufacturing* The quality control division of a motorcycle company classifies the final assembled bike as *passing* or *failing* final inspection. In January, 14,311 vehicles passed whereas 56 failed. In February, 11,077 passed and 158 failed. In March, 12,580 passed and 97 failed.

(a) How many motorcycles passed the inspection during the three months?

(b) In the three months, how many motorcycles were assembled in all?

Use the following facts to solve exercises 49 and 50. It is 87 miles from Springfield to Weston. It is 17 miles from Weston to Boston. Driving directly, it is 98 miles from Springfield to Boston. It is 21 miles from Boston to Hamilton.

49. *Geography* If Melissa drives from Springfield to Weston, then from Weston to Boston, and finally directly home to Springfield, how many miles does she drive?

50. *Geography* If Marcia drives from Hamilton to Boston, then from Boston to Weston, and then from Weston to Springfield, how many miles does she drive?

▲ **51.** *Geometry* Walter Swensen is examining the fences of a farm in Caribou, Maine. One field is in the shape of a four-sided figure with no sides equal. The field is enclosed with 2387 feet of wooden rail fence. The first side is 568 feet long, while the second side is 682 feet long. The third side is 703 feet long. How long is the fourth side?

▲ **52.** *Geometry* Carlos Sontera is riding to examine the fences of a ranch in El Paso, Texas. The field he is examining is in the shape of a rectangle. The perimeter of the rectangle is 3456 feet. One side of the rectangle is 930 feet long. How long are the other sides? (*Hint:* The opposite sides of a rectangle are equal.)

53. *Personal Finance* Answer using the information in the following Western University expense chart for the current academic year.

Western University Yearly Expenses	In-State Student, U.S. Citizen	Out-of-State Student, U.S. Citizen	Foreign Student
Tuition	$3640	$5276	$8352
Room	1926	2437	2855
Board	1753	1840	1840

How much is the total cost for tuition, room, and board for

(a) an out-of-state U.S. citizen?

(b) an in-state U.S. citizen?

(c) a foreign student?

To Think About *In exercises 54–55, add.*

54. 2,368,521,788 + 5,721,368,701 + 4,027,399,206

55. 89 + 166 + 23 + 45 + 72 + 190 + 203 + 77 + 18 + 93 + 46 + 73 + 66

56. What would happen if addition were not commutative?

57. What would happen if addition were not associative?

Cumulative Review *Write the word name for each number.*

58. [1.1.3] 76,208,941

59. [1.1.3] 121,000,374

Write each number in standard notation.

60. [1.1.3] eight million, seven hundred twenty-four thousand, three hundred ninety-six

61. [1.1.3] nine million, fifty-one thousand, seven hundred nineteen

62. [1.1.3] twenty-eight million, three hundred eighty-seven thousand, eighteen

---

Quick Quiz 1.2 *Add.*

1.
```
   56
   38
   92
   17
 +  9
```

2.
```
  831
  276
+ 508
```

3.
```
 681,302
   5,126
  18,371
+ 300,012
```

4. Concept Check Explain how you would use carrying when performing the following calculation: 4567 + 3189 + 895.

1.3 Subtracting Whole Numbers

① Mastering Basic Subtraction Facts

Subtraction is used day after day in the business world. The owner of a bakery placed an ad for his cakes in a local newspaper to see if this might increase his profits. To learn how many cakes had been sold, at closing time he subtracted the number of cakes remaining from the number of cakes the bakery had when it opened. To figure his profits, he subtracted his costs (including the cost of the ad) from his sales. Finally, to see if the ad paid off, he subtracted the profits he usually made in that period from the profits after advertising. He needed subtraction to see whether it paid to advertise.

What is subtraction? We do subtraction when we take objects away from a group. If you have 12 objects and take away 3 of them, 9 objects remain.

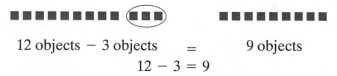

12 objects − 3 objects = 9 objects

$$12 - 3 = 9$$

If you earn \$400 per month, but have \$100 taken out for taxes, how much do you have left?

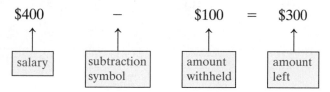

$$\$400 \quad - \quad \$100 \quad = \quad \$300$$

salary	subtraction symbol	amount withheld	amount left

We can use addition to help with a subtraction problem.

To subtract: $200 - 196 =$ what number

We can think: $196 +$ what number $= 200$

Usually when we subtract numbers, we put one number under the other in a column. When we subtract one number from another, the answer is called the **difference.**

$$
\begin{array}{cccc}
9 & 8 & 12 & 17 \\
-2 & -3 & -6 & -9 \\
\hline
7 & 5 & 6 & 8
\end{array}
$$

Each of these is called the difference of the two numbers.

The other two parts of a subtraction problem have labels, although you will not often come across them. The number being subtracted is called the **subtrahend.** The number being subtracted from is called the **minuend.**

$$
\begin{array}{ll}
17 & \text{minuend} \\
-\ 9 & \text{subtrahend} \\
\hline
8 & \text{difference}
\end{array}
$$

In this case, the number 17 is called the *minuend.* The number 9 is called the *subtrahend.* The number 8 is called the *difference.*

QUICK RECALL OF SUBTRACTION FACTS It is helpful if you can subtract quickly. See if you can do Example 1 correctly in 15 seconds or less. Repeat again with Student Practice 1. Strive to obtain all answers correctly in 15 seconds or less.

EXAMPLE 1 Subtract.

(a) $8 - 2$ **(b)** $13 - 5$ **(c)** $12 - 4$
(d) $15 - 8$ **(e)** $16 - 0$

Solution

(a) $\begin{array}{r} 8 \\ -\ 2 \\ \hline 6 \end{array}$ **(b)** $\begin{array}{r} 13 \\ -\ 5 \\ \hline 8 \end{array}$ **(c)** $\begin{array}{r} 12 \\ -\ 4 \\ \hline 8 \end{array}$

(d) $\begin{array}{r} 15 \\ -\ 8 \\ \hline 7 \end{array}$ **(e)** $\begin{array}{r} 16 \\ -\ 0 \\ \hline 16 \end{array}$

NOTE TO STUDENT: Fully worked-out solutions to all of the Student Practice problems can be found at the back of the text starting at page SP-1.

Student Practice 1 Subtract.

(a) $\begin{array}{r} 9 \\ -\ 6 \end{array}$ **(b)** $\begin{array}{r} 12 \\ -\ 5 \end{array}$ **(c)** $\begin{array}{r} 17 \\ -\ 8 \end{array}$ **(d)** $\begin{array}{r} 14 \\ -\ 0 \end{array}$ **(e)** $\begin{array}{r} 18 \\ -\ 9 \end{array}$

② Subtracting Whole Numbers When Borrowing Is Not Necessary

When we subtract numbers with more than two digits, in order to keep track of our work, we line up the ones column, the tens column, the hundreds column, and so on. Note that we begin with the ones column, and move from right to left.

EXAMPLE 2 Subtract. $9867 - 3725$

Solution
$$\begin{array}{r} 9\ 8\ 6\ 7 \\ -\ 3\ 7\ 2\ 5 \\ \hline 6\ 1\ 4\ 2 \end{array}$$

7 ones − 5 ones = 2 ones
6 tens − 2 tens = 4 tens
8 hundreds − 7 hundreds = 1 hundred
9 thousands − 3 thousands = 6 thousands

Student Practice 2 Subtract. $7695 - 3481$

③ **Subtracting Whole Numbers When Borrowing Is Necessary**

In the subtraction that we have looked at so far, each digit in the upper number (the minuend) has been greater than the digit in the lower number (the subtrahend) for each place value. Many times, however, a digit in the lower number is greater than the digit in the upper number for that place value.

$$
\begin{array}{r}
42 \\
- 28 \\
\end{array}
$$

The digit in the ones place in the lower number, the 8 of 28, is greater than the number in the ones place in the upper number, the 2 of 42. To subtract, we must *rename* 42, using place values. This is called **borrowing.**

EXAMPLE 3 Subtract. 42 − 28

Solution

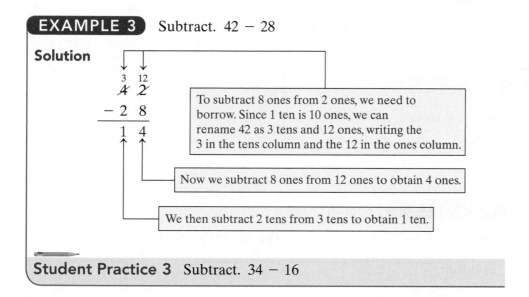

To subtract 8 ones from 2 ones, we need to borrow. Since 1 ten is 10 ones, we can rename 42 as 3 tens and 12 ones, writing the 3 in the tens column and the 12 in the ones column.

Now we subtract 8 ones from 12 ones to obtain 4 ones.

We then subtract 2 tens from 3 tens to obtain 1 ten.

Student Practice 3 Subtract. 34 − 16

EXAMPLE 4 Subtract. 864 − 548

Solution

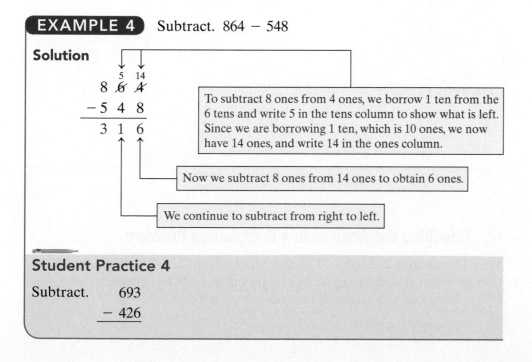

To subtract 8 ones from 4 ones, we borrow 1 ten from the 6 tens and write 5 in the tens column to show what is left. Since we are borrowing 1 ten, which is 10 ones, we now have 14 ones, and write 14 in the ones column.

Now we subtract 8 ones from 14 ones to obtain 6 ones.

We continue to subtract from right to left.

Student Practice 4

Subtract.
$$
\begin{array}{r}
693 \\
- 426 \\
\end{array}
$$

Mc **EXAMPLE 5**   Subtract. 8040 − 6375

Solution

To subtract 5 from 0, we borrow 1 ten from the 4 tens to make 3 tens and 10 ones. 10 − 5 = 5

$$
\begin{array}{r}
\overset{9}{} \ \ \overset{13}{} \\
\overset{7}{} \ \overset{10}{\cancel{10}} \ \overset{3}{\cancel{3}} \ \overset{10}{} \\
\cancel{8} \ \cancel{0} \ \cancel{4} \ \cancel{0} \\
-\ 6 \ 3 \ 7 \ 5 \\
\hline
1 \ 6 \ 6 \ 5
\end{array}
$$

To subtract 7 tens from the 3 tens, we need to borrow 1 hundred to make 10 tens. Since we find a 0 in the hundreds column, first we borrow 1 thousand to make 10 hundreds. We show the number of thousands that are left, and write the 10 in the hundreds column. Now we borrow 1 hundred, show the number of hundreds that are left, and add the 10 tens to the 3 tens. We now do the subtraction. 13 tens − 7 tens = 6 tens

9 hundreds − 3 hundreds = 6 hundreds

7 thousands − 6 thousands = 1 thousand

Student Practice 5   Subtract. 9070 − 5886

EXAMPLE 6   Subtract.

(a) 9521 − 943

(b) 40,000 − 29,056

Solution

(a)
$$
\begin{array}{r}
\overset{14}{} \ \overset{11}{} \\
\overset{8}{} \ \overset{4}{\cancel{5}} \ \overset{1}{\cancel{2}} \ \overset{11}{} \\
\cancel{9} \ \cancel{5} \ \cancel{2} \ \cancel{1} \\
-\ \ \ 9 \ 4 \ 3 \\
\hline
8 \ 5 \ 7 \ 8
\end{array}
$$

(b)
$$
\begin{array}{r}
\overset{3}{} \ \overset{9}{} \ \overset{9}{} \ \overset{9}{} \ \overset{10}{} \\
\cancel{4} \ \cancel{0}, \cancel{0} \ \cancel{0} \ \cancel{0} \\
-\ 2 \ 9, 0 \ 5 \ 6 \\
\hline
1 \ 0, 9 \ 4 \ 4
\end{array}
$$

Student Practice 6   Subtract.

(a)   8964
        − 985

(b)   50,000
        − 32,508

---

④ **Checking the Answer to a Subtraction Problem**

We observe that when 9 − 7 = 2 it follows that 7 + 2 = 9. Each subtraction problem is equivalent to a corresponding addition problem. This gives us a convenient way to check our answers to subtraction.

EXAMPLE 7 Check this subtraction problem.

$$5829 - 3647 = 2182$$

Solution

$$
\begin{array}{r}
5\ 8\ 2\ 9 \\
-\ 3\ 6\ 4\ 7 \\
\hline
2\ 1\ 8\ 2
\end{array}
\quad \text{then} \quad
\begin{array}{r}
3\ 6\ 4\ 7 \\
+\ 2\ 1\ 8\ 2 \\
\hline
5\ 8\ 2\ 9
\end{array}
$$

The sum should equal 5829, which it does. We have checked our work, and it is correct.

Student Practice 7 Check this subtraction problem.

$$9763 - 5732 = 4031$$

EXAMPLE 8 Subtract and check your answers.

(a) $156{,}000 - 29{,}326$ **(b)** $1{,}264{,}308 - 1{,}057{,}612$

Solution

(a)
$$
\begin{array}{r}
156{,}000 \\
-\ 29{,}326 \\
\hline
126{,}674
\end{array}
\qquad
\begin{array}{r}
29{,}326 \\
+\ 126{,}674 \\
\hline
156{,}000
\end{array}
$$
It checks.

(b)
$$
\begin{array}{r}
1{,}264{,}308 \\
-\ 1{,}057{,}612 \\
\hline
206{,}696
\end{array}
\qquad
\begin{array}{r}
1{,}057{,}612 \\
+\ \ \ 206{,}696 \\
\hline
1{,}264{,}308
\end{array}
$$
It checks.

Student Practice 8 Subtract and check your answers.

(a)
$$
\begin{array}{r}
284{,}000 \\
-\ \ 96{,}327
\end{array}
$$

(b)
$$
\begin{array}{r}
8{,}526{,}024 \\
-\ 6{,}397{,}518
\end{array}
$$

Subtraction can be used to solve word problems. Some problems can be expressed (and solved) with an **equation.** An equation is a number sentence with an equals sign, such as

$$10 = 4 + x$$

Here we use the letter x to represent a number we do not know. When we write $10 = 4 + x$, we are stating that 10 is equal to 4 added to some other number. Since $10 - 4 = 6$, we would assume that the number is 6. If we substitute 6 for x in the equation, we have two values that are the same.

$$
\begin{aligned}
10 &= 4 + x \\
10 &= 4 + 6 \quad \text{Substitute 6 for } x. \\
10 &= 10 \quad\ \ \text{Both sides of the equation are the same.}
\end{aligned}
$$

We can write an equation when one of the addends is not known, then use subtraction to solve for the unknown.

EXAMPLE 9   The librarian knows that he has eight world atlases and that five of them are in full color. How many are not in full color?

Solution   We represent the number that we don't know as x and write an equation, or mathematical sentence.

$$8 = 5 + x$$

To solve an equation means to find those values that will make the equation true. We solve this equation by reasoning and by a knowledge of the relationship between addition and subtraction.

$$8 = 5 + x \text{ is equivalent to } 8 - 5 = x$$

We know that $8 - 5 = 3$. Then $x = 3$. We can check the answer by substituting 3 for x in the original equation.

$$8 = 5 + x$$
$$8 = 5 + 3 \quad \text{True } \checkmark$$

We see that $x = 3$ checks, so our answer is correct. There are three atlases not in full color.

Student Practice 9   Form an equation for each of the following problems. Solve the equation in order to answer the question.

(a) The Salem Harbormaster's daily log noted that seventeen fishing vessels left the harbor yesterday during daylight hours. Walter was at the harbor all morning and saw twelve fishing vessels leave in the morning. How many vessels left in the afternoon? (Assume that sunset was at 6 P.M.)

(b) The Appalachian Mountain Club noted that twenty-two hikers left to climb Mount Washington during the morning. By 4 P.M., ten of them had returned. How many of the hikers were still on the mountain?

⑤ Applying Subtraction to Real-Life Situations

We use subtraction in all kinds of situations. There are several key words in word problems that imply subtraction. Words that involve comparison, such as *how much more, how much greater,* or how much a quantity *increased* or *decreased,* all imply subtraction. The *difference* between two numbers implies subtraction.

EXAMPLE 10   Look at the following population table.

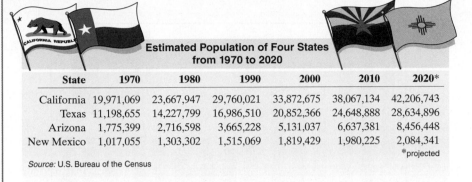

State	1970	1980	1990	2000	2010	2020*
California	19,971,069	23,667,947	29,760,021	33,872,675	38,067,134	42,206,743
Texas	11,198,655	14,227,799	16,986,510	20,852,366	24,648,888	28,634,896
Arizona	1,775,399	2,716,598	3,665,228	5,131,037	6,637,381	8,456,448
New Mexico	1,017,055	1,303,302	1,515,069	1,819,429	1,980,225	2,084,341

*projected

Source: U.S. Bureau of the Census

(a) In 1980, how much greater was the population of Texas than that of Arizona?

(b) How much did the population of California increase from 1970 to 2010?

(c) How much greater was the population of California in 2010 than that of the other three states combined?

Solution

(a)   14,227,799   1980 population of Texas
   − 2,716,598   1980 population of Arizona
   11,511,201   difference

The population of Texas was greater by 11,511,201.

(b)   38,067,134   2010 population of California
   − 19,971,069   1970 population of California
   18,096,065   difference

The population of California increased by 18,096,065 in those 40 years.

(c) First we need to find the total population in 2010 of Texas, Arizona, and New Mexico.

   24,648,888   2010 population of Texas
   6,637,381   2010 population of Arizona
   + 1,980,225   2010 population of New Mexico
   33,266,494

We use subtraction to compare this total with the population of California.

   38,067,134   2010 population of California
   − 33,266,494
   4,800,640

The population of California in 2010 was 4,800,640 more than the population of the other three states combined.

Student Practice 10

(a) In 2020, the projected population of California will be how much greater than the projected population of Texas?

(b) How much did the population of Texas increase from 1970 to 1980?

EXAMPLE 11   The number of real estate transfers in several towns during the years 2009 to 2011 is given in the following bar graph.

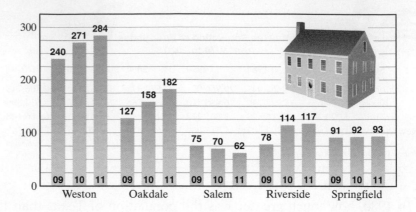

(a) What was the increase in homes sold in Weston from 2010 to 2011?

(b) What was the decrease in homes sold in Salem from 2009 to 2011?

(c) Between what two years did Oakdale have the greatest increase in sales?

Solution

(a) From the labels on the bar graph we see that 284 homes were sold in 2011 in Weston and 271 homes were sold in 2010. Thus the increase can be found by subtracting 284 − 271 = 13. There was an increase of 13 homes sold in Weston from 2010 to 2011.

(b) In 2009, 75 homes were sold in Salem. In 2011, 62 homes were sold in Salem. The decrease in the number of homes sold is 75 − 62 = 13. There was a decrease of 13 homes sold in Salem from 2009 to 2011.

(c) Here we will need to make two calculations in order to decide where the greatest increase occurs.

$$
\begin{array}{rl}
158 & \text{2010 sales} \\
- 127 & \text{2009 sales} \\
\hline
31 & \text{Sales increase} \\
& \text{from 2009 to 2010}
\end{array}
\qquad
\begin{array}{rl}
182 & \text{2011 sales} \\
- 158 & \text{2010 sales} \\
\hline
24 & \text{Sales increase} \\
& \text{from 2010 to 2011}
\end{array}
$$

The greatest increase in sales in Oakdale occurred from 2009 to 2010.

Student Practice 11   Based on the preceding bar graph, answer the following questions.

(a) What was the increase in homes sold in Riverside from 2009 to 2010?

(b) How many more homes were sold in Springfield in 2009 than in Riverside in 2009?

(c) Between what two years did Weston have the greatest increase in sales?

1.3 Exercises MyMathLab®

 Watch the videos in MyMathLab

 Download the MyDashBoard App

Verbal and Writing Skills, Exercises 1–4

1. Explain how you can check a subtraction problem.

2. Explain how you use borrowing to calculate $107 - 88$.

3. Explain what number should be used to replace the question mark in the subtraction equation $32?5 - 1683 = 1592$.

4. Explain what steps need to be done to calculate 7 feet − 11 inches.

Try to do exercises 5–20 in one minute or less with no errors.

Subtract.

5. $\begin{array}{r} 8 \\ -3 \\ \hline \end{array}$

6. $\begin{array}{r} 17 \\ -8 \\ \hline \end{array}$

7. $\begin{array}{r} 15 \\ -9 \\ \hline \end{array}$

8. $\begin{array}{r} 14 \\ -5 \\ \hline \end{array}$

9. $\begin{array}{r} 16 \\ -0 \\ \hline \end{array}$

10. $\begin{array}{r} 17 \\ -9 \\ \hline \end{array}$

11. $\begin{array}{r} 18 \\ -9 \\ \hline \end{array}$

12. $\begin{array}{r} 12 \\ -7 \\ \hline \end{array}$

13. $\begin{array}{r} 11 \\ -4 \\ \hline \end{array}$

14. $\begin{array}{r} 15 \\ -8 \\ \hline \end{array}$

15. $\begin{array}{r} 13 \\ -7 \\ \hline \end{array}$

16. $\begin{array}{r} 16 \\ -9 \\ \hline \end{array}$

17. $\begin{array}{r} 11 \\ -8 \\ \hline \end{array}$

18. $\begin{array}{r} 10 \\ -7 \\ \hline \end{array}$

19. $\begin{array}{r} 15 \\ -6 \\ \hline \end{array}$

20. $\begin{array}{r} 12 \\ -5 \\ \hline \end{array}$

Subtract. Check your answers by adding.

21. $\begin{array}{r} 47 \\ -26 \\ \hline \end{array}$

22. $\begin{array}{r} 96 \\ -51 \\ \hline \end{array}$

23. $\begin{array}{r} 85 \\ -73 \\ \hline \end{array}$

24. $\begin{array}{r} 77 \\ -36 \\ \hline \end{array}$

25. $\begin{array}{r} 379 \\ -36 \\ \hline \end{array}$

26. $\begin{array}{r} 189 \\ -65 \\ \hline \end{array}$

27. $\begin{array}{r} 869 \\ -548 \\ \hline \end{array}$

28. $\begin{array}{r} 659 \\ -247 \\ \hline \end{array}$

29. $\begin{array}{r} 4799 \\ -596 \\ \hline \end{array}$

30. $\begin{array}{r} 5780 \\ -530 \\ \hline \end{array}$

31. $\begin{array}{r} 155,835 \\ -12,600 \\ \hline \end{array}$

32. $\begin{array}{r} 243,951 \\ -12,400 \\ \hline \end{array}$

33. $\begin{array}{r} 986,302 \\ -433,201 \\ \hline \end{array}$

34. $\begin{array}{r} 807,965 \\ -304,214 \\ \hline \end{array}$

Check each subtraction. If the problem has not been done correctly, find the correct answer.

35.   129
       − 19
       ─────
        110

36.   186
       − 45
       ─────
        141

37.   8596
       − 3215
       ─────
        5781

38.   9956
       − 7254
       ─────
        2702

39.   6030
       − 5020
       ─────
        1020

40.   7890
       − 3200
       ─────
        7670

41.   47,869
       − 33,846
       ───────
        13,023

42.   99,583
       − 41,181
       ───────
        58,402

Subtract. Use borrowing if necessary.

43.   98
       − 52
       ────

44.   86
       − 33
       ────

45.   174
       − 82
       ────

46.   136
       − 95
       ────

47.   647
       − 263
       ─────

48.   706
       − 435
       ─────

49.   955
       − 237
       ─────

50.   861
       − 345
       ─────

51.   20,000
       − 9285
       ──────

52.   50,000
       − 7338
       ──────

53.   152,000
       − 117,908
       ────────

54.   361,000
       − 121,520
       ────────

55.   45,312
       − 37,865
       ───────

56.   64,381
       − 29,997
       ───────

57.   2,378,862
       − 1,469,932
       ──────────

58.   3,554,830
       − 1,710,913
       ──────────

Solve.

59. $x + 14 = 19$

60. $x + 35 = 50$

61. $28 = x + 20$

62. $25 = x + 18$

63. $100 + x = 127$

64. $140 + x = 200$

Applications

65. ***Current Events*** In the 2008 presidential election, a total of 1,060,327 votes were cast in Arkansas for either Democrat Barack Obama or Republican John McCain. If McCain received 638,017 votes, how many did Obama receive?

66. ***Current Events*** In the 2008 presidential election, a total of 701,360 votes were cast in New Hampshire for either Democrat Barack Obama or Republican John McCain. If Obama received 384,826 votes, how may did McCain receive?

67. ***Population Trends*** In 2010, the population of Ireland was approximately 4,470,510. In the same year, the population of Portugal was approximately 10,707,924. How much less than the population of Portugal was the population of Ireland?

68. ***Geography*** The Nile River, the longest river in the world, is approximately 22,070,400 feet long. The Yangtze Kiang River, which is the longest river in China, is approximately 19,018,560 feet long. How much longer is the Nile River than the Yangtze Kiang River?

69. ***Personal Finance*** Michaela's gross pay on her last paycheck was $1280. Her deductions totaled $318 and she deposited $200 into her savings account. She put the remaining amount into her checking account to pay bills. How much did Michaela put into her checking account?

70. ***Personal Finance*** Adam earned $3450 last summer at his construction job. He owed his brother $375 and saved $2300 to pay for his college tuition. He used the remaining amount as a down payment for a car. How much did Adam have for the down payment?

Population Trends *In answering exercises 71–78, consider the following population table.*

	1970	1980	1990	2000	2010	2020*
Illinois	11,110,285	11,427,409	11,430,602	12,051,683	12,916,894	13,236,720
Michigan	8,881,826	9,262,044	9,295,297	9,679,052	10,428,683	10,695,993
Indiana	5,195,392	5,490,212	5,544,159	6,045,521	6,392,139	6,627,008
Minnesota	3,806,103	4,075,970	4,375,099	4,830,784	5,420,636	5,900,769

Source: U.S. Census Bureau
*estimated

71. How much did the population of Minnesota increase from 1970 to 2010?

72. How much did the population of Michigan increase from 1970 to 2010?

73. In 1970, how much greater was the population of Illinois than the populations of Indiana and Minnesota combined?

74. In 2000, how much greater was the population of Illinois than the populations of Indiana and Minnesota combined?

75. How much did the population of Illinois increase from 1970 to 1990?

76. How much did the population of Michigan increase from 1970 to 1990?

77. Compare your answers to exercises 75 and 76. How much greater was the population increase of Michigan than the population increase of Illinois from 1970 to 1990?

78. In 2020, what will be the difference in population between the state with the highest population and the state with the lowest population?

Real Estate *The number of real estate transfers in several towns during the years 2009 to 2011 is given in the following bar graph. Use the bar graph to answer exercises 79–86. The figures in the bar graph reflect sales of single-family detached homes only.*

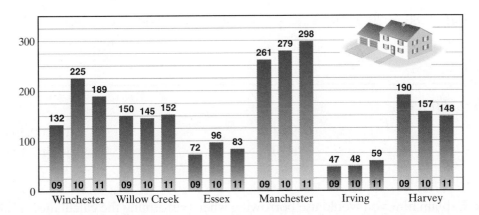

79. What was the increase in the number of homes sold in Winchester from 2009 to 2010?

80. What was the increase in the number of homes sold in Irving from 2010 to 2011?

81. What was the decrease in the number of homes sold in Essex from 2010 to 2011?

82. What was the decrease in the number of homes sold in Harvey from 2009 to 2010?

83. Between what two years did the greatest change occur in the number of homes sold in Willow Creek?

84. Between what two years did the greatest change occur in the number of homes sold in Manchester?

85. A real estate agent was trying to determine which two towns were closest to having the same number of sales in 2011. Which two towns should she select?

86. A real estate agent was trying to determine which two towns were closest to having the same number of sales in 2009. Which two towns should he select?

To Think About

87. In general, subtraction is not commutative. If a and b are whole numbers, $a - b \neq b - a$. For what types of numbers would it be true that $a - b = b - a$?

88. In general, subtraction is not associative. For example, $8 - (4 - 3) \neq (8 - 4) - 3$. In general, $a - (b - c) \neq (a - b) - c$. Can you find some numbers a, b, c for which $a - (b - c) = (a - b) - c$? (Remember, do operations inside the parentheses first.)

89. *Consumer Mathematics* Walter Swensen wants to replace some of the fences on a farm in Caribou, Maine. The wooden rail fence costs about $60 for wood and $50 for labor to install a fence that is 12 feet long. His son estimates he would need 276 feet of new fence. However, when he measures it he realizes he only needs 216 new feet of fence. What is the difference in cost of his son's estimate versus his estimate with regard to how many feet of fence are needed?

90. *Consumer Mathematics* Carlos Sontera is replacing an expensive barbed-wire fence on a ranch in El Paso, Texas. The barbed wire and poles for 12 feet of fence cost about $80. The labor cost to install 12 feet of fence is about $40. A ranch hand reports that 300 new feet of fence are needed. However, when Carlos actually rides out there and measures it, he finds that only 228 new feet of fence are needed. What is the difference in cost of the ranch hand's estimate versus Carlos's estimate of how many feet of fence are needed?

Cumulative Review

91. **[1.1.3]** Write in standard notation: eight million, four hundred sixty-six thousand, eighty-four

92. **[1.1.3]** Write a word name for 296,308.

93. **[1.2.4]** Add. $25 + 75 + 80 + 20 + 18$

94. **[1.2.4]** Add.
$$278{,}563$$
$$+\ 896{,}187$$

Quick Quiz 1.3 *Subtract.*

1.
$$5392$$
$$-\ \ 938$$

2.
$$609{,}240$$
$$-\ 386{,}307$$

3.
$$17{,}200{,}300$$
$$-\ 11{,}562{,}178$$

4. Concept Check Explain how you would use borrowing when performing the calculation $12{,}345 - 11{,}976$.

1.4 Multiplying Whole Numbers

① Mastering Basic Multiplication Facts

Like subtraction, multiplication is related to addition. Suppose that the pastry chef at the Gourmet Restaurant bakes croissants on a sheet that holds four croissants across, with room for three rows. How many croissants does the sheet hold?

We can add $4 + 4 + 4$ to get the total, or we can use a shortcut: three rows of four is the same as 3 times 4, which equals 12. This is **multiplication,** a shortcut for repeated addition.

The numbers that we multiply are called **factors.** The answer is called the **product.** For now, we will use $\times$ to show multiplication. 3×4 is read "three times four."

$$\underbrace{3}_{\text{factor}} \times \underbrace{4}_{\text{factor}} = \underbrace{12}_{\text{product}} \qquad \begin{array}{r} 3 \\ \times\, 4 \\ \hline 12 \end{array} \begin{array}{l} \text{factor} \\ \text{factor} \\ \text{product} \end{array}$$

Your skill in multiplication depends on how well you know the basic multiplication facts. Look at the table on page 36. You should learn these facts well enough to quickly and correctly give the products of any two factors in the table. To check your knowledge, try Exercises 1.4, exercises 3 and 4.

Study the table to see if you can discover any properties of multiplication. What do you see as results when you multiply zero by any number? When you multiply any number times zero, the result is zero. That is the **multiplication property of zero.**

$$2 \times 0 = 0 \qquad 5 \times 0 = 0 \qquad 0 \times 6 = 0 \qquad 0 \times 0 = 0$$

You may recall that zero plays a special role in addition. Zero is the *identity element* for addition. When we add any number to zero, that number does not change. Is there an identity element for multiplication? Look at the table. What is the identity element for multiplication? Do you see that it is 1? The **identity element for multiplication** is 1.

$$5 \times 1 = 5 \qquad 1 \times 5 = 5$$

What other properties of addition hold for multiplication? Is multiplication commutative? Does the order in which you multiply two numbers change the results? Find the product of 3×4. Then find the product of 4×3.

$$3 \times 4 = 12$$
$$4 \times 3 = 12$$

The **commutative property of multiplication** tells us that when we multiply two numbers, changing the order of the numbers gives the same result.

Student Learning Objectives

After studying this section, you will be able to:

① Master basic multiplication facts.

② Multiply a single-digit number by a several-digit number.

③ Multiply a whole number by a power of 10.

④ Multiply a several-digit number by a several-digit number.

⑤ Use the properties of multiplication to perform calculations.

⑥ Apply multiplication to real-life situations.

Basic Multiplication Facts

×	0	1	2	3	4	5	6	7	8	9	10	11	12
0	0	0	0	0	0	0	0	0	0	0	0	0	0
1	0	1	2	3	4	5	6	7	8	9	10	11	12
2	0	2	4	6	8	10	12	14	16	18	20	22	24
3	0	3	6	9	12	15	18	21	24	27	30	33	36
4	0	4	8	12	16	20	24	28	32	36	40	44	48
5	0	5	10	15	20	25	30	35	40	45	50	55	60
6	0	6	12	18	24	30	36	42	48	54	60	66	72
7	0	7	14	21	28	35	42	49	56	63	70	77	84
8	0	8	16	24	32	40	48	56	64	72	80	88	96
9	0	9	18	27	36	45	54	63	72	81	90	99	108
10	0	10	20	30	40	50	60	70	80	90	100	110	120
11	0	11	22	33	44	55	66	77	88	99	110	121	132
12	0	12	24	36	48	60	72	84	96	108	120	132	144

QUICK RECALL OF MULTIPLICATION FACTS It is helpful if you can multiply quickly. See if you can do Example 1 correctly in 15 seconds or less. Repeat again with Student Practice 1. Strive to obtain all answers correctly in 15 seconds or less.

EXAMPLE 1 Multiply.

(a) 5×7 **(b)** 8×9 **(c)** 6×8

(d) 9×3 **(e)** 7×8

Solution

(a)   5   **(b)**   8   **(c)**   6
     $\times$ 7 $\times$ 9 $\times$ 8
     ────   ────   ────
       35   72   48

(d)   9   **(e)**   7
     $\times$ 3 $\times$ 8
     ────   ────
       27   56

NOTE TO STUDENT: *Fully worked-out solutions to all of the Student Practice problems can be found at the back of the text starting at page SP-1.*

Student Practice 1 Multiply.

(a)   8   **(b)**   7   **(c)**   5
     $\times$ 8 $\times$ 6 $\times$ 8

(d)   9   **(e)**   9
     $\times$ 7 $\times$ 9

② Multiplying a Single-Digit Number by a Several-Digit Number

EXAMPLE 2   Multiply. 4312×2

Solution   We first multiply the ones column, then the tens column, and so on, moving right to left.

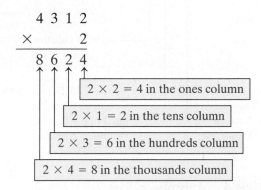

$$
\begin{array}{r}
4\ 3\ 1\ 2 \\
\times \qquad 2 \\
\hline
8\ 6\ 2\ 4
\end{array}
$$

$2 \times 2 = 4$ in the ones column

$2 \times 1 = 2$ in the tens column

$2 \times 3 = 6$ in the hundreds column

$2 \times 4 = 8$ in the thousands column

Student Practice 2   Multiply. 3021×3

Usually, we will have to carry one digit of the result of some of the multiplication into the next column on the left.

EXAMPLE 3   Multiply. 36×7

Solution

$$
\begin{array}{r}
{}^{4} \\
3\ 6 \\
\times \quad 7 \\
\hline
2\ 5\ 2
\end{array}
$$

carry number = 4

$7 \times 6 = 42$. Leave the 2 in the ones column and carry the 4 to the tens column.

7×3 tens = 21 tens. Add 21 tens + 4 tens to obtain 25 tens or 2 hundreds + 5 tens.

Student Practice 3   Multiply. 43×8

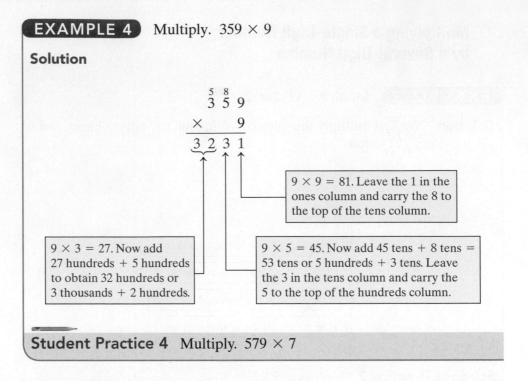

EXAMPLE 4 Multiply. 359×9

Solution

$$
\begin{array}{r}
\overset{5}{}\overset{8}{} \\
3\;5\;9 \\
\times\qquad 9 \\
\hline
3\;2\;3\;1
\end{array}
$$

$9 \times 9 = 81$. Leave the 1 in the ones column and carry the 8 to the top of the tens column.

$9 \times 3 = 27$. Now add 27 hundreds + 5 hundreds to obtain 32 hundreds or 3 thousands + 2 hundreds.

$9 \times 5 = 45$. Now add 45 tens + 8 tens = 53 tens or 5 hundreds + 3 tens. Leave the 3 in the tens column and carry the 5 to the top of the hundreds column.

Student Practice 4 Multiply. 579×7

③ Multiplying a Whole Number by a Power of 10

Observe what happens when a number is multiplied by 10, 100, 1000, 10,000, and so on.

$$56 \times 10 = 560 \quad \text{(one zero)}$$
$$56 \times 100 = 5600 \quad \text{(two zeros)}$$
$$56 \times 1000 = 56,000 \quad \text{(three zeros)}$$
$$56 \times 10,000 = 560,000 \quad \text{(four zeros)}$$

A **power of 10** is a whole number that begins with 1 and ends in one or more zeros. The numbers 10, 100, 1000, 10,000, and so on are powers of 10.

> To multiply a whole number by a power of 10:
>
> 1. Count the number of zeros in the power of 10.
>
> 2. Attach that number of zeros to the right side of the other whole number to obtain the answer.

EXAMPLE 5 Multiply 358 by each number.

(a) 10   **(b)** 100   **(c)** 1000   **(d)** 100,000

Solution

(a) $358 \times 10 = 3580$ (one zero)   **(b)** $358 \times 100 = 35{,}800$ (two zeros)

(c) $358 \times 1000 = 358{,}000$ (three zeros)

(d) $358 \times 100{,}000 = 35{,}800{,}000$ (five zeros)

Student Practice 5 Multiply 1267 by each number.

(a) 10   **(b)** 1000   **(c)** 10,000   **(d)** 1,000,000

How can we handle zeros in multiplication involving a number that is not 10, 100, 1000, or any other power of 10? Consider 32×400. We can rewrite 400 as 4×100, which gives us $32 \times 4 \times 100$. We can simply multiply 32×4 and then attach two zeros for the factor 100. We find that $32 \times 4 = 128$. Attaching two zeros gives us 12,800, or $32 \times 400 = 12,800$.

EXAMPLE 6 Multiply.

(a) 12×3000 **(b)** 25×600 **(c)** 430×260

Solution

(a) $12 \times 3000 = 12 \times 3 \times 1000 = 36 \times 1000 = 36,000$
(b) $25 \times 600 = 25 \times 6 \times 100 = 150 \times 100 = 15,000$
(c) $430 \times 260 = 43 \times 26 \times 10 \times 10 = 1118 \times 100 = 111,800$

Student Practice 6 Multiply.

(a) $9 \times 60,000$ **(b)** 15×400 **(c)** 270×800

④ **Multiplying a Several-Digit Number by a Several-Digit Number**

EXAMPLE 7 Multiply. 234×21

Solution   We can consider 21 as 2 tens (20) and 1 one (1). First we multiply 234 by 1.

We also multiply 234×20. This gives us two **partial products.**

$$
\begin{array}{r}
234 \\
\times\ 1 \\
\hline
234
\end{array}
\qquad
\begin{array}{r}
234 \\
\times\ 20 \\
\hline
4680
\end{array}
$$

Now we combine these two operations together by adding the two partial products to reach the final product, which is the solution.

$$
\begin{array}{r}
2\ 3\ 4 \\
\times\ \ 2\ 1 \\
\hline
2\ 3\ 4 \quad \longleftarrow \text{Multiply } 234 \times 1. \\
4\ 6\ 8\ 0 \quad \longleftarrow \text{Multiply } 234 \times 20. \\
\hline
4\ 9\ 1\ 4 \quad \longleftarrow \text{Add the two partial products.}
\end{array}
$$

Student Practice 7 Multiply. 323×32

EXAMPLE 8   Multiply. 671×35

Solution

```
      6 7 1
    ×   3 5
      3 3 5 5  ←——— First multiply 671 × 5.
    2 0 1 3 0  ←——— Now multiply 671 × 30.
    2 3 4 8 5  ←——— Now add the two partial products.
```

Note: We could omit zero on this line and leave the ones place blank.

Student Practice 8   Multiply. 385×69

EXAMPLE 9   Multiply. 14×20

Solution

```
          1 4
        × 2 0
            0  ←——— Multiply 14 by 0.
        2 8 0  ←——— Multiply 14 by 2 tens.
Now add the ——→ 2 8 0
partial products.
```

Place 28 with the 8 in the tens column. To line up the digits for adding, we can insert a 0 in the ones column.

Notice that you will also get this result if you multiply $14 \times 2 = 28$ and then attach a zero to multiply it by 10: 280.

Student Practice 9   Multiply. 34×20

EXAMPLE 10   Multiply. 120×40

Solution

```
          1 2 0
        ×   4 0
              0  ←——— Multiply 120 × 0.
        4 8 0 0  ←——— Multiply 120 by 4 tens.
Now add the ——→ 4 8 0 0
partial products.
```

The answer is 480 tens. We place the 0 of the 480 in the tens column. To line up the digits for adding, we can insert a 0 in the ones column.

Notice that this result is the same as $12 \times 4 = 48$ with two zeros attached: 4800.

Student Practice 10   Multiply. 130×50

Mᴄ **EXAMPLE 11**   Multiply. 684×763

Solution

$$
\begin{array}{r}
6\ 8\ 4 \\
\times 7\ 6\ 3 \\
\hline
2\ 0\ 5\ 2 \\
4\ 1\ 0\ 4 \\
4\ 7\ 8\ 8 \\
\hline
5\ 2\ 1\ 8\ 9\ 2
\end{array}
$$

← Multiply 684×3.

← Multiply 684×60. Note that we omit the final zero.

← Multiply 684×700. Note that we omit the final two zeros.

Student Practice 11

Multiply. 923×675

---

⑤ **Using the Properties of Multiplication to Perform Calculations**

When we add three numbers, we use the associative property. Recall that the associative property allows us to group the three numbers in different ways. Thus to add $9 + 7 + 3$, we can group the numbers as $9 + (7 + 3)$ because it is easier to find the sum. $9 + (7 + 3) = 9 + 10 = 19$. We can demonstrate that multiplication is also associative.

Is this true? $2 \times (5 \times 3) = (2 \times 5) \times 3$
$$2 \times (15) = (10) \times 3$$
$$30 = 30$$

The final product is the same in both cases.

The way we group numbers to be multiplied does not change the product. This property is called the **associative property of multiplication.**

EXAMPLE 12   Multiply. $14 \times 2 \times 5$

Solution Since we can group any two numbers together, let's take advantage of the ease of multiplying by 10.

$$14 \times 2 \times 5 = 14 \times (2 \times 5) = 14 \times 10 = 140$$

Student Practice 12   Multiply. $25 \times 4 \times 17$

For convenience, we list the properties of multiplication that we have discussed in this section.

1. **Associative Property of Multiplication.** When we multiply three numbers, the multiplication can be grouped in any way.	$(7 \times 3) \times 2 = 7 \times (3 \times 2)$ $21 \times 2 = 7 \times 6$ $42 = 42$

2. **Commutative Property of Multiplication.** Two numbers can be multiplied in either order with the same result.

$$9 \times 8 = 8 \times 9$$
$$72 = 72$$

3. **Identity Property of One.** When one is multiplied by a number, the result is that number.

$$7 \times 1 = 7$$
$$1 \times 15 = 15$$

4. **Multiplication Property of Zero.** The product of any number and zero yields zero as a result.

$$0 \times 14 = 0$$
$$2 \times 0 = 0$$

Sometimes you can use several properties in one problem to make the calculation easier.

EXAMPLE 13   Multiply. $7 \times 20 \times 5 \times 6$

Solution $7 \times 20 \times 5 \times 6 = 7 \times (20 \times 5) \times 6$ Associative property
$$= 7 \times 6 \times (20 \times 5) \quad \text{Commutative property}$$
$$= 42 \times 100$$
$$= 4200$$

Student Practice 13   Multiply. $8 \times 4 \times 3 \times 25$

Thus far we have discussed the properties of addition and the properties of multiplication. There is one more property that links both operations.

Before we discuss that property, we will illustrate several different ways of showing multiplication. The following are all the ways to show "3 times 4."

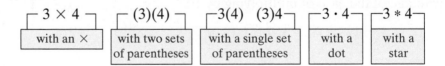

3×4	$(3)(4)$	$3(4)$ $(3)4$	$3 \cdot 4$	$3 * 4$
with an $\times$	with two sets of parentheses	with a single set of parentheses	with a dot	with a star

We will use parentheses to mean multiplication when we use the **distributive property.**

SIDELIGHT: The Distributive Property
Why does our method of multiplying several-digit numbers work? Why can we say that 234×21 is the same as $234 \times 1 + 234 \times 20$?

The *distributive property of multiplication over addition* allows us to distribute the multiplication and then add the results. To illustrate, 234×21 can be written as $234(20 + 1)$. By the distributive property

$$234(20 + 1) = (234 \times 20) + (234 \times 1)$$
$$= \quad 4680 \quad + \quad 234$$
$$= \quad 4914$$

This is what we actually do when we multiply.

$$
\begin{array}{r}
234 \\
\times\ \ 21 \\
\hline
234 \\
4680 \\
\hline
4914
\end{array}
$$

DISTRIBUTIVE PROPERTY OF MULTIPLICATION OVER ADDITION

Multiplication can be distributed over addition without changing the result.

$$5 \times (10 + 2) = (5 \times 10) + (5 \times 2)$$

---

⑥ Applying Multiplication to Real-Life Situations

To use multiplication in word problems, the number of items or the value of each item must be the same. Recall that multiplication is a quick way to do repeated addition where *each addend is the same.* In the beginning of the section we showed three rows of four croissants to illustrate 3×4. The number of croissants in each row was the same, 4. Look at another example. If we had six nickels, we could use multiplication to find the total value of the coins because the value of each nickel is the same, 5¢. Since $6 \times 5 = 30$, six nickels are worth 30¢.

In the following example the word *average* is used. The word *average* has several different meanings. In this example, we are told that the *average annual salary* of an employee at Software Associates is $42,132. This means that we can calculate the total payroll as if each employee made $42,132 even though we know that the president probably makes more than any other employee.

EXAMPLE 14   The average annual salary of an employee at Software Associates is $42,132. There are 38 employees. What is the annual payroll?

Solution

$$
\begin{array}{r}
\$42{,}132 \\
\times\ \ \ \ \ \ \ \ 38 \\
\hline
337\ 056 \\
1\ 263\ 96 \\
\hline
1{,}601{,}016
\end{array}
$$

The total annual payroll is $1,601,016.

Student Practice 14   The average cost of a new car sold last year at Westover Chevrolet was $17,348. The dealership sold 378 cars. What were the total sales of cars at the dealership last year?

Another useful application of multiplication is area. The following example involves the area of a rectangle.

▲ **EXAMPLE 15** What is the area of a rectangular hallway that measures 4 feet by 8 feet?

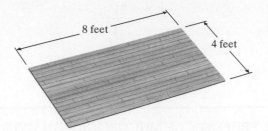

Solution The **area** of a rectangle is the product of the length times the width. Thus for this hallway

$$\text{Area} = 8 \text{ feet} \times 4 \text{ feet} = 32 \text{ square feet.}$$

The area of the hallway is 32 square feet.

Note: All measurements for area are given in square units such as square feet, square meters, square yards, and so on.

▲ **Student Practice 15** What is the area of a rectangular rug that measures 5 yards by 7 yards?

 STEPS TO SUCCESS Doing Homework for Each Class Is Critical.

Usually every student in the course has to ask the question, "Is homework really that important? Do I actually have to do it?"

You learn by doing. It really makes a difference. Mathematics involves mastering a set of skills that you learn by practicing, not by watching someone else do it. Your instructor may make solving a mathematics problem look very easy, but for you to learn the necessary skills, you must practice them over and over.

The key to success is practice. Learning mathematics is like learning to play a musical instrument, to type, or to play a sport. No matter how much you watch someone else do mathematical calculations, no matter how many books you read on "how to" do it, no matter how easy it appears to be, the key to success in mathematics is practice on each homework set.

Do each kind of problem. Some exercises in a homework set are more difficult than others. Some stress different concepts. Usually you need to work at least all the odd-numbered problems in the exercise set. This allows you to cover the full range of skills in the problem set. Remember, the more exercises you do, the better you will become in your mathematical skills.

Making it personal: Write down your personal reason for why you think doing the homework in each section is very important for success. Which of the three points made do you find is the most convincing? ▼

1.4 Exercises

MyMathLab®

Watch the videos
in MyMathLab

Download the
MyDashBoard App

Verbal and Writing Skills, Exercises 1 and 2

1. Explain in your own words.

   (a) the commutative property of multiplication

   (b) the associative property of multiplication

2. How does the distributive property of multiplication over addition help us to multiply 4×13?

Complete the multiplication facts for each table. Strive for total accuracy, but work quickly. (Allow a maximum of six minutes for each table.)

3.

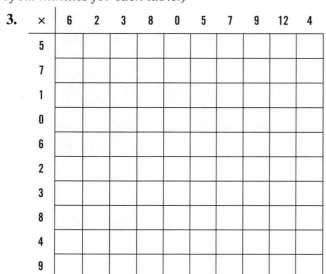

×	6	2	3	8	0	5	7	9	12	4
5										
7										
1										
0										
6										
2										
3										
8										
4										
9										

4.

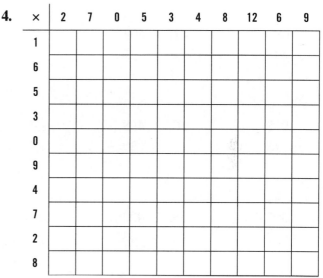

×	2	7	0	5	3	4	8	12	6	9
1										
6										
5										
3										
0										
9										
4										
7										
2										
8										

Multiply.

5.   32
  × 3

6.   21
  × 4

7.   14
  × 5

8.   15
  × 6

9.   87
  × 6

10.   95
  × 7

11.   231
  × 3

12.   313
  × 3

13.   276
  × 7

14.   538
  × 8

15.   6102
  × 3

16.   5203
  × 2

17.   12,203
  × 3

18.   31,206
  × 3

19.   5218
  × 6

20.   3215
  × 6

21.   12,526
  × 8

22.   48,761
  × 7

23.   344,601
  × 9

24.   257,021
  × 9

Multiply by powers of 10.

25.   156
  × 10

26.   278
  × 10

27.   27,158
  × 100

28.   89,361
  × 100

29.   482
  × 1000

30.   579
  × 1000

31.   37,256
  × 10,000

32.   614,260
  × 10,000

Multiply by multiples of 10.

33. 423
    × 20

34. 332
    × 30

35. 2210
    × 40

36. 4230
    × 20

37. 14,000
    × 4000

38. 62,000
    × 3000

Multiply.

39. 514
    × 12

40. 432
    × 13

41. 146
    × 54

42. 163
    × 35

43. 89
    × 64

44. 68
    × 49

45. 607
    × 25

46. 780
    × 24

47. 544
    × 38

48. 652
    × 92

49. 912
    × 76

50. 498
    × 39

51. 5123
    × 29

52. 1268
    × 38

53. 9053
    × 91

54. 3078
    × 72

55. 4326
    × 435

56. 3725
    × 546

57. 458
    × 314

58. 629
    × 512

Mixed Practice

59. 2076
    × 105

60. 5092
    × 302

61. 1324
    × 2004

62. 2074
    × 1003

63. 12,000
    × 60

64. 15,200
    × 30

65. 250
    × 40

66. 302
    × 30

67. 302
    × 300

68. 3000
    × 302

69. $7 \cdot 2 \cdot 5$

70. $8 \cdot 3 \cdot 2$

71. $11 \cdot 7 \cdot 4$

72. $16 \cdot 5 \cdot 6$

73. 412×33

74. 526×21

75. $5 \cdot 8 \cdot 4 \cdot 10$

76. $4 \cdot 5 \cdot 5 \cdot 15$

77. What is x if
$x = 8 \cdot 7 \cdot 6 \cdot 0$?

78. What is x if
$x = 3 \cdot 12 \cdot 0 \cdot 5$?

Applications

▲ **79.** *Geometry* Find the area of a patio that is 16 feet wide and 24 feet long.

▲ **80.** *Geometry* Find the area of the Apple iPhone that has a height of 115 millimeters and a width of 59 millimeters.

▲ **81.** *Consumer Mathematics* Ken Thompson and his wife want to put down new carpet in the living room and the hallway of their house on Cape Cod. The living room measures 12 feet by 14 feet. The hallway measures 9 feet by 3 feet. If the living room and the hallway are rectangular in shape, how many square feet of new carpet do Don and his wife need?

▲ **82.** *Wildlife Management* Robert Tobey in Copper Center, Alaska, wants to put a field under helicopter surveillance because of a roving pack of wolves that is killing other wildlife in the area. The field consists of two rectangular regions. The first one is 4 miles by 5 miles. The second one is 12 miles by 8 miles. How many square miles does he want to place under surveillance?

83. *Business Decisions* The Student Commons Coffee Shop needs to purchase espresso coffee. Find the cost of purchasing 240 pounds of espresso coffee beans at $5 per pound.

84. *Business Decisions* The music department of Wheaton College wishes to purchase 345 sets of headphones at the music supply store at a cost of $8 each. What will be the total amount of the purchase?

85. *Personal Finance* Helen pays $266 per month for her car payment on her new Honda Civic. What is her automobile payment cost for a one-year period?

86. *Personal Finance* A company rents a Ford Escort for a salesman at $276 per month for eight months. What is the cost for the car rental during this time?

87. *Environmental Studies* Marcos has a Toyota Corolla that gets 34 miles per gallon during highway driving. Approximately how far can he travel if he has 18 gallons of gas in the tank?

88. *Environmental Studies* Cheryl has a subcompact car that gets 48 miles per gallon on highway driving. Approximately how far can she travel if she has 12 gallons of gas in the tank?

89. *Personal Finance* Sylvia worked as a camp counselor for 12 weeks during the summer. She earned $420 per week. What is the total amount Sylvia earned during the summer?

90. *Personal Finance* Each time Jorge receives a paycheck, $125 is put into his IRA retirement savings account. If he gets paid twice a month, how much does Jorge contribute to his IRA in one year?

91. *International Relations* In 2010, the country of Morocco had an average per capita (per person) income of $1060. If the approximate population of Morocco was 32,000,000, what was the approximate total yearly income of the entire country?

92. *International Relations* In 2010, the country of Austria had an approximate population of 8,200,000. The average per capita (per person) income was $24,000. What was the approximate total yearly income of the entire country?

To Think About *Use the following information to answer exercises 93–96. There are 98 puppies in a room, with an assortment of black and white ears and paws. 18 puppies have totally black ears and 2 white paws; 26 puppies have 1 black ear and 4 white paws; and 54 puppies have no black ears and 1 white paw.*

93. How many black paws are in the room?

94. How many white paws are in the room?

95. How many black ears are in the room?

96. How many white ears are in the room?

In exercises 97–100, find the value of x in each equation.

97. $5(x) = 40$

98. $7(x) = 56$

99. $72 = 8(x)$

100. $63 = 9(x)$

101. Would the distributive property of multiplication be true for Roman numerals such as (XII) × (IV)? Why or why not?

102. We saw that multiplication is distributive over addition. Is it distributive over subtraction? Why or why not? Give examples.

Cumulative Review

103. **[1.3.3]** Subtract.

$$\begin{array}{r} 34{,}084 \\ -\ 27{,}328 \\ \hline \end{array}$$

104. **[1.2.4]** Add.

$$\begin{array}{r} 263 \\ 27 \\ 891 \\ 5 \\ +\ 63 \\ \hline \end{array}$$

105. **[1.3.5]** *Personal Finance* Adam Goulet has $1278 in his checking account. After writing checks for $345 and $128, how much is left in his account?

106. **[1.3.5]** *Personal Finance* Katrina Feldman was earning $1772 each month at her job. She recently received a raise and now her paychecks are $1932. What was the increase?

107. **[1.3.5]** *Population Studies* The population of Richfield in 2000 was 42,667. In 2010, the population had increased to 45,918. By how many people did the net population increase?

108. **[1.3.5]** *International Relations* In 2000, the gross domestic product of Spain was $720,800,000,000. In 2008, it had grown to $1,466,000,000,000. How much did the gross domestic product increase from 2000 to 2008?

Quick Quiz 1.4 *Multiply.*

1.
$$\begin{array}{r} 34{,}986 \\ \times\quad\ 5 \\ \hline \end{array}$$

2.
$$\begin{array}{r} 79 \\ \times\ 64 \\ \hline \end{array}$$

3.
$$\begin{array}{r} 698 \\ \times\ 297 \\ \hline \end{array}$$

4. Concept Check Explain what you do with the zeros when you multiply 3457 × 2008.

1.5 Dividing Whole Numbers

① Mastering Basic Division Facts

Suppose that we have eight quarters and want to divide them into two equal piles. We would discover that each pile contains four quarters.

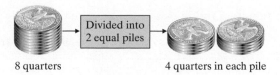

8 quarters   4 quarters in each pile

In mathematics we would express this thought by saying that

$$8 \div 2 = 4.$$

We know that this answer is right because two piles of four quarters is the same dollar amount as eight quarters. In other words, we know that $8 \div 2 = 4$ because $2 \times 4 = 8$. These two mathematical sentences are called **related sentences.** The division sentence $8 \div 2 = 4$ is related to the multiplication sentence $2 \times 4 = 8$.

In fact, in mathematics we usually define **division** in terms of multiplication. The answer to the division problem $12 \div 3$ is that number which when multiplied by 3 yields 12. Thus

$$12 \div 3 = 4 \quad \text{because } 3 \times 4 = 12.$$

Suppose that a surplus of $30 in the French Club budget at the end of the year is to be equally divided among the five club members. We would want to divide the $30 into five equal parts. We would write $30 \div 5 = 6$ because $5 \times 6 = 30$. Thus each of the five people would get $6 in this situation.

$30   5 piles of $6 each

As a mathematical sentence, $30 \div 5 = 6$.

The division problem $30 \div 5 = 6$ could also be written $\frac{30}{5} = 6$ or $5\overline{)30}$.

When referring to division, we sometimes use the words **divisor, dividend,** and **quotient** to identify the three parts.

$$\text{divisor} \overline{)\text{dividend}}^{\text{quotient}}$$

With $30 \div 5 = 6$, 30 is the dividend, 5 is the divisor, and 6 is the quotient.

$$\text{divisor} \rightarrow 5\overline{)30}^{\,6 \; \leftarrow \text{quotient}} \; \leftarrow \text{dividend}$$

So the quotient is the answer to a division problem. It is important that you be able to do short problems involving basic division facts quickly.

Student Learning Objectives

After studying this section, you will be able to:

① Master basic division facts.

② Perform division by a one-digit number.

③ Perform division by a two- or three-digit number.

④ Apply division to real-life situations.

EXAMPLE 1 Divide.

(a) $12 \div 4$ **(b)** $81 \div 9$ **(c)** $56 \div 8$ **(d)** $54 \div 6$

Solution

(a) $4\overline{)12}$ with 3

(b) $9\overline{)81}$ with 9

(c) $8\overline{)56}$ with 7

(d) $6\overline{)54}$ with 9

Student Practice 1 Divide.

(a) $36 \div 4$ **(b)** $25 \div 5$ **(c)** $72 \div 9$ **(d)** $30 \div 6$

NOTE TO STUDENT: Fully worked-out solutions to all of the Student Practice problems can be found at the back of the text starting at page SP-1.

Zero can be divided by any nonzero number, but division by zero is not possible. Why is this?

Suppose that we could divide by zero. Then $7 \div 0 =$ some number. Let us represent "some number" by the letter a.

$$\text{If } 7 \div 0 = a, \text{ then } 7 = 0 \times a,$$

because every division problem has a related multiplication problem. But zero times any number is zero, $0 \times a = 0$. Thus

$$7 = 0 \times a = 0.$$

That is, $7 = 0$, which we know is not true. Therefore, our assumption that $7 \div 0 = a$ is wrong. Thus we conclude that we cannot divide by zero. Mathematicians state this by saying, "Division by zero is **undefined**."

It is helpful to remember the following basic concepts:

DIVISION PROBLEMS INVOLVING THE NUMBER 1 AND THE NUMBER 0

1. Any nonzero number divided by itself is 1 ($7 \div 7 = 1$).

2. Any number divided by 1 remains unchanged ($29 \div 1 = 29$).

3. Zero may be divided by any nonzero number; the result is always zero ($0 \div 4 = 0$).

4. Zero can never be the divisor in a division problem ($3 \div 0$ is undefined).

EXAMPLE 2 Divide, if possible. If it is not possible, state why.

(a) $8 \div 8$ **(b)** $9 \div 1$ **(c)** $0 \div 6$ **(d)** $20 \div 0$

Solution

(a) $\dfrac{8}{8} = 1$ Any number divided by itself is 1.

(b) $\dfrac{9}{1} = 9$ Any number divided by 1 remains unchanged.

(c) $\dfrac{0}{6} = 0$ Zero divided by any nonzero number is zero.

(d) $\dfrac{20}{0}$ cannot be done Division by zero is undefined.

Student Practice 2   Divide, if possible.

(a) $7 \div 1$ **(b)** $\dfrac{9}{9}$ **(c)** $\dfrac{0}{5}$ **(d)** $12 \div 0$

② Performing Division by a One-Digit Number

Our accuracy with division is improved if we have a checking procedure. For each division fact, there is a related multiplication fact.

If $20 \div 4 = 5$,   then $20 = 4 \times 5$.
If $36 \div 9 = 4$,   then $36 = 9 \times 4$.

We will often use multiplication to check our answers.

When two numbers do not divide exactly, a number called the **remainder** is left over. For example, 13 cannot be divided exactly by 2. The number 1 is left over. We call this 1 the *remainder*.

$$\begin{array}{r} 6 \\ 2\overline{)13} \\ \underline{12} \\ 1 \end{array} \leftarrow \text{remainder}$$

Thus $13 \div 2 = 6$ with a remainder of 1. We can abbreviate this answer as

6 R 1.

To check this division, we multiply $2 \times 6 = 12$ and add the remainder: $12 + 1 = 13$. That is, $(2 \times 6) + 1 = 13$. The result will be the dividend if the division was done correctly. The following box shows you how to check a division that has a remainder.

(divisor $\times$ quotient) + remainder = dividend

EXAMPLE 3   Divide and check your answer. $33 \div 4$

Solution $\begin{array}{r} 8 \\ 4\overline{)33} \end{array}$ $\rightarrow$ How many times can 4 be divided into 33?   8.

$\begin{array}{r} 32 \\ \hline 1 \end{array}$ $\leftarrow$ What is 8×4?   32.
$\leftarrow$ 32 subtracted from 33 is 1.

The answer is 8 with a remainder of 1. We abbreviate this as 8 R 1.

CHECK.
$$\begin{array}{r} 8 \\ \times\ 4 \\ \hline 32 \\ +\ 1 \\ \hline 33 \end{array}$$

Multiply. $8 \times 4 = 32$.
Add the remainder. $32 + 1 = 33$.
Because the dividend is 33, the answer is correct.

Student Practice 3   Divide and check your answer. $45 \div 6$

EXAMPLE 4   Divide and check your answer. 158 ÷ 5

Solution
$$\begin{array}{r} 31 \\ 5\overline{)158} \end{array}$$ 5 divided into 15?   3.

$\underline{15}$ ← What is 3 × 5?   15.

08 ← 15 subtract 15?   0.   Bring down 8.

$\underline{5}$ ← 5 divided into 8?   1.   What is 1 × 5?   5.

3 ← 8 subtract 5?   3.

The answer is 31 R 3.

CHECK.
$$\begin{array}{r} 31 \\ \times\ 5 \\ \hline 155 \end{array}$$ Multiply. 31 × 5 = 155.

$$\begin{array}{r} +\ \ \ 3 \\ \hline 158 \end{array}$$ Add the remainder 3.

Because the dividend is 158, the answer is correct.

Student Practice 4   Divide and check your answer. 129 ÷ 6

EXAMPLE 5   Divide. 3672 ÷ 7

Solution
$$\begin{array}{r} 524 \\ 7\overline{)3672} \end{array}$$ How many times can 7 be divided into 36?   5.

$\underline{35}$ ← What is 5 × 7?   35.

17 ← 36 subtract 35?   1.   Bring down 7.

$\underline{14}$ ← 7 divided into 17?   2.   What is 2 × 7?   14.

32 ← 17 subtract 14?   3.   Bring down 2.

$\underline{28}$ ← 7 divided into 32?   4.   What is 4 × 7?   28.

4 ← 32 subtract 28?   4.

The answer is 524 R 4.

Student Practice 5   Divide. 4237 ÷ 8

③ Performing Division by a Two- or Three-Digit Number

When the divisor has more than one digit, an estimation technique may help. Figure how many times the first digit of the divisor goes into the first two digits of the dividend. Try this answer as the first number in the quotient.

EXAMPLE 6   Divide. 283 ÷ 41

Solution

First guess:
$$\begin{array}{r} 7 \\ 41\overline{)283} \\ 287 \end{array}$$ too large

How many times can the first digit of the divisor (4) be divided into the first two digits of the dividend (28)? 7. We try the answer 7 as the first number of the quotient. We multiply 7 × 41 = 287. We see that 287 is larger than 283.

Second guess:
$$\begin{array}{r} 6 \\ 41\overline{)283} \\ \underline{246} \\ 37 \end{array}$$

Because 7 is slightly too large, we try 6.

$246 \leftarrow$ 6×41?   246.

$37 \leftarrow$ 283 subtract 246?   37.

The answer is 6 R 37. (Note that the remainder must always be less than the divisor.)

Student Practice 6 Divide. $243 \div 32$

EXAMPLE 7 Divide. $33{,}897 \div 56$

Solution

First guess:
$$\begin{array}{r} 60 \\ 56\overline{)33897} \\ \underline{336} \\ 29 \end{array}$$

How many times can 33 be divided by 5?   6.

$336 \leftarrow$ What is 6×56?   336.

338 subtract 336?   2.   Bring down 9.

56 cannot be divided into 29.   Write 0 in quotient.

Second set of steps:
$$\begin{array}{r} 605 \\ 56\overline{)33897} \\ \underline{336} \\ 297 \\ \underline{280} \\ 17 \end{array}$$

Bring down 7.

How many times can 5 be divided into 29?   5.

$280 \leftarrow$ What is 5×56?   280.   Subtract $297 - 280$.

Remainder is 17.

The answer is 605 R 17.

Student Practice 7 Divide. $42{,}183 \div 33$

EXAMPLE 8 Divide. $5629 \div 134$

Solution
$$\begin{array}{r} 42 \\ 134\overline{)5629} \\ \underline{536} \\ 269 \\ \underline{268} \\ 1 \end{array}$$

How many times does 134 divide into 562?
We guess by saying that 1 divides into 5 five times, but this is too large. ($5 \times 134 = 670$!)
So we try 4. What is 4×134?   536.
Subtract $562 - 536$. We obtain 26. Bring down 9.
How many times does 134 divide into 269?
We guess by saying that 1 divided into 2 goes 2 times.
What is 2×134?   268. Subtract $269 - 268$.
The remainder is 1.

The answer is 42 R 1.

Student Practice 8 Divide. $3227 \div 128$

④ Applying Division to Real-Life Situations

When you solve a word problem that requires division, you will be given the total number and asked to calculate the number of items in each group or to calculate the number of groups. In the beginning of this section we showed eight quarters (the total number) and we divided them into two equal piles (the number of groups). Division was used to find how many quarters were in each pile (the number in each group). That is, $8 \div 2 = 4$. There were four quarters in each pile.

Let's look at another example. Suppose that $30 is to be divided equally among the members of a group. If each person receives $6, how many people are in the group? We use division, $30 \div 6 = 5$, to find that there are five people in the group.

You will find many real-world examples where you know the total cost of several identical items, and you need to find the cost per item. You will encounter this situation in the following example.

EXAMPLE 9 City Service Realty just purchased nine identical computers for the real estate agents in the office. The total cost for the nine computers was $13,932. What was the cost of one computer? Check your answer.

Solution To find the cost of one computer, we need to divide the total cost by 9. Thus we will calculate $13,932 \div 9$.

$$
\begin{array}{r}
1548 \\
9\overline{)13932} \\
\underline{9} \\
49 \\
\underline{45} \\
43 \\
\underline{36} \\
72 \\
\underline{72} \\
0
\end{array}
$$

Therefore, the cost of one computer was $1548. In order to check our work we will need to see if nine computers each costing $1548 will in fact result in a total of $13,932. We use multiplication to check division.

$$
\begin{array}{r}
1548 \\
\times \quad 9 \\
\hline
13932 \quad \checkmark
\end{array}
$$

We did obtain 13,932. Our answer is correct.

Student Practice 9 The Dallas police department purchased seven identical used police cars at a total cost of $117,964. Find the cost of one used car. Check your answer.

In the following example you will see the word *average* used as it applies to division. The problem states that a car traveled 1144 miles in 22 hours. The problem asks you to find the average speed in miles per hour. This means that we will treat the problem as if the speed of the car were the same during each hour of the trip. We will use division to solve.

EXAMPLE 10   A car traveled from California to Texas, a distance of 1144 miles, in 22 hours. What was the average speed in miles per hour?

Solution   When doing distance problems, it is helpful to remember that distance ÷ time = rate. We need to divide 1144 miles by 22 hours to obtain the rate or speed in miles per hour.

$$
\begin{array}{r}
52 \\
22\overline{)1144} \\
\underline{110} \\
44 \\
\underline{44} \\
0
\end{array}
$$

The car traveled an average of 52 miles per hour.

Student Practice 10   An airplane traveled 5138 miles in 14 hours. What was the average speed in miles per hour?

👣 STEPS TO SUCCESS Taking Good Notes Each Class Session

Don't copy down everything the teacher says. You will get overloaded with facts. Instead write down the important ideas and examples as the instructor lectures. Be sure to include any helpful hints or suggestions that your instructor gives you. You will be amazed at how easily you forget these if you do not write them down.

Be an active listener. Keep your mind on what the instructor is saying. Be ready with questions whenever you do not understand something. Stay alert in class. Keep your mind on mathematics.

Preview the lesson material. Before class glance over the topics that will be covered. You can take much better notes if you know what the topics of the lecture will be.

Look back at your notes. Try to review them the same day sometime after class. You will find the content of your notes much easier to understand if you read them again within a few hours of class.

Making it personal: Which of these suggestions are the most helpful to you? How can you improve your skills in note taking? ▼

1.5 Exercises   MyMathLab®

Verbal and Writing Skills, Exercise 1

1. Explain in your own words what happens when you
   (a) divide a nonzero number by itself.   (b) divide a number by 1.

   (c) divide zero by a nonzero number.   (d) divide a nonzero number by 0.

Divide. See if you can work exercises 2–30 in three minutes or less.

2. $5\overline{)35}$	3. $6\overline{)42}$	4. $4\overline{)32}$	5. $8\overline{)24}$	6. $9\overline{)27}$	7. $5\overline{)25}$
8. $7\overline{)49}$	9. $9\overline{)36}$	10. $4\overline{)16}$	11. $7\overline{)21}$	12. $9\overline{)81}$	13. $5\overline{)30}$
14. $6\overline{)54}$	15. $7\overline{)63}$	16. $4\overline{)28}$	17. $8\overline{)72}$	18. $8\overline{)64}$	19. $6\overline{)36}$
20. $9\overline{)72}$	21. $1\overline{)9}$	22. $1\overline{)8}$	23. $10\overline{)0}$	24. $7\overline{)0}$	25. $9 \div 0$
26. $12 \div 0$	27. $\dfrac{0}{8}$	28. $\dfrac{0}{7}$	29. $6 \div 6$	30. $5 \div 5$	

Divide. In exercises 31–42, check your answer.

31. $29 \div 6$	32. $42 \div 8$	33. $76 \div 8$	34. $75 \div 9$	35. $128 \div 5$
36. $6\overline{)103}$	37. $9\overline{)196}$	38. $8\overline{)427}$	39. $9\overline{)288}$	40. $7\overline{)294}$
41. $5\overline{)185}$	42. $8\overline{)224}$	43. $4\overline{)1289}$	44. $3\overline{)758}$	45. $6\overline{)763}$

46. 7)403   **47.** 7)6083   **48.** 9)6417   **49.** 4)4954   **50.** 5)7139

51. 8)16,450   **52.** 6)18,127   **53.** 5)12,813   **54.** 8)32,223   **55.** 185 ÷ 6

56. 202 ÷ 5   **57.** 267 ÷ 52   **58.** 315 ÷ 45   **59.** 544 ÷ 68

Mixed Practice, Exercises 60–74

60. 72)432   **61.** 12)5024   **62.** 13)6810   **63.** 30)1452   **64.** 40)1125

65. 7)5915   **66.** 8)6144   **67.** 36)7568   **68.** 32)3527   **69.** 182)2550

70. 19)1982   **71.** 174)700   **72.** 128)896   **73.** 224)28,000   **74.** 235)31,490

Solve.

75. $518 \div 14 = x$. What is the value of x?

76. $1572 \div 131 = x$. What is the value of x?

Applications

77. *Sports* A *run* in skiing is going from the top of the ski lift to the bottom. If over seven days, 431,851 runs were made, what was the average number of ski runs per day?

78. *Farming* Western Saddle Stable uses 36,500 pounds of feed per year to feed its 25 horses. How much does each horse eat per year?

79. *Sports* Coach Deno Johnson purchased 9 pairs of cross-country skis for his team. He spent a total of $2592. How much did each pair of skis cost?

80. *Business Finances* A state's department of transportation spent $8,400,000 on 70 new snowplows. How much did each snowplow cost?

81. *Business Finances* A horse and carriage company in New York City bought seven new carriages at exactly the same price each. The total bill was $147,371. How much did each carriage cost?

82. *Real Estate* A group of eight friends invested the same amount each in a beach property that sold for $369,432. How much did each friend pay?

83. *Business Finances* Bridgewater Community College spent $10,360 to equip their math tutoring lab with 56 new flat-panel monitors. How much did each monitor cost?

84. *Business Finances* Metropolitan College spent $13,020 on new bookcases for their faculty offices. If 70 faculty members received new bookcases, how much did each bookcase cost?

85. *Business Planning* The 2nd Avenue Delicatessen is making bagel sandwiches for a New York City Marathon party. The sandwich maker has 360 bagel halves, 340 slices of turkey, and 330 slices of Swiss cheese. If he needs to make sandwiches each consisting of two bagel halves, two slices of turkey, and two slices of Swiss cheese, what is the greatest number of sandwiches he can make?

▲ **86.** *Geometry* Ace Landscaping is mowing a rectangular lawn that has an area of 2652 square feet. The company keeps a record of all lawn mowed in terms of length, width, square feet, and number of minutes it takes to mow the lawn. The width of the lawn is 34 feet. However, the page that lists the length of the lawn is soiled and the number cannot be read. Determine the length of the lawn.

87. *Business Management* Dick Wightman is managing a company that is manufacturing and shipping modular homes in Canada. He has a truck that has made the trip from Toronto, Ontario, to Halifax, Nova Scotia, 12 times and has made the return run from Halifax to Toronto 12 times. The distance from Toronto to Halifax is 1742 kilometers.

(a) How many kilometers has the truck traveled on these 12 trips from Toronto to Halifax and back?

(b) If Dick wants to limit the truck to a total of 50,000 kilometers driven this year, how many more kilometers can the truck be driven?

88. *Space Travel* The space shuttle has recently gone through a number of repairs and improvements. NASA recently approved the use of a shuttle control panel that has an area of 3526 square centimeters. The control panel is rectangular. The width of the panel is 43 centimeters. What is the length of the panel?

To Think About

89. Division is not commutative. For example, $12 \div 4 \neq 4 \div 12$. If $a \div b = b \div a$, what must be true of the numbers a and b besides the fact that $b \neq 0$ and $a \neq 0$?

90. You can think of division as repeated subtraction. Show how $874 \div 138$ is related to repeated subtraction.

Cumulative Review *Solve.*

91. [1.4.4]
$$\begin{array}{r} 108 \\ \times\ \ 50 \\ \hline \end{array}$$

92. [1.4.4]
$$\begin{array}{r} 7162 \\ \times\ \ 145 \\ \hline \end{array}$$

93. [1.2.4] $316{,}214 + 89{,}981$

94. [1.3.3] $1{,}360{,}000 - 1{,}293{,}156$

Quick Quiz 1.5 *Divide. If there is a remainder, be sure to state it as part of your answer.*

1. $9\overline{)4203}$ **2.** $8\overline{)26{,}299}$ **3.** $76\overline{)24{,}928}$

4. Concept Check When performing the division problem $2956 \div 43$, you need to decide how many times 43 goes into 295. Explain how you would decide this.

How Am I Doing?   Sections 1.1–1.5

How are you doing with your homework assignments in Sections 1.1 to 1.5? Do you feel you have mastered the material so far? Do you understand the concepts you have covered? Before you go further in the textbook, take some time to do each of the following problems.

1.1

1. Write in words. 78,310,436

2. Write in expanded notation. 38,247

3. Write in standard notation.   five million, sixty-four thousand, one hundred twenty-two

Use the following table to answer questions 4 and 5.

Enrollment in Degree-Granting Colleges and Universities (in thousands)	
2000	15,312
2005	17,487
2008	18,200
2012*	19,048
2017*	20,080

*estimated
Source: U.S. Department of Education.

4. How many students were enrolled in colleges and universities in 2005?

5. How many students are expected to be enrolled in colleges and universities in 2017?

1.2   *Add.*

6.
$$\begin{array}{r} 13 \\ 31 \\ 88 \\ 43 \\ + 69 \end{array}$$

7.
$$\begin{array}{r} 28,318 \\ 5,039 \\ + 17,213 \end{array}$$

8.
$$\begin{array}{r} 833,576 \\ + 517,885 \end{array}$$

1.3   *Subtract.*

9.
$$\begin{array}{r} 5728 \\ - 1735 \end{array}$$

10.
$$\begin{array}{r} 100,450 \\ - 24,139 \end{array}$$

11.
$$\begin{array}{r} 45,861,413 \\ - 43,879,761 \end{array}$$

1.4   *Multiply.*

12. $9 \times 6 \times 1 \times 2$

13. $50 \times 10 \times 200$

14.
$$\begin{array}{r} 2658 \\ \times \quad 7 \end{array}$$

15.
$$\begin{array}{r} 68 \\ \times 55 \end{array}$$

16.
$$\begin{array}{r} 365 \\ \times 908 \end{array}$$

1.5   *Divide. If there is a remainder, be sure to state it as part of the answer.*

17. $8)\overline{84,840}$ **18.** $7)\overline{51,633}$ **19.** $76)\overline{1984}$ **20.** $42)\overline{5838}$

Now turn to page SA-2 for the answer to each of these problems. Each answer also includes a reference to the objective in which the problem is first taught. If you missed any of these problems, you should stop and review the Examples and Student Practice problems in the referenced objective. A little review now will help you master the material in the upcoming sections of the text.

1. _____
2. _____
3. _____
4. _____
5. _____
6. _____
7. _____
8. _____
9. _____
10. _____
11. _____
12. _____
13. _____
14. _____
15. _____
16. _____
17. _____
18. _____
19. _____
20. _____

1.6 Exponents and the Order of Operations

Student Learning Objectives

After studying this section, you will be able to:

① Evaluate expressions with whole-number exponents.

② Perform several arithmetic operations in the proper order.

① Evaluating Expressions with Whole-Number Exponents

Sometimes a simple math idea comes "disguised" in technical language. For example, an **exponent** is just a "shorthand" number that saves writing multiplication of the same number.

10^3 The exponent 3 means $10 \times 10 \times 10$
(which takes longer to write).

The product 5×5 can be written as 5^2. The small number 2 is called the *exponent*. The exponent tells us how many factors are in the multiplication. The number 5 is called the **base.** The base is the number that is multiplied.

$$3 \times 3 \times 3 \times 3 = 3^4 \longleftarrow \text{exponent}$$
$$\underset{\text{base}}{\uparrow}$$

In 3^4 the base is 3 and the exponent is 4. (The 4 is sometimes called the *superscript.*) 3^4 is read as " three to the fourth power."

EXAMPLE 1   Write each product in exponent form.

(a) $15 \times 15 \times 15$

(b) $7 \times 7 \times 7 \times 7 \times 7$

Solution

(a) $15 \times 15 \times 15 = 15^3$

(b) $7 \times 7 \times 7 \times 7 \times 7 = 7^5$

Student Practice 1   Write each product in exponent form.

(a) $12 \times 12 \times 12 \times 12$

(b) $2 \times 2 \times 2 \times 2 \times 2 \times 2$

NOTE TO STUDENT: Fully worked-out solutions to all of the Student Practice problems can be found at the back of the text starting at page SP-1.

EXAMPLE 2   Find the value of each expression.

(a) 3^3 (b) 7^2 (c) 2^5 (d) 1^8

Solution

(a) To find the value of 3^3, multiply the base 3 by itself 3 times.

$$3^3 = 3 \times 3 \times 3 = 27$$

(b) To find the value of 7^2, multiply the base 7 by itself 2 times.

$$7^2 = 7 \times 7 = 49$$

(c) $2^5 = 2 \times 2 \times 2 \times 2 \times 2 = 32$

(d) $1^8 = 1 \times 1 \times 1 \times 1 \times 1 \times 1 \times 1 \times 1 = 1$

Student Practice 2   Find the value of each expression.

(a) 12^2 (b) 6^3 (c) 2^6 (d) 1^{10}

If a whole number does not have a visible exponent, the exponent is understood to be 1. Thus

$$3 = 3^1 \quad \text{and} \quad 10 = 10^1.$$

Large numbers are often expressed as a power of 10.

$10^1 = 10 = 1 \text{ ten}$ $10^4 = 10{,}000 = 1 \text{ ten thousand}$
$10^2 = 100 = 1 \text{ hundred}$ $10^5 = 100{,}000 = 1 \text{ hundred thousand}$
$10^3 = 1000 = 1 \text{ thousand}$ $10^6 = 1{,}000{,}000 = 1 \text{ million}$

What does it mean to have an exponent of zero? What is 10^0? Any whole number that is not zero can be raised to the zero power. The result is 1. Thus $10^0 = 1$, $3^0 = 1$, $5^0 = 1$, and so on. Why is this? Let's reexamine the powers of 10. As we go down one line at a time, notice the pattern that occurs.

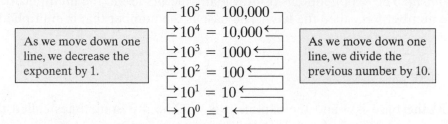

As we move down one line, we decrease the exponent by 1.

$10^5 = 100{,}000$
$10^4 = 10{,}000$
$10^3 = 1000$
$10^2 = 100$
$10^1 = 10$
$10^0 = 1$

As we move down one line, we divide the previous number by 10.

Therefore, we present the following definition.

> For any whole number a other than zero, $a^0 = 1$.

If numbers with exponents are added to other numbers, it is first necessary to **evaluate,** or find the value of, the number that is raised to a power. Then we may combine the results with another number.

EXAMPLE 3   Find the value of each expression.

(a) $3^4 + 2^3$ **(b)** $5^3 + 7^0$ **(c)** $6^3 + 6$

Solution

(a) $3^4 + 2^3 = (3)(3)(3)(3) + (2)(2)(2) = 81 + 8 = 89$
(b) $5^3 + 7^0 = (5)(5)(5) + 1 = 125 + 1 = 126$
(c) $6^3 + 6 = (6)(6)(6) + 6 = 216 + 6 = 222$

Student Practice 3   Find the value of each expression.

(a) $7^3 + 8^2$ **(b)** $9^2 + 6^0$ **(c)** $5^4 + 5$

② Performing Several Arithmetic Operations in the Proper Order

Sometimes the order in which we do things is not important. The order in which chefs hang up their pots and pans probably does not matter. The order in which they add and mix the elements in preparing food, however, makes all the difference in the world! If various cooks follow a recipe, though, they will get similar results. The recipe ensures that the results will be consistent. It shows the **order of operations.**

In mathematics the order of operations is a list of priorities for working with the numbers in computational problems. This mathematical "recipe" tells how to handle certain indefinite computations. For example, how does a person find the value of $5 + 3 \times 2$?

A problem such as $5 + 3 \times 2$ sometimes causes students difficulty. Some people think $(5 + 3) \times 2 = 8 \times 2 = 16$. Some people think $5 + (3 \times 2) = 5 + 6 = 11$. Only one answer is right, 11. To obtain the right answer, follow the steps outlined in the following box.

ORDER OF OPERATIONS

In the absence of grouping symbols:

Do first   **1.** Simplify any expressions with exponents.

↓   **2.** Multiply or divide from left to right.

Do last   **3.** Add or subtract from left to right.

EXAMPLE 4 Evaluate. $3^2 + 5 - 4 \times 2$

Solution

$$3^2 + 5 - 4 \times 2 = 9 + 5 - 4 \times 2 \quad \text{Evaluate the expression with exponents.}$$
$$= 9 + 5 - 8 \quad \text{Multiply from left to right.}$$
$$= 14 - 8 \quad \text{Add from left to right.}$$
$$= 6 \quad \text{Subtract.}$$

Student Practice 4 Evaluate. $7 + 4^3 \times 3$

EXAMPLE 5 Evaluate. $5 + 12 \div 2 - 4 + 3 \times 6$

Solution There are no numbers to raise to a power, so we first do any multiplication or division in order from *left to right*.

$$5 + 12 \div 2 - 4 + 3 \times 6 \quad \text{Multiply or divide from left to right.}$$
$$= 5 + 6 - 4 + 3 \times 6 \quad \text{Divide.}$$
$$= 5 + 6 - 4 + 18 \quad \text{Multiply. Add or subtract from left to right.}$$
$$= 11 - 4 + 18 \quad \text{Add.}$$
$$= 7 + 18 \quad \text{Subtract.}$$
$$= 25 \quad \text{Add.}$$

Student Practice 5 Evaluate. $37 - 20 \div 5 + 2 - 3 \times 4$

EXAMPLE 6 Evaluate. $2^3 + 3^2 - 7 \times 2$

Solution

$$2^3 + 3^2 - 7 \times 2 = 8 + 9 - 7 \times 2 \quad \text{Evaluate exponent expressions } 2^3 = 8 \text{ and } 3^2 = 9.$$
$$= 8 + 9 - 14 \quad \text{Multiply.}$$
$$= 17 - 14 \quad \text{Add.}$$
$$= 3 \quad \text{Subtract.}$$

Student Practice 6 Evaluate. $4^3 - 2 + 3^2$

You can change the order in which you compute by using grouping symbols. Place the numbers you want to calculate first within parentheses. This tells you to do those calculations first.

ORDER OF OPERATIONS

With grouping symbols:

Do first   1. Perform operations inside parentheses.

↓   2. Simplify any expressions with exponents.

            3. Multiply or divide from left to right.

Do last   4. Add or subtract from left to right.

EXAMPLE 7   Evaluate. $2 \times (7 + 5) \div 4 + 3 - 6$

Solution   First, we combine numbers inside the parentheses by adding the 7 to the 5. Next, because multiplication and division have equal priority, we work from left to right doing whichever of these operations comes first.

$$2 \times (7 + 5) \div 4 + 3 - 6$$
$$= 2 \times 12 \div 4 + 3 - 6 \quad \text{Parentheses.}$$
$$= 24 \div 4 + 3 - 6 \quad \text{Multiply.}$$
$$= 6 + 3 - 6 \quad \text{Divide.}$$
$$= 9 - 6 \quad \text{Add.}$$
$$= 3 \quad \text{Subtract.}$$

Student Practice 7   Evaluate. $(17 + 7) \div 6 \times 2 + 7 \times 3 - 4$

$\mathbb{M}_{\mathbb{C}}$ **EXAMPLE 8**   Evaluate.
$4^3 + 18 \div 3 - 2^4 - 3 \times (8 - 6)$

Solution

$$4^3 + 18 \div 3 - 2^4 - 3 \times (8 - 6)$$
$$= 4^3 + 18 \div 3 - 2^4 - 3 \times 2 \quad \text{Work inside the parentheses.}$$
$$= 64 + 18 \div 3 - 16 - 3 \times 2 \quad \text{Evaluate exponents.}$$
$$= 64 + 6 - 16 - 3 \times 2 \quad \text{Divide.}$$
$$= 64 + 6 - 16 - 6 \quad \text{Multiply.}$$
$$= 70 - 16 - 6 \quad \text{Add.}$$
$$= 54 - 6 \quad \text{Subtract.}$$
$$= 48 \quad \text{Subtract.}$$

Student Practice 8   Evaluate.
$5^2 - 6 \div 2 + 3^4 + 7 \times (12 - 10)$

Verbal and Writing Skills, Exercises 1–6

1. Explain what the expression 5^3 means. Evaluate 5^3.

2. In exponent form, the _____ tells how many times to multiply the base.

3. In exponent form, the ____ is the number that is multiplied.

4. 10^5 is read as _____.

5. Explain the order in which we perform mathematical operations to ensure consistency.

6. Use the order of operations to evaluate $12 \times 5 + 3 \times 5 + 7 \times 5$. Is this the same as $5(12 + 3 + 7)$? Why or why not?

Write each number in exponent form.

7. $6 \times 6 \times 6 \times 6$

8. $2 \times 2 \times 2 \times 2 \times 2 \times 2 \times 2 \times 2$

9. $4 \times 4 \times 4 \times 4 \times 4 \times 4 \times 4$

10. $3 \times 3 \times 3 \times 3 \times 3 \times 3$

11. $9 \times 9 \times 9 \times 9$

12. $1 \times 1 \times 1 \times 1 \times 1 \times 1 \times 1$

13. 7

14. 27

Find the value of each expression.

15. 2^4 **16.** 3^3 **17.** 4^3 **18.** 5^2 **19.** 6^2

20. 10^3 **21.** 10^4 **22.** 1^{20} **23.** 1^{17} **24.** 2^5

25. 2^6 **26.** 4^2 **27.** 3^5 **28.** 12^2 **29.** 15^2

30. 3^4 **31.** 7^3 **32.** 5^4 **33.** 4^4 **34.** 7^2

35. 9^0 **36.** 8^0 **37.** 25^2 **38.** 20^3 **39.** 10^6

40. 8^1 **41.** 13^2 **42.** 11^2 **43.** 9^1 **44.** 14^2

45. 8^2 **46.** 5^3 **47.** $3^2 + 1^2$ **48.** $7^0 + 4^3$ **49.** $2^3 + 10^2$

50. $7^3 + 4^2$ **51.** $8^3 + 8$ **52.** $9^2 + 9$

Work each exercise, using the correct order of operations.

53. $8 \times 7 - 20$

54. $6 \times 9 + 32$

55. $3 \times 9 - 10 \div 2$

56. $4 \times 6 - 24 \div 4$

57. $48 \div 2^3 + 4$

58. $4^3 \div 4 - 11$

59. $3 \times 6^2 - 50$

60. $2 \times 12^2 - 80$

61. $10^2 + 3 \times (8 - 3)$

62. $4^3 - 5 \times (9 + 1)$

63. $(400 \div 20) \div 20$

64. $(600 \div 30) \div 20$

65. $950 \div (25 \div 5)$

66. $875 \div (35 \div 7)$

67. $(14)(4) - (14 + 4)$

68. $5(30) - (30 + 5)$

69. $3^2 + 4^2 \div 2^2$

70. $7^2 + 9^2 \div 3^2$

71. $(6)(7) - (12 - 8) \div 4$

72. $(8)(9) - (15 - 5) \div 5$

73. $100 - 3^2 \times 4$

74. $130 - 4^2 \times 5$

75. $5^2 + 2^2 + 3^3$

76. $2^3 + 3^2 + 4^3$

77. $72 \div 9 \times 3 \times 1 \div 2$

78. $120 \div 30 \times 2 \times 5 \div 8$

79. $12^2 - 2 \times 0 \times 5 \times 6$

80. $8^2 - 4 \times 3 \times 0 \times 7$

Mixed Practice *Work each exercise, using the correct order of operations.*

81. $4^2 \times 6 \div 3$

82. $7^2 \times 3 \div 3$

83. $60 - 2 \times 4 \times 5 + 10$

84. $75 - 3 \times 5 \times 2 + 15$

85. $3 + 3^2 \times 6 + 4$

86. $5 + 4^3 \times 2 + 7$

87. $32 \div 2 \times (3 - 1)^4$

88. $24 \div 3 \times (5 - 3)^2$

89. $3^2 \times 6 \div 9 + 4 \times 3$

90. $5^2 \times 3 \div 25 + 7 \times 6$

91. $4^0 + 5^3 + 9^1$

92. $10^1 + 6^0 + 4^3$

93. $1200 - 2^3(3) \div 6$

94. $2150 - 3^4(2) \div 9$

95. $120 \div (30 + 10) - 1$

96. $100 - 48 \div (2 \times 3)$

97. $120 \div 30 + 10 - 1$

98. $100 - 48 \div 2 \times 3$

99. $7 \times 3 - (9 - 7)^3 + 3^0$

100. $8 \times 6 - 5^0 + (6 - 2)^3$

To Think About

101. *Astronomy* The planet Earth rotates once every 23 hours, 56 minutes, 4 seconds. How many seconds is that?

102. *Astronomy* The planet Saturn rotates once every 10 hours, 12 minutes. How many minutes is that? How many seconds?

Cumulative Review

103. [1.1.1] In the number 2,038,754

    (a) What digit tells the number of ten thousands?

    (b) What is the value of the digit 2?

104. [1.1.3] Write in standard notation. two hundred million, seven hundred sixty-five thousand, nine hundred nine

105. [1.1.3] Write in words. 261,763,002

▲ **106.** [1.2.6 and 1.4.6] *Geometry* New Boston High School has an athletic field that needs to be enclosed by fencing. The rectangular field is 250 feet wide and 480 feet long. How many feet of fencing are needed to surround the field? Grass needs to be planted for a new playing field for next year. What is the area in square feet of the amount of grass that must be planted?

Quick Quiz 1.6

1. Write in exponent form.

    $12 \times 12 \times 12 \times 12 \times 12$

2. Find the value of 6^4.

3. Perform each operation in the proper order.

    $42 - 2^5 + 3 \times (9 - 6)^3$

4. Concept Check Explain in what order you would do the steps to evaluate the expression $7 \times 6 \div 3 \times 4^2 - 2$.

1.7 Rounding and Estimating

Student Learning Objectives

After studying this section, you will be able to:

① Round whole numbers.

② Estimate the answer to a problem involving whole numbers.

① Rounding Whole Numbers

Large numbers are often expressed to the nearest hundred or to the nearest thousand, because an approximate number is "good enough" for certain uses.

Distances from Earth to other galaxies are measured in light-years. Although light really travels at 5,865,696,000,000 miles a year, we usually **round** this number to the nearest trillion and say it travels at 6,000,000,000,000 miles a year. To round a number, we first determine the place we are rounding to—in this case, trillion. Then we find which value is closest to the number that we are rounding. In this case, the number we want to round is closer to 6 trillion than to 5 trillion. How do we know the number is closer to 6 trillion than to 5 trillion?

To see which is the closest value, we may picture a **number line,** where whole numbers are represented by points on a line. To show how to use a number line in rounding, we will round 368 to the nearest hundred. 368 is between 300 and 400. When we round, we pick the hundred 368 is "closest to." We draw a number line to show 300 and 400. We also show the point midway between 300 and 400 to help us to determine which hundred 368 is closest to.

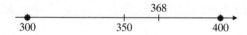

We find that the number 368 is closer to 400 than to 300, so we round 368 *up to* 400.

Let's look at another example. We will round 129 to the nearest hundred. 129 is between 100 and 200. We show this on the number line. We include the midpoint 150 as a guide.

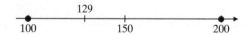

We find that the number 129 is closer to 100 than to 200, so we round 129 *down to* 100.

This leads us to the following simple rule for rounding.

ROUNDING A WHOLE NUMBER

1. If the first digit to the right of the round-off place is
   (a) *less than 5,* we make no change to the digit in the round-off place. (We know it is closer to the smaller number, so we round down.)
   (b) *5 or more,* we increase the digit in the round-off place by 1. (We know it is closer to the larger number, so we round up.)

2. Then we replace the digits to the right of the round-off place by zeros.

EXAMPLE 1 Round 37,843 to the nearest thousand.

Solution   3 7, 8 4 3   According to the directions, the thousands will be the round-off place. We locate the thousands place.

3 7, (8) 4 3   We see that the first digit to the right of the round-off place is 8, which is 5 or more. We increase the thousands digit by 1, and replace all digits to the right by zero.

3 8, 0 0 0

We have rounded 37,843 to the nearest thousand: 38,000. This means that 37,843 is closer to 38,000 than to 37,000.

Student Practice 1 Round 65,528 to the nearest thousand.

NOTE TO STUDENT: Fully worked-out solutions to all of the Student Practice problems can be found at the back of the text starting at page SP-1.

EXAMPLE 2 Round 2,445,360 to the nearest hundred thousand.

Solution   2, 4 4 5, 3 6 0   Locate the hundred thousands round-off place.

2, 4 (4) 5, 3 6 0   The first digit to the right of this is less than 5, so round down. Do not change the hundred thousands digit.

2, 4 0 0, 0 0 0   Replace all digits to the right by zero.

Student Practice 2 Round 172,963 to the nearest ten thousand.

EXAMPLE 3 Round as indicated.

(a) 561,328 to the nearest ten   **(b)** 3,798,152 to the nearest hundred
(c) 51,362,523 to the nearest million

Solution

(a)   ↓   First locate the digit in the tens place.
561,328   The digit to the right of the tens place is greater than 5.
561,330   Round up.

561,328 rounded to the nearest ten is 561,330.

(b) 3,798,152   The digit to the right of the hundreds place is 5.
3,798,200   Round up.

3,798,152 rounded to the nearest hundred is 3,798,200.

(c) 51,362,523   The digit to the right of the millions place is less than 5.
51,000,000   Round down.

51,362,523 rounded to the nearest million is 51,000,000.

Student Practice 3 Round as indicated.

(a) 53,282 to the nearest ten
(b) 164,485 to the nearest thousand
(c) 1,365,273 to the nearest hundred thousand

EXAMPLE 4   Round 763,571.

(a) To the nearest thousand
(b) To the nearest ten thousand
(c) To the nearest million

Solution

(a) 763,571 = 764,000 to the nearest thousand. The digit to the right of the thousands place is 5. We rounded up.

(b) 763,571 = 760,000 to the nearest ten thousand. The digit to the right of the ten thousands place is less than 5. We rounded down.

(c) 763,571 does not have any digits for millions. If it helps, you can think of this number as 0,763,571. Since the digit to the right of the millions place is 7, we round up to obtain one million or 1,000,000.

Student Practice 4   Round 935,682 as indicated.

(a) To the nearest thousand
(b) To the nearest hundred thousand
(c) To the nearest million

EXAMPLE 5   Astronomers use the parsec as a measurement of distance. One parsec is approximately 30,900,000,000,000 kilometers. Round 1 parsec to the nearest trillion kilometers.

Solution   30,900,000,000,000 km is 31,000,000,000,000 km or 31 trillion km to the nearest trillion kilometers.

Student Practice 5   One light-year is approximately 9,460,000,000,000,000 meters. Round to the nearest hundred trillion meters.

② Estimating the Answer to a Problem Involving Whole Numbers

Often we need to quickly check the answer of a calculation to be reasonably assured that the answer is correct. If you expected your bill to be "around $40" for the groceries you had selected and the cashier's total came to $41.89, you would probably be confident that the bill is correct and pay it. If, however, the cashier rang up a bill of $367, you would not just assume that it is correct. You would know an error had been made. If the cashier's total came to $60, you might not be certain, but you would probably suspect an error and check the calculation.

In mathematics we often **estimate,** or determine the approximate value of a calculation, if we need to do a quick check. There are many ways to estimate, but in this book we will use one simple principle of estimation. We use the symbol ≈ to mean **approximately equal to.**

PRINCIPLE OF ESTIMATION

1. Round the numbers so that there is one nonzero digit in each number.

2. Perform the calculation with the rounded numbers.

EXAMPLE 6   Estimate the sum. $163 + 237 + 846 + 922$

Solution   We first determine where to round each number in our problem to leave only one nonzero digit in each. In this case, we round all numbers to the nearest hundred. Then we perform the calculation with the rounded numbers.

Actual Sum	*Estimated Sum*
163	200
237	200
846	800
+ 922	+ 900
	2100

We estimate the answer to be 2100. We say the sum ≈ 2100. If we calculate using the exact numbers, we obtain a sum of 2168, so our estimate is quite close to the actual sum.

Student Practice 6   Estimate the sum. $3456 + 9876 + 5421 + 1278$

When we use the principle of estimation, we will not always round each number in a problem to the same place.

EXAMPLE 7   Phil and Melissa bought their first car last week. The selling price of this compact car was $8980. The dealer preparation charge was $289 and the sales tax was $449. Estimate the total cost that Phil and Melissa had to pay.

Solution   We round each number to have only one nonzero digit, and add the rounded numbers.

8980	9000
289	300
+ 449	+ 400
	9700

The total cost $\approx \$9700$. (The exact answer is $9718, so we see that our answer is quite close.)

Student Practice 7   Greg and Marcia purchased a new sofa for $697, plus $35 sales tax. The store also charged them $19 to deliver the sofa. Estimate their total cost.

Now we turn to a case where an estimate can help us discover an error.

EXAMPLE 8   Roberto added together four numbers and obtained the following result. Estimate the sum and determine if the answer seems reasonable.

$$12{,}456 + 17{,}976 + 18{,}452 + 32{,}128 \stackrel{?}{=} 61{,}012$$

Solution   We round each number so that there is one nonzero digit. In this case, we round them all to the nearest ten thousand.

12,456	10,000
17,976	20,000
18,452	20,000
+ 32,128	+ 30,000
	80,000 Our estimate is 80,000.

This is significantly different from 61,012, so we would suspect that an error has been made. In fact, Roberto did make an error. The exact sum is actually 81,012!

Student Practice 8   Ming did the following calculation. Estimate to see if her sum appears to be correct or incorrect.

$$11{,}849 + 14{,}376 + 16{,}982 + 58{,}151 = 81{,}358$$

Next we look at a subtraction example where estimation is used.

EXAMPLE 9   The profit from Techno Industries for the first quarter of the year was $642,987,000. The profit for the second quarter was $238,890,000. Estimate how much less the profit was for the second quarter than for the first quarter.

Solution   We round each number so that there is one nonzero digit. Then we subtract, using the two rounded numbers.

642,987,000	600,000,000
− 238,890,000	− 200,000,000
	400,000,000

We estimate that the profit was $400,000,000 less for the second quarter.

Student Practice 9   The 2010 population of Texas was 24,648,888. The 2010 population of California was 38,067,134. Estimate how many more people lived in California in 2010 than in Texas.

We also use this principle to estimate results of multiplication and division.

Mc **EXAMPLE 10**   Estimate the product. 56,789 × 529

Solution   We round each number so that there is one nonzero digit. Then we multiply the rounded numbers to obtain our estimate.

$$
\begin{array}{r}
56{,}789 \\
\times\ \ \ \ \ 529 \\
\end{array}
\qquad
\begin{array}{r}
60{,}000 \\
\times\ \ \ \ \ 500 \\
\hline
30{,}000{,}000
\end{array}
$$

Therefore the product is ≈30,000,000. (This is reasonably close to the exact answer of 30,041,381.)

Student Practice 10
Estimate the product. 8945 × 7317

EXAMPLE 11   Estimate the answer for the following division problem.

$$23\overline{)148{,}902}$$

Solution   We round each number to a number with one nonzero digit. Then we perform the division, using the two rounded numbers.

$$23\overline{)148{,}902} \qquad \overset{5000}{20\overline{)100{,}000}}$$

Our estimate is 5000. (The exact answer is 6474. We see that our estimate is "in the ballpark" but is not very close to the exact answer. Remember, an estimate is just a rough approximation of the exact answer.)

Student Practice 11   Estimate the answer for the following division problem.

$$39\overline{)75{,}342}$$

Not all division estimates come out so easily. In some cases, you may need to carry out a long-division problem of several steps just to obtain the estimate. Do not be in a hurry. Students often want to rush the steps of division. It is better to take your time and carefully do each step. This approach will be very worthwhile in the long run.

EXAMPLE 12   John and Stephanie drove their Honda CR-V a distance of 778 miles. They used 25 gallons of gas. Estimate how many miles they can travel on 1 gallon of gas.

Solution   In order to solve this problem, we need to divide 778 by 25 to obtain the number of miles John and Stephanie get with 1 gallon of gas. We round each number to a number with one nonzero digit and then perform the division, using the rounded numbers.

$$25\overline{)778} \qquad
\begin{array}{r}
26\ \ \ \ \\
30\overline{)800} \\
\underline{60}\ \ \ \ \\
200 \\
\underline{180} \\
20 \quad \text{Remainder}
\end{array}$$

Continued on next page

We obtain an answer of 26 with a remainder of 20. For our estimate we will use the whole number 27. Thus we estimate that the number of miles their CR-V obtained on 1 gallon of gas was 27 miles. (This is reasonably close to the exact answer, which is just slightly more than 31 miles per gallon of gas.)

Student Practice 12 The highway department purchased 58 identical trucks at a total cost of $1,864,584. Estimate the cost for one truck.

STEPS TO SUCCESS Getting the Greatest Value from Your Homework

Read the textbook first before doing the homework. Take some time to read the text and study the sample examples. Try working out the Student Practice problems. You will be amazed at the amount of understanding you will obtain by studying the book before jumping into the homework exercises.

Take your time. Read the directions carefully. Be sure you understand what is being asked. Check your answers with those given in the back of the textbook. If your answer is incorrect, study similar sample examples in the text. Then redo the problem, watching for errors.

Make a schedule. You will need to allow two hours outside of class for each hour of actual class time. Make a weekly schedule of the times you have class. Now write down the times each day you will devote to doing math homework. Then write down the times you will spend doing homework for your other classes. If you have a job be sure to write down all your work hours.

Making it personal: Write down your own schedule of class, work, and study time. ▼

Sunday	Monday	Tuesday	Wednesday	Thursday	Friday	Saturday

MyMathLab®

Watch the videos
in MyMathLab

Download the
MyDashBoard App

Verbal and Writing Skills, Exercises 1 and 2

1. Explain the rule for rounding and provide examples.

2. What happens when you round 98 to the nearest ten?

Round to the nearest ten.

3. 83   **4.** 45   **5.** 65   **6.** 57   **7.** 168   **8.** 132

9. 7438   **10.** 2834   **11.** 2961   **12.** 4355

Round to the nearest hundred.

13. 247   **14.** 661   **15.** 2781   **16.** 1249   **17.** 7692   **18.** 1643

Round to the nearest thousand.

19. 7621   **20.** 3754   **21.** 1489   **22.** 515   **23.** 27,863   **24.** 94,489

Applications, Exercises 25–32

25. *History* The worst death toll from an earthquake was in Shaanxi, China, in 1556. That earthquake killed an estimated 832,400 people. Round this number to the nearest hundred thousand.

26. *Astronomy* One light-year (the distance light travels in 1 year) measures 5,878,612,843,000 miles. Round this figure to the nearest hundred million.

27. *Astronomy* The Hubble Space Telescope's *Guide Star Catalogue* lists 15,169,873 stars. Round this figure to the nearest million.

28. *Geography* One of the highest active volcanoes in the world is Popocatépetl in Mexico, with a height of 17,802 feet. Round this figure to the nearest ten thousand.

29. *Population Studies* The population of the United States in 2030 is projected to be 373,504,000. Round this figure to

(a) the nearest ten thousand.

(b) the nearest hundred million.

30. *Population Studies* In 2030, the number of children in the United States 5 years old or younger is expected to be 24,271,894. Round this figure to

(a) the nearest thousand.

(b) the nearest ten million.

▲ **31.** *Geography* The total area of mainland China is 3,705,392 square miles, or 9,596,960 square kilometers. For *both* square miles and square kilometers, round these figures to

(a) the nearest hundred thousand.

(b) the nearest ten thousand.

▲ **32.** *Geography* The area of the Atlantic Ocean is 31,830,000 square miles. Round this figure to

(a) the nearest hundred thousand.

(b) the nearest million.

Use the principle of estimation to find an estimate for each calculation.

33. 772 + 324 + 225   **34.** 186 + 509 + 872   **35.** 42 + 69 + 95 + 18

36. 62 + 27 + 54 + 98

37. 158,270 + 53,441 + 8701

38. 238,271 + 77,304 + 9551

39. 324,230 − 70,290

40. 975,935 − 593,228

41. 842,512 − 78,234

42. 382,140 − 56,117

43. 33,261,378 − 18,199,276

44. 89,263,000 − 54,198,635

45. 47 × 62

46. 43 × 95

47. 1324 × 8

48. 5926 × 3

49. 565,382 × 275

50. 441,630 × 632

51. 6368 ÷ 38

52. 7813 ÷ 22

53. 362,881 ÷ 39

54. 596,450 ÷ 64

55. 3,885,720 ÷ 831

56. 12,447,312 ÷ 497

Estimate the result of each calculation. Some results are correct and some are incorrect. Which results appear to be correct? Which results appear to be incorrect?

57.
```
   361
   522
   873
 + 164
  1320
```

58.
```
   476
   124
   516
 + 389
  1505
```

59.
```
  97,635
  52,123
+ 41,986
 291,744
```

60.
```
  26,181
  47,998
+ 63,271
 137,450
```

61.
```
  302,360
 − 89,518
  212,842
```

62.
```
  735,128
 − 116,733
  518,395
```

63. 78,126,345
  − 48,972,103
  19,154,242

64. 42,765,317
  − 29,318,274
  23,447,043

65. 378
  × 32
  21,096

66. 512
  × 46
  20,552

67. 5896
  × 72
  424,512

68. 8076
  × 89
  718,764

69. 36)82,116   (2281)

70. 52)28,912   (556)

71. 423)161,163   (381)

72. 781)477,972   (612)

Applications

▲ **73. Geometry** Victor and Shannon just purchased a new home with a two-car garage measuring 17 feet wide and 22 feet long. Estimate the number of square feet in the garage.

▲ **74. Geometry** A huge restaurant in New York City is 43 yards wide and 112 yards long. Estimate the number of square yards in the restaurant.

75. International Relations In 2009, the populations of the three largest cities in Canada were Toronto with 5,623,450 people, Montreal with 3,814,738 people, and Vancouver with 2,328,007 people. Estimate the total population of these three cities.

76. Financial Management The highway departments in four towns in northwestern New York had the following budgets for snow removal for the year: $329,560, $672,940, $199,734, and $567,087. Estimate the total amount that the four towns spend for snow removal in one year.

77. Business Management The local pizzeria makes 267 pizzas on an average day. Estimate how many pizzas were made in the last 134 days.

78. Personal Finance Darcy makes $68 for each shift she works. She is scheduled for 33 shifts during the next two months. Estimate how much she will earn in the next two months.

79. Transportation In 2009, the two busiest airports in the United States were Atlanta with 90,039,280 passengers and Chicago O'Hare with 69,353,876 passengers. Round each figure to the nearest hundred thousand. Then estimate the difference.

80. Sports In 2009, the average attendance at a Dallas Cowboys football game was 89,756 people. This ranked first in the National Football League (NFL) for attendance. The lowest-ranking average attendance in the NFL was for the Oakland Raiders, with 44,284 people. Round each figure to the nearest thousand. Then estimate the difference.

▲ **81.** *Geography* The largest state of the United States is Alaska, with a land area of 586,412 square miles. The second largest state is Texas, with an area of 267,339 square miles. Round each figure to the nearest ten thousand. Then estimate how many square miles larger Alaska is than Texas.

▲ **82.** *International Relations* The largest country in Africa is Sudan, measuring 966,757 square miles. South America's largest country is Brazil, measuring 3,286,470 square miles. Round each figure to the nearest hundred thousand, then estimate how many square miles larger Brazil is than Sudan.

To Think About

83. *Space Travel* A space probe travels at 23,560 miles per hour for a distance of 7,824,560,000 miles.

  (a) How many *hours* will it take the space probe to travel that distance? (Estimate.)

  (b) How many *days* will it take the space probe to travel that distance? (Estimate.)

84. *Space Travel* A space probe travels at 28,367 miles per hour for a distance of 9,348,487,000 miles.

  (a) Estimate the number of *hours* it will take the space probe to travel that distance.

  (b) Estimate the number of *days* it will take the space probe to travel that distance.

Cumulative Review *Evaluate.*

85. **[1.6.2]** $26 \times 3 + 20 \div 4$

86. **[1.6.2]** $5^2 + 3^2 - (17 - 10)$

87. **[1.6.2]** $3 \times (16 \div 4) + 8 \times 2$

88. **[1.6.2]** $126 + 4 - (20 \div 5)^3$

89. **[1.4.4]** $\begin{array}{r} 5489 \\ \times 67 \\ \hline \end{array}$

90. **[1.5.3]** $52\overline{)4524}$

Quick Quiz 1.7

  1. Round to the nearest hundred.   92,354

  2. Round to the nearest ten thousand.   2,342,786

  3. Use the principle of estimation to find an estimation for this calculation. $7862 \times 329,182$

  4. **Concept Check** Explain how to round 682,496,934 to the nearest million.

1.8 Solving Applied Problems Involving Whole Numbers

① Solving Problems Involving One Operation

When a builder constructs a new home or office building, he or she usually has a *blueprint*. This accurate drawing shows the basic structure of the building. It also shows the dimensions of the structure to be built. This blueprint serves as a useful reference throughout the construction process.

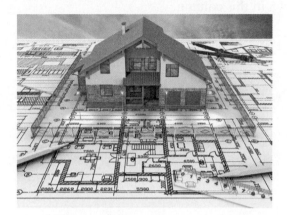

Similarly, when solving applied problems, it is helpful to have a "mathematics blueprint." This is a simple way to organize the information provided in a word problem. You record the facts you need to use and specify what you are solving for. You also record any other information that you feel will be helpful. We will use a Mathematics Blueprint for Problem Solving in this section.

Sometimes people feel totally lost when trying to solve a word problem. They sometimes say, "Where do I begin?" or "How in the world do you do this?" When you have this type of feeling, it sometimes helps to have a formal strategy or plan. Here is a plan you may find helpful:

---

1. **Understand the problem.**
   - (a) Read the problem carefully.
   - (b) Draw a picture if this helps you see the relationships more clearly.
   - (c) Fill in the Mathematics Blueprint so that you have the facts and a method of proceeding in this situation.

2. **Solve and state the answer.**
   - (a) Perform the calculations.
   - (b) State the answer, including the unit of measure.

3. **Check.**
   - (a) Estimate the answer.
   - (b) Compare the exact answer with the estimate to see if your answer is reasonable.

Now exactly what does the Mathematics Blueprint for Problem Solving look like? It is a simple sheet of paper with four columns. Each column tells you something to do.

Gather the Facts—Find the numbers that you will need to use in your calculations.

What Am I Asked to Do?—Are you finding an area, a volume, a cost, the total number of people? What is it that you need to find?

How Do I Proceed?—Do you need to add items together? Do you need to multiply or divide? What types of calculations are required?

Key Points to Remember—Write down things you might forget. The length is in feet. The area is in square feet. We need the total number of something, not the intermediate totals. Whatever you need to help you, write it down in this column.

Mathematics Blueprint for Problem Solving

Gather the Facts	What Am I Asked to Do?	How Do I Proceed?	Key Points to Remember

EXAMPLE 1 Gerald made deposits of $317, $512, $84, and $161 into his checking account. He also made out checks for $100 and $125. What was the total of his deposits?

Solution

1. *Understand the problem.* First we read over the problem carefully and fill in the Mathematics Blueprint.

Mathematics Blueprint for Problem Solving

Gather the Facts	What Am I Asked to Do?	How Do I Proceed?	Key Points to Remember
We need only deposits—not checks. The **deposits** are $317, $512, $84, and $161.	Find the total of Gerald's four deposits.	I must add the four deposits to obtain the total.	Watch out! Don't use the **checks** of $100 and $125 in the calculation. We only want the total of the **deposits**.

2. Solve and state the answer. We need to *add* to find the sum of the deposits.

$$
\begin{array}{r}
317 \\
512 \\
84 \\
+\ 161 \\
\hline
1074
\end{array}
$$

The total of the four deposits is $1074.

3. Check. Reread the problem. Be sure you have answered the question that was asked. Did it ask for the total of the deposits? Yes. ✓

Is the calculation correct? You can use estimation to check. Here we round each of the deposits so that we have one nonzero digit.

$$
\begin{array}{rr}
317 & 300 \\
512 & 500 \\
84 & 80 \\
+\ 161 & +\ 200 \\
\hline
 & 1080
\end{array}
$$

Our estimate is $1080. $1074 is close to our estimated answer of $1080. Our answer is reasonable. ✓

Thus we conclude that the total of the four deposits is $1074.

Student Practice 1 Use the Mathematics Blueprint to solve the following problem. Diane's paycheck shows deductions of $135 for federal taxes, $28 for state taxes, $13 for FICA, and $34 for health insurance. Her gross pay (amount before deductions) is $1352. What is the total amount that is taken out of Diane's paycheck?

NOTE TO STUDENT: Fully worked-out solutions to all of the Student Practice problems can be found at the back of the text starting at page SP-1.

Mathematics Blueprint for Problem Solving

Gather the Facts	What Am I Asked to Do?	How Do I Proceed?	Key Points to Remember

Portland   028,353.0

Kansas   030,162.0
City

EXAMPLE 2 Theofilos looked at his odometer before he began his trip from Portland, Oregon, to Kansas City, Missouri. He checked his odometer again when he arrived in Kansas City. The two readings are shown in the figure. How many miles did Theofilos travel?

Solution

1. *Understand the problem.* Determine what information is given.
   The mileage reading before the trip began and when the trip was over.
   What do you need to find?
   The number of miles traveled.

Mathematics Blueprint for Problem Solving

Gather the Facts	What Am I Asked to Do?	How Do I Proceed?	Key Points to Remember
At the start of the trip, the odometer read 28,353 miles. At the end of the trip, the odometer read 30,162 miles.	Find out how many miles Theofilos traveled.	I must subtract the two mileage readings.	Subtract the mileage at the start of the trip from the mileage at the end of the trip.

2. *Solve and state the answer.* We need to subtract the two mileage readings to find the difference in the number of miles. This will give us the number of miles the car traveled on this trip alone.

$$30{,}162 - 28{,}353 = 1809 \quad \text{The trip totaled 1809 miles.}$$

3. *Check.* We estimate to the nearest thousandth and compare the estimate with the preceding answer.

Kansas City	30,162 $\longrightarrow$	30,000	We subtract
Portland	28,353 $\longrightarrow$	28,000	our rounded values.
		2000	

Our estimate is 2000 miles. We compare this estimate with our answer. Our answer is reasonable. ✓

2008 Presidential Race, Popular Votes

Candidate	Number of Votes
Obama (D)	69,456,897
McCain (R)	59,934,814
Nader	738,475

Source: Federal Election Commission

Student Practice 2 The table on the left shows the results of the 2008 presidential race in the United States. By how many popular votes did the Democratic candidate beat the Republican candidate in that year?

Mathematics Blueprint for Problem Solving

Gather the Facts	What Am I Asked to Do?	How Do I Proceed?	Key Points to Remember

EXAMPLE 3   One horsepower is the power needed to lift 550 pounds a distance of 1 foot in 1 second. How many pounds can be lifted 1 foot in 1 second by 7 horsepower?

Solution

1. *Understand the problem.* Simplify the problem. If 1 horsepower can lift 550 pounds, how many pounds can be lifted by 7 horsepower? We draw and label a diagram.

7 Horsepower

550   550   550   550   550   550   550

We use the Mathematics Blueprint to organize the information.

Mathematics Blueprint for Problem Solving

Gather the Facts	What Am I Asked to Do?	How Do I Proceed?	Key Points to Remember
One horsepower will lift 550 pounds.	Find how many pounds can be lifted by 7 horsepower.	I need to multiply 550 by 7.	I do not use the information about moving one foot in one second.

2. *Solve and state the answer.* To solve the problem we multiply the 7 horsepower by 550 pounds for each horsepower.

$$\begin{array}{r} 550 \\ \times\ \ \ 7 \\ \hline 3850 \end{array}$$

We find that 7 horsepower moves 3850 pounds 1 foot in 1 second. We include 1 foot in 1 second in our answer because it is part of the unit of measure.

3. *Check.* We estimate our answer. We round 550 to 600 pounds.

$$600 \times 7 = 4200 \text{ pounds}$$

Our estimate is 4200 pounds. Our calculations in step 2 gave us 3850. Is this reasonable? This answer is close to our estimate. Our answer is reasonable. ✓

Student Practice 3 In a measure of liquid capacity, 1 gallon is 1024 fluid drams. How many fluid drams would be in 9 gallons?

Mathematics Blueprint for Problem Solving

Gather the Facts	What Am I Asked to Do?	How Do I Proceed?	Key Points to Remember

EXAMPLE 4 Laura can type 35 words per minute. She has to type an English theme that has 5180 words. How many minutes will it take her to type the theme? How many hours and how many minutes will it take her to type the theme?

Solution

1. *Understand the problem.* We draw a picture. Each "package" of 1 minute is 35 words. We want to know how many packages make up 5180 words.

We use the Mathematics Blueprint to organize the information.

Mathematics Blueprint for Problem Solving

Gather the Facts	What Am I Asked to Do?	How Do I Proceed?	Key Points to Remember
Laura can type 35 words per minute. She must type a paper with 5180 words.	Find out how many 35-word units are in 5180 words.	I need to divide 5180 by 35.	In converting minutes to hours, I will use the fact that 1 hour = 60 minutes.

2. *Solve and state the answer.*

$$
\begin{array}{r}
148 \\
35\overline{)5180} \\
\underline{35} \\
168 \\
\underline{140} \\
280 \\
\underline{280} \\
0
\end{array}
$$

It will take 148 minutes.

We will change this answer to hours and minutes. Since 60 minutes = 1 hour, we divide 148 by 60. The quotient will tell us how many hours. The remainder will tell us how many minutes.

$$
\begin{array}{r}
2 \text{ R } 28 \\
60\overline{)148} \\
\underline{-120} \\
28
\end{array}
$$

Laura can type the theme in 148 minutes or 2 hours, 28 minutes.

3. Check. The theme has 5180 words; she can type 35 words per minute. 5180 words is approximately 5000 words.

5180 words → 5000 words rounded to nearest thousand.

$$40\overline{)5000}^{\,125}$$

35 words per minute → 40 words per minute rounded to nearest ten. We divide our estimated values.

Our estimate is 125 minutes. This is close to our calculated answer. Our answer is reasonable. ✓

Student Practice 4 Donna bought 45 shares of stock for $1620. How much did the stock cost her per share?

Mathematics Blueprint for Problem Solving

Gather the Facts	What Am I Asked to Do?	How Do I Proceed?	Key Points to Remember

② Solving Problems Involving More Than One Operation

Sometimes a chart, table, or bill of sale can be used to help us organize the data in an applied problem. In such cases, a blueprint may not be needed.

EXAMPLE 5 Cleanway Rent-A-Car bought four used luxury sedans at $21,000 each, three compact sedans at $14,000 each, and seven subcompact sedans at $8000 each. What was the total cost of the purchase?

Solution

1. **Understand the problem.** We will make an imaginary bill of sale to help us to visualize the problem.

2. **Solve and state the answer.** We do the calculation and enter the results in the bill of sale.

Car Fleet Sales, Inc. Hamilton, Massachusetts

Customer: Cleanway Rent-A-Car

Quantity	Type of Car	Cost per Car	Amount for This Type of Car
4	Luxury sedans	$21,000	$84,000 (4 × $21,000 = $84,000)
3	Compact sedans	$14,000	$42,000 (3 × $14,000 = $42,000)
7	Subcompact sedans	$8000	$56,000 (7 × $8,000 = $56,000)
		TOTAL	$182,000 (sum of the three amounts)

The total cost of all 14 cars is $182,000.

Continued on next page

3. *Check.* You may use estimation to check. The check is left to the student.

Student Practice 5   Anderson Dining Commons purchased 50 tables at $200 each, 180 chairs at $40 each, and six moving carts at $65 each. What was the total cost of the purchase?

EXAMPLE 6   Dawn had a balance of $410 in her checking account last month. She made deposits of $46, $18, $150, $379, and $22. She made out checks for $316, $400, and $89. What is her balance?

Solution

1. *Understand the problem.* We want to *add* to get a total of all deposits and *add* to get a total of all checks.

$$\boxed{\text{Old balance}} \ + \ \boxed{\text{total of deposits}} \ - \ \boxed{\text{total of checks}} \ = \ \boxed{\text{new balance}}$$

Mathematics Blueprint for Problem Solving

Gather the Facts	What Am I Asked to Do?	How Do I Proceed?	Key Points to Remember
Old balance: $410. New deposits: $46, $18, $150, $379, and $22. New checks: $316, $400, and $89.	Find the amount of money in the checking account after deposits are made and checks are withdrawn.	**(a)** I need to calculate the total of the deposits and the total of the checks. **(b)** I add the total deposits to the old balance. **(c)** Then I subtract the total of the checks from that result.	Deposits are added to a checking account. Checks are subtracted from a checking account.

2. *Solve and state the answer.*

First we find the total of deposits:

$$\begin{array}{r} 46 \\ 18 \\ 150 \\ 379 \\ + \ 22 \\ \hline 615 \end{array}$$

Then the total of checks:

$$\begin{array}{r} 316 \\ 400 \\ + \ 89 \\ \hline 805 \end{array}$$

Add the deposits to the old balance and subtract the amount of the checks.

Old balance	410
+ total deposits	+ 615
	1025
– total checks	– 805
New balance	220

The new balance of the checking account is $220.

3. *Check.* Work backward. You can add the total checks to the new balance and then subtract the total deposits. The result should be the old balance. Try it.

$$
\begin{array}{r}
410 \quad \text{Old balance} \;\checkmark \\
-\ 615 \\
\hline
1025 \\
+\ 805 \\
\hline
220 \quad \text{Work backward.}
\end{array}
$$

Student Practice 6   Last month Bridget had $498 in a savings account. She made two deposits: one for $607 and one for $163. The bank credited her with $36 interest. Since last month, she has made four withdrawals: $19, $158, $582, and $74. What is her balance this month?

Mathematics Blueprint for Problem Solving

Gather the Facts	What Am I Asked to Do?	How Do I Proceed?	Key Points to Remember

---

EXAMPLE 7   When Lorenzo began his car trip, his gas tank was full and the odometer read 76,358 miles. He ended his trip at 76,668 miles and filled the gas tank with 10 gallons of gas. How many miles per gallon did he get with his car?

Solution

1. *Understand the problem.*

Mathematics Blueprint for Problem Solving

Gather the Facts	What Am I Asked to Do?	How Do I Proceed?	Key Points to Remember
Odometer reading at end of trip: 76,668 miles. Odometer reading at start of trip: 76,358 miles. Used on trip: 10 gallons of gas	Find the number of miles per gallon that the car obtained on the trip.	**(a)** I need to subtract the two odometer readings to obtain the number of miles traveled. **(b)** I divide the number of miles driven by the number of gallons of gas used to get the number of miles obtained per gallon of gas.	The gas tank was full at the beginning of the trip. 10 gallons fills the tank at the end of the trip.

Continued on next page

2. *Solve and state the answer.* First we subtract the odometer readings to obtain the miles traveled.

$$\begin{array}{r} 76{,}668 \\ -\ 76{,}358 \\ \hline 310 \end{array}$$

The trip was 310 miles.

Next we divide the miles driven by the number of gallons.

$$\begin{array}{r} 31 \\ 10\overline{)310} \\ \underline{30} \\ 10 \\ \underline{10} \\ 0 \end{array}$$

Thus Lorenzo obtained 31 miles per gallon on the trip.

3. *Check.* We do not want to round to one nonzero digit here because, if we do, the result will be zero when we subtract. Thus we will round to the nearest hundred for the values of mileage.

$$76{,}668 \longrightarrow 76{,}700$$

$$76{,}358 \longrightarrow 76{,}400$$

Now we subtract the estimated values.

$$\begin{array}{r} 76{,}700 \\ -\ 76{,}400 \\ \hline 300 \end{array}$$

Thus we estimate the trip to be 300 miles.

Then we divide.

$$\begin{array}{r} 30 \\ 10\overline{)300} \end{array}$$

We obtain 30 miles per gallon for our estimate. This is very close to our calculated value of 31 miles per gallon. ✓

Student Practice 7 Deidre took a car trip starting with a full tank of gas. Her trip began with the odometer at 50,698 and ended at 51,118 miles. She then filled the tank with 12 gallons of gas. How many miles per gallon did her car get on the trip?

Mathematics Blueprint for Problem Solving

Gather the Facts	What Am I Asked to Do?	How Do I Proceed?	Key Points to Remember

In general, most students find that they are more successful at solving applied problems if they take extra time to understand the problem. This requires careful reading and thinking about what the problem is asking you to do. Use a colored pen or pencil and underline the most important facts. Draw a picture or sketch if it will help you visualize the situation. Remember, if you understand what you are solving for, your work will go much more quickly.

STEPS TO SUCCESS

If you attend a traditional mathematics class that meets one or more times each week:

Faithful Class Attendance Is Well Worth It.

Get started in the right direction. Make a personal commitment to attend class every day, beginning with the first day of class. Teachers and students all over the country have discovered that faithful class attendance and good grades go together.

The vital content of class. What goes on in class is designed to help you learn more quickly. Each day significant information is given that will truly help you to understand concepts. There is no substitute for this firsthand learning experience.

Meet a friend. You will soon discover that other students are also coming to class every single class period. It is easy to strike up a friendship with those students who share this common commitment. They will usually be available to answer a question after class and give you an additional source of help when you encounter difficulty.

Making it personal: Write down what you think is the most compelling reason to attend every class meeting. Make that commitment and see how much it helps you. ▼

If you are enrolled in an online mathematics class, self-paced mathematics class taught in a math lab, or some other type of nontraditional class:

Keeping Yourself on Schedule

The key to success is to keep on schedule. In a class where you determine your own pace, you will need to commit yourself to following the suggested pace provided in your course materials. Check off each assignment as you do it so you can see your progress.

Make sure all your class materials are organized and available. Keep all course schedules and assignments right where you can quickly find them. Review them often to be sure you are doing everything that you should.

Discipline yourself to follow the detailed schedule exactly for the first six weeks. Professor Tobey and Professor Slater have both taught online classes for several years. They have found that students usually succeed in the course as long as they do every suggested activity for the first six weeks!

Making it personal: Are you good at following schedules and keeping track of details? Which of the suggestions above do you find the most helpful? ▼

Applications

You may want to use the Mathematics Blueprint for Problem Solving to help you to solve the applied problems in exercises 1–34.

1. *Real Estate* Donna and Miguel want to buy a cabin for $31,500. After repairs, the total cost will be $40,300. How much will the repairs cost?

▲ **2.** *Geography* China has a total area of 9,596,960 square kilometers. Bodies of water account for 270,550 square kilometers. How many square kilometers of land does China have?

3. *Business Management* Paula is organizing a large two-day convention. Bert's Bagels is providing the breakfast bagels. If Paula orders 120 bakers' dozen, how many bagels is that? (There are 13 in a bakers' dozen.)

4. *Business Management* There are 144 pencils in a gross. Mr. Jim Weston ordered 14 gross of pencils for the office. How many pencils did he order?

5. *Consumer Affairs* A 12-ounce can of Hunts tomato sauce costs 84¢. What is the unit cost of the tomato sauce? (How much does the tomato sauce cost per ounce?)

6. *Consumer Affairs* A 15-ounce can of Del Monte pears costs 90¢. What is the unit cost of the pears? (How much do the pears cost per ounce?)

7. *Sports* Kimberly began running 3 years ago. She has spent $832 on 13 pairs of running shoes during this time. How much on average did each pair of shoes cost?

8. *Wildlife Management* There are approximately 50,000 bison living in the United States. If Northwest Trek, the animal preserve located in Mt. Rainier National Park, has 103 bison, how many bison are living elsewhere?

9. *Population Studies* In 2010, the population of the United States was approximately 309,163,000. By the year 2043, it is projected that the population will reach 400,000,000. What is the expected increase in population from 2010 to 2043?

▲ **10.** *Geometry* Valleyfair, an amusement park in Minnesota, covers 26 acres. If there are 44,010 square feet in 1 acre, how many square feet of land does Valleyfair cover?

11. *Business Management* A games arcade has recently opened in a West Chicago neighborhood. The owners were nervous about whether it would be a success. Fortunately, the gross revenues over the last four weeks were $7356, $3257, $4777, and $4992. What was the gross revenue for the arcade over these four weeks?

12. *International Relations* In 2010, the two largest cities in France were Paris, with 2,203,817 people, and Marseille, with 839,043 people. What is the difference in population between these two cities?

13. *Wildlife Management* The Federal Nigeria game preserve has 24,111 animals, 327 full-time staff, and 793 volunteers. What is the total of these three groups? How many more volunteers are there than full-time staff?

14. *Geography* The longest rivers in the world are the Nile River, the Amazon River, and the Mississippi River. Their lengths are 4132 miles, 3915 miles, and 3741 miles, respectively. How many total miles do these three rivers run? What is the difference in the lengths of the Nile and the Mississippi?

15. *World History* Every 60 minutes, the world population increases by 100,000 people. How many people will be born during the next 480 minutes?

16. *Personal Finance* Roberto had $2158 in his savings account six months ago. In the last six months he made four deposits: $156, $238, $1119, and $866. The bank deposited $136 in interest over the six-month period. How much does he have in the savings account at present?

In exercises 17–34, more than one type of operation is required.

17. *Sports* Carmen gives golf lessons every Saturday. She charges $15 for adults, $9 for children, and $5 for club rental. Last Saturday she taught six adults and eight children, and six people needed to rent clubs. How much money did Carmen make on that day?

18. *Business Management* Whale Watch Excursions charges $10 for adults, $6 for children, and $7 for senior citizens. On the last trip of the day, there were five adults, seven children, and three senior citizens. How much money did the company make on this trip?

19. *Personal Finance* Sue Li had a balance in her checking account of $132. During the last two months she has deposited four paychecks of $715 each. She wrote two rent checks for $575 each, and wrote checks totaling $482 for other bills. When all the deposits are recorded and the checks clear, what will be the balance in her checking account?

20. *Space Travel* From 1957 to 2005, the number of successful space launches totaled 4361. Of these, 2746 were launched by the Soviet Union/Russia, and 1305 were launched by the United States. How many launches were completed by other countries?

21. *Real Estate* Diana owns 85 acres of forest land in Oregon. She rents it to a timber grower for $250 per acre per year. Her property taxes are $57 per acre. How much profit does she make on the land each year?

22. *Real Estate* Todd owns 13 acres of commercially zoned land in the city of Columbus, Ohio. He rents it to a construction company for $12,350 per acre per year. His property taxes to the city are $7362 per acre per year. How much profit does he make on the land each year?

23. *Environmental Studies* Hanna wants to determine the miles-per-gallon rating of her Chevrolet Cavalier. She filled the tank when the odometer read 14,926 miles. She then drove her car on a trip. At the end of the trip, the odometer read 15,276 miles. It took 14 gallons to fill the tank. How many miles per gallon does her car deliver?

24. *Environmental Studies* Gary wants to determine the miles-per-gallon rating of his Geo Metro. He filled the tank when the odometer read 28,862 miles. After ten days, the odometer read 29,438 miles and the tank required 18 gallons to be filled. How many miles per gallon did Gary's car achieve?

25. *Forestry* A beautiful piece of land in the Wilmot Nature Preserve has three times as many oak trees as birches, two times as many maples as oaks, and seven times as many pine trees as maples. If there are 18 birches on the land, how many of each of the other trees are there? How many trees are there in all?

26. *Business Management* The Cool Coffee Lounge in Albuquerque, New Mexico, has 27 tables, and each table has either two or four chairs. If there are a total of 94 chairs accompanying the 27 tables, how many tables have four chairs? How many tables have two chairs?

Radio Stations

The following is a partial list of the different formats of commercial radio stations in the United States in 2009. Use this list to answer exercises 27–30.

Primary Format	Number of Stations
Country	1997
News/Talk	1401
Spanish	800
Oldies	669
Sports	635
Adult Contemporary	626
Top 40	483
Classic Rock	479
Hot Adult Contemporary	407
Classic Hits	367
Adult Standards	330
Religion (Teaching, Variety)	328
Rock	294
Black Gospel	241
Southern Gospel	212
Soft Adult Contemporary	207
Adult Hits	207
Urban Adult Contemporary	161
Christian	153
R&B	129

Source: World Almanac and Book of Facts, 2010

27. How many radio stations play primarily some kind of adult contemporary music?

28. How many radio stations are categorized as religion, gospel, or Christian stations?

29. How many more country stations are there than classic rock or classic hits stations?

30. How many more Spanish or oldies stations are there than news/talk stations?

Cable Television

Use the following bar graph to answer exercises 31–34.

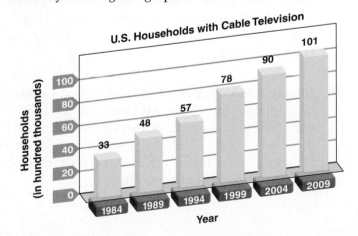

31. How many more households had cable television in 2004 than in 1984?

32. How many more households had cable television in 2009 than in 1989?

33. If the same increase occurs between 2009 and 2014 as occurred between 2004 and 2009, how many households will have cable television in 2014?

34. If the same increase occurs between 2009 and 2019 as occurred between 1999 and 2009, how many households will have cable television in 2019?

Cumulative Review

35. [1.6.1] Evaluate. 7^3

36. [1.6.2] Perform in the proper order.
$3 \times 2^3 + 15 \div 3 - 4 \times 2$

37. [1.4.4] Multiply. 126×38

38. [1.5.3] Divide. $12\overline{)3096}$

39. [1.2.4] Add. $96 + 123 + 57 + 526$

40. [1.3.3] Subtract. $509{,}263 - 485{,}978$

41. [1.7.1] Round to the nearest thousand.
526,195,726

42. [1.1.3] Write this number in standard notation. Three billion, four hundred million, six hundred three thousand, twenty-five.

Quick Quiz 1.8

1. Sixteen people in a travel club chartered a bus to go to Vermont to see the fall foliage. The bill for the bus charter was $4304. How much did each club member pay if the cost is shared equally?

2. Maria had a balance of $471 in her checking account last month. She then deposited $198, $276, and $347. She made out checks for $49, $227, and $158. What will her new balance be?

3. The entire Tobey family went on a fishing charter. The cost was $11 for people 60 or older, $14 for people age 12 to 59, and $5 for children under 12. The captain counted two people over 60, six people age 12 to 59, and four children under 12. What was the total cost for the Tobey family members to go on the fishing charter?

4. **Concept Check** A company has purchased 38 new cars for the sales department for $836,543. Assuming that each car cost the same, explain how you would estimate the cost of each car.

Did You Know...
That Managing Your Debts Can Save You Money Over Time?

LIVE DEBT FREE

Understanding the Problem:

Tracy and Max are in debt.

- Each of their three credit cards is maxed out to the limit of $8000.
- They owe $12,000 in hospital bills and $2000 for their car loan.
- After borrowing from friends, they still owe $100 to one friend and $300 to another.

Before they can save for a much-needed vacation, they must pay off their debts.

Making a Plan:

Tracy and Max come up with a plan for tackling their debt.

Step 1: They list all of their debts, ordered from smallest to largest.

Task 1: *Complete Step 1 for Tracy and Max. Remember that there are three credit cards.*

Step 2: They make minimum monthly payments on each debt.

- Each credit card has a minimum monthly payment of $25.
- The hospital expects a payment of $50 per month.
- The monthly car payment is $200.
- Tracy and Max arrange to pay $20 per month for each loan from friends.

Task 2: *What is the total amount of their minimum monthly payments?*

Step 3: They pay off their three smallest debts first.

Task 3: *What are their three smallest debts?*

Finding a Solution:

Step 4: Tracy and Max decide to eliminate any unnecessary spending until the three smallest debts are paid off. By doing this, they can pay off the two smallest debts in only two months while still making the minimum payments on the other debts.

Task 4: *What is the total amount of the minimum monthly payments for the two smallest debts?*

Step 5: Each month, Tracy and Max take the amount that they would have used to pay the two smallest debts and apply it toward the third smallest, all while paying the minimum monthly payments on the remaining debts.

Task 5: *How many more months will it take Tracy and Max to pay off the third smallest debt if they follow Step 5? Round your answer to the nearest whole number when you perform division operations.*

Step 6: After the three smallest debts are paid off, Tracy and Max take the money that they would have spent per month to pay those debts and use it on the principal of the remaining debts. To pay the debts more quickly, Tracy and Max decide to stop using credit cards for new purchases. After a few years of careful budgeting, not using credit cards, and paying more than the minimum payment on their credit card debt, they finally pay off their debts.

Task 6: *Besides avoiding credit cards for new purchases, can you think of other ways that Tracy and Max could have budgeted their money and paid off the debt more quickly?*

Applying the Situation to Your Life:

Debt counselors often provide this simple, practical plan for people in debt:

- Arrange debts in order.
- Pay off the smallest debt first.
- Let the consequences of paying off the smallest debt help you to pay off the rest of the debts more quickly.
- Avoid unnecessary spending and incorporate budgeting strategies into your daily life.

Chapter 1 Organizer

Topic and Procedure	Examples	✏️ You Try It
Place value of numbers, p. 2 Each digit has a value depending on location. millions | hundred thousands | ten thousands | thousands | hundreds | tens | ones	In the number 2,896,341, what place value does 9 have? ten thousands	1. In the number 731,244, what place value does 7 have?
Writing expanded notation, p. 2 Take each digit and multiply it by one, ten, hundred, thousand, . . . according to its place.	Write in expanded notation. 46,235 $40{,}000 + 6000 + 200 + 30 + 5$	2. Write in expanded notation. 132,259
Writing whole numbers in words, p. 4 Take the number in each period and indicate if it is (millions) (thousands) (ones) xxx, xxx, xxx	Write in words. 134,718,216 one hundred thirty-four million, seven hundred eighteen thousand, two hundred sixteen	3. Write in words. 58,872,150
Adding whole numbers, p. 12 Starting with the right column, add each column separately from right to left. If a two-digit sum occurs, "carry" the left digit over to the next column to the left.	Add. $\begin{array}{r} {}^{2\ 1}\ \\ 2\ 5\ 8 \\ 3\ 6\ 7 \\ 2\ 9\ 1 \\ +\ 4\ 5\ 3 \\ \hline 1\ 3\ 6\ 9 \end{array}$	4. Add. $\begin{array}{r} 478 \\ 134 \\ 260 \\ +\ 73 \\ \hline \end{array}$
Subtracting whole numbers, p. 23 Starting with the right column, subtract each column separately from right to left. If necessary, borrow a unit from the column to the left and bring it to the right as a "10."	Subtract. $\begin{array}{r} {}^{13} \\ {}^{6\ \ 3\ 12} \\ 1\ 6{,}\ 7\ 4\ 2 \\ -\ 1\ 2{,}\ 3\ 9\ 5 \\ \hline 4\ 3\ 4\ 7 \end{array}$	5. Subtract. $\begin{array}{r} 23{,}495 \\ -19{,}297 \\ \hline \end{array}$
Multiplying several-digit numbers, p. 39 Multiply the top factor by the ones digit, then by the tens digit, then by the hundreds digit. Add the partial products together.	Multiply. $\begin{array}{r} 5\ 6\ 7 \\ \times\ 2\ 3\ 8 \\ \hline 4\ 5\ 3\ 6 \\ 1\ 7\ 0\ 1\ \ \\ 1\ 1\ 3\ 4\ \ \ \\ \hline 1\ 3\ 4{,}9\ 4\ 6 \end{array}$	6. Multiply. $\begin{array}{r} 532 \\ \times\ 167 \\ \hline \end{array}$
Multiplying several factors, p. 41 Keep multiplying from left to right. Take each product and multiply by the next factor to the right. Continue until all factors are used once. (Since multiplication is commutative and associative, the factors can be multiplied in any order.)	Multiply. $\begin{aligned} 2 \times 9 \times 7 \times 6 \times 3 &= 18 \times 7 \times 6 \times 3 \\ &= 126 \times 6 \times 3 \\ &= 756 \times 3 \\ &= 2268 \end{aligned}$	7. Multiply. $5 \times 2 \times 4 \times 6 \times 8$

Topic and Procedure	Examples	✏️ You Try It
Dividing by a two- or three-digit number, p. 52 Figure how many times the first digit of the divisor goes into the first two digits of the dividend. To try this answer, multiply it back to see if it is too large or small. Continue each step of long division until finished.	Divide. $$\begin{array}{r} 589 \\ 238\overline{)140{,}182} \\ \underline{1190} \\ 2118 \\ \underline{1904} \\ 2142 \\ \underline{2142} \\ 0 \end{array}$$	8. Divide. $135\overline{)84{,}780}$
Exponent form, p. 61 To show in short form the repeated multiplication of the same number, write the number being multiplied. (This is the base.) Write in smaller print above the line the number of times it appears as a factor. (This is the exponent.) To evaluate the exponent form, write the factor the number of times shown in the exponent. Then multiply.	**(a)** Write in exponent form. $10 \times 10 \times 10 \times 10 \times 10 \times 10 \times 10 \times 10$ 10^8 **(b)** Evaluate. 6^3 $6 \times 6 \times 6 = 216$	9. **(a)** Write in exponent form. $9 \times 9 \times 9 \times 9 \times 9$ **(b)** Evaluate. 7^4
Order of operations, p. 64 1. Perform operations inside parentheses. 2. Simplify exponents. 3. Then do multiplication and division in order from left to right. 4. Then do addition and subtraction in order from left to right.	Evaluate. $$2^3 + 16 \div 4^2 \times 5 - 3$$ Evaluate exponents first. $$8 + 16 \div 16 \times 5 - 3$$ Then do multiplication or division from left to right. $$8 + 1 \times 5 - 3$$ $$8 + 5 - 3$$ Then do addition and subtraction. $$13 - 3 = 10$$	10. Evaluate. $6^2 \div 2 \times 3 - (10 - 8)^2$
Rounding, p. 68 1. If the first digit to the right of the round-off place is less than 5, the digit in the round-off place is unchanged. 2. If the first digit to the right of the round-off place is 5 or more, the digit in the round-off place is increased by 1. 3. Digits to the right of the round-off place are replaced by zeros.	**(a)** Round to the nearest hundred. 56,743 $\downarrow$ 5 6,7④3 The digit 4 is less than 5. 56,700 **(b)** Round to the nearest thousand. 128,517 $\downarrow$ 1 2 8,⑤17 The digit 5 is obviously 5 or more. We increase the thousands digit by 1. 129,000	11. **(a)** Round to the nearest thousand. 338,912 **(b)** Round to the nearest hundred thousand. 745,830
Estimating the answer to a calculation, p. 70 1. Round each number so that there is one nonzero digit. 2. Perform the calculation with the rounded numbers.	Estimate the answer. $$45{,}780 \times 9453$$ First we round. $$50{,}000 \times 9000$$ Then we multiply. $$\begin{array}{r} 50{,}000 \\ \times\ \ \ \ 9{,}000 \\ \hline 450{,}000{,}000 \end{array}$$ We estimate the answer to be 450,000,000.	12. Estimate the answer. $92{,}785 \times 1685$

Procedure for Solving Applied Problems

Using the Mathematics Blueprint for Problem Solving, p. 79

In solving an applied problem, students may find it helpful to complete the following steps. You will not use all the steps all the time. Choose the steps that best fit the conditions of the problem.

1. Understand the problem.
   (a) Read the problem carefully.
   (b) Draw a picture if this helps you to visualize the situation. Think about what facts you are given and what you are asked to find.
   (c) Use the Mathematics Blueprint for Problem Solving to organize your work. Follow these four parts.
      1. Gather the facts (Write down specific values given in the problem.)
      2. What am I asked to do? (Identify what you must obtain for an answer.)
      3. Decide what calculations need to be done.
      4. Key points to remember. (Record any facts, warnings, formulas, or concepts you think will be important as you solve the problem.)

2. Solve and state the answer.
   (a) Perform the necessary calculations.
   (b) State the answer, including the unit of measure.

3. Check.
   (a) Estimate the answer to the problem. Compare this estimate to the calculated value. Is your answer reasonable?
   (b) Repeat your calculations.
   (c) Work backward from your answer. Do you arrive at the original conditions of the problem?

EXAMPLE The Manchester highway department has just purchased two pickup trucks and three dump trucks. The cost of a pickup truck is $17,920. The cost of a dump truck is $48,670. What was the cost to purchase these five trucks?

1. **Understand the problem.**

2. **Solve and state the answer.**
   Calculate cost of pickup trucks

   $$\begin{array}{r} \$17,920 \\ \times \quad\quad 2 \\ \hline \$35,840 \end{array}$$

   Calculate cost of dump trucks

   $$\begin{array}{r} \$48,670 \\ \times \quad\quad 3 \\ \hline \$146,010 \end{array}$$

   Find total cost. $35,840 + $146,010 = $181,850
   The total cost of the five trucks is $181,850.

3. **Check.**
   Estimate cost of pickup trucks

   $$20,000 \times 2 = 40,000$$

   Estimate cost of dump trucks

   $$50,000 \times 3 = 150,000$$

   Total estimate

   $$40,000 + 150,000 = 190,000$$

   This is close to our calculated answer of $181,850. We determine that our answer is reasonable. ✓

Mathematics Blueprint for Problem Solving

Gather the Facts	What Am I Asked to Do?	How Do I Proceed?	Key Points to Remember
Buy 2 pickup trucks 3 dump trucks Cost Pickup: $17,920 Dump: $48,670	Find the total cost of the 5 trucks.	Find the cost of 2 pickup trucks. Find the cost of 3 dump trucks. Add to get final cost of all 5 trucks.	Multiply 2 times pickup truck cost. Multiply 3 times dump truck cost.

Chapter 1 Review Problems

If you have trouble with a particular type of exercise, review the examples in the section indicated for that group of exercises. Answers to all exercises are located in the answer key.

Section 1.1
Write in words.

1. 892

2. 109,276

Write in expanded notation.

3. 4364

4. 42,166,037

Write in standard notation.

5. five thousand three hundred two

6. one million, three hundred twenty-eight thousand, eight hundred twenty-eight

Section 1.2
Add.

7. $76 + 39$

8. $235 + 165$

9. $12 + 28 + 34 + 76$

10.
$$\begin{array}{r} 123 \\ 61 \\ 9 \\ 84 \\ + 123 \\ \hline \end{array}$$

11.
$$\begin{array}{r} 226 \\ 134 \\ + 647 \\ \hline \end{array}$$

12.
$$\begin{array}{r} 52{,}134 \\ + 7966 \\ \hline \end{array}$$

13.
$$\begin{array}{r} 1356 \\ 2892 \\ 561 \\ 89 \\ + 9805 \\ \hline \end{array}$$

Section 1.3
Subtract.

14.
$$\begin{array}{r} 36 \\ - 19 \\ \hline \end{array}$$

15.
$$\begin{array}{r} 126 \\ - 99 \\ \hline \end{array}$$

16.
$$\begin{array}{r} 543 \\ - 372 \\ \hline \end{array}$$

17.
$$\begin{array}{r} 7000 \\ - 845 \\ \hline \end{array}$$

18.
$$\begin{array}{r} 201{,}340 \\ - 120{,}618 \\ \hline \end{array}$$

19.
$$\begin{array}{r} 6{,}325{,}034 \\ - 89{,}023 \\ \hline \end{array}$$

20.
$$\begin{array}{r} 5{,}412{,}022 \\ - 79{,}031 \\ \hline \end{array}$$

Section 1.4
Multiply.

21. $8 \times 1 \times 9 \times 2$

22. $7 \times 6 \times 0 \times 4$

23. $2 \cdot 5 \cdot 10 \cdot 8$

24. 621×100

25. $84{,}312 \times 1000$

26. $78 \times 10{,}000$

27.
$$\begin{array}{r} 3492 \\ \times \quad 7 \\ \hline \end{array}$$

28.
$$\begin{array}{r} 6257 \\ \times \quad 8 \\ \hline \end{array}$$

29. 58 ⨉ 32	**30.** 73 ⨉ 24	**31.** 709 ⨉ 36	**32.** 123 ⨉ 714

33. 431 ⨉ 623	**34.** 1782 ⨉ 305	**35.** 2057 ⨉ 124	**36.** 3182 ⨉ 35

37. 1200 ⨉ 6000	**38.** 100,000 ⨉ 20,000

Section 1.5

Divide, if possible.

39. $20 \div 10$ **40.** $40 \div 8$ **41.** $0 \div 8$ **42.** $7 \div 1$

43. $\dfrac{81}{9}$ **44.** $\dfrac{42}{6}$ **45.** $\dfrac{5}{0}$ **46.** $\dfrac{24}{6}$

Divide. Be sure to indicate the remainder, if one exists.

47. $6\overline{)750}$ **48.** $9\overline{)1863}$ **49.** $6\overline{)15,024}$ **50.** $8\overline{)24,512}$

51. $6\overline{)221,748}$ **52.** $5\overline{)184,605}$ **53.** $8\overline{)120,371}$ **54.** $67\overline{)490}$

55. $21\overline{)666}$ **56.** $53\overline{)3202}$ **57.** $45\overline{)8775}$ **58.** $132\overline{)7128}$

59. $204\overline{)3876}$

Section 1.6

Write in exponent form.

60. $21 \times 21 \times 21$ **61.** $8 \times 8 \times 8 \times 8 \times 8$

62. $10 \times 10 \times 10 \times 10 \times 10 \times 10$

Evaluate.

63. 2^6 **64.** 3^4 **65.** 5^3 **66.** 7^2

67. 9^2 **68.** 6^3

Perform each operation in proper order.

69. $7 + 2 \times 3 - 5$ **70.** $2^5 + 4 - (5 + 3^2)$

71. $34 - 9 \div 9 \times 12$ **72.** $2^3 \times 5 \div 8 + 3 \times 4$

73. $2^3 + 4 \times 5 - 32 \div (1 + 3)^2$ **74.** $6 \times 3 + 3 \times 5^2 - 63 \div (5 - 2)^2$

Section 1.7

Round to the nearest ten.

75. 3364

76. 5895

77. 42,644

In exercises 78–80, round to the nearest thousand.

78. 12,350

79. 22,986

80. 202,498

81. Round to the nearest hundred thousand. 4,649,320

82. Round to the nearest ten thousand. 9,995,312

Use the principle of estimation to find an estimate for each calculation.

83. $18,702 + 8331 + 36,612$

84. $34,950 - 15,439$

85. $2,965,372 \times 893$

86. $83,421 \div 24$

Section 1.8

Solve.

87. *Computer Applications* Ward can type 25 words per minute on his computer. He typed for seven minutes at that speed. How many words did he type?

88. *Travel* In June, 2462 people visited the Renaissance Festival. There were 1997 visitors in July, and 2561 in August. How many people visited the festival during these three months?

89. *Aviation* A plane was flying at 14,630 feet. It flew over a mountain 4329 feet high. How many feet was it from the plane to the top of the mountain?

90. *Personal Finance* Gerardo was billed $4330 for tuition, and he needs to spend $268 on books. He received a $1250 scholarship. How much will he have to pay for tuition and books after the scholarship is deducted?

91. *Business Management* Middlebury College ordered 112 dormitory beds for $8288. What was the cost per bed?

92. *Personal Finance* Melissa's savings account balance last month was $810. The bank added $24 interest. Melissa deposited $105, $36, and $177. She made withdrawals of $18, $145, $250, and $461. What will be her balance this month?

93. *Environmental Studies* Ali began a trip on a full tank of gas with the car odometer at 56,320 miles. He ended the trip at 56,720 miles and added 16 gallons of gas to refill the tank. How many miles per gallon did he get on the trip?

94. *Business Management* Anita is opening a new café in town. She bought 15 tables at $65 each, 60 chairs for $12 each, and eight ceiling fans for $42 each. What was the total purchase price for these items?

Environmental Protection *Use the following bar graph to answer exercises 95–97.*

95. How many more tons of solid waste were recovered and recycled in 2000 than in 1985?

96. What was the greatest increase in tons of solid waste recovered and recycled in a five-year period?

97. If the exact same increase in the number of tons recovered occurs from 2010 to 2020 as occurred from 2000 to 2010, how many tons of solid waste will be recovered and recycled in 2020?

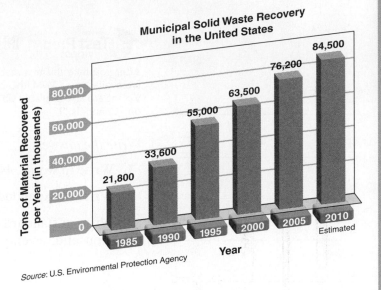

Source: U.S. Environmental Protection Agency

Mixed Practice

Perform each calculation.

98. $205 + 36 + 1983 + 60$

99.
$$\begin{array}{r} 56{,}793 \\ -\ 48{,}926 \\ \hline \end{array}$$

100. 396×28

101. $37\overline{)4773}$

102. Evaluate. $4 \times 12 - (12 + 9) + 2^3 \div 4$

103. ***Personal Finance*** Michael Evans has $3000 in his checking account. He buys 3 computers at $699 each and 2 printers at $78 each. How much does he have remaining after the purchases?

▲ **104.** ***Geometry*** Milton is building a rectangular patio in his backyard. The patio measures 22 feet by 15 feet.

(a) How many square feet is the patio?

(b) If Milton wanted to fence in the patio, how many feet of fence would he need?

How Am I Doing?   Chapter 1 Test

Write the answers.

1. Write in words. 44,007,635

2. Write in expanded notation. 26,859

3. Write in standard notation. three million, five hundred eighty-one thousand, seventy-six

Add.

4.	189	5.	763	6.	135,484
	26		220		2,376
	12		+ 508		81,004
	528				+ 100,113
	+ 76				

Subtract.

7.	8961	ᴹᴄ 8.	501,760	9.	18,400,100
	− 894		− 328,902		− 13,174,332

Multiply.

10. $1 \times 6 \times 9 \times 7$

11.	45
	× 96

ᴹᴄ 12.	326
	× 592

13.	18,491
	× 7

In problems 14–16, divide. If there is a remainder, be sure to state it as part of your answer.

14. $5\overline{)15,071}$

15. $6\overline{)14,148}$

16. $37\overline{)13,024}$

17. Write in exponent form. $14 \times 14 \times 14$

18. Evaluate. 2^6

1. _____ ☐

2. _____ ☐

3. _____ ☐

4. _____ ☐

5. _____ ☐

6. _____ ☐

7. _____ ☐

8. _____ ☐

9. _____ ☐

10. _____ ☐

11. _____ ☐

12. _____ ☐

13. _____ ☐

14. _____ ☐

15. _____ ☐

16. _____ ☐

17. _____ ☐

18. _____ ☐

In problems 19–21, perform each operation in proper order.

$\mathbb{M}_{\mathbb{C}}$ 19. $5 + 6^2 - 2 \times (9 - 6)^2$ **20.** $2^4 + 3^3 + 28 \div 4$

21. $4 \times 6 + 3^3 \times 2 + 23 \div 23$

22. Round to the nearest hundred.   94,768

23. Round to the nearest ten thousand.   6,462,431

24. Round to the nearest hundred thousand.   5,278,963

Estimate the answer.

$\mathbb{M}_{\mathbb{C}}$ 25. $4,867,010 \times 27,058$ **26.** $1423 + 3298 + 4103 + 7614$

Solve.

27. A cruise for 15 people costs $32,220. If each person paid the same amount, how much will it cost each individual?

28. The river is 602 feet wide at Big Bend Corner. A boy is in the shallow water, 135 feet from the shore. How far is the boy from the other side of the river?

29. At the bookstore, Hector bought three notebooks at $2 each, one textbook for $45, two lamps at $21 each, and two sweatshirts at $17 each. What was his total bill?

30. Patricia is looking at her checkbook. She had a balance last month of $31. She deposited $902 and $399. She made out checks for $885, $103, $26, $17, and $9. What will be her new balance?

▲ 31. The runway at Beverly Airport needs to be resurfaced. The rectangular runway is 6800 feet long and 110 feet wide. What is the area of the runway that needs to be resurfaced?

▲ 32. Nancy Tobey planted a vegetable garden in the backyard. However, the deer and raccoons have been stealing all the vegetables. She asked John to fence in the garden. The rectangular garden measures 8 feet by 15 feet. How many feet of fence should John purchase if he wants to enclose the garden?

19. _____ ☐

20. _____ ☐

21. _____ ☐

22. _____ ☐

23. _____ ☐

24. _____ ☐

25. _____ ☐

26. _____ ☐

27. _____ ☐

28. _____ ☐

29. _____ ☐

30. _____ ☐

31. _____ ☐

32. _____ ☐

Total Correct: ☐

MATH COACH

Mastering the skills you need to do well on the test.

Students often make the same types of errors when they do the Chapter 1 Test. Here are some helpful hints to keep you from making these common errors on test problems.

Subtract Whole Numbers with Borrowing—Problem 8

$$501,760$$
$$-\ 328,902$$

> **Helpful Hint** It is wise to show the borrowing steps. This will help you avoid a borrowing error.

Look at your work for Problem 8. Examine your steps. Do your borrowing steps match the solution below?

Yes ____ No ____

Write out the borrowing steps:

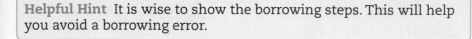

$$\begin{array}{r} {\scriptstyle 4\ \ 9\ \ 10\ 17\ \ 5\ \ 10}\\ \cancel{5}\ \cancel{0}\ \cancel{1},\cancel{7}\ \cancel{6}\ \cancel{0}\\ -\ 3\ 2\ 8,9\ 0\ 2 \end{array}$$

If you answered No, be sure to write your borrowing steps carefully and then check each subtraction step for errors.

Multiply Whole Numbers with Several Digits—Problem 12

$$326$$
$$\times\,592$$

> **Helpful Hint** When multiplying, students sometimes make errors in alignment. Take the extra time to line up each column carefully. Your calculation accuracy will improve if you write your numbers about 50% larger than normal.

Did you line up each column accurately with the numbers shown below?

Yes ____ No ____

$$\begin{array}{r} 326\\ \times\,592\\ \hline 652\\ 2934\\ 1630 \end{array}$$

If you answered No, rework the problem using correct alignment. Then, be sure to check each step for multiplication and addition errors.

Need help? Watch the MATH COACH videos in MyMathLab® or on YouTube™.

104

Order of Operations—Problem 19 $5 + 6^2 - 2 \times (9 - 6)^2$

> **Helpful Hint** Be sure to show each step. Students often skip steps, which leads to errors.

The first step is to combine numbers inside the parentheses.

$$5 + 6^2 - 2 \times (3)^2$$

The second step is to evaluate exponents in two places.

$$5 + 36 - 2 \times 9$$

Did you do the first step correctly?

Yes ____   No ____

Did you do the second step correctly?

Yes ____   No ____

Try to do Problem 19 correctly now. Go slowly and write each step.

Estimate the Product—Problem 25 $4,867,010 \times 27,058$

> **Helpful Hint** Make sure to round each number so that it has only one nonzero digit. This will make the calculation easier to complete.

Did you round each number so that it has only one non-zero digit?

Yes ____   No ____

Did you write down the correct number of zeros?

Yes ____   No ____

$$5,000,000 \times 30,000 = 150,000,000,000$$

If you answered Problem 25 incorrectly, try to rework it now.

Need more help? Look for section examples marked with $\mathbb{MC}$ to review.

105

CHAPTER 2

All of us have seen pictures of the Pyramids of Egypt. These amazing structures were built very carefully. Measurements had to be made that were very precise. The ancient Egyptians used an elaborate system of fractions that allowed them to make highly accurate measurements. As you master the topics of this chapter, you will master the basic skills used by the designers of the Pyramids of Egypt.

Fractions

2.1 Understanding Fractions

① Using a Fraction to Represent Part of a Whole

In Chapter 1 we studied whole numbers. In this chapter we will study a fractional part of a whole number. One way to represent parts of a whole is with **fractions.** The word *fraction* (like the word *fracture*) suggests that something is being broken. In mathematics, fractions represent the part that is "broken off" from a whole. The whole can be a single object (like a whole pie) or a group (the employees of a company). Here are some examples.

Student Learning Objectives

After studying this section, you will be able to:

① Use a fraction to represent part of a whole.

② Draw a sketch to illustrate a fraction.

③ Use fractions to represent real-life situations.

Single object

$\frac{1}{3}$

The whole is the pie on the left. The fraction $\frac{1}{3}$ represents the shaded part of the pie, 1 of 3 pieces. $\frac{1}{3}$ is read "one-third."

A group: ACE company employs 150 men, 200 women.

$$\frac{150}{350}$$

The whole is the company of 350 people (150 men plus 200 women). The fraction $\frac{150}{350}$ represents that part of the company consisting of men.

Recipe: Applesauce
4 apples
1/2 cup sugar
1 teaspoon cinnamon

The whole is 1 whole cup of sugar. This recipe calls for $\frac{1}{2}$ cup of sugar. Notice that in many real-life situations $\frac{1}{2}$ is written as 1/2.

When we say "$\frac{3}{8}$ of a pizza has been eaten," we mean 3 of 8 equal parts of a pizza have been eaten. (See the figure.) When we write the fraction $\frac{3}{8}$, the number on the top, 3, is the **numerator,** and the number on the bottom, 8, is the **denominator.**

The numerator specifies how many parts → 3
The denominator specifies the total number of parts → 8

When we say, "$\frac{2}{3}$ of the marbles are red," we mean 2 marbles out of a total of 3 are red marbles.

Part we are interested in → 2 numerator
Total number in the group → 3 denominator

EXAMPLE 1   Use a fraction to represent the shaded or completed part of the whole shown.

(a)   **(b)**

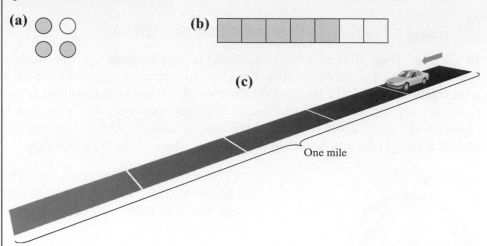

(c)

One mile

Solution

(a) Three out of four circles are shaded. The fraction is $\dfrac{3}{4}$.

(b) Five out of seven equal parts are shaded. The fraction is $\dfrac{5}{7}$.

(c) The mile is divided into five equal parts. The car has traveled 1 part out of 5 of the one-mile distance. The fraction is $\dfrac{1}{5}$.

NOTE TO STUDENT: Fully worked-out solutions to all of the Student Practice problems can be found at the back of the text starting at page SP-1.

Student Practice 1   Use a fraction to represent the shaded part of the whole.

(a)   **(b)**   **(c)**

We can also think of a fraction as a division problem.

$$\frac{1}{3} = 1 \div 3 \quad \text{and} \quad 1 \div 3 = \frac{1}{3}$$

The division way of looking at fractions asks the question:

What is the result of dividing one whole into three equal parts?

Thus we can say the fraction $\frac{a}{b}$ means the same as $a \div b$. However, special care must be taken with the number 0.

Suppose that we had four equal parts and we wanted to take none of them. We would want $\frac{0}{4}$ of the parts. Since $\frac{0}{4} = 0 \div 4 = 0$, we see that $\frac{0}{4} = 0$. Any fraction with a 0 numerator equals zero.

$$\frac{0}{8} = 0 \qquad \frac{0}{5} = 0 \qquad \frac{0}{13} = 0$$

What happens when zero is in the denominator? $\frac{4}{0}$ means 4 out of 0 parts. Taking 4 out of 0 does not make sense. We say $\frac{4}{0}$ is **undefined.**

$$\frac{3}{0}, \ \frac{7}{0}, \ \frac{4}{0} \quad \text{are } \textbf{undefined.}$$

We cannot have a fraction with 0 in the denominator. Since $\frac{4}{0} = 4 \div 0$, we say division by zero is *undefined*. We cannot divide by 0.

② Drawing a Sketch to Illustrate a Fraction

Drawing a sketch of a mathematical situation is a powerful problem-solving technique. The picture often reveals information not always apparent in the words.

EXAMPLE 2   Draw a sketch to illustrate.

(a) $\frac{7}{11}$ of an object

(b) $\frac{2}{9}$ of a group

Solution

(a) The easiest figure to draw is a rectangular bar.

We divide the bar into 11 equal parts. We then shade in 7 parts to show $\frac{7}{11}$.

(b) We draw 9 circles of equal size to represent a group of 9.

We shade in 2 of the 9 circles to show $\frac{2}{9}$.

Student Practice 2   Draw a sketch to illustrate.

(a) $\frac{4}{5}$ of an object

(b) $\frac{3}{7}$ of a group

Recall these facts about division problems involving the number 1 and the number 0.

DIVISION INVOLVING THE NUMBER 1 AND THE NUMBER 0

1. Any nonzero number divided by itself is 1.

$$\frac{7}{7} = 1$$

2. Any number divided by 1 remains unchanged. $\frac{29}{1} = 29$

3. Zero may be divided by any nonzero number; the result is always zero.

$$\frac{0}{4} = 0$$

4. Division by zero is undefined. $\frac{3}{0}$ is undefined

③ Using Fractions to Represent Real-Life Situations

Many real-life situations can be described using fractions.

EXAMPLE 3   Use a fraction to describe each situation.

(a) A baseball player gets a hit 5 out of 12 times at bat.

(b) There are 156 men and 185 women taking psychology this semester. Describe the part of the class that consists of women.

(c) Robert Tobey found in the Alaska moose count that five-eighths of the moose observed were female.

Solution

(a) The baseball player got a hit $\frac{5}{12}$ of his times at bat.

(b) The total class is $156 + 185 = 341$. The fractional part that is women is 185 out of 341. Thus $\frac{185}{341}$ of the class is women.

(c) Five-eighths of the moose observed were female. The fraction is $\frac{5}{8}$.

156 men	185 women

Total class
341 students

Student Practice 3   Use a fraction to describe each situation.

(a) 9 out of the 17 players on the basketball team are on the dean's list.

(b) The senior class has 382 men and 351 women. Describe the part of the class consisting of men.

(c) John needed seven-eighths of a yard of material.

EXAMPLE 4   Wanda made 13 calls, out of which she made five sales. Albert made 17 calls, out of which he made six sales. Write a fraction that describes for both people together the number of calls in which a sale was made compared with the total number of calls.

Solution   There are $5 + 6 = 11$ calls in which a sale was made.

There were $13 + 17 = 30$ total calls.

Thus $\dfrac{11}{30}$ of the calls resulted in a sale.

Student Practice 4   An inspector found that one out of seven belts was defective. She also found that two out of nine shirts were defective. Write a fraction that describes what part of all the objects examined were defective.

👣 STEPS TO SUCCESS Look Ahead to See What Is Coming.

You will find that learning new material is much easier if you know what is coming. Take a few minutes at the end of your study time to glance over the next section of the book. If you quickly look over the topics and ideas in this new section, it will help you get your bearings when the instructor presents new material. Students find that when they preview new material, it enables them to see what is coming. It helps them to be able to grasp new ideas much more quickly.

Making it personal: Do this right now. Look ahead to the next section of the book. Glance over the ideas and concepts. Write down a couple of facts about the next section. ▼

2.1 Exercises

MyMathLab®

 Watch the videos in MyMathLab

 Download the MyDashBoard App

Verbal and Writing Skills, Exercises 1–4

1. A _____ can be used to represent part of a whole or part of a group.

2. In a fraction, the _____ tells the number of parts we are interested in.

3. In a fraction, the _____ tells the total number of parts in the whole or in the group.

4. Describe a real-life situation that involves fractions.

Name the numerator and the denominator in each fraction.

5. $\dfrac{3}{5}$ 6. $\dfrac{9}{11}$ 7. $\dfrac{7}{8}$ 8. $\dfrac{9}{10}$ 9. $\dfrac{1}{17}$ 10. $\dfrac{1}{15}$

In exercises 11–30, use a fraction to represent the shaded part of the object or the shaded portion of the set of objects.

11.   12.   13.   14.

15.   16.   17.   18.

19.   20.   21.  22.

23.   24.   25. ○○○○○○○ 26.

27.   28.   29.   30.

Draw a sketch to illustrate each fractional part.

31. $\dfrac{1}{5}$ of an object 32. $\dfrac{3}{7}$ of an object 33. $\dfrac{3}{8}$ of an object

34. $\dfrac{5}{12}$ of an object 35. $\dfrac{7}{10}$ of an object 36. $\dfrac{5}{9}$ of an object

111

Applications

37. *Literature Class* Professor Lundeen has 95 students in his British Literature class. Fifty-one of the students are women. What fractional part of the class consists of women?

38. *Personal Finance* Miguel bought a notebook with a total purchase price of 98¢. Of this amount, 7¢ was sales tax. What fractional part of the total purchase price was sales tax?

39. *Personal Finance* Theo bought a used pool table for $950. Part of it was paid for with the $329 he earned helping his cousin finish a construction job. What fractional part of the cost of the pool table was paid for by helping his cousin?

40. *Personal Finance* Jillian earned $167 over the weekend at her waitressing job. She used $48 of it to repay a loan to her sister. What fractional part of her earnings did Jillian use to repay her sister?

41. *Political Campaigns* The Democratic National Committee fundraising event served 122 chicken dinners and 89 roast beef dinners to its contributors. What fractional part of the guests ate roast beef?

42. *Education* Dowling Community College will have 407 students and 64 faculty attend the graduation ceremony. What fractional part of the attendees will be students?

43. *Selling Trees* Boy Scout Troop #33 had a Christmas tree sale to raise money for a summer camping trip. In one afternoon, they sold 9 balsam firs, 12 Norwegian pines, and 5 Douglas firs. What fractional part of the trees sold were balsam firs?

44. *Animal Shelters* At the local animal shelter there are 12 puppies, 25 adult dogs, 14 kittens, and 31 adult cats. What fractional part of the animals are either puppies or adult dogs?

45. *DVD Collection* Jamie has 12 comedies, 15 action films, 8 drama movies, and 9 romance films in her DVD collection. What fractional part of her collection consists of either comedies or romance films?

46. *Music Collection* A box of compact discs contains 5 classical CDs, 6 jazz CDs, 4 soundtracks, and 24 blues CDs. What fractional part of the total CDs is either jazz or blues?

47. *Manufacturing* The West Peabody Engine Company manufactured two items last week: 101 engines and 94 lawn mowers. It was discovered that 19 engines and 3 lawn mowers were defective. Of the engines that were not defective, 40 were properly constructed but 42 were not of the highest quality. Of the lawn mowers that were not defective, 50 were properly constructed but 41 were not of the highest quality.

  (a) What fractional part of all items manufactured was of the highest quality?

  (b) What fractional part of all items manufactured was defective?

48. *Garden Planning* Shaunna bought 18 perennial plants and 35 annual plants for her new garden. Eight of the perennial plants and 24 of the annual plants need to be planted in full sun. Of the plants that need full sun, 4 perennials and 19 annuals have yellow flowers.

  (a) What fractional part of the plants need to be planted in full sun?

  (b) What fractional part of the plants that need full sun have yellow flowers?

To Think About

49. Illustrate a real-life example of the fraction $\frac{0}{6}$.

50. What happens when we try to illustrate a real-life example of the fraction $\frac{6}{0}$? Why?

Cumulative Review

51. [1.2.4] Add.
18
27
34
16
125
+ 21

52. [1.3.3] Subtract.
56,203
− 42,987

53. [1.4.4] Multiply.   3178
× 46

54. [1.5.3] Divide. 24)30,196

Quick Quiz 2.1

1. Use a fraction to represent the shaded part of the object.

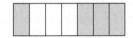

2. Silverstone Community College has 371 students taking classes on Monday night. Of those students, 204 drive a car to campus. Write a fraction that describes the part of the Monday night students who drive a car to class.

3. At the YMCA at 10:00 P.M. last Friday, 8 men were lifting weights and 5 women were lifting weights. At the same time, 7 men were riding stationary bikes and 13 women were riding stationary bikes. No other people were in the gym at that time. What fractional part of the people in the gym were lifting weights?

4. Concept Check One hundred twenty new businesses have opened in Springfield in the last five years. Sixty-five of them were restaurants; the remaining ones were not. Thirty new restaurants went out of business; the other new restaurants did not. Of all the new businesses that were not restaurants, 25 of them went out of business; the others did not. Explain how you can find a fraction that represents the fractional part of the new businesses that did not go out of business.

2.2 Simplifying Fractions

① Writing a Number as a Product of Prime Factors

A **prime number** is a whole number greater than 1 that cannot be evenly divided except by itself and 1. If you examine all the whole numbers from 1 to 50, you will find 15 prime numbers.

> **THE FIRST 15 PRIME NUMBERS**
>
> 2, 3, 5, 7, 11, 13, 17, 19, 23, 29, 31, 37, 41, 43, 47

A **composite number** is a whole number greater than 1 that can be divided by whole numbers other than 1 and itself. The number 12 is a composite number.

$$12 = 2 \times 6 \quad \text{and} \quad 12 = 3 \times 4$$

The number 1 is neither a prime nor a composite number. The number 0 is neither a prime nor a composite number.

Recall that factors are numbers that are multiplied together. Prime factors are prime numbers. To check to see if a number is prime or composite, simply divide the smaller primes (such as 2, 3, 5, 7, 11, . . .) into the given number. If the number can be divided exactly without a remainder by one of the smaller primes, it is a composite and not a prime.

Some students find the following rules helpful when deciding if a number can be divided by 2, 3, or 5.

> **DIVISIBILITY TESTS**
>
> **1.** A number is divisible by 2 if the last digit is 0, 2, 4, 6, or 8.
>
> **2.** A number is divisible by 3 if the sum of the digits is divisible by 3.
>
> **3.** A number is divisible by 5 if the last digit is 0 or 5.

To illustrate:

1. 478 is divisible by 2 since it ends in 8.

2. 531 is divisible by 3 since when we add the digits of 531 (5 + 3 + 1) we get 9, which is divisible by 3.

3. 985 is divisible by 5 since it ends in 5.

EXAMPLE 1 Write each whole number as the product of prime factors.

(a) 12 **(b)** 60 **(c)** 168

Solution

(a) To start, write 12 as the product of any two factors. We will write 12 as 4×3.

$12 = \quad 4 \quad \times 3$ Now check whether the factors are prime. If not, factor these.

$\quad\quad\quad 2 \times 2 \times 3$

$12 = 2 \times 2 \times 3$ Now all factors are prime, so 12 is completely factored.

Instead of writing $2 \times 2 \times 3$, we can write $2^2 \times 3$.

Note: To start, we could write 12 as 2×6. Begin this way and follow the preceding steps. Is the product of prime factors the same? Will this always be true?

(b) We follow the same steps as in (a).

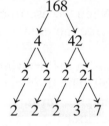

$$60 = \quad 6 \quad \times \quad 10$$

$$3 \times 2 \times 2 \times 5 \qquad \text{Check that all factors are prime.}$$

$$60 = 2 \times 2 \times 3 \times 5$$

Instead of writing $2 \times 2 \times 3 \times 5$, we can write $2^2 \times 3 \times 5$.

Note that in the final answer the prime factors are listed in order from least to greatest.

(c) Some students like to use a **factor tree** to help write a number as a product of prime factors as illustrated below.

168

4   42

2 2 2 21

2 2 2 3 7

$$168 = 2 \times 2 \times 2 \times 3 \times 7$$
$$\text{or} \quad 168 = 2^3 \times 3 \times 7$$

Student Practice 1 Write each whole number as a product of primes.

(a) 18   **(b)** 72   **(c)** 400

NOTE TO STUDENT: Fully worked-out solutions to all of the Student Practice problems can be found at the back of the text starting at page SP-1.

Suppose we started Example 1(c) by writing $168 = 14 \times 12$. Would we get the same answer? Would our answer be correct? Let's compare.

Again we will use a factor tree.

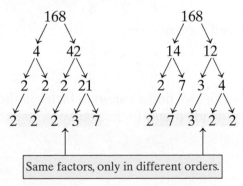

168

4   42

2 2 2 21

2 2 2 3 7

168

14   12

2 7 3 4

2 7 3 2 2

Same factors, only in different orders.

$$\text{Thus} \quad 168 = 2 \times 2 \times 2 \times 3 \times 7$$
$$\text{or} \qquad = 2^3 \times 3 \times 7.$$

The order of prime factors is not important because multiplication is commutative. No matter how we start, when we factor a composite number, we always get exactly the same prime factors.

> **THE FUNDAMENTAL THEOREM OF ARITHMETIC**
>
> Every composite number can be written in exactly one way as a product of prime numbers.

We have seen this in our Solution to Example 1(c).

You will be able to check this theorem again in Exercises 2.2, exercises 7–26. Writing a number as a product of prime factors is also called **prime factorization.**

② Reducing a Fraction to Lowest Terms

You know that $5 + 2$ and $3 + 4$ are two ways to write the same number. We say they are *equivalent* because they are *equal* to the same *value.* They are both ways of writing the value 7.

Like whole numbers, fractions can be written in more than one way. For example, $\frac{2}{4}$ and $\frac{1}{2}$ are two ways to write the same number. The value of the fractions is the same. When we use fractions, we often need to write them in another form. If we make the numerator and denominator smaller, we *simplify* the fractions.

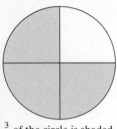

$\frac{3}{4}$ of the circle is shaded.

Compare the two fractions in the drawings on the left. In each picture the shaded part is the same size. The fractions $\frac{3}{4}$ and $\frac{6}{8}$ are called **equivalent fractions.** The fraction $\frac{3}{4}$ is in **simplest form.** To see how we can change $\frac{6}{8}$ to $\frac{3}{4}$, we look at a property of the number 1.

Any nonzero number divided by itself is 1.

$$\frac{5}{5} = \frac{17}{17} = \frac{c}{c} = 1$$

Thus, if we multiply a fraction by $\frac{5}{5}$ or $\frac{17}{17}$ or $\frac{c}{c}$ (remember, c cannot be zero), the value of the fraction is unchanged because we are multiplying by a form of 1. We can use this rule to show that $\frac{3}{4}$ and $\frac{6}{8}$ are equivalent. (We will introduce the concept of multiplication of fractions that will be explained in detail in Section 2.4.)

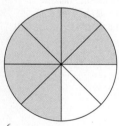

$\frac{6}{8}$ of the circle is shaded.

$$\frac{3}{4} \times \frac{2}{2} = \frac{6}{8}$$

In general, if b and c are not zero,

$$\frac{a}{b} = \frac{a \times c}{b \times c}$$

To reduce a fraction, we find a **common factor** in the numerator and in the denominator and divide it out. In the fraction $\frac{6}{8}$, the common factor is 2.

$$\frac{6}{8} = \frac{3 \times \overset{1}{\cancel{2}}}{4 \times \underset{1}{\cancel{2}}} = \frac{3}{4}$$

$$\frac{6}{8} = \frac{3}{4}$$

For all fractions (where a, b, and c are not zero), if c is a common factor,

$$\frac{a}{b} = \frac{a \div c}{b \div c}$$

A fraction is called **simplified, reduced,** or **in lowest terms** if the numerator and the denominator have only 1 *as a common factor.*

Mc **EXAMPLE 2**   Simplify (write in lowest terms).

(a) $\dfrac{15}{25}$ **(b)** $\dfrac{42}{56}$

Solution

(a) $\dfrac{15}{25} = \dfrac{15 \div 5}{25 \div 5} = \dfrac{3}{5}$ The greatest common factor is 5. Divide the numerator and the denominator by 5.

(b) $\dfrac{42}{56} = \dfrac{42 \div 14}{56 \div 14} = \dfrac{3}{4}$ The greatest common factor is 14. Divide the numerator and the denominator by 14.

Perhaps 14 was not the first common factor you thought of. Perhaps you did see the common factor 2. Divide out 2. Then look for another common factor, 7. Now divide out 7.

$$\frac{42}{56} = \frac{42 \div 2}{56 \div 2} = \frac{21}{28} = \frac{21 \div 7}{21 \div 7} = \frac{3}{4}$$

If we do not see large factors at first, sometimes we can simplify a fraction by dividing both numerator and denominator by a smaller common factor several times, until no common factors are left.

Student Practice 2   Simplify by dividing out common factors.

(a) $\dfrac{30}{42}$ **(b)** $\dfrac{60}{132}$

A second method to reduce or simplify fractions is called the *method of prime factors*. We factor the numerator and the denominator into prime numbers. We then divide the numerator and the denominator by any common prime factors.

EXAMPLE 3   Simplify the fractions by the method of prime factors.

(a) $\dfrac{35}{42}$ **(b)** $\dfrac{22}{110}$

Solution

(a) $\dfrac{35}{42} = \dfrac{5 \times 7}{2 \times 3 \times 7}$ We factor 35 and 42 into prime factors. The common prime factor is 7.

$= \dfrac{5 \times \overset{1}{7}}{2 \times 3 \times \underset{1}{7}}$ Now we divide out 7.

$= \dfrac{5 \times 1}{2 \times 3 \times 1} = \dfrac{5}{6}$ We multiply the factors in the numerator and denominator to write the reduced or simplified form.

Thus $\dfrac{35}{42} = \dfrac{5}{6}$, and $\dfrac{5}{6}$ is the simplified form.

(b) $\dfrac{22}{110} = \dfrac{2 \times 11}{2 \times 5 \times 11} = \dfrac{\overset{1}{2} \times \overset{1}{11}}{\underset{1}{2} \times 5 \times \underset{1}{11}} = \dfrac{1}{5}$

Continued on next page

Student Practice 3   Simplify the fractions by the method of prime factors.

(a) $\dfrac{120}{135}$ (b) $\dfrac{715}{880}$

③ Determining Whether Two Fractions Are Equal

After we simplify, how can we check that a reduced fraction is *equivalent* to the original fraction? If two fractions are equal, their diagonal products or **cross products** are equal. This is called the **equality test for fractions.** If $\frac{3}{4} = \frac{6}{8}$, then

$$\begin{array}{c} 3 \; \underset{?}{\times} \; 6 \longrightarrow 4 \times 6 = 24 \longleftarrow \fbox{Products} \\ 4 \; = \; 8 \longrightarrow 3 \times 8 = 24 \longleftarrow \fbox{are equal.} \end{array}$$

If two fractions are unequal (we use the symbol $\neq$), their *cross products* are unequal. If $\dfrac{5}{6} \neq \dfrac{6}{7}$, then

$$\begin{array}{c} 5 \; \underset{?}{\times} \; 6 \longrightarrow 6 \times 6 = 36 \longleftarrow \fbox{Products} \\ 6 \; = \; 7 \longrightarrow 5 \times 7 = 35 \longleftarrow \fbox{are not equal.} \end{array}$$

Since $36 \neq 35$, we know that $\dfrac{5}{6} \neq \dfrac{6}{7}$. The test can be described in this way.

EQUALITY TEST FOR FRACTIONS

For any two fractions where a, b, c, and d are whole numbers and $b \neq 0$, $d \neq 0$, if $\dfrac{a}{b} = \dfrac{c}{d}$, then $a \times d = b \times c$.

EXAMPLE 4   Are these fractions equal? Use the equality test.

(a) $\dfrac{2}{11} \stackrel{?}{=} \dfrac{18}{99}$ (b) $\dfrac{3}{16} \stackrel{?}{=} \dfrac{12}{62}$

Solution

(a) $\begin{array}{c} 2 \; \underset{?}{\times} \; 18 \longrightarrow 11 \times 18 = 198 \longleftarrow \fbox{Products} \\ 11 \; = \; 99 \longrightarrow 2 \times 99 = 198 \longleftarrow \fbox{are equal.} \end{array}$

Since $198 = 198$, we know that $\dfrac{2}{11} = \dfrac{18}{99}$.

(b) $\begin{array}{c} 3 \; \underset{?}{\times} \; 12 \longrightarrow 16 \times 12 = 192 \longleftarrow \fbox{Products} \\ 16 \; = \; 62 \longrightarrow 3 \times 62 = 186 \longleftarrow \fbox{are not equal.} \end{array}$

Since $192 \neq 186$, we know that $\dfrac{3}{16} \neq \dfrac{12}{62}$.

Student Practice 4   Test whether the following fractions are equal.

(a) $\dfrac{84}{108} \stackrel{?}{=} \dfrac{7}{9}$ (b) $\dfrac{3}{7} \stackrel{?}{=} \dfrac{79}{182}$

2.2 Exercises

MyMathLab®

Watch the videos in MyMathLab

Download the MyDashBoard App

Verbal and Writing Skills, Exercises 1–6

1. Which of these whole numbers are prime?
   4, 12, 11, 15, 6, 19, 1, 41, 38, 24, 5, 46

2. A prime number is a whole number greater than 1 that cannot be evenly _____ except by itself and 1.

3. A _____ is a whole number greater than 1 that can be divided by whole numbers other than itself and 1.

4. Every composite number can be written in exactly one way as a _____ of _____ numbers.

5. Give an example of a composite number written as a product of primes.

6. Give an example of equivalent (equal) fractions.

Write each number as a product of prime factors.

7. 15	**8.** 9	**9.** 35	**10.** 8	**11.** 49
12. 30	**13.** 16	**14.** 81	**15.** 55	**16.** 42
17. 63	**18.** 48	**19.** 84	**20.** 125	**21.** 54
22. 99	**23.** 120	**24.** 135	**25.** 184	**26.** 216

Determine which of these whole numbers are prime. If a number is composite, write it as the product of prime factors.

27. 47	**28.** 31	**29.** 57	**30.** 51
31. 67	**32.** 71	**33.** 62	**34.** 91
35. 89	**36.** 97	**37.** 127	**38.** 119
39. 121	**40.** 95	**41.** 145	**42.** 143

Reduce each fraction by finding a common factor in the numerator and in the denominator and dividing by the common factor.

43. $\dfrac{18}{27}$	**44.** $\dfrac{16}{24}$	**45.** $\dfrac{36}{48}$	**46.** $\dfrac{28}{49}$
47. $\dfrac{54}{84}$	**48.** $\dfrac{45}{75}$	**49.** $\dfrac{260}{290}$	**50.** $\dfrac{110}{140}$

Reduce each fraction by the method of prime factors.

51. $\dfrac{5}{30}$	**52.** $\dfrac{7}{21}$	**53.** $\dfrac{66}{88}$	**54.** $\dfrac{42}{56}$
55. $\dfrac{30}{45}$	**56.** $\dfrac{65}{91}$	**57.** $\dfrac{60}{75}$	**58.** $\dfrac{42}{70}$

Mixed Practice *Reduce each fraction by any method.*

59. $\dfrac{48}{66}$

60. $\dfrac{35}{90}$

61. $\dfrac{63}{108}$

62. $\dfrac{72}{132}$

63. $\dfrac{88}{121}$

64. $\dfrac{125}{200}$

65. $\dfrac{120}{200}$

66. $\dfrac{200}{300}$

67. $\dfrac{220}{260}$

68. $\dfrac{210}{390}$

Are these fractions equal? Why or why not?

69. $\dfrac{4}{16} \overset{?}{=} \dfrac{7}{28}$

70. $\dfrac{10}{65} \overset{?}{=} \dfrac{2}{13}$

71. $\dfrac{12}{40} \overset{?}{=} \dfrac{3}{13}$

72. $\dfrac{24}{72} \overset{?}{=} \dfrac{15}{45}$

73. $\dfrac{23}{27} \overset{?}{=} \dfrac{92}{107}$

74. $\dfrac{70}{120} \overset{?}{=} \dfrac{41}{73}$

75. $\dfrac{27}{57} \overset{?}{=} \dfrac{45}{95}$

76. $\dfrac{18}{24} \overset{?}{=} \dfrac{23}{28}$

77. $\dfrac{60}{95} \overset{?}{=} \dfrac{12}{19}$

78. $\dfrac{21}{27} \overset{?}{=} \dfrac{112}{144}$

Applications *Reduce the fractions in your answers.*

79. *Pizza Delivery* Prinzi's Gourmet Pizza made 128 deliveries on Saturday night. The manager found that 32 of the deliveries were of more than one pizza. He wanted to study the deliveries that consisted of just one pizza. What fractional part of the deliveries consisted of just one pizza?

80. *Medical Students* Medical students frequently work long hours. Susan worked a 16-hour shift, spending 12 hours in the emergency room and 4 hours in surgery. What fractional part of her shift was she in the emergency room? What fractional part of her shift was she in surgery?

81. *Teaching* Professor Nguyen found that 12 out of 96 students in his Aspects of Chemistry course failed the first exam. What fractional part of the class failed the exam? What fractional part of the class passed?

82. *Laptop Computers* Marvin works for a computer manufacturer that makes laptops. He inspected 225 laptops and found that 18 were defective. What fractional part of the laptops was not defective?

83. *Personal Finance* Amelia earned $8400 during her summer vacation. She saved $6000 of her earnings for a trip to New Zealand. What fractional part of her earnings did she save for her trip?

84. *Real Estate* Rachel and her husband have been working two jobs each to put a down payment on a summer cabin. The purchase price is $56,000. They have saved $8400. What fractional part of the cost of the cabin have they saved?

Education *The following data was compiled on the students attending day classes at North Shore Community College.*

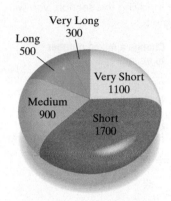

Number of Students	Daily Distance Traveled from Home to College (miles)	Length of Commute
1100	0–6	Very short
1700	7–12	Short
900	13–18	Medium
500	19–24	Long
300	More than 24	Very long

The number of students with each type of commute is displayed in the circle graph to the right.

Answer exercises 85–88 based on the preceding data. Reduce all fractions in your answers.

85. What fractional part of the student body has a short daily commute to the college?

86. What fractional part of the student body has a medium daily commute to the college?

87. What fractional part of the student body has a long or very long daily commute to the college?

88. What fractional part of the student body has a daily commute to the college that is considered less than long?

Cumulative Review

89. **[1.4.4]** Multiply. 386×425

90. **[1.5.3]** Divide. $15,552 \div 12$

91. **[1.4.3]** Multiply. 3200×300

92. **[1.3.5]** *Movies* Released in 2009, *Avatar* became the top-grossing movie worldwide, generating $2,734,603,864 at the box office. The second top-grossing movie of all time, released in 1997, was *Titanic*, generating $1,835,300,000. How much more money did *Avatar* generate than *Titanic*? (*Source:* www.imdb.com)

Quick Quiz 2.2 *Reduce each fraction.*

1. $\dfrac{25}{35}$

2. $\dfrac{14}{84}$

3. $\dfrac{40}{105}$

4. Concept Check Explain how you would determine if the fraction $\frac{195}{231}$ can be reduced.

2.3 Converting Between Improper Fractions and Mixed Numbers

Student Learning Objectives

After studying this section, you will be able to:

① Change a mixed number to an improper fraction.

② Change an improper fraction to a mixed number.

③ Reduce a mixed number or an improper fraction to lowest terms.

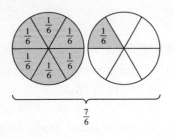

$$\frac{7}{6}$$

① Changing a Mixed Number to an Improper Fraction

We have names for different kinds of fractions. If the value of a fraction is less than 1, we say the fraction is proper.

$$\frac{3}{5}, \frac{5}{7}, \frac{1}{8} \quad \text{are called \textbf{proper fractions.}}$$

Notice that the numerator is less than the denominator. If the numerator is less than the denominator, the fraction is a proper fraction.

If the value of a fraction is greater than or equal to 1, the quantity can be written as an improper fraction or as a mixed number.

Suppose that we have 1 whole pizza and $\frac{1}{6}$ of a pizza. We could write this as $1\frac{1}{6}$. $1\frac{1}{6}$ is called a mixed number. A **mixed number** is the sum of a whole number greater than zero and a proper fraction. The notation $1\frac{1}{6}$ actually means $1 + \frac{1}{6}$. The plus sign is not usually shown.

Another way of writing $1\frac{1}{6}$ pizza is to write $\frac{7}{6}$ pizza. $\frac{7}{6}$ is called an improper fraction. Notice that the numerator is greater than the denominator. If the numerator is greater than or equal to the denominator, the fraction is an improper fraction.

$$\frac{7}{6}, \frac{6}{6}, \frac{5}{4}, \frac{8}{3}, \frac{2}{2} \quad \text{are \textbf{improper fractions.}}$$

The following chart will help you visualize these different fractions and their names.

Because in some cases improper fractions are easier to add, subtract, multiply, and divide than mixed numbers, we often change mixed numbers to improper fractions when we perform calculations with them.

Value Less Than 1	Value Equal To 1	Value Greater Than 1	
Proper Fraction	Improper Fraction	Improper Fraction	or Mixed Number
$\frac{3}{4}$	$\frac{4}{4}$	$\frac{5}{4}$ or $1\frac{1}{4}$	
$\frac{7}{8}$	$\frac{8}{8}$	$\frac{17}{8}$ or $2\frac{1}{8}$	
$\frac{3}{100}$	$\frac{100}{100}$	$\frac{109}{100}$ or $1\frac{9}{100}$	

CHANGING A MIXED NUMBER TO AN IMPROPER FRACTION

1. Multiply the whole number by the denominator of the fraction.

2. Add the numerator of the fraction to the product found in step 1.

3. Write the sum found in step 2 over the denominator of the fraction.

EXAMPLE 1   Change each mixed number to an improper fraction.

(a) $3\dfrac{2}{5}$ **(b)** $5\dfrac{4}{9}$ **(c)** $18\dfrac{3}{5}$

Solution

Multiply the whole number by the denominator.

Add the numerator to the product.

(a) $3\dfrac{2}{5} = \dfrac{3 \times 5 + 2}{5} = \dfrac{15 + 2}{5} = \dfrac{17}{5}$

Write the sum over the denominator.

(b) $5\dfrac{4}{9} = \dfrac{5 \times 9 + 4}{9} = \dfrac{45 + 4}{9} = \dfrac{49}{9}$

(c) $18\dfrac{3}{5} = \dfrac{18 \times 5 + 3}{5} = \dfrac{90 + 3}{5} = \dfrac{93}{5}$

Student Practice 1   Change the mixed numbers to improper fractions.

(a) $4\dfrac{3}{7}$ **(b)** $6\dfrac{2}{3}$ **(c)** $19\dfrac{4}{7}$

NOTE TO STUDENT: Fully worked-out solutions to all of the Student Practice problems can be found at the back of the text starting at page SP-1.

② Changing an Improper Fraction to a Mixed Number

We often need to change an improper fraction to a mixed number.

CHANGING AN IMPROPER FRACTION TO A MIXED NUMBER

1. Divide the numerator by the denominator.

2. Write the quotient followed by the fraction with the remainder over the denominator.

$$\text{quotient} \; \dfrac{\text{remainder}}{\text{denominator}}$$

EXAMPLE 2   Write each improper fraction as a mixed number.

(a) $\dfrac{13}{5}$ **(b)** $\dfrac{29}{7}$ **(c)** $\dfrac{105}{31}$ **(d)** $\dfrac{85}{17}$

Continued on next page

Solution

(a) We divide the denominator 5 into 13.

$$5)\overline{13} \quad \leftarrow \text{quotient} = 2$$

$$\underline{10}$$

$$3 \quad \leftarrow \text{remainder}$$

The answer is in the form quotient $\dfrac{\text{remainder}}{\text{denominator}}$.

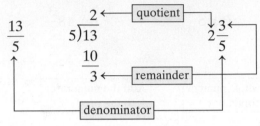

Thus $\dfrac{13}{5} = 2\dfrac{3}{5}$.

(b) $7)\overline{29}$ $\quad \dfrac{29}{7} = 4\dfrac{1}{7}$ $\qquad$ **(c)** $31)\overline{105}$ $\quad \dfrac{105}{31} = 3\dfrac{12}{31}$

$\quad\underline{28}$ $\qquad\qquad\qquad\qquad\qquad\qquad \underline{93}$

$\quad\ 1$ $\qquad\qquad\qquad\qquad\qquad\qquad\ 12$

(d) $17)\overline{85}$

$\quad\underline{85}$

$\quad\ 0 \quad$ The remainder is 0, so $\dfrac{85}{17} = 5$, a whole number.

Student Practice 2 Write as a mixed number or a whole number.

(a) $\dfrac{17}{4}$ $\qquad$ **(b)** $\dfrac{36}{5}$ $\qquad$ **(c)** $\dfrac{116}{27}$ $\qquad$ **(d)** $\dfrac{91}{13}$

③ Reducing a Mixed Number or an Improper Fraction to Lowest Terms

Mixed numbers and improper fractions may need to be reduced if they are not in simplest form. Recall that we write the fraction in terms of prime factors. Then we look for common factors in the numerator and the denominator of the fraction. Then we divide the numerator and the denominator by the common factor.

EXAMPLE 3 Reduce the improper fraction. $\dfrac{22}{8}$

Solution

$$\frac{22}{8} = \frac{\overset{1}{\cancel{2}} \times 11}{\underset{1}{\cancel{2}} \times 2 \times 2} = \frac{11}{4}$$

Student Practice 3 Reduce the improper fraction.

$$\frac{51}{15}$$

EXAMPLE 4   Reduce the mixed number. $4\dfrac{21}{28}$

Solution   We cannot reduce the whole number 4, only the fraction $\dfrac{21}{28}$.

$$\frac{21}{28} = \frac{3 \times \overset{1}{\cancel{7}}}{4 \times \underset{1}{\cancel{7}}} = \frac{3}{4}$$

Therefore, $4\dfrac{21}{28} = 4\dfrac{3}{4}$.

Student Practice 4   Reduce the mixed number.

$$3\frac{16}{80}$$

If an improper fraction contains a very large numerator and denominator, it is best to change the fraction to a mixed number before reducing.

EXAMPLE 5   Reduce $\dfrac{945}{567}$ by first changing to a mixed number.

Solution
$$\begin{array}{r} 1 \\ 567\overline{)945} \\ \underline{567} \\ 378 \end{array} \qquad \text{so } \frac{945}{567} = 1\frac{378}{567}$$

To reduce the fraction we write

$$\frac{378}{567} = \frac{2 \times 3 \times 3 \times 3 \times 7}{3 \times 3 \times 3 \times 3 \times 7} = \frac{2 \times \overset{1}{\cancel{3}} \times \overset{1}{\cancel{3}} \times \overset{1}{\cancel{3}} \times \overset{1}{\cancel{7}}}{3 \times \underset{1}{\cancel{3}} \times \underset{1}{\cancel{3}} \times \underset{1}{\cancel{3}} \times \underset{1}{\cancel{7}}} = \frac{2}{3}$$

So $\dfrac{945}{567} = 1\dfrac{378}{567} = 1\dfrac{2}{3}$.

Problems like Example 5 can be done in several different ways. It is not necessary to follow these exact steps when reducing this fraction.

Student Practice 5   Reduce $\dfrac{1001}{572}$ by first changing to a mixed number.

TO THINK ABOUT: When a Denominator Is Prime   A student concluded that just by looking at the denominator he could tell that the fraction $\frac{1655}{97}$ cannot be reduced unless $1655 \div 97$ is a whole number. How did he come to that conclusion?

Note that 97 is a prime number. The only factors of 97 are 97 and 1. Therefore, *any* fraction with 97 in the denominator can be reduced only if 97 is a factor of the numerator. Since $1655 \div 97$ is not a whole number (see the following division), it is therefore impossible to reduce $\frac{1655}{97}$.

$$\begin{array}{r} 17 \\ 97\overline{)1655} \\ \underline{97} \\ 685 \\ \underline{679} \\ 6 \end{array}$$

You may explore this idea in Exercises 2.3, exercises 83 and 84.

Verbal and Writing Skills, Exercises 1 and 2

1. Describe in your own words how to change a mixed number to an improper fraction.

2. Describe in your own words how to change an improper fraction to a mixed number.

Change each mixed number to an improper fraction.

3. $2\frac{1}{3}$ **4.** $2\frac{3}{4}$ **5.** $2\frac{3}{7}$ **6.** $3\frac{3}{8}$ **7.** $9\frac{2}{9}$ **8.** $8\frac{3}{8}$

9. $10\frac{2}{3}$ **10.** $15\frac{3}{4}$ **11.** $11\frac{3}{5}$ **12.** $15\frac{4}{5}$ **13.** $7\frac{2}{7}$ **14.** $9\frac{5}{8}$

15. $20\frac{1}{6}$ **16.** $6\frac{6}{7}$ **17.** $10\frac{11}{12}$ **18.** $13\frac{5}{7}$ **19.** $7\frac{9}{10}$ **20.** $4\frac{1}{50}$

21. $8\frac{1}{25}$ **22.** $12\frac{5}{6}$ **23.** $5\frac{5}{12}$ **24.** $207\frac{2}{3}$ **25.** $164\frac{2}{3}$ **26.** $33\frac{1}{3}$

27. $8\frac{11}{15}$ **28.** $5\frac{19}{20}$ **29.** $6\frac{19}{30}$ **30.** $4\frac{3}{22}$

Change each improper fraction to a mixed number or a whole number.

31. $\frac{4}{3}$ **32.** $\frac{13}{4}$ **33.** $\frac{11}{4}$ **34.** $\frac{9}{5}$ **35.** $\frac{15}{7}$ **36.** $\frac{23}{6}$

37. $\frac{27}{8}$ **38.** $\frac{80}{5}$ **39.** $\frac{100}{4}$ **40.** $\frac{42}{13}$ **41.** $\frac{86}{9}$ **42.** $\frac{47}{2}$

43. $\frac{70}{3}$ **44.** $\frac{54}{17}$ **45.** $\frac{25}{4}$ **46.** $\frac{19}{3}$ **47.** $\frac{57}{10}$ **48.** $\frac{83}{10}$

49. $\frac{35}{2}$ **50.** $\frac{132}{11}$ **51.** $\frac{91}{7}$ **52.** $\frac{183}{7}$ **53.** $\frac{210}{15}$ **54.** $\frac{196}{9}$

55. $\frac{102}{17}$ **56.** $\frac{104}{8}$ **57.** $\frac{180}{25}$ **58.** $\frac{200}{30}$

Reduce each mixed number.

59. $5\frac{3}{6}$ **60.** $4\frac{6}{8}$ **61.** $4\frac{11}{66}$ **62.** $3\frac{15}{90}$ **63.** $15\frac{18}{72}$ **64.** $10\frac{15}{75}$

Reduce each improper fraction.

65. $\frac{24}{6}$ **66.** $\frac{36}{4}$ **67.** $\frac{36}{15}$ **68.** $\frac{63}{45}$ **69.** $\frac{105}{28}$ **70.** $\frac{112}{21}$

Change to a mixed number and reduce.

71. $\frac{340}{126}$ **72.** $\frac{390}{360}$ **73.** $\frac{580}{280}$

74. $\frac{764}{328}$ **75.** $\frac{508}{296}$ **76.** $\frac{2150}{1000}$

Applications

77. ***Banner Display*** The Science Museum is hanging banners all over the building to commemorate the Apollo astronauts. The art department is using $360\frac{2}{3}$ yards of starry-sky parachute fabric. Change this number to an improper fraction.

78. ***Sculpture*** Emily is a well-known sculptor who works with metal to create art for museums. She just ordered a shipment of a metal alloy and the total weight of the package is $178\frac{3}{4}$ pounds. Change this number to an improper fraction.

79. ***Environmental Studies*** A Cape Cod cranberry bog was contaminated by waste from abandoned oil storage tanks at Otis Air Force Base. Damage was done to $\frac{151}{3}$ acres of land. Write this as a mixed number.

80. ***Theater*** Waite Auditorium needs new velvet stage curtains. The manufacturer took measurements and calculated he would need $\frac{331}{4}$ square yards of fabric. Write this as a mixed number.

81. ***Cooking*** The cafeteria workers at Ipswich High School cafeteria used $\frac{1131}{8}$ pounds of flour while cooking for the students last week. Write this as a mixed number.

82. ***Shelf Construction*** The new Danvers Main Building at North Shore Community College has several new offices for the faculty and staff. Shelving was constructed for these offices. A total of $\frac{1373}{8}$ feet of shelving was used in the construction. Write this as a mixed number.

To Think About

83. Can $\frac{5687}{101}$ be reduced? Why or why not?

84. Can $\frac{9810}{157}$ be reduced? Why or why not?

Cumulative Review

85. **[1.3.3]** Subtract. $1,398,210 - 1,137,963$

86. **[1.7.2]** Estimate the answer. $17,335 \times 114,576$

87. **[1.7.2]** Estimate the answer. $328,515 \div 966$

88. **[2.1.3]** ***E-mail*** When Eric checked his e-mail after being on vacation, he had 156 new e-mails in his inbox. Of these, 98 were spam. What fractional part of his new e-mails was not spam?

Quick Quiz 2.3

1. Change to an improper fraction.

$4\dfrac{7}{13}$

2. Change to a mixed number.

$\dfrac{89}{12}$

3. Reduce the improper fraction.

$\dfrac{42}{14}$

4. **Concept Check** Explain how you change the mixed number $5\frac{6}{13}$ to an improper fraction.

2.4 Multiplying Fractions and Mixed Numbers

Student Learning Objectives

After studying this section, you will be able to:

1. Multiply two fractions that are proper or improper.

2. Multiply a whole number by a fraction.

3. Multiply mixed numbers.

① Multiplying Two Fractions That Are Proper or Improper

FUDGE SQUARES

Ingredients:

2 cups sugar	1/4 teaspoon salt
4 oz chocolate	1 teaspoon vanilla
1/2 cup butter	1 cup all-purpose flour
4 eggs	1 cup nutmeats

Suppose you want to make an amount equal to half of what the recipe shown will produce. You would multiply the measure given for each ingredient by $\frac{1}{2}$.

$\frac{1}{2}$ of 2 cups sugar $\frac{1}{2}$ of $\frac{1}{4}$ teaspoon salt

$\frac{1}{2}$ of 4 oz chocolate $\frac{1}{2}$ of 1 teaspoon vanilla

$\frac{1}{2}$ of $\frac{1}{2}$ cup butter $\frac{1}{2}$ of 1 cup all-purpose flour

$\frac{1}{2}$ of 4 eggs $\frac{1}{2}$ of 1 cup nutmeats

We often use multiplication of fractions to describe taking a fractional part of something. To find $\frac{1}{2}$ of $\frac{3}{7}$, we multiply

$$\frac{1}{2} \times \frac{3}{7} = \frac{3}{14}.$$

We begin with a bar that is $\frac{3}{7}$ shaded. To find $\frac{1}{2}$ of $\frac{3}{7}$ we divide the bar in half and take $\frac{1}{2}$ of the shaded section. $\frac{1}{2}$ of $\frac{3}{7}$ yields 3 out of 14 squares.

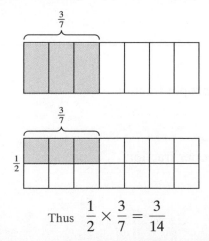

Thus $\frac{1}{2} \times \frac{3}{7} = \frac{3}{14}$

When you multiply two proper fractions together, you get a smaller fraction.

To multiply two fractions, we multiply the numerators and multiply the denominators.

$$\frac{2}{3} \times \frac{5}{7} = \frac{10}{21} \quad \begin{matrix} \leftarrow 2 \times 5 = 10 \\ \leftarrow 3 \times 7 = 21 \end{matrix}$$

MULTIPLICATION OF FRACTIONS

In general, for all positive whole numbers a, b, c, and d,

$$\frac{a}{b} \times \frac{c}{d} = \frac{a \times c}{b \times d} \quad b \neq 0, d \neq 0$$

EXAMPLE 1 Multiply.

(a) $\dfrac{3}{8} \times \dfrac{5}{7}$ **(b)** $\dfrac{1}{11} \times \dfrac{2}{13}$

Solution

(a) $\dfrac{3}{8} \times \dfrac{5}{7} = \dfrac{3 \times 5}{8 \times 7} = \dfrac{15}{56}$ **(b)** $\dfrac{1}{11} \times \dfrac{2}{13} = \dfrac{1 \times 2}{11 \times 13} = \dfrac{2}{143}$

Student Practice 1 Multiply.

(a) $\dfrac{6}{7} \times \dfrac{3}{13}$ **(b)** $\dfrac{1}{5} \times \dfrac{11}{12}$

NOTE TO STUDENT: Fully worked-out solutions to all of the Student Practice problems can be found at the back of the text starting at page SP-1.

Some products may be reduced. $\dfrac{12}{35} \times \dfrac{25}{18} = \dfrac{300}{630} = \dfrac{10}{21}$

By simplifying before multiplication, the reducing can be done more easily. For a multiplication problem, a factor in the numerator can be paired with a common factor in the denominator of the same or a different fraction. We can begin by finding the prime factors in the numerators and denominators. We then divide numerator and denominator by their common prime factors.

EXAMPLE 2 Simplify first and then multiply. $\dfrac{12}{35} \times \dfrac{25}{18}$

Solution

$\dfrac{12}{35} \times \dfrac{25}{18} = \dfrac{2 \cdot 2 \cdot 3}{5 \cdot 7} \times \dfrac{5 \cdot 5}{2 \cdot 3 \cdot 3}$ First we find the prime factors.

$\qquad = \dfrac{2 \cdot 2 \cdot 3 \cdot 5 \cdot 5}{5 \cdot 7 \cdot 2 \cdot 3 \cdot 3}$ Write the product as one fraction.

$\qquad = \dfrac{\overset{1}{\cancel{2}} \cdot 2 \cdot \overset{1}{\cancel{3}} \cdot \overset{1}{\cancel{5}} \cdot 5}{\underset{1}{\cancel{2}} \cdot \underset{1}{\cancel{3}} \cdot 3 \cdot \underset{1}{\cancel{5}} \cdot 7}$ Arrange the factors in order and divide the numerator and denominator by the common factors.

$\qquad = \dfrac{10}{21}$ Multiply the remaining factors.

Student Practice 2 Simplify first and then multiply.

$$\frac{55}{72} \times \frac{16}{33}$$

Note: Although finding the prime factors of the numerators and denominators will help you avoid errors, you can also begin these problems by dividing the numerators and denominators by larger common factors. This method will be used for the remainder of the Examples in this section of the text.

② Multiplying a Whole Number by a Fraction

When multiplying a fraction by a whole number, it is more convenient to express the whole number as a fraction with a denominator of 1. We know that $5 = \frac{5}{1}, 7 = \frac{7}{1}$, and so on.

EXAMPLE 3 Multiply.

(a) $5 \times \dfrac{3}{8}$ **(b)** $\dfrac{22}{7} \times 14$

Solution

(a) $5 \times \dfrac{3}{8} = \dfrac{5}{1} \times \dfrac{3}{8} = \dfrac{15}{8}$ or $1\dfrac{7}{8}$ **(b)** $\dfrac{22}{7} \times 14 = \dfrac{22}{7} \times \dfrac{\overset{2}{\cancel{14}}}{1} = \dfrac{44}{1} = 44$

Student Practice 3 Multiply.

(a) $7 \times \dfrac{5}{13}$ **(b)** $\dfrac{13}{4} \times 8$

EXAMPLE 4 Mr. and Mrs. Jones found that $\frac{2}{7}$ of their income went to pay federal income taxes. Last year they earned \$37,100. How much did they pay in taxes?

Solution We need to find $\frac{2}{7}$ of \$37,100. So we must multiply $\frac{2}{7} \times 37,100$.

$$\dfrac{2}{\underset{1}{\cancel{7}}} \times \overset{5300}{\cancel{37,100}} = \dfrac{2}{1} \times 5300 = 10,600$$

They paid \$10,600 in federal income taxes.

Student Practice 4 Fred and Linda own 98,400 square feet of land. They found that $\frac{3}{8}$ of the land is in a wetland area and cannot be used for building. How many square feet of land are in the wetland area?

---

③ Multiplying Mixed Numbers

To multiply a fraction by a mixed number or to multiply two mixed numbers, first change each mixed number to an improper fraction.

EXAMPLE 5 Multiply.

(a) $\dfrac{5}{7} \times 3\dfrac{1}{4}$ **(b)** $20\dfrac{2}{5} \times 6\dfrac{2}{3}$ **(c)** $\dfrac{3}{4} \times 1\dfrac{1}{2} \times \dfrac{4}{7}$ **(d)** $4\dfrac{1}{3} \times 2\dfrac{1}{4}$

Solution

(a) $\dfrac{5}{7} \times 3\dfrac{1}{4} = \dfrac{5}{7} \times \dfrac{13}{4} = \dfrac{65}{28}$　or　$2\dfrac{9}{28}$

(b) $20\dfrac{2}{5} \times 6\dfrac{2}{3} = \dfrac{\overset{34}{\cancel{102}}}{\underset{1}{\cancel{5}}} \times \dfrac{\overset{4}{\cancel{20}}}{\underset{1}{\cancel{3}}} = \dfrac{136}{1} = 136$

(c) $\dfrac{3}{4} \times 1\dfrac{1}{2} \times \dfrac{4}{7} = \dfrac{3}{\underset{1}{\cancel{4}}} \times \dfrac{3}{2} \times \dfrac{\overset{1}{\cancel{4}}}{7} = \dfrac{9}{14}$

(d) $4\dfrac{1}{3} \times 2\dfrac{1}{4} = \dfrac{13}{\underset{1}{\cancel{3}}} \times \dfrac{\overset{3}{\cancel{9}}}{4} = \dfrac{39}{4}$　or　$9\dfrac{3}{4}$

Student Practice 5　Multiply.

(a) $2\dfrac{1}{6} \times \dfrac{4}{7}$　　　　　　(b) $10\dfrac{2}{3} \times 13\dfrac{1}{2}$

(c) $\dfrac{3}{5} \times 1\dfrac{1}{3} \times \dfrac{5}{8}$　　　　　(d) $3\dfrac{1}{5} \times 2\dfrac{1}{2}$

▲ **EXAMPLE 6**　Find the area in square miles of a rectangle with width $1\dfrac{1}{3}$ miles and length $12\dfrac{1}{4}$ miles.

Length $= 12\dfrac{1}{4}$ miles

Width $= 1\dfrac{1}{3}$ miles

Solution　We find the area of a rectangle by multiplying the width times the length.

$$1\dfrac{1}{3} \times 12\dfrac{1}{4} = \dfrac{\overset{1}{\cancel{4}}}{3} \times \dfrac{49}{\underset{1}{\cancel{4}}} = \dfrac{49}{3}\quad \text{or}\quad 16\dfrac{1}{3}$$

The area is $16\dfrac{1}{3}$ square miles.

▲ **Student Practice 6**　Find the area in square meters of a rectangle with width $1\dfrac{1}{5}$ meters and length $4\dfrac{5}{6}$ meters.

EXAMPLE 7　Find the value of x if

$$\dfrac{3}{7} \cdot x = \dfrac{15}{42}.$$

Solution　The variable x represents a fraction. We know that 3 times one number equals 15 and 7 times another equals 42.

Since $3 \cdot 5 = 15$
and　$7 \cdot 6 = 42$　we know that　$\dfrac{3}{7} \cdot \dfrac{5}{6} = \dfrac{15}{42}.$

Therefore, $x = \dfrac{5}{6}.$

Student Practice 7　Find the value of x if $\dfrac{8}{9} \cdot x = \dfrac{80}{81}.$

Watch the videos
in MyMathLab

Download the
MyDashBoard App

Multiply. Make sure all fractions are simplified in the final answer.

1. $\dfrac{3}{5} \times \dfrac{7}{11}$

2. $\dfrac{1}{8} \times \dfrac{5}{11}$

3. $\dfrac{3}{4} \times \dfrac{5}{13}$

4. $\dfrac{4}{7} \times \dfrac{3}{5}$

5. $\dfrac{6}{5} \times \dfrac{10}{12}$

6. $\dfrac{7}{8} \times \dfrac{16}{21}$

7. $\dfrac{5}{36} \times \dfrac{9}{20}$

8. $\dfrac{22}{45} \times \dfrac{5}{11}$

9. $\dfrac{12}{25} \times \dfrac{5}{11}$

10. $\dfrac{9}{4} \times \dfrac{13}{27}$

11. $\dfrac{9}{10} \times \dfrac{35}{12}$

12. $\dfrac{12}{17} \times \dfrac{3}{24}$

13. $8 \times \dfrac{3}{7}$

14. $\dfrac{8}{9} \times 6$

15. $\dfrac{5}{12} \times 8$

16. $5 \times \dfrac{7}{25}$

17. $\dfrac{4}{9} \times \dfrac{3}{7} \times \dfrac{7}{8}$

18. $\dfrac{8}{7} \times \dfrac{5}{12} \times \dfrac{3}{10}$

19. $\dfrac{5}{4} \times \dfrac{9}{10} \times \dfrac{8}{3}$

20. $\dfrac{5}{7} \times \dfrac{15}{2} \times \dfrac{28}{15}$

Multiply. Change any mixed number to an improper fraction before multiplying.

21. $2\dfrac{5}{6} \times \dfrac{3}{17}$

22. $\dfrac{5}{6} \times 3\dfrac{3}{5}$

23. $10 \times 3\dfrac{1}{10}$

24. $12 \times 5\dfrac{7}{12}$

25. $1\dfrac{3}{16} \times 0$

26. $0 \times 6\dfrac{2}{3}$

27. $3\dfrac{7}{8} \times 1$

28. $\dfrac{5}{5} \times 11\dfrac{5}{7}$

29. $1\dfrac{1}{4} \times 3\dfrac{2}{3}$

30. $2\dfrac{3}{5} \times 1\dfrac{4}{7}$

31. $2\dfrac{3}{10} \times \dfrac{3}{5}$

32. $4\dfrac{3}{5} \times \dfrac{1}{10}$

33. $4\dfrac{1}{5} \times 8\dfrac{1}{3}$

34. $5\dfrac{1}{4} \times 4\dfrac{4}{7}$

35. $6\dfrac{2}{5} \times \dfrac{1}{4}$

36. $\dfrac{8}{9} \times 4\dfrac{1}{11}$

Mixed Practice *Multiply. Make sure all fractions are simplified in the final answer.*

37. $\dfrac{11}{15} \times \dfrac{35}{33}$

38. $\dfrac{13}{12} \times \dfrac{96}{65}$

39. $2\dfrac{3}{8} \times 5\dfrac{1}{3}$

40. $2\dfrac{2}{9} \times 4\dfrac{1}{2}$

Solve for x.

41. $\dfrac{4}{9} \cdot x = \dfrac{28}{81}$

42. $\dfrac{12}{17} \cdot x = \dfrac{144}{85}$

43. $\dfrac{7}{13} \cdot x = \dfrac{56}{117}$

44. $x \cdot \dfrac{11}{15} = \dfrac{77}{225}$

Applications

▲ **45.** *Geometry* A spy is running from his captors in a forest that is $8\frac{3}{4}$ miles long and $4\frac{1}{3}$ miles wide. Find the area of the forest where he is hiding. (*Hint:* The area of a rectangle is the product of the length times the width.)

▲ **46.** *Geometry* An area in the Midwest is a designated tornado danger zone. The land is $22\frac{5}{8}$ miles long and $16\frac{1}{2}$ miles wide. Find the area of the tornado danger zone. (*Hint:* The area of a rectangle is the product of the length times the width.)

47. *Airplane Travel* A Lear jet airplane has 360 gallons of fuel. The plane averages $4\frac{1}{3}$ miles per gallon. How far can the plane go?

48. *Real Estate* Mike and Ellen Horton bought their house in 1986 for $63,400. Twenty-five years later, in 2011, their house was worth $5\frac{1}{2}$ times what they paid for it. How much was Mike and Ellen's house worth in 2011?

49. *Cooking* A recipe from Nanette's French cookbook for a scalloped potato tart requires $90\frac{1}{2}$ grams of grated cheese. How many grams of cheese would she need if she made one tart for each of her 18 cousins?

▲ **50.** *Geometry* The dormitory rooms in Selkirk Hall are being carpeted. Each room requires $20\frac{1}{2}$ square feet of carpet. If there are 30 rooms, how much carpet is needed?

51. *College Students* Of the 7998 students at Normandale Community College, $\frac{2}{3}$ of them are under 25 years of age. How many students are under 25 years of age?

52. *Health Care* A nurse finds that of the 225 rooms at Dover Area Hospital, $\frac{1}{15}$ of them are occupied by surgery patients. How many rooms contain surgery patients?

53. *Job Search* Carlos has sent his resume to 12,064 companies through an Internet job search service. If $\frac{1}{32}$ of the companies e-mail him with an invitation for an interview, how many companies will he have heard from?

54. *Car Purchase* Keiko purchased a new 2011 Toyota Camry Hybrid for $24,800. After a year, the car was worth $\frac{7}{8}$ of the purchase price. What was the car worth after one year?

55. *Jogging* Mary jogged $4\frac{1}{4}$ miles per hour for $1\frac{1}{3}$ hours. During $\frac{1}{3}$ of her jogging time, she was jogging in the rain. How many miles did she jog in the rain?

56. *College Students* There were 1340 students at the Danvers campus of North Shore Community College during the spring 2011 semester. The registrar discovered that $\frac{2}{5}$ of these students live in the city of Beverly. He further discovered that $\frac{1}{4}$ of the students living in Beverly attend classes only on Monday, Wednesday, and Friday. How many students at the Danvers campus live in the city of Beverly and attend classes only on Monday, Wednesday, and Friday?

To Think About

57. When we multiply two fractions, we look for opportunities to divide a numerator and a denominator by the same number. Why do we bother with that step? Why don't we just multiply the two numerators and the two denominators?

58. Suppose there is an unknown fraction that has *not* been simplified (it is not reduced). You multiply this unknown fraction by $\frac{2}{5}$ and you obtain a simplified answer of $\frac{6}{35}$. How many possible values could this unknown fraction have? Give at least three possible answers.

Cumulative Review

59. **[1.5.4]** *Toll Bridge* A total of 16,399 cars used a toll bridge in January (31 days). What is the average number of cars using the bridge in one day?

60. **[1.5.4]** *Sales* The Office of Investors Services has 15,456 calls made per month by the sales personnel. There are 42 sales personnel in the office. What is the average number of calls made per month by one salesperson?

61. **[2.1.3]** *Sales* Of the 78 cars sold this month at Midtown Used Cars, 41 of them were foreign cars. What fractional part of the cars sold was made in the United States?

62. **[2.1.3]** *Grades* Of the 96 students in Professor Chang's biology lecture, 15 failed the first exam. What fractional part of the class passed the first exam?

Quick Quiz 2.4 *Multiply.*

1. $32 \times \dfrac{5}{16}$

2. $\dfrac{11}{13} \times \dfrac{4}{5}$

3. $4\dfrac{1}{3} \times 2\dfrac{3}{4}$

4. **Concept Check** Explain how you would multiply the whole number 6 times the mixed number $4\frac{3}{5}$.

2.5 Dividing Fractions and Mixed Numbers

① Dividing Two Proper or Improper Fractions

Why would you divide fractions? Consider this problem.

A copper pipe that is $\frac{3}{4}$ of a foot long is to be cut into $\frac{1}{4}$-foot pieces. How many pieces will there be?

To find how many $\frac{1}{4}$'s are in $\frac{3}{4}$, we divide $\frac{3}{4} \div \frac{1}{4}$. We draw a sketch.

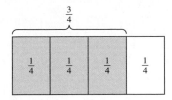

Notice that there are three $\frac{1}{4}$'s in $\frac{3}{4}$.

How do we divide two fractions? We **invert** the second fraction and multiply.

$$\frac{3}{4} \div \frac{1}{4} = \frac{3}{\cancel{4}} \times \frac{\cancel{4}^{1}}{1} = \frac{3}{1} = 3$$

When we invert a fraction, we interchange the numerator and the denominator. If we invert $\frac{5}{9}$, we obtain $\frac{9}{5}$. If we invert $\frac{6}{1}$, we obtain $\frac{1}{6}$. Numbers such as $\frac{5}{9}$ and $\frac{9}{5}$ are called **reciprocals** of each other.

RULE FOR DIVISION OF FRACTIONS

To divide two fractions, we invert the second fraction and multiply.

$$\frac{a}{b} \div \frac{c}{d} = \frac{a}{b} \times \frac{d}{c}$$

(when b, c, and d are not zero).

EXAMPLE 1 Divide.

(a) $\dfrac{3}{11} \div \dfrac{2}{5}$ 　　　　(b) $\dfrac{5}{8} \div \dfrac{25}{16}$

Solution

(a) $\dfrac{3}{11} \div \dfrac{2}{5} = \dfrac{3}{11} \times \dfrac{5}{2} = \dfrac{15}{22}$ 　　(b) $\dfrac{5}{8} \div \dfrac{25}{16} = \dfrac{\cancel{5}^{1}}{\cancel{8}_{1}} \times \dfrac{\cancel{16}^{2}}{\cancel{25}_{5}} = \dfrac{2}{5}$

Student Practice 1 Divide.

(a) $\dfrac{7}{13} \div \dfrac{3}{4}$ 　　　　(b) $\dfrac{16}{35} \div \dfrac{24}{25}$

NOTE TO STUDENT: *Fully worked-out solutions to all of the Student Practice problems can be found at the back of the text starting at page SP-1.*

Student Learning Objectives

After studying this section, you will be able to:

① **Divide two proper or improper fractions.**

② **Divide a whole number and a fraction.**

③ **Divide mixed numbers.**

② Dividing a Whole Number and a Fraction

When dividing with whole numbers, it is helpful to remember that for any whole number a, $a = \dfrac{a}{1}$.

EXAMPLE 2 Divide.

(a) $\dfrac{3}{7} \div 2$

(b) $5 \div \dfrac{10}{13}$

Solution

(a) $\dfrac{3}{7} \div 2 = \dfrac{3}{7} \div \dfrac{2}{1} = \dfrac{3}{7} \times \dfrac{1}{2} = \dfrac{3}{14}$

(b) $5 \div \dfrac{10}{13} = \dfrac{5}{1} \div \dfrac{10}{13} = \dfrac{\overset{1}{\cancel{5}}}{1} \times \dfrac{13}{\underset{2}{\cancel{10}}} = \dfrac{13}{2}$ or $6\dfrac{1}{2}$

Student Practice 2 Divide.

(a) $\dfrac{3}{17} \div 6$

(b) $14 \div \dfrac{7}{15}$

EXAMPLE 3 Divide, if possible.

(a) $\dfrac{23}{25} \div 1$

(b) $1 \div \dfrac{7}{5}$

(c) $0 \div \dfrac{4}{9}$

(d) $\dfrac{3}{17} \div 0$

Solution

(a) $\dfrac{23}{25} \div 1 = \dfrac{23}{25} \times \dfrac{1}{1} = \dfrac{23}{25}$

(b) $1 \div \dfrac{7}{5} = \dfrac{1}{1} \times \dfrac{5}{7} = \dfrac{5}{7}$

(c) $0 \div \dfrac{4}{9} = \dfrac{0}{1} \times \dfrac{9}{4} = \dfrac{0}{4} = 0$ Zero divided by any nonzero number is zero.

(d) $\dfrac{3}{17} \div 0$ Division by zero is undefined.

Student Practice 3 Divide, if possible.

(a) $1 \div \dfrac{11}{13}$

(b) $\dfrac{14}{17} \div 1$

(c) $\dfrac{3}{11} \div 0$

(d) $0 \div \dfrac{9}{16}$

SIDELIGHT: Invert and Multiply

Why do we divide by inverting the second fraction and multiplying? What is really going on when we do this? We are actually multiplying by 1. To see why, consider the following.

$$\frac{3}{7} \div \frac{2}{3} = \frac{\dfrac{3}{7}}{\dfrac{2}{3}}$$ We write the division by using another fraction bar.

$$= \frac{\dfrac{3}{7}}{\dfrac{2}{3}} \times 1$$ Any fraction can be multiplied by 1 without changing the value of the fraction. This is the multiplicative identity.

$$= \frac{\dfrac{3}{7}}{\dfrac{2}{3}} \times \frac{\dfrac{3}{2}}{\dfrac{3}{2}}$$ Any nonzero number divided by itself equals 1.

$$= \frac{\dfrac{3}{7} \times \dfrac{3}{2}}{\dfrac{2}{3} \times \dfrac{3}{2}}$$ Definition of multiplication of fractions.

$$= \frac{\dfrac{3}{7} \times \dfrac{3}{2}}{1} = \frac{3}{7} \times \frac{3}{2}$$ Any number can be written as a fraction with a denominator of 1 without changing its value.

Thus

$$\frac{3}{7} \div \frac{2}{3} = \frac{3}{7} \times \frac{3}{2} = \frac{9}{14}.$$

③ Dividing Mixed Numbers

If one or more mixed numbers are involved in the division, they should be converted to improper fractions first.

EXAMPLE 4   Divide.

(a) $3\dfrac{7}{15} \div 1\dfrac{1}{25}$ **(b)** $\dfrac{3}{5} \div 2\dfrac{1}{7}$

Solution

(a) $3\dfrac{7}{15} \div 1\dfrac{1}{25} = \dfrac{52}{15} \div \dfrac{26}{25} = \dfrac{\overset{2}{\cancel{52}}}{\underset{3}{\cancel{15}}} \times \dfrac{\overset{5}{\cancel{25}}}{\underset{1}{\cancel{26}}} = \dfrac{10}{3}$ or $3\dfrac{1}{3}$

(b) $\dfrac{3}{5} \div 2\dfrac{1}{7} = \dfrac{3}{5} \div \dfrac{15}{7} = \dfrac{\overset{1}{\cancel{3}}}{5} \times \dfrac{7}{\underset{5}{\cancel{15}}} = \dfrac{7}{25}$

Student Practice 4   Divide.

(a) $1\dfrac{1}{5} \div \dfrac{7}{10}$ **(b)** $2\dfrac{1}{4} \div 1\dfrac{7}{8}$

The division of two fractions may be indicated by a wide fraction bar.

EXAMPLE 5　Divide.

(a) $\dfrac{10\frac{2}{9}}{2\frac{1}{3}}$

(b) $\dfrac{1\frac{1}{15}}{3\frac{1}{3}}$

Solution

(a) $\dfrac{10\frac{2}{9}}{2\frac{1}{3}} = 10\frac{2}{9} \div 2\frac{1}{3} = \dfrac{92}{9} \div \dfrac{7}{3} = \dfrac{92}{\cancel{9}_{3}} \times \dfrac{\cancel{3}^{1}}{7} = \dfrac{92}{21}$　or　$4\dfrac{8}{21}$

(b) $\dfrac{1\frac{1}{15}}{3\frac{1}{3}} = 1\frac{1}{15} \div 3\frac{1}{3} = \dfrac{16}{15} \div \dfrac{10}{3} = \dfrac{\cancel{16}^{8}}{\cancel{15}_{5}} \times \dfrac{\cancel{3}^{1}}{\cancel{10}_{5}} = \dfrac{8}{25}$

Student Practice 5　Divide.

(a) $\dfrac{5\frac{2}{3}}{7}$

(b) $\dfrac{1\frac{2}{5}}{2\frac{1}{3}}$

Some students may find Example 6 difficult at first. Read it slowly and carefully. It may be necessary to read it several times before it becomes clear.

EXAMPLE 6　Find the value of x if $x \div \frac{8}{7} = \frac{21}{40}$.

Solution　First we will change the division problem to an equivalent multiplication problem.

$$x \div \dfrac{8}{7} = \dfrac{21}{40}$$

$$x \cdot \dfrac{7}{8} = \dfrac{21}{40}$$

x represents a fraction.

In the numerator, we want to know what times 7 equals 21. In the denominator, we want to know what times 8 equals 40.

$$\dfrac{3}{5} \cdot \dfrac{7}{8} = \dfrac{21}{40}$$

Thus $x = \frac{3}{5}$.

Student Practice 6　Find the value of x if $x \div \frac{3}{2} = \frac{22}{36}$.

EXAMPLE 7 There are 117 milligrams of cholesterol in $4\frac{1}{3}$ cups of milk. How much cholesterol is in 1 cup of milk?

Solution We want to divide the 117 by $4\frac{1}{3}$ to find out how much is in 1 cup.

$$117 \div 4\frac{1}{3} = 117 \div \frac{13}{3} = \frac{\overset{9}{\cancel{117}}}{1} \times \frac{3}{\underset{1}{\cancel{13}}} = \frac{27}{1} = 27$$

Thus there are 27 milligrams of cholesterol in 1 cup of milk.

Student Practice 7 A copper pipe that is $19\frac{1}{4}$ feet long will be cut into 14 equal pieces. How long will each piece be?

Take a little time to review Examples 1–7 and Student Practice 1–7. This is important material. It is crucial to understand how to do each of these problems. Some extra time spent reviewing here will make the homework exercises go much more quickly.

STEPS TO SUCCESS Why Does Reviewing Make Such a Big Difference?

Students are often amazed that reviewing makes such a huge difference in helping them to learn. It is one of the most powerful tools that a math student can have.

Mathematics involves learning concepts one step at a time. Then the concepts are put together in a chapter. At the end of the chapter you need to know each of these concepts. Therefore, to succeed in each chapter you need to be able to put together all the pieces of each chapter. Reviewing in each section and reviewing at the end of each chapter are the amazing tools that help you to master the mathematical concepts.

As you review, if you find you cannot work out a problem be sure to study the Examples and the Student Practice problems very carefully. Then things will become more clear.

Making it personal: Start with this section. Do each Cumulative Review problem at the end of the exercise set. Check your answer for each one in the back of the book. If you miss any problem, then go back to the appropriate section of the book for help.

For example, if you miss Problem 81 in Exercises 2.5 you see it is coded [1.1.3]. This means to go back to Section 1.1 of the book and look at objective 3. There you will find similar examples that explain this kind of problem.

Now think about what was said in this section. Why do you think reviewing is so important in mathematics? Write your answer below. ▼

Make sure all fractions are simplified in the final answer.

Verbal and Writing Skills, Exercises 1 and 2

1. In your own words explain how to remember that when you divide two fractions you invert the *second* fraction and multiply by the first. How can you be sure that you don't invert the *first* fraction by mistake?

2. Explain why $2 \div \frac{1}{3}$ is a larger number than $2 \div \frac{1}{2}$.

Divide, if possible.

3. $\dfrac{7}{16} \div \dfrac{3}{4}$

4. $\dfrac{3}{13} \div \dfrac{9}{26}$

5. $\dfrac{2}{3} \div \dfrac{4}{27}$

6. $\dfrac{25}{49} \div \dfrac{5}{7}$

7. $\dfrac{7}{18} \div \dfrac{21}{6}$

8. $\dfrac{8}{15} \div \dfrac{24}{35}$

9. $\dfrac{5}{9} \div \dfrac{1}{5}$

10. $\dfrac{3}{4} \div \dfrac{2}{3}$

11. $\dfrac{4}{15} \div \dfrac{4}{15}$

12. $\dfrac{2}{7} \div \dfrac{2}{7}$

13. $\dfrac{3}{7} \div \dfrac{7}{3}$

14. $\dfrac{9}{14} \div \dfrac{1}{3}$

15. $\dfrac{4}{5} \div 1$

16. $1 \div \dfrac{3}{7}$

17. $\dfrac{3}{11} \div 4$

18. $3 \div \dfrac{5}{6}$

19. $1 \div \dfrac{7}{27}$

20. $\dfrac{9}{16} \div 1$

21. $0 \div \dfrac{3}{17}$

22. $0 \div \dfrac{5}{16}$

23. $\dfrac{18}{19} \div 0$

24. $\dfrac{24}{29} \div 0$

25. $8 \div \dfrac{4}{5}$

26. $16 \div \dfrac{8}{11}$

27. $\dfrac{7}{8} \div 4$

28. $\dfrac{5}{6} \div 12$

29. $\dfrac{9}{16} \div \dfrac{3}{4}$

30. $\dfrac{3}{4} \div \dfrac{9}{16}$

31. $3\dfrac{1}{4} \div 2\dfrac{1}{4}$

32. $2\dfrac{2}{3} \div 4\dfrac{1}{3}$

33. $6\dfrac{2}{5} \div 3\dfrac{1}{5}$

34. $9\dfrac{1}{3} \div 3\dfrac{1}{9}$

35. $6000 \div \dfrac{6}{5}$

36. $12{,}000 \div \dfrac{3}{8}$

37. $\dfrac{\dfrac{4}{5}}{200}$

38. $\dfrac{\dfrac{5}{9}}{100}$

39. $\dfrac{\dfrac{5}{8}}{\dfrac{25}{7}}$

40. $\dfrac{\dfrac{3}{16}}{\dfrac{5}{8}}$

Mixed Practice *Multiply or divide.*

41. $3\dfrac{1}{5} \div \dfrac{1}{5}$

42. $4\dfrac{3}{4} \div \dfrac{1}{4}$

43. $2\dfrac{1}{3} \times \dfrac{1}{6}$

44. $7\dfrac{5}{6} \times \dfrac{1}{2}$

45. $5\dfrac{1}{4} \div 2\dfrac{5}{8}$

46. $1\dfrac{7}{8} \div 3\dfrac{3}{4}$

47. $5 \div 1\dfrac{1}{4}$

48. $7 \div 1\dfrac{2}{5}$

49. $5\dfrac{2}{3} \div 2\dfrac{1}{4}$

50. $14\dfrac{2}{3} \div 3\dfrac{1}{2}$

51. $\dfrac{7}{2} \div 3\dfrac{1}{2}$

52. $\dfrac{16}{3} \div 5\dfrac{1}{3}$

53. $\dfrac{13}{25} \times 2\dfrac{1}{3}$

54. $\dfrac{11}{20} \times 4\dfrac{1}{2}$

55. $3\dfrac{3}{4} \div 9$

56. $5\dfrac{5}{6} \div 7$

57. $\dfrac{5}{3\dfrac{1}{6}}$

58. $\dfrac{10}{2\dfrac{1}{2}}$

59. $\dfrac{0}{4\dfrac{3}{8}}$

60. $\dfrac{5\dfrac{2}{5}}{0}$

61. $\dfrac{\dfrac{7}{12}}{3\dfrac{2}{3}}$

62. $\dfrac{4\dfrac{1}{2}}{\dfrac{8}{9}}$

63. $4\dfrac{2}{5} \times 2\dfrac{8}{11}$

64. $4\dfrac{2}{3} \times 5\dfrac{1}{7}$

Review Example 6. Then find the value of x in each of the following.

65. $x \div \dfrac{4}{3} = \dfrac{21}{20}$

66. $x \div \dfrac{2}{5} = \dfrac{15}{16}$

67. $x \div \dfrac{10}{7} = \dfrac{21}{100}$

68. $x \div \dfrac{11}{6} = \dfrac{54}{121}$

Applications *Answer each question.*

69. *Leather Factory* A leather factory in Morocco tans leather. In order to make the leather soft, it has to soak in a vat of uric acid and other ingredients. The main holding tank holds $20\dfrac{1}{4}$ gallons of the tanning mixture. If the mixture is distributed evenly into nine vats of equal size for the different colored leathers, how much will each vat hold?

70. *Marine Biology* A specially protected stretch of beach bordering the Great Barrier Reef in Australia is used for marine biology and ecological research. The beach, which is $7\dfrac{1}{2}$ miles long, has been broken up into 20 equal segments for comparison purposes. How long is each segment of the beach?

71. *Vehicle Travel* Bruce drove in a snowstorm to get to his favorite mountain to do some snowboarding. He traveled 125 miles in $3\dfrac{1}{3}$ hours. What was his average speed (in miles per hour)?

72. *Vehicle Travel* Roberto drove his truck to Cedarville, a distance of 200 miles, in $4\dfrac{1}{6}$ hours. What was his average speed (in miles per hour)?

73. *Cooking* The school cafeteria is making hamburgers for the annual Senior Day Festival. The cooks have decided that because hamburger shrinks on the grill, they will allow $\frac{2}{3}$ pound of meat for each student. If the kitchen has $38\frac{2}{3}$ pounds of meat, how many students will be fed?

74. *Making Costumes* Costumes are needed for the junior high school's *Wizard of Oz* performance. Each costume requires $4\frac{1}{3}$ yards of fabric and $151\frac{2}{3}$ yards are available. How many costumes can be made?

75. *Cooking* A coffee pot that holds 150 cups of coffee is being used at a company meeting. Each large Styrofoam cup holds $1\frac{1}{2}$ cups of coffee. How many large Styrofoam cups can be filled?

76. *Medicine Dosage* A small bottle of eye drops contains 16 milliliters. If the recommended use is $\frac{2}{3}$ milliliter, how many times can a person use the drops before the bottle is empty?

77. *Time Capsule* In 1911, a time capsule was placed behind a steel wall measuring $4\frac{3}{4}$ inches thick. On December 22, 2011, a special drill was used to bore through the wall and extricate the time capsule. The drill could move only $\frac{5}{6}$ inch at a time. How many drill attempts did it take to reach the other side of the steel wall?

78. *Ink Production* Imagination Ink supplies different colored inks for highlighter pens. Vat 1 has yellow ink, holds 150 gallons, and is $\frac{4}{5}$ full. Vat 2 has green ink, holds 50 gallons, and is $\frac{5}{8}$ full. One gallon of ink will fill 1200 pens. How many pens can be filled with the existing ink from Vats 1 and 2?

To Think About *When multiplying or dividing mixed numbers it is wise to estimate your answer by rounding each mixed number to the nearest whole number.*

79. Estimate your answer to $14\frac{2}{3} \div 5\frac{1}{6}$ by rounding each mixed number to the nearest whole number. Then find the exact answer. How close was your estimate?

80. Estimate your answer to $18\frac{1}{4} \times 27\frac{1}{2}$ by rounding each mixed number to the nearest whole number. Then find the exact answer. How close was your estimate?

Cumulative Review

81. **[1.1.3]** Write in words. 39,576,304

82. **[1.1.1]** Write in expanded form. 509,270

83. **[1.2.4]** Add. $126 + 34 + 9 + 891 + 12 + 27$

84. **[1.1.3]** Write in standard notation. eighty-seven million, five hundred ninety-five thousand, six hundred thirty-one

---

Quick Quiz 2.5 *Divide.*

1. $\dfrac{15}{24} \div \dfrac{5}{6}$

2. $6\dfrac{1}{3} \div 2\dfrac{5}{12}$

3. $7\dfrac{3}{4} \div 4$

4. **Concept Check** Explain how you would divide the whole number 7 by the mixed number $3\frac{3}{5}$.

How Am I Doing?   Sections 2.1–2.5

How are you doing with your homework assignments in Sections 2.1 to 2.5? Do you feel you have mastered the material so far? Do you understand the concepts you have covered? Before you go further in the textbook, take some time to do each of the following problems.

2.1

1. Use a fraction to represent the shaded part of the object.

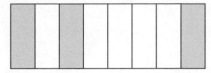

2. Frederich University had 3500 students from inside the state, 2600 students from outside the state but inside the country, and 800 students from outside the country. Write a fraction that describes the part of the student body from outside the country. Reduce the fraction.

3. An inspector tested 224 MP3 players. Of these, 10 were defective. Write a fraction that describes the part that was defective.

2.2

Reduce each fraction.

4. $\dfrac{4}{28}$ **5.** $\dfrac{13}{39}$ **6.** $\dfrac{16}{112}$ **7.** $\dfrac{175}{200}$ **8.** $\dfrac{44}{121}$

2.3

Change to an improper fraction.

9. $3\dfrac{2}{3}$ **10.** $15\dfrac{1}{3}$

Change to a mixed number.

11. $\dfrac{81}{4}$ **12.** $\dfrac{29}{5}$ **13.** $\dfrac{36}{17}$

2.4

Multiply.

14. $\dfrac{5}{11} \times \dfrac{1}{4}$ **15.** $\dfrac{3}{7} \times \dfrac{14}{9}$ **16.** $3\dfrac{1}{3} \times 5\dfrac{1}{3}$

2.5

Divide.

17. $\dfrac{3}{7} \div \dfrac{3}{7}$ **18.** $\dfrac{7}{16} \div \dfrac{7}{8}$ **19.** $6\dfrac{4}{7} \div 1\dfrac{5}{21}$ **20.** $12 \div \dfrac{4}{7}$

Now turn to page SA-4 for the answer to each of these problems. Each answer also includes a reference to the objective in which the problem is first taught. If you missed any of these problems, you should stop and review the Examples and Student Practice in the referenced objective. A little review now will help you master the material in the upcoming sections of the text.

1. _____
2. _____
3. _____
4. _____
5. _____
6. _____
7. _____
8. _____
9. _____
10. _____
11. _____
12. _____
13. _____
14. _____
15. _____
16. _____
17. _____
18. _____
19. _____
20. _____

How Am I Doing? Test on Sections 2.1–2.5

Solve. Make sure all fractions are simplified in the final answer.

1. Norah answered 33 out of 40 questions correctly on her chemistry exam. Write a fraction that describes the part of the exam she answered correctly.

2. Carlos inspected the boxes that were shipped from the central warehouse. He found that 340 were the correct weight and 112 were not. Write a fraction that describes what part of the total number of the boxes were at the correct weight.

Reduce each fraction.

3. $\dfrac{19}{38}$

4. $\dfrac{40}{56}$

5. $\dfrac{24}{66}$

6. $\dfrac{125}{155}$

7. $\dfrac{50}{140}$

8. $\dfrac{84}{36}$

Change each mixed number to an improper fraction.

9. $12\dfrac{2}{3}$

10. $4\dfrac{1}{8}$

Change each improper fraction to a mixed number.

11. $\dfrac{45}{7}$

12. $\dfrac{75}{9}$

Multiply.

13. $\dfrac{3}{8} \times \dfrac{7}{11}$

14. $\dfrac{35}{16} \times \dfrac{4}{5}$

15. $18 \times \dfrac{5}{6}$

16. $\dfrac{3}{8} \times 44$

17. $2\dfrac{1}{3} \times 5\dfrac{3}{4}$

18. $24 \times 3\dfrac{1}{3}$

Divide.

19. $\dfrac{4}{7} \div \dfrac{3}{4}$

20. $\dfrac{8}{9} \div \dfrac{1}{6}$

21. $5\dfrac{1}{4} \div \dfrac{3}{4}$

22. $5\dfrac{3}{5} \div 2\dfrac{1}{3}$

1. _____

2. _____

3. _____

4. _____

5. _____

6. _____

7. _____

8. _____

9. _____

10. _____

11. _____

12. _____

13. _____

14. _____

15. _____

16. _____

17. _____

18. _____

19. _____

20. _____

21. _____

22. _____

Mixed Practice

Perform the indicated operations. Simplify your answers.

23. $2\frac{1}{4} \times 3\frac{1}{2}$

24. $6 \times 2\frac{1}{3}$

25. $12 \div 2\frac{1}{4}$

26. $5\frac{3}{4} \div 2$

27. $\frac{13}{20} \div \frac{4}{5}$

28. $\frac{5}{3} \div 5$

29. $\frac{7}{10} \times \frac{20}{23}$

30. $\frac{14}{25} \times \frac{65}{42}$

Solve. Simplify your answer.

▲ **31.** A garden measures $5\frac{1}{4}$ feet by $8\frac{3}{4}$ feet. What is the area of the garden in square feet?

32. A recipe for two loaves of bread calls for $2\frac{2}{3}$ cups of flour. Lexi wants to make $1\frac{1}{2}$ times as much bread. How many cups of flour will she need?

33. Lisa drove $62\frac{1}{2}$ miles to visit a friend. Three-fourths of her trip was on the highway. How many miles did she drive on the highway?

34. The butcher prepared $12\frac{3}{8}$ pounds of lean ground round. He placed it in packages that held $\frac{3}{4}$ of a pound. How many full packages did he have? How much lean ground round was left over?

35. The college computer center has 136 computers. Samuel found that $\frac{3}{8}$ of them have Windows 7 installed on them. How many computers have Windows 7 installed on them?

36. The average household uses 82,000 gallons of water each year. About $\frac{3}{10}$ of this amount is used for showers and baths. How many gallons of water are used each year for showers and baths in an average household?

37. Yung Kim's rectangular kitchen measures 102 square feet. If the length of the kitchen is $12\frac{3}{4}$ feet, how wide is it?

38. The Outdoor Shop is making some custom tents that are very light but totally waterproof. Each tent requires $8\frac{1}{4}$ yards of cloth. How many tents can be made from $56\frac{1}{2}$ yards of cloth? How much cloth will be left over?

39. A container of vanilla-flavored syrup holds $32\frac{4}{5}$ ounces. Nate uses $\frac{4}{5}$ ounce every morning in his coffee. How many days will it take Nate to use up the container?

23. _____

24. _____

25. _____

26. _____

27. _____

28. _____

29. _____

30. _____

31. _____

32. _____

33. _____

34. _____

35. _____

36. _____

37. _____

38. _____

39. _____

2.6 The Least Common Denominator and Creating Equivalent Fractions

Student Learning Objectives

After studying this section, you will be able to:

① Find the least common multiple (LCM) of two numbers.

② Find the least common denominator (LCD) given two or three fractions.

③ Create equivalent fractions with a least common denominator.

NOTE TO STUDENT: *Fully worked-out solutions to all of the Student Practice problems can be found at the back of the text starting at page SP-1.*

① Finding the Least Common Multiple (LCM) of Two Numbers

The idea of a multiple of a number is fairly straightforward.

The **multiples** of a number are the products of that number and the numbers 1, 2, 3, 4, 5, 6, 7, . . .

For example, the multiples of 4 are   4, 8, 12, 16, 20, 24, 28, . . .

The multiples of 5 are   5, 10, 15, 20, 25, 30, 35, . . .

The **least common multiple,** or **LCM,** of two natural numbers is the smallest number that is a multiple of both.

EXAMPLE 1   Find the least common multiple of 10 and 12.

Solution

The multiples of 10 are   10,   20,   30,   40,   50,   60 ,   70, . . .

The multiples of 12 are   12,   24,   36,   48,   60 ,   72,   84, . . .

The first multiple that appears on both lists is the least common multiple. Thus the number 60 is the least common multiple of 10 and 12.

Student Practice 1   Find the least common multiple of 14 and 21.

EXAMPLE 2   Find the least common multiple of 6 and 8.

Solution

The multiples of 6 are   6,   12,   18,   24 ,   30,   36,   42, . . .

The multiples of 8 are   8,   16,   24 ,   32,   40,   48,   56, . . .

The first multiple that appears on both lists is the least common multiple. Thus the number 24 is the least common multiple of 6 and 8.

Student Practice 2   Find the least common multiple of 10 and 15.

Now of course we can do the problem immediately if the larger number is a multiple of the smaller number. In such cases the larger number is the least common multiple.

EXAMPLE 3   Find the least common multiple of 7 and 35.

Solution   Because $7 \times 5 = 35$, 35 is a multiple of 7.

So we can state immediately that the least common multiple of 7 and 35 is 35.

Student Practice 3   Find the least common multiple of 6 and 54.

② Finding the Least Common Denominator (LCD) Given Two or Three Fractions

We need some way to determine which of two fractions is larger. Suppose that Marcia and Melissa each have some leftover pizza.

Marcia's Pizza
$\frac{1}{3}$ of a pizza left

Melissa's Pizza
$\frac{1}{4}$ of a pizza left

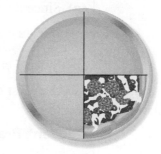

Who has more pizza left? How much more? Comparing the amounts of pizza left would be easy if each pizza had been cut into equal-sized pieces. If the original pizzas had each been cut into 12 pieces, we would be able to see that Marcia had $\frac{1}{12}$ of a pizza more than Melissa had.

Marcia's Pizza

$\left(\begin{array}{c}\text{We know that} \\ \dfrac{4}{12} = \dfrac{1}{3} \text{ by reducing.}\end{array}\right)$

Melissa's Pizza

$\left(\begin{array}{c}\text{We know that} \\ \dfrac{3}{12} = \dfrac{1}{4} \text{ by reducing.}\end{array}\right)$

The denominator 12 appears in the fractions $\frac{4}{12}$ and $\frac{3}{12}$. We call the smallest denominator that allows us to compare fractions directly the *least common denominator,* abbreviated LCD. The number 12 is the least common denominator for the fractions $\frac{1}{3}$ and $\frac{1}{4}$.

Notice that 12 is the least common multiple of 3 and 4.

LEAST COMMON DENOMINATOR

The **least common denominator (LCD)** of two or more fractions is the smallest number that can be divided evenly by each of the fractions' denominators.

How does this relate to least common multiples? The LCD of two fractions is the least common multiple of the two denominators.

In some problems you may be able to guess the LCD quite quickly. With practice, you can often find the LCD mentally. For example, you now know that if the denominators of two fractions are 3 and 4, the LCD is 12. For the fractions $\frac{1}{2}$ and $\frac{1}{4}$, the LCD is 4; for the fractions $\frac{1}{3}$ and $\frac{1}{6}$, the LCD is 6. We can see that if the denominator of one fraction divides without remainder into the denominator of another, the LCD of the two fractions is the larger of the denominators.

EXAMPLE 4   Determine the LCD for each pair of fractions.

(a) $\dfrac{7}{15}$ and $\dfrac{4}{5}$ **(b)** $\dfrac{2}{3}$ and $\dfrac{5}{27}$

Solution

(a) Since 5 can be divided into 15, the LCD of $\dfrac{7}{15}$ and $\dfrac{4}{5}$ is 15. (Notice that the least common multiple of 5 and 15 is 15.)

(b) Since 3 can be divided into 27, the LCD of $\dfrac{2}{3}$ and $\dfrac{5}{27}$ is 27. (Notice that the least common multiple of 3 and 27 is 27.)

Student Practice 4   Determine the LCD for each pair of fractions.

(a) $\dfrac{3}{4}$ and $\dfrac{11}{12}$ **(b)** $\dfrac{1}{7}$ and $\dfrac{8}{35}$

Sometimes, the LCD is the product of the two denominators.

EXAMPLE 5   Find the LCD for $\dfrac{1}{4}$ and $\dfrac{3}{5}$.

Solution   We see that $4 \times 5 = 20$. Also, 20 is the *smallest* number that can be divided without remainder by 4 and by 5. We know this because the least common multiple of 4 and 5 is 20. So the LCD = 20.

Student Practice 5   Find the LCD for $\dfrac{3}{7}$ and $\dfrac{5}{6}$.

In cases where the LCD is not obvious, the following procedure will help us find the LCD.

THREE-STEP PROCEDURE FOR FINDING THE LEAST COMMON DENOMINATOR

1. Write each denominator as a product of prime factors.

2. List all the prime factors that appear in either product.

3. Form a product of those prime factors, using each factor the greatest number of times it appears in any one denominator.

EXAMPLE 6 Find the LCD by the three-step procedure.

(a) $\dfrac{5}{6}$ and $\dfrac{4}{15}$ (b) $\dfrac{7}{18}$ and $\dfrac{7}{30}$ (c) $\dfrac{10}{27}$ and $\dfrac{5}{18}$

Solution

(a) Step 1 Write each denominator as a product of prime factors.

$$6 = 2 \times 3 \qquad 15 = 5 \times 3$$

Step 2 The LCD will contain the factors 2, 3, and 5.

$$6 = 2 \times 3 \qquad 15 = 5 \times 3$$

Step 3 LCD $= 2 \times 3 \times 5$ We form a product.

$$= 30$$

(b) Step 1 Write each denominator as a product of prime factors.

$$18 = 2 \times 9 = 2 \times 3 \times 3$$
$$30 = 3 \times 10 = 2 \times 3 \times 5$$

Step 2 The LCD will be a product containing 2, 3, and 5.

Step 3 The LCD will contain the factor 3 twice since it occurs twice in the denominator 18.

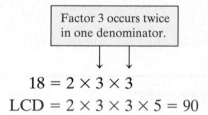

Factor 3 occurs twice in one denominator.

$$18 = 2 \times 3 \times 3$$
$$\text{LCD} = 2 \times 3 \times 3 \times 5 = 90$$

(c) Write each denominator as a product of prime factors.

$$27 = 3 \times 3 \times 3 \qquad 18 = 3 \times 3 \times 2$$

Factor 3 occurs three times.

The LCD will contain the factor 2 once but the factor 3 three times.

$$\text{LCD} = 2 \times 3 \times 3 \times 3 = 54$$

Student Practice 6 Find the LCD for each pair of fractions.

(a) $\dfrac{3}{14}$ and $\dfrac{1}{10}$

(b) $\dfrac{1}{15}$ and $\dfrac{7}{50}$

(c) $\dfrac{3}{16}$ and $\dfrac{5}{12}$

A similar procedure can be used for three fractions.

EXAMPLE 7 Find the LCD of $\dfrac{7}{12}, \dfrac{1}{15},$ and $\dfrac{11}{30}.$

Solution

$$12 = 2 \times 2 \times 3$$
$$15 = \qquad 3 \times 5$$
$$30 = \qquad 2 \times 3 \times 5$$

$$\text{LCD} = 2 \times 2 \times 3 \times 5$$
$$= 60$$

Continued on next page

Student Practice 7   Find the LCD of $\dfrac{3}{49}$, $\dfrac{5}{21}$, and $\dfrac{6}{7}$.

③ Creating Equivalent Fractions with a Least Common Denominator

In Section 2.7, we will discuss how to add fractions. We cannot add fractions that have different denominators. To change denominators, we must (1) find the LCD and (2) build up the addends—the fractions being added—into equivalent fractions that have the LCD as the denominator. We know now how to find the LCD. Let's look at how we build fractions. We know, for example, that

$$\frac{1}{2} = \frac{2}{4} = \frac{50}{100} \qquad \frac{1}{4} = \frac{25}{100} \quad \text{and} \quad \frac{3}{4} = \frac{75}{100}.$$

In these cases, we have mentally multiplied the given fraction by 1, in the form of a certain number, c, in the numerator and that same number, c, in the denominator.

$$\frac{1}{2} \times \boxed{\frac{c}{c}} = \frac{2}{4} \qquad \text{Here } c = 2, \frac{2}{2} = 1.$$

$$\frac{1}{2} \times \boxed{\frac{c}{c}} = \frac{50}{100} \qquad \text{Here } c = 50, \frac{50}{50} = 1.$$

This property is called the *building fraction property*.

BUILDING FRACTION PROPERTY

For whole numbers a, b, and c where $b \neq 0$, $c \neq 0$,

$$\frac{a}{b} = \frac{a}{b} \times 1 = \frac{a}{b} \times \boxed{\frac{c}{c}} = \frac{a \times c}{b \times c}.$$

EXAMPLE 8   Build each fraction to an equivalent fraction with the given LCD.

(a) $\dfrac{3}{4}$, LCD = 28   **(b)** $\dfrac{4}{5}$, LCD = 45   **(c)** $\dfrac{1}{3}$ and $\dfrac{4}{5}$, LCD = 15

Solution

(a) $\dfrac{3}{4} \times \boxed{\dfrac{c}{c}} = \dfrac{?}{28}$ We know that $4 \times 7 = 28$, so the value c that we multiply numerator and denominator by is 7.

$$\frac{3}{4} \times \frac{7}{7} = \frac{21}{28}$$

(b) $\dfrac{4}{5} \times \boxed{\dfrac{c}{c}} = \dfrac{?}{45}$ We know that $5 \times 9 = 45$, so $c = 9$.

$$\frac{4}{5} \times \frac{9}{9} = \frac{36}{45}$$

(c) $\dfrac{1}{3} = \dfrac{?}{15}$ We know that $3 \times 5 = 15$, so we multiply numerator and denominator by 5.

$$\dfrac{1}{3} \times \boxed{\dfrac{5}{5}} = \dfrac{5}{15}$$

$\dfrac{4}{5} = \dfrac{?}{15}$ We know that $5 \times 3 = 15$, so we multiply numerator and denominator by 3.

$$\dfrac{4}{5} \times \boxed{\dfrac{3}{3}} = \dfrac{12}{15}$$

Thus $\dfrac{1}{3} = \dfrac{5}{15}$ and $\dfrac{4}{5} = \dfrac{12}{15}$.

Student Practice 8 Build each fraction to an equivalent fraction with the given LCD.

(a) $\dfrac{3}{5}$, LCD = 40   **(b)** $\dfrac{7}{11}$, LCD = 44   **(c)** $\dfrac{2}{7}$ and $\dfrac{3}{4}$, LCD = 28

EXAMPLE 9

(a) Find the LCD of $\dfrac{1}{32}$ and $\dfrac{7}{48}$.

(b) Build the fractions to equivalent fractions that have the LCD as their denominators.

Solution

(a) First we find the prime factors of 32 and 48.

$$32 = 2 \times 2 \times 2 \times 2 \times 2$$
$$48 = 2 \times 2 \times 2 \times 2 \times 3$$

Thus the LCD will require a factor of 2 five times and a factor of 3 one time.

$$\text{LCD} = 2 \times 2 \times 2 \times 2 \times 2 \times 3 = 96$$

(b) $\dfrac{1}{32} = \dfrac{?}{96}$ Since $32 \times 3 = 96$ we multiply by the fraction $\dfrac{3}{3}$.

$$\dfrac{1}{32} = \dfrac{1}{32} \times \boxed{\dfrac{3}{3}} = \dfrac{3}{96}$$

$\dfrac{7}{48} = \dfrac{?}{96}$ Since $48 \times 2 = 96$, we multiply by the fraction $\dfrac{2}{2}$.

$$\dfrac{7}{48} = \dfrac{7}{48} \times \boxed{\dfrac{2}{2}} = \dfrac{14}{96}$$

Continued on next page

Student Practice 9

(a) Find the LCD of $\dfrac{3}{20}$ and $\dfrac{11}{15}$.

(b) Build the fractions to equivalent fractions that have the LCD as their denominators.

EXAMPLE 10

(a) Find the LCD of $\dfrac{2}{125}$ and $\dfrac{8}{75}$.

(b) Build the fractions to equivalent fractions that have the LCD as their denominators.

Solution

(a) First we find the prime factors of 125 and 75.

$$125 = 5 \times 5 \times 5$$
$$75 = 5 \times 5 \times 3$$

Thus the LCD will require a factor of 5 three times and a factor of 3 one time.

$$\text{LCD} = 5 \times 5 \times 5 \times 3 = 375$$

(b) $\dfrac{2}{125} = \dfrac{?}{375}$ Since $125 \times 3 = 375$, we multiply by the fraction $\dfrac{3}{3}$.

$$\frac{2}{125} = \frac{2}{125} \times \boxed{\frac{3}{3}} = \frac{6}{375}$$

$\dfrac{8}{75} = \dfrac{?}{375}$ Since $75 \times 5 = 375$, we multiply by the fraction $\dfrac{5}{5}$.

$$\frac{8}{75} = \frac{8}{75} \times \boxed{\frac{5}{5}} = \frac{40}{375}$$

Student Practice 10

(a) Find the LCD of $\dfrac{5}{64}$ and $\dfrac{3}{80}$.

(b) Build the fractions to equivalent fractions that have the LCD as their denominators.

MyMathLab®

Watch the videos
in MyMathLab

Download the
MyDashBoard App

Find the least common multiple (LCM) for each pair of numbers.

1. 8 and 12   **2.** 6 and 9   **3.** 20 and 50   **4.** 22 and 55   **5.** 12 and 15

6. 18 and 30   **7.** 10 and 15   **8.** 8 and 60   **9.** 21 and 49   **10.** 25 and 35

Find the LCD for each pair of fractions.

11. $\dfrac{1}{5}$ and $\dfrac{3}{10}$ **12.** $\dfrac{2}{9}$ and $\dfrac{11}{18}$ **13.** $\dfrac{3}{7}$ and $\dfrac{1}{4}$ **14.** $\dfrac{5}{6}$ and $\dfrac{3}{5}$ **15.** $\dfrac{2}{5}$ and $\dfrac{3}{7}$

16. $\dfrac{1}{16}$ and $\dfrac{2}{3}$ **17.** $\dfrac{1}{6}$ and $\dfrac{5}{9}$ **18.** $\dfrac{7}{8}$ and $\dfrac{5}{12}$ **19.** $\dfrac{7}{12}$ and $\dfrac{14}{15}$ **20.** $\dfrac{7}{15}$ and $\dfrac{9}{25}$

21. $\dfrac{7}{32}$ and $\dfrac{3}{4}$ **22.** $\dfrac{2}{11}$ and $\dfrac{1}{44}$ **23.** $\dfrac{5}{10}$ and $\dfrac{11}{45}$ **24.** $\dfrac{13}{20}$ and $\dfrac{17}{30}$ **25.** $\dfrac{7}{16}$ and $\dfrac{17}{80}$

26. $\dfrac{5}{6}$ and $\dfrac{19}{30}$ **27.** $\dfrac{5}{21}$ and $\dfrac{8}{35}$ **28.** $\dfrac{1}{20}$ and $\dfrac{5}{70}$ **29.** $\dfrac{11}{24}$ and $\dfrac{7}{30}$ **30.** $\dfrac{23}{30}$ and $\dfrac{37}{50}$

Find the LCD for each set of three fractions.

31. $\dfrac{2}{3}, \dfrac{1}{2}, \dfrac{5}{6}$ **32.** $\dfrac{1}{5}, \dfrac{1}{3}, \dfrac{7}{10}$ **33.** $\dfrac{1}{4}, \dfrac{11}{12}, \dfrac{5}{6}$ **34.** $\dfrac{21}{48}, \dfrac{1}{12}, \dfrac{3}{8}$

35. $\dfrac{5}{11}, \dfrac{7}{12}, \dfrac{1}{6}$ **36.** $\dfrac{11}{16}, \dfrac{3}{20}, \dfrac{2}{5}$ **37.** $\dfrac{7}{12}, \dfrac{1}{21}, \dfrac{3}{14}$ **38.** $\dfrac{28}{45}, \dfrac{4}{15}, \dfrac{17}{30}$

39. $\dfrac{7}{15}, \dfrac{11}{12}, \dfrac{7}{8}$ **40.** $\dfrac{5}{36}, \dfrac{2}{48}, \dfrac{1}{24}$

Build each fraction to an equivalent fraction with the specified denominator. State the numerator.

41. $\dfrac{1}{3} = \dfrac{?}{9}$

42. $\dfrac{1}{6} = \dfrac{?}{30}$

43. $\dfrac{5}{7} = \dfrac{?}{49}$

44. $\dfrac{7}{9} = \dfrac{?}{81}$

45. $\dfrac{4}{11} = \dfrac{?}{55}$

46. $\dfrac{5}{14} = \dfrac{?}{42}$

47. $\dfrac{5}{12} = \dfrac{?}{96}$

48. $\dfrac{3}{50} = \dfrac{?}{100}$

49. $\dfrac{8}{9} = \dfrac{?}{108}$

50. $\dfrac{6}{7} = \dfrac{?}{147}$

51. $\dfrac{7}{20} = \dfrac{?}{180}$

52. $\dfrac{3}{25} = \dfrac{?}{175}$

The LCD of each pair of fractions is listed. Build each fraction to an equivalent fraction with the given LCD.

53. LCD $= 36, \dfrac{7}{12}$ and $\dfrac{5}{9}$

54. LCD $= 20, \dfrac{9}{10}$ and $\dfrac{3}{4}$

55. LCD $= 80, \dfrac{5}{16}$ and $\dfrac{17}{20}$

56. LCD $= 72, \dfrac{5}{24}$ and $\dfrac{7}{36}$

57. LCD $= 20, \dfrac{9}{10}$ and $\dfrac{19}{20}$

58. LCD $= 150, \dfrac{19}{25}$ and $\dfrac{7}{30}$

Find the LCD. Build the fractions to equivalent fractions having the LCD as the denominator.

59. $\dfrac{2}{5}$ and $\dfrac{9}{35}$

60. $\dfrac{7}{9}$ and $\dfrac{35}{54}$

61. $\dfrac{5}{24}$ and $\dfrac{3}{8}$

62. $\dfrac{19}{42}$ and $\dfrac{6}{7}$

63. $\dfrac{8}{15}$ and $\dfrac{1}{6}$

64. $\dfrac{19}{20}$ and $\dfrac{7}{8}$

65. $\dfrac{4}{15}$ and $\dfrac{5}{12}$

66. $\dfrac{9}{10}$ and $\dfrac{3}{25}$

67. $\dfrac{5}{18}, \dfrac{11}{36}, \dfrac{7}{12}$

68. $\dfrac{1}{30}, \dfrac{7}{15}, \dfrac{1}{45}$

69. $\dfrac{3}{56}, \dfrac{7}{8}, \dfrac{5}{7}$

70. $\dfrac{5}{9}, \dfrac{1}{6}, \dfrac{3}{54}$

71. $\dfrac{5}{63}, \dfrac{4}{21}, \dfrac{8}{9}$

72. $\dfrac{3}{8}, \dfrac{5}{14}, \dfrac{13}{16}$

Applications

73. *Door Repair* Suppose that you wish to compare the lengths of the three portions of the given stainless steel bolt that came out of a door.

(a) What is the LCD for the three fractions?

(b) Build each fraction to an equivalent fraction that has the LCD as a denominator.

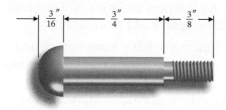

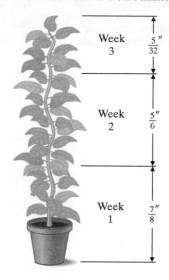

74. *Plant Growth* Suppose that you want to prepare a report on the growth of a plant. The total height of the plant in the pot is recorded for each week of a three-week experiment.

(a) What is the LCD for the three fractions?

(b) Build each fraction to an equivalent fraction that has the LCD for a denominator.

Week 3	$\frac{5}{32}''$
Week 2	$\frac{5}{6}''$
Week 1	$\frac{7}{8}''$

Cumulative Review

75. [1.6.2] Evaluate. $(5 - 3)^2 + 4 \times 6 - 3$ **76. [2.4.3]** Multiply. $4\frac{3}{4} \times \frac{2}{3}$ **77. [2.5.3]** Divide. $16\frac{1}{2} \div \frac{3}{4}$

Quick Quiz 2.6

1. Find the least common denominator of $\frac{5}{6}$ and $\frac{5}{21}$

2. Find the least common denominator of $\frac{27}{28}, \frac{3}{4}, \frac{19}{20}$

3. Build the fraction to an equivalent fraction with the specified denominator.
$\frac{7}{26} = \frac{?}{78}$

4. Concept Check Explain how you would find the least common denominator of the fractions $\frac{5}{6}, \frac{11}{14}$, and $\frac{2}{15}$.

2.7 Adding and Subtracting Fractions

Student Learning Objectives

After studying this section, you will be able to:

① Add and subtract fractions with a common denominator.

② Add and subtract fractions with different denominators.

① Adding and Subtracting Fractions with a Common Denominator

You must have common denominators (denominators that are alike) to add or subtract fractions.

If your problem has fractions without a common denominator or if it has mixed numbers, you must use what you already know about changing the form of each fraction (how the fraction looks). Only after all the fractions have a common denominator can you add or subtract.

An important distinction: You must have common denominators to add or subtract fractions, but you need not have common denominators to multiply or divide fractions.

To add two fractions that have the same denominator, add the numerators and write the sum over the common denominator.

To illustrate we use $\frac{1}{5} + \frac{2}{5} = \frac{3}{5}$. The figure shows that $\frac{1}{5} + \frac{2}{5} = \frac{3}{5}$.

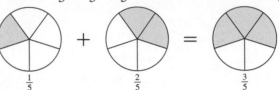

EXAMPLE 1 Add. $\frac{5}{13} + \frac{7}{13}$

Solution

$$\frac{5}{13} + \frac{7}{13} = \frac{12}{13}$$

Student Practice 1 Add.

$$\frac{3}{17} + \frac{12}{17}$$

NOTE TO STUDENT: Fully worked-out solutions to all of the Student Practice problems can be found at the back of the text starting at page SP-1.

The answer may need to be reduced. Sometimes the answer may be written as a mixed number.

EXAMPLE 2 Add.

(a) $\frac{4}{9} + \frac{2}{9}$

(b) $\frac{5}{7} + \frac{6}{7}$

Solution

(a) $\frac{4}{9} + \frac{2}{9} = \frac{6}{9} = \frac{2}{3}$

(b) $\frac{5}{7} + \frac{6}{7} = \frac{11}{7}$ or $1\frac{4}{7}$

Student Practice 2 Add.

(a) $\frac{1}{12} + \frac{5}{12}$

(b) $\frac{13}{15} + \frac{7}{15}$

A similar rule is followed for subtraction, except that the numerators are subtracted and the result placed over the common denominator. Be sure to reduce all answers when possible.

EXAMPLE 3   Subtract.

(a) $\dfrac{5}{13} - \dfrac{4}{13}$

(b) $\dfrac{17}{20} - \dfrac{3}{20}$

Solution

(a) $\dfrac{5}{13} - \dfrac{4}{13} = \dfrac{1}{13}$

(b) $\dfrac{17}{20} - \dfrac{3}{20} = \dfrac{14}{20} = \dfrac{7}{10}$

Student Practice 3 Subtract.

(a) $\dfrac{5}{19} - \dfrac{2}{19}$

(b) $\dfrac{21}{25} - \dfrac{6}{25}$

---

② Adding and Subtracting Fractions with Different Denominators

If the two fractions do not have a common denominator, we follow the procedure in Section 2.6: Find the LCD and then build each fraction so that its denominator is the LCD.

EXAMPLE 4   Add. $\dfrac{7}{12} + \dfrac{1}{4}$

Solution   The LCD is 12. The fraction $\frac{7}{12}$ already has the least common denominator.

$$
\begin{array}{rcl}
\dfrac{7}{12} & = & \boxed{\dfrac{7}{12}} \\[2mm]
+\ \dfrac{1}{4} \times \dfrac{3}{3} & = & +\ \boxed{\dfrac{3}{12}} \\[2mm]
\hline
& & \boxed{\dfrac{10}{12}}
\end{array}
$$

We will need to reduce this fraction. Then we will have

$$\dfrac{7}{12} + \dfrac{1}{4} = \dfrac{7}{12} + \dfrac{3}{12} = \dfrac{10}{12} = \dfrac{5}{6}.$$

It is very important to remember to reduce our final answer.

Student Practice 4   Add.

$$\dfrac{2}{15} + \dfrac{1}{5}$$

EXAMPLE 5 Add. $\dfrac{7}{20} + \dfrac{4}{15}$

Solution LCD = 60.

$$\frac{7}{20} \times \frac{3}{3} = \frac{21}{60} \qquad\qquad \frac{4}{15} \times \frac{4}{4} = \frac{16}{60}$$

Thus

$$\frac{7}{20} + \frac{4}{15} = \frac{21}{60} + \frac{16}{60} = \frac{37}{60}$$

Student Practice 5 Add.

$$\frac{5}{12} + \frac{5}{16}$$

A similar procedure holds for the addition of three or more fractions.

EXAMPLE 6 Add. $\dfrac{3}{8} + \dfrac{5}{6} + \dfrac{1}{4}$

Solution LCD = 24.

$$\frac{3}{8} \times \frac{3}{3} = \frac{9}{24} \qquad \frac{5}{6} \times \frac{4}{4} = \frac{20}{24} \qquad \frac{1}{4} \times \frac{6}{6} = \frac{6}{24}$$

$$\frac{3}{8} + \frac{5}{6} + \frac{1}{4} = \frac{9}{24} + \frac{20}{24} + \frac{6}{24} = \frac{35}{24} \quad \text{or} \quad 1\frac{11}{24}$$

Student Practice 6 Add.

$$\frac{3}{16} + \frac{1}{8} + \frac{1}{12}$$

EXAMPLE 7 Subtract. $\dfrac{17}{25} - \dfrac{3}{35}$

Solution LCD = 175.

$$\frac{17}{25} \times \frac{7}{7} = \frac{119}{175} \qquad\qquad \frac{3}{35} \times \frac{5}{5} = \frac{15}{175}$$

Thus

$$\frac{17}{25} - \frac{3}{35} = \frac{119}{175} - \frac{15}{175} = \frac{104}{175}.$$

Student Practice 7 Subtract.

$$\frac{9}{48} - \frac{5}{32}$$

▲ **EXAMPLE 8**   John and Stephanie and the triplets have a house on $\frac{7}{8}$ acre of land. They have $\frac{1}{3}$ acre of land planted with grass. How much of the land is not planted with grass?

Solution

1. Understand the problem. Draw a picture.

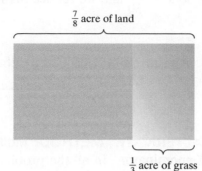

$\frac{7}{8}$ acre of land

$\frac{1}{3}$ acre of grass

We need to subtract. $\frac{7}{8} - \frac{1}{3}$

2. Solve and state the answer. The LCD is 24.

$$\frac{7}{8} \times \frac{3}{3} = \frac{21}{24} \qquad \frac{1}{3} \times \frac{8}{8} = \frac{8}{24}$$

$$\frac{7}{8} - \frac{1}{3} = \frac{21}{24} - \frac{8}{24} = \frac{13}{24}$$

We conclude that $\frac{13}{24}$ acre of the land is not planted with grass.

3. Check. The check is left to the student.

Student Practice 8   Leon had $\frac{9}{10}$ gallon of cleaning fluid in the garage. He used $\frac{1}{4}$ gallon to clean the garage floor. How much cleaning fluid is left?

Some students may find Example 9 difficult. Read it slowly and carefully.

EXAMPLE 9   Find the value of x in the equation $x + \frac{5}{6} = \frac{9}{10}$. Reduce your answer.

Solution   The LCD for the two fractions $\frac{5}{6}$ and $\frac{9}{10}$ is 30.

$$\frac{5}{6} \times \frac{5}{5} = \frac{25}{30} \qquad \frac{9}{10} \times \frac{3}{3} = \frac{27}{30}$$

Thus we can write the equation in the equivalent form.

$$x + \frac{25}{30} = \frac{27}{30}$$

Continued on next page

The denominators are the same. Look at the numerators. We must add 2 to 25 to get 27.

$$\frac{2}{30} + \frac{25}{30} = \frac{27}{30}$$

So $x = \frac{2}{30}$ and we reduce the fraction to obtain $x = \frac{1}{15}$.

Student Practice 9 Find the value of x in the equation. $x + \frac{3}{10} = \frac{23}{25}$

ALTERNATE METHOD: Multiply the Denominators as a Common Denominator In all the problems in this section so far, we have combined two fractions by first finding the least common denominator. However, there is an alternate approach. You are only required to find a common denominator, not necessarily the least common denominator. One way to quickly find a common denominator of two fractions is to multiply the two denominators. However, if you use this method, the numbers will usually be larger and you will often need to simplify the fraction in your final answer.

EXAMPLE 10 Add $\frac{11}{12} + \frac{13}{30}$ by using the product of the two denominators as a common denominator.

Solution Using this method we just multiply the numerator and denominator of each fraction by the denominator of the other fraction. Thus no steps are needed to determine what to multiply by.

$$\frac{11}{12} \times \frac{30}{30} = \frac{330}{360} \qquad \frac{13}{30} \times \frac{12}{12} = \frac{156}{360}$$

Thus $\frac{11}{12} + \frac{13}{30} = \frac{330}{360} + \frac{156}{360} = \frac{486}{360}$

We must reduce the fraction: $\frac{486}{360} = \frac{27}{20}$ or $1\frac{7}{20}$

Student Practice 10 Add $\frac{15}{16} + \frac{3}{40}$ by using the product of the two denominators as a common denominator.

Some students find this alternate method helpful because you do not have to find the LCD or the number each fraction must be multiplied by. Other students find this alternate method more difficult because of errors encountered when working with large numbers or in reducing the final answer. You are encouraged to try a couple of the homework exercises by this method and make up your own mind.

2.7 Exercises

MyMathLab® Watch the videos in MyMathLab Download the MyDashBoard App

Add or subtract. Simplify all answers.

1. $\dfrac{5}{9} + \dfrac{2}{9}$

2. $\dfrac{5}{8} + \dfrac{2}{8}$

3. $\dfrac{7}{18} + \dfrac{15}{18}$

4. $\dfrac{11}{25} + \dfrac{17}{25}$

5. $\dfrac{19}{20} - \dfrac{11}{20}$

6. $\dfrac{19}{45} - \dfrac{4}{45}$

7. $\dfrac{53}{88} - \dfrac{19}{88}$

8. $\dfrac{103}{110} - \dfrac{3}{110}$

Add or subtract. Simplify all answers.

9. $\dfrac{1}{3} + \dfrac{1}{2}$

10. $\dfrac{1}{4} + \dfrac{1}{3}$

11. $\dfrac{3}{10} + \dfrac{3}{20}$

12. $\dfrac{4}{9} + \dfrac{1}{6}$

13. $\dfrac{1}{8} + \dfrac{3}{4}$

14. $\dfrac{11}{18} + \dfrac{1}{9}$

15. $\dfrac{4}{5} + \dfrac{7}{20}$

16. $\dfrac{2}{3} + \dfrac{4}{7}$

17. $\dfrac{3}{10} + \dfrac{7}{100}$

18. $\dfrac{13}{100} + \dfrac{7}{10}$

19. $\dfrac{3}{10} + \dfrac{1}{6}$

20. $\dfrac{8}{15} + \dfrac{3}{10}$

21. $\dfrac{7}{8} + \dfrac{5}{12}$

22. $\dfrac{5}{6} + \dfrac{7}{8}$

23. $\dfrac{3}{8} + \dfrac{3}{10}$

24. $\dfrac{12}{35} + \dfrac{1}{10}$

25. $\dfrac{29}{18} - \dfrac{5}{9}$

26. $\dfrac{37}{20} - \dfrac{2}{5}$

27. $\dfrac{3}{7} - \dfrac{9}{21}$

28. $\dfrac{8}{9} - \dfrac{3}{8}$

29. $\dfrac{5}{9} - \dfrac{5}{36}$

30. $\dfrac{9}{10} - \dfrac{1}{15}$

31. $\dfrac{5}{12} - \dfrac{7}{30}$

32. $\dfrac{9}{24} - \dfrac{3}{8}$

33. $\dfrac{11}{12} - \dfrac{2}{3}$

34. $\dfrac{7}{10} - \dfrac{2}{5}$

35. $\dfrac{17}{21} - \dfrac{1}{7}$

36. $\dfrac{20}{25} - \dfrac{4}{5}$

37. $\dfrac{5}{12} - \dfrac{7}{18}$

38. $\dfrac{7}{8} - \dfrac{1}{12}$

39. $\dfrac{10}{16} - \dfrac{5}{8}$

40. $\dfrac{2}{3} - \dfrac{12}{18}$

41. $\dfrac{23}{36} - \dfrac{2}{9}$

42. $\dfrac{2}{3} - \dfrac{1}{16}$

43. $\dfrac{1}{2} + \dfrac{2}{7} + \dfrac{3}{14}$

44. $\dfrac{7}{8} + \dfrac{5}{6} + \dfrac{7}{24}$

45. $\dfrac{5}{30} + \dfrac{3}{40} + \dfrac{1}{8}$

46. $\dfrac{1}{12} + \dfrac{3}{14} + \dfrac{4}{21}$

47. $\dfrac{7}{30} + \dfrac{2}{5} + \dfrac{5}{6}$

48. $\dfrac{1}{12} + \dfrac{5}{36} + \dfrac{5}{6}$

Study Example 9 carefully. Then find the value of x in each equation.

49. $x + \dfrac{1}{7} = \dfrac{5}{14}$

50. $x + \dfrac{1}{8} = \dfrac{7}{16}$

51. $x + \dfrac{2}{3} = \dfrac{9}{11}$

52. $x + \dfrac{4}{5} = \dfrac{33}{40}$

53. $x - \dfrac{3}{10} = \dfrac{4}{15}$

54. $x - \dfrac{7}{12} = \dfrac{5}{24}$

Applications

55. *Cooking* Fiona is baking a pie and a cake for a bake sale. She needs $\frac{1}{4}$ cup sugar for the pie crust and $\frac{2}{3}$ cup sugar for the cake's frosting. How much sugar will she use to make both?

56. *Fitness Training* Kia is training for a short triathlon. On Monday she swam $\frac{1}{4}$ mile and ran $\frac{5}{6}$ mile. On Tuesday she swam $\frac{1}{2}$ mile and ran $\frac{3}{4}$ mile. How many miles has she swum so far this week? How many miles has she run so far?

57. *Food Purchase* Yasmin wants to make a trail mix of nuts and dried fruit. She has $\frac{2}{3}$ pound peanuts and $\frac{1}{2}$ pound dried cranberries. She purchases $\frac{3}{4}$ pound almonds and $\frac{3}{8}$ pound raisins to mix with the peanuts and cranberries. After mixing the four ingredients how many pounds of nuts and how many pounds of dried fruit will there be in the trail mix?

58. *Automobile Maintenance* Mandy purchased two new steel-belted all-weather radial tires for her car. The tread depth on the new tires measures $\frac{11}{32}$ of an inch. The dealer told her that when the tires have worn down and their tread depth measures $\frac{1}{8}$ of an inch, she should replace the worn tires with new ones. How much will the tread depth decrease over the useful life of the tire?

59. *Power Outage* Travis typed $\frac{11}{12}$ of his research paper on his computer. Then he printed out $\frac{3}{5}$ of his paper on his computer printer. Suddenly, there was a power outage, and he discovered that he hadn't saved his paper before the power went off. What fractional part of the research paper was lost when the power failed?

60. *Childcare* An infant's father knows that straight apple juice is too strong for his daughter. Her bottle is $\frac{1}{2}$ full, and he adds $\frac{1}{3}$ of a bottle of water to dilute the apple juice.

(a) How much is there to drink in the bottle after this addition?

(b) If she drinks $\frac{2}{5}$ of what is in the bottle, how much is left?

61. *Food Purchase* While he was at the grocery store, Raymond purchased a box of candy for himself. On the way back to the dorm he ate $\frac{1}{4}$ of the candy. As he was putting away the groceries he ate $\frac{1}{2}$ of what was left. There are now six chocolates left in the box. How many chocolates were in the box to begin with?

62. *Baking* Peter has $\frac{3}{4}$ cup of cocoa. He needs $\frac{1}{8}$ cup to make brownies, and another $\frac{1}{4}$ cup to make fudge squares. After making the brownies and the fudge, how much cocoa will Peter have left?

63. *Business Management* The manager at Fit Factory Health Club was going through his files for 2007 and discovered that only $\frac{7}{10}$ of the members actually used the club. When he checked the numbers from the previous year of 2006, he found that $\frac{7}{8}$ of the members had used the club. What fractional part of the membership represents the decrease in club usage?

Cumulative Review

64. **[2.2.2]** Reduce to lowest terms. $\frac{15}{85}$

65. **[2.2.2]** Reduce to lowest terms. $\frac{27}{207}$

66. **[2.3.2]** Change to a mixed number. $\frac{125}{14}$

67. **[2.3.1]** Change to an improper fraction. $14\frac{3}{7}$

68. **[2.5.3]** Divide. $4\frac{1}{3} \div 1\frac{1}{2}$

69. **[2.4.3]** Multiply. $5\frac{1}{2} \times 1\frac{3}{11}$

Quick Quiz 2.7 *Simplify all answers.*

1. Add. $\frac{7}{16} + \frac{3}{4}$

2. Add. $\frac{1}{3} + \frac{5}{7} + \frac{10}{21}$

3. Subtract. $\frac{8}{9} - \frac{7}{15}$

4. **Concept Check** Explain how you would subtract the fractions $\frac{8}{9} - \frac{3}{7}$.

2.8 Adding and Subtracting Mixed Numbers and the Order of Operations

Student Learning Objectives

After studying this section, you will be able to:

① Add mixed numbers.

② Subtract mixed numbers.

③ Evaluate fractional expressions using the order of operations.

① Adding Mixed Numbers

When adding mixed numbers, it is best to add the fractions together and then add the whole numbers together.

EXAMPLE 1 Add. $3\frac{1}{8} + 2\frac{5}{8}$

Solution

$$
\begin{array}{r}
3\ \dfrac{1}{8} \\[2mm]
+2\ \dfrac{5}{8} \\[2mm]
\hline
5\ \dfrac{6}{8}
\end{array}
$$

Add the whole numbers. $3 + 2 = 5$ → $5\ \dfrac{6}{8}$ ← Add the fractions. $\dfrac{1}{8} + \dfrac{5}{8} = \dfrac{6}{8}$

$$= 5\ \dfrac{3}{4} \longleftarrow \text{Reduce } \dfrac{6}{8} = \dfrac{3}{4}$$

NOTE TO STUDENT: *Fully worked-out solutions to all of the Student Practice problems can be found at the back of the text starting at page SP-1.*

Student Practice 1 Add. $5\frac{1}{12} + 9\frac{5}{12}$

If the fraction portions of the mixed numbers do not have a common denominator, we must build the fraction parts to obtain a common denominator before adding.

EXAMPLE 2 Add. $1\frac{2}{7} + 5\frac{1}{3}$

Solution The LCD of $\frac{2}{7}$ and $\frac{1}{3}$ is 21.

$$\frac{2}{7} \times \frac{3}{3} = \frac{6}{21} \qquad \frac{1}{3} \times \frac{7}{7} = \frac{7}{21}$$

Thus $1\frac{2}{7} + 5\frac{1}{3} = 1\frac{6}{21} + 5\frac{7}{21}$.

$$
\begin{array}{r}
1\dfrac{2}{7} = \quad 1\ \dfrac{6}{21} \\[3mm]
+\,5\dfrac{1}{3} = +\,5\ \dfrac{7}{21} \\[2mm]
\hline
6\ \dfrac{13}{21}
\end{array}
$$

Add the whole numbers. $1 + 5$ → $6\ \dfrac{13}{21}$ ← Add the fractions. $\dfrac{6}{21} + \dfrac{7}{21}$

Student Practice 2 Add. $6\frac{1}{4} + 2\frac{2}{5}$

If the sum of the fractions is an improper fraction, we convert it to a mixed number and add the whole numbers together.

EXAMPLE 3 Add. $6\frac{5}{6} + 4\frac{3}{8}$

Solution The LCD of $\frac{5}{6}$ and $\frac{3}{8}$ is 24.

$$6 \boxed{\frac{5}{6} \times \frac{4}{4}} = 6 \boxed{\frac{20}{24}}$$

$$+4 \boxed{\frac{3}{8} \times \frac{3}{3}} = +4 \boxed{\frac{9}{24}}$$

Add the whole numbers. $\longrightarrow 10 \boxed{\frac{29}{24}} \leftarrow$ Add the fractions.

$$= 10 + \boxed{1\frac{5}{24}} \quad \text{Since } \frac{29}{24} = 1\frac{5}{24}$$

$$= 11\frac{5}{24} \quad \text{We add the whole numbers } 10 + 1 = 11.$$

Student Practice 3 Add. $7\frac{1}{4} + 3\frac{5}{6}$

② Subtracting Mixed Numbers

Subtracting mixed numbers is like adding.

EXAMPLE 4 Subtract. $8\frac{5}{7} - 5\frac{5}{14}$

Solution The LCD of $\frac{5}{7}$ and $\frac{5}{14}$ is 14.

$$8 \boxed{\frac{5}{7} \times \frac{2}{2}} = 8\frac{10}{14}$$

$$-5\frac{5}{14} = -5\frac{5}{14}$$

$$\boxed{\text{Subtract the whole numbers.}} \longrightarrow 3\frac{5}{14} \longleftarrow \boxed{\text{Subtract the fractions.}}$$

Student Practice 4 Subtract. $12\frac{5}{6} - 7\frac{5}{12}$

Sometimes we must borrow before we can subtract.

EXAMPLE 5 Subtract.

(a) $9\frac{1}{4} - 6\frac{5}{14}$ 　　　　　　**(b)** $15 - 9\frac{3}{16}$

Continued on next page

Solution   This example is fairly challenging. Read through each step carefully. Be sure to have paper and pencil handy and see if you can verify each step.

(a) The LCD of $\frac{1}{4}$ and $\frac{5}{14}$ is 28.

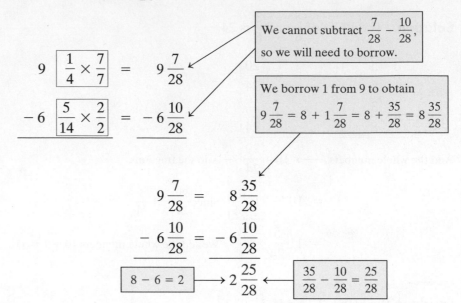

$$9 \begin{array}{|c|} \hline \dfrac{1}{4} \times \dfrac{7}{7} \\ \hline \end{array} = 9\dfrac{7}{28}$$

$$-6 \begin{array}{|c|} \hline \dfrac{5}{14} \times \dfrac{2}{2} \\ \hline \end{array} = -6\dfrac{10}{28}$$

We cannot subtract $\dfrac{7}{28} - \dfrac{10}{28}$, so we will need to borrow.

We borrow 1 from 9 to obtain
$$9\dfrac{7}{28} = 8 + 1\dfrac{7}{28} = 8 + \dfrac{35}{28} = 8\dfrac{35}{28}$$

$$9\dfrac{7}{28} = 8\dfrac{35}{28}$$

$$-6\dfrac{10}{28} = -6\dfrac{10}{28}$$

$$\boxed{8 - 6 = 2} \longrightarrow 2\dfrac{25}{28} \longleftarrow \boxed{\dfrac{35}{28} - \dfrac{10}{28} = \dfrac{25}{28}}$$

(b) The LCD = 16.

$$15 \quad = \quad 14\dfrac{16}{16}$$

We borrow 1 from 15 to obtain
$$15 = 14 + 1 = 14 + \dfrac{16}{16} = 14\dfrac{16}{16}$$

$$-9\dfrac{3}{16} = -9\dfrac{3}{16}$$

$$\boxed{14 - 9 = 5} \longrightarrow 5\dfrac{13}{16} \longleftarrow \boxed{\dfrac{16}{16} - \dfrac{3}{16} = \dfrac{13}{16}}$$

Student Practice 5   Subtract.

(a) $9\dfrac{1}{8} - 3\dfrac{2}{3}$ **(b)** $18 - 6\dfrac{7}{18}$

EXAMPLE 6   A plumber had a pipe $5\frac{3}{16}$ inches long for a fitting under the sink. He needed a pipe that was $3\frac{7}{8}$ inches long, so he cut the pipe down. How much of the pipe did he cut off?

Solution   We will need to subtract $5\frac{3}{16} - 3\frac{7}{8}$ to find the length that was cut off.

$$5\dfrac{3}{16} \quad = \quad 5\dfrac{3}{16}$$

$$-3\dfrac{7}{8} \times \dfrac{2}{2} = -3\dfrac{14}{16}$$

$$4\frac{19}{16} \leftarrow \boxed{\begin{array}{l}\text{We borrow 1 from 5 to obtain}\\[4pt] 5\frac{3}{16} = 4 + 1\frac{3}{16} = 4 + \frac{19}{16}\end{array}}$$

$$-\ 3\frac{14}{16}$$

$$\boxed{4 - 3 = 1} \longrightarrow 1\frac{5}{16} \leftarrow \boxed{\dfrac{19}{16} - \dfrac{14}{16} = \dfrac{5}{16}}$$

The plumber had to cut off $1\frac{5}{16}$ inches of pipe.

Student Practice 6 Hillary and Sam purchased $6\frac{1}{4}$ gallons of paint to paint the inside of their house. They used $4\frac{2}{3}$ gallons of paint. How much paint was left over?

ALTERNATIVE METHOD: Add or Subtract Mixed Numbers as Improper Fractions Can mixed numbers be added and subtracted as improper fractions? Yes. Recall Example 5(a).

$$9\frac{1}{4} - 6\frac{5}{14} = 2\frac{25}{28}$$

If we write $9\frac{1}{4} - 6\frac{5}{14}$ using improper fractions, we have $\frac{37}{4} - \frac{89}{14}$. Now we build each of these improper fractions so that they both have the LCD for their denominators.

$$\begin{array}{r}\dfrac{37}{4}\ \boxed{\times\dfrac{7}{7}} = \dfrac{259}{28}\\[10pt] -\ \dfrac{89}{14}\ \boxed{\times\dfrac{2}{2}} = -\dfrac{178}{28}\\[8pt]\hline\\[-6pt]\dfrac{81}{28} = 2\dfrac{25}{28}\end{array}$$

The same result is obtained as in Example 5(a). This method does not require borrowing. However, you do work with larger numbers. For more practice, see exercises 53–54.

③ Evaluating Fractional Expressions Using the Order of Operations

Recall that in Section 1.6 we discussed the order of operations when we were combining whole numbers. We will now encounter some similar problems involving fractions and mixed numbers. We will repeat here the four-step order of operations that we studied previously:

ORDER OF OPERATIONS

With grouping symbols:

Do first **1.** Perform operations inside parentheses.

2. Simplify any expressions with exponents.

3. Multiply or divide from left to right.

Do last **4.** Add or subtract from left to right.

Mc **EXAMPLE 7** Evaluate. $\dfrac{3}{4} - \dfrac{2}{3} \times \dfrac{1}{8}$

Solution

$\dfrac{3}{4} - \dfrac{2}{3} \times \dfrac{1}{8} = \dfrac{3}{4} - \dfrac{1}{12}$ First we must multiply $\dfrac{2}{3} \times \dfrac{1}{8}$.

$= \dfrac{9}{12} - \dfrac{1}{12}$ Now we subtract, but first we need to build $\dfrac{3}{4}$ to an equivalent fraction with a common denominator of 12.

$= \dfrac{8}{12}$ Now we can subtract $\dfrac{9}{12} - \dfrac{1}{12}$.

$= \dfrac{2}{3}$ Finally we reduce the fraction.

Student Practice 7

Evaluate. $\dfrac{3}{5} - \dfrac{1}{15} \times \dfrac{10}{13}$

EXAMPLE 8 Evaluate. $\dfrac{2}{3} \times \dfrac{1}{4} + \dfrac{2}{5} \div \dfrac{14}{15}$

Solution

$\dfrac{2}{3} \times \dfrac{1}{4} + \dfrac{2}{5} \div \dfrac{14}{15} = \dfrac{1}{6} + \dfrac{2}{5} \div \dfrac{14}{15}$ First we multiply $\dfrac{2}{3} \times \dfrac{1}{4}$.

$= \dfrac{1}{6} + \dfrac{2}{5} \times \dfrac{15}{14}$ We express the division as a multiplication problem. We invert $\dfrac{14}{15}$ and multiply.

$= \dfrac{1}{6} + \dfrac{3}{7}$ Now we perform the multiplication.

$= \dfrac{7}{42} + \dfrac{18}{42}$ We obtain equivalent fractions with the LCD of 42.

$= \dfrac{25}{42}$ We add the two fractions.

Student Practice 8 Evaluate.

$$\dfrac{1}{7} \times \dfrac{5}{6} + \dfrac{5}{3} \div \dfrac{7}{6}$$

👣 STEPS TO SUCCESS Improving Your Accuracy

It is easy to make a mistake. But here are five ways to cut down on errors. Look over each one and think about how each suggestion can help you.

1. **Work carefully, and take your time.** Do not rush through a problem just to get it done.

2. **Concentrate on the problem.** Sometimes your mind starts to wander. Then you get careless and will likely make a mistake.

3. **Check your problem.** Be sure you copied it correctly from the book.

4. **Check your computations from step to step.** Did you do each step correctly?

5. **Check your final answer.** Does it work? Is it reasonable?

Making it personal: Look over these five suggestions. Which one do you think will help you the most? Write down how you can use this suggestion to help you personally as you try to improve your accuracy. ▼

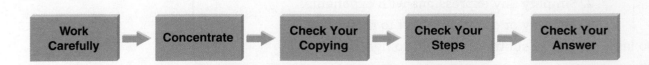

Work Carefully → Concentrate → Check Your Copying → Check Your Steps → Check Your Answer

Add or subtract. Express the answer as a mixed number. Simplify all answers.

1. $7\frac{1}{8} + 2\frac{5}{8}$

2. $6\frac{3}{10} + 4\frac{1}{10}$

3. $15\frac{3}{14} - 11\frac{1}{14}$

4. $8\frac{3}{4} - 3\frac{1}{4}$

5. $12\frac{1}{3} + 5\frac{1}{6}$

6. $20\frac{1}{4} + 3\frac{1}{8}$

7. $4\frac{3}{5} + 8\frac{2}{5}$

8. $8\frac{2}{9} + 7\frac{7}{9}$

9. $1 - \frac{3}{7}$

10. $1 - \frac{9}{11}$

11. $1\frac{3}{4} + \frac{5}{16}$

12. $1\frac{2}{3} + \frac{13}{18}$

13. $5\frac{1}{6} + 4\frac{5}{18}$

14. $6\frac{2}{5} + 7\frac{3}{20}$

15. $8\frac{1}{4} - 8\frac{4}{16}$

16. $9\frac{3}{4} - 5\frac{1}{6}$

17. $12\frac{1}{3} - 7\frac{2}{5}$

18. $10\frac{10}{15} - 10\frac{2}{3}$

19. $30 - 15\frac{3}{7}$

20. $25 - 14\frac{2}{11}$

21. $3 + 4\frac{2}{5}$

22. $8 + 2\frac{3}{4}$

23. $14 - 3\frac{7}{10}$

24. $18 - 4\frac{3}{5}$

Add or subtract. Express the answer as a mixed number. Simplify all answers.

25. $15\frac{4}{15}$
  $+ 26\frac{8}{15}$

26. $22\frac{1}{8}$
  $+ 14\frac{3}{8}$

27. $6\frac{1}{6}$
  $+ 2\frac{1}{4}$

28. $8\frac{1}{5}$
  $+ 7\frac{1}{4}$

29. $3\frac{3}{4}$
  $+ 4\frac{5}{12}$

30. $10\frac{5}{6}$
  $+ 9\frac{2}{3}$

31. $47\frac{3}{10}$
  $+ 26\frac{5}{8}$

32. $34\frac{1}{20}$
  $+ 45\frac{8}{15}$

33. $19\frac{5}{6}$
  $- 14\frac{1}{3}$

34. $22\frac{7}{9}$
  $- 16\frac{1}{4}$

35. $6\frac{1}{12}$
  $- 5\frac{10}{24}$

36. $4\frac{1}{12}$
  $- 3\frac{7}{18}$

37. $12\frac{3}{20}$
  $- 7\frac{7}{15}$

38. $8\frac{5}{12}$
  $- 5\frac{9}{10}$

39. 12
  $- 3\frac{7}{15}$

40. 40
  $- 6\frac{3}{7}$

41. 120
  $- 17\frac{3}{8}$

42. 87
  $- 56\frac{7}{10}$

43. $3\frac{5}{8}$
  $2\frac{2}{3}$
  $+ 7\frac{3}{4}$

44. $4\frac{2}{3}$
  $3\frac{4}{5}$
  $+ 6\frac{3}{4}$

Applications

45. *Mountain Biking* Lee Hong rode his mountain bike through part of the Sangre de Cristo Mountains in New Mexico. On Wednesday he rode $20\frac{3}{4}$ miles. On Thursday he rode $22\frac{3}{8}$ miles. What was his total biking distance during those two days?

46. *Triathlon* Participants in the Lifetime Fitness Triathlon in Minneapolis must complete three events: a $\frac{9}{10}$-mile swim, a $24\frac{4}{5}$-mile bike ride, and a $6\frac{1}{5}$-mile run. How many total miles does the triathlon consist of?

47. *Bicycling* Lake Harriet and Lake Calhoun have paved paths around them for runners, walkers, and bicyclists. The distance around Lake Harriet is $2\frac{4}{5}$ miles, and the distance around Lake Calhoun is $3\frac{1}{10}$ miles. The road connecting the two lakes is $\frac{1}{2}$ mile. If Lola rides her bike around both lakes, and uses the connecting road twice, how long is her bike ride?

48. *World Record Fish* The heaviest muskellunge on record weighed $69\frac{15}{16}$ pounds. The heaviest carp on record weighed $57\frac{13}{16}$ pounds. How much heavier was the muskellunge?

49. *Basketball* Nina and Julie are the two tallest basketball players on their high school team. Nina is $69\frac{3}{4}$ inches tall and Julie is $72\frac{1}{2}$ inches tall. How many inches taller is Julie than Nina?

50. *Food Purchase* Julio bought $3\frac{3}{4}$ pounds of roast turkey and $1\frac{2}{3}$ pounds of salami at the deli. How many more pounds of turkey than salami did he buy?

51. *Food Purchase* Lara needs 8 pounds of haddock for her dinner party. At the grocery store, haddock portions weighing $1\frac{3}{4}$ pounds and $2\frac{1}{6}$ pounds are placed on the scale.

   (a) How many pounds of haddock are on the scale?

   (b) How many more pounds of haddock does Lara need?

52. *Medical Care* A young man has been under a doctor's care to lose weight. His doctor wanted him to lose 46 pounds in the first three months. He lost $17\frac{5}{8}$ pounds the first month and $13\frac{1}{2}$ pounds the second month.

   (a) How much did he lose during the first two months?

   (b) How much would he need to lose in the third month to reach the goal?

To Think About, Exercises 53–56

Use improper fractions and the Alternative Method as discussed in the text to perform each calculation.

53. $\dfrac{379}{8} + \dfrac{89}{5}$

54. $\dfrac{151}{6} - \dfrac{130}{7}$

When adding or subtracting mixed numbers, it is wise to estimate your answer by rounding each mixed number to the nearest whole number.

55. Estimate your answer to $35\frac{1}{6} + 24\frac{5}{12}$ by rounding each mixed number to the nearest whole number. Then find the exact answer. How close was your estimate?

56. Estimate your answer to $102\frac{5}{7} - 86\frac{2}{3}$ by rounding each mixed number to the nearest whole number. Then find the exact answer. How close was your estimate?

Evaluate using the correct order of operations.

57. $\dfrac{6}{7} - \dfrac{4}{7} \times \dfrac{1}{3}$

58. $\dfrac{3}{5} - \dfrac{1}{3} \times \dfrac{6}{5}$

59. $\dfrac{1}{2} + \dfrac{3}{8} \div \dfrac{3}{4}$

60. $\dfrac{3}{4} + \dfrac{1}{4} \div \dfrac{5}{3}$

61. $\dfrac{9}{10} \div \dfrac{3}{8} \times \dfrac{5}{8}$

62. $\dfrac{5}{12} \div \dfrac{3}{10} \times \dfrac{9}{5}$

63. $\dfrac{3}{5} \times \dfrac{1}{2} + \dfrac{1}{5} \div \dfrac{2}{3}$

64. $\dfrac{5}{6} \times \dfrac{1}{2} + \dfrac{2}{3} \div \dfrac{4}{3}$

65. $\left(\dfrac{3}{5} - \dfrac{3}{20}\right) \times \dfrac{4}{5}$

66. $\left(\dfrac{1}{3} + \dfrac{1}{6}\right) \times \dfrac{5}{11}$

67. $\left(\dfrac{1}{3}\right)^2 \div \dfrac{4}{9}$

68. $\left(\dfrac{1}{4}\right)^2 \div \dfrac{3}{4}$

69. $\dfrac{1}{4} \times \left(\dfrac{2}{3}\right)^2$

70. $\dfrac{5}{8} \times \left(\dfrac{2}{5}\right)^2$

71. $\dfrac{5}{6} \div \left(\dfrac{2}{3} + \dfrac{1}{6}\right)^2$

72. $\dfrac{4}{3} \div \left(\dfrac{3}{5} - \dfrac{3}{10}\right)^2$

Cumulative Review *Multiply.*

73. **[1.4.3]** $\begin{array}{r} 1200 \\ \times\ 400 \\ \hline \end{array}$

74. **[1.4.4]** $\begin{array}{r} 4050 \\ \times\,2106 \\ \hline \end{array}$

Quick Quiz 2.8

1. Add. Express the answer as a mixed number.

$3\dfrac{4}{5} + 5\dfrac{3}{8}$

2. Subtract. Express the answer as a mixed number.

$6\dfrac{5}{12} - 4\dfrac{7}{10}$

3. Evaluate using the correct order of operations.

$\dfrac{1}{5} + \dfrac{3}{10} \div \dfrac{11}{20}$

4. **Concept Check** Explain how you would evaluate the following expression using the correct order of operations. $\dfrac{4}{5} - \dfrac{1}{4} \times \dfrac{2}{3}$

2.9 Solving Applied Problems Involving Fractions

Student Learning Objective

After studying this section, you will be able to:

① Solve real-life problems with fractions.

① Solving Real-Life Problems with Fractions

All problem solving requires the same kind of thinking. In this section we will combine problem-solving skills with our new computational skills with fractions. Sometimes the difficulty is in figuring out what must be done. Sometimes it is in doing the computation. Remember that *estimating* is important in problem solving. We may use the following steps.

1. *Understand the problem.*

    (a) Read the problem carefully.

    (b) Draw a picture if this helps you.

    (c) Fill in the Mathematics Blueprint.

2. *Solve.*

    (a) Perform the calculations.

    (b) State the answer, including the units of measure.

3. *Check.*

    (a) Estimate the answer. Round fractions to the nearest whole number.

    (b) Compare the exact answer with the estimate to see if your answer is reasonable.

▲ **EXAMPLE 1** In designing a modern offshore speedboat, the designing engineer has determined that one of the oak frames near the engine housing needs to be $26\frac{1}{8}$ inches long. At the end of the oak frame there will be $2\frac{5}{8}$ inches of insulation. Finally there will be a steel mounting that is $3\frac{3}{4}$ inches long. When all three items are assembled, how long will the oak frame and insulation and steel mounting extend?

Solution

1. *Understand the problem.*

We draw a picture to help us.

Then we fill in the Mathematics Blueprint.

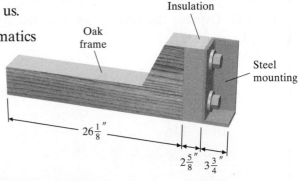

Mathematics Blueprint for Problem Solving

Gather the Facts	What Am I Asked to Do?	How Do I Proceed?	Key Points to Remember
Oak frame: $26\frac{1}{8}''$ Insulation: $2\frac{5}{8}''$ Steel mounting: $3\frac{3}{4}''$	Find the total length.	Add the lengths of the three items.	When adding mixed numbers, add the whole numbers first and then add the fractions.

2. *Solve and state the answer.*

Add the three amounts. $26\frac{1}{8} + 2\frac{5}{8} + 3\frac{3}{4}$

$$\text{LCD} = 8 \quad 26\frac{1}{8} \quad = \quad 26\frac{1}{8}$$

$$2\frac{5}{8} \quad = \quad 2\frac{5}{8}$$

$$+ 3 \;\boxed{\frac{3}{4} \times \frac{2}{2}} = \; + 3\frac{6}{8}$$

$$31\frac{12}{8} = 32\frac{4}{8} = 32\frac{1}{2}$$

The entire assembly will be $32\frac{1}{2}$ inches.

3. *Check.* Estimate the sum by rounding each fraction to the nearest whole number.

$$\text{Thus} \qquad 26\frac{1}{8} + 2\frac{5}{8} + 3\frac{3}{4}$$

$$\text{becomes} \qquad 26 + 3 + 4 = 33$$

This is close to our answer, $32\frac{1}{2}$. Our answer seems reasonable.

One of the most important uses of estimation in mathematics is in the calculation of problems involving fractions. People find it easier to detect significant errors when working with whole numbers. However, the extra steps involved in the calculations with fractions and mixed numbers often distract our attention from an error that we should have detected.

Thus it is particularly critical to take the time to check your answer by estimating the results of the calculation with whole numbers. Be sure to ask yourself, is this answer reasonable? Does this answer seem realistic? Only by estimating our results with whole numbers will we be able to answer those questions. It is this estimating skill that you will find more useful in your own life as a consumer and as a citizen.

Student Practice 1 Nicole required the following amounts of gas for her farm tractor in the last three fill-ups: $18\frac{7}{10}$ gallons, $15\frac{2}{5}$ gallons, and $14\frac{1}{2}$ gallons. How many gallons did she need altogether?

NOTE TO STUDENT: Fully worked-out solutions to all of the Student Practice problems can be found at the back of the text starting at page SP-1.

The word *diameter* has two common meanings. First, it means a line segment that passes through the center of and intersects a circle twice. It has its endpoints on the circle. Second, it means the *length* of this segment.

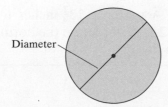

Diameter

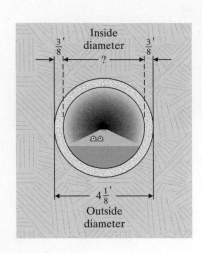

Inside diameter

$\frac{3}{8}'$ $\frac{3}{8}'$

?

$4\frac{1}{8}'$

Outside diameter

▲ **EXAMPLE 2**   What is the inside diameter (distance across) of a concrete storm drain pipe that has an outside diameter of $4\frac{1}{8}$ feet and is $\frac{3}{8}$ foot thick?

Solution

1. ***Understand the problem.*** Read the problem carefully. Draw a picture. The picture is in the margin on the left. Now fill in the Mathematics Blueprint.

Mathematics Blueprint for Problem Solving

Gather the Facts	What Am I Asked to Do?	How Do I Proceed?	Key Points to Remember
Outside diameter is $4\frac{1}{8}$ *feet.* Thickness is $\frac{3}{8}$ foot on both ends of the diameter.	Find the *inside* diameter of the pipe.	Add the two measures of thickness. Then subtract this total from the outside diameter.	Since the LCD = 8, all fractions must have this denominator.

2. ***Solve and state the answer.*** Add the two thickness measurements together. Adding $\frac{3}{8} + \frac{3}{8} = \frac{6}{8}$ gives the total thickness of the pipe, $\frac{6}{8}$ foot. We will not reduce $\frac{6}{8}$ since the LCD is 8.

We subtract the total of the two thickness measurements from the outside diameter.

$$
\begin{array}{rcr}
4\frac{1}{8} & = & 3\frac{9}{8} \\
-\ \frac{6}{8} & = & -\ \frac{6}{8} \\
\hline
& & 3\frac{3}{8}
\end{array}
$$

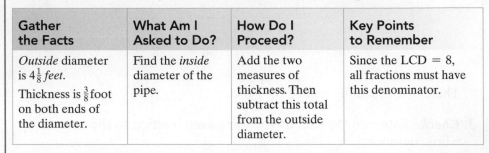

We borrow 1 from 4 to get $3 + 1\frac{1}{8}$ or $3\frac{9}{8}$.

The inside diameter is $3\frac{3}{8}$ feet.

3. ***Check.*** We will work backward to check. We will use the exact values. If we have done our work correctly, $\frac{3}{8}$ foot $+\ 3\frac{3}{8}$ feet $+\ \frac{3}{8}$ foot should add up to the outside diameter, $4\frac{1}{8}$ feet.

$$\frac{3}{8} + 3\frac{3}{8} + \frac{3}{8} \overset{?}{=} 4\frac{1}{8}$$

$$3\frac{9}{8} \overset{?}{=} 4\frac{1}{8}$$

$$4\frac{1}{8} = 4\frac{1}{8} \quad \checkmark$$

Our answer of $3\frac{3}{8}$ feet is correct.

▲ **Student Practice 2**   A poster is $12\frac{1}{4}$ inches long. We want a $1\frac{3}{8}$-inch border on the top and a 2-inch border on the bottom. What is the length of the inside portion of the poster?

EXAMPLE 3 A slipcover manufacturer uses $8\frac{1}{4}$ yards to make a slipcover for a chair. To make a slipcover for a recliner, $1\frac{1}{2}$ times that amount is needed. How many yards of fabric are needed to fill an order of 8 chair slipcovers and 10 recliner slipcovers?

Solution

1. ***Understand the problem.*** We draw a picture of the parts that make up the total number of yards needed.

| Yards needed for chairs | + | Yards needed for recliners | = | Total number of yards needed |

Now fill in the Mathematics Blueprint.

Mathematics Blueprint for Problem Solving

Gather the Facts	What Am I Asked to Do?	How Do I Proceed?	Key Points to Remember
Eight chair slipcovers are needed, each requiring $8\frac{1}{4}$ yards. Ten recliner slipcovers are needed, each requiring $1\frac{1}{2}$ times the amount needed for a chair slipcover.	Find the total number of yards needed to fill the order.	Find out how many yards are needed for the chairs. Find out how many yards are needed for the recliners. Then add the two.	The amount of fabric needed to make a recliner slipcover is $1\frac{1}{2}$ multiplied by the amount needed to make a chair slipcover.

2. ***Solve and state the answer.*** Find the amount of fabric needed for the chair slipcovers.

$$8 \times 8\frac{1}{4} = 8 \times \frac{33}{4} = 66 \text{ yards}$$

Now, find the amount of fabric needed for the recliner slipcovers.
For 1 recliner slipcover,

$$1\frac{1}{2} \times 8\frac{1}{4} = \frac{3}{2} \times \frac{33}{4} = \frac{99}{8} \text{ yards are needed.}$$

For 10 recliner slipcovers,

$$10 \times \frac{99}{8} = \frac{495}{4} = 123\frac{3}{4} \text{ yards are needed.}$$

Now we add to find the total yards needed.

$$\begin{array}{ll} 66 \text{ yards} & \text{Yards needed for chair slipcovers} \\ +\ 123\frac{3}{4} \text{ yards} & \text{Yards needed for recliner slipcovers} \\ \hline 189\frac{3}{4} \text{ yards} & \end{array}$$

To make 8 chair slipcovers and 10 recliner slipcovers, $189\frac{3}{4}$ yards of fabric are needed.

Continued on next page

3. *Check.* We estimate the amount of fabric needed for a chair slipcover as 8 yards.

We estimate the amount of fabric needed for a recliner slipcover as

$$1\frac{1}{2} \times 8 = \frac{3}{2} \times 8 = 12 \text{ yards.}$$

$$8 \text{ chair slipcovers} \times 8 \text{ yards per chair} = 64 \text{ yards}$$

$$10 \text{ recliner slipcovers} \times 12 \text{ yards per recliner} = 120 \text{ yards}$$

Estimated sum: 64 yards + 120 yards = 184 yards

184 yards is close to our calculated value, $189\frac{3}{4}$ yards, so our answer is reasonable.

Student Practice 3 A tent manufacturer uses $8\frac{1}{4}$ yards of waterproof duck cloth to make a regular tent. She uses $1\frac{1}{2}$ times that amount to make a large tent. How many yards of cloth will she need to make 6 regular tents and 16 large tents?

▲ **EXAMPLE 4** Alicia is buying some 8-foot boards for shelving. She wishes to make two bookcases, each with three shelves. Each shelf will be $3\frac{1}{4}$ feet long.

(a) How many boards does she need to buy?

(b) How many linear feet of shelving are actually needed to build the bookcases?

(c) How many linear feet of shelving will be left over?

Solution

1. *Understand the problem.* Draw a sketch of a bookcase. Each bookcase will have three shelves. Alicia is making two such bookcases. (Alicia's boards are for the shelves, not the sides.)

Now fill in the Mathematics Blueprint.

Mathematics Blueprint for Problem Solving

Gather the Facts	What Am I Asked to Do?	How Do I Proceed?	Key Points to Remember
She needs three shelves for each bookcase. Each shelf is $3\frac{1}{4}$ feet long. She will make two bookcases. Shelves are cut from 8-foot boards.	Find out how many boards to buy. Find out how many feet of board are needed for shelves and how many feet will be left over.	First find out how many $3\frac{1}{4}$-foot shelves she can get from one board. Then see how many boards she needs to make all six shelves.	Each time she cuts up an 8-foot board, she will get some shelves and some leftover wood.

2. Solve and state the answer. We want to know how many $3\frac{1}{4}$-foot boards are in an 8-foot board. By drawing a rough sketch, we would probably guess the answer is 2. To find exactly how many $3\frac{1}{4}$-foot-long pieces are in 8 feet, we will use division.

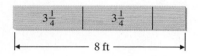

$$8 \div 3\frac{1}{4} = \frac{8}{1} \div \frac{13}{4} = \frac{8}{1} \times \frac{4}{13} = \frac{32}{13} = 2\frac{6}{13}$$

She will get two shelves from each board, and some wood will be left over.

(a) How many boards does Alicia need to build two bookcases? For two bookcases, she needs six shelves. She will get two shelves out of each board. $6 \div 2 = 3$. She will need three 8-foot boards.

(b) How many linear feet of shelving are actually needed to build the bookcases?

She needs 6 shelves at $3\frac{1}{4}$ feet.

$$6 \times 3\frac{1}{4} = \overset{3}{6} \times \frac{13}{\underset{2}{4}} = \frac{39}{2} = 19\frac{1}{2}$$

A total of $19\frac{1}{2}$ linear feet of shelving is needed.

(c) How many linear feet of shelving will be left over?
Each time she uses one board she will have

$$8 - 3\frac{1}{4} - 3\frac{1}{4} = 8 - \left(3\frac{1}{4} + 3\frac{1}{4}\right) = 8 - 6\frac{1}{2} = 1\frac{1}{2}$$

feet left over. Each of the three boards will have $1\frac{1}{2}$ feet left over.

$$3 \times 1\frac{1}{2} = 3 \times \frac{3}{2} = \frac{9}{2} = 4\frac{1}{2}$$

A total of $4\frac{1}{2}$ linear feet of shelving will be left over.

3. Check. Work backward. See if you can check that with three 8-foot boards you
   (a) can make the six shelves for the two bookcases.
   (b) will use exactly $19\frac{1}{2}$ linear feet to make the shelves.
   (c) will have exactly $4\frac{1}{2}$ linear feet left over.

The check is left to you.

Continued on next page

▲ | **Student Practice 4**   Michael is purchasing 12-foot boards for shelving. He wishes to make two bookcases, each with four shelves. Each shelf will be $2\frac{3}{4}$ feet long.

(a) How many boards does he need to buy?

(b) How many linear feet of shelving are actually needed to build the bookcases?

(c) How many linear feet of shelving will be left over?

Another useful method for solving applied problems is called "Do a similar, simpler problem." When a problem seems difficult to understand because of the fractions, change the problem to an easier but similar problem. Then decide how to solve the simpler problem and use the same steps to solve the original problem. For example:

How many gallons of water can a tank hold if its volume is $58\frac{2}{3}$ cubic feet? (1 cubic foot holds about $7\frac{1}{2}$ gallons.)

A similar, easier problem would be: "If 1 cubic foot holds about 8 gallons and a tank holds 60 cubic feet, how many gallons of water does the tank hold?"

The easier problem can be read more quickly and seems to make more sense. Probably we will see how to solve the easier problem right away: "I can find the number of gallons by multiplying 8×60." Therefore we can solve the first problem by multiplying $7\frac{1}{2} \times 58\frac{2}{3}$ to obtain the number of gallons of water. See the next example.

EXAMPLE 5   A fishing boat traveled $69\frac{3}{8}$ nautical miles in $3\frac{3}{4}$ hours. How many knots (nautical miles per hour) did the fishing boat average?

Solution

1. **Understand the problem.** Let us think of a simpler problem. If a boat traveled 70 nautical miles in 4 hours, how many knots did it average? We would divide distance by time.

$$70 \div 4 = \text{average speed}$$

Likewise in our original problem we need to divide distance by time.

$$69\frac{3}{8} \div 3\frac{3}{4} = \text{average speed}$$

Now fill in the Mathematics Blueprint.

Mathematics Blueprint for Problem Solving

Gather the Facts	What Am I Asked to Do?	How Do I Proceed?	Key Points to Remember
Distance is $69\frac{3}{8}$ nautical miles. Time is $3\frac{3}{4}$ hours.	Find the average speed of the boat.	Divide the distance in nautical miles by the time in hours.	You must change the mixed numbers to improper fractions before dividing.

Student Practice 5   Alfonso traveled $199\frac{3}{4}$ miles in his car and used $8\frac{1}{2}$ gallons of gas. How many miles per gallon did he get?

2. *Solve and state the answer.* Divide distance by time to get speed in knots.

$$69\frac{3}{8} \div 3\frac{3}{4} = \frac{555}{8} \div \frac{15}{4} = \frac{\overset{37}{\cancel{555}}}{\cancel{8}} \cdot \frac{\overset{1}{\cancel{4}}}{\cancel{15}}$$

$$= \frac{37}{2} \cdot \frac{1}{1} = \frac{37}{2} = 18\frac{1}{2}$$

The speed of the boat was $18\frac{1}{2}$ knots.

3. *Check.* We estimate $69\frac{3}{8} \div 3\frac{3}{4}$.

$$\text{Use } 70 \div 4 = 17\frac{1}{2} \text{ knots}$$

Our estimate is close to the calculated value.
Our answer is reasonable. ✓

Be sure to allow extra time to read over Examples 1–5 and Student Practice. Many students find it is helpful to study them on two different days. This allows you additional time to really understand the steps of reasoning involved.

👣 STEPS TO SUCCESS What Good Is It to Study Mathematics?

Students often question the value of mathematics. They see little real use for it in their everyday lives. Let us think about three things.

Get a good job. Mathematics is often the key that opens the door to a better-paying job or just to get a job if you are unemployed. In our present-day technological world, many people use mathematics daily. Many vocational and professional areas—such as the fields of business, statistics, economics, psychology, finance, computer science, chemistry, physics, engineering, electronics, nuclear energy, banking, quality control, nursing, medical technology, and teaching—require a certain level of expertise in mathematics. Those who want to work in these fields must be able to function at a given mathematical level. Those who cannot will not be able to enter these job areas.

Save money. These are challenging financial times. We are all looking for ways to save money. The more mathematics you learn, the more you will be able to find ways to save money. Several suggestions for saving money are given in this book. Be sure to read over each one and think how it might apply to your life.

Make decisions. Should I buy a car or lease one? Should I buy a house or rent an apartment? Should I drive to work or take public transportation? What career field should I pick if I want to increase my chances of getting a good job? Mathematics will help you to think more clearly and make better decisions because it will help you collect all the facts.

Making it personal: Which of these three paragraphs is the most relevant to you? Which one helps you to see why mathematics can really help you? Write down what you think is the most important reason for you to study mathematics. ▼

You may benefit from using the Mathematics Blueprint for Problem Solving when solving the following exercises.

Applications

▲ **1.** **Geometry** A triangle has three sides that measure $8\frac{1}{3}$ inches, $5\frac{4}{5}$ inches, and $9\frac{3}{10}$ inches. What is the perimeter (total distance around) of the triangle?

2. **Automobile Travel** On Tuesday, Sally drove $10\frac{1}{2}$ miles while running errands. On Friday and Saturday, she had more errands to run and drove $6\frac{1}{3}$ miles and $12\frac{1}{4}$ miles, respectively. How many total miles did Sally drive this week while running errands?

3. **Wildlife** In 2010, only 700 mountain gorillas remained in the world. Of these, about $\frac{5}{9}$ of them were living along the Virunga Mountain range between Rwanda and Uganda. How many gorillas were living in this mountain range? Round your answer to the nearest whole number.

4. **Consumer Decisions** When the Apple iPod was first sold in 2001, it cost $400. Today, the iPod Nano costs about $\frac{3}{8}$ of the 2001 price. How much does the iPod Nano cost now?

5. **Carpentry** A bolt extends through $\frac{3}{4}$-inch-thick plywood, two washers that are each $\frac{1}{16}$ inch thick, and a nut that is $\frac{3}{16}$ inch thick. The main body of the bolt must be $\frac{1}{2}$ inch longer than the sum of the thicknesses of plywood, washers, and nut. What is the minimum length of the bolt?

6. **Carpentry** A carpenter is using an 8-foot length of wood for a frame. The carpenter needs to cut a notch in the wood that is $4\frac{7}{8}$ feet from one end and $1\frac{2}{3}$ feet from the other end. How long does the notch need to be?

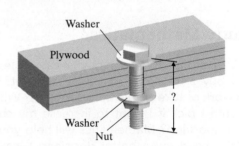

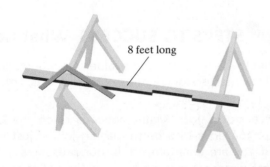

7. **Running a Marathon** Josiah is running the Boston Marathon, which is $26\frac{1}{5}$ miles long. At $6\frac{3}{4}$ miles from the start, he meets his wife, who is cheering him on. $9\frac{1}{2}$ miles further down the marathon course, he sees some friends from his running club volunteering at a water stop. Once he passes his friends, how many more miles does Josiah have left to run?

8. **Carpentry** Norman Olerud makes birdhouses as a hobby. He has a long piece of lumber that measures $14\frac{1}{4}$ feet. He needs to cut it into pieces that are $\frac{3}{4}$ foot long for the birdhouse floors. How many floors will he be able to cut from the long piece?

9. **Baking** Nathaniel works as a baker and makes several batches of muffins every morning. Each batch requires $1\frac{1}{2}$ cups of sugar. For a large meeting, Nathaniel needs $8\frac{1}{2}$ times more muffins than what a regular batch makes. How much sugar will he use? If each batch fills 3 boxes, how many boxes will be needed to pack the muffins for the meeting?

10. **Baking** A brownie requires $1\frac{3}{4}$ cups of flour. Stella wants to make 3 times the recipe. How much flour will she use? If each cup of flour weighs $4\frac{1}{2}$ ounces, what is the weight of the flour she will use?

▲ **11.** *Geometry* How many gallons can a tank hold that has a volume of $36\frac{3}{4}$ cubic feet? (Assume that 1 cubic foot holds about $7\frac{1}{2}$ gallons.)

▲ **12.** *Geometry* A tank can hold a volume of $7\frac{1}{4}$ cubic feet. If it is filled with water, how much does the water weigh? (Assume that 1 cubic foot of water weighs $62\frac{1}{2}$ pounds.)

13. *Titanic Disaster* The night of the *Titanic* cruise ship disaster, the captain decided to run his ship at $22\frac{1}{2}$ knots (nautical miles per hour). The *Titanic* traveled at that speed for $4\frac{3}{4}$ hours before it met its tragic demise. How far did the *Titanic* travel at this excessive speed before the disaster?

14. *Personal Finance* William built a porch for his neighbor and got paid $1200. He gave $\frac{1}{10}$ of this to his brother to pay back a debt. He used $\frac{1}{3}$ of it to pay bills and used $\frac{1}{6}$ to pay his helper. How much of the $1200 did William have left?

15. *Personal Finance* Noriko earns $660 per week. She has $\frac{1}{5}$ of her income withheld for federal taxes, $\frac{1}{15}$ of her income withheld for state taxes, and $\frac{1}{20}$ of her income withheld for medical coverage. How much per week is left for Noriko after these three deductions?

16. *Real Estate* Dan and Estella are saving for a down payment on a house. Their total take-home pay is $960 per week. They have allotted $\frac{1}{4}$ of their weekly income for rent, $\frac{1}{10}$ for car insurance, and $\frac{1}{3}$ for all other expenses including groceries, clothing, entertainment, and monthly bills. How much is left per week to be saved for their down payment?

17. *Making Jewelry* Emily makes children's bracelets. Each bracelet requires $\frac{2}{5}$ foot of wire to make. She has a large piece of wire 22 feet long.

  (a) How many bracelets can Emily make from the large piece of wire?

  (b) To make a necklace, Emily would need $3\frac{1}{2}$ times what is needed for a bracelet. How many feet is this?

▲ **18.** *Home Improvement* The Costellos are having new carpet and molding installed in their sunroom. The room measures $7\frac{1}{2}$ feet by $11\frac{2}{3}$ feet.

  (a) If new carpet costs $3 per square foot to install, how much will the new carpet cost?

  (b) The new molding will be placed around the room where the walls and ceiling meet. How many feet of molding will they need?

19. *Food Purchase* Cecilia bought a loaf of sourdough bread that was made by a local gourmet bakery. The label said that the bread, plus its fancy box, weighed $18\frac{1}{2}$ ounces in total. Of this, $1\frac{1}{4}$ ounces turned out to be the weight of the ribbon. The box weighed $3\frac{1}{8}$ ounces.

  (a) How many ounces of bread did she actually buy?

  (b) The box stated its net weight as $14\frac{3}{4}$ ounces. (This means that she should have found $14\frac{3}{4}$ ounces of gourmet sourdough bread in the box.) How much in error was this measurement?

20. *Cooking* Marnie has $12\frac{1}{2}$ cups of flour. She wants to make two pies, each requiring $1\frac{1}{4}$ cups of flour, and three cakes, each requiring $2\frac{1}{8}$ cups. How much flour will be left after Marnie makes the pies and cakes?

21. *Coast Guard Boat Operation* The largest Coast Guard boat stationed at San Diego can travel $160\frac{1}{8}$ nautical miles in $5\frac{1}{4}$ hours.

(a) At how many knots (nautical miles per hour) is the boat traveling?

(b) At this speed, how long would it take the Coast Guard boat to travel $213\frac{1}{2}$ nautical miles?

22. *Water Ski Boat* Russ and Norma's Mariah water ski boat can travel $72\frac{7}{8}$ nautical miles in $2\frac{3}{4}$ hours.

(a) At how many knots (nautical miles per hour) is the boat traveling?

(b) At this speed, how long would it take their water ski boat to travel $92\frac{3}{4}$ nautical miles?

▲ **23.** *Farming* A Kansas wheat farmer has a storage bin with a capacity of $6856\frac{1}{4}$ cubic feet.

(a) If a bushel of wheat is $1\frac{1}{4}$ cubic feet, how many bushels can the storage bin hold?

(b) If the farmer wants to make a new storage bin $1\frac{3}{4}$ times larger, how many cubic feet will it hold?

(c) How many bushels will the new bin hold?

▲ **24.** *Farming* A Texas wheat farmer has a storage bin with a capacity of $8693\frac{1}{3}$ cubic feet.

(a) If a barrel of wheat is $1\frac{1}{3}$ cubic feet, how many barrels can the storage bin hold?

(b) If the farmer wants to make a new storage bin $1\frac{1}{3}$ times larger, how many cubic feet will it hold?

(c) How many barrels will the new bin hold?

Cumulative Review

25. [2.7.2] Subtract. $\dfrac{17}{36} - \dfrac{2}{9}$

26. [2.8.3] Evaluate. $\dfrac{1}{5} + \dfrac{2}{5} \times \dfrac{3}{2} - \dfrac{1}{10}$

27. [2.4.3] Multiply. $30 \times 4\frac{2}{3}$

28. [2.5.3] Divide. $\dfrac{15}{16} \div 1\frac{1}{4}$

Quick Quiz 2.9

1. Marcia wants to put wall-to-wall carpet in her bedroom. The room measures $15\frac{3}{4}$ feet by $10\frac{2}{3}$ feet. How many square feet of carpeting does she need?

2. Ken Thompson shipped out $41\frac{3}{5}$ pounds of electrical supplies. The supplies are placed in individual packets that weigh $2\frac{3}{5}$ pounds each. How many packets did he ship out?

3. Lexi bicycled $1\frac{1}{8}$ miles from Beverly to Beverly Cove. She then traveled $1\frac{1}{2}$ miles from Beverly Cove to Chapman's Corner. Finally she traveled $2\frac{3}{4}$ miles from Chapman's Corner to Beverly Farms. How far did she travel on her bicycle? Express your answer as a mixed number.

4. **Concept Check** A trail to a peak on Mount Washington is $3\frac{3}{5}$ miles long. Caleb started hiking on the trail and stopped after walking $1\frac{7}{8}$ miles to take a break. Explain how you would find how far he still has to go to get to the peak.

Did You Know...
That You Can Save Money by Changing Your Daily Habits?

FORGET THE COFFEE AND SAVE THE MONEY

Understanding the Problem:
Do you find yourself running short of money each month? Do you wish you could find a little extra cash for yourself? Is there some daily habit that costs money that maybe you could give up? Let's start with buying coffee. Consider the story of Tricia. Tricia used to buy a cup of coffee every day on her way to work, buy another at lunchtime, and also buy two cups every day on the weekends. However, the coffee shop Tricia went to every day was closed for renovations in August, and she was not able to buy coffee every day.

Making a Plan:
Step 1: Tricia noticed she had more money left over at the end of the month. She got to thinking. Where she lives, a large cup of coffee costs $3. Tricia would buy one on the way to work every day, buy another at lunchtime, and also buy two cups every day on the weekends. How much was she spending per month on coffee? (Figure 30 days for an average month.)

Task 1: Find out how much Tricia spent in a month (30 days) on coffee.

Task 2: Use the answer from Task 1 to find out how much Tricia would spend on coffee in 12 months.

Step 2: Tricia thought she could purchase a really nice plasma television for $1000.

Task 3: If she put the money she saved each month by not buying coffee into a savings account, would there be enough money to buy the TV by Tricia's birthday (which is seven months from now)?

Task 4: Would there be extra money for a birthday celebration dinner for Tricia? If so, how much?

Task 5: If Tricia found a plasma television on sale that costs only 3/4 of what the television costs in Step 2, how much money would be available for the birthday dinner?

Finding a Solution:
Step 3: Tricia realizes that she can make her own coffee. She discovers that one pound of coffee will make approximately 20 large-size cups of coffee. She knows that she can buy coffee for $10 per pound.

Task 6: Assuming Tricia drinks the same amount of coffee every day, how much money will she save in the course of a month by making coffee instead of buying it?

Task 7: How much would Tricia save in a year by making her own coffee?

Applying the Situation to Your Life:
Task 8: Do you have any habits that cost you money?

Task 9: Can you think of one extra expense you could cut out or cut back on so you could save money to purchase a big-ticket item? Calculate the savings.

Task 10: Is there anything you could make yourself instead of buying it to help you save money? Calculate your savings.

Chapter 2 Organizer

Topic and Procedure	Examples	✏️ You Try It
Concept of a fractional part, p. 107 The numerator is the number of parts selected. The denominator is the number of total parts.	What part of this sketch is shaded? $\frac{7}{10}$	1. What part of this sketch is shaded?
Using fractions in real-life situations, p. 110 The numerator is the number of items of interest. The denominator is the total number of items.	Of the 15 players on the Chicago Bulls 2011 team, 10 are 6 feet 5 inches or taller. Write a fraction that describes the part of the team that is 6 feet 5 inches or taller. $\underline{10}\leftarrow$ number of players 6 feet 5 inches or taller $15 \leftarrow$ total number of players Reduce the fraction. $\frac{10}{15}=\frac{2}{3}$	2. A baseball team played 115 games one season. They won 85 of the games. Write a fraction that describes the part of the season that the team won.
Prime factorization, p. 114 Prime factorization is the writing of a number as the product of prime numbers.	Write the prime factorization of 36. $36 = \quad 4 \quad \times \quad 9$ $\qquad\quad 2\times 2 \quad 3\times 3$ $\quad = 2\times 2\times 3\times 3 = 2^2\times 3^2$	3. Write the prime factorization of 60.
Reducing fractions, p. 116 1. Factor numerator and denominator into prime factors. 2. Divide out factors common to numerator and denominator.	Reduce. $\frac{54}{90}$ $\frac{54}{90}=\frac{\overset{1}{2}\times\overset{1}{3}\times\overset{1}{3}\times 3}{\underset{1}{2}\times\underset{1}{3}\times\underset{1}{3}\times 5}=\frac{3}{5}$	4. Reduce. $\frac{24}{80}$
Changing a mixed number to an improper fraction, p. 122 1. Multiply whole number by denominator. 2. Add product to numerator. 3. Place sum over denominator.	Write as an improper fraction. $7\frac{3}{4}=\frac{7\times 4+3}{4}=\frac{28+3}{4}=\frac{31}{4}$	5. Write as an improper fraction. $10\frac{2}{3}$
Changing an improper fraction to a mixed number, p. 123 1. Divide denominator into numerator. 2. The quotient is the whole number. 3. The fraction is the remainder over the divisor.	Change to a mixed number. $\frac{32}{5}$ $5\overline{)32}$ with quotient 6 so $\frac{32}{5}=6\frac{2}{5}$ $\quad\underline{30}$ $\quad\ \ 2$	6. Change to a mixed number. $\frac{28}{3}$
Multiplying fractions, p. 128 1. Divide out common factors from the numerators and denominators whenever possible. 2. Multiply numerators. 3. Multiply denominators.	**(a)** Multiply. $\frac{3}{7}\times\frac{5}{13}=\frac{15}{91}$ **(b)** Multiply. $\frac{\overset{1}{5}}{\underset{1}{8}}\times\frac{\overset{2}{16}}{\underset{3}{15}}=\frac{2}{3}$	7. **(a)** Multiply. $\frac{2}{5}\times\frac{2}{9}$ **(b)** Multiply. $\frac{4}{5}\times\frac{25}{28}$
Multiplying mixed and/or whole numbers, p. 130 1. Change any whole numbers to fractions with a denominator of 1. 2. Change any mixed numbers to improper fractions. 3. Use multiplication rule for fractions.	Multiply. $7\times 3\frac{1}{4}$ $\frac{7}{1}\times\frac{13}{4}=\frac{91}{4}$ or $22\frac{3}{4}$	8. Multiply. $2\frac{1}{2}\times 4\frac{2}{5}$
Dividing fractions, p. 135 To divide two fractions, we invert the second fraction and multiply.	Divide. $\frac{3}{7}\div\frac{2}{9}=\frac{3}{7}\times\frac{9}{2}=\frac{27}{14}$ or $1\frac{13}{14}$	9. Divide. $\frac{1}{3}\div\frac{2}{5}$

Topic and Procedure	Examples	You Try It
Dividing mixed numbers and/or whole numbers, p. 136 1. Change any whole numbers to fractions with a denominator of 1. 2. Change any mixed numbers to improper fractions. 3. Use rule for division of fractions.	Divide. $8\frac{1}{3} \div 5\frac{5}{9} = \frac{25}{3} \div \frac{50}{9}$ $= \frac{\overset{1}{\cancel{25}}}{\underset{1}{\cancel{3}}} \times \frac{\overset{3}{\cancel{9}}}{\underset{2}{\cancel{50}}} = \frac{3}{2}$ or $1\frac{1}{2}$	10. Divide. $7\frac{1}{5} \div 2\frac{1}{10}$
Finding the least common denominator, p. 147 1. Write each denominator as the product of prime factors. 2. List all the prime factors that appear in both products. 3. Form a product of those factors, using each factor the greatest number of times it appears in any denominator.	Find the LCD of $\frac{1}{10}, \frac{3}{8}$, and $\frac{7}{25}$. $10 = 2 \times 5$ $8 = 2 \times 2 \times 2$ $25 = 5 \times 5$ $LCD = 2 \times 2 \times 2 \times 5 \times 5 = 200$	11. Find the LCD of $\frac{1}{6}, \frac{7}{10}$, and $\frac{5}{24}$.
Building fractions, p. 150 1. Find how many times the original denominator can be divided into the new denominator. 2. Multiply that value by numerator and denominator of original fraction.	Build $\frac{5}{7}$ to an equivalent fraction with a denominator of 42. First we find $7\overline{)42}^{6}$. Then we multiply the numerator and denominator by 6. $\frac{5}{7} \times \frac{6}{6} = \frac{30}{42}$	12. Build $\frac{4}{9}$ to an equivalent fraction with a denominator of 54.
Adding or subtracting fractions with a common denominator, p. 156 1. Add or subtract the numerators. 2. Keep the common denominator.	**(a)** Add. $\frac{3}{13} + \frac{5}{13} = \frac{8}{13}$ **(b)** Subtract. $\frac{15}{17} - \frac{12}{17} = \frac{3}{17}$	13. **(a)** Add. $\frac{7}{15} + \frac{1}{15}$ **(b)** Substract. $\frac{8}{11} - \frac{7}{11}$
Adding or subtracting fractions without a common denominator, p. 157 1. Find the LCD of the fractions. 2. Build each fraction, if needed, to obtain the LCD in the denominator. 3. Follow the steps for adding and subtracting fractions with the same denominator.	Add. $\frac{1}{4} + \frac{3}{7} + \frac{5}{8}$ $LCD = 56$ $\frac{1}{4} \times \frac{14}{14} + \frac{3}{7} \times \frac{8}{8} + \frac{5}{8} \times \frac{7}{7}$ $= \frac{14}{56} + \frac{24}{56} + \frac{35}{56} = \frac{73}{56}$ or $1\frac{17}{56}$	14. Add. $\frac{1}{3} + \frac{3}{5} + \frac{9}{10}$
Adding mixed numbers, p. 164 1. Change fractional parts to equivalent fractions with LCD as a denominator, if needed. 2. Add whole numbers and fractions separately. 3. If improper fractions occur, change to mixed numbers and simplify.	Add. $6\frac{3}{4} + 2\frac{5}{8}$ $\begin{array}{rcl} 6\boxed{\frac{3}{4} \times \frac{2}{2}} &=& 6\frac{6}{8} \\ + 2\frac{5}{8} &=& + 2\frac{5}{8} \\ \hline && 8\frac{11}{8} = 9\frac{3}{8} \end{array}$	15. Add. $8\frac{5}{6} + 3\frac{1}{3}$
Subtracting mixed numbers, p. 165 1. Change fractional parts to equivalent fractions with LCD as a denominator, if needed. 2. If necessary, borrow from whole number to subtract fractions. 3. Subtract whole numbers and fractions separately.	Subtract. $8\frac{1}{5} - 4\frac{2}{3}$ $\begin{array}{rcccl} 8\boxed{\frac{1}{5} \times \frac{3}{3}} &=& 8\frac{3}{15} &=& 7\frac{18}{15} \\ - 4\boxed{\frac{2}{3} \times \frac{5}{5}} &=& - 4\frac{10}{15} &=& - 4\frac{10}{15} \\ \hline &&&& 3\frac{8}{15} \end{array}$	16. Substract. $10\frac{1}{4} - 3\frac{4}{5}$

185

Topic and Procedure	Examples	✏ You Try It

Order of Operations p. 167
With grouping symbols:
 	
Do first **1.** Perform operations inside parentheses.
        2. Simplify any expressions with exponents.
        3. Multiply or divide from left to right.
Do last **4.** Add or subtract from left to right.
	$\frac{5}{6} \div \left(\frac{4}{5} - \frac{7}{15}\right)$ First combine numbers inside the parentheses.
	
	$\frac{5}{6} \div \left(\frac{12}{15} - \frac{7}{15}\right)$ Transform $\frac{4}{5}$ to equivalent fraction $\frac{12}{15}$
	
	$\frac{5}{6} \div \frac{1}{3}$ Subtract the two fractions inside the parentheses and reduce.
	
	$\frac{5}{6} \times \frac{3}{1}$ Invert the second fraction and multiply.
	
	$\frac{5}{2}$ or $2\frac{1}{2}$ Simplify.
		17. Evaluate. $6 \times \frac{1}{2} + \left(\frac{9}{10} - \frac{2}{5}\right)$

Procedure for Solving Applied Problems

Using the Mathematics Blueprint for Problem Solving, p. 172

In solving an applied problem with fractions, students may find it helpful to complete the following steps. You will not use all the steps all of the time. Choose the steps that best fit the conditions of the problem.

1. **Understand the problem.**
   (a) Read the problem carefully.
   (b) Draw a picture if this helps you to visualize the situation. Think about what facts you are given and what you are asked to find.
   (c) It may help to write a similar, simpler problem to get started and to determine what operation to use.
   (d) Use the Mathematics Blueprint for Problem Solving to organize your work. Follow these four parts.

   1. Gather the facts (Write down specific values given in the problem.)
   2. What am I asked to do? (Identify what you must obtain for an answer.)
   3. How do I proceed? (Decide what calculations need to be done.)
   4. Key points to remember (Record any facts, warnings, formulas, or concepts you think will be important as you solve the problem.)

2. **Solve and state the answer.**
   (a) Perform the necessary calculations.
   (b) State the answer, including the unit of measure.

3. **Check.**
   (a) Estimate the answer to the problem. Compare this estimate to the calculated value. Is your answer reasonable?

   (b) Repeat your calculations.
   (c) Work backward from your answer. Do you arrive at the original conditions of the problem?

EXAMPLE A wire is $95\frac{1}{3}$ feet long. It is cut up into smaller, equal-sized pieces, each $4\frac{1}{3}$ feet long. How many pieces will there be?

1. **Understand the problem.**

   Draw a picture of the situation.

   How will we find the number of pieces?

   Now we will use a simpler problem to clarify the idea. A wire 100 feet long is cut up into smaller pieces each 4 feet long. How many pieces will there be? We readily see that we would divide 100 by 4. Thus in our original problem we should divide $95\frac{1}{3}$ feet by $4\frac{1}{3}$ feet. This will tell us the number of pieces. Now we fill in the Mathematics Blueprint (see below).

2. **Solve and state the answer.**

   We need to divide $95\frac{1}{3} \div 4\frac{1}{3}$.

   $$\frac{286}{3} \div \frac{13}{3} = \frac{\overset{22}{\cancel{286}}}{\underset{1}{\cancel{3}}} \times \frac{\overset{1}{\cancel{3}}}{\underset{1}{\cancel{13}}} = \frac{22}{1} = 22$$

   There will be 22 pieces of wire.

3. **Check.**

   Estimate. Rounded to the nearest ten, $95\frac{1}{3} \approx 100$.

   Rounded to the nearest whole number, $4\frac{1}{3} \approx 4$.

   $$100 \div 4 = 25$$

   This is close to our estimate. Our answer is reasonable. ✓

Mathematics Blueprint for Problem Solving

Gather the Facts	What Am I Asked to Do?	How Do I Proceed?	Key Points to Remember
Wire is $95\frac{1}{3}$ feet. It is cut into equal pieces $4\frac{1}{3}$ feet long.	Determine how many pieces of wire there will be.	Divide $95\frac{1}{3}$ by $4\frac{1}{3}$.	Change mixed numbers to improper fractions before carrying out the division.

Chapter 2 Review Problems

Be sure to simplify all answers.

Section 2.1

Use a fraction to represent the shaded part of each object.

1.

2.

In exercises 3 and 4, draw a sketch to illustrate each fraction.

3. $\dfrac{4}{7}$ of an object

4. $\dfrac{7}{10}$ of a group

5. *Quality Control* An inspector looked at 80 semiconductors and found 9 of them defective. What fractional part of these items was defective?

6. *Education* The dean asked 100 freshmen if they would be staying in the dorm over the holidays. A total of 87 said they would not. What fractional part of the freshmen said they would not?

Section 2.2

Express each number as a product of prime factors.

7. 54

8. 120

9. 168

Determine which of the following numbers are prime. If a number is composite, express it as the product of prime factors.

10. 59

11. 78

12. 167

Reduce each fraction.

13. $\dfrac{12}{42}$

14. $\dfrac{13}{52}$

15. $\dfrac{27}{72}$

16. $\dfrac{168}{192}$

Section 2.3

Change each mixed number to an improper fraction.

17. $4\dfrac{3}{8}$

18. $15\dfrac{3}{4}$

19. $6\dfrac{3}{5}$

Change each improper fraction to a mixed number.

20. $\dfrac{45}{8}$

21. $\dfrac{100}{21}$

22. $\dfrac{53}{7}$

23. Reduce and leave your answer as a mixed number. $3\dfrac{15}{55}$

24. Reduce and leave your answer as an improper fraction. $\dfrac{234}{16}$

25. Change to a mixed number and then reduce. $\dfrac{132}{32}$

Section 2.4

Multiply.

26. $\dfrac{4}{7} \times \dfrac{5}{11}$

27. $\dfrac{7}{9} \times \dfrac{21}{35}$

28. $12 \times \dfrac{3}{7} \times 0$

29. $\dfrac{3}{5} \times \dfrac{2}{7} \times \dfrac{10}{27}$

30. $5\dfrac{1}{8} \times 3\dfrac{1}{5}$

31. $36 \times \dfrac{4}{9}$

32. ***Stock Market*** In 1999, one share of stock cost $37\dfrac{5}{8}$. How much money did 18 shares cost? (In April 2001, stock market prices switched from fractions to decimals.)

▲ **33.** ***Geometry*** The O'Gara's new family room addition measures $13\dfrac{1}{2}$ feet long by $9\dfrac{2}{3}$ feet wide. Find the area of the addition.

Section 2.5

Divide, if possible.

34. $\dfrac{3}{7} \div \dfrac{2}{5}$

35. $900 \div \dfrac{3}{5}$

36. $5\dfrac{3}{4} \div 11\dfrac{1}{2}$

37. $\dfrac{\dfrac{20}{1}}{2\dfrac{1}{2}}$

38. $0 \div 3\dfrac{7}{5}$

39. $4\dfrac{2}{11} \div 3$

▲ **40.** ***Floor Carpeting*** Each roll of carpet covers $28\dfrac{1}{2}$ square yards. The community center has 342 square yards of flooring to carpet. How many rolls are needed?

▲ **41.** There are 420 calories in $2\dfrac{1}{4}$ cans of grape soda. How many calories are in 1 can of soda?

Section 2.6

Find the LCD for each pair of fractions.

42. $\dfrac{7}{14}$ and $\dfrac{3}{49}$

43. $\dfrac{13}{20}$ and $\dfrac{3}{25}$

44. $\dfrac{5}{18}, \dfrac{1}{6}, \dfrac{7}{45}$

Build each fraction to an equivalent fraction with the specified denominator.

45. $\dfrac{3}{7} = \dfrac{?}{56}$

46. $\dfrac{11}{24} = \dfrac{?}{72}$

47. $\dfrac{8}{15} = \dfrac{?}{150}$

Section 2.7

Add or subtract.

48. $\dfrac{9}{14} - \dfrac{5}{14}$

49. $\dfrac{1}{2} + \dfrac{1}{3} + \dfrac{1}{4}$

50. $\dfrac{7}{8} - \dfrac{3}{5}$

51. $\dfrac{7}{30} + \dfrac{2}{21}$

52. $\dfrac{5}{18} + \dfrac{7}{10}$

53. $\dfrac{14}{15} - \dfrac{3}{25}$

Section 2.8

Evaluate using the correct order of operations.

54. $8 - 2\dfrac{3}{4}$

55. $3 + 5\dfrac{2}{3}$

56. $3\dfrac{3}{8} + 2\dfrac{3}{4}$

57. $5\dfrac{11}{16} - 2\dfrac{1}{5}$

58. $\dfrac{3}{5} \times \dfrac{1}{2} + \dfrac{2}{5} \div \dfrac{2}{3}$

59. $\left(\dfrac{4}{5} - \dfrac{1}{2}\right)^2 \times \dfrac{10}{3}$

60. *Jogging* Bob jogged $1\frac{7}{8}$ miles on Monday, $2\frac{3}{4}$ miles on Tuesday, and $4\frac{1}{10}$ miles on Wednesday. How many miles did he jog on these three days?

61. *Fuel Economy* When it was new, Ginny Sue's car got $28\frac{1}{6}$ miles per gallon. It now gets $1\frac{5}{6}$ miles per gallon less. How far can she drive now if the car has $10\frac{3}{4}$ gallons in the tank?

62. *Cooking* A recipe calls for $3\frac{1}{3}$ cups of sugar and $4\frac{1}{4}$ cups of flour. How much sugar and how much flour would be needed for $\frac{1}{2}$ of that recipe?

63. *Fuel Economy* Rafael travels in a car that gets $24\frac{1}{4}$ miles per gallon. He has $8\frac{1}{2}$ gallons of gas in the gas tank. Approximately how far can he drive?

64. *Construction* How many lengths of pipe $3\frac{1}{5}$ inches long can be cut from a pipe 48 inches long?

65. *Automobile Maintenance* A car radiator holds $15\frac{3}{4}$ liters. If it contains $6\frac{1}{8}$ liters of antifreeze and the rest is water, how much is water?

66. *Reading Speed* Tim found that he can read 5 pages of his biology book in $32\frac{1}{2}$ minutes. He has three chapters to read over the weekend. The first is 12 pages, the second is 9 pages, and the third is 14 pages. How long will it take him?

67. *Cooking* A biscuit recipe requires $1\frac{3}{4}$ cups of flour. Marcia wants to make $2\frac{1}{2}$ times the recipe. How much flour will she need? If she uses this amount from a new bag of flour containing 12 cups, how much will be left?

68. *Carpentry* A 3-inch bolt passes through $1\frac{1}{2}$ inches of pine board, a $\frac{1}{16}$-inch washer, and a $\frac{1}{8}$-inch nut. How many inches does the bolt extend beyond the board, washer, and nut if the head of the bolt is $\frac{1}{4}$ inch long?

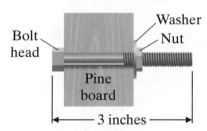

69. *Budgeting* Francine has a take-home pay of $880 per month. She gives $\frac{1}{10}$ of it to her church, spends $\frac{1}{2}$ of it for rent and food, and spends $\frac{1}{8}$ of it on electricity, heat, and telephone. How many dollars per month does she have left for other things?

70. *Cost of Auto Travel* Manuel's new car used $18\frac{2}{5}$ gallons of gas on a 460-mile trip. How many miles can his car travel on 1 gallon of gas?

Mixed Practice

Perform each calculation or each requested operation.

71. Reduce. $\dfrac{27}{63}$

72. $\dfrac{7}{15} + \dfrac{11}{25}$

73. $4\frac{1}{3} - 2\frac{11}{12}$

74. $\dfrac{36}{49} \times \dfrac{14}{33}$

75. $\left(\dfrac{4}{7}\right)^3$

76. $\dfrac{3}{8} \div \dfrac{1}{10}$

77. $5\frac{1}{2} \times 18$

78. $150 \div 3\frac{1}{8}$

How Am I Doing? Chapter 2 Test

 Test Prep VIDEOS   **MATH COACH**   **MyMathLab®**   **You Tube**

After you take this test read through the Math Coach on pages 192 and 193. Math Coach videos are available via MyMathLab and YouTube. Step-by-step test solutions in the Chapter Test Prep Videos are also available via MyMathLab and YouTube. (Search "TobeyBasicCollMath" and click on "Channels.")

Solve.

1. Use a fraction to represent the shaded part of the object.

2. A basketball star shot at the hoop 388 times. The ball went in 311 times. Write a fraction that describes the part of the time that his shots went in.

Reduce each fraction.

3. $\dfrac{18}{42}$

4. $\dfrac{15}{70}$

ᴹᴄ **5.** $\dfrac{225}{50}$

6. Change to an improper fraction. $6\dfrac{4}{5}$

7. Change to a mixed number. $\dfrac{145}{14}$

Multiply.

8. $42 \times \dfrac{2}{7}$

9. $\dfrac{7}{9} \times \dfrac{2}{5}$

10. $2\dfrac{2}{3} \times 5\dfrac{1}{4}$

Divide.

11. $\dfrac{7}{8} \div \dfrac{5}{11}$

12. $\dfrac{12}{31} \div \dfrac{8}{13}$

13. $7\dfrac{1}{5} \div 1\dfrac{1}{25}$

14. $5\dfrac{1}{7} \div 3$

Find the least common denominator of each set of fractions.

15. $\dfrac{5}{12}$ and $\dfrac{7}{18}$

ᴹᴄ **16.** $\dfrac{3}{16}$ and $\dfrac{1}{24}$

17. $\dfrac{1}{4}, \dfrac{3}{8}, \dfrac{5}{6}$

18. Build the fraction to an equivalent fraction with the specified denominator. $\dfrac{5}{12} = \dfrac{?}{72}$

1. _____
2. _____
3. _____
4. _____
5. _____
6. _____
7. _____
8. _____
9. _____
10. _____
11. _____
12. _____
13. _____
14. _____
15. _____
16. _____
17. _____
18. _____

Evaluate using the correct order of operations.

19. $\dfrac{7}{9} - \dfrac{5}{12}$

20. $\dfrac{2}{15} + \dfrac{5}{12}$

21. $\dfrac{1}{4} + \dfrac{3}{7} + \dfrac{3}{14}$

22. $8\dfrac{3}{5} + 5\dfrac{4}{7}$

23. $18\dfrac{6}{7} - 13\dfrac{13}{14}$

24. $\dfrac{2}{9} \div \dfrac{8}{3} \times \dfrac{1}{4}$

ℳ⒞ 25. $\left(\dfrac{1}{2} + \dfrac{1}{3}\right) \times \dfrac{7}{5}$

Answer each question.

▲ **26.** Erin needs to find the area of her kitchen so she knows how much tile to purchase. The room measures $16\frac{1}{2}$ feet by $9\frac{1}{3}$ feet. How many square feet is the kitchen?

ℳ⒞ **27.** A butcher has $18\frac{2}{3}$ pounds of steak that he wishes to place into packages that average $2\frac{1}{3}$ pounds each. How many packages can he make?

28. From central parking it is $\frac{9}{10}$ of a mile to the science building. Bob started at central parking and walked $\frac{1}{5}$ of a mile toward the science building. He stopped for coffee. When he finished, how much farther did he have to walk to reach the science building?

29. Robin jogged $4\frac{1}{8}$ miles on Monday, $3\frac{1}{6}$ miles on Tuesday, and $6\frac{3}{4}$ miles on Wednesday. How far did she jog on those three days?

30. Mr. and Mrs. Samuel visited Florida and purchased 120 oranges. They gave $\frac{1}{4}$ of them to relatives, ate $\frac{1}{12}$ of them in the hotel, and gave $\frac{1}{3}$ of them to friends. They shipped the rest home to Illinois. How many oranges did they ship?

31. A candle company purchased $48\frac{1}{8}$ pounds of wax to make specialty candles. It takes $\frac{5}{8}$ pound of wax to make one specialty candle. How many candles can they make? To make one pillar candle, it takes $2\frac{1}{2}$ times the amount to make one specialty candle. How many pounds is this?

19. _____ ☐

20. _____ ☐

21. _____ ☐

22. _____ ☐

23. _____ ☐

24. _____ ☐

25. _____ ☐

26. _____ ☐

27. _____ ☐

28. _____ ☐

29. _____ ☐

30. _____ ☐

31. _____ ☐

Total Correct: ☐

MATH COACH

Mastering the skills you need to do well on the test.

Students often make the same types of errors when they do the Chapter 2 Test. Here are some helpful hints to keep you from making these common errors on test problems.

Simplifying Fractions—Problem 5 $\dfrac{225}{50}$

> **Helpful Hint** After you complete each step when simplifying a fraction, stop to see if your result can be simplified any further.

Did you divide the numerator and denominator by 25? Yes ____ No ____

If you answered Yes, then you chose the best factor to divide by. But, if you answered No, then check your division.

If you answered Problem 5 incorrectly, rework it now using these suggestions.

Finding the Least Common Denominator—Problem 16 $\dfrac{3}{16}$ and $\dfrac{1}{24}$

> **Helpfule Hint** Write each denominator as the product of prime factors. When forming the LCD, use each factor the greatest number of times that it appears in any one denominator.

Did you factor 16 into $2 \times 2 \times 2 \times 2$? Yes ____ No ____
If you answered No, stop and perform that step.

Did you factor 24 into $2 \times 2 \times 2 \times 3$? Yes ____ No ____
If you answered No, stop and perform that step.

If you answered Problem 16 incorrectly, rework it now using these suggestions. Remember to use the factor 2 four times.

Need more help? Watch the **MATH COACH** videos in MyMathLab® or on YouTube™.

192

Evaluate Using the Correct Order of Operations—Problem 25 $\left(\dfrac{1}{2} + \dfrac{1}{3}\right) \times \dfrac{7}{5}$

> **Helpful Hint** Write out each step separately. Be sure to combine operations inside the parentheses first.

Did you first combine $\dfrac{1}{2} + \dfrac{1}{3}$? Yes _____ No _____

Did you rewrite the fractions as $\dfrac{3}{6} + \dfrac{2}{6}$ before adding them together? Yes _____ No _____

If you answered No to either of these questions, stop and perform these operations first.

If you answered Problem 25 incorrectly, rework it now using these suggestions.

Solve Applied Problems Involving Fractions—Problem 27

A butcher has $18\dfrac{2}{3}$ pounds of steak that he wishes to place into packages that average $2\dfrac{1}{3}$ pounds each. How many packages can he make?

> **Helpful Hint** Remember to use the problem-solving steps and the Mathematics Blueprint for Problem Solving to help you set up the problem correctly. When dividing, remember to invert the second fraction and then multiply.

Did you translate the problem correctly to form the calculation $18\dfrac{2}{3} \div 2\dfrac{1}{3}$? Yes _____ No _____

If you answered No, stop and perform that calculation.

Did you remember to change the mixed numbers to fractions and write $\dfrac{56}{3} \div \dfrac{7}{3}$? Yes _____ No _____

If you answered No, stop and perform that calculation.

If you answered Problem 27 incorrectly, rework it now using these suggestions.

Need more help? Look for section examples marked with $\mathbb{MC}$ to review.

The rising cost of gasoline is of great concern to every driver in the country. But what is the best bargain when you purchase gasoline? How can you use math to save money at the pump? Should you always go to the station with the lowest price? Turn to page 251 and you may be surprised at some of the answers.

Decimals

3.1 Using Decimal Notation

① Writing a Word Name for a Decimal Fraction

In Chapter 2 we discussed *fractions*—the set of numbers such as $\frac{1}{2}, \frac{2}{3}, \frac{1}{10}, \frac{6}{7}, \frac{18}{100}$, and so on. Now we will take a closer look at **decimal fractions**—that is, fractions with 10, 100, 1000, and so on, in the denominator, such as $\frac{1}{10}, \frac{18}{100}$, and $\frac{43}{1000}$.

Why, of all fractions, do we take special notice of these? Our hands have 10 digits. Our U.S. money system is based on the dollar, which has 100 equal parts, or cents. And the international system of measurement called the *metric system* is based on 10 and powers of 10.

As with other numbers, these decimal fractions can be written in different ways (forms). For example, the shaded part of the whole in the following drawing can be written:

in words (one-tenth)
in fractional form $\left(\frac{1}{10}\right)$
in decimal form (0.1)

All mean the same quantity, namely 1 out of 10 equal parts of the whole. We'll see that when we use decimal notation, computations can be easily done based on the old rules for whole numbers and a few new rules about where to place the decimal point. In a world where calculators and computers are commonplace, many of the fractions we encounter are decimal fractions. A decimal fraction is a fraction whose denominator is a power of 10.

$\frac{7}{10}$ is a decimal fraction. $\frac{89}{10^2} = \frac{89}{100}$ is a decimal fraction.

Decimal fractions can be written with numerals in two ways: fractional form or decimal form. Some decimal fractions are shown in decimal form below.

Fractional Form		Decimal Form
$\frac{3}{10}$	=	0.3
$\frac{59}{100}$	=	0.59
$\frac{171}{1000}$	=	0.171

The zero in front of the decimal point is not actually required. We place it there simply to make sure that we don't miss seeing the decimal point. A number written in decimal notation has three parts.

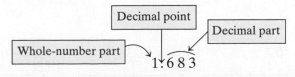

When a number is written in decimal form, the first digit to the right of the decimal point represents tenths, the next digit hundredths, the next digit thousandths, and so on. 0.9 means nine tenths and is equivalent to $\frac{9}{10}$.

0.51 means fifty-one hundredths and is equivalent to $\frac{51}{100}$. Some decimals are larger than 1. For example, 1.683 means one and six hundred eighty-three thousandths. It is equivalent to $1\frac{683}{1000}$. Note that the word *and* is used to indicate the decimal point. A place-value chart is helpful.

Decimal Place Values

Hundreds	Tens	Ones	Decimal point	Tenths	Hundredths	Thousandths	Ten-thousandths
100	10	1	"and"	$\frac{1}{10}$	$\frac{1}{100}$	$\frac{1}{1000}$	$\frac{1}{10,000}$
1	5	6	.	2	8	7	4

So, we can write 156.2874 in words as one hundred fifty-six and two thousand eight hundred seventy-four ten-thousandths. We say ten-thousandths because it is the name of the last decimal place on the right.

EXAMPLE 1   Write the word name for each decimal.

(a) 0.79   **(b)** 0.5308   **(c)** 1.6   **(d)** 23.765

Solution

(a) 0.79 = seventy-nine hundredths
(b) 0.5308 = five thousand three hundred eight ten-thousandths
(c) 1.6 = one and six tenths
(d) 23.765 = twenty-three and seven hundred sixty-five thousandths

---

Student Practice 1   Write the word name for each decimal.

(a) 0.073   **(b)** 4.68   **(c)** 0.0017   **(d)** 561.78

NOTE TO STUDENT: Fully worked-out solutions to all of the Student Practice problems can be found at the back of the text starting at page SP-1.

Sometimes, decimals are used where we would not expect them. For example, we commonly say that there are 365 days in a year, with 366 days in every fourth year (or leap year). However, this is not quite correct. In fact, from time to time further adjustments need to be made to the calendar to adjust for these inconsistencies. Astronomers know that a more accurate measure of a year is called a **tropical year** (measured from one equinox to the next). Rounded to the nearest hundred-thousandth, 1 tropical year = 365.24122 days. This is read "three hundred sixty-five and twenty-four thousand, one hundred twenty-two hundred-thousandths." This approximate value is a more accurate measurement of the amount of time it takes Earth to complete one orbit around the sun.

Note the relationship between fractions and their equivalent numbers' decimal forms.

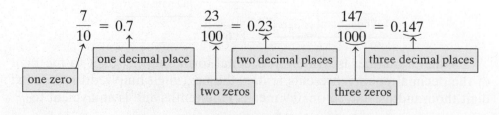

Decimal notation is commonly used with money. When writing a check, we often write the amount that is less than 1 dollar, such as 23¢, as $\frac{23}{100}$ dollar.

```
ALICE J. JENNINGTON                37-86              3680
208 BARTON SPRINGS 512-555-1212    110
AUSTIN, TX 78704
                                 DATE March 18, 2011
PAY TO THE  Rosetta Ramirez                    $ 59.23
ORDER OF
Fifty-nine and 23/100                          DOLLARS
ACB Austin Central Bank
Austin, Texas
MEMO      textbooks
⑉063000420⑉ 800136492811⑉ 3680
```

EXAMPLE 2   Write a word name for the amount on a check made out for $672.89.

Solution   Six hundred seventy-two and $\frac{89}{100}$ dollars

Student Practice 2   Write a word name for the amount of a check made out for $7863.04.

② Changing from Fractional Notation to Decimal Notation

It is helpful to be able to write decimals in both decimal notation and fractional notation. First we illustrate changing a fraction with a denominator of 10, 100, or 1000 into decimal form.

EXAMPLE 3   Write as a decimal.

(a) $\frac{8}{10}$ **(b)** $\frac{74}{100}$ **(c)** $1\frac{3}{10}$ **(d)** $2\frac{56}{1000}$

Solution

(a) $\frac{8}{10} = 0.8$ **(b)** $\frac{74}{100} = 0.74$ **(c)** $1\frac{3}{10} = 1.3$ **(d)** $2\frac{56}{1000} = 2.056$

Note: In part (d), we need to add a zero before the digits 56. Since there are three zeros in the denominator, we need three decimal places in the decimal number.

Student Practice 3   Write as a decimal.

(a) $\frac{9}{10}$ **(b)** $\frac{136}{1000}$ **(c)** $2\frac{56}{100}$ **(d)** $34\frac{86}{1000}$

③ Changing from Decimal Notation to Fractional Notation

EXAMPLE 4   Write in fractional notation.

(a) 0.51   **(b)** 18.1   **(c)** 0.7611   **(d)** 1.363

Solution

(a) $0.51 = \frac{51}{100}$ **(b)** $18.1 = 18\frac{1}{10}$ **(c)** $0.7611 = \frac{7611}{10,000}$ **(d)** $1.363 = 1\frac{363}{1000}$

Student Practice 4   Write in fractional notation.

(a) 0.37   **(b)** 182.3   **(c)** 0.7131   **(d)** 42.019

When we convert from decimal form to fractional form, we reduce whenever possible.

EXAMPLE 5   Write in fractional notation. Reduce whenever possible.

(a) 2.6   **(b)** 0.38   **(c)** 0.525   **(d)** 361.007

Solution

(a) $2.6 = 2\dfrac{6}{10} = 2\dfrac{3}{5}$ **(b)** $0.38 = \dfrac{38}{100} = \dfrac{19}{50}$

(c) $0.525 = \dfrac{525}{1000} = \dfrac{105}{200} = \dfrac{21}{40}$

(d) $361.007 = 361\dfrac{7}{1000}$ (cannot be reduced)

Student Practice 5   Write in fractional notation. Reduce whenever possible.

(a) 8.5   **(b)** 0.58   **(c)** 36.25   **(d)** 106.013

EXAMPLE 6   A chemist found that the concentration of lead in a water sample was 5 parts per million. What fraction would represent the concentration of lead?

Solution   Five parts per million means 5 parts out of 1,000,000. As a fraction, this is $\frac{5}{1,000,000}$. We can reduce this by dividing numerator and denominator by 5. Thus

$$\frac{5}{1,000,000} = \frac{1}{200,000}.$$

The concentration of lead in the water sample is $\frac{1}{200,000}$.

Student Practice 6   A chemist found that the concentration of PCBs in a water sample was 2 parts per billion. What fraction would represent the concentration of PCBs?

Verbal and Writing Skills, Exercises 1–4

1. Describe a decimal fraction and provide examples.

2. What word is used to describe the decimal point when writing the word name for a decimal that is greater than one?

3. What is the name of the last decimal place on the right for the decimal 132.45678?

4. When writing $82.75 on a check, we write 75¢ as

_____.

Write the word name for each decimal.

5. 0.57

6. 0.78

7. 3.8

8. 12.4

9. 7.013

10. 2.056

11. 28.0037

12. 54.0013

Write a word name as you would on a check.

13. $124.20

14. $510.31

15. $1236.08

16. $5304.05

17. $18,045.19

18. $10,540.28

Write in decimal notation.

19. seven tenths

20. six tenths

21. ninety-six hundredths

22. eighteen hundredths

23. four hundred eighty-one thousandths

24. twenty-two thousandths

25. six thousand one hundred fourteen millionths

26. three thousand one hundred nine millionths

Write each fraction or mixed number as a decimal.

27. $\dfrac{7}{10}$

28. $\dfrac{3}{10}$

29. $\dfrac{76}{100}$

30. $\dfrac{84}{100}$

31. $\dfrac{1}{100}$

32. $\dfrac{6}{100}$

33. $\dfrac{53}{1000}$

34. $\dfrac{328}{1000}$

35. $\dfrac{2403}{10,000}$

36. $\dfrac{7794}{10,000}$

37. $10\dfrac{9}{10}$

38. $5\dfrac{3}{10}$

39. $84\dfrac{13}{100}$

40. $52\dfrac{77}{100}$

41. $3\dfrac{529}{1000}$

42. $2\dfrac{23}{1000}$

43. $235\dfrac{104}{10,000}$

44. $116\dfrac{312}{10,000}$

Write in fractional notation. Reduce whenever possible.

45. 0.02

46. 0.05

47. 3.6

48. 8.9

49. 7.41

50. 15.75

51. 12.625

52. 29.875

53. 7.0615

54. 4.0016

55. 8.0108

56. 7.0605

57. 235.1254   **58.** 581.2406   **59.** 0.0125   **60.** 0.3375

Applications

61. *Cigarette Use* Kentucky has one of the highest smoking rates in the United States. In 2008, 26,300 out of every 100,000 men age 18 or older who lived in Kentucky were smokers. That same year, 24,200 out of every 100,000 women age 18 or older who lived in Kentucky were smokers.

(a) What fractional part of the male population in Kentucky were smokers?

(b) What fractional part of the female population in Kentucky were smokers? Be sure to express these fractions in reduced form. (*Source:* www.cdc.gov)

62. *Cigarette Use* Utah has the lowest smoking rate in the United States. In 2008, 10,600 out of every 100,000 men age 18 or older who lived in Utah were smokers. That same year, 8000 out of every 100,000 women age 18 or older who lived in Utah were smokers.

(a) What fractional part of the male population in Utah were smokers?

(b) What fractional part of the female population in Utah were smokers? Be sure to express these fractions in reduced form. (*Source:* www.cdc.gov)

63. *Bald Eagle Eggs* American bald eagles have been fighting extinction due to environmental hazards such as DDT, PCBs, and dioxin. The problem is with the food chain. Fish or rodents consume contaminated food and/or water. Then the eagles ingest the poison, which in turn affects the durability of the eagles' eggs. It takes only 4 parts per million of certain chemicals to ruin an eagle egg; write this number as a fraction in lowest terms. (In 1994 the bald eagle was removed from the endangered species list.)

64. *Turtle Eggs* Every year turtles lay eggs on the islands of South Carolina. Unfortunately, due to illegal polluting, a lot of the eggs are contaminated. If the turtle eggs contain more than 2 parts per one hundred million of chemical pollutants, they will not hatch and the population will continue to head toward extinction. Write the preceding amount of chemical pollutants as a fraction in lowest terms.

Cumulative Review

65. **[1.7.1]** Round to the nearest *hundred*.
56,758

66. **[1.7.1]** Round to the nearest *thousand*.
8,069,482

67. **[2.2.2]** Reduce. $\dfrac{36}{80}$

68. **[2.7.2]** Subtract. $\dfrac{7}{8} - \dfrac{21}{40}$

Quick Quiz 3.1

1. Write the word name for the decimal. 5.367

2. Write as a decimal. $\dfrac{523}{10,000}$

3. Write in fractional notation. Reduce your answer. 12.58

4. **Concept Check** Explain how you know how many zeros to put in your answer if you need to write $\dfrac{953}{100,000}$ as a decimal.

3.2 Comparing, Ordering, and Rounding Decimals

① Comparing Decimals

All of the numbers we have studied have a specific order. To illustrate this order, we can place the numbers on a **number line.** Look at the number line in the margin. Each number has a specific place on it. The arrow points in the direction of increasing value. Thus, if one number is to the right of a second number, it is larger, or greater, than that number. Since 5 is to the right of 2 on the number line, we say that 5 is greater than 2. We write $5 > 2$.

Since 4 is to the left of 6 on the number line, we say that 4 is less than 6. We write $4 < 6$. The symbols ">" and "<" are called **inequality symbols.**

$a < b$ is read "a is less than b."

$a > b$ is read "a is greater than b."

We can assign exactly one point on the number line to each decimal number. When two decimal numbers are placed on a number line, the one farther to the right is the larger. Thus we can say that $3.4 > 2.7$ and $4.3 > 4.0$. We can also say that $0.5 < 1.0$ and $1.8 < 2.2$. Why?

To compare or order decimals, we compare each digit.

> **COMPARING TWO NUMBERS IN DECIMAL NOTATION**
>
> 1. Start at the left and compare corresponding digits. If the digits are the same, move one place to the right.
>
> 2. When two digits are different, the larger number is the one with the larger digit.

EXAMPLE 1 Write an inequality statement with 0.167 and 0.166.

Solution The numbers in the tenths place are the same. They are both 1.

0.1 6 7   0.1 6 6

The numbers in the hundredths place are the same. They are both 6.

0.1 6 7   0.1 6 6

The numbers in the thousandths place differ.

0.1 6 7   0.1 6 6

Since $7 > 6$, we know that $0.167 > 0.166$.

Student Practice 1 Write an inequality statement with 5.74 and 5.75.

Student Learning Objectives

After studying this section, you will be able to:

① Compare decimals.

② Place decimals in order from smallest to largest.

③ Round decimals to a specified decimal place.

NOTE TO STUDENT: *Fully worked-out solutions to all of the Student Practice problems can be found at the back of the text starting at page SP-1.*

Whenever necessary, extra zeros can be written to the right of the last digit—that is, to the right of the decimal point—without changing the value of the decimal. Thus

$$0.56 = 0.56000 \quad \text{and} \quad 0.7768 = 0.77680.$$

The zero to the left of the decimal point is optional. Thus $0.56 = .56$. Both notations are used. You are encouraged to place a zero to the left of the decimal point so that you don't miss the decimal point when you work with decimals.

EXAMPLE 2   Fill in the blank with one of the symbols $<$, $=$, or $>$.

$$0.77 \_\_\_\_ 0.777$$

Solution   We begin by adding a zero to the first decimal.

$$0.77\underline{0} \qquad 0.77\underline{7}$$

We see that the tenths and hundredths digits are equal. But the thousandths digits differ. Since $0 < 7$, we have $0.770 < 0.777$.

Student Practice 2   Fill in the blank with one of the symbols $<$, $=$, or $>$.

$$0.894 \_\_\_\_ 0.89$$

② Placing Decimals in Order from Smallest to Largest

Which is the heaviest—a puppy that weighs 6.2 ounces, a puppy that weighs 6.28 ounces, or a puppy that weighs 6.028 ounces? Did you choose the puppy that weighs 6.28 ounces? You are correct.

You can place two or more decimals in order. If you are asked to order the decimals from smallest to largest, look for the smallest decimal and place it first.

EXAMPLE 3   Place the following five decimal numbers in order from smallest to largest.

$$1.834, \quad 1.83, \quad 1.381, \quad 1.38, \quad 1.8$$

Solution   First we add zeros to make the comparison easier.

$$1.834, \quad 1.830, \quad 1.381, \quad 1.380, \quad 1.800$$

Now we rearrange with smallest first.

$$1.380, \quad 1.381, \quad 1.800, \quad 1.830, \quad 1.834$$

Student Practice 3   Place the following five decimal numbers in order from smallest to largest.

$$2.45, \quad 2.543, \quad 2.46, \quad 2.54, \quad 2.5$$

③ Rounding Decimals to a Specified Decimal Place

Sometimes in calculations involving money, we see numbers like $386.432 and $29.5986. To make these useful, we usually round them to the nearest cent. $386.432 is rounded to $386.43. $29.5986 is rounded to $29.60. A general rule for rounding decimals follows.

ROUNDING DECIMALS

1. Find the decimal place (units, tenths, hundredths, and so on) to which rounding is required.

2. If the first digit to the right of the given place value is less than 5, drop it and all digits to the right of it.

3. If the first digit to the right of the given place value is 5 or greater, increase the number in the given place value by one. Drop all digits to the right of this place.

EXAMPLE 4　Round 156.37 to the nearest tenth.

156.3 7
　　↑
　　└——— We find the tenths place.

Solution　Note that 7, the next place to the right, is greater than 5. We round up to 156.4 and drop the digits to the right. The answer is 156.4.

Student Practice 4　Round 723.88 to the nearest tenth.

EXAMPLE 5　Round to the nearest thousandth.

(a) 0.06358　　　　　　　　**(b)** 128.37448

Solution

(a) 0.06 3 58
　　　↑
　　　└——— We locate the thousandths place.

Note that the digit to the right of the thousandths place is 5. We round up to 0.064 and drop all the digits to the right.

(b) 128.37 4 48
　　　　↑
　　　　└——— We locate the thousandths place.

Note that the digit to the right of the thousandths place is less than 5. We round to 128.374 and drop all the digits to the right.

Student Practice 5　Round to the nearest thousandth.

(a) 12.92647　　　　　　　**(b)** 0.007892

Remember that rounding up to the next digit in a position may result in several digits being changed.

EXAMPLE 6 Round to the nearest hundredth. Fred and Linda used 203.9964 kilowatt-hours of electricity in their house in May.

$$203.9\ 9\ 64$$

We locate the hundredths place.

Solution Since the digit to the right of the hundredths place is greater than 5, we round up. This affects the next two positions. Do you see why? The result is 204.00 kilowatt-hours. Notice that we have the two zeros to the right of the decimal place to show we have rounded to the nearest hundredth.

Student Practice 6 Round to the nearest tenth. Last month the college gymnasium used 15,699.953 kilowatt-hours of electricity.

Sometimes we round a decimal to the nearest whole number. For example, when writing figures on income tax forms, a taxpayer may round all figures to the nearest dollar.

EXAMPLE 7 To complete her income tax return, Marge needs to round these figures to the nearest whole dollar.

Medical bills $779.86   Taxes $563.49
Retirement contributions $674.38   Contributions to charity $534.77

Solution Round the amounts.

	Original Figure	*Rounded to Nearest Dollar*
Medical bills	$779.86	$780
Taxes	$563.49	$563
Retirement	$674.38	$674
Charity	$534.77	$535

Student Practice 7 Round the following figures to the nearest whole dollar.

Medical bills $375.50   Taxes $971.39
Retirement contributions $980.49   Contributions to charity $817.65

CAUTION: Why is it so important to consider only *one* digit to the right of the desired round-off position? What is wrong with rounding in steps? Suppose that Mark rounds 1.349 to the nearest tenth in steps. First he rounds 1.349 to 1.35 (nearest hundredth). Then he rounds 1.35 to 1.4 (nearest tenth). What is wrong with this reasoning?

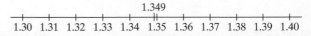

To round 1.349 to the nearest tenth, we ask if 1.349 is closer to 1.3 or to 1.4. It is closer to 1.3. Mark got 1.4, so he is not correct. He "rounded in steps" by first moving to 1.35, thus increasing the error and moving in the wrong direction. To control rounding errors, we consider *only* the first digit to the right of the decimal place to which we are rounding.

3.2 Exercises   MyMathLab®

Watch the videos in MyMathLab
Download the MyDashBoard App

Fill in the blank with one of the symbols <, =, *or* >.

1. 1.3 ___ 1.29   **2.** 2.6 ___ 2.58   **3.** 0.34 ___ 0.340   **4.** 72.54 ___ 72.56

5. 18.92 ___ 18.93   **6.** 0.460 ___ 0.46   **7.** 0.00043 ___ 0.0004   **8.** 0.0037 ___ 0.036

9. 1.002 ___ 1.0021   **10.** 2.0056 ___ 2.006   **11.** 126.34 ___ 125.35   **12.** 406.78 ___ 407.75

13. 0.888 ___ 0.8888   **14.** 0.666 ___ 0.6666   **15.** 0.777 ___ 0.7077   **16.** 0.555 ___ 0.5505

17. $\dfrac{72}{1000}$ ___ 0.072   **18.** $\dfrac{54}{1000}$ ___ 0.054   **19.** $\dfrac{8}{10}$ ___ 0.08   **20.** $\dfrac{5}{100}$ ___ 0.005

Arrange each set of decimals from smallest to largest.

21. 12.6, 12.8, 12.65   **22.** 18.32, 18.038, 18.04   **23.** 0.0071, 0.05, 0.007

24. 0.0025, 0.0052, 0.002   **25.** 8.4, 8.39, 8.41, 8.31   **26.** 5.1, 5.01, 5.23, 5.02

27. 26.034, 26.003, 26.04, 26.033   **28.** 33.082, 33.02, 33.088, 33.079

29. 18.006, 18.060, 18.066, 18.606, 18.065   **30.** 15.020, 15.002, 15.001, 15.018, 15.0019

Round to the nearest tenth.

31. 6.92   **32.** 9.76   **33.** 28.98   **34.** 47.94

35. 578.064   **36.** 115.95   **37.** 2176.83   **38.** 4082.74

Round to the nearest hundredth.

39. 26.032   **40.** 47.071   **41.** 36.997   **42.** 24.999

43. 156.1749   **44.** 283.8441   **45.** 2786.706   **46.** 4609.285

Round to the nearest indicated place.

47. 7.8155; thousandths   **48.** 1.30465; thousandths

49. 0.05951; ten-thousandths   **50.** 0.092156; ten-thousandths

51. 12.0157823; hundred-thousandths   **52.** 22.1126843; hundred-thousandths

53. 135.564; nearest whole number   **54.** 534.675; nearest whole number

Round to the nearest dollar.

55. $788.42   **56.** $912.75   **57.** $15,020.50   **58.** $20,159.48

Round to the nearest cent.

59. $96.3357   **60.** $42.9261   **61.** $5783.716   **62.** $3928.649

Applications

63. *Baseball* During the 2010 baseball season, the winning percentages of the New York Yankees and the Minnesota Twins were 0.63636 and 0.54455, respectively. Round these values to the nearest thousandth.

64. *Sales Tax* Bryan purchased a CD for himself and a toy for his daughter. The sales tax calculated on the CD was $1.2593 and the sales tax on the toy was $1.7143. Round these values to the nearest cent.

65. *Astronomy* The number of days in a year is approximately 365.24122. Round this value to the nearest hundredth.

66. *Mathematics History* The numbers π and e are approximately equal to 3.14159 and 2.71828, respectively. We will be using π later in this textbook. You will encounter e in higher-level mathematics courses. Round these values to the nearest hundredth.

To Think About

67. Arrange in order from smallest to largest.

$$0.61, 0.062, \frac{6}{10}, 0.006, 0.0059,$$
$$\frac{6}{100}, 0.0601, 0.0519, 0.0612$$

68. Arrange in order from smallest to largest.

$$1.05, 1.512, \frac{15}{10}, 1.0513, 0.049,$$
$$\frac{151}{100}, 0.0515, 0.052, 1.051$$

69. A person wants to round 86.23498 to the nearest hundredth. He first rounds 86.23498 to 86.2350. He then rounds to 86.235. Finally, he rounds to 86.24. What is wrong with his reasoning?

70. *Personal Finance* Fred is checking the calculations on his monthly bank statement. An interest charge of $16.3724 was rounded to $16.38. An interest charge of $43.7214 was rounded to $43.73. What rule does the bank use for rounding to the nearest cent?

Cumulative Review

71. **[2.8.1]** Add. $3\frac{1}{4} + 2\frac{1}{2} + 6\frac{3}{8}$

72. **[2.8.2]** Subtract. $27\frac{1}{5} - 16\frac{3}{4}$

73. **[1.7.2]** *Boat Sales* Don's New and Used Watercraft sold four boats one weekend for $18,650, $2490, $835, and $9845. Estimate the total amount of the sales.

74. **[3.1.2]** Write each fraction as a decimal.

(a) $\frac{7}{100}$ (b) $\frac{145}{1000}$

Quick Quiz 3.2

1. Arrange from smallest to largest:

4.56, 4.6, 4.056, 4.559

2. Round to the nearest hundredth. 27.1782

3. Round to the nearest thousandth. 155.52525

4. **Concept Check** Explain how you would round 34.958365 to the nearest ten-thousandth.

3.3 Adding and Subtracting Decimals

① Adding Decimals

We often add decimals when we check the addition of our bill at a restaurant or at a store. We can relate addition of decimals to addition of fractions. For example,

$$\frac{3}{10} + \frac{6}{10} = \frac{9}{10} \quad \text{and} \quad 1\frac{1}{10} + 2\frac{8}{10} = 3\frac{9}{10}.$$

These same problems can be written more efficiently as decimals.

$$\begin{array}{r} 0.3 \\ + 0.6 \\ \hline 0.9 \end{array} \qquad \begin{array}{r} 1.1 \\ + 2.8 \\ \hline 3.9 \end{array}$$

The steps to follow when adding decimals are listed in the following box.

Student Learning Objectives

After studying this section, you will be able to:

① Add decimals.

② Subtract decimals.

NOTE TO STUDENT: *Fully worked-out solutions to all of the Student Practice problems can be found at the back of the text starting at page SP-1.*

ADDING DECIMALS

1. Write the numbers to be added vertically and line up the decimal points. Extra zeros may be placed to the right of the decimal points if needed.

2. Add all the digits with the same place value, starting with the right column and moving to the left.

3. Place the decimal point of the sum in line with the decimal points of the numbers added.

EXAMPLE 1 Add.

(a) $2.8 + 5.6 + 3.2$

(b) $158.26 + 200.07 + 315.98$

(c) $5.3 + 26.182 + 0.0007 + 624$

Solution

(a)
$$\begin{array}{r} \overset{1}{2}.8 \\ 5.6 \\ + 3.2 \\ \hline 11.6 \end{array}$$

(b)
$$\begin{array}{r} \overset{11}{1}\overset{2}{5}8.26 \\ 200.07 \\ + 315.98 \\ \hline 674.31 \end{array}$$

(c)
$$\begin{array}{r} 5.3000 \\ \overset{1}{2}6.1820 \\ 0.0007 \\ + 624.0000 \\ \hline 655.4827 \end{array}$$

In part (c), extra zeros have been added to make the problem easier.
Note: The decimal point is understood to be to the right of the digit 4.

Student Practice 1 Add.

(a)
$$\begin{array}{r} 9.8 \\ 3.6 \\ + 5.4 \end{array}$$

(b)
$$\begin{array}{r} 300.72 \\ 163.75 \\ + 291.08 \end{array}$$

(c) $8.9 + 37.056 + 0.0023 + 945$

SIDELIGHT: Adding in Extra Zeros

When we add decimals like 3.1 + 2.16 + 4.007, we may write in zeros, as shown:

$$
\begin{array}{r}
3.100 \\
2.160 \\
+\ 4.007 \\
\hline
9.267
\end{array}
$$

What are we really doing here? What is the advantage of adding these extra zeros?

"Decimals" means "decimal fractions." If we look at the numbers as fractions, we see that we are actually using the property of multiplying a fraction by 1 in order to obtain common denominators. Look at the problem this way:

$$
\left.
\begin{array}{l}
3.1 \ = 3\dfrac{1}{10} \\[2mm]
2.16 \ = 2\dfrac{16}{100} \\[2mm]
4.007 = 4\dfrac{7}{1000}
\end{array}
\right\}
$$

The least common denominator is 1000. To obtain the common denominator for the first two fractions, we multiply.

$$
\left.
\begin{array}{l}
3 \ \dfrac{1}{10} \times \dfrac{100}{100} = 3\dfrac{100}{1000} \\[2mm]
2 \ \dfrac{16}{100} \times \dfrac{10}{10} = 2\dfrac{160}{1000} \\[2mm]
+ 4 \ \dfrac{7}{1000} \qquad\ = 4\dfrac{7}{1000}
\end{array}
\right\}
$$

Once we obtain a common denominator, we can add the three fractions.

$$
9\dfrac{267}{1000} = 9.267
$$

This is the answer we arrived at earlier using the decimal form for each number. Thus writing in zeros in a decimal fraction is really an easy way to transform fractions to equivalent fractions with a common denominator. Working with decimal fractions is easier than working with other fractions.

The final digit of most odometers measures tenths of a mile. The odometer reading shown in the odometer on the left is 38,516.2 miles.

Calculator

Adding Decimals

The calculator can be used to verify your work. You can use your calculator to add decimals. To find 23.08 + 8.53 + 9.31 enter:

23.08 $\boxed{+}$ 8.53 $\boxed{+}$

9.31 $\boxed{=}$

Display:

$\boxed{40.92}$

EXAMPLE 2   Barbara checked her odometer before the summer began. It read 49,645.8 miles. She traveled 3852.6 miles that summer in her car. What was the odometer reading at the end of the summer?

Solution

$$
\begin{array}{r}
\overset{11}{4}9,6\overset{1}{4}5.8 \\
+\ \ 3852.6 \\
\hline
53,498.4
\end{array}
$$

The odometer read 53,498.4 miles.

Student Practice 2   A car odometer read 93,521.8 miles before a trip of 1634.8 miles. What was the final odometer reading?

EXAMPLE 3 During his first semester at Tarrant County Community College, Kelvey deposited checks into his checking account in the amounts of $98.64, $157.32, $204.81, $36.07, and $229.89. What was the sum of his five deposits?

Solution

$$
\begin{array}{r}
\overset{2\,3\,2\,2}{\$\ 98.64} \\
157.32 \\
204.81 \\
36.07 \\
+\quad 229.89 \\
\hline
\$726.73
\end{array}
$$

Student Practice 3 During the spring semester, Will deposited the following checks into his account: $80.95, $133.91, $256.47, $53.08, and $381.32. What was the sum of his five deposits?

② Subtracting Decimals

It is important to see the relationship between the decimal form of a mixed number and the fractional form of a mixed number. This relationship helps us understand why calculations with decimals are done the way they are. Recall that when we subtract mixed numbers with common denominators, sometimes we must borrow from the whole number.

$$
\begin{array}{rcl}
5\dfrac{1}{10} & = & 4\dfrac{11}{10} \\
-\,2\dfrac{7}{10} & = & -\,2\dfrac{7}{10} \\
\hline
& & 2\dfrac{4}{10}
\end{array}
$$

We could write the same problem in decimal form:

$$
\begin{array}{r}
\overset{4\ 11}{\cancel{5}.\cancel{1}} \\
-\ 2.7 \\
\hline
2.4
\end{array}
$$

Subtraction of decimals is thus similar to subtraction of fractions (we get the same result), but it's usually easier to subtract with decimals than to subtract with fractions.

SUBTRACTING DECIMALS

1. Write the decimals to be subtracted vertically and line up the decimal points. Additional zeros may be placed to the right of the decimal point if not all numbers have the same number of decimal places.

2. Subtract all digits with the same place value, starting with the right column and moving to the left. Borrow when necessary.

3. Place the decimal point of the difference in line with the decimal points of the two numbers being subtracted.

EXAMPLE 4   Subtract.

(a)   84.8
        − 27.3

(b)   1076.320
        − 983.518

Solution

(a)
$$
\begin{array}{r}
\overset{7}{\cancel{8}}\overset{14}{\cancel{4}}.8 \\
-\ 2\ 7\ .3 \\
\hline
5\ 7\ .5
\end{array}
$$

(b)
$$
\begin{array}{r}
\overset{9}{\cancel{1}}\ \overset{}{\cancel{0}}\ 7\ \overset{}{6}\ .\overset{}{3}\ \overset{}{2}\ \overset{}{0} \\
-\ 9\ 8\ 3\ .5\ 1\ 8 \\
\hline
9\ 2\ .8\ 0\ 2
\end{array}
$$

Student Practice 4   Subtract.

(a)   38.8
        − 26.9

(b)   2034.908
        − 1986.325

When the two numbers being subtracted do not have the same number of decimal places, write in zeros as needed.

EXAMPLE 5   Subtract.

(a) $12 - 8.362$

(b) $156.381 - 99.82$

Solution

(a)
$$
\begin{array}{r}
\overset{11}{\ }\overset{9}{\ }\overset{9}{\ } \\
\cancel{1}\ \overset{}{2}\ .\overset{}{0}\ \overset{}{0}\ \overset{}{0} \\
-\ \ \ \ 8\ .3\ 6\ 2 \\
\hline
3\ .6\ 3\ 8
\end{array}
$$

(b)
$$
\begin{array}{r}
\overset{14}{\ }\overset{15}{\ } \\
\cancel{1}\ \overset{}{5}\ \overset{}{6}\ .\overset{}{3}\ 8\ 1 \\
-\ \ 9\ 9\ .8\ 2\ 0 \\
\hline
5\ 6\ .5\ 6\ 1
\end{array}
$$

Student Practice 5   Subtract.

(a) $19 - 12.579$

(b) $283.076 - 96.38$

EXAMPLE 6   On Tuesday, Don Ling filled the gas tank in his car. The odometer read 56,098.5. He drove for four days. The next time he filled the tank, the odometer read 56,420.2. How many miles had he driven?

Solution

$$
\begin{array}{r}
5\ 6\ ,\ 4\ 2\ 0\ .2 \\
-\ 5\ 6\ ,\ 0\ 9\ 8\ .5 \\
\hline
3\ 2\ 1\ .7
\end{array}
$$

He had driven 321.7 miles.

Student Practice 6   Abdul had his car oil changed when his car odometer read 82,370.9 miles. When he changed the oil again, the odometer read 87,160.1 miles. How many miles did he drive between oil changes?

EXAMPLE 7 Find the value of x if $x + 3.9 = 14.6$.

Solution Recall that the letter x is a variable. It represents a number that is added to 3.9 to obtain 14.6. We can find the number x if we calculate $14.6 - 3.9$.

$$
\begin{array}{r}
\overset{3\ 16}{1\cancel{4}.\cancel{6}} \\
-\ \ 3.9 \\
\hline
10.7
\end{array}
$$

Thus $x = 10.7$.

Check. Is this true? If we replace x by 10.7, do we get a true statement?

$$x + 3.9 = 14.6$$
$$10.7 + 3.9 \overset{?}{=} 14.6$$
$$14.6 = 14.6 \ \checkmark$$

Student Practice 7 Find the value of x if $x + 10.8 = 15.3$.

Adding and subtracting decimals is an important part of life. When you are recording deposits at the bank, reconciling your checkbook, or completing your income tax forms, you are adding and subtracting decimals. Be sure to learn to do it accurately. In the homework exercises, always check your answers with the Answers section in the back of the book. Making sure you have the correct answers is very important.

STEPS TO SUCCESS Do You Realize How Valuable Friendship Is?

In a math class a friend is a person of fantastic value. Robert Louis Stevenson once wrote, "A friend is a gift you give yourself." This is especially true when you take a mathematics class and make a friend in the class. You will find that you enjoy sitting together and drawing support and encouragement from each other. You may want to exchange phone numbers or e-mail addresses. You may want to study together or review together before a test.

How do you get started? Try talking to the students seated around you. Ask someone for help about something you did not understand in class. Take the time to listen to them and their interests and concerns. You may discover you have a lot in common. If the first few people you talk to seem uninterested, try sitting in a different part of the room and talk to those students who are seated around you in the new location. Don't force a friendship on anyone but just look for chances to open up a good channel of communication.

Making it personal: What do you think is the best way to make a friend of someone in your class? Which of these suggestions do you find the most helpful? Will you take the time to reach out to someone in your class this week and try to begin a new friendship? ▼

Add.

1. 57.1 + 19.7

2. 78.3 + 29.4

3. 384.25 + 209.65

4. 193.42 + 768.78

5.   13.4
           7.6
      + 275.2

6.   176.5
            8.4
       + 22.5

7.   4.71
      + 8.05

8.   9.284
      + 5.77

9.   4.9637
          28.12
       + 3.645

10.   7.0276
           3.451
        + 16.98

11. 12.
          3.62
       + 51.8

12. 13.
          4.52
       + 63.7

13. 108.36 + 14.3 + 85.12 + 28

14. 215.45 + 48 + 30.77 + 15.8

15. 753.61 + 28.75 + 162.3 + 100.5 + 67

16. 432.51 + 16.08 + 892.1 + 301.2 + 84

Applications *In exercises 17 and 18, calculate the perimeter of each triangle.*

17.

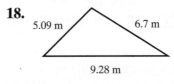

9.28 ft

5.26 ft

6.5 ft

18.

5.09 m   6.7 m

9.28 m

19. *Weight Loss* Lamar is losing weight by walking each evening after dinner. During the first week in February he lost 1.75 pounds. During the second, third, and fourth weeks, he lost 2.5 pounds, 1.55 pounds, and 2.8 pounds, respectively. How many total pounds did Lamar lose in February?

20. *Health* Olivia knows she needs to drink more water while at work. One day during her morning break she drank 7.15 ounces. At lunch she drank 12.45 ounces and throughout the afternoon she drank 10.75 ounces. How many total ounces of water did she drink?

21. *Beach Vacation* Mick and Keith have arrived in Miami and are going to the beach. They buy sunblock for $4.99, beverages for $12.50, sandwiches for $11.85, towels for $28.50, bottled water for $3.29, and two novels for $16.99. After they got what they needed, what was Mick and Keith's bill for their day at the beach?

22. *Consumer Mathematics* Anika bought school supplies at the campus bookstore. She purchased a calculator for $37.25, pens for $5.89, a T-shirt for $13.95, and notebooks for $10.49. The amount of sales tax was $4.05. What was the total of Anika's bill including tax?

23. *Truck Travel* A truck odometer read 46,276.0 miles before a trip of 778.9 miles. What was the final odometer reading?

24. *Car Travel* Jane traveled 1723.1 miles. The car odometer at the beginning of the trip read 23,195.0 miles. What was the final odometer reading?

Personal Banking In exercises 25 and 26, a portion of a bank checking account deposit slip is shown. Add the numbers to determine the total deposit. The line drawn between the dollars and the cents columns serves as the decimal point.

25.
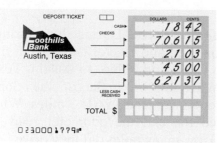

26.
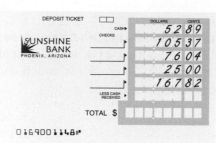

Subtract.

27. $12.8 - 9.3$

28. $18.6 - 7.1$

29. $35.75 - 9.82$

30. $76.22 - 8.17$

31. $126 - 76.22$

32. $209 - 81.54$

33.   586.513
        $- \ 78.2$

34.   243.967
        $- \ 84.2$

35.   220.9
        $- \ 85.47$

36.   181.9
        $- \ 62.23$

37.   24.0079
        $- \ 19.3614$

38.   43.0205
        $- \ 22.7346$

39.   8
        $- \ 1.263$

40.   12
        $- \ 7.981$

41.   7362.14
        $- \ 6173.07$

42.   4986.71
        $- \ 3615.93$

43.   1.5
        $- \ 0.0365$

44.   2.8
        $- \ 0.07763$

Mixed Practice *Add or subtract.*

45. $123.621 + 52.96$

46. $241.983 + 75.48$

47. $98.3 - 56.71$

48. $79.2 - 45.93$

49. $0.0763 + 2 + 3.16$

50. $18 - 2.75$

51. $197.600 - 124.375$

52. $382.700 - 291.927$

Applications

53. *World Records* The heaviest apple on record was grown in Japan in 2005 and weighed 4.0678 pounds. The heaviest lemon was grown in Israel in 2003 and weighed 11.583 pounds. How much heavier was the lemon than the apple? (*Source:* www.guinessworldrecords.com)

54. *Health* At her 4-month checkup, baby Grace weighed 7.675 kilograms. When she was born, she weighed 3.7 kilograms. How much weight has Grace gained since she was born?

55. *Telescope* A child's beginner telescope is priced at $79.49. The price of a certain professional telescope is $37,026.65. How much more does the professional telescope cost?

56. *Automobile Travel* During their spring break vacation, Ted and Hector drove from their college in Cleveland, Ohio, to Austin, Texas, and back. When they began the trip, the odometer of their rental car read 15,675.3 miles. When they returned the car, the odometer read 18,247.1 miles. How many miles did they drive?

57. *Taxi Trip* Malcolm took a taxi from John F. Kennedy Airport in New York to his hotel in the city. His fare was $47.70 and he tipped the driver $7.00. How much change did Malcolm get back if he gave the driver a $100 bill?

58. *Personal Banking* Nathan took $200 out of the ATM. He bought snow boots for $65.49, pet supplies for $27.75, and a bouquet of flowers for $18.95. How much money does he have left?

59. *Electric Wire Construction* An insulated wire measures 12.62 centimeters. The last 0.98 centimeter of the wire is exposed. How long is the part of the wire that is not exposed?

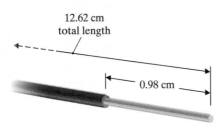

12.62 cm
total length

0.98 cm

60. *Plumbing* The outside radius of a pipe is 9.39 centimeters. The inside radius is 7.93 centimeters. What is the thickness of the pipe?

9.39   7.93

61. *Medical Research* A cancer researcher is involved in an important experiment. She is trying to determine how much of an anticancer drug is necessary for a Stage I (nonhuman or animal) test. She pours 2.45 liters of the experimental anticancer formula in one container and 1.35 liters of a reactive liquid in another. She then pours the contents of one container into the other. If 0.85 liter is expected to evaporate during the process, how much liquid will be left?

62. *Rainforest Loss* Everyone is becoming aware of the rapid loss of Earth's rainforests. Originally, there were 3.968 billion acres of rainforest in the world. By 2010, 1.856 billion acres had been destroyed. How many acres of rainforest remained in 2010? (*Source:* www.wikipedia.org)

The federal water safety standard requires that drinking water contain no more than 0.015 milligram of lead per liter of water. (Source: Environmental Protection Agency)

63. *Well Water Safety* Carlos and Maria had the well that supplies their home analyzed for safety. A sample of well water contained 0.0089 milligram of lead per liter of water. What is the difference between their sample and the federal safety standard? Is it safe for them to drink the water?

64. *City Water Safety* Fred and Donna use water provided by the city for the drinking water in their home. A sample of their tap water contained 0.023 milligram of lead per liter of water. What is the difference between their sample and the federal safety standard? Is it safe for them to drink the water?

Income of Industries *The following table shows the incomes of some major U.S. industries. Use this table for exercises 65–68. Write each answer as a decimal and as a whole number. The table values are recorded in billions of dollars.*

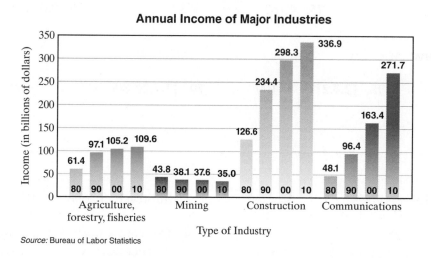

Annual Income of Major Industries

Source: Bureau of Labor Statistics

65. How many more dollars were earned in mining in 1980 than in 2000?

66. How many more dollars were earned in construction in 2000 than in 1980?

67. In 2010, how many more dollars were earned in communications than in agriculture, forestry, and fisheries?

68. In 1990, how many more dollars were earned in communications than in mining?

To Think About *Mr. Jensen made up the following shopping list of items he needs and the cost of each item. Use the list to answer exercises 69 and 70.*

can cranberry sauce	$0.99
hot dog relish	$0.79
ranch salad dressing	$1.47
large can solid white tuna	$2.29
can tomato soup	$0.68
can sliced peaches	$1.26
large jar tomato sauce	$1.65
large box Cheerios	$3.79
medium box Raisin Bran	$2.63
medium jar peanut butter	$2.19

69. *Grocery Shopping* Mr. Jensen goes to the store to buy the following items from his list: Raisin Bran, ranch salad dressing, sliced peaches, hot dog relish, and peanut butter. He has a ten-dollar bill. Estimate the cost of buying these items by first rounding the cost of each item to the nearest ten cents. Does he have enough money to buy all of them? Find the exact cost of these items. How close was your estimate?

70. *Grocery Shopping* The next day the Jensens' daughter, Brenda, goes to the store to buy the following items from the list: Cheerios, tomato sauce, peanut butter, white tuna, tomato soup, and cranberry sauce. She has fifteen dollars. Estimate the cost of buying these items by first rounding the cost of each item to the nearest ten cents. Does she have enough money to buy all of them? Find the exact cost of these items. How close was your estimate?

Find the value of x.

71. $x + 7.1 = 15.5$

72. $x + 4.8 = 23.1$

73. $156.9 + x = 200.6$

74. $185.8 + x = 238.3$

75. $4.162 = x + 2.053$

76. $8.013 = x + 4.708$

Cumulative Review *Multiply.*

77. [1.4.2] $\begin{array}{r} 2536 \\ \times\ \ \ 8 \\ \hline \end{array}$

78. [2.4.2] $\frac{1}{4} \times 100$

79. [2.4.2] $800 \times \frac{1}{2}$

80. [1.7.2] Estimate.
$38{,}125 \div 43$

Quick Quiz 3.3

1. Add. $53.261 + 1.9 + 17.82$

2. Subtract. $5.2608 - 3.0791$

3. Subtract. $59.6 - 3.925$

4. **Concept Check** Explain how you perform the correct borrowing and correct use of the decimal point if you subtract $567.45 - 345.9872$.

👣 STEPS TO SUCCESS Keep Trying! Do Not Quit!

We live in a highly technical world. More and more it depends on mathematics. You cannot afford to give up on the study of mathematics. Dropping mathematics may prevent you from entering some career field that you find really interesting. Learning mathematics can open new doors for you.

Learning mathematics is a process that takes time and effort. Regular study and daily practice are necessary. This will help your level of academic success and lead you toward mastery of mathematics. In hard economic times, this mastery will open a path for you to get a well-paying job. Do not quit! You can do it!

Making it personal: What things can you do to help you master mathematics? Don't let yourself quit or let up on your studies. Your work in mathematics now can lead to financial success in the future! ▼

3.4 Multiplying Decimals

① Multiplying a Decimal by a Decimal or a Whole Number

We learned previously that the product of two fractions is the product of the numerators over the product of the denominators. For example,

$$\frac{3}{10} \times \frac{7}{100} = \frac{21}{1000}$$

In decimal form this product would be written

$$0.3 \times 0.07 = 0.021$$

one decimal place   two decimal places   three decimal places

---

MULTIPLICATION OF DECIMALS

1. Multiply the numbers just as you would multiply whole numbers.

2. Find the sum of the number of decimal places in the two factors.

3. Place the decimal point in the product so that the product has the same number of decimal places as the sum in step 2. You may need to write zeros to the left of the number found in step 1.

Now use these steps to do the preceding multiplication problem.

EXAMPLE 1 Multiply. 0.07×0.3

Solution

0.07	2 decimal places
× 0.3	1 decimal place
0.021	3 decimal places in product $(2 + 1 = 3)$

Student Practice 1 Multiply. 0.09×0.6

When performing the calculation, it is usually easier to place the factor with the smallest number of nonzero digits underneath the other factor.

EXAMPLE 2 Multiply.

(a) 0.38×0.26 **(b)** 12.64×0.572

Solution

(a)

0.38	2 decimal places
× 0.26	2 decimal places
228	
76	
0.0988	4 decimal places $(2 + 2 = 4)$

Note that we need to insert a zero before the 988.

(b)

12.64	2 decimal places
× 0.572	3 decimal places
2528	
8848	
6 320	
7.23008	5 decimal places $(2 + 3 = 5)$

Continued on next page

Student Learning Objectives

After studying this section, you will be able to:

① Multiply a decimal by a decimal or a whole number.

② Multiply a decimal by a power of 10.

Calculator

 Multiplying Decimals

You can use your calculator to multiply a decimal by a decimal. To find 0.08×1.53 enter:

0.08 ☒ 1.53 ☐

Display:

0.1224

NOTE TO STUDENT: *Fully worked-out solutions to all of the Student Practice problems can be found at the back of the text starting at page SP-1.*

Student Practice 2 Multiply.

(a) 0.47×0.28 **(b)** 0.436×18.39

When multiplying a decimal fraction by a whole number, you need to remember that a whole number has no decimal places.

EXAMPLE 3 Multiply. 5.261×45

Solution

5.261	3 decimal places
$\times\quad 45$	0 decimal places
$26\ 305$	
$210\ 44$	
236.745	3 decimal places $(3 + 0 = 3)$

Student Practice 3 Multiply. 0.4264×38

▲ **EXAMPLE 4** Uncle Roger's rectangular front lawn measures 50.6 yards wide and 71.4 yards long. What is the area of the lawn in square yards?

Solution Since the lawn is rectangular, we will use the fact that to find the area of a rectangle we multiply the length by the width.

71.4	1 decimal place
$\times\ 50.6$	1 decimal place
$42\ 84$	
$3570\ 0$	
3612.84	2 decimal places

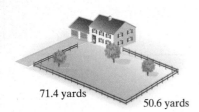

71.4 yards   50.6 yards

The area of the lawn is 3612.84 square yards.

▲ **Student Practice 4** A rectangular computer chip measures 1.26 millimeters wide and 2.3 millimeters long. What is the area of the chip in square millimeters?

② **Multiplying a Decimal by a Power of 10**

Observe the following pattern.

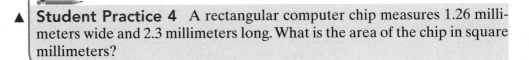

| one zero | Decimal point moved one place to the right. |

$0.035 \times 10^1 = 0.035 \times 10 = 0.35$

| two zeros | Decimal point moved two places to the right. |

$0.035 \times 10^2 = 0.035 \times 100 = 3.5$

| three zeros | Decimal point moved three places to the right. |

$0.035 \times 10^3 = 0.035 \times 1000 = 35.$

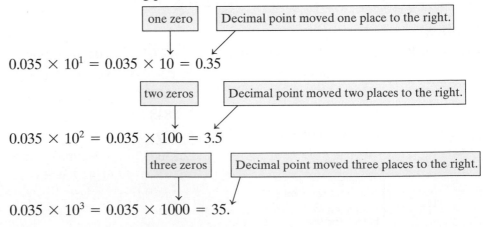

> **MULTIPLICATION OF A DECIMAL BY A POWER OF 10**
>
> To multiply a decimal by a power of 10, move the decimal point to the right the same number of places as the number of zeros in the power of 10.

EXAMPLE 5 Multiply.

(a) 2.671×10 **(b)** 37.85×100

Solution

(a) $2.671 \times 10 \qquad = 26.71$

one zero | Decimal point moved one place to the right.

(b) $37.85 \times 100 \qquad = 3785.$

two zeros | Decimal point moved two places to the right.

Student Practice 5 Multiply.

(a) 0.0561×10 **(b)** 1462.37×100

Sometimes it is necessary to add extra zeros before placing the decimal point in the answer.

EXAMPLE 6 Multiply.

(a) 4.8×1000 **(b)** $0.076 \times 10,000$

Solution

(a) $4.8 \times 1000 \qquad = 4800.$

three zeros | Decimal point moved three places to the right. Two extra zeros were needed.

(b) $0.076 \times 10,000 \qquad = 760.$

four zeros | Decimal point moved four places to the right. One extra zero was needed.

Student Practice 6 Multiply.

(a) 0.26×1000 **(b)** $5862.89 \times 10,000$

If the number that is a power of 10 is in exponent form, move the decimal point to the right the same number of places as the number that is the exponent.

EXAMPLE 7   Multiply. 3.68×10^3

Solution

| Exponent of 3 | | Decimal point moved three places to the right. |

$$3.68 \times 10^3 = 3680.$$

Student Practice 7   Multiply. 7.684×10^4

SIDELIGHT: Moving the Decimal Point

Can you devise a quick rule to use when multiplying a decimal fraction by $\frac{1}{10}, \frac{1}{100}, \frac{1}{1000}$, and so on? How is it like the rules developed in this section? Consider a few examples:

Original Problem	Change Fraction to Decimal	Decimal Multiplication	Observation
$86 \times \dfrac{1}{10}$	86×0.1	$\begin{array}{r} 86 \\ \times\ 0.1 \\ \hline 8.6 \end{array}$	Decimal point moved one place to the left.
$86 \times \dfrac{1}{100}$	86×0.01	$\begin{array}{r} 86 \\ \times\ 0.01 \\ \hline 0.86 \end{array}$	Decimal point moved two places to the left.
$86 \times \dfrac{1}{1000}$	86×0.001	$\begin{array}{r} 86 \\ \times\ 0.001 \\ \hline 0.086 \end{array}$	Decimal point moved three places to the left.

Can you think of a way to describe a rule that you could use in solving this type of problem without going through all the foregoing steps?

You use multiplying by a power of 10 when you convert a larger unit of measure to a smaller unit of measure in the metric system.

EXAMPLE 8   Change 2.96 kilometers to meters.

Solution   Since we are going from a larger unit of measure to a smaller one, we multiply. There are 1000 meters in 1 kilometer. Multiply 2.96 by 1000.

$$2.96 \times 1000 = 2960$$

2.96 kilometers is equal to 2960 meters.

Student Practice 8   Change 156.2 kilometers to meters.

TO THINK ABOUT: Names Used to Describe Large Numbers

Often when reading the newspaper or watching television news shows, we hear words like 3.46 trillion or 67.8 billion. These are abbreviated notations that are used to describe large numbers. When you encounter these numbers, you can change them to standard notation by multiplication of the appropriate value.

For example, if someone says that the population of China is 1.31 billion people, we can write 1.31 billion $= 1.31 \times 1$ billion $= 1.31 \times 1,000,000,000 = 1,310,000,000$. If someone says the population of Chicago is 2.92 million people, we can write

$$2.92 \text{ million} = 2.92 \times 1 \text{ million} = 2.92 \times 1,000,000 = 2,920,000.$$

Verbal and Writing Skills, Exercises 1–4

1. Explain in your own words how to determine where to put the decimal point in the answer when you multiply 0.67×0.08.

2. Explain in your own words how to determine where to put the decimal point in the answer when you multiply 3.45×0.9.

3. Explain in your own words how to determine where to put the decimal point in the answer when you multiply 0.0078×100.

4. Explain in your own words how to determine where to put the decimal point in the answer when you multiply 5.0807 by 1000.

Multiply.

5.
$$\begin{array}{r} 0.6 \\ \times\ 0.2 \\ \hline \end{array}$$

6.
$$\begin{array}{r} 0.9 \\ \times\ 0.3 \\ \hline \end{array}$$

7.
$$\begin{array}{r} 0.12 \\ \times\ 0.5 \\ \hline \end{array}$$

8.
$$\begin{array}{r} 0.17 \\ \times\ 0.4 \\ \hline \end{array}$$

9.
$$\begin{array}{r} 0.0036 \\ \times\ 0.8 \\ \hline \end{array}$$

10.
$$\begin{array}{r} 0.067 \\ \times\ 0.07 \\ \hline \end{array}$$

11.
$$\begin{array}{r} 452 \\ \times\ 0.12 \\ \hline \end{array}$$

12.
$$\begin{array}{r} 316 \\ \times\ 0.24 \\ \hline \end{array}$$

13.
$$\begin{array}{r} 0.043 \\ \times\ 0.012 \\ \hline \end{array}$$

14.
$$\begin{array}{r} 0.037 \\ \times\ 0.011 \\ \hline \end{array}$$

15.
$$\begin{array}{r} 10.97 \\ \times\ 0.06 \\ \hline \end{array}$$

16.
$$\begin{array}{r} 18.07 \\ \times\ 0.05 \\ \hline \end{array}$$

17.
$$\begin{array}{r} 3423 \\ \times\ 0.8 \\ \hline \end{array}$$

18.
$$\begin{array}{r} 5119 \\ \times\ 0.7 \\ \hline \end{array}$$

19.
$$\begin{array}{r} 2.163 \\ \times\ 0.008 \\ \hline \end{array}$$

20.
$$\begin{array}{r} 1.892 \\ \times\ 0.007 \\ \hline \end{array}$$

21.
$$\begin{array}{r} 0.7613 \\ \times\ 1009 \\ \hline \end{array}$$

22.
$$\begin{array}{r} 0.6178 \\ \times\ 5004 \\ \hline \end{array}$$

23.
$$\begin{array}{r} 2350 \\ \times\ 3.6 \\ \hline \end{array}$$

24.
$$\begin{array}{r} 3720 \\ \times\ 8.1 \\ \hline \end{array}$$

25. 4.57×11.8

26. 73.2×2.45

27. 0.001×6523.7

28. 0.01×826.75

Applications, Exercises 29–38

29. *Car Payments* Kenny is making payments on his Ford Escort of $155.40 per month for the next 60 months. How much will he have spent in car payments after he sends in his final payment?

30. *Food Purchase* Each carton of ice cream contains 1.89 liters. Paul stocked his freezer with 25 cartons. How many total liters of ice cream did he buy?

31. *Personal Income* Philippe works as a bank teller and earns $10.50 per hour for a 40-hour week. How much does he earn in one week? (The average wage in 2009 for U.S. bank tellers was $11.91 per hour.) (*Source:* www.bls.gov)

32. *Personal Income* Camilla works as a medical secretary and earns $15.75 per hour for a 40-hour week. How much does she earn in one week? (The average wage in 2009 for U.S. medical secretaries was $15.12 per hour.) (*Source:* www.bls.gov)

▲ **33.** *Geometry* Ralph and Darlene are getting new carpet in their bedroom and need to find how many square feet they need to purchase. The dimensions of their rectangular bedroom are 15.5 feet by 19.2 feet. What is the area of the room in square feet?

▲ **34.** *Geometry* Sal is having his driveway paved by a company that charges by the square yard. Sal's driveway measures 8.6 yards by 17.5 yards. How many square yards is his driveway?

35. *Student Loan* Dwight is paying off a student loan at Westmont College with payments of $36.90 per month for the next 18 months. How much will he pay off during the next 18 months?

36. *Car Payments* Marcia is making car payments to Highfield Center Chevrolet of $230.50 per month for 16 more months. How much will she pay for car payments in the next 16 months?

37. *Fuel Efficiency* Steve's car gets approximately 26.4 miles per gallon. His gas tank holds 19.5 gallons. Approximately how many miles can he travel on a full tank of gas?

38. *Fuel Efficiency* Caleb's 4 × 4 truck gets approximately 18.6 miles per gallon. His gas tank holds 19.5 gallons. Approximately how many miles can he travel on a full tank of gas? Compare this to your answer in exercise 37.

Multiply.

39. 2.86×10

40. 1.98×10

41. 52.125×100

42. 86.375×100

43. 22.615×1000

44. 34.105×1000

45. $5.60982 \times 10,000$

46. $1.27986 \times 10,000$

47. $17,561.44 \times 10^2$

48. 7163.241×10^2

49. 816.32×10^3

50. 763.49×10^4

Applications

51. *Metric Conversion* To convert from meters to centimeters, multiply by 100. How many centimeters are in 5.932 meters?

52. *Metric Conversion* One meter is about 39.36 inches. About how many inches are in 100 meters?

53. *Stock Market* Jeremiah bought 1000 shares of ADPT Corporation stock, each worth $2.71. How much did Jeremiah spend on the shares?

54. *Stock Market* Sara bought 1000 shares of Micron Technology, Inc. stock, each worth $7.64. How much did Sara spend on the shares?

55. *Personal Finance* In April, Ellen received a $925.75 tax refund. She decided to spend the money on some gifts. She spent $95.00 on her parents' anniversary gift, $47.50 on each of her two cousins' graduation gifts, and $39.25 on each of her three nieces' birthday gifts. How much money does she have left over?

56. *Pet Cats* Tomba is a beautiful orange tabby cat. When he was found by the side of the road, he was three weeks old and weighed 0.95 lb. At the age of three months, he weighed 2.85 lb. At the age of nine months, he weighed 6.30 lb; at one year, he weighed 11.7 lb. Today, Tomba the cat is $1\frac{1}{2}$ years old, and weighs 15.75 lb.

(a) How much weight did he gain?

(b) If the veterinarian wants him to lose 0.25 lb per week until he weighs 13.5 lb, how long will it take?

▲ **57.** *Geometry* The college is purchasing new carpeting for the learning center. What is the price of a carpet that is 19.6 yards wide and 254.2 yards long if the cost is $12.50 per square yard?

58. *Jewelry Store Operations* A jewelry store purchased long lengths of gold chain, which will be cut and made into necklaces and bracelets. The store purchased 3220 grams of gold chain at $3.50 per gram.

(a) How much did the jewelry store spend?

(b) If they sell a 28-gram gold necklace for $17.75 per gram, how much profit will they make on the necklace?

To Think About

59. State in your own words a rule for mental multiplication by 0.1, 0.01, 0.001, 0.0001, and so on.

60. State in your own words a rule for mental multiplication by 0.2, 0.02, 0.002, 0.0002, and so on.

Cumulative Review *Divide. Be sure to include any remainder as part of your answer.*

61. **[1.5.3]** $35\overline{)7035}$ **62.** **[1.5.3]** $124\overline{)56,024}$ **63.** **[2.4.2]** Multiply. $\frac{1}{3} \times 1500$ **64.** **[2.5.3]** Divide. $36 \div 3\frac{1}{3}$

Pets in the United States The total number of pets owned in the United States for the year 2010 is given in the bar graph below. Use the graph to answer exercises 65–68.

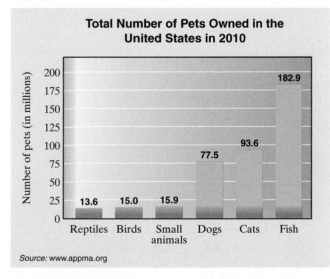

Total Number of Pets Owned in the United States in 2010

Source: www.appma.org

65. **[3.3.2]** How many more pet cats were there in the United States than pet dogs?

66. **[3.3.2]** How many more pet dogs were there in the United States than pet birds?

67. **[3.3.1]** How many more pet fish were there in the United States than pet dogs, small animals, birds, and reptiles combined?

68. **[3.3.1]** How many more pet cats, dogs, and birds combined were there in the United States than pet fish?

Quick Quiz 3.4

1. Multiply. 0.76×0.04

2. Multiply. 25.6×0.128

3. Multiply. 5.162×10^4

4. **Concept Check** Explain how you know where to put the decimal point in the answer when you multiply 3.45×9.236.

How Am I Doing? Sections 3.1–3.4

How are you doing with your homework assignments in Sections 3.1 to 3.4? Do you feel you have mastered the material so far? Do you understand the concepts you have covered? Before you go further in the textbook, take some time to do each of the following problems.

3.1

1. Write the word name for the decimal. 31.903

2. Express as a decimal. $\dfrac{567}{10,000}$

Write as a fraction or a mixed number. Reduce whenever possible.

3. 4.09

4. 0.475

3.2

5. Place the set of numbers in the proper order from smallest to largest.
1.6, 1.59, 1.61, 1.601

6. Round to the nearest tenth. 123.49268

7. Round to the nearest ten thousandth. 8.065447

8. Round to the nearest hundredth. 17.98523

3.3

Add.

9. $5.12 + 4.7 + 8.03 + 1.6$

10. $24.613 + 0.273 + 2.305$

Subtract.

11. $\begin{array}{r} 42.16 \\ -31.57 \\ \hline \end{array}$

12. $26 - 18.329$

3.4

Multiply.

13. $\begin{array}{r} 11.67 \\ \times\ 0.03 \\ \hline \end{array}$

14. 4.7805×1000

15. 0.0003796×10^5

16. 3.14×2.5

17. 982×0.007

18. 0.00052×0.006

Now turn to page SA-7 for the answer to each of these problems. Each answer also includes a reference to the objective in which the problem is first taught. If you missed any of these problems, you should stop and review the Examples and Student Practice problems in the referenced objective. A little review now will help you master the material in the upcoming sections of the text.

1. _____
2. _____
3. _____
4. _____
5. _____
6. _____
7. _____
8. _____
9. _____
10. _____
11. _____
12. _____
13. _____
14. _____
15. _____
16. _____
17. _____
18. _____

3.5 Dividing Decimals

① Dividing a Decimal by a Whole Number

When you divide a decimal by a whole number, place the decimal point for the quotient directly above the decimal point in the dividend. Then divide as if the numbers were whole numbers.

To divide 26.8 by 4, we place the decimal point of our answer (the quotient) directly *above* the decimal point in the dividend.

$$4\overline{)26.8}$$

The decimal points are aligned, one above the other.

Then we divide as if we were dividing whole numbers.

$$
\begin{array}{r}
6.7 \\
4\overline{)26.8} \\
\underline{24} \\
2\,8 \\
\underline{2\,8} \\
0
\end{array}
$$

The quotient is 6.7.

The quotient to a problem may have all digits to the right of the decimal point. In some cases you will have to put a zero in the quotient as a "place holder." Let's divide 0.268 by 4.

$$
\begin{array}{r}
0.067 \\
4\overline{)0.268} \\
\underline{24} \\
28 \\
\underline{28} \\
0
\end{array}
$$

Note that we must have a zero after the decimal point in 0.067.

EXAMPLE 1 Divide.

(a) $9\overline{)0.3204}$ **(b)** $14\overline{)36.12}$

Solution

(a)
$$
\begin{array}{r}
0.0356 \\
9\overline{)0.3204} \\
\underline{27} \\
50 \\
\underline{45} \\
54 \\
\underline{54} \\
0
\end{array}
$$

Note the zero *after* the decimal point.

(b)
$$
\begin{array}{r}
2.58 \\
14\overline{)36.12} \\
\underline{28} \\
8\,1 \\
\underline{7\,0} \\
1\,12 \\
\underline{1\,12} \\
0
\end{array}
$$

Student Practice 1 Divide.

(a) $7\overline{)1.806}$ **(b)** $16\overline{)0.0928}$

Student Learning Objectives

After studying this section, you will be able to:

① Divide a decimal by a whole number.

② Divide a decimal by a decimal.

NOTE TO STUDENT: *Fully worked-out solutions to all of the Student Practice problems can be found at the back of the text starting at page SP-1.*

Some division problems do not yield a remainder of zero. In such cases, we may be asked to round the answer to a specified place. To round when dividing, we carry out the division until our answer contains a digit that is one place to the right of that to which we intend to round. Then we round our answer to the specified place. For example, to round to the nearest thousandth, we carry out the division to the ten-thousandths place. In some division problems, you will need to write in zeros at the end of the dividend so that this division can be carried out.

EXAMPLE 2 Divide and round the quotient to the nearest thousandth.

$$12.67 \div 39$$

Solution We will carry out our division to the ten-thousandths place. Then we will round our answer to the nearest thousandth.

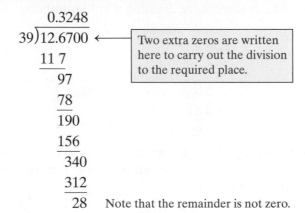

Now we round 0.3248 to 0.325. The answer is rounded to the nearest thousandth.

Student Practice 2 Divide and round the quotient to the nearest hundredth. $23.82 \div 46$

EXAMPLE 3 Maria paid $5.92 for 16 pounds of tomatoes. How much did she pay per pound?

Solution The cost of one pound of tomatoes equals the total cost, $5.92, divided by 16 pounds. Thus we will divide.

$$
\begin{array}{r}
0.37 \\
16\overline{)5.92} \\
4\,8 \\
\hline
1\,12 \\
1\,12 \\
\hline
0
\end{array}
$$

Maria paid
$0.37 per pound
for the tomatoes.

Student Practice 3 Won Lin will pay off his auto loan for $3538.75 over 19 months. If the monthly payments are equal, how much will he pay each month?

② Dividing a Decimal by a Decimal

When the divisor is not a whole number, we can convert the division problem to an equivalent problem that has a whole number as a divisor. Think about the reasons why this procedure will work. We will ask you about it after you study Examples 4 and 5.

DIVIDING A DECIMAL BY A DECIMAL

1. Make the divisor a whole number by moving the decimal point to the right. Mark that position with a caret ($_\wedge$). Count the number of places the decimal point moved.

2. Move the decimal point in the dividend to the right the same number of places. Mark that position with a caret.

3. Place the decimal point of your answer directly above the caret marking the decimal point of the dividend.

4. Divide as with whole numbers.

EXAMPLE 4

(a) Divide. $0.08\overline{)1.632}$ **(b)** Divide. $1.352 \div 0.026$

Solution

(a) $0.08._\wedge\overline{)1.63.2}$ Move each decimal point two places to the right.

Place the decimal point of the answer directly above the caret.

$0.08_\wedge\overline{)1.63_\wedge 2}$ Mark the new position by a caret ($_\wedge$).

$$
\begin{array}{r}
20.4 \\
0.08_\wedge\overline{)1.63_\wedge 2} \\
\underline{1\,6} \\
3\,2 \\
\underline{3\,2} \\
0
\end{array}
$$

The answer is 20.4.

Perform the division.

(b)
$$
\begin{array}{r}
52. \\
0.026_\wedge\overline{)1.352_\wedge} \\
\underline{1\,30} \\
52 \\
\underline{52} \\
0
\end{array}
$$

Move each decimal point three places to the right and mark the new position by a caret.

The answer is 52.

Student Practice 4 Divide.

(a) $0.09\overline{)0.1008}$ **(b)** $1.702 \div 0.037$

TO THINK ABOUT: The Multiplicative Identity Why do we move the decimal point to the right in the divisor and the dividend? What rule allows us to do this? How do we know the answer will be valid? We are actually using the property that multiplication of a fraction by 1 leaves the fraction unchanged. This is called the *multiplicative identity*. Let us examine Example 4(b) again. We will write $1.352 \div 0.026$ as a fraction.

$\dfrac{1.352}{0.026} \times 1$	Multiplication of a fraction by 1 does not change the value of the fraction.
$= \dfrac{1.352}{0.026} \times \dfrac{1000}{1000}$	We know that $\dfrac{1000}{1000} = 1$.
$= \dfrac{1352}{26}$	Multiplication by 1000 can be done by moving the decimal point three places to the right.
$= 52$	Divide the whole numbers.

Thus in Example 4(b) when we moved the decimal point three places to the right in the divisor and the dividend, we were actually creating an equivalent fraction where the numerator and the denominator of the original fraction were multiplied by 1000.

EXAMPLE 5   Divide.

(a) $1.7\overline{)0.0323}$ **(b)** $0.0032\overline{)7.68}$

Solution

(a)
$$
\begin{array}{r}
0.019 \\
1.7_\wedge\overline{)0.0_\wedge323} \\
\underline{17} \\
153 \\
\underline{153} \\
0
\end{array}
$$
Move the decimal point in the divisor and dividend one place to the right and mark that position with a caret.

(b)
$$
\begin{array}{r}
2400. \\
0.0032_\wedge\overline{)7.6800_\wedge} \\
\underline{6\,4} \\
1\,28 \\
\underline{1\,28} \\
000
\end{array}
$$
Note that two extra zeros are needed in the dividend as we move the decimal point four places to the right.

Student Practice 5   Divide.

(a) $1.8\overline{)0.0414}$ **(b)** $0.0036\overline{)8.316}$

EXAMPLE 6

(a) Find $2.9\overline{)431.2}$ rounded to the nearest tenth.
(b) Find $2.17\overline{)0.08}$ rounded to the nearest thousandth.

Solution

(a)
$$\begin{array}{r}14\,8.\,68 \\ 2.9_\wedge\overline{)431.2_\wedge\,00} \\ \underline{29} \\ 141 \\ \underline{116} \\ 25\,2 \\ \underline{23\,2} \\ 2\,0\;0 \\ \underline{1\,7\;\;4} \\ 2\;60 \\ \underline{2\;32} \\ 28 \end{array}$$

Calculate to the hundredths place and round the answer to the nearest tenth.

The answer rounded to the nearest tenth is 148.7.

(b)
$$\begin{array}{r}0.\,0368 \\ 2.17_\wedge\overline{)0.08_\wedge\,0000} \\ \underline{6\;51} \\ 1\;490 \\ \underline{1\;302} \\ 1880 \\ \underline{1736} \\ 144 \end{array}$$

Calculate to the ten-thousandths place and then round the answer. Rounding 0.0368 to the nearest thousandth, we obtain 0.037.

Calculator

Dividing Decimals

You can use your calculator to divide a decimal by a decimal. To find $21.38\overline{)54.53}$ rounded to the nearest hundredth, enter:

54.53 $\boxed{\div}$ 21.38 $\boxed{=}$

Display:

$$\boxed{2.5505145}$$

This is an approximation. Some calculators will round to eight digits. The answer rounded to the nearest hundredth is 2.55.

Student Practice 6

(a) Find $3.8\overline{)521.6}$ rounded to the nearest tenth.

(b) Find $8.05\overline{)0.17}$ rounded to the nearest thousandth.

EXAMPLE 7 John drove his 1997 Cavalier 420.5 miles to Chicago. He used 14.5 gallons of gas on the trip. How many miles per gallon did his car get on the trip?

Solution To find miles per gallon we need to divide the number of miles, 420.5, by the number of gallons, 14.5.

$$\begin{array}{r}2\,9. \\ 14.5_\wedge\overline{)420.5_\wedge} \\ \underline{290} \\ 130\;5 \\ \underline{130\;5} \\ 0 \end{array}$$

John's car achieved 29 miles per gallon on the trip to Chicago.

Student Practice 7 Sarah rented a large truck to move to Boston. She drove 454.4 miles yesterday. She used 28.5 gallons of gas on the trip. How many miles per gallon did the rental truck get? Round to the nearest tenth.

EXAMPLE 8 Find the value of n if $0.8 \times n = 2.68$.

Solution Here 0.8 is multiplied by some number n to obtain 2.68. What is this number n? If we divide 2.68 by 0.8, we will find the value of n.

$$\begin{array}{r}3.35 \\ 0.8_\wedge\overline{)2.6_\wedge80} \\ \underline{2\,4} \\ 2\;8 \\ \underline{2\;4} \\ 40 \\ \underline{40} \\ 0 \end{array}$$

Thus the value of n is 3.35.

Continued on next page

Check. Is this true? Are we sure the value of $n = 3.35$?
We substitute the value of $n = 3.35$ into the equation to see if it makes the statement true.

$$0.8 \times n \quad = 2.68$$

$$0.8 \times 3.35 \stackrel{?}{=} 2.68$$

$$2.68 = 2.68 \quad \checkmark \quad \text{Yes, it is true.}$$

Student Practice 8 Find the value of n if $0.12 \times n = 0.696$.

EXAMPLE 9 The level of sulfur dioxide emissions in the air has slowly been decreasing over the last several decades, as can be seen in the accompanying bar graph. Find the average amount of sulfur dioxide emissions in the air over these five specific years.

Solution

First we take the sum of the five years.

$$
\begin{array}{r}
31.2 \\
25.9 \\
20.1 \\
16.3 \\
+ \ 11.4 \\
\hline
104.9
\end{array}
$$

Then we divide by five to obtain the average.

$$
\begin{array}{r}
20.98 \\
5\overline{)104.90} \\
\underline{10} \\
4 \\
\underline{0} \\
4\ 9 \\
\underline{4\ 5} \\
40 \\
\underline{40} \\
0
\end{array}
$$

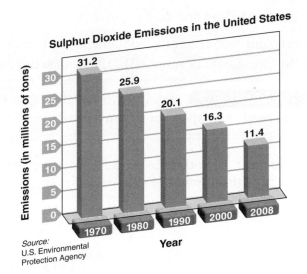

Thus the yearly average is 20.98 million tons of sulfur dioxide emissions in the United States.

Student Practice 9 Use the accompanying bar graph to find the average level of sulfur dioxide for the three years 1980, 1990, and 2000. By how much does the three-year average differ from the five-year average? Round your answers to the nearest thousandth.

3.5 Exercises

MyMathLab®

Watch the videos
in MyMathLab

Download the
MyDashBoard App

Divide until there is a remainder of zero.

1. $6\overline{)12.6}$ **2.** $8\overline{)17.28}$ **3.** $4\overline{)71.32}$ **4.** $6\overline{)83.16}$

5. $7\overline{)73.64}$ **6.** $8\overline{)168.48}$ **7.** $0.6\overline{)81.9}$ **8.** $0.5\overline{)32.15}$

9. $0.2706 \div 0.05$ **10.** $0.6092 \div 0.08$ **11.** $153.7 \div 2.9$ **12.** $76.8 \div 3.2$

13. $68.4 \div 3.8$ **14.** $728 \div 5.6$ **15.** $40.30 \div 0.31$

Divide and round your answer to the nearest tenth.

16. $8\overline{)45}$ **17.** $9\overline{)47.31}$ **18.** $1.3\overline{)4.67}$

19. $1.9\overline{)2.36}$ **20.** $0.95\overline{)32.067}$ **21.** $0.85\overline{)41.901}$

Divide and round your answer to the nearest hundredth.

22. $4\overline{)263.82}$ **23.** $5\overline{)471.03}$ **24.** $1.7\overline{)20.8}$

25. $1.8\overline{)24.41}$ **26.** $24\overline{)3.126}$ **27.** $35\overline{)7.369}$

Divide and round your answer to the nearest thousandth.

28. $8\overline{)0.2019}$ **29.** $7\overline{)0.5681}$ **30.** $0.54\overline{)12.97}$ **31.** $0.87\overline{)79.40}$

Divide and round your answer to the nearest whole number.

32. $12\overline{)1396}$ **33.** $19\overline{)2341}$ **34.** $0.0019\overline{)0.165}$ **35.** $0.0046\overline{)0.981}$

Applications, Exercises 36–45

36. *Travel in Mexico* Rhett and Liza are traveling in Mexico, where distances on the highway are given in kilometers. There are approximately 1.6 kilometers in one mile. They see a sign that reads "Mexico City: 342 km." How many miles is it to Mexico City?

37. *Computer Payments* The Miller family wants to use the latest technology to access the Internet from their home television system. The equipment needed to upgrade their existing equipment will cost $992.76. If the Millers make 12 equal monthly payments, how much will they pay per month?

38. *Lasagna Dinner* Four students sit down to their weekly lasagna dinner. At one end of the table, there is a bottle containing 67.6 ounces of a popular soft drink. At the other end of the table is a bottle that contains 33.6 ounces of water.

  (a) If the students share the soft drink and water equally, how many ounces of liquid will each student drink?

  (b) At the last minute, another student is asked to join the group. How many ounces of liquid will each of the five students share?

39. *Fuel Efficiency* Wally owns a Dodge Caliber that travels 360 miles on 13.2 gallons of gas. How many miles per gallon does it achieve? (Round your answer to the nearest tenth.)

40. ***Costs of a Ski Trip*** A church youth group went on a ski trip. The ski resort charged the group $1840 for 32 lift tickets. How much was each ticket?

41. ***Flower Sales*** Andrea makes Mother's Day bouquets each year for extra income. This year her goal is to make $300. If she sells each bouquet for $12.50, how many bouquets must she sell to reach her goal?

42. ***Outdoor Deck Payments*** Demitri had a contractor build an outdoor deck for his back porch. He now has $1131.75 to pay off, and he agreed to pay $125.75 per month. How many payments on the outdoor deck must he make?

43. ***Wedding Reception Costs*** For their wedding reception, Sharon and Richard spent $1865.50 on food and drinks. If the caterer charged them $10.25 per person, how many guests did they have?

44. ***Record Rainfall***

(a) Using the chart below, find the average amount of precipitation for the months April, May, and June.

(b) On average, how much more precipitation does Mount Waialeale get per day in April than in March? (Use 30 days in a month, and round to the nearest thousandth.)

45. ***Quality Inspection*** Yoshi is working as an inspector for a company that makes snowboards. A Mach 1 snowboard weighs 3.8 kilograms. How many of these snowboards are contained in a box in which the contents weigh 87.40 kilograms? If the box is labeled CONTENTS: 24 SNOWBOARDS, how great an error was made in packing the box?

Month	Average Amount of Precipitation in Mount Waialeale, Hawaii, for January–June
January	24.78 in.
February	24.63 in.
March	27.24 in.
April	47.75 in.
May	28.34 in.
June	30.65 in.

Source: www.wrcc.dri.edu

Find the value of n.

46. $0.5 \times n = 3.55$

47. $0.3 \times n = 9.66$

48. $1.7 \times n = 129.2$

49. $1.3 \times n = 1267.5$

50. $n \times 0.063 = 2.835$

51. $n \times 0.098 = 4.312$

To Think About *Multiply the numerator and denominator of each fraction by 10,000. Then divide the numerator by the denominator. Is the result the same if we divide the original numerator by the original denominator? Why?*

52. $\dfrac{3.8702}{0.0523}$

53. $\dfrac{2.9356}{0.0716}$

Cumulative Review

54. [2.8.1] Add. $\dfrac{3}{8} + 2\dfrac{4}{5}$

55. [2.8.2] Subtract. $2\dfrac{13}{16} - 1\dfrac{7}{8}$

56. [2.4.3] Multiply. $3\dfrac{1}{2} \times 2\dfrac{1}{6}$

57. [2.5.3] $7\dfrac{1}{2} \div \dfrac{1}{2}$

Most Damaging Hurricanes *The amount of property damage for the five most destructive hurricanes to hit the United States is represented in the following bar graph. Use the bar graph to answer exercises 58–61.*

58. [3.3.2] How much more property damage occurred during Hurricane Andrew than Hurricane Hugo?

59. [3.3.2] How much more property damage occurred during Hurricane Hugo than Hurricane Agnes?

60. [3.5.2] How many times more property damage occurred during Hurricane Katrina than Hurricane Andrew?

61. [3.5.2] How many times more property damage occurred during Hurricane Katrina than Hurricane Agnes?

Quick Quiz 3.5

1. Divide. $0.07\overline{)0.04606}$

2. Divide. $0.52\overline{)1.69416}$

3. Divide and round to the nearest hundredth. $8\overline{)52.643}$

4. Concept Check Explain how you would know where to place the decimal point in the answer if you divide $0.173 \div 0.578$.

3.6 Converting Fractions to Decimals and the Order of Operations

Student Learning Objectives

After studying this section, you will be able to:

1. Convert a fraction to a decimal.
2. Use the order of operations with decimals.

① Converting a Fraction to a Decimal

A number can be expressed in two equivalent forms: as a fraction or as a decimal.

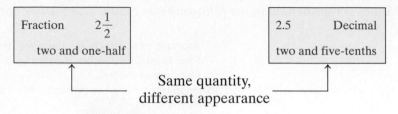

Same quantity, different appearance

Every decimal in this chapter can be expressed as an equivalent fraction or mixed number. For example,

Decimal form $\Rightarrow$ fraction form

$$0.75 = \frac{75}{100} \quad \text{or} \quad \frac{3}{4}$$

$$0.5 = \frac{5}{10} \quad \text{or} \quad \frac{1}{2}$$

$$2.5 = 2\frac{5}{10} = 2\frac{1}{2} \quad \text{or} \quad \frac{5}{2}.$$

And every fraction can be expressed as an equivalent decimal, as we will learn in this section. For example,

Fraction form $\Rightarrow$ decimal form

$$\frac{1}{5} = 0.20 \quad \text{or} \quad 0.2$$

$$\frac{3}{8} = 0.375$$

$$\frac{5}{11} = 0.4545\ldots \ . \ (\text{The "45" keeps repeating.})$$

Some of these decimal equivalents are so common that people find it helpful to memorize them. You would be wise to memorize the following equivalents:

$$\frac{1}{2} = 0.5 \qquad \frac{1}{4} = 0.25 \qquad \frac{1}{5} = 0.2 \qquad \frac{1}{10} = 0.1.$$

We previously studied how to convert some fractions with a denominator of 10, 100, 1000, and so on to decimal form. For example, $\frac{3}{10} = 0.3$ and $\frac{7}{100} = 0.07$. We need to develop a procedure to write other fractions, such as $\frac{3}{8}$ and $\frac{5}{16}$, in decimal form.

> **CONVERTING A FRACTION TO AN EQUIVALENT DECIMAL**
>
> Divide the denominator into the numerator until
>
> **(a)** the remainder becomes zero, or
>
> **(b)** the remainder repeats itself, or
>
> **(c)** the desired number of decimal places is achieved.

EXAMPLE 1   Write as an equivalent decimal.

(a) $\dfrac{3}{8}$

(b) $\dfrac{31}{40}$ of a second

Divide the denominator into the numerator until the remainder becomes zero.

Solution

(a)
$$\begin{array}{r} 0.375 \\ 8\overline{)3.000} \\ \underline{2\,4} \\ 60 \\ \underline{56} \\ 40 \\ \underline{40} \\ 0 \end{array}$$

(b)
$$\begin{array}{r} 0.775 \\ 40\overline{)31.000} \\ \underline{28\,0} \\ 3\,00 \\ \underline{2\,80} \\ 200 \\ \underline{200} \\ 0 \end{array}$$

Therefore, $\dfrac{3}{8} = 0.375$.

Therefore, $\dfrac{31}{40} = 0.775$ of a second.

Student Practice 1 Write as an equivalent decimal.

(a) $\dfrac{5}{16}$

(b) $\dfrac{11}{80}$

NOTE TO STUDENT: Fully worked-out solutions to all of the Student Practice problems can be found at the back of the text starting at page SP-1.

Athletes' times in Olympic events, such as the 100-meter dash, are measured to the nearest hundredth of a second. Future Olympic athletes' times will be measured to the nearest thousandth of a second.

Decimals such as 0.375 and 0.775 are called **terminating decimals.** When converting $\frac{3}{8}$ to 0.375 or $\frac{31}{40}$ to 0.775, the division operation eventually yields a remainder of zero. Other fractions yield a repeating pattern. For example, $\frac{1}{3} = 0.3333\ldots$ and $\frac{2}{3} = 0.6666\ldots$ have a pattern of repeating digits. Decimals that have a digit or a group of digits that repeats are called **repeating decimals.** We often indicate the repeating pattern with a bar over the repeating group of digits:

$$0.\,3333\ldots = 0.\overline{3} \qquad 0.\,74\ 74\ 74\ldots = 0.\overline{74}$$
$$0.\,218\ 218\ 218\ldots = 0.\overline{218} \qquad 0.\,8942\ 8942\ldots = 0.\overline{8942}$$

If when converting fractions to decimal form the remainder repeats itself, we know that we have a repeating decimal.

EXAMPLE 2   Write as an equivalent decimal.

(a) $\dfrac{5}{11}$

(b) $\dfrac{13}{22}$

(c) $\dfrac{5}{37}$

Continued on next page

Solution

(a) $11\overline{)5.0000}$ → 0.4545

$$\begin{array}{r} 44 \\ \hline 60 \\ 55 \\ \hline 50 \\ 44 \\ \hline 60 \\ 55 \\ \hline 5 \end{array}$$

repeating remainders

Thus $\dfrac{5}{11} = 0.4545\ldots = 0.\overline{45}$.

(b) $22\overline{)13.00000}$ → 0.59090

$$\begin{array}{r} 11\,0 \\ \hline 2\,00 \\ 1\,98 \\ \hline 2\,00 \\ 1\,98 \\ \hline 20 \end{array}$$

repeating remainders

Thus $\dfrac{13}{22} = 0.5909090\ldots = 0.5\overline{90}$.

Notice that the bar is over the digits 9 and 0 but *not* over the digit 5.

(c) $37\overline{)5.0000}$ → 0.1351

$$\begin{array}{r} 37 \\ \hline 130 \\ 111 \\ \hline 190 \\ 185 \\ \hline 50 \\ 37 \\ \hline 13 \end{array}$$

repeating remainders

Thus $\dfrac{5}{37} = 0.135135\ldots = 0.\overline{135}$.

Student Practice 2   Write as an equivalent decimal.

(a) $\dfrac{7}{11}$

(b) $\dfrac{8}{15}$

(c) $\dfrac{13}{44}$

Calculator

 Fraction to Decimal

You can use a calculator to change $\dfrac{5}{8}$ to a decimal.

Enter:

$5 \;\boxed{\div}\; 8 \;\boxed{=}$

The display should read

$\boxed{0.625}$

Try the following.

(a) $\dfrac{17}{25}$ (b) $\dfrac{2}{9}$

(c) $\dfrac{13}{10}$ (d) $\dfrac{15}{19}$

Note: 0.78947368 is an approximation for $\dfrac{15}{19}$. Some calculators round to only eight places.

EXAMPLE 3   Write as an equivalent decimal.

(a) $3\dfrac{7}{15}$

(b) $\dfrac{20}{11}$

Solution

(a) $3\dfrac{7}{15}$ means $3 + \dfrac{7}{15}$

$$15\overline{)7.000} \;\to\; 0.466$$
$$\begin{array}{r} 60 \\ \hline 100 \\ 90 \\ \hline 100 \\ 90 \\ \hline 10 \end{array}$$

Thus $\dfrac{7}{15} = 0.4\overline{6}$ and $3\dfrac{7}{15} = 3.4\overline{6}$.

(b) $11\overline{)20.000} \;\to\; 1.818$
$$\begin{array}{r} 11 \\ \hline 9\,0 \\ 8\,8 \\ \hline 20 \\ 11 \\ \hline 90 \\ 88 \\ \hline 2 \end{array}$$

Thus $\dfrac{20}{11} = 1.818181\ldots = 1.\overline{81}$.

Student Practice 3   Write as an equivalent decimal.

(a) $2\dfrac{11}{18}$

(b) $\dfrac{28}{27}$

In some cases, the pattern of repeating is quite long. For example,

$$\frac{1}{7} = 0.142857142857\ldots = 0.\overline{142857}$$

Such problems are often rounded to a certain value.

EXAMPLE 4 Express $\frac{5}{7}$ as a decimal rounded to the nearest thousandth.

Solution

$$
\begin{array}{r}
0.7142 \\
7)\overline{5.0000} \\
\underline{4\,9} \\
10 \\
\underline{7} \\
30 \\
\underline{28} \\
20 \\
\underline{14} \\
6
\end{array}
$$

Rounding to the nearest thousandth, we round 0.7142 to 0.714. (In repeating form, $\frac{5}{7} = 0.714285714285\ldots = 0.\overline{714285}$.)

Student Practice 4 Express $\frac{19}{24}$ as a decimal rounded to the nearest thousandth.

Recall that we studied placing two decimals in order in Section 3.2. If we are required to place a fraction and a decimal in order, it is usually easiest to change the fraction to decimal form and then compare the two decimals.

EXAMPLE 5 Fill in the blank with one of the symbols $<$, $=$, or $>$.

Solution

$$\frac{7}{16} \underline{\hspace{1cm}} 0.43$$

Now we divide to find the decimal equivalent of $\frac{7}{16}$.

$$
\begin{array}{r}
0.4375 \\
16)\overline{7.0000} \\
\underline{64} \\
60 \\
\underline{48} \\
120 \\
\underline{112} \\
80 \\
\underline{80} \\
0
\end{array}
$$

Continued on next page

Now in the thousandths place $7 > 0$, so we know

$$0.43\,7\,5 > 0.43\,0\,0.$$

Therefore, $\dfrac{7}{16} > 0.43$.

Student Practice 5 Fill in the blank with one of the symbols $<, =,$ or $>$.

$$\dfrac{5}{8} \underline{\hspace{1cm}} 0.63$$

② Using the Order of Operations with Decimals

The rules for order of operations that we discussed in Section 1.6 and Section 2.8 apply to operations with decimals.

ORDER OF OPERATIONS

Do first   1. Perform operations inside parentheses.

                2. Simplify any expressions with exponents.

                3. Multiply or divide from left to right.

Do last   4. Add or subtract from left to right.

Sometimes exponents are used with decimals. In such cases, we merely evaluate using repeated multiplication.

$$(0.2)^2 = 0.2 \times 0.2 = 0.04$$
$$(0.2)^3 = 0.2 \times 0.2 \times 0.2 = 0.008$$
$$(0.2)^4 = 0.2 \times 0.2 \times 0.2 \times 0.2 = 0.0016$$

EXAMPLE 6 Evaluate. $(0.3)^3 + 0.6 \times 0.2 + 0.013$

Solution First we need to evaluate $(0.3)^3 = 0.3 \times 0.3 \times 0.3 = 0.027$. Thus

$(0.3)^3 + 0.6 \times 0.2 + 0.013$

$= 0.027 + 0.6 \times 0.2 + 0.013$

$= 0.027 + 0.12 + 0.013 \longleftarrow$ When addends have a different number of decimal places, writing the problem in column form makes adding easier.

$$\begin{array}{r} 0.027 \\ 0.120 \\ + \ 0.013 \\ \hline 0.160 \end{array}$$

$= 0.16$

Student Practice 6 Evaluate. $0.3 \times 0.5 + (0.4)^3 - 0.036$

In the next example all four steps of the rules for order of operations will be used.

 EXAMPLE 7 Evaluate. $(8 - 0.12) \div 2^3 + 5.68 \times 0.1$

Student Practice 7 Evaluate.
$6.56 \div (2 - 0.36) + (8.5 - 8.3)^2$

Solution

$(8 - 0.12) \div 2^3 + 5.68 \times 0.1$

$= 7.88 \div 2^3 + 5.68 \times 0.1$ First do subtraction inside the parentheses.

$= 7.88 \div 8 + 5.68 \times 0.1$ Simplify the expression with an exponent.

$= 0.985 + 0.568$ From left to right do division and multiplication.

$= 1.553$ Add the final two numbers.

Take the time to review these seven Examples and seven Student Practice problems. This is an important skill to master. Some careful review will help you to work the homework exercises much more quickly and accurately.

👣 STEPS TO SUCCESS What Is the Best Way to Review Before a Test?

Here is what students have found.

1. Read over your textbook again. Make a list of any terms, rules, or formulas you need to know for the exam. Make sure you understand them all.

2. Go over your notes. Look back at your homework and quizzes. Redo the problems you missed. Make sure you can get the right answer.

3. Practice some of each type of problem covered in the chapter(s) you are to be tested on.

4. At the end of the chapter are special sections to help you review. Be sure to do the Chapter Review Problems. Study each part of the Chapter Organizer and do the You Try It problems.

5. When you think you are ready, take the How Am I Doing? Chapter Test. Make sure you study the Math Coach notes right after the test.

6. Get help for those concepts that are giving you difficulty. Don't be afraid to ask for help. Teachers, tutors, friends in class, and other friends are ready to help you. Don't wait. Get help now.

Making it personal: Which of these six steps do you most need to follow? Take some time today to start doing these things. These methods of review have helped thousands of students! They can help you—NOW! ▼

3.6 Exercises

MyMathLab®

Watch the videos in MyMathLab

Download the MyDashBoard App

Verbal and Writing Skills, Exercises 1–4

1. 0.75 and $\frac{3}{4}$ are different ways to express the _____.

2. To convert a fraction to an equivalent decimal, divide the _____ into the numerator.

3. Why is $0.\overline{8942}$ called a repeating decimal?

4. The order of operations for decimals is the same as the order of operations for whole numbers. Write the steps for the order of operations.

Write as an equivalent decimal. If a repeating decimal is obtained, use notation such as $0.\overline{7}$, $0.\overline{16}$, or $0.\overline{245}$.

5. $\frac{1}{4}$

6. $\frac{3}{4}$

7. $\frac{4}{5}$

8. $\frac{2}{5}$

9. $\frac{1}{8}$

10. $\frac{3}{8}$

11. $\frac{7}{20}$

12. $\frac{3}{40}$

13. $\frac{31}{50}$

14. $\frac{23}{25}$

15. $\frac{9}{4}$

16. $\frac{14}{5}$

17. $2\frac{7}{8}$

18. $3\frac{13}{16}$

19. $5\frac{3}{16}$

20. $2\frac{5}{12}$

21. $\frac{2}{3}$

22. $\frac{5}{6}$

23. $\frac{5}{11}$

24. $\frac{7}{11}$

25. $3\frac{7}{12}$

26. $7\frac{1}{3}$

27. $4\frac{2}{9}$

28. $8\frac{7}{9}$

Write as an equivalent decimal or a decimal approximation. Round your answer to the nearest thousandth if needed.

29. $\frac{4}{13}$

30. $\frac{8}{17}$

31. $\frac{19}{21}$

32. $\frac{20}{21}$

33. $\frac{7}{48}$

34. $\frac{5}{48}$

35. $\frac{57}{28}$

36. $\frac{15}{7}$

37. $\frac{21}{52}$

38. $\frac{3}{38}$

39. $\frac{17}{18}$

40. $\frac{7}{13}$

41. $\frac{22}{7}$

42. $\frac{17}{14}$

43. $3\frac{9}{19}$

44. $5\frac{3}{17}$

Fill in the blank with one of the symbols $<$, $=$, or $>$.

45. $\frac{7}{8}$ ___ 0.88

46. $\frac{10}{11}$ ___ 0.9

47. 0.07 ___ $\frac{1}{16}$

48. 0.08 ___ $\frac{1}{12}$

240

Applications, Exercises 49–54

49. *Carpentry* A carpenter is using several bolts that measure $\frac{5}{16}$ of an inch while building a bookcase. Write this measurement as a decimal.

50. *Car Repair* A machinist needs to use a drill that measures $\frac{7}{16}$ of an inch while repairing a car engine. Write this measurement as a decimal.

51. *U.S. Women's Shoe Sizes* A size 7 women's shoe measures 9.31 inches and a size $7\frac{1}{2}$ measures $9\frac{1}{2}$ inches. What is the difference in length between a size 7 and a size $7\frac{1}{2}$ shoe?

52. *U.S. Men's Shoe Sizes* A size $9\frac{1}{2}$ men's shoe measures $10\frac{1}{2}$ inches and a size 10 measures 10.69 inches. What is the difference in length between a size $9\frac{1}{2}$ and a size 10 shoe?

53. *Safety Regulations* Federal safety regulations specify that the slots between the bars on a baby's crib must not be more than $2\frac{3}{8}$ inches. One crib's slots measured 2.4 inches apart. Is this too wide? If so, by how much?

54. *Manufacturing* To manufacture a circuit board, Rick must program a computer to place a piece of thin plastic atop a circuit board. For the current to flow through the circuit, the top plastic piece must form a border of exactly $\frac{1}{16}$ inch with the circuit board. A few circuit boards were made with a border of 0.055 inch by accident. Is this border too small or too large? By how much?

Evaluate.

55. $2.4 + (0.5)^2 - 0.35$

56. $8.5 + 2.4 - (0.3)^2$

57. $2.3 \times 3.2 - 5 \times 0.8$

58. $1.5 \times 8 + 12.4 \div 4$

59. $12 \div 0.03 - 50 \times (0.5 + 1.5)^3$

60. $61.95 \div 1.05 - 2 \times (1.7 + 1.3)^3$

61. $(1.1)^3 + 2.6 \div 0.13 + 0.083$

62. $(1.1)^3 + 8.6 \div 2.15 - 0.086$

63. $(14.73 - 14.61)^2 \div (1.18 + 0.82)$

64. $(32.16 - 32.02)^2 \div (2.24 + 1.76)$

65. $(0.5)^3 + (3 - 2.6) \times 0.5$

66. $(0.6)^3 + (7 - 6.3) \times 0.07$

67. $(0.76 + 4.24) \div 0.25 + 8.6$

68. $(2.4)^2 + 3.6 \div (1.2 - 0.7)$

Evaluate.

69. $(1.6)^3 + (2.4)^2 + 18.666 \div 3.05 + 4.86$

70. $5.9 \times 3.6 \times 2.4 - 0.1 \times 0.2 \times 0.3 \times 0.4$

Write as a decimal. Round your answer to six decimal places.

71. $\dfrac{5236}{8921}$

72. $\dfrac{17,359}{19,826}$

To Think About

73. Subtract. $0.\overline{16} - 0.00\overline{16}$

    (a) What do you obtain?

    (b) Now subtract $0.\overline{16} - 0.01\overline{6}$. What do you obtain?

    (c) What is different about these results?

74. Subtract. $1.\overline{89} - 0.01\overline{89}$

    (a) What do you obtain?

    (b) Now subtract $1.\overline{89} - 0.18\overline{9}$. What do you obtain?

    (c) What is different about these results?

Cumulative Review

75. **[2.9.1]** *Boating Dock* John and Nancy put in a new dock at the end of Tobey Lane. A pipe at the end of the dock supports the dock and is driven deep into the mud and sand at the bottom of Eel Pond. The pipe is 25 feet long. Half of the pipe is above the surface of the water at low tide. The pipe is driven $6\frac{3}{4}$ feet deep into the mud and sand. How deep is the water at the end of the dock at low tide?

76. **[2.9.1]** *Tidal Fluctuation* Fisherman's Wharf in Digby, Nova Scotia, has an average tidal range of $25\frac{4}{5}$ feet. These huge tidal ranges require considerable ingenuity in the design of docks and ramps for boats. If the water is $6\frac{1}{2}$ feet deep at low tide at the end of Fisherman's Wharf during an average low tide, how deep is the water at the same location during an average high tide? (*Source:* Nova Scotia Board of Tourism)

Quick Quiz 3.6

1. Write as an equivalent decimal. $3\dfrac{9}{16}$

2. Write as an equivalent decimal. Round your answer to the nearest hundredth. $\dfrac{5}{17}$

3. Perform the operations in the proper order.
$(0.7)^2 + 1.92 \div 0.3 - 0.79$

4. **Concept Check** Explain how you would perform the operations in the calculation
$45.78 - (3.42 - 2.09)^2 \times 0.4$.

3.7 Estimating and Solving Applied Problems Involving Decimals

1 Estimating Sums, Differences, Products, and Quotients of Decimals

When we encounter real-life applied problems, it is important to know if an answer is reasonable. A car may get 21.8 miles per gallon. However, a car will not get 218 miles per gallon. Neither will a car get 2.18 miles per gallon. To avoid making an error in solving applied problems, it is wise to make an estimate. The most useful time to make an estimate is at the end of solving the problem, in order to see if the answer is reasonable.

There are several different rules for estimating. Not all mathematicians agree on the best method in each case. Most students find that a quick and simple method is to round each number so that there is one nonzero digit. Then perform the calculation. We will use that approach in this section of the book. However, you should be aware that there are other valid approaches. Your instructor may wish you to use another method.

Student Learning Objectives

After studying this section, you will be able to:

① Estimate sums, differences, products, and quotients of decimals.

② Solve applied problems using operations with decimals.

EXAMPLE 1 Estimate.

(a) $184{,}987.09 + 676{,}393.95$

(b) $0.00782 - 0.00358$

(c) 145.87×78.323

(d) $138.85 \div 5.887$

Solution In each case we will round to one nonzero digit to estimate.

(a) $184{,}987.09 + 676{,}393.95 \approx 200{,}000 + 700{,}000 = 900{,}000$

(b) $0.00782 - 0.00358 \approx 0.008 - 0.004 = 0.004$

(c) $145.87 \times 78.323 \approx$

$$
\begin{array}{r}
100 \\
\times\ \ 80 \\
\hline
8000
\end{array}
$$

Thus $145.87 \times 78.323 \approx 8000$

(d) $138.85 \div 5.887 \approx 6\overline{)100} = 16\dfrac{4}{6} \approx 17$

$$
\begin{array}{r}
16\ \ \ \ \ \\
6\overline{)100} \\
\underline{6}\ \ \ \ \\
40 \\
\underline{36} \\
4
\end{array}
$$

Thus $138.85 \div 5.887 \approx 17$ (Rounding to the nearest whole number.)

Student Practice 1 Round to one nonzero digit. Then estimate the result of the indicated calculation.

(a) $385.98 + 875.34$

(b) $0.0932 - 0.0579$

(c) 5876.34×0.087

(d) $46{,}873 \div 8.456$

NOTE TO STUDENT: Fully worked-out solutions to all of the Student Practice problems can be found at the back of the text starting at page SP-1.

Take a few minutes to review Example 1. Be sure you can perform these estimation steps. We will use this type of estimation to check our work in the applied problems in this section.

② Solving Applied Problems Using Operations with Decimals

We use the basic plan of solving applied problems that we discussed in Section 1.8 and Section 2.9. Let us review how we analyze applied-problem situations.

1. *Understand the problem.*
2. *Solve and state the answer.*
3. *Check.*

In the United States for almost all jobs where you are paid an hourly wage, if you work more than 40 hours in one week, you should be paid overtime. The overtime rate is 1.5 times the normal hourly rate for the extra hours worked in that week. The next problem deals with overtime wages.

EXAMPLE 2 A laborer is paid $10.38 per hour for a 40-hour week and 1.5 times that wage for any hours worked beyond the standard 40. If he works 47 hours in a week, what will he earn?

Solution

1. *Understand the problem.*

Mathematics Blueprint for Problem Solving

Gather the Facts	What Am I Asked to Do?	How Do I Proceed?	Key Points to Remember
He works 47 hours. He gets paid $10.38 per hour for 40 hours. He gets paid 1.5 × $10.38 per hour for 7 hours.	Find the earnings of the laborer if he works 47 hours in one week.	Add the earnings of 40 hours at $10.38 per hour to the earnings of 7 hours at overtime pay.	Multiply 1.5 × $10.38 to find the pay he earns for overtime.

2. *Solve and state the answer.*

We compute his regular pay and his overtime pay and add the results.

$$\text{Regular pay} + \text{Overtime pay} = \text{Total pay}$$

Regular pay: Calculate his pay for 40 hours of work.

$$
\begin{array}{r}
10.38 \\
\times\ \ \ 40 \\
\hline
415.20
\end{array}
$$
He earns $415.20 at $10.38 per hour.

Overtime pay: Calculate his overtime pay rate. This is 10.38×1.5.

$$
\begin{array}{r}
10.38 \\
\times\ \ 1.5 \\
\hline
5\,190 \\
10\,38\ \ \\
\hline
15.570
\end{array}
$$
He earns $15.57 per hour for overtime.

Calculate how much he earned doing 7 hours of overtime work.

$$
\begin{array}{r}
15.57 \\
\times\ \ \ \ \ 7 \\
\hline
108.99
\end{array}
$$
For 7 overtime hours he earns $108.99.

Total pay: Add the two amounts.

$415.20   Regular 40-hour-week earnings
+   108.99   Overtime earnings
$524.19   Total earnings

The total earnings of the laborer for a 47-hour workweek will be $524.19.

3. Check. Estimate his regular pay.

$$40 \times \$10 = \$400$$

Estimate his overtime rate of pay, and then his overtime pay.

$$2 \times \$10 = \$20$$
$$7 \times \$20 = \$140$$

Then add.

$$\$400$$
$$+ \quad 140$$
$$\$540$$

Our estimate of $540 is close to our answer of $524.19. Our answer is reasonable. ✓

Student Practice 2   Melinda works for the phone company as a line repair technician. She earns $12.36 per hour. She worked 51 hours last week. If she gets time and a half (1.5 times the hourly wage) for all hours worked above 40 hours per week, how much did she earn last week?

EXAMPLE 3   A chemist is testing 36.85 liters of cleaning fluid. She wishes to pour it into several smaller containers that each hold 0.67 liter of fluid. (a) How many containers will she need? (b) If each liter of this fluid costs $3.50, how much does the cleaning fluid in one container cost? (Round your answer to the nearest cent.)

Mathematics Blueprint for Problem Solving

Gather the Facts	What Am I Asked to Do?	How Do I Proceed?	Key Points to Remember
The total amount of cleaning fluid is 36.85 liters. Each small container holds 0.67 liter. Each liter of fluid costs $3.50.	**(a)** Find out how many containers the chemist needs. **(b)** Find the cost of cleaning fluid in each small container.	**(a)** Divide the total, 36.85 liters, by the amount in each small container, 0.67 liter, to find the number of containers. **(b)** Multiply the cost of one liter, $3.50, by the amount of liters in one container, 0.67.	If you are not clear as to what to do at any stage of the problem, then do a similar, simpler problem.

Solution

(a) How many containers will the chemist need?

She has 36.85 liters of cleaning fluid and she wants to put it into several equal-sized containers each holding 0.67 liter. Suppose we are not sure what to do. Let's do a similar, simpler problem. If we had 40 liters of

Continued on next page

cleaning fluid and we wanted to put it into little containers each holding 2 liters, what would we do? Since the little containers would only hold 2 liters, we would need 20 containers. We know that $40 \div 2 = 20$. So we see that, in general, we divide the total number of liters by the amount the small container holds. Thus $36.85 \div 0.67$ will give us the number of containers in this case.

$$
\begin{array}{r}
55. \\
0.67_\wedge)\overline{36.85_\wedge} \\
\underline{33\ 5} \\
3\ 35 \\
\underline{3\ 35} \\
\end{array}
$$

The chemist will need 55 containers to hold this amount of cleaning fluid.

(b) How much does the cleaning fluid in each container cost? Each container will hold only 0.67 liter. If one liter costs $3.50, then to find the cost of one container we multiply $0.67 \times \$3.50$.

$$
\begin{array}{r}
3.50 \\
\times\ 0.67 \\
\hline
2450 \\
\underline{2100} \\
2.3450 \\
\end{array}
$$

We round our answer to the nearest cent. Thus each container would cost $2.35.

Check.

(a) Is it really true that 55 containers each holding 0.67 liter will hold a total of 36.85 liters? To check, we multiply.

$$
\begin{array}{r}
55 \\
\times\ 0.67 \\
\hline
385 \\
\underline{330} \\
36.85\ \checkmark \\
\end{array}
$$

(b) One liter of cleaning fluid costs $3.50. We would expect the cost of 0.67 liter to be less than $3.50. $2.35 is less than $3.50. ✓
    We use estimation to check more closely.

$$
\begin{array}{ccc}
\$3.50 & \longrightarrow & \$4.00 \\
\times\quad 0.67 & \longrightarrow & \times\quad 0.7 \\
\hline
 & & \$2.800 \\
\end{array}
$$

$2.80 is fairly close to $2.35. Our answer is reasonable. ✓

Student Practice 3 A butcher divides 17.4 pounds of prime steak into small, equal-sized packages. Each package contains 1.45 pounds of prime steak. **(a)** How many packages of steak will he have? **(b)** Prime steak sells for $7.60 per pound. How much will each package of prime steak cost?

In exercises 1–10, first round each number to one nonzero digit. Then perform the calculation using the rounded numbers to obtain an estimate.

1. 238,598,980 + 487,903,870

2. 5,927,000 + 9,983,000

3. 56,789.345 − 33,875.125

4. 6949.45 − 1432.88

5. 12,638 × 0.7892

6. 47,225 × 0.463

7. 879.654 ÷ 56.82

8. 34.5684 ÷ 0.55

9. *Car Sales* Last year the sales of Honda Accords at Hopkins Honda totaled $11,760,770. If this represented a purchase of 483 Accords, estimate the average price per car.

10. *Boat Sales* Last year the sales of boats in Massachusetts totaled $865,987,273.45. If this represented a purchase of 55,872 boats, estimate the average price per boat.

Applications *Estimate an answer to each of the following by rounding each number first, then perform the actual calculation.*

11. *Currency Conversion* Kristy is taking a trip to Denmark. Before she leaves, she checks the newspaper and finds that every U.S. dollar is equal to 5.68 kroner (Danish currency). If Kristy takes $525 on her trip, how many kroner will she receive when she does the exchange?

▲ **12.** *Football Field Dimensions* The dimensions of a professional football field, including the end zones, are about 48.8 meters wide by 109.7 meters long. What is the area of a professional football field?

▲ **13.** *Geometry* Juan and Gloria are having their roof reshingled and need to determine its area in square feet. The dimensions of the roof are 48.3 feet by 56.9 feet. What is the area of the roof in square feet?

14. *Baby Formula* A large can of infant formula contains 808 grams of powder. To prepare a bottle, 35.2 grams are needed. How many bottles can be prepared from the can? Round to the nearest whole number.

15. *Cooking* Hans is making gourmet chocolate in Switzerland. He has 11.52 liters of liquid white chocolate that will be poured into molds that hold 0.12 liter each. How many individual molds can Hans make with his 11.52 liters of liquid white chocolate?

16. *Food Purchase* David bought MacIntosh apples and Anjou pears at the grocery store for a fruit salad. At the checkout counter, the apples weighed 2.7 pounds and the pears weighed 1.8 pounds. If the apples cost $1.29 per pound and the pears cost $1.49 per pound, how much did David spend on fruit? (Round your answer to the nearest cent.)

17. *Hawaii Rainfall* One year in Mount Waialeale, Hawaii, considered the "rainiest place in the world," the yearly rainfall totaled 11.68 meters. The next year, the yearly rainfall on this mountain totaled 10.42 meters. The third year it was 12.67 meters. On average, how much rain falls on Mount Waialeale, Hawaii, per year?

18. *Auto Travel* Emma and Jennie took a trip in their Ford Taurus from Saskatoon, Saskatchewan, to Calgary, Alberta, in Canada to check out the glacier lakes. When they left, their odometer read 54,089. When they returned home, the odometer read 55,401. They used 65.6 gallons of gas. How many miles per gallon did they get on the trip?

19. *Food Portions* A jumbo bag of potato chips contains 18 ounces of chips. The recommended serving is 0.75 ounce. How many servings are in the jumbo bag?

20. *Telephone Costs* Sylvia's telephone company offers a special rate of $0.23 per minute on calls made to the Philippines during certain parts of the day. If Sylvia makes a 28.5-minute call to the Philippines at this special rate, how much will it cost?

21. *Consumer Mathematics* The local Police Athletic League raised money to renovate the local youth hall and turn it into a coffeehouse/activity center so that there is a safe place to hang out. The room that holds the Ping-Pong table needs 43.9 square yards of new carpeting. The entryway needs 11.3 square yards, and the stage/seating area needs 63.4 square yards. The carpeting will cost $10.65 per square yard. What will be the total bill for carpeting these three areas of the coffeehouse?

22. *Painting Costs* Kevin has a job as a house painter. One family needs its kitchen, family room, and hallway painted. The respective amounts needed are 2.7 gallons, 3.3 gallons, and 1.8 gallons. If paint costs $7.40 per gallon, how much will Kevin need to spend on paint to do the job?

23. *Overtime Pay* Lucy earns $8.50 per hour at the neighborhood café. She earns time and a half for each hour she works on a holiday. Lucy worked eight hours each day for six days, then worked eight hours on New Year's Day. How much did she earn for that week?

24. *Electrician's Pay* An electrician is paid $18.30 per hour for a 40-hour week. She is paid time and a half for every hour more than 40 hours worked in the same week. If she works 48 hours in one week, what will she earn for that week?

25. *Rainforest Loss* In 2005, Brazil had 3.413 million square kilometers of rainforest. Each year, approximately 0.011 million square kilometers are lost to deforestation and development. By 2015, how many square kilometers of rainforest will remain in Brazil? (*Source:* www.mongabay.com)

26. *Consumer Mathematics* At the beginning of each month, Raul withdraws $100 for small daily purchases. This month he spent $18.50 on bus fares, $42.75 on coffee and snacks, and $21.25 on news magazines. How much did Raul have left at the end of the month?

27. *Car Payments* Charlie borrowed $11,500 to purchase a new car. His loan requires him to pay $288.65 each month over the next 60 months (five years). How much will he pay over the five years? How much more will he pay back than the amount of the loan?

28. *House Payments* Mel and Sally borrowed $140,000 to buy their new home. They make monthly payments to the bank of $764.35 to repay the loan. They will be making these payments for the next 30 years. How much money will they pay to the bank in the next 30 years? How much more will they pay back than they borrowed?

29. *Drinking Water Safety* The EPA standard for safe drinking water is a maximum of 1.3 milligrams of copper per liter of water. A study was conducted on a sample of 7 liters of water drawn from Jeff Slater's house. The analysis revealed 8.06 milligrams of copper in the sample. Is the water safe or not? By how much?

30. *Drinking Water Safety* The EPA standard for safe drinking water is a maximum of 0.015 milligram of lead per liter of water. A study was conducted on 6 liters of water from West Towers Dormitory. The analysis revealed 0.0795 milligram of lead in the sample. Is the water safe or not? By how much?

31. *Jet Travel* A jet fuel tank containing 17,316.8 gallons is being emptied at the rate of 126.4 gallons per minute. How many minutes will it take to empty the tank?

32. *Monopoly Game* In a New Jersey mall, the average price of a Parker Brothers Monopoly game is $11.50. The Alfred Dunhill Company made a special commemorative set for $25,000,000.00. Instead of plastic houses and hotels, you can buy and trade gold houses and silver hotels! How many regular Monopoly games could you purchase for the price of one special commemorative set?

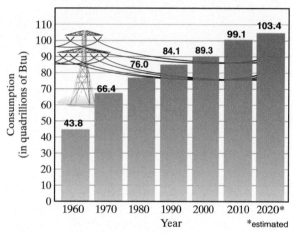

Total Yearly U.S. Consumption of Energy

Source: U.S. Department of Energy

Energy Consumption *Use the bar graph above to answer exercises 33–36.*

33. How many more Btu were consumed in the United States during 2000 than in 1970?

34. What was the greatest increase in consumption of energy in a 10-year period? When did it occur?

35. What was the average consumption of energy per year in the United States for the years 1960, 1970, and 1980? Write your answer in quadrillion Btu and then write your answer in Btu. (Remember that a quadrillion is 1000 trillion.)

36. What will be the average consumption of energy per year in the United States for the years 2000, 2010, and 2020? Write your answer in quadrillion Btu and then write your answer in Btu. (Remember that a quadrillion is 1000 trillion.)

Cumulative Review

37. [2.3.1] Write as an improper fraction. $7\frac{5}{6}$

38. [2.3.2] Write as a mixed number. $\frac{37}{5}$

39. [2.4.1] Calculate. $\frac{7}{25} \times \frac{15}{42}$

40. [2.8.3] Calculate. $\frac{4}{7} + \frac{1}{2} \times \frac{2}{3}$

Quick Quiz 3.7

1. The rainfall for Springfield last year was 1.23 inches in March, 2.58 inches in April, and 3.67 inches in May. Normally that city gets 8.5 inches during those three months. How much less rain was received during those three months compared to the normal rainfall amount?

2. Melissa and Phil started on a trip to the mountains with their Honda CR-V. Their odometer read 87,569.2 miles at the start of the trip and 87,929.2 miles at the end of the trip. They used 15.5 gallons of gas on the trip. How many miles per gallon did they achieve with their car? (Round to the nearest tenth.)

3. Chris Smith is making car payments of $275.50 for the next 36 months to pay off a car loan for a new Saturn. He borrowed $8000 from a bank to purchase the car. How much will he make in car payments over the next three years? How much more will he pay back than the original amount of the loan?

4. **Concept Check** Explain how you would solve the following problem. The Classic Chocolate Company has 24.7 pounds of chocolate. They wish to place it in individual boxes that each hold 1.3 pounds of chocolate. How many boxes will they need?

 STEPS TO SUCCESS Why Do We Have to Learn to Solve Word Problems?

Many students ask that question. It is a fair question and it deserves an honest answer.

Applications or word problems are the very life of mathematics! They are the reason for doing mathematics. They teach you how to put into use the mathematical skills you have developed.

Almost all branches of mathematics are studied because they solve problems in real life. You will find in this textbook that studying the Use Math to Save Money problems will help you manage your finances. Many decisions in daily life can be made more easily if mathematics is used to evaluate the possible options.

How can I learn to solve word problems more easily? Many students ask that question also. The key to success is practice. Make yourself do as many problems as you can. You may not be able to do them all correctly at first, but keep trying. If you cannot solve a problem, try another one. Ask for help from your teacher or someone in the tutoring lab. Ask other classmates how they solved the problem. Soon you will see great progress in your own problem-solving ability.

Making it personal: Which of these statements best answers the question "Why do we have to learn to solve word problems?" What suggestion do you think is the most helpful to you in learning to solve word problems more easily? Write the suggestion in your own words. ▼

Did You Know...
That You May Be Paying More Than You Think?

SEE THROUGH THE HIDDEN FEES

Understanding the Problem:

Sam lives in California. Before taking a trip one weekend in March 2011 he needed to put gas in his car.

* On his street is a SHELL gas station, which charged $4.55 per gallon for gas.
* There is also an ARCO gas station, which charged $4.43 per gallon for gas.
* The ARCO station also charged an "ATM Transaction Fee" of $0.45.

Sam needed to decide which gas station to go to in order to pay the least amount for gas.

Making a Plan:

Sam needed to know how much gas would cost at each station to make a choice that would save him money.

Step 1: Sam needed to find out how much filling up his car at the gas station really cost after any hidden fees.

Task 1: Determine how much it cost Sam to buy one gallon of gas at each station.

Task 2: Determine how much it cost Sam to buy three gallons of gas at each station.

Task 3: Determine how much it cost Sam to buy four gallons of gas at each station.

Task 4: Determine how much it cost Sam to buy 10 gallons of gas at each station.

Step 2: Notice that when Sam bought more gas, it was less expensive to buy at the ARCO station than at the SHELL station. The price at SHELL rose faster than the price at ARCO.

Task 5: Find the number of gallons for which the cost was the same.

Making a Decision:

Step 3: Sam needed to take into consideration the price of gas and any fees involved when he made his decision about where to buy gas.

Task 6: Which station was less expensive if Sam only needed a small amount of gas, say less than four gallons?

Task 7: Which station was less expensive if Sam needed more than four gallons of gas?

Applying the Situation to Your Life:

Some gas stations have different prices depending on whether you pay with cash or a credit card.

Task 8: Does the station where you normally buy gas charge the same price for cash or credit?

Task 9: Do you know if the gas station charges an ATM transaction fee?

Task 10: Have the increases in gas prices caused you to change your driving habits? If so, please explain.

There is a simple plan to help you save money by avoiding hidden fees:

* Always check for any hidden fees.
* Find the real total cost for your purchases, including fees, taxes, shipping costs, and tips.
* Watch for different prices for cash or credit.

Chapter 3 Organizer

Topic and Procedure	Examples	✏ You Try It
Word names for decimals, p. 195 Hundreds / Tens / Ones / Decimal point / Tenths / Hundredths / Thousandths / Ten-thousandths 3 4 1 . 6 7 8 3	The word name for 341.6783 is three hundred forty-one and six thousand seven hundred eighty-three ten-thousandths.	1. Write the word name for 332.194.
Writing a decimal fraction as a decimal, p. 197 $\frac{8}{1000}$ is read "eight thousandths." The decimal form must have three decimal places to end in the thousandths place. Two zeros must be added before the 8.	Write $\frac{8}{1000}$ as a decimal. $\frac{8}{1000} = 0.008$	2. Write $\frac{54}{1000}$ as a decimal.
Writing a decimal as a fraction, p. 197 1. Read the decimal in words. 2. Write it in fraction form. 3. Reduce if possible.	Write 0.36 as a fraction. 0.36 is read "thirty-six hundredths." Write the fractional form. $\frac{36}{100}$ Reduce. $\frac{36}{100} = \frac{9}{25}$	3. Write 0.844 as a fraction.
Writing decimals in order from smallest to largest, p. 202 1. Add zeros so all decimals have the same number of decimal places. 2. Since all the numbers start with 3, look at the numbers to the right of the decimal point. Write these numbers from smallest to largest. 3. Write the corresponding decimals in order.	Arrange from smallest to largest: 3.12, 3.125, 3.1, 3.01 3.120, 3.125, 3.100, 3.010 010, 100, 120, 125 3.010, 3.100, 3.120, 3.125	4. Arrange from smallest to largest: 5.73, 5.713, 5.735, 5.7
Rounding decimals, p. 203 1. Locate the place (units, tenths, hundredths, etc.) to which rounding is required. 2. If the first digit to the right of the given place value is less than 5, drop it and all the digits to the right of it. 3. If the first digit to the right of the given place value is 5 or greater, increase the number in the given place value by one. Drop all digits to the right.	(a) Round to the nearest hundredth: 0.8652 0.87 (b) Round to the nearest thousandth: 0.21648 0.216	5. (a) Round to the nearest hundredth: 1.354 (b) Round to the nearest ten-thousandth: 9.077641
Adding and subtracting decimals, p. 207 1. Write the numbers vertically and line up the decimal points. Extra zeros may be written to the right of the decimal points after the nonzero digits if needed. 2. Add or subtract all the digits with the same place value, starting with the right column, moving to the left. Use carrying or borrowing as needed. 3. Place the decimal point of the result in line with the decimal points of all the numbers added or subtracted.	Add or subtract. (a) $36.3 + 8.007 + 5.26$ $\begin{array}{r} \overset{1}{3}6.300 \\ 8.007 \\ + 5.260 \\ \hline 49.567 \end{array}$ (b) $82.5 - 36.843$ $\begin{array}{r} 8\,2\,.\,5\,0\,0 \\ -3\,6\,.\,8\,4\,3 \\ \hline 4\,5\,.\,6\,5\,7 \end{array}$	6. Add or subtract. (a) $24.35 + 0.017 + 9.44$ (b) $172.2 - 42.186$

Topic and Procedure	Examples	✏ You Try It
Multiplying decimals, p. 217 1. Multiply the numbers just as you would multiply whole numbers. 2. Find the sum of the number of decimal places in the two factors. 3. Place the decimal point in the product so that the product has the same number of decimal places as the sum in step 2. You may need to insert zeros to the left of the number found in step 1.	Multiply. **(a)** $\begin{array}{r} 0.2 \\ \times\ 0.6 \\ \hline 0.12 \end{array}$ **(b)** $\begin{array}{r} 0.3174 \\ \times\quad 0.8 \\ \hline 0.25392 \end{array}$ **(c)** $\begin{array}{r} 0.0064 \\ \times\quad 0.21 \\ \hline 64 \\ 128 \\ \hline 0.001344 \end{array}$ **(d)** $\begin{array}{r} 1364 \\ \times\quad 0.7 \\ \hline 954.8 \end{array}$	7. Multiply. **(a)** $\begin{array}{r} 0.95 \\ \times\ 0.3 \end{array}$ **(b)** $\begin{array}{r} 1.225 \\ \times\ 2.8 \end{array}$
Multiplying a decimal by a power of 10, p. 218 Move the decimal point to the right the same number of places as there are zeros in the power of 10 or the same number of places as the exponent on the 10. (Sometimes it is necessary to write extra zeros before placing the decimal point in the answer.)	Multiply. **(a)** $5.623 \times 10 = 56.23$ **(b)** $0.0082 \times 1000 = 8.2$ **(c)** $28.93 \times 10^2 = 2893$ **(d)** $0.597 \times 10^4 = 5970$ **(e)** $0.075 \times 10^6 = 75{,}000$	8. Multiply. **(a)** 7.93×100 **(b)** 0.00015×1000 **(c)** 0.3125×10^3 **(d)** 0.4119×10^4 **(e)** 2.375×10^5
Dividing by a decimal, p. 227 1. Make the divisor a whole number by moving the decimal point to the right. Mark that position with a caret ($\wedge$). 2. Move the decimal point in the dividend to the right the same number of places. Mark that position with a caret. 3. Place the decimal point of your answer directly above the caret in the dividend. 4. Divide as with whole numbers.	Divide. **(a)** $0.06\overline{)0.162}$ **(b)** $0.003\overline{)85.8}$ **(a)** $\begin{array}{r} 2.\,7 \\ 0.06_\wedge\overline{)0.16_\wedge2} \\ \underline{12} \\ 4\ 2 \\ \underline{4\ 2} \\ 0 \end{array}$ **(b)** $\begin{array}{r} 28\ 600. \\ 0.003_\wedge\overline{)85.800_\wedge} \\ \underline{6} \\ 25 \\ \underline{24} \\ 1\ 8 \\ \underline{1\ 8} \\ 0 \end{array}$	9. Divide. **(a)** $0.04\overline{)0.152}$ **(b)** $0.007\overline{)13.3}$
Converting a fraction to a decimal, p. 234 Divide the denominator into the numerator until 1. the remainder is zero, or 2. the decimal repeats itself, or 3. the desired number of decimal places is achieved.	Find the decimal equivalent. **(a)** $\dfrac{13}{22}$ **(b)** $\dfrac{5}{7}$, rounded to the nearest ten-thousandth **(a)** $\begin{array}{r} 0.5909 \\ 22\overline{)13.0000} \\ \underline{11\ 0} \\ 2\ 00 \\ \underline{1\ 98} \\ 200 \\ \underline{198} \\ 2 \end{array}$ **(b)** $\begin{array}{r} 0.71428 \\ 7\overline{)5.00000} \\ \underline{4\ 9} \\ 10 \\ \underline{7} \\ 30 \\ \underline{28} \\ 20 \\ \underline{14} \\ 60 \\ \underline{56} \\ 4 \end{array}$ 0.71428 rounded to the nearest ten-thousandth is 0.7143. $\dfrac{13}{22} = 0.5\overline{90}$ or $0.5909090\ldots$	10. Find the decimal equivalent. **(a)** $\dfrac{11}{25}$ **(b)** $\dfrac{9}{14}$, rounded to the nearest thousandth
Order of operations with decimal numbers, p. 238 Same as order of operations of whole numbers. 1. Perform operations inside parentheses. 2. Simplify any expressions with exponents. 3. Multiply or divide from left to right. 4. Add or subtract from left to right.	Evaluate. $(0.4)^3 + 1.26 \div 0.12 - 0.12 \times (1.3 - 1.1)$ $= (0.4)^3 + 1.26 \div 0.12 - 0.12 \times 0.2$ $= 0.064 + 1.26 \div 0.12 - 0.12 \times 0.2$ $= 0.064 + 10.5 - 0.024$ $= 10.564 - 0.024$ $= 10.54$	11. Evaluate. $3.5 - (0.3)^2 \div 0.45 + (9.5 - 8.1)$

Procedure for Solving Applied Problems

Using the Mathematics Blueprint for Problem Solving, p. 244

In solving a real-life problem with decimals, students may find it helpful to complete the following steps. You will not use all the steps all of the time. Choose the steps that best fit the conditions of the problem.

1. Understand the problem.
   (a) Read the problem carefully.
   (b) Draw a picture if it helps you visualize the situation. Think about what facts you are given and what you are asked to find.
   (c) It may help to write a similar, simpler problem to get started and to determine what operation to use.
   (d) Use the Mathematics Blueprint for Problem Solving to organize your work. Follow these four parts.

   1. Gather the Facts (Write down specific values given in the problem.)
   2. What Am I Asked to Do? (Identify what you must obtain for an answer.)
   3. How Do I Proceed? (Determine what calculations need to be done.)
   4. Key Points to Remember (Record any facts, warnings, formulas, or concepts you think will be important as you solve the problem.)

2. Solve and state the answer.
   (a) Perform the necessary calculations.
   (b) State the answer, including the units of measure.

3. Check.
   (a) Estimate the answer to the problem. Compare this estimate to the calculated value. Is your answer reasonable?
   (b) Repeat your calculations.
   (c) Work backward from your answer. Do you arrive at the original conditions of the problem?

▲ **EXAMPLE**   Fred has a rectangular living room that measures 3.5 yards by 6.8 yards. He has a hallway that measures 1.8 yards by 3.5 yards. He wants to carpet each area using carpeting that costs $12.50 per square yard. What will the carpeting cost him?

Understand the problem.

It is helpful to draw a sketch.

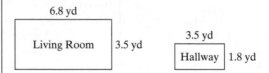

Mathematics Blueprint for Problem Solving

Gather the Facts	What Am I Asked to Do?	How Do I Proceed?	Key Points to Remember
Living room: 6.8 yards by 3.5 yards Hallway: 3.5 yards by 1.8 yards Cost of carpet: $12.50 per square yard	Find out what the carpeting will cost Fred.	Find the area of each room. Add the two areas. Multiply the total area by $12.50.	Multiply the length by the width to get the area of the room. Remember, area is measured in square yards.

To find the area of each room, we multiply the dimensions for each room.

   Living room 6.8 × 3.5 = 23.8 square yards
   Hallway 3.5 × 1.8 = 6.3 square yards

Add the two areas.

$$\begin{array}{r} 23.8 \\ +\ 6.3 \\ \hline 30.1 \text{ square yards} \end{array}$$

Multiply the total area by the cost per square yard.

$$30.1 \times \$12.50 = \$376.25$$

Estimate to check. You may be able to do some of this mentally.

   7 × 4 = 28 square yards   4 × 2 = 8 square yards

$$\begin{array}{r} 28 \\ +\ 8 \\ \hline 36 \text{ square yards} \end{array}$$

36 × 10 = $360 $360 is close to $376.25.   ✓

Chapter 3 Review Problems

Section 3.1

Write a word name for each decimal.

1. 13.672

2. 0.00084

Write as a decimal.

3. $\dfrac{7}{10}$

4. $\dfrac{81}{100}$

5. $1\dfrac{523}{1000}$

6. $\dfrac{79}{10,000}$

Write as a fraction or a mixed number.

7. 0.17

8. 0.036

9. 34.24

10. 1.00025

Section 3.2

Fill in the blank with $<$, $=$, or $>$.

11. $2\dfrac{9}{100}$ ___ 2.09

12. 0.716 ___ 0.706

13. $\dfrac{65}{100}$ ___ 0.655

In exercises 14–16, arrange each set of decimal numbers from smallest to largest.

14. 0.981, 0.918, 0.98, 0.901

15. 5.62, 5.2, 5.6, 5.26, 5.59

16. 2.36, 2.3, 2.362, 2.302

17. Round to the nearest tenth. 0.613

18. Round to the nearest hundredth. 19.2076

19. Round to the nearest ten-thousandth. 9.85215

20. Round to the nearest dollar. $156.48

Section 3.3

21. Add.
$$\begin{array}{r} 9.6 \\ 11.5 \\ 21.8 \\ +\ 34.7 \end{array}$$

22. Add.
$$\begin{array}{r} 2.5 \\ 32.7 \\ 116.94 \\ +\ \ \ 0.67 \end{array}$$

23. Subtract.
$$\begin{array}{r} 17.03 \\ -\ 2.448 \end{array}$$

24. Subtract.
$$\begin{array}{r} 182.422 \\ -\ 68.55 \end{array}$$

Section 3.4

In exercises 25–29, multiply.

25.
$$\begin{array}{r} 0.098 \\ \times\ 0.032 \end{array}$$

26.
$$\begin{array}{r} 126.83 \\ \times\ \ \ \ \ 7 \end{array}$$

27.
$$\begin{array}{r} 7053 \\ \times\ 0.34 \end{array}$$

28. 0.000613×10^3

29. 1.2354×10^5

30. *Food Cost* Roast beef was on sale for $3.49 per pound. How much would 2.5 pounds cost? Round to the nearest cent.

Section 3.5

In exercises 31 and 32, divide until there is a remainder of zero.

31. $5.2\overline{)191.36}$

32. $8\overline{)1863.2}$

33. Divide and round your answer to the nearest tenth.

$$1.3\overline{)746.75}$$

34. Divide and round your answer to the nearest thousandth.

$$0.06\overline{)0.003539}$$

Section 3.6

Write as an equivalent decimal.

35. $\dfrac{11}{12}$

36. $\dfrac{17}{20}$

37. $1\dfrac{5}{6}$

Write as a decimal rounded to the nearest thousandth.

38. $\dfrac{11}{14}$

39. $\dfrac{10}{29}$

40. $3\dfrac{9}{23}$

Evaluate by doing the operations in proper order.

41. $2.3 \times 1.82 + 3 \times 5.12$

42. $3.57 - (0.4)^3 \times 2.5 \div 5$

43. $2.4 \div (2 - 1.6)^2 + 8.13$

Mixed Practice

Calculate.

44. $2398.26 - 1959.07$

45. $32.15 \times 0.02 \times 10^2$

46. $1.809 - 0.62 + 3.27$

47. $2.0792 \div 2.3$

48. $8 \div 0.4 + 0.1 \times (0.2)^2$

49. $(3.8 - 2.8)^3 \div (0.5 + 0.3)$

Applications
Section 3.7

Solve each problem.

50. *Football Tickets* At a large football stadium there are 2600 people in line for tickets. In the first two minutes the computer is running slowly and tickets 228 people. Then the computer stops. For the next 2.5 minutes, the computer runs at medium speed and tickets 388 people per minute. For the next three minutes the computer runs at full speed and tickets 430 people per minute. Then the computer stops. How many people still have not received their tickets?

51. *Fuel Efficiency* Phil drove to the mountains. His odometer read 26,005.8 miles at the start, and 26,325.8 miles at the end of the trip. He used 12.9 gallons of gas on the trip. How many miles per gallon did his car get? (Round your answer to the nearest tenth.)

52. *Drinking Water Safety* The EPA standard for safe drinking water is a maximum of 0.002 milligram of mercury in one liter of water. The town wells at Winchester were tested. The test was done on 12 liters of water. The entire 12-liter sample contained 0.03 milligram of mercury. Is the water safe or not? By how much does it differ from the standard?

53. *Infant Head Size* It is common for infants to have their heads measured during the first year of life. At two months, Will's head measured 40 centimeters. There are 2.54 centimeters in one inch. How many inches was this measurement? Round to the nearest hundredth.

54. *Geometry* Dick Wright's new rectangular garden measures 18.3 feet by 9.6 feet. He needs to install wire fence on all four sides.

(a) How many feet of fence does he need?

(b) The number of bags of wood chips Dick buys depends on the area of the garden. What is the area?

55. *Geometry* Bill Tupper's rectangular driveway needs to be resurfaced. It measures 75.5 feet by 18.5 feet. How large is the area of the driveway?

56. *Travel Distances* The following strip map shows the distances in miles between several local towns in Pennsylvania. How much longer is the distance from Coudersport to Gaines than the distance from Galeton to Wellsboro?

57. *Geometry* A farmer in Vermont has a field with an irregular shape. The distances are marked on the diagram. There is no fence but there is a path on the edge of the field. How long is the walking path around the field?

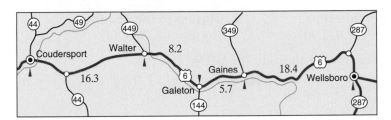

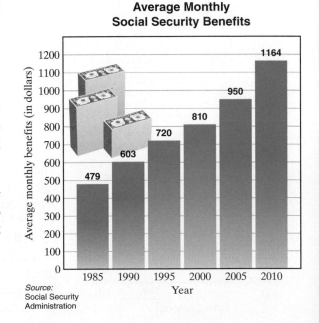

58. *Car Payments* Marcia and Greg purchased a new car. For the next five years they will be making monthly payments of $212.50. Their bank has offered to give them a loan at a smaller interest rate so that they would make monthly payments of only $199.50. The bank would charge them $285.00 to reissue their car loan. How much would it cost them to keep their original loan? How much would it cost them if they took the new loan from the bank? Should they make the change or keep the original loan?

Social Security Benefits Use the following bar graph to answer exercises 59–62. Round all answers to the nearest cent.

59. How much did the average monthly social security benefit increase from 1995 to 2005?

60. What was the average daily social security benefit in 2005? (Assume 30 days in a month.)

61. If the average daily social security benefit increases by the same amount from 2010 to 2025 as it did from 1995 to 2010, what will be the average daily social security benefit in 2025? (Assume 30 days in a month.)

62. If the average daily social security benefit increases by the same amount from 2010 to 2020 as it did from 2000 to 2010, what will be the average daily social security benefit in 2020? (Assume 30 days in a month.)

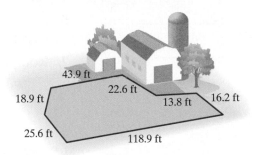

Average Monthly Social Security Benefits

Source: Social Security Administration

How Am I Doing?   Chapter 3 Test

 Test Prep VIDEOS   **MATH COACH**   **MyMathLab**   **You Tube**

After you take this test read through the Math Coach on pages 259–260. Math Coach videos are available via MyMathLab and YouTube. Step-by-step test solutions on the Chapter Test Prep Videos are also available via MyMathLab and YouTube. (Search "TobeyBasicCollMath" and click on "Channels.")

1. Write a word name for the decimal. 12.043

2. Write as a decimal. $\dfrac{3977}{10,000}$

In questions 3 and 4, write in fractional notation. Reduce whenever possible.

3. 7.15

4. 0.261

5. Arrange from smallest to largest. 2.19, 2.91, 2.9, 2.907

6. Round to the nearest hundredth. 78.6562

7. Round to the nearest ten-thousandth. 0.0341752

Add.
8.
$$\begin{array}{r} 96.2 \\ 1.348 \\ +\ 2.15 \\ \hline \end{array}$$

MC 9. $17 + 2.1 + 16.8 + 0.04 + 1.59$

Subtract.
10.
$$\begin{array}{r} 1.0075 \\ -\ 0.9096 \\ \hline \end{array}$$

MC 11. $72.3 - 1.145$

Multiply.
12.
$$\begin{array}{r} 8.31 \\ \times\ 0.07 \\ \hline \end{array}$$

13. 2.189×10^3

Divide.
14. $0.08\overline{)0.01028}$

15. $0.69\overline{)32.43}$

Write as a decimal.

16. $\dfrac{11}{9}$

MC 17. $\dfrac{7}{8}$

In questions 18 and 19, perform the operations in the proper order.

18. $(0.3)^3 + 1.02 \div 0.5 - 0.58$

MC 19. $19.36 \div (0.24 + 0.26) \times (0.4)^2$

20. Peter put 8.5 gallons of gas in his car. The price per gallon is $3.17. How much did Peter spend on gas? Round to the nearest cent.

21. Frank traveled from the city to the shore. His odometer read 42,620.5 miles at the start and 42,780.5 at the end of the trip. He used 8.5 gallons of gas. How many miles per gallon did his car achieve? Round to the nearest tenth.

22. The rainfall for March in Central City was 8.01 centimeters; for April, 5.03 centimeters; and for May, 8.53 centimeters. The normal rainfall for these three months is 25 centimeters. How much less rain fell during these three months than usual—that is, how does this year's figure compare with the figure for normal rainfall?

23. Wendy is earning $7.30 per hour in her new job as a teller trainee at the Springfield National Bank. She earns 1.5 times that amount for every hour over 40 hours she works in one week. She was asked to work 49 hours last week. How much did she earn last week?

1. _____
2. _____
3. _____
4. _____
5. _____
6. _____
7. _____
8. _____
9. _____
10. _____
11. _____
12. _____
13. _____
14. _____
15. _____
16. _____
17. _____
18. _____
19. _____
20. _____
21. _____
22. _____
23. _____

Total Correct: ____

MATH COACH

Mastering the skills you need to do well on the test.

Students often make the same types of errors when they do the Chapter 3 Test. Here are some helpful hints to keep you from making these common errors on test problems.

Adding Decimals—Problem 9 $17 + 2.1 + 16.8 + 0.04 + 1.59$

> **Helpful Hint**   Place additional zeros at the end of each number so that each decimal has two digits to the right of each decimal point. Include the decimal point before adding zeros to any whole numbers.

Did you change the problem to
$17.00 + 2.10 + 16.80 + 0.04 + 1.59$?

Yes ▭ No ▭

If you answered No, stop and make those changes now.

Did you line up the decimal points carefully when you wrote the numbers in a column?

Yes ▭ No ▭

$$
\begin{array}{r}
17.00 \\
2.10 \\
16.80 \\
0.04 \\
+\ 1.59 \\
\end{array}
$$

If you answered No, stop and write the numbers carefully in a column with the decimal points exactly aligned.

Subtracting Decimals—Problem 11 $72.3 - 1.145$

> **Helpful Hint**   Place additional zeros at the end of each number so that each decimal has three digits to the right of the decimal point. Be sure to align the decimal points when you write the subtraction in column form.

Did you change the problem to $72.300 - 1.145$?

Yes ▭ No ▭

If you answered No, stop and make those changes now.

Did you line up the decimal points carefully when you wrote the numbers one beneath the other?

Yes ▭ No ▭

$$
\begin{array}{r}
72.300 \\
-\ 1.145 \\
\end{array}
$$

If you answered No, stop and write the numbers carefully in a column with the decimal points exactly aligned.

Need help? Watch the MATH COACH **videos in** MyMathLab® **or on** YouTube™.

259

Writing Fractions in Decimal Form—Problem 17 $\frac{7}{8}$

> **Helpful Hint** Divide the denominator into the numerator until the remainder becomes zero. Be careful with each division step.

In your first step of division, was the result a quotient of 8?

Yes ____   No ____

When you multiplied $8 \cdot 8$, was the result a product of 64?

Yes ____   No ____

When you subtracted 64 from 70, was the result a difference of 6?

Yes ____   No ____

If you answered No to any of these questions, stop and do the first part of the division problem again.

Did you write the problem as $8\overline{)7.000}$ and complete a total of three division steps?

Yes ____   No ____

If you answered No, try to go back and complete Problem 17 correctly now.

Order of Operations with Decimals—Problem 19 $19.36 \div (0.24 + 0.26) \times (0.4)^2$

> **Helpful Hint** Be sure to write out each step. Combine operations inside the parentheses first. Then raise numbers to a power.

After the first step, your result should be $19.36 \div 0.5 \times (0.4)^2$. Did you get that result?

Yes ____   No ____

If you answered No, stop and perform that step correctly.

After the second step, your result should be $19.36 \div 0.5 \times 0.16$.

Did you get that result?

Yes ____   No ____

If you answered No, stop and perform that step correctly.

Now try to answer the entire Problem 19 correctly using these suggestions.

Need more help? Look for section examples marked with $\mathbb{MC}$ to review.

Cumulative Test for Chapters 1–3

This test provides a comprehensive review of the key objectives for Chapters 1–3.

1. Write in words. 38,056,954

2. Add.   156,028
        301,579
    + 21,980

3. Subtract.   1,091,000
        −1,036,520

4. Multiply.   589
       × 67

5. Multiply.   1200
      × 40

6. Divide. $15\overline{)4740}$

7. Evaluate. $20 \div 4 + 2^5 - 7 \times 3$

8. Round to the nearest thousand.
236,813

9. Estimate. $58,216 \times 438,207$

10. Reduce. $\dfrac{18}{45}$

11. Change to an improper fraction.
$4\dfrac{3}{8}$

12. Multiply. $2\dfrac{1}{5} \times 3\dfrac{1}{3}$

13. Add. $5\dfrac{3}{8} + 2\dfrac{11}{12}$

14. Subtract. $\dfrac{23}{35} - \dfrac{2}{5}$

15. Evaluate. $\dfrac{5}{16} \times \dfrac{4}{5} + \dfrac{9}{10} \times \dfrac{2}{3}$

16. Divide. $52 \div 3\dfrac{1}{4}$

17. Divide. $1\dfrac{3}{8} \div \dfrac{5}{12}$

18. Write as a decimal. $\dfrac{39}{1000}$

19. Arrange from smallest to largest.
2.1, 20.1, 2.01, 2.12, 2.11

20. Round to the nearest thousandth.
26.07984

21. Add.   3.126
      8.4
    10.33
   + 0.09

22. Subtract.   28.1
       −14.982

23. Multiply. 28.7×0.05

24. Multiply. 0.1823×1000

25. Divide. $0.06\overline{)0.06348}$

26. Write as a decimal. $\dfrac{13}{16}$

▲ 27. Dr. Bob Wells has a small square garden that measures 10.5 feet on each side.

    (a) What is the area of this garden?

    (b) What is the perimeter of this garden?

28. Russ and Norma Camp borrowed some money from the bank to purchase a new car. They are paying off the car loan at the rate of $320.50 per month. At the end of the loan period they will have paid $19,230.00 to the bank. How many months will it take to pay off this car loan?

1. _____

2. _____

3. _____

4. _____

5. _____

6. _____

7. _____

8. _____

9. _____

10. _____

11. _____

12. _____

13. _____

14. _____

15. _____

16. _____

17. _____

18. _____

19. _____

20. _____

21. _____

22. _____

23. _____

24. _____

25. _____

26. _____

27. (a) _____

     (b) _____

28. _____

We all know that too much fast food is not good for us. Many people eat fast food nearly every day, but they may be unaware of just how many calories they are consuming. Do you know how many calories are in some common fast foods? How much exercise is necessary to burn off these calories? The mathematics you learn in this chapter will enable you to answer those questions.

Ratio and Proportion

4.1 Ratios and Rates

① Using a Ratio to Compare Two Quantities with the Same Units

Assume that you earn 13 dollars an hour and your friend earns 10 dollars per hour. The *ratio* 13 : 10 compares what you and your friend make. This ratio means that for every 13 dollars you earn, your friend earns 10. The *rate* you are paid is 13 dollars per hour, which compares 13 dollars to 1 hour. In this section we see how to use both ratios and rates to solve many everyday problems.

Suppose that we want to compare an object weighing 20 pounds to an object weighing 23 pounds. The ratio of their weights would be 20 to 23. We may also write this as $\frac{20}{23}$. A **ratio** is the comparison of two quantities that have the *same units*.

A commonly used video display for a computer has a horizontal dimension of 14 inches and a vertical dimension of 10 inches. The ratio of the horizontal dimension to the vertical dimension is 14 to 10. In reduced form we would write that as 7 to 5. We can express the ratio three ways.

We can write "the ratio of 7 to 5."

We can write 7 : 5 using a colon.

We can write $\frac{7}{5}$ using a fraction.

All three notations are valid ways to compare 7 to 5. Each is read as "7 to 5."

> We always want to write a ratio in simplest form. A ratio is in **simplest form** when the two numbers do not have a common factor and both numbers are whole numbers.

EXAMPLE 1 Write in simplest form. Express your answer as a fraction.

(a) the ratio of 15 hours to 20 hours

(b) the ratio of 36 hours to 30 hours

(c) 125 : 150

Solution

(a) $\dfrac{15}{20} = \dfrac{3}{4}$ **(b)** $\dfrac{36}{30} = \dfrac{6}{5}$ **(c)** $\dfrac{125}{150} = \dfrac{5}{6}$

Notice that in each case the two numbers *do* have a common factor. When we form the fraction—that is, the ratio—we take the extra step of *reducing* the fraction. However, improper fractions are *not* changed to mixed numbers.

Student Practice 1 Write in simplest form. Express your answer as a fraction.

(a) the ratio of 36 feet to 40 feet

(b) the ratio of 18 feet to 15 feet

(c) 220 : 270

Student Learning Objectives

After studying this section, you will be able to:

① Use a ratio to compare two quantities with the same units.

② Use a rate to compare two quantities with different units.

CLICK & READ
WebNews
The latest news updated by the minute!
Heatwave Continues in Texas

HOME
NATIONAL
LOCAL
SPORTS
STOCKS
WEATHER

NOTE TO STUDENT: Fully worked-out solutions to all of the Student Practice problems can be found at the back of the text starting at page SP-1.

EXAMPLE 2   Martin earns $350 weekly. However, he takes home only $250 per week in his paycheck.

$350.00 gross pay   (what Martin earns)

45.00 withheld for federal tax ⎫
20.00 withheld for state tax   ⎬   (what is taken out of Martin's earnings)
35.00 withheld for retirement   ⎭

$250.00 take-home pay   (what Martin has left)

(a) What is the ratio of the amount withheld for federal tax to gross pay?

(b) What is the ratio of the amount withheld for state tax to the amount withheld for federal tax?

Solution

(a) The ratio of the amount withheld for federal tax to gross pay is

$$\frac{45}{350} = \frac{9}{70}.$$

(b) The ratio of the amount withheld for state tax to the amount withheld for federal tax is

$$\frac{20}{45} = \frac{4}{9}.$$

Student Practice 2 Recently President Burton conducted a survey of students at North Shore Community College who use the Internet. He wanted to determine how many of the students use the college Internet provider versus how many use AOL, MSN, or other commercial Internet providers. The results of his survey are shown in the circle graph.

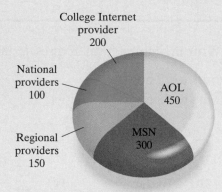

(a) Write the ratio of the number of students who use the college Internet provider to the number of students who use AOL.

(b) Write the ratio of the number of students who use MSN to the total number of students who use the Internet.

TO THINK ABOUT: Mach Numbers Perhaps you have heard statements like "a certain jet plane travels at Mach 2.2." What does that mean? A Mach number is a ratio that compares the velocity (speed) of an object to the velocity of sound. Sound travels at about 330 meters per second. The Mach number is written in decimal form.

What is the Mach number of a jet traveling at 690 meters per second?

$$\text{Mach number of jet} = \frac{690 \text{ meters per second}}{330 \text{ meters per second}} = \frac{69}{33} = \frac{23}{11}$$

Dividing this out, we obtain

$$2.09090909 \ldots \text{ or } 2.\overline{09}$$

Rounded to the nearest tenth, the Mach number of the jet is 2.1.
Exercises 73 and 74 in Exercises 4.1 deal with Mach numbers.

② Using a Rate to Compare Two Quantities with Different Units

A **rate** is a comparison of two quantities with *different units*. Usually, to avoid misunderstanding, we express a rate as a reduced or simplified fraction with the units included.

EXAMPLE 3   Recently an automobile manufacturer spent $946,000 for a 48-second television commercial shown on a national network. What is the rate of dollars spent to seconds of commercial time?

Solution

$$\text{The rate is } \frac{946{,}000 \text{ dollars}}{48 \text{ seconds}} = \frac{59{,}125 \text{ dollars}}{3 \text{ seconds}}.$$

Student Practice 3 A farmer is charged a $44 storage fee for every 900 tons of grain he stores. What is the rate of the storage fee in dollars to tons of grain?

Often we want to know the rate for a single unit, which is the unit rate. A **unit rate** is a rate in which the denominator is the number 1. Often we need to divide the numerator by the denominator to obtain this value. Finding the unit rate may result in a whole number or a decimal.

EXAMPLE 4   A car traveled 301 miles in seven hours. Find the unit rate.

Solution $\frac{301}{7}$ can be simplified. We find $301 \div 7 = 43$.

Thus

$$\frac{301 \text{ miles}}{7 \text{ hours}} = \frac{43 \text{ miles}}{1 \text{ hour}}$$

The denominator is 1. We write our answer as 43 miles/hour. The fraction line is read as the word *per,* so our answer here is read "43 miles per hour." *Per* means "for every," so a rate of 43 miles per hour means 43 miles traveled for every hour traveled.

Student Practice 4 A car traveled 212 miles in four hours. Find the unit rate.

EXAMPLE 5 A grocer purchased 200 pounds of apples for $68. He sold the 200 pounds of apples for $86. How much profit did he make per pound of apples?

Solution

$$\begin{array}{r} \$86 \quad \text{selling price} \\ - \quad 68 \quad \text{cost} \\ \hline \$18 \quad \text{profit} \end{array}$$

The rate that compares profit to pounds of apples sold is $\dfrac{18 \text{ dollars}}{200 \text{ pounds}}$. We will find $18 \div 200$.

$$\begin{array}{r} 0.09 \\ 200\overline{)18.00} \\ \underline{18\ 00} \\ 0 \end{array}$$

The unit rate of profit is $0.09 per pound.

Student Practice 5 A retailer purchased 120 nickel-cadmium batteries for flashlights for $129.60. She sold them for $170.40. What was her profit per battery?

EXAMPLE 6 Hamburger at a local butcher shop is packaged in large and extra-large packages. A large package costs $7.86 for 6 pounds and an extra-large package is $10.08 for 8 pounds.

(a) What is the unit rate in dollars per pound for each size package?
(b) How much per pound does a consumer save by buying the extra-large package?

Solution

(a) $\dfrac{7.86 \text{ dollars}}{6 \text{ pounds}} = \$1.31/\text{pound for the large package}$

$\dfrac{10.08 \text{ dollars}}{8 \text{ pounds}} = \$1.26/\text{pound for the extra-large package}$

(b)

$$\begin{array}{r} \$1.31 \\ - \quad 1.26 \\ \hline \$0.05 \end{array}$$ A person saves $0.05/pound by buying the extra-large package.

Student Practice 6 A 12-ounce package of Fred's favorite cereal costs $2.04. A 20-ounce package of the same cereal costs $2.80.

(a) What is the unit rate in cost per ounce of each size of cereal package?
(b) How much per ounce would Fred save by buying the larger size?

Verbal and Writing Skills, Exercises 1–4

1. A _____ is a comparison of two quantities that have the same units.

2. A rate is a comparison of two quantities that have _____ units.

3. The ratio 5 : 8 is read _____.

4. Marion compares the number of loaves of bread she bakes to the number of pounds of flour she needs to make the bread. Is this a ratio or a rate? Why?

Write in simplest form. Express your answer as a fraction.

5. 6 : 18   **6.** 8 : 20   **7.** 21 : 18   **8.** 50 : 35

9. 150 : 225   **10.** 360 : 480   **11.** 165 to 90   **12.** 135 to 120

13. 60 to 72   **14.** 55 to 77   **15.** 28 to 42   **16.** 21 to 98

17. 32 to 20   **18.** 90 to 54   **19.** 8 ounces to 12 ounces

20. 50 years to 85 years   **21.** 39 kilograms to 26 kilograms   **22.** 255 meters to 15 meters

23. $75 to $95   **24.** $54 to $78   **25.** 312 yards to 24 yards

26. 91 tons to 133 tons   **27.** $2\frac{1}{2}$ pounds to $4\frac{1}{4}$ pounds   **28.** $4\frac{1}{3}$ feet to $5\frac{2}{3}$ feet

Personal Finance *Use the following table to answer exercises 29–32.*

ROBIN'S WEEKLY PAYCHECK

Total (Gross) Pay	Federal Withholding	State Withholding	Retirement	Insurance	Savings Contribution	Take-Home Pay
$285	$35	$20	$28	$16	$21	$165

29. What is the ratio of take-home pay to total (gross) pay?

30. What is the ratio of retirement to insurance?

31. What is the ratio of federal withholding to take-home pay?

32. What is the ratio of retirement to total (gross) pay?

Useful Life of an Automobile   An automobile insurance company prepared the following analysis for its clients. Use this table for exercises 33–36.

33. What is the ratio of sedans that lasted two years or less to the total number of sedans?

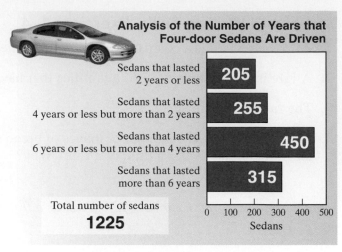

34. What is the ratio of sedans that lasted more than six years to the total number of sedans?

35. What is the ratio of the number of sedans that lasted six years or less but more than four years to the number of sedans that lasted two years or less?

36. What is the ratio of the number of sedans that lasted more than six years to the number of sedans that lasted four years or less but more than two years?

37. *Basketball* A basketball team scored a total of 704 points during one season. Of these, 44 points were scored by making one-point free throws. What is the ratio of free-throw points to total points?

38. *Sales Tax* When Michael bought his home theater sound system, he paid $34 in tax. The total cost was $714. What is the ratio of tax to total cost?

Write as a rate in simplest form.

39. $42 for 12 pairs of socks

40. $50 for 15 deli sandwiches

41. $170 for 12 bushes

42. 168 people for 20 vans

43. $114 for 12 CDs

44. $150 for 12 house plants

45. 6150 revolutions for every 15 miles

46. 9540 revolutions for every 18 miles

47. $330,000 for 12 employees

48. $156,000 for 24 people

Write as a unit rate.

49. Earn $600 in 40 hours

50. Earn $315 in 35 hours

51. Travel 308 miles on 11 gallons of gas

52. Travel 221 miles on 13 gallons of gas

▲ **53.** 1120 people in 16 square miles

▲ **54.** 3600 people in 24 square miles

55. 840 books for 12 libraries

56. 930 points in 15 games

57. Travel 297 miles in 4.5 hours

58. Travel 374 miles in 5.5 hours

59. 475 patients for 25 doctors

60. 248 children on 4 buses

61. 84 geraniums planted in 28 pots

62. 78 children in 26 families

Applications

63. *Stock Market* $2970 was spent for 135 shares of Mattel, Inc. stock. Find the cost per share.

64. *Stock Market* $5440 was spent for 170 shares of Sony Corporation stock. Find the cost per share.

65. *Toy Store Profit* A toy store owner purchased 90 puppets for $1080. She sold them for $1485. How much profit did she make per puppet?

66. *Clothing Store Profit* The manager of an outdoor clothing store ordered 40 pairs of hiking boots for $2400. The boots will sell for a total of $3560. How much profit will the store make per pair?

67. *Food Cost* A 16-ounce box of dry pasta costs $1.28. A 24-ounce box of the same pasta costs $1.68.

(a) What is the cost per ounce of each box of pasta?

(b) How much does the educated consumer save by buying the larger box?

(c) How much does the consumer save by buying 2 large boxes instead of 3 small boxes?

68. *Food Cost* A 16-ounce can of beef stew costs $2.88. A 26-ounce can of the same beef stew costs $4.16.

(a) What is the cost per ounce of each can of stew?

(b) How much does the consumer save per ounce by buying the larger can?

69. *Moose Population Density* Dr. Robert Tobey completed a count of two herds of moose in the central regions of Alaska. He recorded 3978 moose on the North Slope and 5520 moose on the South Slope. There are 306 acres on the North Slope and 460 acres on the South Slope.

(a) How many moose per acre were found on the North Slope?

(b) How many moose per acre were found on the South Slope?

(c) In which region are the moose more closely crowded together?

70. *Australia Population Density* In Melbourne, Australia, 27,900 people live in the suburb of St. Kilda and 38,700 live in the suburb of Caulfield. The area of St. Kilda is 6500 acres. The area of Caulfield is 9200 acres. Round your answers to the nearest tenth.

(a) How many people per acre live in St. Kilda?

(b) How many people per acre live in Caulfield?

(c) Which suburb is more crowded?

71. *Stock Market*

   (a) Ms. Handley bought 350 shares of Home Depot stock for $14,332.50. How much did she pay per share?

   (b) Mr. Johnston bought 210 shares of Office Max stock for $11,088. How much did he pay per share?

   (c) How much more per share did Mr. Johnston pay than Ms. Handley?

72. *Baseball Statistics* In 2010, the top professional baseball players with the highest batting averages were Josh Hamilton of the Texas Rangers and Justin Morneau of the Minnesota Twins. Morneau had 102 hits out of 296 "at-bats." Hamilton had 148 hits out of 414 at-bats.

   (a) What are the rates of at-bats per hit for Morneau and Hamilton? Round to the nearest tenth.

   (b) Which player got a hit more often?

To Think About *For exercises 73 and 74, recall that the speed of sound is about 330 meters per second. (See the To Think About discussion on pages 264 and 265.) Round your answers to the nearest tenth.*

73. *Jet Speed* A jet plane was originally designed to fly at 750 meters per second. It was modified to fly at 810 meters per second. By how much was its Mach number increased?

74. *Rocket Speed* A new rocket was first flown at 1960 meters per second. It proved unstable and unreliable at that speed. It is now flown at a maximum of 1920 meters per second. By how much was its Mach number decreased?

Cumulative Review *Calculate.*

75. [2.8.1] $2\frac{1}{4} + \frac{3}{8}$

76. [2.5.1] $\frac{5}{7} \div \frac{3}{21}$

77. [2.8.3] $\frac{3}{5} \times \frac{5}{8} - \frac{2}{3} \times \frac{1}{4}$

78. [2.8.2] $3\frac{1}{16} - 2\frac{1}{24}$

▲ **79.** [3.7.2] *Geometry* A rectangular room 12 yards by 5.2 yards had a carpet installed. The bill was $764.40. What was the cost of the installed carpet per square yard?

80. [1.8.2] *Electronic Game Store Profit* An electronics superstore bought 1050 computer games for $23 each. How much did the store pay in all for these games? The store sold the games for $39 each. How much profit did the store make?

Quick Quiz 4.1

1. Write as a ratio in simplest form.

   51 to 85

2. Write as a rate in simplest form. Express your answer as a fraction.

   1700 square feet for 55 pounds

3. Write as a unit rate. Round to the nearest hundredth if necessary.

   462 trees planted on 17 acres

4. **Concept Check** At a company picnic, there were 663 cans of soda for 231 people. Explain how you would write that as a rate in simplest form.

4.2 The Concept of Proportions

① Writing a Proportion

A **proportion** states that two ratios or two rates are equal. For example, $\frac{5}{8} = \frac{15}{24}$ is a proportion and $\frac{7 \text{ feet}}{8 \text{ dollars}} = \frac{35 \text{ feet}}{40 \text{ dollars}}$ is also a proportion. A proportion can be read two ways. The proportion $\frac{5}{8} = \frac{15}{24}$ can be read "five eighths equals fifteen twenty-fourths," or it can be read "five *is to* eight *as* fifteen *is to* twenty-four."

Student Learning Objectives

After studying this section, you will be able to:

① Write a proportion.

② Determine whether a statement is a proportion.

EXAMPLE 1 Write the proportion 5 is to 7 as 15 is to 21.

Solution

$$\frac{5}{7} = \frac{15}{21}$$

Student Practice 1 Write the proportion 6 is to 8 as 9 is to 12.

NOTE TO STUDENT: Fully worked-out solutions to all of the Student Practice problems can be found at the back of the text starting at page SP-1.

EXAMPLE 2 Write a proportion to express the following: If four rolls of wallpaper measure 300 feet, then eight rolls of wallpaper will measure 600 feet.

Solution When you write a proportion, order is important. Be sure that the similar units for the rates are in the same position in the fractions.

$$\frac{4 \text{ rolls}}{300 \text{ feet}} = \frac{8 \text{ rolls}}{600 \text{ feet}}$$

Student Practice 2 Write a proportion to express the following: If it takes two hours to drive 72 miles, then it will take three hours to drive 108 miles.

② Determining Whether a Statement Is a Proportion

By definition, a proportion states that two ratios are equal. $\frac{2}{7} = \frac{4}{14}$ is a proportion because $\frac{2}{7}$ and $\frac{4}{14}$ are equivalent fractions. You might say that $\frac{2}{7} = \frac{4}{14}$ is a *true* statement. It is easy enough to see that $\frac{2}{7} = \frac{4}{14}$ is true. However, is $\frac{4}{14} = \frac{6}{21}$ true? Is $\frac{4}{14} = \frac{6}{21}$ a proportion? To determine whether a statement is a proportion, we use the equality test for fractions.

EQUALITY TEST FOR FRACTIONS

For any two fractions where $b \neq 0$ and $d \neq 0$,

$$\frac{a}{b} = \frac{c}{d} \text{ if and only if } a \times d = b \times c.$$

Thus, to see if $\dfrac{4}{14} = \dfrac{6}{21}$, we can multiply.

$$\dfrac{4}{14} \diagdown \dfrac{6}{21} \qquad \begin{array}{l} 14 \times 6 = 84 \leftarrow \\ 4 \times 21 = 84 \leftarrow \end{array} \boxed{\begin{array}{l}\text{The cross}\\\text{products}\\\text{are equal.}\end{array}}$$

$\dfrac{4}{14} = \dfrac{6}{21}$ is true. $\dfrac{4}{14} = \dfrac{6}{21}$ is a proportion.

This method is called finding **cross products.**

EXAMPLE 3   Determine which equations are proportions.

(a) $\dfrac{14}{18} \stackrel{?}{=} \dfrac{35}{45}$ 　　　　　　**(b)** $\dfrac{16}{21} \stackrel{?}{=} \dfrac{174}{231}$

Solution

(a) $\dfrac{14}{18} \stackrel{?}{=} \dfrac{35}{45}$

$$18 \times 35 = 630$$
$$\dfrac{14}{18} \diagdown \dfrac{35}{45} \qquad \text{The cross products are equal.} \qquad \text{Thus } \dfrac{14}{18} = \dfrac{35}{45}. \text{ This is a proportion.}$$
$$14 \times 45 = 630$$

(b) $\dfrac{16}{21} \stackrel{?}{=} \dfrac{174}{231}$

$$21 \times 174 = 3654$$
$$\dfrac{16}{21} \diagdown \dfrac{174}{231} \qquad \text{The cross products are not equal.} \qquad \text{Thus } \dfrac{16}{21} \neq \dfrac{174}{231}. \text{ This is not a proportion.}$$
$$16 \times 231 = 3696$$

Student Practice 3   Determine which equations are proportions.

(a) $\dfrac{10}{18} \stackrel{?}{=} \dfrac{25}{45}$ 　　　　　　**(b)** $\dfrac{42}{100} \stackrel{?}{=} \dfrac{22}{55}$

Proportions may involve fractions or decimals.

EXAMPLE 4   Determine which equations are proportions.

(a) $\dfrac{5.5}{7} \stackrel{?}{=} \dfrac{33}{42}$ 　　　　　　**(b)** $\dfrac{5}{8\frac{3}{4}} \stackrel{?}{=} \dfrac{40}{72}$

Solution

(a) $\dfrac{5.5}{7} \stackrel{?}{=} \dfrac{33}{42}$

$$7 \times 33 = 231$$
$$\dfrac{5.5}{7} \diagdown \dfrac{33}{42} \qquad \text{The cross products are equal.} \qquad \text{Thus } \dfrac{5.5}{7} = \dfrac{33}{42}. \text{ This is a proportion.}$$
$$5.5 \times 42 = 231$$

(b) $\dfrac{5}{8\frac{3}{4}} \overset{?}{=} \dfrac{40}{72}$

First we multiply $8\dfrac{3}{4} \times 40 = \dfrac{35}{\cancel{4}_1} \times \cancel{40}^{10} = 35 \times 10 = 350$

$$8\dfrac{3}{4} \times 40 = 350$$

$\dfrac{5}{8\frac{3}{4}} \overset{\diagup\diagdown}{} \dfrac{40}{72}$ The cross products are not equal. Thus $\dfrac{5}{8\frac{3}{4}} \neq \dfrac{40}{72}$. This is not a proportion.

$$5 \times 72 = 360$$

Student Practice 4 Determine which equations are proportions.

(a) $\dfrac{2.4}{3} \overset{?}{=} \dfrac{12}{15}$ **(b)** $\dfrac{2\frac{1}{3}}{6} \overset{?}{=} \dfrac{14}{38}$

M℃ EXAMPLE 5 **(a)** Is the rate $\dfrac{\$86}{13 \text{ tons}}$ equal to the rate $\dfrac{\$79}{12 \text{ tons}}$?

(b) Is the rate $\dfrac{3 \text{ American dollars}}{2 \text{ British pounds}}$ equal to the rate $\dfrac{27 \text{ American dollars}}{18 \text{ British pounds}}$?

Solution

(a) We want to know whether $\dfrac{86}{13} = \dfrac{79}{12}$.

$$13 \times 79 = 1027$$

$\dfrac{86}{13} \overset{\diagup\diagdown}{} \dfrac{79}{12}$ The cross products are not equal. Thus the two rates are not equal. This is not a proportion.

$$86 \times 12 = 1032$$

(b) We want to know whether $\dfrac{3}{2} = \dfrac{27}{18}$.

$$2 \times 27 = 54$$

$\dfrac{3}{2} \overset{\diagup\diagdown}{} \dfrac{27}{18}$ The cross products are equal. Thus the two rates are equal. This is a proportion.

$$3 \times 18 = 54$$

Student Practice 5

(a) Is the rate $\dfrac{1260 \text{ words}}{7 \text{ pages}}$ equal to the rate $\dfrac{3530 \text{ words}}{20 \text{ pages}}$?

(b) Is the rate $\dfrac{2 \text{ U.S. dollars}}{11 \text{ Egyptian pounds}}$ equal to the rate $\dfrac{16 \text{ U.S. dollars}}{88 \text{ Egyptian pounds}}$?

Verbal and Writing Skills, Exercises 1 and 2

1. A proportion states that two ratios or rates are _____ .

2. Explain in your own words how we use the equality test for fractions to determine if a statement is a proportion. Give an example.

Write a proportion.

3. 6 is to 8 as 3 is to 4.

4. 18 is to 15 as 6 is to 5.

5. 20 is to 36 as 5 is to 9.

6. 120 is to 15 as 160 is to 20.

7. 220 is to 11 as 400 is to 20.

8. $2\frac{1}{2}$ is to 10 as $7\frac{1}{2}$ is to 30.

9. $4\frac{1}{3}$ is to 13 as $5\frac{2}{3}$ is to 17.

10. 8.5 is to 11 as 17 is to 22.

11. 6.5 is to 14 as 13 is to 28.

Applications, Exercises 12–21 *Write a proportion.*

12. *Cooking* When Jenny makes rice in her steamer, she mixes 2 cups of rice with 3 cups of water. To make 8 cups of rice, she needs 12 cups of water.

13. *Cartography* A cartographer (a person who makes maps) uses a scale of 3 inches to represent 40 miles. 27 inches would then represent 360 miles.

14. *Reading Speed* If Mallory can read 16 pages of her biology book in 2 hours, she can read 40 pages in 5 hours.

15. *Tips for Valets* Stephen works as a valet parker. If he earns $40 in tips for parking 12 cars, he should earn $60 for parking 18 cars.

16. *Food Cost* If 20 pounds of pistachio nuts cost $75, then 30 pounds will cost $112.50.

17. *Education* If three credit hours at El Paso Community College cost $525, then seven credit hours should cost $1225.

18. *Education* When Ridgewood Community College had 1200 students enrolled, 24 mathematics sections were offered. This year there are 1450 students enrolled, so 29 mathematics sections should be offered.

19. *Teaching Ratio* There are 3 teaching assistants for every 40 children in the elementary school. If we have 280 children, then we will have 21 teaching assistants.

20. *Lawn Care* If 16 pounds of fertilizer cover 1520 square feet of lawn, then 19 pounds of fertilizer should cover 1805 square feet of lawn.

21. *Restaurants* When New City had 4800 people, it had three restaurants. Now New City has 11,200 people, so it should have seven restaurants.

Determine which equations are proportions.

22. $\dfrac{8}{6} \overset{?}{=} \dfrac{20}{15}$

23. $\dfrac{10}{25} \overset{?}{=} \dfrac{6}{15}$

24. $\dfrac{15}{12} \overset{?}{=} \dfrac{16}{13}$

25. $\dfrac{8}{10} \overset{?}{=} \dfrac{13}{15}$

26. $\dfrac{99}{100} \overset{?}{=} \dfrac{49}{50}$

27. $\dfrac{17}{75} \overset{?}{=} \dfrac{22}{100}$

28. $\dfrac{315}{2100} \overset{?}{=} \dfrac{15}{100}$

29. $\dfrac{102}{120} \overset{?}{=} \dfrac{85}{100}$

30. $\dfrac{6}{14} \overset{?}{=} \dfrac{4.5}{10.5}$

31. $\dfrac{2.5}{4} \overset{?}{=} \dfrac{7.5}{12}$

32. $\dfrac{11}{12} \overset{?}{=} \dfrac{9.5}{10}$

33. $\dfrac{3}{17} \overset{?}{=} \dfrac{4.5}{24.5}$

34. $\dfrac{7}{1\frac{1}{2}} \overset{?}{=} \dfrac{14}{3}$

35. $\dfrac{6}{2\frac{1}{2}} \overset{?}{=} \dfrac{12}{5}$

36. $\dfrac{2\frac{1}{3}}{3} \overset{?}{=} \dfrac{7}{15}$

37. $\dfrac{7\frac{1}{3}}{3} \overset{?}{=} \dfrac{23}{9}$

38. $\dfrac{2.5}{\frac{1}{2}} \overset{?}{=} \dfrac{21}{5}$

39. $\dfrac{\frac{1}{4}}{2} \overset{?}{=} \dfrac{\frac{7}{20}}{2.8}$

40. $\dfrac{90 \text{ miles}}{6 \text{ hours}} \overset{?}{=} \dfrac{135 \text{ miles}}{9 \text{ hours}}$

41. $\dfrac{135 \text{ miles}}{3 \text{ hours}} \overset{?}{=} \dfrac{225 \text{ miles}}{5 \text{ hours}}$

42. $\dfrac{286 \text{ gallons}}{12 \text{ acres}} \overset{?}{=} \dfrac{429 \text{ gallons}}{18 \text{ acres}}$

43. $\dfrac{166 \text{ gallons}}{14 \text{ acres}} \overset{?}{=} \dfrac{249 \text{ gallons}}{21 \text{ acres}}$

44. $\dfrac{34 \text{ three-point shots}}{60 \text{ attempts}} \overset{?}{=} \dfrac{30 \text{ three-point shots}}{55 \text{ attempts}}$

45. $\dfrac{21 \text{ home runs}}{96 \text{ games}} \overset{?}{=} \dfrac{18 \text{ home runs}}{81 \text{ games}}$

Applications

46. *Concert Audiences* At the Michael W. Smith concert on Friday there were 9600 female fans and 8200 male fans. The concert on Saturday had 12,480 female fans and 10,660 male fans. Is the ratio of female fans to male fans the same for both nights of the concert?

47. *Baseball Team Wins* Since Harding High School opened, they have won 132 baseball games and have lost 22. Derry High School has won 160 games and lost 32 since it opened. Is the ratio of lost games to won games the same for both schools?

48. *Machine Operating Rate* A machine folds 650 boxes in five hours. Another machine folds 580 boxes in 4 hours.

(a) Do they fold boxes at the same rate?

(b) Which machine folds more boxes in 24 hours?

49. *Speed of Vehicle* A bus traveled 675 miles in 18 hours. A passenger van traveled 820 miles in 20 hours.

(a) Did they travel at the same rate?

(b) Which vehicle traveled at a faster rate?

▲ **50.** *Television Screen Size* A common size for a television screen is 22 inches wide by 16 inches tall. Does a smaller television screen that is 11 inches wide by 8.5 inches tall have the same ratio of width to height?

▲ **51.** *Driveway Size* A common size for a driveway in suburban Wheaton, Illinois, is 75 feet long by 20 feet wide. Does a larger driveway that is 105 feet long by 28 feet wide have the same ratio of width to length?

To Think About

52. Determine whether $\dfrac{63}{161} = \dfrac{171}{437}$

   (a) by reducing each side to lowest terms.

   (b) by using the equality test for fractions. (This is the cross-product method.)

   (c) Which method was faster? Why?

53. Determine whether $\dfrac{169}{221} = \dfrac{247}{323}$

   (a) by reducing each side to lowest terms.

   (b) by using the equality test for fractions. (This is the cross-product method.)

   (c) Which method was faster? Why?

Cumulative Review *Calculate.*

54. [3.3.1] $9.6 + 7.8 + 2.56 + 3.004 + 0.1765$

55. [3.4.1] 5.92×3.04

56. [3.3.2]
$$\begin{array}{r} 29{,}366.215 \\ -28{,}963.807 \end{array}$$

57. [3.5.2] $7.03\overline{)181.374}$

58. [2.9.1] *Walking Distance* Jamie has a goal of walking 20 miles this week. On Monday she walked $3\frac{1}{4}$ miles and on Tuesday she walked $4\frac{3}{8}$ miles. How many more miles does she need to walk to reach her goal?

Quick Quiz 4.2

1. Write as a proportion.

  8 is to 18 as 28 is to 63.

2. Write as a proportion.

  13 is to 32 as $3\frac{1}{4}$ is to 8.

3. Determine if this equation is a proportion.

$$\frac{15 \text{ shots}}{4 \text{ goals}} \overset{?}{=} \frac{75 \text{ shots}}{22 \text{ goals}}$$

4. Concept Check Explain how to determine if

$$\frac{33 \text{ chairs}}{45 \text{ employees}} \overset{?}{=} \frac{165 \text{ chairs}}{225 \text{ employees}} \text{ is a}$$

proportion.

How Am I Doing?   Sections 4.1–4.2

*How are you doing with your homework assignments in Sections 4.1 and 4.2?
Do you feel you have mastered the material so far? Do you understand the
concepts you have covered? Before you go further in the textbook, take some
time to do each of the following problems.*

4.1 *In questions 1–4, write each ratio in simplest form.*

1. 13 to 18

2. 44 to 220

3. $72 to $16

4. 135 meters to 165 meters

5. Eleanor's take-home pay is $330 per week. $60 per week is withheld for
federal taxes and $32 per week is withheld for state taxes.

   (a) Find the ratio of federal withholding to take-home pay.

   (b) Find the ratio of state withholding to take-home pay.

Write each rate in simplest form.

6. 9 flight attendants for 300 passengers

7. 620 gallons of water for each 840 square feet of lawn

Write as a unit rate. Round to the nearest tenth if necessary.

8. A professional bicyclist travels 65 miles in 4 hours. What is the rate in
miles per hour?

9. 18 MP3 players are purchased for $630. What is the cost per MP3 player?

10. In a certain recipe, 2400 cookies are made with 15 pounds of cookie
dough. How many cookies can be made with 1 pound of cookie dough?

4.2 *Write a proportion.*

11. 13 is to 40 as 39 is to 120

12. 116 is to 148 as 29 is to 37

13. If a speedboat can travel 33 nautical miles in 2 hours, then it can travel
49.5 nautical miles in 3 hours.

14. If the cost to manufacture 3000 athletic shoes is $370, then the cost to
manufacture 7500 athletic shoes is $925.

Determine whether each equation is a proportion.

15. $\dfrac{14}{31} = \dfrac{42}{93}$

16. $\dfrac{17}{33} = \dfrac{19}{45}$

17. $\dfrac{6.5}{4.8} = \dfrac{120}{96}$

18. $\dfrac{15}{24} = \dfrac{1\frac{5}{8}}{2\frac{3}{5}}$

19. The Pine Street Inn can produce 670 servings of a chicken dinner for
homeless people at a cost of $1541. It will therefore cost $1886 to produce 820 servings of the same chicken dinner.

20. For every 60 flights that arrive at Logan Airport in December approximately 8 of them are more than 15 minutes late. Therefore, for every
3000 flights that arrive at Logan Airport in December approximately
400 of them will be more than 15 minutes late.

*Now turn to page SA-8 for the answer to each of these problems. Each answer
also includes a reference to the objective in which the problem is first taught. If
you missed any of these problems, you should stop and review the Examples
and Student Practice problems in the referenced objective. A little review now
will help you master the material in the upcoming sections of the text.*

1. _____

2. _____

3. _____

4. _____

5. (a) _____ (b) _____

6. _____

7. _____

8. _____

9. _____

10. _____

11. _____

12. _____

13. _____

14. _____

15. _____

16. _____

17. _____

18. _____

19. _____

20. _____

4.3 Solving Proportions

Student Learning Objectives

After studying this section, you will be able to:

① Solve for the variable *n* in an equation of the form $a \times n = b$.

② Find the missing number in a proportion.

① Solving for the Variable *n* in an Equation of the Form $a \times n = b$

Consider this expression: "3 times a number yields 15. What is the number?" We could write this as

$$3 \times \boxed{?} = 15$$

and guess that the number $\boxed{?} = 5$. There is a better way of solving this problem, a way that eliminates the guesswork. We will begin by using a **variable.** That is, we will use a letter to represent a number we do not yet know. We briefly used variables in Chapters 1–3. Now we use them more extensively.

Let the letter *n* represent the unknown number. We write

$$3 \times n = 15.$$

This is called an **equation.** An equation has an equals sign. This indicates that the values on each side of it are equivalent. We want to find the number *n* in this equation without guessing. We will not change the value of *n* in the equation if we divide both sides of the equation by 3. Thus if

$$3 \times n = 15,$$

we can say $\quad \dfrac{3 \times n}{3} = \dfrac{15}{3},$

which is $\quad \dfrac{3}{3} \times n = 5$

or $\quad 1 \times n = 5.$

Since 1 × any number is the same number, we know that $n = 5$. Any equation of the form $a \times n = b$ can be solved in this way. We divide both sides of an equation of the form $a \times n = b$ by the number that is multiplied by *n*. (We do this because division is the inverse operation of multiplication. This method will not work for $3 + n = 15$, since here the 3 is added to *n* and not multiplied by *n*.)

EXAMPLE 1 Solve for *n*.

(a) $16 \times n = 80$ **(b)** $24 \times n = 240$

Solution

(a) $16 \times n = 80$

$\dfrac{16 \times n}{16} = \dfrac{80}{16}$ Divide each side by 16.

$\qquad n = 5$ because $16 \div 16 = 1$ and $80 \div 16 = 5$.

(b) $24 \times n = 240$

$\dfrac{24 \times n}{24} = \dfrac{240}{24}$ Divide each side by 24.

$\qquad n = 10$ because $24 \div 24 = 1$ and $240 \div 24 = 10$.

Student Practice 1 Solve for *n*.

(a) $5 \times n = 45$ **(b)** $7 \times n = 84$

NOTE TO STUDENT: Fully worked-out solutions to all of the Student Practice problems can be found at the back of the text starting at page SP-1.

The same procedure is followed if the variable n is on the right side of the equation.

EXAMPLE 2 Solve for n.

(a) $66 = 11 \times n$ **(b)** $143 = 13 \times n$

Solution

(a) $66 = 11 \times n$

$\dfrac{66}{11} = \dfrac{11 \times n}{11}$ Divide each side by 11.

$6 = n$

(b) $143 = 13 \times n$

$\dfrac{143}{13} = \dfrac{13 \times n}{13}$ Divide each side by 13.

$11 = n$

Student Practice 2 Solve for n.

(a) $108 = 9 \times n$ **(b)** $210 = 14 \times n$

The numbers in the equations are not always whole numbers, and the answer to an equation is not always a whole number.

EXAMPLE 3 Solve for n.

(a) $16 \times n = 56$ **(b)** $18.2 = 2.6 \times n$

Solution

(a) $16 \times n = 56$

$\dfrac{16 \times n}{16} = \dfrac{56}{16}$ Divide each side by 16.

$n = 3.5$

$$\begin{array}{r} 3.5 \\ 16\overline{)56.0} \\ 48 \\ \hline 8\,0 \\ 8\,0 \\ \hline 0 \end{array}$$

(b) $18.2 = 2.6 \times n$

$\dfrac{18.2}{2.6} = \dfrac{2.6 \times n}{2.6}$ Divide each side by 2.6

$7 = n$

$$\begin{array}{r} 7. \\ 2.6_\wedge\overline{)18.2_\wedge} \\ 18\,2 \\ \hline 0 \end{array}$$

Student Practice 3 Solve for n.

(a) $15 \times n = 63$ **(b)** $39.2 = 5.6 \times n$

② Finding the Missing Number in a Proportion

Sometimes one of the pieces of a proportion is unknown. We can use an equation such as $a \times n = b$ and solve for n to find the unknown quantity. Suppose we want to know the value of n in the proportion

$$\frac{5}{12} = \frac{n}{144}.$$

Since this is a proportion, we know that $5 \times 144 = 12 \times n$. Simplifying, we have

$$720 = 12 \times n.$$

Next we divide both sides by 12.

$$\frac{720}{12} = \frac{12 \times n}{12}$$

$$60 = n$$

We check to see if this is correct. Do we have a true proportion?

$$\frac{5}{12} \overset{?}{=} \frac{60}{144}$$

$$\frac{5}{12} \bowtie \frac{60}{144} \qquad \begin{array}{l} 12 \times 60 = 720 \\ 5 \times 144 = 720 \end{array} \quad \text{The cross products are equal.}$$

Thus $\dfrac{5}{12} = \dfrac{60}{144}$ is true. We have checked our answer.

TO SOLVE FOR A MISSING NUMBER IN A PROPORTION

1. Find the cross products.

2. Divide each side of the equation by the number being multiplied by n.

3. Simplify the result.

4. Check your answer.

EXAMPLE 4 Find the value of n in $\dfrac{25}{4} = \dfrac{n}{12}$.

Solution
$$25 \times 12 = 4 \times n \qquad \text{Find the cross products.}$$
$$300 = 4 \times n$$
$$\frac{300}{4} = \frac{4 \times n}{4} \qquad \text{Divide each side by 4.}$$
$$75 = n$$

Check. *Is this a proportion?*

$$\frac{25}{4} \overset{?}{=} \frac{75}{12}$$

$$25 \times 12 \overset{?}{=} 4 \times 75$$

$$300 = 300 \quad \checkmark$$

It is a proportion. The answer $n = 75$ is correct.

Student Practice 4 Find the value of n in

$$\frac{24}{n} = \frac{3}{7}.$$

The answer to the next problem is not a whole number.

EXAMPLE 5   Find the value of n in $\dfrac{125}{2} = \dfrac{150}{n}$.

Solution

$125 \times n = 2 \times 150$ Find the cross products.

$125 \times n = 300$

$\dfrac{125 \times n}{125} = \dfrac{300}{125}$ Divide each side by 125.

$n = 2.4$

Check. $\dfrac{125}{2} \stackrel{?}{=} \dfrac{150}{2.4}$

$125 \times 2.4 \stackrel{?}{=} 2 \times 150$

$300 = 300$ ✓

Student Practice 5   Find the value of n in

$$\frac{176}{4} = \frac{286}{n}.$$

EXAMPLE 6   Find the value of n in

$$\frac{n}{20} = \frac{\frac{3}{4}}{5}.$$

Solution $5 \times n = 20 \times \dfrac{3}{4}$ Find the cross products.

$5 \times n = 15$ Simplify.

$\dfrac{5 \times n}{5} = \dfrac{15}{5}$ Divide each side by 5.

$n = 3$

Check.   *Can you verify that this is a proportion?*

Student Practice 6   Find the value of n in

$$\frac{n}{30} = \frac{\frac{2}{3}}{4}.$$

In real-life situations it is helpful to write the units of measure in the proportion. Remember, order is important. The same units should be in the same position in the fractions.

EXAMPLE 7   If 5 grams of a non-icing additive are placed in 8 liters of diesel fuel, how many grams n should be added to 12 liters of diesel fuel?

Solution   We need to find the value of n in $\dfrac{n \text{ grams}}{12 \text{ liters}} = \dfrac{5 \text{ grams}}{8 \text{ liters}}$.

$8 \times n = 12 \times 5$

$8 \times n = 60$

$\dfrac{8 \times n}{8} = \dfrac{60}{8}$

$n = 7.5$

Continued on next page

The answer is 7.5 grams. 7.5 grams of the additive should be added to 12 liters of the diesel fuel.

Check.

$$\frac{7.5 \text{ grams}}{12 \text{ liters}} \overset{?}{=} \frac{5 \text{ grams}}{8 \text{ liters}}$$

$$7.5 \times 8 \overset{?}{=} 12 \times 5$$

$$60 = 60 \qquad \checkmark$$

Student Practice 7 If 2.5 tablespoons of a lawn fertilizer is to be mixed with 3 gallons of water, how many tablespoons of fertilizer should be mixed with 24 gallons of water?

Some answers will be exact values. In other cases we will obtain answers that are rounded to a certain decimal place. Recall that we sometimes use the $\approx$ symbol, which means "is approximately equal to."

EXAMPLE 8 Find the value of n in $\dfrac{141 \text{ miles}}{4.5 \text{ hours}} = \dfrac{67 \text{ miles}}{n \text{ hours}}$. Round to the nearest tenth.

Solution

$$141 \times n = 67 \times 4.5$$

$$141 \times n = 301.5$$

$$\frac{141 \times n}{141} = \frac{301.5}{141}$$

If we calculate to four decimal places, we have $n = 2.1382$. Rounding to the nearest tenth, $n \approx 2.1$.

The answer to the nearest tenth is $n = 2.1$. The check is up to you.

Student Practice 8 Find the value of n in $\dfrac{264 \text{ meters}}{3.5 \text{ seconds}} = \dfrac{n \text{ meters}}{2 \text{ seconds}}$. Round to the nearest tenth.

TO THINK ABOUT: Proportions with Mixed Numbers or Fractions
Suppose that the proportion contains many fractions or mixed numbers. Could you still follow all the steps? For example, find n when

$$\frac{n}{3\frac{1}{4}} = \frac{5\frac{1}{6}}{2\frac{1}{3}}$$

We have $2\frac{1}{3} \times n = 5\frac{1}{6} \times 3\frac{1}{4}$.

This can be written as

$$\frac{7}{3} \times n = \frac{31}{6} \times \frac{13}{4}$$

$$\boxed{\frac{7}{3} \times n = \frac{403}{24}} \qquad \text{equation (1)}$$

Now we divide each side of equation (1) by $\dfrac{7}{3}$. Why?

$$\frac{\dfrac{7}{3} \times n}{\dfrac{7}{3}} = \frac{\dfrac{403}{24}}{\dfrac{7}{3}}$$

Be careful here. The right-hand side means $\dfrac{403}{24} \div \dfrac{7}{3}$, which we evaluate by *inverting* the second fraction and multiplying.

$$\frac{403}{\underset{8}{\cancel{24}}} \times \frac{\overset{1}{\cancel{3}}}{7} = \frac{403}{56}$$

Thus $n = \frac{403}{56}$ or $7\frac{11}{56}$. Think about all the steps to solving this problem. Can you follow them? There is another way to do the problem. We could multiply each side of equation (1) by $\frac{3}{7}$.

$$\frac{7}{3} \times n = \frac{403}{24} \qquad \text{equation (1)}$$

$$\frac{3}{7} \times \frac{7}{3} \times n = \frac{3}{7} \times \frac{403}{24}$$

$$n = \frac{403}{56}$$

Why does this work? Now try exercises 56–59 in Exercises 4.3.

Accuracy is especially important in this section. When you do Exercises 4.3, be sure to verify your answers on pages SA-8 and SA-9. Take a little extra time with those problems that result in fraction or decimal answers. It is very important that you learn how to do this type of problem.

Verbal and Writing Skills, Exercises 1 and 2

1. Suppose you have an equation of the form $a \times n = b$, where the letters a and b represent whole numbers and $a \neq 0$. Explain in your own words how you would solve the equation.

2. Suppose you have an equation of the form $\frac{n}{a} = \frac{b}{c}$, where a, b, and c represent whole numbers and $a, c \neq 0$. Explain in your own words how you would solve the equation.

Solve for n.

3. $8 \times n = 72$

4. $6 \times n = 72$

5. $3 \times n = 16.8$

6. $4 \times n = 33.2$

7. $n \times 6.7 = 134$

8. $n \times 2.9 = 58$

9. $50.4 = 6.3 \times n$

10. $40.6 = 5.8 \times n$

11. $\frac{4}{9} \times n = 22$ (*Hint:* Divide each side by $\frac{4}{9}$.)

12. $\frac{6}{7} \times n = 26$ (*Hint:* Divide each side by $\frac{6}{7}$.)

Find the value of n. Check your answer.

13. $\frac{n}{20} = \frac{3}{4}$

14. $\frac{n}{28} = \frac{3}{7}$

15. $\frac{6}{n} = \frac{3}{8}$

16. $\frac{10}{n} = \frac{2}{5}$

17. $\frac{12}{40} = \frac{n}{25}$

18. $\frac{13}{30} = \frac{n}{15}$

19. $\frac{50}{100} = \frac{2.5}{n}$

20. $\frac{40}{160} = \frac{1.5}{n}$

21. $\frac{n}{6} = \frac{150}{12}$

22. $\frac{n}{22} = \frac{25}{11}$

23. $\frac{15}{4} = \frac{n}{6}$

24. $\frac{18}{12} = \frac{n}{9}$

25. $\frac{240}{n} = \frac{5}{4}$

26. $\frac{180}{n} = \frac{4}{3}$

Solve for n. Round your answer to the nearest tenth when necessary.

27. $\frac{21}{n} = \frac{2}{3}$

28. $\frac{62}{n} = \frac{5}{4}$

29. $\frac{9}{26} = \frac{n}{52}$

30. $\frac{12}{8} = \frac{21}{n}$

31. $\frac{15}{12} = \frac{10}{n}$

32. $\frac{n}{16} = \frac{2.5}{1}$

33. $\frac{n}{36} = \frac{4.5}{1}$

34. $\frac{2.5}{n} = \frac{0.5}{10}$

35. $\frac{1.8}{n} = \frac{0.7}{12}$

36. $\frac{4}{15} = \frac{n}{25.6}$

37. $\frac{11}{12} = \frac{n}{32.8}$

38. $\frac{10.5}{22} = \frac{n}{9}$

39. $\frac{13.8}{15} = \frac{n}{6}$

40. $\frac{5}{n} = \frac{12\frac{1}{2}}{100}$

41. $\frac{3}{n} = \frac{6\frac{1}{4}}{100}$

Applications *Find the value of n. Round to the nearest hundredth when necessary. If fractions are involved, express your answer as a fraction or mixed number.*

42. $\dfrac{n \text{ grams}}{10 \text{ liters}} = \dfrac{7 \text{ grams}}{25 \text{ liters}}$

43. $\dfrac{n \text{ pounds}}{20 \text{ ounces}} = \dfrac{2 \text{ pounds}}{32 \text{ ounces}}$

44. $\dfrac{190 \text{ kilometers}}{3 \text{ hours}} = \dfrac{n \text{ kilometers}}{5 \text{ hours}}$

45. $\dfrac{145 \text{ kilometers}}{2 \text{ hours}} = \dfrac{220 \text{ kilometers}}{n \text{ hours}}$

46. $\dfrac{50 \text{ gallons}}{12 \text{ acres}} = \dfrac{36 \text{ gallons}}{n \text{ acres}}$

47. $\dfrac{32 \text{ meters}}{5 \text{ yards}} = \dfrac{24 \text{ meters}}{n \text{ yards}}$

48. $\dfrac{3 \text{ kilograms}}{6.6 \text{ pounds}} = \dfrac{n \text{ kilograms}}{10 \text{ pounds}}$

49. $\dfrac{36.4 \text{ feet}}{5 \text{ meters}} = \dfrac{n \text{ feet}}{12 \text{ meters}}$

50. $\dfrac{12 \text{ quarters}}{3 \text{ dollars}} = \dfrac{87 \text{ quarters}}{n \text{ dollars}}$

51. $\dfrac{35 \text{ dimes}}{3.5 \text{ dollars}} = \dfrac{n \text{ dimes}}{8 \text{ dollars}}$

52. $\dfrac{2\frac{1}{2} \text{ acres}}{3 \text{ people}} = \dfrac{n \text{ acres}}{5 \text{ people}}$

53. $\dfrac{3\frac{1}{4} \text{ feet}}{8 \text{ pounds}} = \dfrac{n \text{ feet}}{12 \text{ pounds}}$

▲ **54.** *Photography* You have a photographic negative that is 3.5 centimeters wide and 2.5 centimeters tall. If you want to make a color print that is 6 centimeters tall, how wide will the print be?

▲ **55.** *Photography* A color photograph is 5 inches wide and 3 inches tall. If you want to make an enlargement of this photograph that is 6.6 inches tall, how wide will the enlargement be?

To Think About *Study the "To Think About" example in the text on page 282. Then solve for n in exercises 56–59. Express n as a mixed number.*

56. $\dfrac{n}{7\frac{1}{4}} = \dfrac{2\frac{1}{5}}{4\frac{1}{8}}$

57. $\dfrac{n}{2\frac{1}{3}} = \dfrac{4\frac{5}{6}}{3\frac{1}{9}}$

58. $\dfrac{9\frac{3}{4}}{n} = \dfrac{8\frac{1}{2}}{4\frac{1}{3}}$

59. $\dfrac{8\frac{1}{6}}{n} = \dfrac{5\frac{1}{2}}{7\frac{1}{3}}$

Cumulative Review *Evaluate by doing each operation in the proper order.*

60. [1.6.2] $4^3 + 20 \div 5 + 6 \times 3 - 5 \times 2$

61. [1.6.2] $(3 + 1)^3 - 30 \div 6 - 144 \div 12$

62. [3.1.1] Write a word name for the decimal 0.563.

63. [3.1.1] Write thirty-four ten-thousandths in decimal notation.

64. [1.8.2] *Profit on Cell Phones* If a man purchases 156 cell phones for $32 each and sells half of them for $45 and half of them for $39, how much profit will he make?

65. [1.8.2] *Soccer Team* The North Bend Women's Soccer League has eight teams. If each team plays all the others twice, how many games will have been played? (Think carefully. This is a challenging question.)

Quick Quiz 4.3

1. Solve.

$$\frac{n}{26} = \frac{9}{130}$$

2. Solve.

$$\frac{8}{6} = \frac{2\frac{2}{3}}{n}$$

3. Solve. Round to the nearest tenth.

$$\frac{17 \text{ hits}}{93 \text{ pitches}} = \frac{n \text{ hits}}{62 \text{ pitches}}$$

4. Concept Check Explain how you would solve the proportion $\dfrac{2\frac{1}{2}}{3\frac{3}{4}} = \dfrac{16\frac{1}{2}}{n}$.

4.4 Solving Applied Problems Involving Proportions

① Solving Applied Problems Using Proportions

Let us examine a variety of applied problems that can be solved by proportions.

Student Learning Objective

After studying this section, you will be able to:

① Solve applied problems using proportions.

EXAMPLE 1 A company that makes eyeglasses conducted a recent survey using a quality control test. It was discovered that 37 pairs of eyeglasses in a sample of 120 pairs of eyeglasses were defective. If this rate remains the same each year, how many of the 36,000 pairs of eyeglasses made by this company each year are defective?

Solution

Mathematics Blueprint for Problem Solving

Gather the Facts	What Am I Asked to Do?	How Do I Proceed?	Key Points to Remember
Sample: 37 defective pairs in a total of 120 pairs 36,000 pairs were made by the company.	Find how many of the 36,000 pairs of eyeglasses are defective.	Set up a proportion comparing defective eyeglasses to total eyeglasses.	Make sure one fraction represents the sample and one fraction represents the total number of eyeglasses made by the company.

We will use the letter n to represent the number of defective eyeglasses in the total.

$$\underbrace{\frac{37 \text{ defective pairs}}{120 \text{ total pairs of eyeglasses}}}_{\text{We compare the sample}} = \underbrace{\frac{n \text{ defective pairs}}{36,000 \text{ total pairs of eyeglasses}}}_{\text{to the total number}}$$

$$37 \times 36,000 = 120 \times n \quad \text{Find the cross products.}$$
$$1,332,000 = 120 \times n \quad \text{Simplify.}$$
$$\frac{1,332,000}{120} = \frac{120 \times n}{120} \quad \text{Divide each side by 120.}$$
$$11,100 = n$$

Thus, if the rate of defective eyeglasses holds steady, there are about 11,100 defective pairs of eyeglasses made by the company each year.

Student Practice 1 Yesterday an automobile assembly line produced 243 engines, of which 27 were defective. If the same rate is true each day, how many of the 4131 engines produced this month are defective?

NOTE TO STUDENT: Fully worked-out solutions to all of the Student Practice problems can be found at the back of the text starting at page SP-1.

Looking back at Example 1, perhaps it occurred to you that the fractions in the proportion could be set up in an alternative way. **You can set up this problem in several different ways as long as the units are in correctly**

corresponding positions. It would be correct to set up the problem in the form

$$\frac{\text{defective pairs in sample}}{\text{total defective pairs}} = \frac{\text{total glasses in sample}}{\text{total glasses made by company}}$$

or

$$\frac{\text{total glasses in sample}}{\text{defective pairs in sample}} = \frac{\text{total glasses made by company}}{\text{total defective pairs}}$$

But we **cannot** set up the problem this way.

$$\frac{\text{defective pairs in sample}}{\text{total glasses made by company}} = \frac{\text{total defective pairs}}{\text{total glasses in sample}}$$

This is *not* correct. Do you see why?

EXAMPLE 2 Ted's car can go 245 miles on 7 gallons of gas. Ted wants to take a trip of 455 miles. Approximately how many gallons of gas will this take?

Solution Let n = the unknown number of gallons.

$$\frac{245 \text{ miles}}{7 \text{ gallons}} = \frac{455 \text{ miles}}{n \text{ gallons}}$$

$245 \times n = 7 \times 455$ Find the cross products.

$245 \times n = 3185$ Simplify.

$$\frac{245 \times n}{245} = \frac{3185}{245}$$ Divide both sides by 245.

$n = 13$

Ted will need approximately 13 gallons of gas for the trip.

Student Practice 2 Cindy's car travels 234 miles on 9 gallons of gas. How many gallons of gas will Cindy need to take a 312-mile trip?

EXAMPLE 3 In a certain gear, Alice's 18-speed bicycle has a gear ratio of three revolutions of the pedal for every two revolutions of the bicycle wheel. If her bicycle wheel is turning at 65 revolutions per minute, how many times must she pedal per minute?

Solution Let n = the number of revolutions of the pedal.

$$\frac{3 \text{ revolutions of the pedal}}{2 \text{ revolutions of the wheel}} = \frac{n \text{ revolutions of the pedal}}{65 \text{ revolutions of the wheel}}$$

$3 \times 65 = 2 \times n$ Find the cross products.

$195 = 2 \times n$ Simplify.

$$\frac{195}{2} = \frac{2 \times n}{2}$$ Divide both sides by 2.

$97.5 = n$

Alice will pedal at the rate of 97.5 revolutions per 1 minute.

Student Practice 3   Alicia must pedal at 80 revolutions per minute to ride her bicycle at 16 miles per hour. If she pedals at 90 revolutions per minute, how fast will she be riding?

EXAMPLE 4   Tim operates a bicycle rental center during the summer months on the island of Martha's Vineyard. He discovered that when the ferryboats brought 8500 passengers a day to the island, his center rented 340 bicycles a day. Next summer the ferryboats plan to bring 10,300 passengers a day to the island. How many bicycles a day should Tim plan to rent?

Solution   Two important cautions are necessary before we solve the proportion. We need to be sure that the bicycle rentals are directly related to the number of people on the ferryboat. (Presumably, people who fly to the island or who take small pleasure boats to the island also rent bicycles.)

Next we need to be sure that the people who represent the increase in passengers per day would be as likely to rent bicycles as the present number of passengers do. For example, if the new visitors to the island are all senior citizens, they are not as likely to rent bicycles as younger people. If we assume those two conditions are satisfied, then we can solve the problem as follows.

$$\frac{8500 \text{ passengers per day now}}{340 \text{ bike rentals per day now}} = \frac{10{,}300 \text{ passengers per day later}}{n \text{ bike rentals per day later}}$$

$$8500 \times n = 340 \times 10{,}300$$
$$8500 \times n = 3{,}502{,}000$$
$$\frac{8500 \times n}{8500} = \frac{3{,}502{,}000}{8500}$$
$$n = 412$$

If the two conditions are satisfied, we would predict 412 bicycle rentals.

Student Practice 4   For every 4050 people who walk into Tom's Souvenir Shop, 729 make a purchase. Assuming the same conditions, if 5500 people walk into Tom's Souvenir Shop, how many people may be expected to make a purchase?

Wildlife Population Counting   Biologists and others who observe or protect wildlife sometimes use the capture-mark-recapture method to determine how many animals are in a certain region. In this approach some animals are caught and tagged in a way that does not harm them. They are then released into the wild, where they mix with their kind.

It is assumed (usually correctly) that the tagged animals will mix throughout the entire population in that region, so that when they are recaptured in a future sample, the biologists can use them to make reasonable estimates about the total population. We will employ the capture-mark-recapture method in the next example.

EXAMPLE 5 A biologist catches 42 fish in a lake and tags them. She then quickly returns them to the lake. In a few days she catches a new sample of 50 fish. Of those 50 fish, 7 have her tag. Approximately how many fish are in the lake?

Solution

$$\frac{42 \text{ fish tagged in 1st sample}}{n \text{ fish in lake}} = \frac{7 \text{ fish tagged in 2nd sample}}{50 \text{ fish caught in 2nd sample}}$$

$$42 \times 50 = 7 \times n$$
$$2100 = 7 \times n$$
$$\frac{2100}{7} = \frac{7 \times n}{7}$$
$$300 = n$$

Assuming that no tagged fish died and that the tagged fish mixed throughout the population of fish in the lake, we estimate that there are 300 fish in the lake.

Student Practice 5 A park ranger in Alaska captures and tags 50 bears. He then releases them to range through the forest. Sometime later he captures 50 bears. Of the 50, 4 have tags from the previous capture. Estimate the number of bears in the forest.

STEPS TO SUCCESS Help! When Do I Get It? Where Do I Get It?

Getting the right kind of help at the right time in this course can be the key ingredient in being successful in Basic College Mathematics. When you have made every effort to learn the math on your own and you still need assistance it is time to go for help.

When should you go for help? As soon as you discover that you don't understand something. Don't wait until the night before a test. When you try the homework and have trouble and you are unable to clear up the difficulty in the next class period *that is the time to immediately seek help.*

Where do you go for help? The best source is your instructor. Make an appointment to see your instructor and explain exactly what you are having trouble with. If that is not possible use the tutoring services at your college, or visit the mathematics lab, watch the videotapes, use MyMathLab and let it help you, or call the 1-800 phone number for phone tutoring. You may want to talk with a classmate and see if you can help one another.

Making it personal: Look over these suggestions. Pick the one that you think is the best source of help and then use it this week. Make it a priority in your life to get help immediately in this math course whenever you do not understand something. We all need a little help from time to time. ▼

Verbal and Writing Skills

1. Dan saw 12 people on the beach on Friday night. He counted 5 dogs on the beach at that time. On Saturday night he saw 60 people on the same beach. He is trying to estimate how many dogs might have been on the beach Saturday night. He started by writing the equation

$$\frac{12 \text{ people}}{5 \text{ dogs}} = \underline{\qquad}.$$

Explain how he should set up the rest of the proportion.

2. Connie drove to the top of Mount Washington. As she drove up the roadway she observed 15 cars. On the same trip she observed 17 people walking. Later that afternoon she drove down the mountain. On that trip she observed 60 cars. She is trying to estimate how many people she might have seen walking. She started by writing the equation

$$\frac{15 \text{ cars}}{17 \text{ people}} = \underline{\qquad}.$$

Explain how she should set up the rest of the proportion.

Applications

3. **Car Repairs** An automobile dealership has found that for every 140 cars sold, 23 will be brought back to the dealer for major repairs. If the dealership sells 980 cars this year, approximately how many cars will be brought back for major repairs?

4. **Hotel Management** The policy at the Colonnade Hotel is to have 19 desserts for every 16 people if a buffet is being served. If the Saturday buffet has 320 people, how many desserts must be available?

5. **Consumer Product Use** The directions on a bottle of bleach say to use $\frac{3}{4}$ cup bleach for every 1 gallon of soapy water. Ron needs a 4-gallon mixture to mop his floors. How many cups of bleach will he need?

6. **Food Preparation** To make an 8-ounce serving of hot chocolate, $1\frac{1}{2}$ tablespoons of cocoa are needed. How much cocoa is needed to make 12 ounces of hot chocolate?

7. **Measurement** There are approximately $1\frac{1}{2}$ kilometers in one mile. Approximately how many kilometers are in 5 miles?

8. **Measurement** There are approximately $2\frac{1}{2}$ centimeters in one inch. Approximately how many centimeters are in one foot?

9. **Exchange Rate** When Anna flew to India in 2010, the exchange rate was 230 Indian rupees for every 5 U.S. dollars. If Anna brought 120 U.S. dollars for spending money, how many Indian rupees did she receive?

10. **Exchange Rate** One day in August 2010, one U.S. dollar was worth 0.627 British pounds. Huang exchanged 240 U.S. dollars when he arrived in London. How many British pounds did he receive?

Shadow Length *In exercises 11 and 12 two nearby objects cast shadows at the same time of day. The ratio of the height of one of the objects to the length of its shadow is equal to the ratio of the height of the other object to the length of its shadow.*

▲ **11.** A pro football offensive tackle who stands 6.5 feet tall casts a 5-foot shadow. At the same time, the football stadium, which he is standing next to, casts a shadow of 152 feet. How tall is the stadium? If necessary, round your answer to the nearest tenth.

▲ **12.** In Copper Center, Alaska, at 2:00 P.M., a boulder that is 7 feet high casts a shadow that is 11 feet long. At that same time, Melinda Tobey is standing by a tree that is on the bank of the river. She measured the tree and found it is exactly 22 feet tall. The tree has a shadow that crosses the entire width of the river. How wide is the river? Round your answer to the nearest tenth.

13. ***Map Scale*** On a tour guide map of Madagascar, the scale states that 3 inches represent 125 miles. Two beaches are 5.2 inches apart on the map. What is the approximate distance in miles between the two beaches? Round your answer to the nearest mile.

14. ***Map Scale*** On Dr. Jennings's map of Antarctica, the scale states that 4 inches represent 250 miles of actual distance. Two Antarctic mountains are 5.7 inches apart on the map. What is the approximate distance in miles between the two mountains? Round your answer to the nearest mile.

15. ***Food Preparation*** In his curried chicken recipe, Deepak uses 3 cups of curry sauce for every 8 people. How many cups of curry sauce will he need to make curry chicken for a dinner party of 34 people? Write your answer as a mixed number.

16. ***Food Preparation*** Bianca's chocolate fondue recipe uses 4 cups of chocolate chips to make fondue sauce for 6 people. How many cups of chocolate chips are needed to make fondue for 26 people? Write your answer as a mixed number.

17. ***Basketball*** During a basketball game against the Miami Heat, the Denver Nuggets made 17 out of 25 free throws attempted. If they attempt 150 free throws in the remaining games of the season, how many will they make if their success rate remains the same?

18. ***Baseball*** A baseball pitcher gave up 52 earned runs in 260 innings of pitching. At that rate, how many runs would he give up in a 9-inning game? (This decimal is called the pitcher's *earned run average.*)

19. ***Fuel Efficiency*** In her Honda Civic, Claire can drive 192 miles on 6 gallons of gas. During spring break, she plans to drive 600 miles. How many gallons of gas will she use?

20. ***Fuel Efficiency*** In her Nissan Pathfinder, Juanita can drive 75 miles on 5 gallons of gas. She drove 318 miles for a business trip. How many gallons of gas did she use?

21. ***Wildlife Population Counting*** An ornithologist is studying hawks in the Adirondack Mountains. She catches 24 hawks over a period of one month, tags them, and releases them back into the wild. The next month, she catches 20 hawks and finds that 12 are already tagged. Estimate the number of hawks in this part of the mountains.

22. ***Wildlife Population Counting*** At a game preserve in Kenya, a worker captures 26 giraffes, tags them, and then releases them back into the preserve. The next month, he captures 18 giraffes and finds that 6 of them have already been tagged. Estimate the number of giraffes on the preserve.

23. **Farming** Bill and Shirley Grant are raising tomatoes to sell at the local co-op. The farm has a yield of 425 pounds of tomatoes for every 3 acres. The farm has 14 acres of good tomatoes. The crop this year should bring $1.80 per pound for each pound of tomatoes. How much will the Grants get from the sale of the tomato crop?

▲ 24. **Painting** A paint manufacturer suggests 2 gallons of flat latex paint for every 750 square feet of wall. A painter is going to paint 7875 square feet of wall in a Dallas office building with paint that costs $8.50 per gallon. How much will the painter spend for the paint?

25. **Manufacturing Quality** A company that manufactures computer chips expects 5 out of every 100 made to be defective. In a shipment of 5400 chips, how many are expected to be defective?

26. **Customer Satisfaction** The editor of a small-town newspaper conducted a survey to find out how many customers are satisfied with their delivery service. Of the 100 people surveyed, 88 customers said they were satisfied. If 1700 people receive the newspaper, how many are satisfied?

To Think About

Cooking *The following chart is used for several brands of instant mashed potatoes. Use this chart in answering exercises 27–30.*

TO MAKE	WATER	MARGARINE OR BUTTER	SALT (optional)	MILK	FLAKES
2 servings	2/3 cup	1 tablespoon	1/8 teaspoon	1/4 cup	2/3 cup
4 servings	1-1/3 cups	2 tablespoons	1/4 teaspoon	1/2 cup	1-1/3 cups
6 servings	2 cups	3 tablespoons	1/2 teaspoon	3/4 cup	2 cups
Entire box	5 cups	1/2 cup	1 teaspoon	2-1/2 cups	Entire box

27. How many cups of water and how many cups of milk are needed to make enough mashed potatoes for three people?

28. How many cups of water and how many cups of milk are needed to make enough mashed potatoes for five people?

29. Phil and Melissa went camping near Denver. They found that the instructions on the box say that at high altitudes (above 5000 feet) the amount of water should be increased by $\frac{1}{4}$. If you had to make enough mashed potatoes for eight people in a high-altitude city, how many cups of water and how many cups of milk would be needed?

30. Noah and Olivia went camping near Denver. They found that the instructions on the box say that at high altitudes (above 5000 feet) the amount of water should be increased by $\frac{1}{4}$.
    (a) If you had to make two boxes of mashed potatoes at a high altitude, how many cups of water and how many cups of milk would be used?
    (b) How many servings would be obtained?

Baseball Salaries *During the 2009 baseball season, Albert Pujols of the St. Louis Cardinals hit 47 home runs and was paid an annual salary of $14,427,326. During the same season, Ryan Howard of the Philadelphia Phillies hit 45 home runs and was paid an annual salary of $15,000,000. (Sources: www.usatoday.com and www.espn.com)*

31. Express the salary of each player as a unit rate in terms of dollars paid to home runs hit.

32. Which player hit more home runs per dollar?

Basketball Salaries During the 2009–2010 basketball season, Kobe Bryant of the Los Angeles Lakers made 617 two-point shots, and he was paid an annual salary of $23,034,375. During the same season, Tim Duncan of the San Antonio Spurs made 559 two-point shots, and he was paid an annual salary of $22,183,218. (Sources: www.usatoday.com and www.espn.com)

33. Express the salary of each player as a unit rate in terms of dollars paid to shots made.

34. Which player made more two-point shots per dollar?

Cumulative Review

35. **[3.7.2]** Calvin purchased 3 pounds of wild rice salad from the local deli as his contribution for the office picnic. The salad cost $5.75 per pound. How much did Calvin spend?

36. **[3.7.2]** Kate put 12.5 gallons of gasoline in her car. If the price per gallon was $2.80, how much did she pay for the gasoline?

37. **[3.2.3]** Round to the nearest tenth. 56.148

38. **[3.2.3]** Round to the nearest ten-thousandth. 2.74895

▲ 39. **[2.9.1]** *Sunglass Production* An eyewear company makes very expensive carbon fiber sunglasses. The material is made into long sheets, and the basic eyeglass frame is punched out by a machine. It takes a section measuring $1\frac{3}{16}$ feet by $\frac{4}{5}$ foot to make one pair of glasses.

(a) How many square feet of this material are needed for one frame?

(b) How many square feet of this material are needed for 1500 frames?

Quick Quiz 4.4 *Solve using a proportion. Round your answer to the nearest hundredth when necessary.*

1. A copper cable 36 feet long weighs 160 pounds. How much will 54 feet of this cable weigh?

2. If 11 inches on a map represent a distance of 64 miles, what distance does 5 inches represent?

3. During the first few games of the basketball season, Caleb shot 16 free throws and made 7 of them. If Caleb shoots free throws at the same rate as the first few games, he expects to shoot 100 more during the rest of the season. How many of these additional free throws should he expect to make? Round to the nearest whole number.

4. **Concept Check** When Fred went to France he discovered that 70 euros were worth 104 American dollars. He brought 400 American dollars on his trip. Explain how he would find what that is worth in euros.

Did You Know...
That You Can Save Money Each Semester by Choosing the Right Meal Plan?

CHOOSING A MEAL PLAN

Understanding the Problem:

Jake is a college student who lives in a college dormitory. He is going over the meal plans and calculating his food budget for the semester. He is interested in choosing a meal plan that will cost him the least while still meeting his needs.

Making a Plan:

First he needs to compare the prices and options for the meal plans. Then he needs to determine how many meals he will eat in the dining halls so he can determine which plan is right for him.

Step 1: There are three dining plans that cover the 15-week semester.

- The Gold plan costs $2205 a semester and covers 21 meals a week.
- The Silver plan costs $1764 a semester and covers 16 meals a week, with extra meals costing $10.
- The Bronze plan costs $1386 a semester and covers 12 meals a week, with extra meals costing $10.

Which plan is best for the following situations:

Task 1: *Jake plans on eating 20 meals per week in the dining halls?*

Task 2: *Jake plans on eating 18 meals per week in the dining halls?*

Task 3: *Jake plans on eating 16 meals per week in the dining halls?*

Task 4: *Jake plans on eating 14 meals per week in the dining halls?*

Step 2: Jake also has the option of buying "dining dollars" to use at other eating establishments on campus. With dining dollars Jake will get $\frac{1}{10}$ off the cost of all food purchases.

How much will Jake save over the course of the semester by using dining dollars in the following situations:

Task 5: *if he spends $20 a week at these establishments?*

Task 6: *if he spends $40 a week at these establishments?*

Task 7: *if he spends $60 a week at these establishments?*

Finding a Solution:

Step 3: Jake plans on eating 17 meals a week in the dining halls and spending $30 a week at other establishments.

Task 8: *Which plan should he buy?*

Task 9: *How much will Jake spend on food for the semester?*

Applying the Situation to Your Life:

You can do these same calculations for your own circumstances if your college has a meal plan.

- See what plans are offered.
- Determine how often you will eat in the dining halls.
- Decide which plan would be right for you.
- Calculate how much you will need to spend on food for the semester.

Chapter 4 Organizer

Topic and Procedure	Examples	✏ You Try It
Forming a ratio, p. 263 A *ratio* is the comparison of two quantities that have the same units. A ratio is usually expressed as a fraction. The fraction should be in reduced form.	**(a)** Find the ratio of 20 books to 35 books. $$\frac{20}{35} = \frac{4}{7}$$ **(b)** Find the ratio in simplest form of 88 : 99. $$\frac{88}{99} = \frac{8}{9}$$ **(c)** Bob earns \$250 each week, but \$15 is withheld for medical insurance. Find the ratio of medical insurance to total pay. $$\frac{\$15}{\$250} = \frac{3}{50}$$	1. **(a)** Find the ratio of 15 cars to 45 cars. **(b)** Find the ratio in simplest form of 64 : 80. **(c)** Maya earns \$320 each week, but \$70 is withheld for tax. Find the ratio of tax to total pay.
Forming a rate, p. 265 A *rate* is a comparison of two quantities that have different units. A rate is usually expressed as a fraction in reduced form.	A college has 2520 students with 154 faculty. What is the rate of students to faculty? $$\frac{2520 \text{ students}}{154 \text{ faculty}} = \frac{180 \text{ students}}{11 \text{ faculty}}$$	2. An elementary school has 27 teachers and 720 students. What is the rate of teachers to students?
Forming a unit rate, p. 265 A *unit rate* is a rate with a denominator of 1. Divide the denominator into the numerator to obtain the unit rate.	**(a)** A car traveled 416 miles in 8 hours. Find the unit rate. $$\frac{416 \text{ miles}}{8 \text{ hours}} = 52 \text{ miles/hour}$$ **(b)** Bob spread 50 pounds of fertilizer over 1870 square feet of land. Find the unit rate of square feet per pound. $$\frac{1870 \text{ square feet}}{50 \text{ pounds}} = 37.4 \text{ square feet/pound}$$	3. **(a)** A car traveled 756 miles in 12 hours. Find the unit rate. **(b)** At a new playground, 1550 pounds of sand were spread over 620 square feet. Find the unit rate of pounds per square feet.
Writing proportions, p. 271 A *proportion* is a statement that two rates or two ratios are equal. The proportion statement *a* is to *b* as *c* is to *d* can be written $$\frac{a}{b} = \frac{c}{d}.$$	Write a proportion for 17 is to 34 as 13 is to 26. $$\frac{17}{34} = \frac{13}{26}$$	4. Write a proportion for 15 is to 60 as 13 is to 52.
Determining whether a relationship is a proportion, p. 271 For any two fractions where $b \neq 0$ and $d \neq 0$, $\frac{a}{b} = \frac{c}{d}$ if and only if $a \times d = b \times c$. A proportion is a statement that two rates or two ratios are equal.	**(a)** Is this a proportion? $\frac{7}{56} \stackrel{?}{=} \frac{3}{24}$ $$7 \times 24 \stackrel{?}{=} 56 \times 3$$ $$168 = 168$$ It is a proportion. **(b)** Is this a proportion? $$\frac{64 \text{ gallons}}{5 \text{ acres}} \stackrel{?}{=} \frac{89 \text{ gallons}}{7 \text{ acres}}$$ $$64 \times 7 \stackrel{?}{=} 5 \times 89$$ $$448 \neq 445$$ It is not a proportion.	5. **(a)** Is this a proportion? $\frac{6}{70} \stackrel{?}{=} \frac{2}{24}$ **(b)** Is this a proportion? $$\frac{45 \text{ pounds}}{30 \text{ square feet}} \stackrel{?}{=} \frac{54 \text{ pounds}}{36 \text{ square feet}}$$
Solving a proportion, p. 278 To solve a proportion where the value *n* is not known: **1.** Find the cross products. **2.** Divide both sides of the equation by the number being multiplied by *n*.	Solve for *n*. $$\frac{17}{n} = \frac{51}{9}$$ $$17 \times 9 = 51 \times n \quad \text{Find the cross products.}$$ $$153 = 51 \times n \quad \text{Simplify.}$$ $$\frac{153}{51} = \frac{51 \times n}{51} \quad \text{Divide by 51.}$$ $$3 = n$$	6. Solve for *n*. $$\frac{n}{4} = \frac{49}{28}$$

Topic and Procedure	Examples	✏ You Try It
Solving applied problems, p. 287 **1.** Write a proportion with n representing the unknown value. **2.** Solve the proportion.	Bob purchased eight notebooks for $19. How much would 14 notebooks cost? $$\frac{8 \text{ notebooks}}{\$19} = \frac{14 \text{ notebooks}}{n}$$ $$8 \times n = 19 \times 14$$ $$8 \times n = 266$$ $$\frac{8 \times n}{8} = \frac{266}{8}$$ $$n = 33.25$$ The 14 notebooks would cost $33.25.	**7.** Emily purchased ten mechanical pencils for $24. How much would 15 pencils cost?

Chapter 4 Review Problems

Section 4.1

Write in simplest form. Express your answer as a fraction.

1. $88 : 40$

2. $28 : 35$

3. $250 : 475$

4. $2\frac{1}{3}$ to $4\frac{1}{4}$

5. 168 to 300

6. 26 tons to 65 tons

Personal Finance Bob earns $480 per week and has $60 withheld for federal taxes and $45 withheld for state taxes.

7. Write the ratio of federal taxes withheld to earned income.

8. Write the ratio of total withholdings to earned income.

Write as a rate in simplest form.

9. $75 donated by every 6 people

10. 44 revolutions every 121 minutes

11. 75 heartbeats every 60 seconds

In exercises 12–15, write as a unit rate. Round to the nearest tenth when necessary.

12. $2125 was paid for 125 shares of stock. Find the cost per share.

13. $1344 was paid for 12 credit-hours. Find the cost per credit-hour.

▲ **14.** $742.50 was spent for 55 square yards of carpet. Find the cost per square yard.

15. *Food Costs* A 4-ounce jar of instant coffee costs $2.96. A 9-ounce jar of the same brand of instant coffee costs $5.22.

    (a) What is the cost per ounce of the 4-ounce jar?

    (b) What is the cost per ounce of the 9-ounce jar?

    (c) How much per ounce do you save by buying the larger jar?

Section 4.2

Write as a proportion.

16. 12 is to 48 as 7 is to 28

17. $1\frac{1}{2}$ is to 5 as 4 is to $13\frac{1}{3}$

18. ***Bus Capacity*** If three buses can transport 138 passengers, then five buses can transport 230 passengers.

19. ***Cost of Products*** If 15 pounds cost $4.50, then 27 pounds will cost $8.10.

Determine whether each equation is a proportion.

20. $\frac{16}{48} \stackrel{?}{=} \frac{2}{12}$

21. $\frac{36}{30} \stackrel{?}{=} \frac{60}{50}$

22. $\frac{37}{33} \stackrel{?}{=} \frac{22}{19}$

23. $\frac{84 \text{ miles}}{7 \text{ gallons}} \stackrel{?}{=} \frac{108 \text{ miles}}{9 \text{ gallons}}$

24. $\frac{156 \text{ revolutions}}{6 \text{ minutes}} \stackrel{?}{=} \frac{181 \text{ revolutions}}{7 \text{ minutes}}$

Section 4.3

Solve for n.

25. $9 \times n = 162$

26. $5 \times n = 38$

27. $442 = 20 \times n$

Solve. Round to the nearest tenth when necessary.

28. $\frac{3}{11} = \frac{9}{n}$

29. $\frac{n}{28} = \frac{6}{24}$

30. $\frac{n}{32} = \frac{15}{20}$

31. $\frac{3\frac{1}{3}}{2\frac{2}{3}} = \frac{7}{n}$

32. $\frac{42}{50} = \frac{n}{6}$

33. $\frac{2.25}{9} = \frac{4.75}{n}$

34. $\frac{36}{n} = \frac{109}{18}$

35. $\frac{35 \text{ miles}}{28 \text{ gallons}} = \frac{15 \text{ miles}}{n \text{ gallons}}$

36. $\frac{8 \text{ defective parts}}{100 \text{ perfect parts}} = \frac{44 \text{ defective parts}}{n \text{ perfect parts}}$

Section 4.4

Solve using a proportion. Round your answer to the nearest hundredth when necessary.

37. ***Painting*** The school volunteers used 3 gallons of paint to paint two rooms. How many gallons would they need to paint 10 rooms of the same size?

38. ***Coffee Consumption*** Several recent surveys show that 49 out of every 100 adults in America drink coffee. If a computer company employs 3450 people, how many of those employees would you expect would drink coffee? Round to the nearest whole number.

39. ***Exchange Rate*** When Marguerite traveled as a child, the rate of French francs to American dollars was 24 francs to 5 dollars. How many francs did Marguerite receive for 420 dollars?

40. ***Exchange Rate*** When John and Nancy traveled to Sweden in 2010, the rate of Swedish kronor to U.S. dollars was 6 kronor to 0.83 dollar. How many Swedish kronor would they have received for 125 U.S. dollars?

41. *Map Scale* Two cities located 225 miles apart appear 3 inches apart on a map. If two other cities appear 8 inches apart on the map, how many miles apart are the cities?

▲ **42.** *Shadow Length* In the setting sun, a 6-foot man casts a shadow 16 feet long. At the same time, a building casts a shadow of 320 feet. How tall is the building?

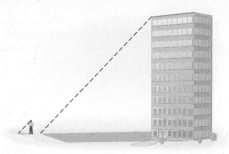

43. *Gasoline Consumption* During the first 680 miles of a trip, Johnny and Stephanie used 26 gallons of gas. They need to travel 200 more miles. Assume that the car will have the same rate of gas consumption.

(a) How many more gallons of gas will they need?

(b) If gas costs $4.20 per gallon, what will fuel cost them for the last 200 miles?

45. The dosage of a certain medication is 3 grams for every 50 pounds of body weight. If a person weighs 125 pounds, how many grams of this medication should she take?

▲ **47.** When Carlos was painting his apartment he found that he used 3 gallons of paint to cover 500 square feet of wall space. He is planning to paint his sister's apartment. She said there are 1400 square feet of wall space that need to be painted. How many gallons of paint will Carlos need?

▲ **49.** A scale model of a new church sanctuary has a length of 14 centimeters. When the church is built, the actual length will be 145 feet. In the scale model, the width measures 11 centimeters. What will be the actual width of the church sanctuary?

51. A recent survey showed that 3 out of every 10 people in Massachusetts read the *Boston Globe*. In a Massachusetts town of 45,600 people, how many people would you expect read the *Boston Globe*?

▲ **44.** *Photography* A film negative is 3.5 centimeters wide and 2.5 centimeters tall. If you want to make a color print that is 8 centimeters wide, how tall will the print be?

46. Greta did a study at her community college for her sociology class. Of the 35 students she interviewed, 21 of them said they eat in the campus cafeteria at least once a week. If there are a total of 2800 students at the college, how many of them eat in the cafeteria at least once a week?

48. From previous experience, the directors of a large running race know they need 2 liters of water for every 3 runners. This year 1250 runners will participate in the race. How many liters of water do they need?

50. Stella, the top soccer player for the Springfield Comets, scored a total of 68 goals during the season. During the season the team played 32 games, but Stella played in only 27 of them due to a leg injury. The league has been expanded and next season the team will play 34 games. If Stella scores goals at the same rate and is able to play in every game, how many goals might she be expected to score? Round your answer to the nearest whole number.

52. Greg and Marcia are managing a boat for Eastern Whale Watching Tours this summer. For every 16 trips out to the ocean, the passengers spotted at least one whale during 13 trips. If Greg and Marcia send out 240 trips this month, how many trips will have the passengers spotting at least one whale?

How Am I Doing? Chapter 4 Test

After you take this test read through the Math Coach on pages 302 and 303. Math Coach videos are available via MyMathLab and YouTube. Step-by-step test solutions on the Chapter Test Prep Videos are also available via MyMathLab and YouTube. (Search "TobeyBasicCollMath" and click on "Channels.")

CHAPTER Test Prep VIDEOS   MATH COACH   MyMathLab®   You Tube™

Write as a ratio in simplest form.

1. $18 : 52$

2. 70 to 185

Write as a rate in simplest form. Express your answer as a fraction.

3. 784 miles per 24 gallons

4. 2100 square feet per 45 pounds

Write as a unit rate. Round to the nearest hundredth when necessary.

5. 19 tons in five days

6. $57.96 for seven hours

Mc 7. 5400 feet per 22 telephone poles

8. $9373 for 110 shares of stock

Write as a proportion.

9. 17 is to 29 as 51 is to 87

10. $2\frac{1}{2}$ is to 10 as 6 is to 24

11. 490 miles is to 21 gallons as 280 miles is to 12 gallons

Mc 12. 3 hours is to 180 miles as 5 hours is to 300 miles

Determine whether each equation is a proportion.

13. $\frac{50}{24} \overset{?}{=} \frac{34}{16}$

14. $\frac{3\frac{1}{2}}{14} \overset{?}{=} \frac{5}{20}$

15. $\frac{32 \text{ smokers}}{46 \text{ nonsmokers}} \overset{?}{=} \frac{160 \text{ smokers}}{230 \text{ nonsmokers}}$

Mc 16. $\frac{\$0.74}{16 \text{ ounces}} \overset{?}{=} \frac{\$1.84}{40 \text{ ounces}}$

1. _____ ☐
2. _____ ☐
3. _____ ☐
4. _____ ☐
5. _____ ☐
6. _____ ☐
7. _____ ☐
8. _____ ☐
9. _____ ☐
10. _____ ☐
11. _____ ☐
12. _____ ☐
13. _____ ☐
14. _____ ☐
15. _____ ☐
16. _____ ☐

Find the value of n. Round to the nearest tenth when necessary.

17. $\dfrac{n}{20} = \dfrac{4}{5}$

18. $\dfrac{8}{3} = \dfrac{60}{n}$

19. $\dfrac{2\frac{2}{3}}{8} = \dfrac{6\frac{1}{3}}{n}$

20. $\dfrac{4.2}{11} = \dfrac{n}{77}$

21. $\dfrac{45 \text{ women}}{15 \text{ men}} = \dfrac{n \text{ women}}{40 \text{ men}}$

22. $\dfrac{5 \text{ kilograms}}{11 \text{ pounds}} = \dfrac{32 \text{ kilograms}}{n \text{ pounds}}$

23. $\dfrac{n \text{ inches of snow}}{14 \text{ inches of rain}} = \dfrac{12 \text{ inches of snow}}{1.4 \text{ inches of rain}}$

24. $\dfrac{5 \text{ pounds of coffee}}{\$n} = \dfrac{1/2 \text{ pound of coffee}}{\$5.20}$

Solve using a proportion. Round your answer to the nearest hundredth when necessary.

25. Josiah's recipe for pancakes calls for three eggs and will serve 11 people. If he wants to feed 22 people, how many eggs will he need?

26. A steel cable 42 feet long weighs 170 pounds. How much will 20 feet of this cable weigh?

MC 27. If 9 inches on a map represent 57 miles, what distance does 3 inches represent?

28. Dan and Connie found it would cost $240 per year to fertilize their front lawn of 4000 square feet. How much would it cost to fertilize 6000 square feet?

29. Nathaniel Tobey knows that 1 mile is approximately 1.61 kilometers. While he is driving in Canada, a sign reads "Montreal 220 km." How many miles is Nathaniel from Montreal? Round to the nearest tenth of a mile.

30. Caleb traveled 570 kilometers in 9 hours. At this rate, how far could he go in 11 hours?

31. During the first few games of the basketball season, Tyler shot 15 free throws and made 11 of them. If Tyler shoots free throws at the same rate as the first few games, he expects to shoot 120 more during the rest of the season. How many of these free throws should he expect to make?

32. On the tri-city softball league, Lexi got seven hits in 34 times at bat last week. During the entire playing season, she was at bat 155 times. If she gets hits at the same rate all season as during last week's game, how many hits would she have for the entire season? Round to the nearest whole number.

17. _____ ☐

18. _____ ☐

19. _____ ☐

20. _____ ☐

21. _____ ☐

22. _____ ☐

23. _____ ☐

24. _____ ☐

25. _____ ☐

26. _____ ☐

27. _____ ☐

28. _____ ☐

29. _____ ☐

30. _____ ☐

31. _____ ☐

32. _____ ☐

Total Correct: ☐

MATH COACH

Mastering the skills you need to do well on the test.

Students often make the same types of errors when they do the Chapter 4 Test. Here are some helpful hints to keep you from making these common errors on test problems.

Write as a Unit Rate—Problem 7   5400 feet per 22 telephone poles

> **Helpful Hint** Write the word phrase as a ratio. Then divide the denominator into the numerator to find the unit rate. Round your answer to the nearest hundredth.

Did you divide the denominator (22) into the numerator (5400)?

Yes ____ No ____

If you answered No, please perform the correct division.

Did you stop your division steps with 245 as your answer?

Yes ____ No ____

If you answered Yes, you need to continue you calculation further since the problem requires an answer rounded to the nearest hundredth. Remember to include the units with your answer.

If you did not solve Problem 7 correctly, please rework the problem using these suggestions.

Write as a Proportion—Problem 12   3 hours is to 180 miles as 5 hours is to 300 miles

> **Helpful Hint** Write the appropriate units in the top and bottom of each fraction. Compare the two fractions to make sure that the same unit appears in both numerators and the same unit appears in both denominators.

Did you write 3 hours on the top of one fraction and 5 hours on the top of the other fraction?

Yes ____ No ____

If you answered No, then stop and make this correction.

Did you write 180 miles in the denominator of the fraction that has 3 hours as the numerator?

Yes ____ No ____

If you answered No, then stop and make this correction.

If you answered Problem 12 incorrectly, then please rework the problem using these suggestions.

Need help? Watch the **MATH COACH** videos in MyMathLab® or on YouTube™.

302

Determine If an Equation Is a Proportion—Problem 16 $\dfrac{\$0.74}{16 \text{ ounces}} \overset{?}{=} \dfrac{\$1.84}{40 \text{ ounces}}$

> **Helpful Hint** Find the cross products and see if they are equal.

Did you multiply 0.74×40 to obtain 29.60?

Yes ____   No ____

If you answered No, perform this calculation.

Did you multiply 16×1.84 to obtain 29.44?

Yes ____   No ____

If you answered No, perform this calculation.

If you did not solve Problem 16 correctly, please rework the problem using these suggestions.

Now go back and state whether or not the equation is a proportion.

Solving Applied Problems Using Proportions—Problem 27   If 9 inches on a map represent 57 miles, what distance does 3 inches represent?

> **Helpful Hint** Take the time to follow all the rules of calculations with fractions.

Did you translate the problem into the proportion
$\dfrac{9 \text{ inches}}{57 \text{ miles}} = \dfrac{3 \text{ inches}}{n \text{ miles}}$?

Yes ____   No ____

If you answered No, reread the problem and try to translate the situation again.

Did you find the cross products to obtain the equation $9 \times n = 57 \times 3$?

Yes ____   No ____

If you answered No, please find the cross products again.

Did you multiply 57 and 3 to obtain 171 on the right side of the equation?

Yes ____   No ____

If you answered No, perform the multiplication again.

Did you divide both sides of the equation by 9 to find the value for n?

Yes ____   No ____

If you answered No, perform this step to solve for n.

If you did not solve Problem 27 correctly, please rework the problem using these suggestions. (Remember to include the proper units with your final answer.)

Need more help? Look for section examples marked with $\mathbb{MC}$ to review.

303

Compared to thirty years ago, today more high school students take a modern foreign language rather than an ancient language such as Latin. This increase is partly due to a greater demand from businesses in our own country for more employees who are bilingual. However, there is a huge, growing need for companies that place employees overseas to have modern foreign language skills. Many companies require that a certain percent of their employees be bilingual. Many percent calculations require the knowledge of the mathematics of this chapter.

Percent

5.1 Understanding Percent

① Writing a Fraction with a Denominator of 100 as a Percent

"My raise came through. I got a 6% increase!"

"The leading economic indicators show inflation rising at a rate of 1.3%."

"Mark McGwire and Babe Ruth each hit quite a few home runs. But I wonder who has the higher percentage of home runs per at-bat?"

We use percents often in our everyday lives. In business, in sports, in shopping, and in many areas of life, percentages play an important role. In this section we introduce the idea of percent, which means "*per centum*" or "per hundred." We then show how to use percentages.

In previous chapters, when we described parts of a whole, we used fractions or decimals. Using a percent is another way to describe a part of a whole. Percents can be described as ratios whose denominators are 100. The word **percent** means per 100. The sketch to the right has 100 rectangles.

Of the 100 rectangles, 23 are shaded. We can say that 23 percent of the whole is shaded. We use the symbol % for percent. It means "parts per 100." When we write 23 percent as 23%, we understand that it means 23 parts per one hundred, or, as a fraction, $\frac{23}{100}$.

Student Learning Objectives

After studying this section, you will be able to:

① Write a fraction with a denominator of 100 as a percent.

② Write a percent as a decimal.

③ Write a decimal as a percent.

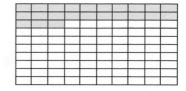

EXAMPLE 1 Recently 100 college students were surveyed about their intentions for voting in the next presidential election. 39 students intended to vote for the Republican candidate, 28 students intended to vote for the Democratic candidate, and 22 students were undecided about which candidate to vote for. The remaining 11 students admitted that they were not planning to vote.

(a) What percent of the students intended to vote for the Democratic candidate?

(b) What percent of the students intended to vote for the Republican candidate?

(c) What percent of the students were undecided as to which candidate they would vote for?

(d) What percent of the students were not planning to vote?

Solution

(a) $\frac{28}{100} = 28\%$ **(b)** $\frac{39}{100} = 39\%$

(c) $\frac{22}{100} = 22\%$ **(d)** $\frac{11}{100} = 11\%$

Percent notation is often used in circle graphs or pie charts.

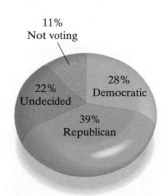

Student Practice 1 Write as a percent.

(a) 51 out of 100 students in a class were women.

(b) 68 out of 100 cars in a parking lot have front-wheel drive.

(c) 7 out of 100 students in a dorm quit smoking.

(d) 26 out of 100 students did not vote in class elections.

Some percents are larger than 100%. When you see expressions like 140% or 400%, you need to understand what they represent. Consider the following situations.

EXAMPLE 2

(a) Write $\dfrac{386}{100}$ as a percent.

(b) Twenty years ago, four car tires for a full-size car cost $100. Now the average price for four car tires for a full-size car is $270. Write the present cost as a percent of the cost 20 years ago.

Solution

(a) $\dfrac{386}{100} = 386\%$

(b) The ratio is $\dfrac{\$270 \text{ for four tires now}}{\$100 \text{ for four tires then}}$. $\dfrac{270}{100} = 270\%$

The present cost of four car tires for a full-size car is 270% of the cost 20 years ago.

Student Practice 2

(a) Write $\dfrac{238}{100}$ as a percent.

(b) Last year 100 students tried out for varsity baseball. This year 121 students tried out. Write this year's number as a percent of last year's number.

Some percents are smaller than 1%.

$\dfrac{0.7}{100}$ can be written as 0.7%. $\dfrac{0.3}{100}$ can be written as 0.3%.

$\dfrac{0.04}{100}$ can be written as 0.04%.

EXAMPLE 3   Write as a percent.

(a) $\dfrac{0.9}{100}$ **(b)** $\dfrac{0.002}{100}$ **(c)** $\dfrac{0.07}{100}$

Solution

(a) $\dfrac{0.9}{100} = 0.9\%$ **(b)** $\dfrac{0.002}{100} = 0.002\%$ **(c)** $\dfrac{0.07}{100} = 0.07\%$

Student Practice 3   Write as a percent.

(a) $\dfrac{0.5}{100}$ **(b)** $\dfrac{0.06}{100}$ **(c)** $\dfrac{0.003}{100}$

Remember: Whenever the denominator of a fraction is 100, the numerator is the percent.

② Writing a Percent as a Decimal

Suppose we have a percent such as 59%. What would be the equivalent in decimal form? Using our definition of percent, $59\% = \frac{59}{100}$. This fraction could be written in decimal form as 0.59. In a similar way, we could write 21% as $\frac{21}{100} = 0.21$. This pattern allows us to quickly change the form of a number from a percent to a fraction whose denominator is 100 to a decimal.

EXAMPLE 4   Write as a decimal.

(a) 38%   **(b)** 6%

Solution

(a) $38\% = \dfrac{38}{100} = 0.38$ **(b)** $6\% = \dfrac{6}{100} = 0.06$

Student Practice 4   Write as a decimal.

(a) 47%   **(b)** 2%

The results of Example 4 suggest that **when you remove a percent sign (%), you are dividing by 100.** When we divide by 100 this moves the decimal point of a number two places to the left. Now that you understand this process we can abbreviate it with the following rule.

CHANGING A PERCENT TO A DECIMAL

1. Drop the % symbol.

2. Move the decimal point two places to the left.

EXAMPLE 5   Write as a decimal.

(a) 26.9%   **(b)** 7.2%   **(c)** 0.13%   **(d)** 158%

Solution   In each case, we drop the percent symbol and move the decimal point two places to the left.

(a) $26.9\% = 0.269 = 0.269$

(b) $7.2\% = 0.072 = 0.072$ Note that we need to add an extra zero to the left of the seven.

(c) $0.13\% = 0.0013 = 0.0013$ Here we added zeros to the left of the 1.

(d) $158\% = 1.58 = 1.58$

Student Practice 5   Write as a decimal.

(a) 80.6%   **(b)** 2.5%   **(c)** 0.29%   **(d)** 231%

③ **Writing a Decimal as a Percent**

In Example 4(a) we changed 38% to $\frac{38}{100}$ to 0.38. We can start with 0.38 and reverse the process. We obtain $0.38 = \frac{38}{100} = 38\%$. Study all the parts of Examples 4 and 5. You will see that the steps are reversible. Thus $0.38 = 38\%$, $0.06 = 6\%$, $0.269 = 26.9\%$, $0.072 = 7.2\%$, $0.0013 = 0.13\%$, and $1.58 = 158\%$.

In each part we are multiplying by 100. **To change a decimal number to a percent, we are multiplying the number by 100.** In each part the decimal point is moved two places to the right. Then the percent symbol is written after the number.

CHANGING A DECIMAL TO A PERCENT

1. Move the decimal point two places to the right.

2. Then write the % symbol at the end of the number.

 EXAMPLE 6 Write as a percent.

(a) 0.47   **(b)** 0.08   **(c)** 6.31

(d) 0.055   **(e)** 0.001

Solution In each part we move the decimal point two places to the right and write the percent symbol at the end of the number.

(a) $0.47 = 47\%$ **(b)** $0.08 = 8\%$ **(c)** $6.31 = 631\%$

(d) $0.055 = 5.5\%$ **(e)** $0.001 = 0.1\%$

Student Practice 6 Write as a percent.

(a) 0.78   **(b)** 0.02

(c) 5.07   **(d)** 0.029

(e) 0.006

Calculator

Percent to Decimal

You can use a calculator to change 52% to a decimal. Enter

52 [%]

The display should read

[0.52]

Try the following.

(a) 46%   **(b)** 137%

(c) 9.3%   **(d)** 6%

Note: The calculator divides by 100 when the percent key is pressed. If you do not have a [%] key then you can use the keystrokes [÷] 100 [=].

TO THINK ABOUT: The Meaning of Percent What is really happening when we change a decimal to a percent? Suppose that we wanted to change 0.59 to a percent.

$$0.59 = \frac{59}{100} \qquad \text{Definition of a decimal.}$$

$$= 59 \times \frac{1}{100} \qquad \text{Definition of multiplying fractions.}$$

$$= 59 \text{ percent} \qquad \text{Because ``per 100'' means percent.}$$

$$= 59\% \qquad \text{Writing the symbol for percent.}$$

Can you see why each step is valid? Since we know the reason behind each step, we know we can always move the decimal point two places to the right and write the percent symbol. See Exercises 5.1, exercises 81 and 82.

5.1 Exercises

MyMathLab®

Watch the videos
in MyMathLab

Download the
MyDashBoard App

Verbal and Writing Skills, Exercises 1–4

1. In this section we introduced percent, which means "per centum" or "per _____."

2. The number 1 written as a percent is _____.

3. To change a percent to a decimal, move the decimal point _____ places to the _____, _____ the % symbol.

4. To change a decimal to a percent, move the decimal point _____ places to the _____, _____ the % symbol at the end of the number.

Write as a percent.

5. $\frac{59}{100}$

6. $\frac{67}{100}$

7. $\frac{4}{100}$

8. $\frac{7}{100}$

9. $\frac{80}{100}$

10. $\frac{90}{100}$

11. $\frac{245}{100}$

12. $\frac{110}{100}$

13. $\frac{12.5}{100}$

14. $\frac{15.8}{100}$

15. $\frac{4\frac{1}{3}}{100}$

16. $\frac{10\frac{2}{5}}{100}$

Applications, Exercises 17–20 *Write a percent to express each of the following.*

17. 13 out of 100 loaves of bread had gone stale.

18. 54 out of 100 dog owners have attended an obedience class.

19. 9 out of 100 customers ordered black coffee.

20. 7 out of 100 students majored in exercise science.

Write as a decimal.

21. 51%

22. 42%

23. 7%

24. 3%

25. 20%

26. 40%

27. 43.6%

28. 72.1%

29. 0.03%

30. 0.09%

31. 0.72%

32. 0.61%

33. 1.25%

34. 9.6%

35. 275%

36. 189%

Write as a percent.

37. 0.74

38. 0.66

39. 0.50

40. 0.10

41. 0.08

42. 0.03

43. 0.563

44. 0.408

45. 0.002

46. 0.009

47. 0.0057

48. 0.0015

49. 1.35

50. 1.86

51. 5.16

52. 9.17

53. ***Income Taxes*** Robert Tansill paid $\frac{27}{100}$ of his income for federal income taxes. This means that 0.27 of his income was paid for federal taxes. Express this as a percent.

54. ***Housing Costs*** Sally LeBlanc spends $\frac{37}{100}$ of her income for housing. This means that 0.37 of her income is spent for housing. Express this as a percent.

55. ***Grade Distribution*** Professor Harlin gave $\frac{2}{10}$ of his students a grade of A for the semester. This means that 0.2 of his students got an A. Express this as a percent.

56. ***Checking Account*** Tomás Garcia puts $\frac{8}{10}$ of his income into his checking account each month. This means that 0.8 of his income goes into his checking account. Express this as a percent.

Mixed Practice *Write as a percent.*

57. 0.94

58. 0.25

59. 2.31

60. 1.48

61. $\frac{10}{100}$

62. $\frac{40}{100}$

63. 0.089

64. 0.055

Write as a decimal.

65. 62%

66. 49%

67. 138%

68. 210%

69. $\frac{0.3}{100}$

70. $\frac{0.8}{100}$

71. $\frac{75}{100}$

72. $\frac{35}{100}$

Applications *Write as a percent.*

73. ***Presidential Election*** Out of every 100 people in Oregon who voted in the 2008 presidential election, 57 voted for Barack Obama.

74. ***Presidential Election*** Out of every 100 people in Idaho who voted in the 2008 presidential election, 61 voted for John McCain.

The following are statements that might be found in newspapers or online. In each case write the percent as a decimal.

75. ***House Value*** The value of the Sanchez's home increased by 115 percent during the last five years.

76. ***Charitable Donations*** According to the Gallup Poll, 1 percent of Americans receiving money from a federal tax cut plan to donate the money.

77. ***Home Value*** 0.6 percent of homes in the United States are valued at more than $1 million.

78. ***Vitamins*** Americans get 30 percent of their vitamin A from carrots.

79. ***College Costs*** In 2010, first-year college students were expected to spend 19% less in back-to-school spending than first-year students in 2009. Overall back-to-college spending was expected to drop only 0.32%. (*Source:* www.nrf.com)

80. ***Baseball*** In 2010, the Minnesota Twins baseball team began playing in a new stadium that has 15.2% fewer seats and 53% fewer private suites than the old stadium.

To Think About

81. Suppose that we want to change 36% to 0.36 by moving the decimal point two places to the left and dropping the % symbol. Explain the steps to show what is really involved in changing 36% to 0.36. Why does the rule work?

82. Suppose that we want to change 10.65 to 1065%. Give a complete explanation of the steps.

Write the given value **(a)** *as a decimal,* **(b)** *as a fraction with a denominator of 100, and* **(c)** *as a reduced fraction.*

83. 1562%

84. 3724%

Cumulative Review *Write as a fraction in simplest form.*

85. **[3.1.3]** 0.56

86. **[3.1.3]** 0.78

Write as a decimal.

87. **[3.6.1]** $\dfrac{11}{16}$

88. **[3.6.1]** $\dfrac{7}{8}$

89. **[1.8.2]** *Ceramics Studio* A very successful commercial ceramics studio makes beautiful vases for gift stores. In one corner of the warehouse, the storage area has 24 shelves. Three shelves have 246 vases each, seven shelves have 380 vases each, five shelves have 168 vases each, and nine shelves have 122 vases each. How many vases are there in this corner of the studio?

Quick Quiz 5.1 *Write as a percent.*

1. 0.007

2. $\dfrac{4.5}{100}$

3. Write as a decimal. 1.25%

4. **Concept Check** Explain how you would change 0.00072% to a decimal.

5.2 Changing Between Percents, Decimals, and Fractions

Student Learning Objectives

After studying this section, you will be able to:

① Change a percent to a fraction.

② Change a fraction to a percent.

③ Change a percent, a decimal, or a fraction to equivalent forms.

NOTE TO STUDENT: Fully worked-out solutions to all of the Student Practice problems can be found at the back of the text starting at page SP-1.

① Changing a Percent to a Fraction

By using the definition of percent, we can write any percent as a fraction whose denominator is 100. Thus when we change a percent to a fraction, we remove the percent symbol and write the number over 100. To write a number over 100 means that we are dividing by 100. If possible, we then simplify the fraction.

EXAMPLE 1 Write as a fraction in simplest form.

(a) 37%   **(b)** 75%   **(c)** 2%

Solution

(a) $37\% = \dfrac{37}{100}$ **(b)** $75\% = \dfrac{75}{100} = \dfrac{3}{4}$ **(c)** $2\% = \dfrac{2}{100} = \dfrac{1}{50}$

Student Practice 1 Write as a fraction in simplest form.

(a) 71%   **(b)** 25%   **(c)** 8%

In some cases, it may be helpful to write the percent as a decimal before you write it as a fraction in simplest form.

EXAMPLE 2 Write as a fraction in simplest form.

(a) 43.5%   **(b)** 36.75%

Solution

(a) $43.5\% = 0.435$ Change the percent to a decimal.

$= \dfrac{435}{1000}$ Change the decimal to a fraction.

$= \dfrac{87}{200}$ Reduce the fraction.

(b) $36.75\% = 0.3675 = \dfrac{3675}{10,000} = \dfrac{147}{400}$

Student Practice 2 Write as a fraction in simplest form.

(a) 8.4%   **(b)** 28.5%

If the percent is greater than 100%, the simplified fraction is usually changed to a mixed number.

EXAMPLE 3 Write as a mixed number.

(a) 225%   **(b)** 138%

Solution

(a) $225\% = 2.25 = 2\dfrac{25}{100} = 2\dfrac{1}{4}$ **(b)** $138\% = 1.38 = 1\dfrac{38}{100} = 1\dfrac{19}{50}$

Student Practice 3 Write as a mixed number.

(a) 170%   **(b)** 288%

Sometimes a percent is not a whole number, such as 9% or 10%. Instead, it contains a fraction, such as $9\frac{1}{12}\%$ or $9\frac{3}{8}\%$. Extra steps will be needed to write such a percent as a simplified fraction.

EXAMPLE 4 Convert $3\frac{3}{8}\%$ to a fraction in simplest form.

Solution

$$3\frac{3}{8}\% = \frac{3\frac{3}{8}}{100}$$ Change the percent to a fraction.

$$= 3\frac{3}{8} \div \frac{100}{1}$$ Write the division horizontally. $\frac{3\frac{3}{8}}{100}$ means $3\frac{3}{8}$ divided by 100.

$$= \frac{27}{8} \div \frac{100}{1}$$ Write $3\frac{3}{8}$ as an improper fraction.

$$= \frac{27}{8} \times \frac{1}{100}$$ Use the definition of division of fractions.

$$= \frac{27}{800}$$ Simplify.

Student Practice 4 Convert $7\frac{5}{8}\%$ to a fraction in simplest form.

EXAMPLE 5 In the fiscal 2011 budget of the United States, approximately $19\frac{1}{6}\%$ of the budget was designated for social security. (*Source:* www.whitehouse.gov) Write this percent as a fraction.

Solution $19\frac{1}{6}\% = \dfrac{19\frac{1}{6}}{100} = 19\frac{1}{6} \div 100 = \dfrac{115}{6} \times \dfrac{1}{100} = \dfrac{115}{600} = \dfrac{23}{120}$

Thus we could say $\frac{23}{120}$ of the fiscal 2011 budget was designated for social security. That is, for every \$120 in the budget, \$23 was spent on social security.

Student Practice 5 In the fiscal 2011 budget of the United States, approximately $12\frac{5}{6}\%$ was designated for Medicare. (*Source:* www.whitehouse.gov) Write this percent as a fraction.

Certain percents occur very often, especially in money matters. Here are some common equivalents that you may already know. If not, be sure to memorize them.

$$25\% = \frac{1}{4} \qquad 33\frac{1}{3}\% = \frac{1}{3} \qquad 10\% = \frac{1}{10}$$

$$50\% = \frac{1}{2} \qquad 66\frac{2}{3}\% = \frac{2}{3}$$

$$75\% = \frac{3}{4}$$

② Changing a Fraction to a Percent

A convenient way to change a fraction to a percent is to write the fraction in decimal form first and then convert the decimal to a percent.

EXAMPLE 6 Write $\frac{3}{8}$ as a percent.

Solution We see that $\frac{3}{8} = 0.375$ by calculating $3 \div 8$.

$$
\begin{array}{r}
0.375 \\
8)\overline{3.000} \\
\underline{2\,4} \\
60 \\
\underline{56} \\
40 \\
\underline{40} \\
0
\end{array}
$$

Thus $\frac{3}{8} = 0.375 = 37.5\%$.

Student Practice 6 Write $\frac{5}{8}$ as a percent.

EXAMPLE 7 Write as a percent.

(a) $\frac{7}{40}$ (b) $\frac{39}{50}$

Solution

(a) $\frac{7}{40} = 0.175 = 17.5\%$ (b) $\frac{39}{50} = 0.78 = 78\%$

Student Practice 7 Write as a percent.

(a) $\frac{21}{25}$ (b) $\frac{7}{16}$

Changing some fractions to decimal form results in infinitely repeating decimals. In such cases, we usually round to the nearest hundredth of a percent.

EXAMPLE 8 Write as a percent. Round to the nearest hundredth of a percent.

(a) $\frac{1}{6}$ (b) $\frac{15}{33}$

Solution

(a) We find that $\dfrac{1}{6} = 0.1666\ldots$ by calculating $1 \div 6$.

$$
\begin{array}{r}
0.1666 \\
6\overline{)1.0000} \\
\underline{6} \\
40 \\
\underline{36} \\
40 \\
\underline{36} \\
40 \\
\underline{36} \\
4
\end{array}
$$

We will need a four-place decimal so that we will obtain a percent to the nearest hundredth. If we round the decimal to the nearest ten-thousandth, we have $\frac{1}{6} \approx 0.1667$. If we change this to a percent, we have

$$\frac{1}{6} \approx 16.67\%.$$

This is correct to the nearest hundredth of a percent.

(b) By calculating $15 \div 33$, we see that $\dfrac{15}{33} = 0.45454545\ldots$. We will need a four-place decimal so that we will obtain a percent to the nearest hundredth. If we round to the nearest ten-thousandth, we have

$$\frac{15}{33} \approx 0.4545 = 45.45\%.$$

This rounded value is correct to the nearest hundredth of a percent.

Student Practice 8 Write as a percent. Round to the nearest hundredth of a percent.

(a) $\dfrac{7}{9}$ **(b)** $\dfrac{19}{30}$

Recall that sometimes percents are written with fractions.

EXAMPLE 9 Express $\dfrac{11}{12}$ as a percent containing a fraction.

Solution We will stop the division after two steps and write the remainder in fraction form.

$$
\begin{array}{r}
0.91 \\
12\overline{)11.00} \\
\underline{10\ 8} \\
20 \\
\underline{12} \\
8
\end{array}
$$

This division tells us that we can write

Continued on next page

$$\frac{11}{12} \quad \text{as} \quad 0.91\frac{8}{12} \quad \text{or} \quad 0.91\frac{2}{3}.$$

We now have a decimal with a fraction. When we express this decimal as a percent, we move the decimal point two places to the right. We do not write the decimal point in front of the fraction.

$$0.91\frac{2}{3} = 91\frac{2}{3}\%$$

Note that our answer in Example 9 is an *exact answer*. We have not rounded off or approximated in any way.

Student Practice 9 Express $\frac{7}{12}$ as a percent containing a fraction.

③ Changing a Percent, a Decimal, or a Fraction to Equivalent Forms

We have seen so far that a fraction, a decimal, and a percent are three different forms (notations) for the same number. We can illustrate this in a chart.

EXAMPLE 10 Complete the following table of equivalent notations. Round decimals to the nearest ten-thousandth. Round percents to the nearest hundredth of a percent.

Fraction	Decimal	Percent
$\frac{11}{16}$		
	0.265	
		$17\frac{1}{5}\%$

Solution Begin with the first row. The number is written as a fraction. We will change the fraction to a decimal and then to a percent.

The fraction is changed to a decimal, which is changed to a percent.

$$\frac{11}{16} \longrightarrow 16\overline{)11.0000}^{0.6875} \longrightarrow 68.75\%$$

In the second row the number is written as a decimal. This can easily be written as a percent.

$$0.265 \longrightarrow 26.5\%$$

Now write 0.265 as a fraction and simplify.

$$0.265$$
$$\downarrow$$
$$\frac{53}{200} \longleftarrow \frac{265}{1000}$$

In the third row the number is written as a percent. Proceed from right to left—that is, write the number as a decimal and then as a fraction.

$$\frac{17\frac{1}{5}}{100} \leftarrow 17\frac{1}{5}\%$$

$$\downarrow$$

$$\boxed{\frac{86}{5} \times \frac{1}{100}}$$

$$\downarrow$$

$$0.172 \leftarrow \frac{86}{500} \quad \text{Divide.} \quad \frac{0.172}{500)\overline{86.000}}$$

and

$$\frac{43}{250} \leftarrow \frac{86}{500}$$

Thus the completed table is as follows.

Fraction	Decimal	Percent
$\frac{11}{16}$	0.6875	68.75%
$\frac{53}{200}$	0.265	26.5%
$\frac{43}{250}$	0.172	$17\frac{1}{5}\%$

Student Practice 10 Complete the following table of equivalent notations. Round decimals to the nearest ten-thousandth. Round percents to the nearest hundredth of a percent.

Fraction	Decimal	Percent
$\frac{23}{99}$		
	0.516	
		$38\frac{4}{5}\%$

Calculator

Fraction to Decimal

You can use a calculator to change $\frac{3}{5}$ to a decimal. Enter

$$3 \;\boxed{\div}\; 5 \;\boxed{=}$$

The display should read

$$\boxed{0.6}$$

Try the following.

(a) $\frac{17}{25}$ (b) $\frac{2}{9}$

(c) $\frac{13}{10}$ (d) $\frac{15}{19}$

Note: 0.78947368 is an approximation for $\frac{15}{19}$. Some calculators round to only eight places.

ALTERNATIVE METHOD: Using Proportions to Convert from Fraction to Percent Another way to convert a fraction to a percent is to use a proportion. To change $\frac{7}{8}$ to a percent, write the proportion

$$\frac{7}{8} = \frac{n}{100}$$

$$7 \times 100 = 8 \times n \qquad \text{Cross-multiply.}$$

$$700 = 8 \times n \qquad \text{Simplify.}$$

$$\frac{700}{8} = \frac{8 \times n}{8} \qquad \text{Divide each side by 8.}$$

$$87.5 = n \qquad \text{Simplify.}$$

Thus $\frac{7}{8} = 87.5\%$. You will use this approach in Exercises 5.2, exercises 85 and 86.

Verbal and Writing Skills, Exercises 1 and 2

1. Explain in your own words how to change a percent to a fraction.

2. Explain in your own words how to change a fraction to a percent.

Write as a fraction or as a mixed number.

3. 6%

4. 8%

5. 33%

6. 47%

7. 55%

8. 35%

9. 75%

10. 25%

11. 20%

12. 40%

13. 9.5%

14. 6.5%

15. 22.5%

16. 72.5%

17. 64.8%

18. 16.4%

19. 71.25%

20. 38.75%

21. 168%

22. 256%

23. 340%

24. 420%

25. 1200%

26. 2700%

27. $3\frac{5}{8}\%$

28. $6\frac{2}{5}\%$

29. $12\frac{1}{2}\%$

30. $37\frac{1}{2}\%$

31. $8\frac{4}{5}\%$

32. $9\frac{1}{5}\%$

Applications

33. ***Crime Rates*** Between 2004 and 2008, the number of robbery crimes in the United States increased by 10.1%. Write the percent as a fraction. (*Source:* www.ojp.usdoj.gov)

34. ***Crime Rates*** Between 1999 and 2008, the number of motor vehicle thefts in the United States decreased by 16.9%. Write the percent as a fraction. (*Source:* www.ojp.usdoj.gov)

35. ***Gasoline Prices*** On August 9, 2010, the average price in the United States for regular gasoline was $2.783 per gallon. This was a $4\frac{4}{5}\%$ increase from the average price of the previous week. Write this percent as a fraction. (*Source:* www.aaa.com)

36. ***Gasoline Prices*** On August 9, 2010, the average price in the United States for regular gasoline was $2.783. This was a $13\frac{3}{5}\%$ increase from the average price of one year ago. Write this percent as a fraction. (*Source:* www.aaa.com)

Write as a percent. Round to the nearest hundredth of a percent when necessary.

37. $\dfrac{3}{4}$ **38.** $\dfrac{1}{4}$ **39.** $\dfrac{7}{10}$ **40.** $\dfrac{9}{10}$ **41.** $\dfrac{7}{20}$ **42.** $\dfrac{11}{20}$

43. $\dfrac{18}{25}$ **44.** $\dfrac{22}{25}$ **45.** $\dfrac{11}{40}$ **46.** $\dfrac{13}{40}$ **47.** $\dfrac{18}{5}$ **48.** $\dfrac{7}{4}$

49. $2\dfrac{1}{2}$ **50.** $3\dfrac{3}{4}$ **51.** $4\dfrac{1}{8}$ **52.** $2\dfrac{5}{8}$ **53.** $\dfrac{1}{3}$ **54.** $\dfrac{2}{3}$

55. $\dfrac{5}{12}$ **56.** $\dfrac{8}{15}$ **57.** $\dfrac{17}{4}$ **58.** $\dfrac{12}{5}$ **59.** $\dfrac{26}{50}$ **60.** $\dfrac{43}{50}$

Applications *Round to the nearest hundredth of a percent.*

61. *Human Brain* The brain represents approximately $\frac{1}{40}$ of an average person's weight. Express this fraction as a percent.

62. *Monthly House Payments* To calculate your maximum monthly house payment, a real estate agent multiplies your monthly income by $\frac{7}{25}$. Express this fraction as a percent.

63. *Size of Africa* Africa is the second largest continent on Earth, measuring 30,301,596 sq km. However, it comprises only $\frac{119}{2000}$ of the earth's total surface area. Express this fraction as a percent.

64. *Size of Antarctica* The continent of Antarctica takes up $\frac{11}{400}$ of the earth's total surface area. Express this fraction as a percent.

Express as a percent containing a fraction. (See Example 9.)

65. $\dfrac{3}{8}$ **66.** $\dfrac{5}{8}$ **67.** $\dfrac{3}{40}$ **68.** $\dfrac{11}{90}$

69. $\dfrac{4}{15}$ **70.** $\dfrac{11}{15}$ **71.** $\dfrac{2}{9}$ **72.** $\dfrac{8}{9}$

Mixed Practice *In exercises 73–82, complete the table of equivalents. Round decimals to the nearest ten-thousandth. Round percents to the nearest hundredth of a percent.*

	Fraction	Decimal	Percent
73.	$\dfrac{11}{12}$		
75.		0.56	
77.		0.005	
79.	$\dfrac{5}{9}$		
81.			$3\dfrac{1}{8}\%$

	Fraction	Decimal	Percent
74.	$\dfrac{1}{12}$		
76.		0.85	
78.		0.085	
80.	$\dfrac{7}{9}$		
82.			$2\dfrac{5}{8}\%$

83. Write $28\frac{15}{16}\%$ as a fraction.

84. Write $18\frac{7}{12}\%$ as a fraction.

Change each fraction to a percent by using a proportion.

85. $\dfrac{123}{800}$

86. $\dfrac{417}{600}$

Cumulative Review *Find the value of n.*

87. [4.3.2] $\dfrac{15}{n} = \dfrac{8}{3}$

88. [4.3.2] $\dfrac{32}{24} = \dfrac{n}{3}$

89. [1.8.1] *Law Firm* The law firm of Dewey, Cheatham, & Howe was required to review 54 years of documents of one of its clients. The first file contained 10,041 documents. The second file contained 986 documents. The third file contained 4,283 documents. The last file contained 533,855 documents. How many total documents were there?

▲ **90.** [2.9.1] *Restaurant Size* A small neighborhood café has an area of 1800 square feet. A new steak house across the street is $2\frac{1}{2}$ times the size of the café. How many square feet is the new steak house?

Quick Quiz 5.2 *Write as a fraction or as a mixed number in simplified form.*

1. 45%

2. $7\frac{3}{5}\%$

3. Change to a percent. $\dfrac{23}{25}$

4. **Concept Check** Explain how you would change $8\frac{3}{8}\%$ to a decimal.

5.3A Solving Percent Problems Using Equations

① Translating a Percent Problem into an Equation

In word problems like the ones in this section, we can translate from words to mathematical symbols and back again. After we have the mathematical symbols arranged in an *equation,* we solve the equation. When we find the values that make the equation true, we have also found the answer to our word problem.

To solve a percent problem, we express it as an equation with an unknown quantity. We use the letter n to represent the number we do not know. The following table is helpful when translating from a percent problem to an equation.

Word	Mathematical Symbol
of	Any multiplication symbol: $\times$ or ( ) or $\cdot$
is	=
what	Any letter; for example, n
find	$n =$

In Examples 1–5 we show how to translate words into an equation. Please do **not** solve the problem. Translate into an equation only.

EXAMPLE 1 Translate into an equation.

$$\text{What is } 5\% \text{ of } 19.00?$$
$$\downarrow \quad \downarrow \quad \downarrow \quad \downarrow \quad \downarrow$$

Solution $\qquad n \ = 5\% \times 19.00$

Student Practice 1 Translate into an equation. What is 26% of 35?

EXAMPLE 2 Translate into an equation.

$$\text{Find } 0.6\% \text{ of } 400.$$

Solution Notice here that the words *what is* are missing. The word *find* is equivalent to *what is.*

$$\text{Find } \ 0.6\% \ \text{ of } 400.$$
$$\downarrow \qquad \downarrow \qquad \downarrow \ \downarrow$$
$$n \ = \ 0.6\% \ \times 400$$

Student Practice 2 Translate into an equation. Find 0.08% of 350.

The unknown quantity, n, does not always stand alone in an equation.

Student Learning Objectives

After studying this section, you will be able to:

① Translate a percent problem into an equation.

② Solve a percent problem by solving an equation.

NOTE TO STUDENT: *Fully worked-out solutions to all of the Student Practice problems can be found at the back of the text starting at page SP-1.*

EXAMPLE 3 Translate into an equation.

(a) 35% of what is 60?   **(b)** 7.2 is 120% of what?

Solution

(a) 35% of what is 60?   **(b)** 7.2 is 120% of what?

$$35\% \times n = 60$$ $$7.2 = 120\% \times n$$

Student Practice 3 Translate into an equation.

(a) 58% of what is 400?   **(b)** 9.1 is 135% of what?

EXAMPLE 4 Translate into an equation.

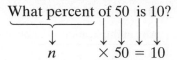

What percent of 50 is 10?

Solution $$n \times 50 = 10$$

We see here that the words *what percent* are represented by the letter *n*.

Student Practice 4 Translate into an equation. What percent of 250 is 36?

EXAMPLE 5 Translate into an equation.

(a) 30 is what percent of 16?   **(b)** What percent of 3000 is 2.6?

Solution

(a) 30 is what percent of 16?   **(b)** What percent of 3000 is 2.6?

$$30 = n \times 16$$ $$n \times 3000 = 2.6$$

Student Practice 5 Translate into an equation.

(a) 50 is what percent of 20?   **(b)** What percent of 2000 is 4.5?

② Solving a Percent Problem by Solving an Equation

The percent problems we have translated are of three types. Consider the equation $60 = 20\% \times 300$. This problem has the form

$$\text{amount} = \text{percent} \times \text{base}$$

Any one of these quantities—amount, percent, or base—may be unknown.

  1. When *we do not know the amount,* we have an equation like

$$n = 20\% \times 300.$$

  2. When *we do not know the base,* we have an equation like

$$60 = 20\% \times n.$$

3. When *we do not know the percent,* we have an equation like

$$60 = n \times 300.$$

We will study each type separately. It is not necessary to memorize the three types, but it is helpful to look carefully at the examples we give of each. In each example, do the computation in a way that is easiest for you. This may be using a pencil and paper, using a calculator, or, in some cases, doing the problem mentally.

Solving Percent Problems When the Amount Is Unknown In solving these equations we will need to change the percent number to decimal form.

EXAMPLE 6 What is 45% of 590?
$$\downarrow \quad \downarrow \quad \downarrow \quad \downarrow \quad \downarrow$$

Solution

$n = 45\% \times 590$	Translate into an equation.
$n = (0.45)(590)$	Change the percent to decimal form.
$n = 265.5$	Multiply 0.45×590.

Student Practice 6 What is 82% of 350?

EXAMPLE 7 Find 160% of 500.

Find 160% of 500.   When you translate, remember that the
$\downarrow \quad \downarrow \quad \downarrow \quad \downarrow$ word *find* is equivalent to *what is.*

Solution $n = 160\% \times 500$

$n = (1.60)(500)$ Change the percent to decimal form.

$n = 800$ Multiply 1.6 by 500.

Student Practice 7 Find 230% of 400.

EXAMPLE 8 When Rick bought a new Toyota Yaris, he had to pay a sales tax of 5% on the cost of the car, which was $12,000. What was the sales tax?

Solution This problem is asking

What is 5% of 12,000?
$$\downarrow \quad \downarrow \quad \downarrow \quad \downarrow \quad \downarrow$$

$n = 5\% \times \$12,000$

$n = 0.05 \times \$12,000$

$n = \$600$

The sales tax was $600.

Student Practice 8 When Oprah bought an airplane ticket, she had to pay a tax of 8% on the cost of the ticket, which was $350. What was the tax?

Solving Percent Problems When the Base Is Unknown If a number is multiplied by the letter n, this can be indicated by a multiplication sign, parentheses, a dot, or placing the number in front of the letter. Thus $3 \times n = 3(n) = 3 \cdot n = 3n$.

Calculator

 Percent of a Number

You can use a calculator to find 12% of 48.
Enter

12 $\boxed{\%}$ $\boxed{\times}$ 48 $\boxed{=}$

The display should read

$\boxed{5.76}$

If your calculator does not have a percent key, use the keystrokes

0.12 $\boxed{\times}$ 48 $\boxed{=}$

What is 54% of 450?

In this section we use equations like $3n = 9$ and $0.5n = 20$. To solve these equations we use the procedures developed in Chapter 4. We divide each side by the number being multiplied by n.

In solving these equations we will need to change the percent number to decimal form.

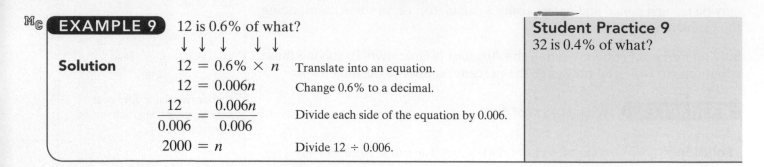

EXAMPLE 9 12 is 0.6% of what?

Solution

$12 = 0.6\% \times n$ Translate into an equation.

$12 = 0.006n$ Change 0.6% to a decimal.

$\dfrac{12}{0.006} = \dfrac{0.006n}{0.006}$ Divide each side of the equation by 0.006.

$2000 = n$ Divide $12 \div 0.006$.

Student Practice 9
32 is 0.4% of what?

EXAMPLE 10 Dave and Elsie went out to dinner. They gave the waiter a tip that was 15% of the total bill. The tip the waiter received was $6. What was the total bill (not including the tip)?

Solution This problem is asking

15% of what is $6?

$15\% \times n = 6$

$0.15n = 6$

$\dfrac{0.15n}{0.15} = \dfrac{6}{0.15}$ $n = 40$

The total bill for the meal (not including the tip) was $40.

Student Practice 10 The coach of the university baseball team said that 30% of the players on his team are left-handed. Six people on the team are left-handed. How many people are on the team?

Solving Percent Problems When the Percent Is Unknown In solving these problems, we notice that there is no % symbol in the problem. The percent is what we are trying to find. Therefore, our answer for this type of problem will always have a percent symbol.

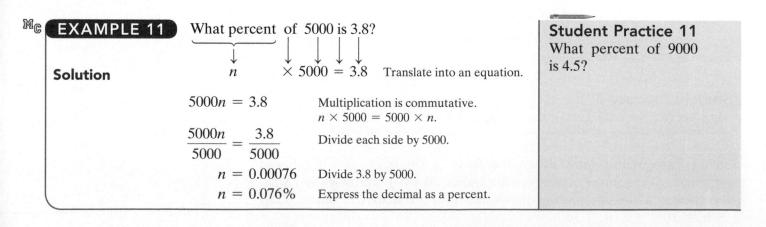

EXAMPLE 11 What percent of 5000 is 3.8?

Solution

$n \times 5000 = 3.8$ Translate into an equation.

$5000n = 3.8$ Multiplication is commutative. $n \times 5000 = 5000 \times n$.

$\dfrac{5000n}{5000} = \dfrac{3.8}{5000}$ Divide each side by 5000.

$n = 0.00076$ Divide 3.8 by 5000.

$n = 0.076\%$ Express the decimal as a percent.

Student Practice 11
What percent of 9000 is 4.5?

EXAMPLE 12   90 is what percent of 20?

Solution

$$90 = n \times 20 \qquad \text{Translate into an equation.}$$

$$90 = 20n \qquad \text{Multiplication is commutative. } n \times 20 = 20 \times n.$$

$$\frac{90}{20} = \frac{20n}{20} \qquad \text{Divide each side by 20.}$$

$$4.5 = n \qquad \text{Divide 90 by 20.}$$

$$450\% = n \qquad \text{Express the decimal as a percent.}$$

Student Practice 12   198 is what percent of 33?

EXAMPLE 13   In a basketball game for the Atlanta Hawks, Jamal Crawford made 10 of his 24 shots. What percent of his shots did he make? (Round to the nearest tenth of a percent.)

Solution   This is equivalent to

10 is what percent of 24?

$$10 = n \times 24$$

$$10 = 24n$$

$$\frac{10}{24} = \frac{24n}{24}$$

$$0.41666\ldots = n$$

To the nearest tenth of a percent we have

$$n = 41.7\%$$

Jamal Crawford made 41.7% of his shots in this game.

Student Practice 13   In a basketball game for the Los Angeles Lakers, Kobe Bryant made 5 of his 16 shots. What percent of his shots did he make? (Round to the nearest tenth of a percent.)

Calculator

 Finding the Percent

You can use a calculator to find a missing percent. What percent of 95 is 19?

1. Enter as a fraction. Enter the number after "is," and then the division key. Then enter the number after the word "of."

   19 $\boxed{\div}$ 95

2. Change to a percent.

   19 $\boxed{\div}$ 95 $\boxed{\times}$ 100 $\boxed{=}$

   The display should read

   $\boxed{20}$

   This means 20%.

What percent of 625 is 250?

Verbal and Writing Skills, Exercises 1–6

1. Give an example of a percent problem when we do not know the amount.

2. Give an example of a percent problem when we do not know the base.

3. Give an example of a percent problem when we do not know the percent.

4. When you encounter a problem like "What is 65% of $600?" what type of percent problem is this? How would you solve such a problem?

5. When you encounter a problem like "108 is 18% of what number?" what type of percent problem is this? How would you solve such a problem?

6. When you encounter a problem like "What percent of 35 is 14?" what type of percent problem is this? How would you solve such a problem?

*Translate into a mathematical equation in exercises 7–12. Use the letter n for the unknown quantity. Do **not** solve, but rather just obtain the equation.*

7. What is 5% of 90?

8. What is 8% of 75?

9. 30% of what is 5?

10. 65% of what is 28?

11. 17 is what percent of 85?

12. 16 is what percent of 128?

Solve.

13. What is 20% of 140?

14. What is 30% of 210?

15. Find 40% of 140.

16. Find 80% of 240.

Applications, Exercises 17 and 18

17. *Sales Tax* Malik bought a new flat-screen television. The price before the 6% sales tax was added on was $850. How much tax did Malik have to pay?

18. *Coin-Counting Service* At the local bank coins can be placed into a machine to be counted. You can then receive bills for the amount the coins are worth. However, the bank charges a fee that is 8% of the coins' value. How much would the service fee be if someone put $215 worth of coins into the machine?

Solve.

19. 2% of what is 26?

20. 3% of what is 18?

21. 52 is 4% of what?

22. 36 is 6% of what?

Applications, Exercises 23 and 24

23. *Australia Tax* In Australia, all general sales (except for food) have a hidden tax of 22% built into the final price. Walter is planning to purchase a camera while in Australia. He wants to know the before-tax price that the dealer is charging before he adds on the hidden tax of $33. Can you determine the before-tax price?

24. *Opinion Poll* A newspaper states that 522 of its residents are in favor of building a new high school. This is 12% of the town's population. What is the population of the town?

Solve.

25. What percent of 200 is 168?

26. What percent of 300 is 135?

27. 33 is what percent of 300?

28. 78 is what percent of 200?

Applications, Exercises 29 and 30

29. *Basketball* The total number of points scored in a basketball game was 120. The winning team scored 78 of those points. What percent of the points were scored by the winning team?

30. *Car Repairs* Randy's bill for car repairs was $140. Of this amount, $28 was charged for labor and $112 was charged for parts. What percent of the bill was for labor?

Mixed Practice *Solve.*

31. 20% of 155 is what?

32. 40% of 310 is what?

33. 170% of what is 144.5?

34. 160% of what is 152?

35. 84 is what percent of 700?

36. 72 is what percent of 900?

37. Find 0.4% of 820.

38. Find 0.3% of 540.

39. What percent of 35 is 22.4?

40. What percent of 45 is 16.2?

41. 15 is 20% of what?

42. 10 is 25% of what?

43. 8 is what percent of 1000?

44. 6 is what percent of 800?

45. What is 10.5% of 180?

46. What is 14.5% of 280?

47. Scoring 44 problems out of 55 problems correctly on a test is what percent?

48. Scoring 27 problems out of 45 problems correctly on a test is what percent?

Applications

49. *Oil Pollution* Each decade 800 million gallons of oil are spilled in North American waters. Of this number, 493 million gallons are spills from natural undersea seepage. (The rest, 307 million gallons, are spills caused by human activity.) What percent of all oil spilled is caused by natural seepage? Round your answer to the nearest hundredth of a percent. (*Source*: www.nationalacademies.org)

50. *Equestrian Rider* An Olympic equestrian rider practiced jumping over a water hazard. In 400 attempts, she and her horse touched the water 15 times. What percent of her jump attempts were not perfect?

51. *College Courses* At Monroe State College, 62% of the freshman class is enrolled in a composition course. There are 1070 freshmen this year. How many of them are taking a composition course? Round your answer to the nearest whole number.

52. *Student Health* A recent study indicates that 15% of all middle school students do not eat a proper breakfast. If Pineridge Middle School has 420 students, how many do not eat a proper breakfast?

53. *Swim Team* The swim team at Stonybrook College has gone on to the state championships 24 times over the years. If that translates to 60% of the time in which the team has qualified for the finals, how many years has the swim team qualified for the finals?

54. *Higher Education* North Shore Community College found that 60% of its graduates go on for further education. Last year 570 of the graduates went on for further education. How many students graduated from the college last year?

55. Find 12% of 30% of $1600.

56. Find 90% of 15% of 2700.

Cumulative Review *Multiply or divide.*

57. [3.4.1] $\begin{array}{r} 1.36 \\ \times\ 1.8 \\ \hline \end{array}$

58. [3.4.1] $\begin{array}{r} 5.06 \\ \times\ 0.82 \\ \hline \end{array}$

59. [3.5.2] $0.06\overline{)170.04}$

60. [3.5.2] $0.9\overline{)2.124}$

Quick Quiz 5.3A

1. What is 152% of 84?

2. 72 is 0.8% of what number?

3. 68 is what percent of 400?

4. Concept Check Explain how to solve the following problem using an equation. Jason found that 85% of all people who purchased Mustangs at Danvers Ford were previous Mustang owners. Last year 120 people purchased Mustangs at Danvers Ford. How many of them were previous Mustang owners?

5.3B Solving Percent Problems Using Proportions

① Identifying the Parts of the Percent Proportion

In Section 5.3A we showed you how to use an equation to solve a percent problem. Some students find it easier to use proportions to solve percent problems. We will show you how to use proportions in this section. The two methods work equally well. Using percent proportions allows you to see another of the many uses of the proportions that we studied in Chapter 4.

Suppose your math class of 25 students has 19 right-handed students and 6 left-handed students. You could say that $\frac{19}{25}$ of the class or 76% is right-handed. Consider the following relationship.

$$\frac{19}{25} = 76\%$$

This can be written as

$$\frac{19}{25} = \frac{76}{100}$$

As a rule, we can write this relationship using the **percent proportion**

$$\frac{\text{amount}}{\text{base}} = \frac{\text{percent number}}{100}.$$

To use this equation effectively, we need to find the amount, base, and percent number in a word problem. The easiest of these three parts to find is the percent number. We use the letter p (a variable) to represent the **percent number.**

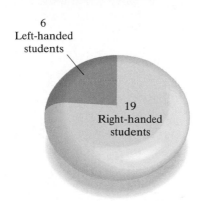

6
Left-handed students

19
Right-handed students

EXAMPLE 1 Identify the percent number p.

(a) Find 16% of 370. **(b)** 28% of what is 25?
(c) What percent of 18 is 4.5?

Solution
(a) Find 16% of 370. **(b)** 28% of what is 25?
The value of p is 16. The value of p is 28.
(c) What percent of 18 is 4.5?

$$p$$

We let p represent the unknown percent number.

Student Practice 1 Identify the percent number p.

(a) Find 83% of 460. **(b)** 18% of what number is 90?
(c) What percent of 64 is 8?

NOTE TO STUDENT: Fully worked-out solutions to all of the Student Practice problems can be found at the back of the text starting at page SP-1.

We use the letter b to represent the base number. The **base** is the entire quantity or the total involved. The number that is the base usually appears after the word *of.* The **amount,** which we represent by the letter a, is the part being compared to the whole.

EXAMPLE 2   Identify the base *b* and the amount *a*.

(a) 20% of 320 is 64.   **(b)** 12 is 60% of what?

Solution

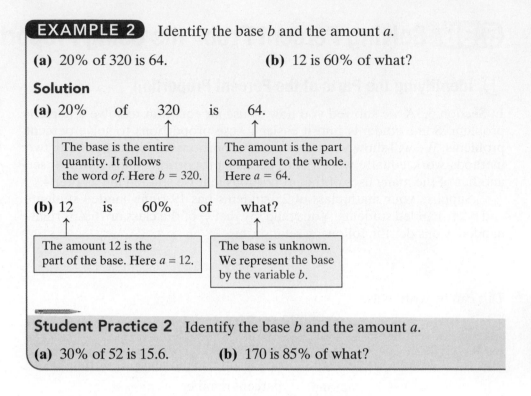

(a) 20%   of   320   is   64.

> The base is the entire quantity. It follows the word *of*. Here *b* = 320.

> The amount is the part compared to the whole. Here *a* = 64.

(b) 12   is   60%   of   what?

> The amount 12 is the part of the base. Here *a* = 12.

> The base is unknown. We represent the base by the variable *b*.

Student Practice 2   Identify the base *b* and the amount *a*.

(a) 30% of 52 is 15.6.   **(b)** 170 is 85% of what?

When identifying *p, b*, and *a* in a problem, it is easiest to identify *p* and *b* first. The remaining quantity or variable is *a*.

EXAMPLE 3   Find *p, b*, and *a*.

(a) What is 52% of 300?   **(b)** What percent of 30 is 18?

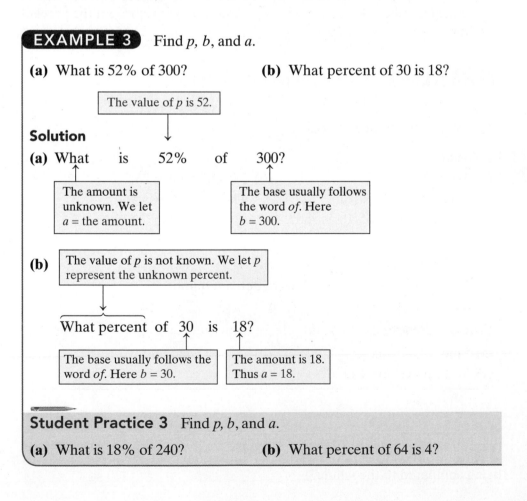

> The value of *p* is 52.

Solution

(a) What   is   52%   of   300?

> The amount is unknown. We let *a* = the amount.

> The base usually follows the word *of*. Here *b* = 300.

(b)

> The value of *p* is not known. We let *p* represent the unknown percent.

What percent of 30 is 18?

> The base usually follows the word *of*. Here *b* = 30.

> The amount is 18. Thus *a* = 18.

Student Practice 3   Find *p, b*, and *a*.

(a) What is 18% of 240?   **(b)** What percent of 64 is 4?

② Using the Percent Proportion to Solve Percent Problems

When we solve the percent proportion, we will have enough information to state the numerical value for two of the three variables a, b, p in the equation

$$\frac{a}{b} = \frac{p}{100}$$

We first identify those two values, and then substitute those values into the equation. Then we will use the skills that we acquired for solving proportions in Chapter 4 to find the value we do not know. Here and throughout the entire chapter we assume that $b \neq 0$.

When solving each problem it is a good idea to look at your answer and see if it is reasonable. Ask yourself, "Does my answer make sense?"

EXAMPLE 4 Find 260% of 40.

Solution The percent $p = 260$. The number that is the base usually appears after the word *of*. The base $b = 40$. The amount is unknown. We use the variable a. Thus

$$\frac{a}{b} = \frac{p}{100} \quad \text{becomes} \quad \frac{a}{40} = \frac{260}{100}.$$

If we reduce the fraction on the right-hand side, we have

$$\frac{a}{40} = \frac{13}{5}$$

$5a = (40)(13)$ Cross-multiply.

$5a = 520$ Simplify.

$\dfrac{5a}{5} = \dfrac{520}{5}$ Divide each side of the equation by 5.

$a = 104$

Thus 260% of 40 is 104.

Student Practice 4 Find 340% of 70.

EXAMPLE 5 85% of what is 221?

Solution The percent $p = 85$. The base is unknown. We use the variable b. The amount a is 221. Thus

$$\frac{a}{b} = \frac{p}{100} \quad \text{becomes} \quad \frac{221}{b} = \frac{85}{100}.$$

Continued on next page

If we reduce the fraction on the right-hand side, we have

$$\frac{221}{b} = \frac{17}{20}$$

$$(221)(20) = 17b \quad \text{Cross-multiply.}$$

$$4420 = 17b \quad \text{Simplify.}$$

$$\frac{4420}{17} = \frac{17b}{17} \quad \text{Divide each side by 17.}$$

$$260 = b. \quad \text{Divide 4420 by 17.}$$

Thus 85% of 260 is 221.

Student Practice 5 68% of what is 476?

EXAMPLE 6 George and Barbara purchased some no-load mutual funds. The account manager charged a service fee of 0.2% of the value of the mutual funds. George and Barbara paid this fee, which amounted to $53. When they got home they could not find the receipt that showed the exact value of the mutual funds that they purchased. Can you find the value of the mutual funds that they purchased?

Solution The basic situation here is that 0.2% of some number is $53. This is equivalent to saying $53 is 0.2% of what? If we want to answer the question "53 is 0.2% of what?", we need to identify a, b, and p.

The percent $p = 0.2$. The base is unknown. We use the variable b. The amount $a = 53$. Thus

$$\frac{a}{b} = \frac{p}{100} \quad \text{becomes} \quad \frac{53}{b} = \frac{0.2}{100}.$$

When we cross-multiply, we obtain

$$(53)(100) = 0.2b$$

$$5300 = 0.2b$$

$$\frac{5300}{0.2} = \frac{0.2b}{0.2}$$

$$26{,}500 = b.$$

Thus $53 is 0.2% of $26,500. Therefore the value of the mutual funds was $26,500.

Student Practice 6 Everett Hatfield recently exchanged U.S. dollars to Canadian dollars for his company, Nova Scotia Central Trucking, Ltd. The bank charged a fee of 0.3% of the total U.S. dollars exchanged. The fee amounted to $216 in U.S. money. How many U.S. dollars were exchanged?

EXAMPLE 7 What percent of 4000 is 160?

Solution The percent is unknown. We use the variable p. The base $b = 4000$. The amount $a = 160$. Thus

$$\frac{a}{b} = \frac{p}{100} \quad \text{becomes} \quad \frac{160}{4000} = \frac{p}{100}.$$

If we reduce the fraction on the left-hand side, we have

$$\frac{1}{25} = \frac{p}{100}$$

$$100 = 25p \qquad \text{Cross-multiply.}$$

$$\frac{100}{25} = \frac{25p}{25} \qquad \text{Divide each side by 25.}$$

$$4 = p \qquad \text{Divide 100 by 25.}$$

Thus 4% of 4000 is 160.

Student Practice 7 What percent of 3500 is 105?

👣 STEPS TO SUCCESS What Are the Absolute Essentials to Succeed in This Course?

If you are in a traditional class:

Students who are successful in this course find there are six absolute essentials.

Here they are:

1. Attend every class session.
2. Read the textbook for every assigned section.
3. Take notes in class.
4. Do the assigned homework for every class.
5. Get help immediately when you need assistance.
6. Review what you are learning.

Making it personal: Which of the six suggestions above is the one you have the greatest trouble actually doing? Will you make a personal commitment to doing that one thing faithfully for the next two weeks? You will be amazed at the results. ▼

If you are in an online class or a nontraditional class:

Students in an online class or a self-paced class find it really helps to spread the homework and the reading over five different days each week.

Making it personal: Will you try to do your homework for five days each week over the next two weeks? You will be amazed at the results. ▼

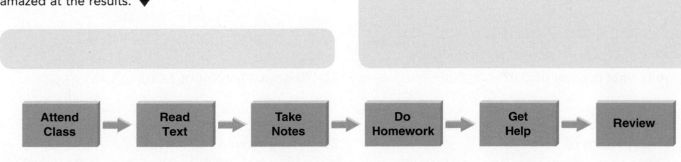

Identify p, b, and a. Do not solve for the unknown.

	p	*b*	*a*
1. 75% of 660 is 495.	_____	_____	_____
2. 65% of 820 is 532.	_____	_____	_____
3. What is 22% of 60?	_____	_____	_____
4. What is 35% of 95?	_____	_____	_____
5. 49% of what is 2450?	_____	_____	_____
6. 38% of what is 2280?	_____	_____	_____
7. 30 is what percent of 50?	_____	_____	_____
8. 50 is what percent of 250?	_____	_____	_____

Solve using the percent proportion

$$\frac{a}{b} = \frac{p}{100}.$$

In exercises 9–14, the amount a is not known.

9. 40% of 70 is what?   **10.** 80% of 90 is what?   **11.** Find 210% of 40.

12. Find 150% of 80.   **13.** 0.7% of 8000 is what?   **14.** 0.8% of 9000 is what?

In exercises 15–20, the base b is not known.

15. 20 is 25% of what?   **16.** 45 is 60% of what?   **17.** 250% of what is 200?

18. 120% of what is 90?   **19.** 3000 is 0.5% of what?   **20.** 6000 is 0.4% of what?

In exercises 21–24, the percent p is not known.

21. 56 is what percent of 280?   **22.** 70 is what percent of 1400?

23. What percent of 90 is 18?   **24.** What percent of 120 is 18?

Mixed Practice

25. 25% of 88 is what?

26. 20% of 75 is what?

27. 300% of what is 120?

28. 200% of what is 120?

29. 82 is what percent of 500?

30. 75 is what percent of 600?

31. Find 0.7% of 520.

32. Find 0.4% of 650.

33. What percent of 66 is 16.5?

34. What percent of 49 is 34.3?

35. 68 is 40% of what?

36. 52 is 40% of what?

Applications *When solving each applied problem, examine your answer and see if it is reasonable. Ask yourself, "Does my answer make sense?"*

37. *Paycheck Deposit* Each time Lowell gets paid, 5% of his paycheck is deposited in his retirement account. Last week, $48 was put into his retirement account. What was the amount of Lowell's paycheck?

38. *Income Tax* Last year, Rachel had 24% of her salary withheld for taxes. If the total amount withheld was $6300 for the year, what was Rachel's annual salary?

39. *Eating Out* Ed and Suzie went out to eat at Pizzeria Uno. The dinner check was $26.00. They left a tip of $3.90. What percent of the check was the tip?

40. *Baseball* During the baseball season, Damon was up to bat 60 times. Of these at-bats, 12 were home runs. What percent of Damon's at-bats resulted in a home run?

41. *Food Expiration Date* The Super Shop and Save store had 120 gallons of milk placed on the shelf one night. During the next morning's inspection, the manager found that 15% of the milk had passed the expiration date. How many gallons of milk had passed the expiration date?

42. *Police Arrests* During June the Wenham police stopped 250 drivers for speed violations. It was found that 8% of the people who were stopped had outstanding warrants for their arrest. How many people had outstanding warrants for their arrest?

43. *Car Purchase* Janelle purchased a used car for $10,500. She made a down payment of 18% of the purchase price. How much was her down payment?

44. *Education* Trudy took a biology test with 40 problems. She got 8 of the problems wrong and 32 of the problems right. What percent of the test problems did she do incorrectly?

To Think About

The Cost of Children A study was conducted to determine the amount of money middle-income families in the United States spend per child, ages 6 to 14, for each of the following: housing, food, transportation, clothing, health care, child care and education, and miscellaneous (including personal care items, entertainment, and reading materials). The amount of money in each of these categories is given in the pie chart below. Use the pie chart to answer questions 45–48. Round all answers to the nearest tenth.

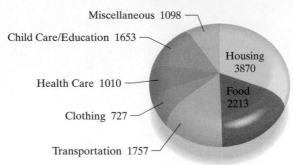

Middle-Income Families' Annual Expenditures per Child in 2008

Miscellaneous 1098
Child Care/Education 1653
Health Care 1010
Clothing 727
Transportation 1757
Housing 3870
Food 2213

Source: www.census.gov

45. What percent of the total expenditures in these seven categories was spent on food?

46. What percent of the total expenditures in these seven categories was spent on health care?

47. Suppose that compared to 2008, expenditures in 2013 for housing were 25% larger, expenditures for child care/education were 10% larger, and the other five categories remained at the same dollar amount. What percent of the total expenditures in 2013 in these seven categories would be used for food?

48. Suppose that compared to 2008, expenditures in 2018 for health care were 25% larger, expenditures for food were 15% larger, and the other five categories remained at the same dollar amount. What percent of the total expenditures in 2018 in these seven categories would be used for miscellaneous items?

Cumulative Review *Simplify.*

49. [2.7.2] $\dfrac{4}{5} + \dfrac{8}{9}$

50. [2.7.2] $\dfrac{7}{13} - \dfrac{1}{2}$

51. [2.4.3] $\left(2\dfrac{4}{5}\right)\left(1\dfrac{1}{2}\right)$

52. [2.5.3] $1\dfrac{2}{5} \div \dfrac{3}{4}$

Quick Quiz 5.3B

1. What is 0.09% of 17,000?　　**2.** 64.8 is 54% of what number?　　**3.** 132 is what percent of 600?

Explain how to solve the following problem using a proportion.

4. Concept Check Alice purchased some stock and was charged a service fee of 0.7% of the value of the stock. The fee she paid was $140. What was the value of the stock that she purchased?

How Am I Doing?   Sections 5.1–5.3

*How are you doing with your homework assignments in Sections 5.1 to 5.3?
Do you feel you have mastered the material so far? Do you understand the
concepts you have covered? Before you go further in the textbook, take some
time to do each of the following problems.*

5.1

Write as a percent.

1. 0.17　　　**2.** 0.387　　　**3.** 7.95　　　**4.** 12.25

5. 0.006　　　**6.** 0.0004　　　**7.** $\dfrac{17}{100}$　　　**8.** $\dfrac{89}{100}$

9. $\dfrac{13.4}{100}$　　　**10.** $\dfrac{19.8}{100}$　　　**11.** $\dfrac{6\frac{1}{2}}{100}$　　　**12.** $\dfrac{3\frac{5}{8}}{100}$

5.2

Change to a percent. Round to the nearest hundredth of a percent when necessary.

13. $\dfrac{8}{10}$　　　**14.** $\dfrac{15}{30}$　　　**15.** $\dfrac{52}{20}$　　　**16.** $\dfrac{17}{16}$

17. $\dfrac{5}{7}$　　　**18.** $\dfrac{2}{7}$　　　**19.** $\dfrac{18}{24}$　　　**20.** $\dfrac{9}{36}$

21. $4\dfrac{2}{5}$　　　**22.** $2\dfrac{3}{4}$　　　**23.** $\dfrac{1}{300}$　　　**24.** $\dfrac{1}{400}$

Write as a fraction or a mixed number in simplified form.

25. 22%　　　**26.** 53%　　　**27.** 150%　　　**28.** 160%

29. $6\dfrac{1}{3}\%$　　　**30.** $3\dfrac{1}{8}\%$　　　**31.** $51\dfrac{1}{4}\%$　　　**32.** $43\dfrac{3}{4}\%$

5.3

Solve. Round to the nearest hundredth when necessary.

33. What is 70% of 60?　　　**34.** Find 15% of 140.

35. 68 is what percent of 72?　　　**36.** What percent of 80 is 64?

37. 8% of what number is 240?　　　**38.** 354 is 40% of what number?

*Now turn to page SA-10 for the answer to each of these problems. Each answer
also includes a reference to the objective in which the problem is first taught. If
you missed any of these problems, you should stop and review the Examples
and Student Practice problems in the referenced objective. A little review now
will help you master the material in the upcoming sections of the text.*

1. _____
2. _____
3. _____
4. _____
5. _____
6. _____
7. _____
8. _____
9. _____
10. _____
11. _____
12. _____
13. _____
14. _____
15. _____
16. _____
17. _____
18. _____
19. _____
20. _____
21. _____
22. _____
23. _____
24. _____
25. _____
26. _____
27. _____
28. _____
29. _____
30. _____
31. _____
32. _____
33. _____
34. _____
35. _____
36. _____
37. _____
38. _____

5.4 Solving Applied Percent Problems

Student Learning Objectives

After studying this section, you will be able to:

1. Solve general applied percent problems.

2. Solve applied problems when percents are added.

3. Solve discount problems.

① Solving General Applied Percent Problems

In Sections 5.3A and 5.3B, we learned the three types of percent problems. Some problems ask you to find a percent of a number. Some problems give you an amount and a percent and ask you to find the base (or whole). Other problems give an amount and a base and ask you to find the percent. We will now see how the three types of percent problems occur in real life.

Of all the sections of this textbook, this one covers some types of problems you will encounter in real life. You see discounts at the mall, sales tax is added to your purchases, and our daily news is filled with statistics reported in terms of percentages. A complete understanding of percents is essential in your daily life.

EXAMPLE 1 Of all the computers manufactured last month, an inspector found 18 that were defective. This is 2.5% of all the computers manufactured last month. How many computers were manufactured last month?

Solution

Method A Translate to an equation.

The problem is equivalent to: 2.5% of the number of computers is 18.

Let n = the number of computers.

$$2.5\% \quad \text{of} \quad \underbrace{\text{the number of computers}} \quad \text{is} \quad 18$$
$$2.5\% \quad \times \qquad\qquad n \qquad\qquad = 18$$

$$0.025n = 18$$

$$\frac{0.025n}{0.025} = \frac{18}{0.025}$$

$$n = 720$$

720 computers were manufactured last month.

Method B Use the percent proportion $\dfrac{a}{b} = \dfrac{p}{100}$.

The percent $p = 2.5$. The base is unknown. We will use the variable b. The amount $a = 18$. Thus

$$\frac{a}{b} = \frac{p}{100} \quad \text{becomes} \quad \frac{18}{b} = \frac{2.5}{100}.$$

Using cross multiplication, we have

$$(18)(100) = 2.5b$$
$$1800 = 2.5b$$
$$\frac{1800}{2.5} = \frac{2.5b}{2.5}$$
$$720 = b.$$

720 computers were manufactured last month.

Student Practice 1 4800 people, or 12%, of all passengers holding tickets for American Airlines flights in one month did not show up for their flights. How many people held tickets that month?

338

By either Method A or Method B, we obtain the same number of computers, 720.

Substitute 720 into the original problem to check.

2.5% of 720 computers are defective.

$$(0.025)(720) = 18 \;\checkmark$$

EXAMPLE 2 How much sales tax will you pay on a plasma HDTV television priced at $499 if the sales tax is 5%?

Solution

Method A Translate to an equation.

$$\text{What is } 5\% \text{ of } \$499?$$
$$\downarrow \;\; \downarrow \;\; \downarrow \;\; \downarrow \;\; \downarrow$$
$$n \; = \; 5\% \; \times \; 499$$
$$n = (0.05)(499)$$
$$n = 24.95 \qquad \text{The tax is } \$24.95.$$

Method B Use the percent proportion $\dfrac{a}{b} = \dfrac{p}{100}$.

The percent $p = 5$. The base $b = 499$. The amount is unknown. We use the variable a. Thus

$$\frac{a}{b} = \frac{p}{100} \quad \text{becomes} \quad \frac{a}{499} = \frac{5}{100}.$$

If we reduce the fraction on the right-hand side, we have

$$\frac{a}{499} = \frac{1}{20}.$$

We then cross-multiply to obtain

$$20a = 499$$
$$\frac{20a}{20} = \frac{499}{20}$$
$$a = 24.95$$

The tax is $24.95.

Thus, by either method, the amount of the sales tax is $24.95.

Let's see if our answer is reasonable. Is 5% of $499 really $24.95? If we round $499 to one nonzero digit, we have $500. Thus we have 5% of $500 = 25$. Since 25 is quite close to our value of $24.95, our answer seems reasonable.

Student Practice 2 A salesperson rented a hotel room for $62.30 per night. The tax in her state is 8%. How much tax did she pay for one night at the hotel? Round to the nearest cent.

NOTE TO STUDENT: Fully worked-out solutions to all of the Student Practice problems can be found at the back of the text starting at page SP-1.

EXAMPLE 3   A failing student attended class 39 times out of the 45 times the class met last semester. What percent of the classes did he attend? Round to the nearest tenth of a percent.

Solution

Method A   Translate to an equation.

This problem is equivalent to:

39 is what percent of 45?

$$39 = n \times 45$$

$$39 = 45n$$

$$\frac{39}{45} = \frac{45n}{45}$$

$$0.8666\ldots = n.$$

To the nearest tenth of a percent we have $n = 86.7\%$.

Method B   Use the percent proportion $\frac{a}{b} = \frac{p}{100}$.

The percent is unknown. We use the variable p. The base b is 45. The amount a is 39. Thus

$$\frac{a}{b} = \frac{p}{100} \quad \text{becomes} \quad \frac{39}{45} = \frac{p}{100}.$$

When we cross-multiply, we get

$$(39)(100) = 45p$$

$$3900 = 45p$$

$$\frac{3900}{45} = \frac{45p}{45}$$

$$86.666\ldots = p.$$

To the nearest tenth, the answer is 86.7%.

By using either method, we discover that the failing student attended approximately 86.7% of the classes.

Verify by estimating that the answer is reasonable.

Student Practice 3   Of the 130 flights at Orange County Airport yesterday, only 105 of them were on time. What percent of the flights were on time? (Round to the nearest tenth of a percent.)

Now you have some experience solving the three types of percent problems in real-life applications. You can use either Method A or Method B to solve applied percent problems. In the following pages we will present more percent applications. We will not list all the steps of Method A or Method B. Most students will find after a careful study of Examples 1–3 that they do not need to write out all the steps of Method A or Method B when solving applied percent problems.

② Solving Applied Problems When Percents Are Added

Percents can be added if the base (whole) is the same. For example, 50% of your salary added to 20% of your salary = 70% of your salary. 100% of your cost added to 15% of your cost = 115% of your cost. Problems like this are often called **markup problems.** If we add 15% of the cost of an item to the original cost, the markup is 15%. We will add percents in some applied situations.

The following example is interesting, but it is a little challenging. So please read it very carefully. A lot of students find it difficult at first.

EXAMPLE 4 Walter and Mary Ann are going out to a restaurant. They have a limit of $63.25 to spend for the evening. They want to tip the waitress 15% of the cost of the meal. How much money can they afford to spend on the meal itself? (Assume there is no tax.)

Solution In some of the problems in this section, it may help you to use the Mathematics Blueprint. We will use it here for Example 4.

Mathematics Blueprint for Problem Solving

Gather the Facts	What Am I Asked to Do?	How Do I Proceed?	Key Points to Remember
They have a spending limit of $63.25. They want to tip the waitress 15% of the cost of the meal.	Find the amount of money that the meal can cost.	Separate the $63.25 into two parts: the cost of the meal and the tip. Add these two parts to get $63.25.	We are not taking 15% of $63.25, but rather 15% of the cost of the meal.

Let n = the cost of the meal. 15% of the cost = the amount of the tip. We want to add the percents of the meal.

$$\boxed{\text{Cost of meal } n} \; + \; \boxed{\begin{array}{c}\text{tip of 15\%}\\\text{of the cost}\end{array}} \; = \; \boxed{\$63.25}$$

$$100\% \text{ of } n \; + \; 15\% \text{ of } n \; = \; \$63.25$$

Note that 100% of n added to 15% of n is 115% of n.

$$115\% \text{ of } n = \$63.25$$
$$1.15 \times n = 63.25$$
$$\frac{1.15 \times n}{1.15} = \frac{63.25}{1.15} \qquad \text{Divide both sides by 1.15.}$$
$$n = 55$$

They can spend up to $55.00 on the meal itself.
Does this answer seem reasonable?

Student Practice 4 Sue and Sam have $46.00 to spend at a restaurant, including a 15% tip. How much can they spend on the meal itself? (Assume there is no tax.)

③ **Solving Discount Problems**

Frequently, we see signs urging us to buy during a sale when the list price is discounted by a certain percent. The amount of a **discount** is the product of the discount rate and the list price.

$$\text{Discount} = \text{discount rate} \times \text{list price}$$

EXAMPLE 5 Jeff purchased a flat-panel LCD TV on sale at a 35% discount. The list price was $430.00.

(a) What was the amount of the discount?
(b) How much did Jeff pay for the flat-panel LCD TV?

Solution

(a) Discount = discount rate × list price

$$= 35\% \times 430$$
$$= 0.35 \times 430$$
$$= 150.5$$

The discount was $150.50.

(b) We subtract the discount from the list price to get the selling price.

$430.00	list price
− $150.50	discount
$279.50	selling price

Jeff paid $279.50 for the flat-panel LCD TV.

Student Practice 5 Betty bought a car that lists for $13,600 at a 7% discount.

(a) What was the discount?
(b) What did she pay for the car?

STEPS TO SUCCESS What Happens When You Read a Mathematics Textbook?

Reading a math book can give you amazing insight. Always take the time at the start of a homework section to read the section(s) assigned in your textbook.

Remember this book was written to help you become successful in this mathematics class. However, you have to read it if you want to take advantage of all its benefits.

Always read your book with a pen or a highlighter in hand. Watch for helpful hints and suggestions as you look over the sample explanations. Underline any step that you think is hard or not clear to you. Put question marks by words you do not understand. You may want to write down a list of these in your notebook. Ask your instructor about things you do not understand.

When you come to a sample example, make your mind work it through step by step. Underline steps that you think are especially important.

Make sure you are thinking about what you are reading. If your mind is wandering, get up and get a drink of water or walk around the room—anything to help you get your mind back on track.

Making it personal: Look over these suggestions. Pick one or two ideas that you think will be the most helpful to you. Use them this week as you read the book carefully before doing homework assignments. ▼

5.4 Exercises

MyMathLab®

Watch the videos
in MyMathLab

Download the
MyDashBoard App

Applications *Exercises 1–18 present the three types of percent problems. They are similar to Examples 1–3. Take the time to master exercises 1–18 before going on to the next ones. Round to the nearest hundredth when necessary.*

1. **Education** No graphite was found in 4500 pencils shipped to Sureway School Supplies. This was 2.5% of the total number of pencils received by Sureway. How many pencils in total were in the order?

2. **Track and Field** A high-jumper on the track and field team hit the bar 58 times last week. This means that he did not succeed in 29% of his jump attempts. How many total attempts did he make last week?

3. **Cable Television Bill** Under Todd's new cable television plan, his bill averages $63 per month. This is 140% of his average monthly bill last year when he had the basic cable package. What was his average monthly cable bill last year?

4. **Salary Changes** Renata now earns $9.50 per hour. This is 125% of what she earned last year. What did she earn per hour last year?

5. **Square Footage of Home** A 2100-square-foot home is for sale. The finished basement has an area of 432 square feet. The basement accounts for what percent of the total square footage?

6. **Coffee Bar** Every day this year, Sam ordered either cappuccino or espresso from the coffee bar downstairs. He had 85 espressos and 280 cappuccinos. What percent of the coffees were espressos?

7. **Sales Tax** Elizabeth bought new towels and sheets for $65. How much tax did she pay if the sales tax is 6%?

8. **Sales Tax** Leon bought new clothes for his bank job. Before tax was added on, his total was $180. How much tax did he pay if the sales tax is 5%?

9. **Mountain Bike** Malia bought a new mountain bike. The sales tax in her state is 7%, and she paid $38.50 in tax. What was the price of the mountain bike before the tax?

10. **Sales Tax** Hiro bought some artwork and paid $10.75 in tax. The sales tax in his state is 5%. What was the price of the artwork?

11. **Mortgage Payment** Paul and Sue Yin together earn $4180 per month. Their mortgage payment is $1254 per month. What percent of their household income goes toward paying the mortgage?

12. **Car Payment** Cora puts aside $52.50 per week for her monthly car payment. She earns $350 per week. What percent of her income is set aside for car payments?

13. **Charities** The Children's Wish Charity raised 75% of its funds from sporting promotions. Last year the charity received $7,200,000 from its sporting promotions. What was the charity's total income last year?

14. **Taxes** Last year, Michael paid $9100 in federal and state taxes as a veterinary assistant, which amounted to 26% of his annual income. What was his annual income last year?

15. **Baseball** 10,001 home runs have been hit in Boston's Fenway Park since it opened in 1912. Ted Williams of the Boston Red Sox hit 248 of these. What percent of the total home runs did Ted Williams hit?

16. **Baseball** During the 2010 baseball season, Jose Bautista was up to bat 397 times. Of these, 35 resulted in a home run. What percent of Bautista's at-bats resulted in a home run?

17. **Population Studies** In 2010, 0.6% of the world's population lived in Kenya. If the population of the world was approximately 6,862,600,000 in 2010, how many people lived in Kenya?

18. **Population Studies** In 2010, 0.95% of the world's population lived in France. If the population of the world was approximately 6,862,600,000 in 2010, how many people lived in France?

Exercises 19–30 include percents that are added and discounts. Solve. Round to the nearest hundredth when necessary.

19. *Sales Tax* Henry has $800 total to spend on a new dining room table and chairs. If the sales tax is 5%, how much can he afford to spend on the table and chairs?

20. *Eating Out* Belinda asked Martin out to dinner. She has $47.50 to spend. She wants to tip the waitress 15% of the cost of their meal. How much money can she afford to spend on the meal itself?

21. *Building Costs* John and Chris Maney are building a new house. When finished, the house will cost $163,500. The price of the house is 9% higher than the price when the original plans were made. What was the price of the house when the original plans were made?

22. *SUV Purchase* Dan and Connie Lacorazza purchased a new Honda Pilot. The purchase price was $25,440. The price was 6% higher than the price of a similar Honda Pilot three years ago. What was the price of the Honda Pilot three years ago?

23. *Manufacturing* When a new computer case is made, there is some waste of the plastic material used for the front of the computer case. Approximately 3% of the plastic that is delivered is of poor quality and is thrown away. Furthermore, 8% of the plastic that is delivered is waste material that is thrown away as the front pieces are created by a giant stamping machine. If 20,000 pounds of plastic are delivered to make the fronts of computer cases each month, how many pounds are thrown away?

24. *Airport Operations* In a recent survey of planes landing at Logan Airport in Boston, it was observed that 12% of the flights were delayed less than an hour and 7% of the flights were delayed an hour or more but less than two hours. If 9000 flights arrive at Logan Airport in a day, how many flights are delayed less than two hours?

25. *Political Parties* The Democratic National Committee has a budget of $33,000,000 to spend on the inauguration of the new president. 15% of the costs will be paid to personnel, 12% of the costs will go toward food, and 10% will go to decorations.

  (a) How much money will go for personnel, food, and decorations?

  (b) How much will be left over to cover security, facility rental, and all other expenses?

26. *Medical Research* A major research facility has developed an experimental drug to treat Alzheimer's disease. Twenty percent of the research costs was paid to the staff. Sixteen percent of the research costs was paid to rent the building where the research was conducted. The rest of the money was used for research. The company has spent $6,000,000 in research on this new drug.

  (a) How much was paid to cover the cost of staff and rental of the building?

  (b) How much was left over for research?

27. *Clothes Purchase* Melinda purchased a new blouse, jeans, and a sweater in Naperville, Illinois. All of the clothes were discounted 35%. Before the sale, the total purchase price would have been $190 for these three items. How much did she pay for them with the discount?

28. *Tire Purchase* Juan went to purchase two new radial tires for his Honda Accord in Austin, Texas. The set of two tires normally costs $130. However, he bought them on sale at a discount of 30%. How much did he pay for the tires with the discount?

29. *Snowmobile Purchase* Jack bought his first Polaris snowmobile in Rice Lake, Wisconsin. The price was $8800, but the dealer gave him a discount of 15%.

   (a) What was the discount?

   (b) How much did he pay for the snowmobile?

30. *Appliance Purchase* Charlotte bought a stainless steel refrigerator and stove that had been used as floor models at Home Depot. The list price of the set was $1150, but the store manager gave her a discount of 25%.

   (a) What was the discount?

   (b) How much did she pay for the refrigerator and stove?

Cumulative Review

31. **[1.7.1]** Round to the nearest thousand. 1,698,481

32. **[1.7.1]** Round to the nearest hundred. 2,452,399

33. **[3.2.3]** Round to the nearest hundredth. 1.63474

34. **[3.2.3]** Round to the nearest thousandth. 0.7995

35. **[3.2.3]** Round to the nearest ten-thousandth. 0.055613

36. **[3.2.3]** Round to the nearest ten-thousandth. 0.079152

Quick Quiz 5.4

1. Chris Smith bought a new laptop computer. The list price was $596. He got a 28% discount.

   (a) What was the discount?

   (b) How much did Chris pay for the laptop?

2. Laurie left on a trip for a week. When she returned she had 87 e-mail messages. 56 of these messages were "spam" junk mail. What percent of her e-mail was spam? Round your answer to the nearest tenth if necessary.

3. A total of 4500 people in the city bought take-out pizza at least once during the week. This was 30% of all the people who live in the city. How many people live in the city?

Explain how to solve the following problem.

4. **Concept Check** Sam works in sales for a pharmaceutical company. He can spend 23% of his budget for travel expenses. He can spend 14% for entertainment of clients. He can spend 17% of his budget for advertising. Last year he had a total budget of $80,000. Last year he spent a total of $48,000 for travel expenses, entertainment, and advertising. Did he stay within his budget allowance for those items?

5.5 Solving Commission, Percent of Increase or Decrease, and Interest Problems

Student Learning Objectives

After studying this section, you will be able to:

1. Solve commission problems.
2. Solve percent-of-increase or percent-of-decrease problems.
3. Solve simple-interest problems.

NOTE TO STUDENT: *Fully worked-out solutions to all of the Student Practice problems can be found at the back of the text starting at page SP-1.*

1 Solving Commission Problems

If you work as a salesperson, your earnings may be in part or in total a certain percentage of the sales you make. The amount of money you get that is a percentage of the value of your sales is called your **commission.** It is calculated by multiplying the percentage (called the **commission rate**) by the value of the sales.

$$\text{Commission} = \text{commission rate} \times \text{value of sales}$$

EXAMPLE 1 A salesperson has a commission rate of 17%. She sells $32,500 worth of goods in a department store in two months. What is her commission?

Solution
$$\text{Commission} = \text{commission rate} \times \text{value of sales}$$
$$\text{Commission} = 17\% \times \$32,500$$
$$= 0.17 \times 32,500$$
$$= 5525$$

Her commission is $5525.00.
Does this answer seem reasonable? Check by estimating.

Student Practice 1 A real estate salesperson earns a commission rate of 6% when he sells a $156,000 home. What is his commission?

In some problems, the unknown quantity will be the commission rate or the value of sales. However, the same equation is used:

$$\text{Commission} = \text{commission rate} \times \text{value of sales}$$

2 Solving Percent-of-Increase or Percent-of-Decrease Problems

We sometimes need to find the percent by which a number increases or decreases. If a car costs $7000 and the price decreases $1750, we say that the percent of decrease is $\frac{1750}{7000} = 0.25 = 25\%$.

$$\text{Percent of decrease} = \frac{\text{amount of decrease}}{\text{original amount}}$$

Similarly, if a population of 12,000 people increases by 1920 people, we say that the percent of increase is $\frac{1920}{12,000} = 0.16 = 16\%$.

$$\text{Percent of increase} = \frac{\text{amount of increase}}{\text{original amount}}$$

Note that for these types of problems the base is always the *original amount.*

The most important thing to remember is that we must **first** find the amount of increase or decrease.

EXAMPLE 2   The population of Center City increased from 50,000 to 59,500. What was the percent of increase?

Solution   For this problem as well as others in this section, you may find it helpful to use the Mathematics Blueprint.

Mathematics Blueprint for Problem Solving

Gather the Facts	What Am I Asked to Do?	How Do I Proceed?	Key Points to Remember
The population increased from 50,000 to 59,500.	We must find the percent of increase.	First subtract to find the amount of increase. Then divide the amount of increase by the original amount.	Always divide by the original amount.

Amount of increase

$$\begin{array}{r} 59{,}500 \\ -\ 50{,}000 \\ \hline 9500 \end{array}$$

$$\text{Percent of increase} = \frac{\text{amount of increase}}{\text{original amount}} = \frac{9500}{50{,}000}$$

$$= 0.19 = 19\%$$

The percent of increase is 19%.

Student Practice 2   A new car is sold for $15,000. A year later its price had decreased to $10,500. What is the percent of decrease?

③ Solving Simple-Interest Problems

Interest is money paid for the use of money. If you deposit money in a bank, the bank uses that money and pays you interest. If you borrow money, you pay the bank interest for the use of that money.

The **principal** is the amount deposited or borrowed. Interest is usually expressed as a percent rate of the principal. The **interest rate** is assumed to be per year, unless otherwise stated. The formula used in business to compute simple interest is

$$\text{Interest} = \text{principal} \times \text{rate} \times \text{time}$$

$$I = P \times R \times T$$

If the interest rate is *per year,* the time *T must* be in *years.*

Calculator

 Interest

You can use a calculator to find simple interest. Find the interest on $450 invested at 6.5% for 15 months. Notice the time is in months. Since the interest formula $I = P \times R \times T$, is in years, you need to change 15 months to years by dividing 15 by 12.

Enter

$$15\ \boxed{\div}\ 12\ \boxed{=}$$

Display

$$\boxed{1.25}$$

Leave this on the display and multiply as follows:

$$1.25\ \boxed{\times}\ 450\ \boxed{\times}$$

$$6.5\ \boxed{\%}\ \boxed{=}$$

The display should read

$$\boxed{36.5625}$$

which would round to $36.56.

Try the following.

(a) $9516 invested at 12% for 30 months

(b) $593 borrowed at 8% for 5 months

EXAMPLE 3 Find the simple interest on a loan of $7500 borrowed at 13% for one year.

Solution $I = P \times R \times T$

P = principal = $7500 R = rate = 13% T = time = 1 year
$I = 7500 \times 13\% \times 1 = 7500 \times 0.13 = 975$

The interest is $975.

Student Practice 3 Find the simple interest on a loan of $5600 borrowed at 12% for one year.

Our formula is based on a yearly interest rate. Time periods of more than one year or a fractional part of a year are sometimes needed.

EXAMPLE 4 Find the simple interest on a loan of $2500 that is borrowed at 9% for

(a) three years.   **(b)** three months.

Solution

(a) $$I = P \times R \times T$$

$$P = \$2500 \qquad R = 9\% \qquad T = 3 \text{ years}$$
$$I = 2500 \times 0.09 \times 3 = 225 \times 3 = 675$$

The interest for three years is $675.

(b) Three months $= \dfrac{1}{4}$ year. The time period must be in years to use the formula.

Since $T = \dfrac{1}{4}$ year, we have

$$I = 2500 \times 0.09 \times \frac{1}{4}$$

$$= 225 \times \frac{1}{4}$$

$$= \frac{225}{4} = 56.25$$

The interest for three months is $56.25.

Student Practice 4 Find the simple interest on a loan of $1800 that is borrowed at 11% for

(a) four years.   **(b)** six months.

Many loans today are based on **compound interest.** This topic is covered in more advanced mathematics courses. The calculations for compound interest are tedious to do by hand. Usually people use a computer or a compound interest table to do compound interest problems.

Applications *Exercises 1–18 are problems involving commissions, percent of increase or decrease, and simple interest.*

1. ***Appliance Sales*** Walter works as an appliance salesman in a department store. Last month he sold $170,000 worth of appliances. His commission rate is 2%. How much money did he earn in commission last month?

2. ***Car Sales*** Susan works at the Acura dealership in Winchester. Last month she had car sales totaling $230,000. Her commission rate is 3%. How much money did she earn in commission last month?

3. ***Mobile Phone Sales*** Allison works in the local Verizon office selling mobile phones. She is paid $300 per month plus 4% of her total sales in mobile phones. Last month she sold $96,000 worth of mobile phones. What was her total income for the month?

4. ***Stockbroker*** Matthew is a stockbroker. He is paid $500 per month plus 0.5% of the total sales of stocks that he sells. Last month he sold $340,000 worth of stock. What was his total income for the month?

5. ***Airline Tickets*** Dawn is searching online for airline tickets. Two weeks ago the cost to fly from San Francisco to Minneapolis was $275. The price now is $330. What is the percent of increase?

6. ***Weight Loss*** Tim weighed 267 pounds before starting an exercise routine. Two years later, he weighed a healthy 183 pounds. What was the percent of decrease in Tim's weight? Round your answer to the nearest tenth of a percent.

7. ***Computer Sales*** In 2003, 10.9 million desktop computers were sold in the United States. This dropped to 9.2 million sold in 2009. What was the percent of decrease in the number of desktop computers sold? Round your answer to the nearest tenth of a percent. (*Source:* www.ce.org)

8. ***Camera Sales*** In 2003, 33.2 million digital cameras were sold in the United States. This dropped to 30.5 million sold in 2009. What was the percent of decrease in the number of digital cameras sold? Round your answer to the nearest tenth of a percent. (*Source:* www.ce.org)

9. ***CD Interest*** Phil placed $2000 in a one-year CD at the bank. The bank is paying simple interest of 7% for one year on the CD. How much interest will Phil earn in one year?

10. ***Checking Account Interest*** Charlotte has a checking account that pays her simple interest of 1.2% on the average balance in her checking account. Last year her average balance was $450. How much interest did she earn in her checking account?

11. ***Credit Card Expenses*** Melinda has a MasterCard account with Centerville Bank. She has to pay a monthly interest rate of 1.5% on the average daily balance of the amount she owes on her credit card. Last month her average daily balance was $500. How much interest was she charged last month? (*Hint:* The formula $I = P \times R \times T$ can be used if the interest rate is *per month* and the time is in *months*.)

12. ***Student Loan*** Walter borrowed $3000 for a student loan to finish college this year. Next year he will need to pay 7% simple interest on the amount he borrowed. How much interest will he need to pay next year?

13. ***House Construction Loan*** James had to borrow $26,000 for a house construction loan for four months. The interest rate was 12% per year. How much interest did he have to pay for borrowing the money for four months?

14. ***Small Business Loan*** Maya needed to borrow $9200 for three months to finance some renovations to her gift shop. The interest rate was 15% per year. How much interest did she have to pay for borrowing the money for three months?

15. ***Life Insurance*** Robert sells life insurance for a major insurance company for a commission. Last year he sold $12,000,000 worth of insurance. He earned $72,000 in commissions. What was his commission rate?

16. ***Medical Supplies*** Hillary sells medical supplies to doctors' offices for a major medical supply company. Last year she sold $9,000,000 worth of medical supplies. She works on a commission basis and last year she earned $63,000 in commissions. What was her commission rate?

17. ***Furniture Sales*** Jennifer sells furniture for a major department store. Last year she was paid $48,000 for commissions. If her commission rate is 3%, what was the sales total of the furniture that she sold last year?

18. ***Auto Sales*** Michael sells used cars for Beltway Motors. Last year he was paid $42,000 in commissions. Beltway Motors pays the salespeople a commission rate of 6%. What was the sales total of the cars that Michael sold last year?

Mixed Applications
Exercises 19–36 are a variety of percent problems. They involve commissions, percent of increase or decrease, and simple interest. There are also some of each kind of percent problem encountered in the chapter. Unless otherwise directed, round to the nearest hundredth.

19. ***Entertainment Expenses*** Ted is trying to decrease his spending on entertainment. He earns $265 per week and is allowing himself to spend only 15% per week on movies, dining out, and so on. How much can Ted spend per week on entertainment?

20. ***Biology*** The maximum capacity of your lungs is 4.58 liters of air. In a typical breath, you breathe in 12% of the maximum capacity. How many liters of air do you breathe in a typical breath?

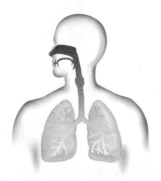

21. ***Girl Scout Cookies*** Of all the boxes of Girl Scout cookies sold, 25% are Thin Mints. If a Girl Scout troop sells 156 boxes of cookies, how many are Thin Mints?

22. ***Scotland*** 11% of the Scottish population has red hair. If the population of Scotland is 5,600,000, how many people have red hair?

23. **Food Stamps** In March 2008, 28 million Americans were using food stamps. By March 2010, this number had gone up to 40.2 million people. What was the percent of increase in the number of people using food stamps? (*Source:* www.usda.gov)

24. **Carbon Dioxide Emissions** In 2005, the amount of energy-related carbon dioxide emissions in the United States was 5955 million metric tons. In 2006, it was estimated that this number had fallen to 5877 million metric tons. What was the percent of decrease in carbon dioxide emissions? (*Source:* www.energy.gov)

25. **Sporting Goods** A sporting goods store buys cross-training shoes for $40, and sells them for $72. What is the percent of increase in the price of the shoes?

26. **Jewelry Costs** A gift store buys earrings from an artist for $20 and sells them for $29. What is the percent of increase in the price of the earrings?

27. **Savings Account** Adam deposited $3700 in his savings account for one year. His savings account earns 2.3% interest annually. He did not add any more money within the year, and at the end of that time, he withdrew all funds.
   (a) How much interest did he earn?
   (b) How much money did he withdraw from the bank?

28. **Credit Card** Nikki had $1258 outstanding on her MasterCard, which charges 2% monthly interest. At the end of this month Nikki paid off the account.
   (a) How much interest did Nikki pay for one month?
   (b) How much did it cost to pay off the account totally?

29. **Shopping Trip** Bryce went shopping and bought a pair of sandals for $52, swimming trunks for $38, and sunglasses for $26. The tax in Bryce's city is 6%.
   (a) What was the total sales tax?
   (b) What was the total of the purchases?

30. **Automobile Purchase** Jin bought a used Toyota Camry for $12,600. The tax in her state is 7%.
   (a) What was the sales tax?
   (b) What was the final price of the Camry?

31. **Property Taxes** Smithville Kitchen Cabinetry Inc. is late in paying $9500 in property taxes to the city of Springfield. It will be assessed 14% interest for being late in property tax payment. Find the one total amount that will pay off both the taxes and the interest charge.

32. **Property Taxes** Raymond and Elsie Ostram are late in paying $1600 in property taxes to the city of New Boston. They will be assessed 12% interest for being late in property tax payment. Find the one total amount that will pay off both the taxes and the interest charge.

33. **Home Purchase** Betty and Michael Bently purchased a new home for $349,000. They paid a down payment of 8% of the cost of the home. They took out a mortgage for the rest of the purchase price of the home.
   (a) What was the amount of their down payment?
   (b) What was the amount of their mortgage?

34. **Home Purchase** Marcia and Dan Perkins purchased a condominium for $188,000. They paid a down payment of 11% of the cost of the condominium. They took out a mortgage for the rest of the purchase price of the condominium.
   (a) What was the amount of their down payment?
   (b) What was the amount of their mortgage?

35. *Interest Charges on a Mortgage* Richard is making monthly mortgage payments of $840 for his home mortgage. He noticed on his monthly statement that $814 is used to pay off the interest charge. Only $26 is used to pay off the principal. What percent of his monthly mortgage payment is used to pay off the interest charge? Round to the nearest tenth of a percent.

36. *Interest Charges on a Mortgage* Alicia is making monthly mortgage payments of $960 for her home mortgage. She noticed on her monthly statement that $917 is used to pay off the interest charge. Only $43 is used to pay off the principal. What percent of her monthly mortgage payment is used to pay off the interest charge? Round to the nearest tenth of a percent.

Solve. Round to the nearest cent.

37. *Sales Tax* How much sales tax would you pay to purchase a new Honda Accord that costs $20,456.82 if the sales tax rate is 4.6%?

38. *Living Room Set Purchase* The Hartling family purchased a new living room set. The list price was $1249.95. However, they got a discount of 29%. How much did they pay for the new living room set?

Cumulative Review *Perform the following calculations using the correct order of operations.*

39. **[1.6.2]** $3(12 - 6) - 4(12 \div 3)$

40. **[1.6.2]** $7 + 4^3 \times 2 - 15$

41. **[2.8.3]** $\left(\dfrac{5}{2}\right)\left(\dfrac{1}{3}\right) - \left(\dfrac{2}{3} - \dfrac{1}{3}\right)^2$

42. **[3.6.2]** $(6.8 - 6.6)^2 + 2(1.8)$

Quick Quiz 5.5

1. A construction contractor working for a real estate developer builds a house that sells for $325,000. He gets a commission of 8% on the house. What is his commission?

2. Susanne runs a bakery. Five years ago the bakery sold 160 loaves of bread each day. Today they sell 275 loaves each day. What is the percent of increase in the number of loaves sold each day?

3. Find the simple interest on a loan of $4600 borrowed at 13% for six months.

4. **Concept Check** Explain how to find simple interest on a loan of $5800 borrowed at an annual rate of 16% for a period of three months.

Did You Know ...
That It May Be Better to Buy a Car Than Lease It?

DECIDING TO BUY OR LEASE A CAR

Understanding the Problem:

Louvy has his eye on a brand new car. He thinks he should lease the car because his best friend Tranh has a car lease and says he can get the same deal for Louvy. On the other hand, Louvy's girlfriend Allie says it is always better to buy the car and finance it by taking out a loan.

Louvy does some research and finds that it is not at all simple. While a lease offers lower monthly payments, at the end of the lease period you are left with nothing.

	Lease	Purchase
Automobile price	$23,000	$23,000
Interest rate	6%	6%
Length of loan/lease	36 months	36 months
Down payment	$1000	$1000
Residual value (the value of the car you are turning in, or the price you would pay if you want to buy it)	$11,000	Not applicable
Monthly payment	$388.06	$669.28

Task 5: How much will he save in payments each month if he leases the car?

Task 6: Which option should Louvy choose if he wants the best overall price? If he is concerned about his monthly payments?

Making a Plan:

Step 1: Louvy needs to compare the total amounts he would pay for the entire loan.

Task 1: Determine how much Louvy would pay to lease the car for three years.

Task 2: Determine how much Louvy would pay to buy the car with a three-year loan.

Step 2: If Louvy wanted to buy the car at the end of the lease, he would have to pay an additional $11,000.

Task 3: Determine the total cost if Louvy wants to buy the car at the end of the lease.

Task 4: Determine the total overall savings if Louvy buys the car instead of leasing it.

Making a Decision:

Step 3: Louvy is unsure what to do. He likes the fact that buying the car would give him an overall savings, but a lower monthly payment is also important to him.

Applying the Situation to Your Life:

Task 7: How much can you afford to pay for a car payment each month?

Task 8: How much would you pay in insurance, gas, and taxes?

Task 9: Would you rather buy or lease a car?

Facts You Should Know:

You may wish to lease a car if

• you don't drive more than the specified number of miles in the lease on a yearly basis.

• you want a new vehicle every two to three years.

You may wish to buy a car if

• you intend to keep it a long time.

• you want to be debt-free after a time.

Chapter 5 Organizer

Topic and Procedure	Examples	✏ You Try It
Converting a fraction with a denominator of 100 to a percent, p. 305 1. Use the numerator only. 2. Add the percent sign.	**(a)** $\frac{29}{100} = 29\%$ **(d)** $\frac{5.6}{100} = 5.6\%$ **(b)** $\frac{3}{100} = 3\%$ **(e)** $\frac{7\frac{1}{3}}{100} = 7\frac{1}{3}\%$ **(c)** $\frac{231}{100} = 231\%$	1. Write each fraction as a percent. **(a)** $\frac{78}{100}$ **(d)** $\frac{3\frac{1}{4}}{100}$ **(b)** $\frac{125}{100}$ **(e)** $\frac{18.5}{100}$ **(c)** $\frac{9}{100}$
Changing a percent to a decimal, p. 307 1. Drop the percent sign. 2. Move the decimal point two places to the left.	**(a)** $49\% = 0.49$ **(d)** $196\% = 1.96$ **(b)** $2\% = 0.02$ **(e)** $1.36\% = 0.0136$ **(c)** $0.5\% = 0.005$	2. Write each percent as a decimal. **(a)** 86% **(d)** 335% **(b)** 1% **(e)** 7.54% **(c)** 1.8%
Converting a decimal to a percent, p. 308 1. Move the decimal point two places to the right. 2. Add the percent sign.	**(a)** $0.19 = 19\%$ **(b)** $0.516 = 51.6\%$ **(c)** $0.04 = 4\%$ **(d)** $1.53 = 153\%$ **(e)** $0.006 = 0.6\%$	3. Write each decimal as a percent. **(a)** 0.12 **(d)** 6.5 **(b)** 2.35 **(e)** 1.005 **(c)** 0.0071
Changing a percent to a fraction, p. 312 1. If the percent does not contain a decimal point, remove the % and write the number over a denominator of 100. Reduce the fraction if possible. 2. If the percent contains a decimal point, change the percent to a decimal by removing the % and moving the decimal point two places to the left. Then write the decimal as a fraction and reduce if possible. 3. If the percent contains a fraction, remove the % and write the number over a denominator of 100. If the numerator is a mixed number, change the numerator to an improper fraction. Next simplify by the "invert and multiply" rule. Then reduce the fraction if possible.	**(a)** $25\% = \frac{25}{100} = \frac{1}{4}$ **(b)** $38\% = \frac{38}{100} = \frac{19}{50}$ **(c)** $130\% = \frac{130}{100} = \frac{13}{10} = 1\frac{3}{10}$ **(d)** $5.8\% = 0.058$ $= \frac{58}{1000} = \frac{29}{500}$ **(e)** $2.72\% = 0.0272$ $= \frac{272}{10{,}000} = \frac{17}{625}$ **(f)** $7\frac{1}{8}\% = \frac{7\frac{1}{8}}{100}$ $= 7\frac{1}{8} \div \frac{100}{1}$ $= \frac{57}{8} \times \frac{1}{100} = \frac{57}{800}$	4. Write each percent as a fraction or a mixed number. **(a)** 20% **(d)** 1.5% **(b)** 45% **(e)** 5.84% **(c)** 260% **(f)** $9\frac{2}{3}\%$
Changing a percent to a mixed number, p. 312 1. Drop the percent sign and move the decimal point two places to the left. 2. Write the decimal part of the number as a fraction. 3. Reduce the fraction if possible.	**(a)** $275\% = 2.75 = 2\frac{75}{100} = 2\frac{3}{4}$ **(b)** $324\% = 3.24 = 3\frac{24}{100} = 3\frac{6}{25}$ **(c)** $107\% = 1.07 = 1\frac{7}{100}$	5. Write each percent as a mixed number. **(a)** 360% **(c)** 208% **(b)** 185%
Changing a fraction (whose denominator is not 100) to a percent, p. 314 1. Divide the numerator by the denominator and obtain a decimal. 2. Change the decimal to a percent.	**(a)** $\frac{13}{50} = 0.26 = 26\%$ **(b)** $\frac{1}{20} = 0.05 = 5\%$ **(c)** $\frac{3}{800} = 0.00375 = 0.375\%$ **(d)** $\frac{312}{200} = 1.56 = 156\%$ **(e)** $2\frac{1}{2} = 2.5 = 250\%$	6. Write each fraction as a percent. **(a)** $\frac{34}{50}$ **(d)** $\frac{480}{300}$ **(b)** $\frac{8}{25}$ **(e)** $3\frac{1}{4}$ **(c)** $\frac{4}{500}$

Topic and Procedure	Examples	✏— You Try It
Solving percent problems by translating to equations, p. 321 **1.** Translate by replacing "of" with $\times$ "is" with $=$ "what" with n "find" with $n =$ **2.** Solve the resulting equation.	**(a)** What is 3% of 56? $n = 3\% \times 56$ $n = (0.03)(56)$ $n = 1.68$ **(b)** 16% of what is 208? $16\% \times n = 208$ $0.16n = 208$ $\dfrac{0.16n}{0.16} = \dfrac{208}{0.16}$ $n = 1300$ **(c)** What percent of 70 is 30? $n \times 70 = 30$ $70n = 30$ $\dfrac{70n}{70} = \dfrac{30}{70}$ $n = 0.428571\ldots$ $n \approx 42.86\%$	**7.** Solve for n by translating to an equation. **(a)** What is 12% of 300? **(b)** 22% of what number is 165? **(c)** What percent of 90 is 32?
Solving percent problems by using proportions, p. 329 and p. 331 **1.** Identify the parts of the percent proportion. a = the amount b = the base (the whole; it usually appears after the word "of") p = the percent number **2.** Write the percent proportion $\dfrac{a}{b} = \dfrac{p}{100}$ using the values obtained in step 1 and solve.	**(a)** What is 28% of 420? The percent $p = 28$. The base $b = 420$. The amount a is unknown. We use the variable a. $\dfrac{a}{b} = \dfrac{p}{100}$ becomes $\dfrac{a}{420} = \dfrac{28}{100}$ If we reduce the fraction on the right-hand side, we have $\dfrac{a}{420} = \dfrac{7}{25}$ $25a = (7)(420)$ $25a = 2940$ $\dfrac{25a}{25} = \dfrac{2940}{25}$ $a = 117.6$ Thus, 28% of 420 is 117.6. **(b)** 64% of what is 320? The percent $p = 64$. The base is unknown. We use the variable b. The amount $a = 320$. $\dfrac{a}{b} = \dfrac{p}{100}$ becomes $\dfrac{320}{b} = \dfrac{64}{100}$ If we reduce the fraction on the right-hand side, we have $\dfrac{320}{b} = \dfrac{16}{25}$ $(320)(25) = 16b$ $8000 = 16b$ $\dfrac{8000}{16} = \dfrac{16b}{16}$ $500 = b$ Thus, 64% of 500 is 320. **(c)** What percent of 140 is 105? The percent is unknown. The base $b = 140$, the amount $a = 105$. $\dfrac{a}{b} = \dfrac{p}{100}$ becomes $\dfrac{105}{140} = \dfrac{p}{100}$ If we reduce the fraction on the left side, we have $\dfrac{3}{4} = \dfrac{p}{100}$ $(100)(3) = 4p$ $300 = 4p$ $\dfrac{300}{4} = \dfrac{4p}{4}$ $75 = p$ Thus 105 is 75% of 140.	**8.** Solve by using proportions. **(a)** What is 32% of 180? **(b)** 48% of what number is 15? **(c)** What percent of 300 is 120?

Continued on next page

Topic and Procedure	Examples	✏ You Try It
Solving discount problems, p. 342 Discount = discount rate × list price	Carla purchased a color TV set that lists for $350 at an 18% discount. **(a)** How much was the discount? **(b)** How much did she pay for the TV set? **(a)** Discount = (0.18)(350) = $63 **(b)** 350 − 63 = 287 She paid $287 for the color TV set.	9. Tony purchased a DVD player that lists for $225 at a 15% discount. **(a)** How much was the discount? **(b)** How much did he pay for the DVD player?
Solving commission problems, p. 346 Commission = commission rate × value of sales	A housewares salesperson gets a 16% commission on sales he makes. How much commission does he earn if he sells $12,000 in housewares? $$\text{Commission} = (0.16)(12,000)$$ $$= 1920$$ He earns a commission of $1920.	10. An electronics salesperson earns 8% commission on sales she makes. How much commission does she earn if she sells $15,500 in electronics?
Percent-of-increase or percent-of-decrease problems, p. 346 Percent of increase or decrease = amount of increase or decrease ÷ base	A car that costs $16,500 now cost only $15,000 last year. What is the percent of increase? $$\begin{array}{r} 16,500 \\ -\ 15,000 \\ \hline 1500 \ \text{increase} \end{array}$$ $$\frac{1500}{15,000} = 0.10$$ Percent of increase = 10%	11. A house that was worth $270,000 five years ago is now worth $310,000. What is the percent of increase?
Solving simple-interest problems, p. 347 Interest = principal × rate × time $$I = P \times R \times T$$ I = interest P = principal R = rate T = time	Hector borrowed $3000 for 4 years at a simple interest rate of 12%. How much interest did he owe after 4 years? $$I = P \times R \times T$$ $$I = (3000)(0.12)(4)$$ $$= (360)(4)$$ $$= 1440$$ Hector owed $1440 in interest.	12. Olivia borrowed $8000 for 3 years at a simple interest rate of 6%. How much interest did she owe after 3 years?

Chapter 5 Review Problems

Section 5.1

Write as a percent.

1. 0.62

2. 0.43

3. 0.372

4. 2.2

5. 2.52

6. 1.036

7. 0.006

8. $\dfrac{62.5}{100}$

9. $\dfrac{4\frac{1}{12}}{100}$

10. $\dfrac{317}{100}$

Write as a decimal.

11. 32%

12. 15.75%

13. 236%

14. $32\frac{1}{8}\%$

Section 5.2

Write as a percent. Round to the nearest hundredth of a percent when necessary.

15. $\dfrac{19}{25}$ **16.** $\dfrac{11}{20}$ **17.** $\dfrac{9}{40}$ **18.** $\dfrac{7}{12}$ **19.** $\dfrac{14}{15}$

20. $2\dfrac{1}{4}$ **21.** $3\dfrac{3}{4}$ **22.** $2\dfrac{7}{9}$ **23.** $\dfrac{200}{80}$ **24.** $\dfrac{5}{800}$

Write as a fraction or mixed number.

25. 72%   **26.** 175%   **27.** 16.4%

28. $31\dfrac{1}{4}\%$ **29.** 0.08%   **30.** 0.04%

Complete the following chart.

	Fraction	Decimal	Percent
31.	$\dfrac{3}{5}$		
32.	$\dfrac{7}{10}$		
33.			37.5%

	Fraction	Decimal	Percent
34.			56.25%
35.		0.008	
36.		0.45	

Section 5.3

Solve. Round to the nearest hundredth when necessary.

37. What is 20% of 85?

38. What is 25% of 92?

39. Find 162% of 60.

40. 15 is 25% of what number?

41. 30 is 75% of what number?

42. 92% of what number is 147.2?

43. 50 is what percent of 130?

44. 70 is what percent of 180?

45. What percent of 70 is 14?

Sections 5.4 and 5.5

Solve. Round your answer to the nearest hundredth when necessary.

46. *Education* Professor Padron found that 35% of his World History course is sophomores. He has 140 students in his class. How many are sophomores?

47. *Truck Dealer* A Vermont truck dealer found that 64% of all the trucks he sold had four-wheel drive. If he sold 150 trucks, how many had four-wheel drive?

48. *Car Depreciation* Today Yvonne's car has 61% of the value that it had two years ago. Today it is worth $6832. What was it worth two years ago?

49. *Administrative Expenses* A charity organization spent 12% of its budget for administrative expenses. It spent $9624 on administrative expenses. What was the total budget?

50. *Rain in Seattle* In Seattle it rained 20 days in February, 18 days in March, and 16 days in April. What percent of those three months did it rain? (Assume it was a leap year.)

51. *Job Applications* Moorehouse Industries received 600 applications and hired 45 of the applicants. What percent of the applicants obtained a job?

52. *Appliance Purchase* Nathan bought new appliances for $3670. The sales tax in his state is 5%. What did he pay in sales tax?

53. *Boat Purchase* Chris and Annette bought a boat for $12,600. The sales tax is 6% in their state. What did they pay in sales tax?

54. *Encyclopedia Sales* Adam sold encyclopedias last summer to raise tuition money. He sold $83,500 worth of encyclopedias and was paid a commission at the rate of 6%. How much commission did he earn?

55. *Commission Sales* Roberta earns a commission at the rate of 7.5%. Last month she sold $16,000 worth of goods. How much commission did she make last month?

56. *Furniture Set* Amy purchased a table and chairs set for her new patio at a 25% discount. The list price was $1450.

(a) What was the discount?

(b) What did she pay for the set?

57. *Medical School Admissions* Between 2008 and 2009, medical school admissions in the United States increased from 18,040 students to 18,400 students. What was the percent of increase in the number of medical school admissions? (*Source:* www.chronicle.com)

58. *Cost of Education* For the 2004–2005 academic year, the average cost of tuition at a four-year public college in the United States was about $5000. By the 2008–2009 academic year, the cost had risen to $6585. Find the percent of increase. Round to the nearest whole percent. (*Source:* www.wikipedia.org)

59. *Log Cabin* Mark and Julie wanted to buy a prefabricated log cabin to put on their property in the Colorado Rockies. The price of the kit is listed at $24,000. At the after-holiday cabin sale, a discount of 14% was offered.

(a) What was the discount?

(b) How much did they pay for the cabin?

60. *Mutual Funds* Sally invested $6000 in mutual funds earning 11% simple interest in one year. How much interest will she earn in

(a) six months?

(b) two years?

61. *College Loan* Reed took out a college loan of $3000. He will be charged 8% simple interest on the loan.

(a) How much interest will be due on the loan in three months?

(b) How much interest will be due on the loan in three years?

How Am I Doing? Chapter 5 Test

 MATH COACH MyMathLab® You Tube™

After you take this test read through the Math Coach on pages 361 and 362. Math Coach videos are available via MyMathLab and YouTube. Step-by-step test solutions in the Chapter Test Prep Videos are also available via MyMathLab and YouTube. (Search "TobeyBasicCollMath" and click on "Channels.")

Write as a percent.

1. 0.57

2. 0.01

3. 0.008

4. 12.8

5. 3.56

6. $\dfrac{71}{100}$

7. $\dfrac{1.8}{100}$

8. $\dfrac{3\frac{1}{7}}{100}$

Write as a percent. Round to the nearest hundredth of a percent when necessary.

9. $\dfrac{19}{40}$

10. $\dfrac{27}{36}$

11. $\dfrac{225}{75}$

12. $1\dfrac{3}{4}$

Write as a percent.

13. 0.0825

ᴹᴄ **14.** 3.024

Write as a fraction or a mixed number in simplified form.

15. 152%

16. $7\dfrac{3}{4}\%$

Solve. Round to the nearest hundredth if necessary.

17. What is 40% of 50?

18. 33.8 is 26% of what number?

19. What percent of 72 is 40?

20. Find 0.8% of 25,000.

1. _____ ☐
2. _____ ☐
3. _____ ☐
4. _____ ☐
5. _____ ☐
6. _____ ☐
7. _____ ☐
8. _____ ☐
9. _____ ☐
10. _____ ☐
11. _____ ☐
12. _____ ☐
13. _____ ☐
14. _____ ☐
15. _____ ☐
16. _____ ☐
17. _____ ☐
18. _____ ☐
19. _____ ☐
20. _____ ☐

21. _____ ☐

ᴹ꜀ **21.** 16% of what number is 800?

22. _____ ☐

22. 92 is what percent of 200?

23. _____ ☐

23. 132% of 530 is what number?

24. _____ ☐

ᴹ꜀ **24.** What percent is 15 of 75?

Solve. Round to the nearest hundredth if necessary.

25. _____ ☐

25. A real estate agent sells a house for $152,300. She gets a commission of 4% on the sale. What is her commission?

26. (a) _____ ☐

26. Julia and Charles bought a new dishwasher at a 33% discount. The list price was $457.

   (b) _____ ☐

   (a) What was the discount?
   (b) How much did they pay for the dishwasher?

27. _____ ☐

27. An inspector found that 75 out of 84 parts were not defective. What percent of the parts were not defective?

28. _____ ☐

28. Last year Charlotte was the top player on the basketball team, scoring 185 points. This year, she scored 228 points. What is the percent of increase in the number of points?

29. _____ ☐

ᴹ꜀ **29.** A total of 5160 people voted in the city election. This was 43% of the registered voters. How many registered voters are in the city?

30. (a) _____ ☐

   (b) _____ ☐

30. Wanda borrowed $3000 at a simple interest rate of 16%.

   (a) How much interest did she pay in six months?
   (b) How much interest did she pay in two years?

Total Correct: _____

MATH COACH

Mastering the skills you need to do well on the test.

Students often make the same types of errors when they do the Chapter 5 Test. Here are some helpful hints to keep you from making these common errors on test problems.

Write a Decimal as a Percent—Problem 14 3.024

> **Helpful Hint** Remember to move the decimal point two places to the right and then write the % symbol at the end of the number.

Did you move the decimal point to the right and add the % symbol?

Yes [ ] No [ ]

If you answered No, please stop and complete this process correctly.

Note that when writing a decimal as a percent, we are multiplying the decimal by 100.

That is why the decimal point moves to the right two places.

Solving Percent Problems When the Base Is Unknown—Problem 21 16% of what number is 800?

> **Helpful Hint** Translate the problem into an equation first. This will show you that division is a necessary step in solving the problem.

Were you able to translate the problem into the equation $16\% \times n = 800$?

Yes [ ] No [ ]

If you answered No, please stop and complete that step correctly first.

Were you able to go to the second step and write $0.16 \times n = 800$?

Yes [ ] No [ ]

If you answered No, go back and complete that step.

Were you able to divide both sides of the equation by 0.16?

Yes [ ] No [ ]

If you answered No, please go back and complete the division carefully to solve for n.

Need help? Watch the MATH COACH videos in MyMathLab® or on YouTube™.

361

Solving Percent Problems When the Percent Is Unknown—Problem 24   What percent is 15 of 75?

> **Helpful Hint** Translate the problem into an equation first. Remember that you must add the % symbol to your answer.

Were you able to translate the problem into the equation $15 = n \times 75$?

Yes _____ No _____

If you answered No, stop and reason through the steps to see if you can obtain that equation.

Were you able to divide both sides of the equation by 75?

Yes _____ No _____

If you answered No, go back and complete the division carefully to solve for n.

Do you see why it is necessary to change the answer from a decimal to a percent?

Yes _____ No _____

If you answered Problem 24 incorrectly, please go back and rework the problem using these suggestions.

Solving Applied Percent Problems—Problem 29   A total of 5160 people voted in the city election. This was 43% of the registered voters. How many registered voters are in the city?

> **Helpful Hint** Write a percent statement that describes the situation in the applied problem first. Then consider whether you want to use percent proportions or percent equations to solve the problem.

Were you able to write a percent statement to describe the problem situation, such as 43% of the registered voters equals 5160?

Yes _____ No _____

If you answered No, please read the problem again and try to write a similar statement.

Were you able to translate the statement into the percent equation $43\% \times n = 5160$ or the percent proportion

$$\frac{5160}{b} = \frac{43}{100}?$$

Yes _____ No _____

If you answered No, please reread your statement and see if you can rewrite it to describe this equation or proportion.

If you answered Problem 29 incorrectly, please go back and rework the problem using these suggestions.

Need more help? Look for section examples marked with ^Mℂ to review.

362

About fifty years ago, a major change in the way land is measured occurred in Canada. All new deeds recording the amount of land area had to be changed from recording the area in acres to recording the area in hectares. Farmers, forestry personnel, and all landowners had to do some extensive calculations. When you have mastered the content of this chapter you will be able to do these types of calculations.

Measurement

6.1 American Units

Student Learning Objectives

After studying this section, you will be able to:

① Identify the basic unit equivalencies in the American system.

② Convert from one unit of measure to another.

① Identifying the Basic Unit Equivalencies in the American System

We often ask questions about measurements. How far is it to work? How much does this bottle hold? What is the weight of this box? How long will it be until exams? To answer these questions we need to agree on a unit of measure for each type of measurement.

At present there are two main systems of measurement, each with its own set of units: the **metric system** and the **American system.** Nearly all countries in the world use the metric system. In the United States, however, except in science, most measurements are made in American units.

The United States is using the metric system more and more frequently, however, and may eventually convert to metric units as the standard. But for now we need to be familiar with both systems. We cover American units in this section.

One of the most familiar measuring devices is a ruler that measures lengths as great as 1 foot. It is divided into 12 inches; that is,

$$1 \text{ foot} = 12 \text{ inches}.$$

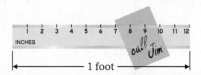

There are several other important relationships you need to know. Your instructor may require you to memorize the following facts.

Length	Time
12 inches = 1 foot	60 seconds = 1 minute
36 inches = 1 yard	60 minutes = 1 hour
3 feet = 1 yard	24 hours = 1 day
5280 feet = 1 mile	7 days = 1 week
1760 yards = 1 mile	

Note that time is measured in the same units in both the metric and the American systems.

Weight	Volume
16 ounces = 1 pound	8 fluid ounces = 1 cup
2000 pounds = 1 ton	2 cups = 1 pint
	2 pints = 1 quart
	4 quarts = 1 gallon

We can choose to measure an object—say, a bridge—using a small unit (an inch), a larger unit (a foot), or a still larger unit (a mile). We may say that the bridge spans 7920 inches, 660 feet, or an eighth of a mile. Although we probably would not choose to express our measurement in inches because this is not a convenient measurement to work with for an object as long as a bridge, the bridge length is the same whatever unit of measurement we use.

Notice that the smaller the measuring unit, the larger the number of those units in the final measurement. The inch is the smallest unit in our example, and the inch measurement has the greatest number of units (7920). The mile is the largest unit, and it has the smallest number of units (an eighth equals 0.125). Whatever measuring system you use, and whatever you measure (length, volume, and so on), the smaller the unit of measurement you use, the greater the number of those units.

After studying the values in the length, time, weight, and volume tables, see if you can quickly do Example 1.

EXAMPLE 1 Answer rapidly the following questions.

(a) How many inches in a foot?   **(b)** How many yards in a mile?
(c) How many seconds in a minute?   **(d)** How many hours in a day?
(e) How many pounds in a ton?   **(f)** How many cups in a pint?

Solution

(a) 12   **(b)** 1760   **(c)** 60   **(d)** 24   **(e)** 2000   **(f)** 2

Student Practice 1 Answer rapidly the following questions.

(a) How many feet in a yard?   **(b)** How many feet in a mile?
(c) How many minutes in an hour?   **(d)** How many days in a week?
(e) How many ounces in a pound?   **(f)** How many pints in a quart?
(g) How many quarts in a gallon?

NOTE TO STUDENT: Fully worked-out solutions to all of the Student Practice problems can be found at the back of the text starting at page SP-1.

② Converting from One Unit of Measure to Another

To convert or change one measurement to another, we simply multiply by 1 since multiplying by 1 does not change the value of a quantity. For example, to convert 180 inches to feet, we look for a name for 1 that has inches and feet.

$$1 = \frac{1 \text{ foot}}{12 \text{ inches}}$$

A ratio of measurements for which the measurement in the numerator is equivalent to the measurement in the denominator is called a **unit fraction.** We now use the unit fraction $\dfrac{1 \text{ foot}}{12 \text{ inches}}$ to convert 180 inches to feet.

$$180 \text{ inches} \times \frac{1 \text{ foot}}{12 \text{ inches}} = \frac{180 \text{ feet}}{12} = 15 \text{ feet}$$

Notice that when we multiplied, the inches divided out. We are left with the unit feet.

What name for 1 (that is, unit fraction) should we choose if we want to change from feet to inches? Convert 4 feet to inches.

$$1 = \frac{12 \text{ inches}}{1 \text{ foot}}$$

$$4 \text{ feet} \times \frac{12 \text{ inches}}{1 \text{ foot}} = \frac{48 \text{ inches}}{1} = 48 \text{ inches}$$

When multiplying by a unit fraction, the unit we want to change to should be in the *numerator*. The unit we start with should be in the *denominator*. This unit will divide out.

EXAMPLE 2 Convert. 8800 yards to miles

Solution $8800 \text{ yards} \times \dfrac{1 \text{ mile}}{1760 \text{ yards}} = \dfrac{8800}{1760} \text{ miles} = 5 \text{ miles}$

Student Practice 2 Convert. 15,840 feet to miles

Some conversions involve fractions or decimals. Whether you want to measure the area of a living room or the dimensions of a piece of property, it is helpful to be able to make conversions like these.

EXAMPLE 3 Convert.

(a) 26.48 miles to yards   **(b)** $3\dfrac{2}{3}$ feet to yards

Solution

(a) $26.48 \text{ miles} \times \dfrac{1760 \text{ yards}}{1 \text{ mile}} = 46{,}604.8 \text{ yards}$

(b) $3\dfrac{2}{3} \text{ feet} \times \dfrac{1 \text{ yard}}{3 \text{ feet}} = \dfrac{11}{3} \times \dfrac{1}{3} \text{ yard} = \dfrac{11}{9} \text{ yards} = 1\dfrac{2}{9} \text{ yards}$

Student Practice 3 Convert.

(a) 18.93 miles to feet

(b) $16\dfrac{1}{2}$ inches to yards

EXAMPLE 4 Lynda's new car weighs 2.43 tons. How many pounds is that?

Solution $2.43 \text{ tons} \times \dfrac{2000 \text{ pounds}}{1 \text{ ton}} = 4860 \text{ pounds}$

Student Practice 4 A package weighs 760.5 pounds. How many ounces does it weigh?

EXAMPLE 5 The chemistry lab has 34 quarts of weak hydrochloric acid. How many gallons of this acid are in the lab? (Express your answer as a decimal.)

Solution $34 \text{ quarts} \times \dfrac{1 \text{ gallon}}{4 \text{ quarts}} = \dfrac{34}{4} \text{ gallons} = 8.5 \text{ gallons}$

Student Practice 5 19 pints of milk is the same as how many quarts? (Express your answer as a decimal.)

EXAMPLE 6   A window is 4 feet 5 inches wide. How many inches is that? 4 feet 5 inches means 4 feet and 5 inches. Change the 4 feet to inches and add the 5 inches.

Solution $4 \text{ feet} \times \dfrac{12 \text{ inches}}{1 \text{ foot}} = 48 \text{ inches}$

               48 inches + 5 inches = 53 inches   The window is 53 inches wide.

Student Practice 6   A path through Dr. Sherf's property measures 26 yards 2 feet in length. How many feet long is the path?

EXAMPLE 7   The Charlotte all-night garage charges $1.50 per hour for parking both day and night. A businessman left his car there for $2\frac{1}{4}$ days. How much was he charged?

Solution

1. *Understand the problem.* Here it might help to look at a simpler problem. If the businessman had left his car for two hours, we would multiply.

   The fraction bar means "per." $\rightarrow$ $\dfrac{1.50 \text{ dollars}}{1 \text{ hour}} \times 2 \text{ hours} = 3.00 \text{ dollars or } \$3.00$

   Thus, if the businessman had left his car for two hours, he would have been charged $3. We see that we need to multiply $1.50 by the number of hours the car was in the garage to solve the problem.

   Since the original problem gave the time in days, not hours, we will need to change the days to hours.

2. *Solve and state the answer.* Now that we know that the way to solve the problem is to multiply by hours, we will begin. To make our calculations easier we will write $2\frac{1}{4}$ as 2.25. Change days to hours. Then multiply by $1.50 per hour.

   $$2.25 \text{ days} \times \frac{24 \text{ hours}}{1 \text{ day}} \times \frac{1.50 \text{ dollars}}{1 \text{ hour}} = 81 \text{ dollars or } \$81$$

   The businessman was charged $81.

3. *Check.* Is our answer in the desired unit? Yes. The answer is in dollars and we would expect it to be in dollars. ✓

   The check is up to you.

Student Practice 7   A businesswoman parked her car at a garage for $1\frac{3}{4}$ days. The garage charges $1.50 per hour. How much did she pay to park the car?

ALTERNATIVE METHOD: Using Proportions How did people first come up with the idea of multiplying by a unit fraction? What mathematical principles are involved here? Actually, this is the same as solving a proportion. Consider Example 5, where we changed 34 quarts to 8.5 gallons by multiplying.

$$34 \text{ quarts} \times \frac{1 \text{ gallon}}{4 \text{ quarts}} = \frac{34}{4} \text{ gallons} = 8.5 \text{ gallons}$$

What we were actually doing is setting up the proportion:

1 gallon is to 4 quarts as n gallons is to 34 quarts.

$$\frac{1 \text{ gallon}}{4 \text{ quarts}} = \frac{n \text{ gallons}}{34 \text{ quarts}}$$

$1 \text{ gallon} \times 34 \text{ quarts} = 4 \text{ quarts} \times n \text{ gallons}$ Cross-multiply.

$$\frac{1 \text{ gallon} \times 34 \text{ quarts}}{4 \text{ quarts}} = \frac{4 \text{ quarts} \times n \text{ gallons}}{4 \text{ quarts}}$$ Divide both sides of the equation by 4 quarts.

$$1 \text{ gallon} \times \frac{34}{4} = n \text{ gallons}$$ Simplify.

$$8.5 \text{ gallons} = n \text{ gallons}$$

Thus the number of gallons is 8.5. Using proportions takes a little longer, so multiplying by a fractional name for 1 is the more popular method.

You will find the methods covered in Section 6.1 to be quite useful in everyday life, so the topic is worth a little extra study. Take a few minutes to read over Examples 1–7. Think about the steps that are shown. Now work out each of the Student Practice problems 1–7. Turn to page SP-18 and make sure you have done them correctly. You will find this review is very worthwhile.

👣 STEPS TO SUCCESS How Important Is the Quick Quiz?

At the end of each homework exercise section is a Quick Quiz. Please be sure to complete that section. You will be amazed at how it will help you prove to yourself that you have mastered the concepts discussed in that section. Make sure you do the Quick Quiz each time as a necessary part of your homework. It will really help you gain confidence in what you have learned. The Quick Quiz is essential.

Here is a fun way you can use the Quick Quiz. Ask a friend in the class if he or she will "quiz" you about five minutes before class by asking you to work out (without using your book) the solution to one of the problems on the Quick Quiz. Tell your friend you can do the same thing for him or her. This will force you to "be ready for anything" when you come to class. This little trick will keep you sharp and ready for class.

Making it personal: Which of these suggestions do you find most helpful? Use those suggestions as you do each section of Chapter 6. ▼

Verbal and Writing Skills, Exercises 1 and 2

1. Explain in your own words how you would use a unit fraction to change 23 miles to inches.

2. Explain in your own words how you would use a unit fraction to change 27 days to minutes.

From memory, write the equivalent value.

3. _____ yards = 1 mile

4. _____ feet = 1 mile

5. 1 ton = _____ pounds

6. 1 pound = _____ ounces

7. _____ quarts = 1 gallon

8. _____ cups = 1 pint

9. 1 quart = _____ pints

10. _____ minutes = 1 hour

Convert. When necessary, express your answer as a decimal.

11. 21 feet = _____ yards

12. 63 feet = _____ yards

13. 108 inches = _____ feet

14. 204 inches = _____ feet

15. 9 feet = _____ inches

16. 11 feet = _____ inches

17. 10,560 feet = _____ miles

18. 5 miles = _____ feet

19. 7 miles = _____ yards

20. 21,120 feet = _____ miles

21. 12 gallons = _____ quarts

22. 15 gallons = _____ quarts

23. 48 quarts = _____ gallons

24. 24 pints = _____ quarts

25. 16 cups = _____ fluid ounces

26. 40 fluid ounces = _____ cups

27. $8\frac{1}{2}$ gallons = _____ pints

28. $6\frac{1}{2}$ gallons = _____ pints

29. 77 days = _____ weeks

30. 56 days = _____ weeks

31. 960 seconds = _____ minutes

32. 1500 seconds = _____ minutes

33. 8 ounces = _____ pound

34. 12 ounces = _____ pound

35. 12,500 pounds = _____ tons

36. $4\frac{3}{4}$ tons = _____ pounds

37. 15 pints = _____ cups

38. 27 pints = _____ cups

39. 2.25 pounds = _____ ounces

40. 3.75 pounds = _____ ounces

41. 66 inches = _____ feet

42. 102 inches = _____ feet

Applications

43. *Wheelchair Racer* Ernst Van Dyke of South Africa has dominated wheelchair racing since 2000. His fastest marathon time was at the 2004 Boston Marathon. He covered the 26.2-mile course in 1:18 (1 hour and 18 minutes). How many feet did he travel in the race?

44. *Marathon Record* In April 2003, Paula Radcliffe of the United Kingdom set the women's world record for the marathon, winning the London Marathon with a time of 2:15:25 (hours:minutes:seconds). How many seconds is that?

45. *Ocean Depth* The deepest part of the Pacific Ocean is known as The Mariana Trench and is 35,840 feet. How many miles is that? Round your answer to the nearest hundredth.

46. *Shot Put* Stacy threw the shot put $36\frac{1}{4}$ feet at a college track meet. How many inches is that?

47. *Food Purchase* Judy is making a wild mushroom sauce for a pasta dinner with a large group of friends. She bought 26 ounces of wild mushrooms at $6.00 per pound. How much were the mushrooms?

48. *Food Purchase* Kurt is trying to eat a low-fat diet. He finds a store that sells 1% fat ground white-meat turkey breast at $6.00 per pound. He buys one packet weighing 18 ounces and another weighing 22 ounces. How much does he pay?

▲ **49.** *Geometry* A window in Lynn and Jorge's house needs to be replaced. The rectangular window is 3 feet 6 inches tall and 2 feet 5 inches wide. Change each of the measurements to inches.

(a) Find the perimeter of the window in inches.

(b) If the perimeter needs to be sealed with insulation that is $0.60 per inch, how much will it cost to insulate the perimeter?

▲ **50.** *Geometry* The cellar in Jeff and Shelley's house needs to be sealed along the edge of the concrete floor. The rectangular floor measures 7 yards 2 feet wide and 12 yards 1 foot long. Change each of the measurements to feet.

(a) Find the perimeter of the cellar in feet.

(b) If the perimeter (edge) of the cellar floor needs to be sealed with waterproof sealer that costs $1.75 per foot, how much will it cost to seal the perimeter?

51. *Heart Capacity* Every day, your heart pumps 7200 quarts of blood through your body. How many cups is that?

52. *Subway System* The total length of all routes of the London subway system is 258 miles. How many feet is that?

Use estimating and rounding to complete exercises 53–56.

53. *Boating Map* A local boater's map shows a marker buoy 6 miles out in the ocean from the harbor at Woods Hole. Estimate the number of yards this distance is. (*Hint:* First round the number of yards in one mile to the nearest thousand yards. Then finish the calculation.)

54. *Plant Growth* Dr. Russ Camp is examining a new experimental type of wheat in his laboratory at Gordon College. The plant is 618 days old. Estimate the age of the plant in months. (*Hint:* First round 618 to the nearest hundred. Then finish the calculation.)

55. **Altitude of a Plane** Greg Salzman is flying from Chicago. The pilot announced that the plane was flying at an altitude of 33,000 feet. The man seated next to Greg asked him to estimate how many miles high the plane was at that point. How should Greg answer the man? (*Hint:* First round the altitude to the nearest ten thousand feet. Then round the conversion equivalent for the number of feet in a mile to the nearest thousand feet. Then perform the calculation.)

56. **Plane Flight Time** Melissa LaBelle is flying from Paris. The pilot announced that the plane had 3170 miles to go to complete the flight. Earlier he had said that the plane was flying at 640 miles per hour. Estimate the number of hours left in the flight. (*Hint:* First round the distance to the nearest thousand miles. Then round the speed to the nearest hundred. Then perform the calculation.) The pilot then announced that it would take 318 minutes to complete the flight. How close was our estimate?

Cumulative Review

57. **[1.8.2] House Payments** Melinda and Robert are refinancing their house. They were making payments of $560 per month for the next 20 years. They obtained a new mortgage for which their payments will be $515 per month for the next 20 years. Starting with the date they began payments on the new mortgage, how much will they save with the new monthly payments during this entire period?

58. **[5.4.1] Hard Drive Storage** Lexi has a hard drive that stores 60 GB of data. She stored three programs that each require 1.5 GB of storage space. She then stored two programs that each require 2.25 GB of storage space. What percent of her hard drive is still free for storage of future programs?

59. **[4.4.1] Bicycle Trip** John and Phil took a bicycle trip covering 115 miles in five days last summer. This summer they hope to take a similar bicycle trip lasting seven days. How many miles are they likely to cover?

60. **[4.4.1] Education** The dean observed that 150 students enrolled in mathematics courses at the University of Hawaii during the last summer semester and that 12 of them dropped the course during the first two weeks. If this ratio has remained consistent over the past seven summers and if 1300 students enrolled in the summer mathematics courses during this time, how many of them dropped the course during the first two weeks of the semester?

Quick Quiz 6.1 *Convert. Express your answer as a decimal rounded to the nearest hundredth when necessary.*

1. Convert 3.5 tons to pounds.

2. Convert 4.5 yards to feet.

3. Convert 24 ounces to pounds.

4. **Concept Check** Explain how you would convert 250 pints to quarts.

6.2 Metric Measurements: Length

Student Learning Objectives

After studying this section, you will be able to:

1. Understand prefixes in metric units.

2. Convert from one metric unit of length to another.

① Understanding Prefixes in Metric Units

The **metric system** of measurement is used in most industrialized nations of the world. As we move toward a global economy, it is important to be familiar with the metric system. The metric system is designed for ease in calculating and in converting from one unit to another.

In the metric system, the basic unit of length measurement is the **meter.** A meter is just slightly longer than a yard. To be more precise, the meter is approximately 39.37 inches long.

1 meter — Approximately 39.37 inches

1 yard — Exactly 36 inches

Units that are larger or smaller than the meter are based on the meter and powers of 10. For example, the unit *deka*meter is *ten* meters. The unit *deci*meter is *one-tenth* of a meter. The prefix *deka* means 10. What does the prefix *deci* mean? All the prefixes in the metric system are names for multiples of 10. A list of metric prefixes and their meanings follows.

Prefix	Meaning
kilo-	thousand
hecto-	hundred
deka-	ten
deci-	tenth
centi-	hundredth
milli-	thousandth

The most commonly used prefixes are *kilo-, centi-,* and *milli-. Kilo-* means thousand, so a *kilo*meter is a thousand meters. Similarly, *centi-* means one hundredth, so a *centi*meter is one hundredth of a meter. And *milli-* means one thousandth, so a *milli*meter is one thousandth of a meter.

The kilometer is used to measure distances much larger than the meter. How far did you travel in a car? The centimeter is used to measure shorter lengths. What are the dimensions of this textbook? The millimeter is used to measure very small lengths. What is the width of the lead in a pencil?

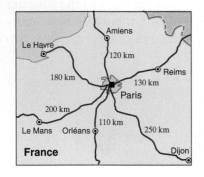

Amiens
Le Havre
120 km
180 km
Reims
130 km
Paris
200 km
110 km
Le Mans Orléans
250 km
France
Dijon

EXAMPLE 1 Write the prefixes that mean **(a)** thousand and **(b)** tenth.

Solution

(a) The prefix *kilo-* is used for thousand.

(b) The prefix *deci-* is used for tenth.

Student Practice 1 Write the prefixes that mean **(a)** ten and **(b)** thousandth.

NOTE TO STUDENT: Fully worked-out solutions to all of the Student Practice problems can be found at the back of the text starting at page SP-1.

② Converting from One Metric Unit of Length to Another

How do we convert from one metric unit to another? For example, how do we change 5 kilometers into an equivalent number of meters?

Recall from Chapter 3 that when we multiply by 10 we move the decimal point one place to the right. When we divide by 10 we move the decimal point one place to the left. Let's see how we use this idea to change from one metric unit to another.

Changing from Larger Metric Units to Smaller Ones

When you change from one metric prefix to another by moving to the **right** on this prefix chart, move the decimal point to the **right** the same number of places.

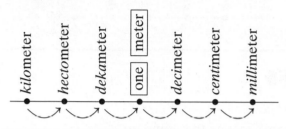

Thus 1 meter = 100 centimeters because we move two places to the right on the chart of prefixes and we also move the decimal point (1.00) two places to the right.

Now let us examine the four most commonly used metric measurements of length and their abbreviations.

COMMONLY USED METRIC LENGTHS

1 kilometer (km) = 1000 meters

1 meter (m) (the basic unit of length in the metric system)

1 centimeter (cm) = 0.01 meter

1 millimeter (mm) = 0.001 meter

Now let us see how we can change a measurement stated in a larger unit to an equivalent measurement stated in smaller units.

EXAMPLE 2

(a) Change 5 kilometers to meters.   **(b)** Change 20 meters to centimeters.

Solution

(a) To go from *kilometer* to meter, we move three places to the right on the prefix chart. So we move the decimal point three places to the right.

5 kilometers = 5.000. meters (move three places) = 5000 meters

(b) To go from meter to *centimeter*, we move two places to the right on the prefix chart. Thus we move the decimal point two places to the right.

20 meters = 20.00. centimeters (move two places) = 2000 centimeters

Student Practice 2

(a) Change 4 meters to centimeters.

(b) Change 30 centimeters to millimeters.

Changing from Smaller Metric Units to Larger Ones

When you change from one metric prefix to another by moving to the **left** on this prefix chart, move the decimal point to the **left** the same number of places.

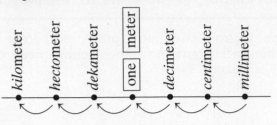

EXAMPLE 3

(a) Change 163 centimeters to meters.
(b) Change 56 millimeters to kilometers.

Solution

(a) To go from *centi*meter to meter, we move *two* places to the left on the prefix chart. Thus we move the decimal point two places to the left.

$$163 \text{ centimeters} = 1.\underset{\curvearrowleft}{63}. \text{ meters (move two places to the left)}$$
$$= 1.63 \text{ meters}$$

(b) To go from *milli*meter to *kilo*meter, we move six places to the left on the prefix chart. Thus we move the decimal point six places to the left.

$$56 \text{ millimeters} = 0.\underset{\curvearrowleft}{000056}. \text{ kilometer} \quad \text{(move six places to the left)}$$
$$= 0.000056 \text{ kilometer}$$

Student Practice 3

(a) Change 3 *milli*meters to meters.
(b) Change 47 *centi*meters to *kilo*meters.

Thinking Metric Long distances are customarily measured in kilometers. A kilometer is about 0.62 mile. It takes about 1.6 kilometers to make a mile. The following drawing shows the relationship between the kilometer and the mile.

1 kilometer

1 mile (about 1.6 kilometers)

Many small distances are measured in centimeters. A centimeter is about 0.394 inch. It takes 2.54 centimeters to make an inch. You can get a good idea of their size by looking at a ruler marked in both inches and centimeters.

1 centimeter

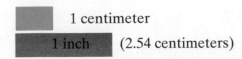

1 inch   (2.54 centimeters)

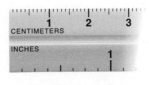

Try to visualize how many centimeters long or wide this book is. It is about 21 centimeters wide and 27.5 centimeters long.

A millimeter is a very small unit of measurement, often used in manufacturing.

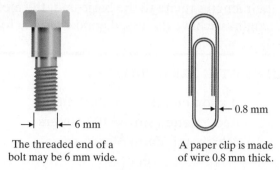

The threaded end of a
bolt may be 6 mm wide.

A paper clip is made
of wire 0.8 mm thick.

Now that you have an understanding of the size of these metric units, let's try to select the most convenient metric unit for measuring the length of an object.

EXAMPLE 4   Bob measured the width of a doorway in his house. He wrote down "73." What unit of measurement did he use?

(a) 73 kilometers   **(b)** 73 meters   **(c)** 73 centimeters

Solution   The most reasonable choice is **(c)**, 73 centimeters. The other two units would be much too long. A meter is close to a yard, and a doorway would not be 73 yards wide! A kilometer is even larger than a meter.

Student Practice 4   Joan measured the length of her car. She wrote down 3.8. Which unit of measurement did she use?

(a) 3.8 kilometers   **(b)** 3.8 meters   **(c)** 3.8 centimeters

The most frequent metric conversions in length are done between kilometers, meters, centimeters, and millimeters. Their abbreviations are km, m, cm, and mm, respectively, and you should be able to use them correctly.

EXAMPLE 5   Convert.   **(a)** 982 cm to m   **(b)** 5.2 m to mm

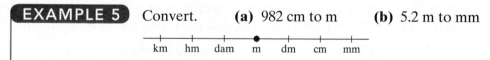

Solution

(a) In the first case, we move the decimal point to the left because we are going from a smaller unit to a larger unit.

$$982 \text{ cm} = 9.82 \text{ m} \qquad \text{(two places to left)}$$
$$= 9.82 \text{ m}$$

(b) In the next case, we need to move the decimal point to the right because we are going from a larger unit to a smaller unit.

$$5.2 \text{ m} = 5200. \text{ mm} \qquad \text{(three places to right)}$$
$$= 5200 \text{ mm}$$

Student Practice 5   Convert.

(a) 375 cm to m   **(b)** 46 m to mm

The other metric units of length are the hectometer, the dekameter, and the decimeter. These are not used very frequently, but it is good to understand how their lengths relate to the basic unit, the meter. A complete list of the metric lengths we have discussed appears in the following table.

METRIC LENGTHS WITH ABBREVIATIONS

1 kilometer (km) = 1000 meters
1 hectometer (hm) = 100 meters
1 dekameter (dam) = 10 meters
1 meter (m)
1 decimeter (dm) = 0.1 meter
1 centimeter (cm) = 0.01 meter
1 millimeter (mm) = 0.001 meter

EXAMPLE 6 Convert.

(a) 426 decimeters to kilometers
(b) 9.47 hectometers to meters

Solution

(a) We are converting from a smaller unit, dm, to a larger one, km. Therefore, there will be fewer kilometers than decimeters. (The number we get will be smaller than 426.) We move the decimal point four places to the left.

$$426 \text{ dm} = 0.0426. \text{ km} \quad \text{(four places to left)}$$
$$= 0.0426 \text{ km}$$

(b) We are converting from a larger unit, hm, to a smaller one, m. Therefore, there will be more meters than hectometers. (The number will be larger than 9.47.) We move the decimal point two places to the right.

$$9.47 \text{ hm} = 9.47. \text{ m} \quad \text{(two places to right)}$$
$$= 947 \text{ m}$$

Student Practice 6 Convert.

(a) 389 millimeters to dekameters
(b) 0.48 hectometer to centimeters

When several metric measurements are to be added, we change them to a convenient common unit.

EXAMPLE 7 Add. 125 m + 1.8 km + 793 m

Solution First we change the kilometer measurement to a measurement in meters.

$$1.8 \text{ km} = 1800 \text{ m}$$

Then we add.

$$
\begin{array}{r}
125 \text{ m} \\
1800 \text{ m} \\
+ \quad 793 \text{ m} \\
\hline
2718 \text{ m}
\end{array}
$$

Student Practice 7 Add. 782 cm + 2 m + 537 m

SIDELIGHT: Extremely Large Metric Distances

Is the biggest length in the metric system a kilometer? Is the smallest length a millimeter? No. The system extends to very large units and very small ones. Usually, only scientists use these units.

1 gigameter = 1,000,000,000 meters
1 megameter = 1,000,000 meters
1 kilometer = 1000 meters
1 meter
1 millimeter = 0.001 meter
1 micrometer = 0.000001 meter
1 nanometer = 0.000000001 meter

For example, 26 megameters equals a length of 26,000,000 meters. A length of 31 micrometers equals a length of 0.000031 meter.

SIDELIGHT: Metric Measurements for Computers

A **byte** is the amount of computer memory needed to store one alphanumeric character. When referring to computers you may hear the following words: **kilobytes, megabytes,** and **gigabytes.** The following chart may help you.

1 terabyte (TB) = one trillion bytes = 1,000,000,000,000 bytes
1 gigabyte (GB) = one billion bytes =   1,000,000,000 bytes
1 megabyte (MB) = one million bytes* =   1,000,000 bytes
1 kilobyte (KB) = one thousand bytes† =   1000 bytes

TO THINK ABOUT: Converting Computer Measurements Before we move on to the exercises, see if you can use the large or small metric distance and use computer memory size to convert the following measurements.

1. 18 megameters = _____ kilometers
2. 26 millimeters = _____ micrometers
3. 17 nanometers = _____ millimeter
4. 38 meters = _____ megameter
5. 1.2 gigabytes = _____ bytes
6. 528 megabytes = _____ bytes
7. 78.9 kilobytes = _____ bytes
8. 24.9 gigabytes = _____ bytes

*Sometimes in computer science 1 megabyte is considered to be 1,048,576 bytes.
†Sometimes in computer science 1 kilobyte is considered to be 1024 bytes.

6.2 Exercises

MyMathLab®

Watch the videos
in MyMathLab

Download the
MyDashBoard App

Verbal and Writing Skills, Exercises 1–6

Write the prefix for each of the following.

1. Hundred

2. Hundredth

3. Tenth

4. Thousandth

5. Thousand

6. Ten

The following conversions involve metric units that are very commonly used. You should be able to perform each conversion without any notes and without consulting your text.

7. 46 centimeters = _____ millimeters

8. 79 centimeters = _____ millimeters

9. 2.61 kilometers = _____ meters

10. 8.3 kilometers = _____ meters

11. 12,500 millimeters = _____ meters

12. 10,600 millimeters = _____ meters

13. 7.32 centimeters = _____ meters

14. 9.14 centimeters = _____ meters

15. 2 kilometers = _____ centimeters

16. 7 kilometers = _____ centimeters

17. 78,000 millimeters = _____ kilometer

18. 840 millimeters = _____ kilometer

Abbreviations are used in the following. Fill in the blanks with the correct values.

19. 35 mm = _____ cm = _____ m

20. 6300 mm = _____ cm = _____ m

21. 4.5 km = _____ m = _____ cm

22. 6.8 km = _____ m = _____ cm

Applications, Exercises 23–32

23. *Driving in Spain* Amanda is driving in Spain and sees a road sign which gives the distance to the next city. Choose the most reasonable measurement.

(a) 24 m   (b) 24 km   (c) 24 cm

24. *Picture Frame Construction* Blair is making a picture frame. She needs to measure the length to know how much decorative border to buy. Choose the most reasonable measurement.

(a) 20 m   (b) 20 cm   (c) 20 mm

25. *Compact Disc Case* Eddie measured the width of a compact disc case. Choose the most appropriate measurement.

(a) 80.5 km   (b) 80.5 m   (c) 80.5 mm

26. *Botanical Gardens* The Botanical Gardens are located in the center of the city. If it takes approximately 10 minutes to walk directly from the east entrance to the west entrance, which would be the most likely measurement of the width of the Gardens?

(a) 1.4 km   (b) 1.4 m   (c) 1.4 mm

27. *Computer Monitor* Julie measured her monitor to see if it would fit under a shelf on her computer desk. Choose the most reasonable measurement.

(a) 45 cm   **(b)** 45 m   **(c)** 45 mm

28. *Jewelry Box Construction* Rich is making a jewelry box and must use very small screws. Which would be the most likely measurement of the screws?

(a) 8 cm   **(b)** 8 m   **(c)** 8 mm

29. *Street Length* Brenda and Stanley live on Brookwood Street. It is a dead end street with about 12 houses on it. Which would be the most likely measurement of the length of the street?

(a) 0.5 km   **(b)** 0.5 m   **(c)** 0.5 cm

30. *Bookcase* Gabe has a bookcase for all of his college textbooks. It is fairly full but one shelf has room for about 5 or 6 more textbooks. Which would be the most likely measurement of the available space on his bookcase?

(a) 32 km   **(b)** 32 cm   **(c)** 32 mm

31. *House Insulation* Bob and Debbie found that one corner of their townhouse is chilly in the winter. Bob examined the wall and found one portion of the wall does not have any insulation. He needed to install insulation between the wallboard and the outer wall. What would be the most likely measurement of the thickness of the insulation?

(a) 11.9 mm   **(b)** 11.9 cm   **(c)** 11.9 m

32. *Baseball* John and Stephanie got excellent tickets to see a Boston Red Sox game at Fenway Park. Which would be the most likely measurement of the distance from their seats to the pitcher's mound on the baseball field?

(a) 320 cm   **(b)** 320 km   **(c)** 320 m

The following conversions involve metric units that are not used extensively. You should be able to perform each conversion, but it is not necessary to do it from memory.

33. 390 decimeters = _____ meters

34. 270 decimeters = _____ meters

35. 800 dekameters = _____ meters

36. 530 dekameters = _____ meters

37. 48.2 meters = _____ hectometer

38. 435 hectometers = _____ kilometers

Change to a convenient unit of measure and add.

39. 243 m + 2.7 km + 312 m

40. 845 m + 5.79 km + 701 m

41. 225 mm + 12.7 cm + 148 cm

42. 305 mm + 45.4 cm + 318 cm

43. 15 mm + 2 dm + 42 cm

44. 8 dm + 21 mm + 38 cm

45. *Stereo Cabinet* The outside casing of a stereo cabinet is built of plywood 0.95 centimeter thick attached to plastic 1.35 centimeters thick and a piece of mahogany veneer 2.464 millimeters thick. How thick is the stereo casing?

46. *House Construction* A plywood board is 2.2 centimeters thick. A layer of tar paper is 3.42 millimeters thick. A layer of false brick siding is 2.7 centimeters thick. A house wall consists of these three layers. How thick is the wall?

Mixed Practice

47. 65 cm + 80 mm + 2.5 m   **48.** 82 m + 471 cm + 0.32 km   **49.** 46 m + 986 cm + 0.884 km

50. 56.3 centimeters = _____ meter   **51.** 96.4 centimeters = _____ meter

Write true *or* false *for each statement.*

52. 0.001 kilometer = 1 meter   **53.** 1 kilometer = 0.001 meter

54. 1000 meters = 1 kilometer   **55.** 10 millimeters = 1 centimeter

56. An airport runway might be 2 kilometers long.   **57.** A man might be 2 meters tall.

58. A kilometer is longer than a mile.   **59.** A yard is longer than a meter.

Applications

60. *Trans-Siberian Train* The Trans-Siberian train in Russia starts in Moscow and ends in Vladivostok, a city on the Pacific Ocean. The length of the run, which has 26 stations, measures 9259 kilometers.

(a) How many meters is the run?

(b) How many centimeters is the run?

61. *Chinese Train* The Qingzang railway in China includes the Tanggula Pass. At 5072 meters high, it is the world's highest track.

(a) How many centimeters high is the track?

(b) How many kilometers high is the track?

62. *Dam Construction* The world's highest dam, the Rogun dam in Tajikistan, is 335 meters high.

(a) How many kilometers high is the dam?

(b) How many centimeters high is the dam?

63. *Virus Size* A typical virus of the human body measures just 0.000000254 centimeter in diameter. How many meters in diameter is a typical virus?

To Think About

World's Longest Subway Systems   The lengths of some of the world's longest subway systems are given in the following bar graph. Use the graph to answer exercises 64–69.

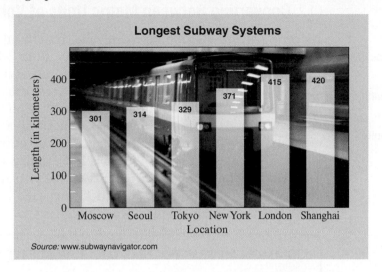

64. How many meters longer is the subway system in Tokyo, Japan, than that of Moscow, Russia?

65. How many meters longer is the Shanghai, China, subway system than that of Seoul, South Korea?

66. How many megameters long is the subway system in New York City?

67. How many megameters long is the subway system in London?

To convert from kilometers to miles, multiply the number of kilometers by 0.62.

68. How many miles long is Tokyo's subway system?

69. How many miles long is Shanghai's subway system?

Cumulative Review

70. [5.3.2] 57% of what number is 2850?

71. [5.3.2] Find 0.03% of 5900.

72. [5.4.3] Chloe purchased a dining room set at a 20% discount. The list price was $660. How much did Chloe pay?

73. [5.5.1] Frank sells jewelry and has a commission rate of 8%. If he sells $27,000 worth of jewelry, what is his commission?

Quick Quiz 6.2

1. Convert 45.9 meters to centimeters.

2. Convert 0.0283 centimeter to millimeters.

3. Convert 5160 meters to kilometers.

4. Concept Check Explain how you would convert 5643 centimeters to kilometers.

After studying this section, you will be able to:

① Convert between metric units of volume.

② Convert between metric units of weight.

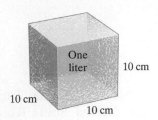

One liter
10 cm
10 cm
10 cm

1 liter   2 liters

① Converting Between Metric Units of Volume

As products are distributed worldwide, more and more of them are being sold in metric units. Soft drinks come in 1-, 2-, or 3-liter bottles. Labels often contain these amounts in both American units and metric units. Try looking at these labels to gain a sense of the size of metric units of volume.

The basic metric unit for volume is the liter. A **liter** is defined as the volume of a box 10 cm × 10 cm × 10 cm, or 1000 cm^3. A cubic centimeter may be written as cc, so we sometimes see 1000 cc = 1 liter. A liter is slightly larger than a quart; 1 liter of liquid is 1.057 quarts of that liquid.

The most common metric units of volume are the milliliter, the liter, and the kiloliter. Sometimes deciliters are used in recipes. Often a capital letter L is used as an abbreviation for *liter*.

COMMON METRIC VOLUME MEASUREMENTS

1 kiloliter (kL) = 1000 liters

1 liter (L)

1 milliliter (mL) = 0.001 liter

We know that 1000 cc = 1 liter. Dividing each side of that equation by 1000, we get 1 cc = 1 mL.

Use this prefix chart as a guide when you change one metric prefix to another. Move the decimal point in the same direction and the same number of places.

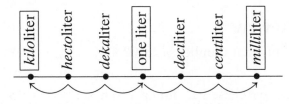

kiloliter   hectoliter   dekaliter   one liter   deciliter   centiliter   milliliter

---

The prefixes for liter follow the pattern we have seen for meter. Note that *kilo-* is three places to the left of the liter and *milli-* is three places to the right. Because the kiloliter, the liter, and the milliliter are the most commonly used units of volume, we will focus exclusively on them in this text.

EXAMPLE 1   Convert.

(a) 3 L = _____ mL   **(b)** 24 kL = _____ L   **(c)** 0.084 L = _____ mL

Solution

(a) The prefix *milli-* is three places to the right. We move the decimal point three places to the right.

3 L = 3.000 = 3000 mL

(b) 24 kL = 24.000 = 24,000 L   **(c)** 0.084 L = 0.084 = 84 mL

Student Practice 1 Convert.

(a) $5 \text{ L} = $ ____ mL **(b)** $84 \text{ kL} = $ _____ L **(c)** $0.732 \text{ L} = $ ____ mL

NOTE TO STUDENT: Fully worked-out solutions to all of the Student Practice problems can be found at the back of the text starting at page SP-1.

EXAMPLE 2 Convert.

(a) $26.4 \text{ mL} = $ ____ L **(b)** $5982 \text{ mL} = $ ____ L **(c)** $6.7 \text{ L} = $ ____ kL

Solution

(a) The unit L is three places to the left of the unit mL. We move the decimal three places to the left. $26.4 \text{ mL} = 0.0264 \text{ L}$

(b) $5982 \text{ mL} = 5.982 \text{ L}$

(c) $6.7 \text{ L} = 0.0067 \text{ kL}$

Student Practice 2 Convert.

(a) $15.8 \text{ mL} = $ _____ L **(b)** $12,340 \text{ mL} = $ _____ L

(c) $86.3 \text{ L} = $ _____ kL

The cubic centimeter is often used in medicine. Recall $1 \text{ mL} = 1 \text{ cm}^3$ or 1 cc.

EXAMPLE 3 Convert.

(a) $26 \text{ mL} = $ _____ cm^3 **(b)** $0.82 \text{ L} = $ _____ cc

Solution

(a) A milliliter and a cubic centimeter are equivalent. $26 \text{ mL} = 26 \text{ cm}^3$

(b) We use the same rule to convert liters to cubic centimeters as we do to convert liters to milliliters.
Since $1 \text{ cm}^3 = 1 \text{ cc}$, $0.82 \text{ L} = 820 \text{ cm}^3 = 820 \text{ cc}$.

Student Practice 3 Convert.

(a) $396 \text{ mL} = $ _____ cm^3 **(b)** $0.096 \text{ L} = $ _____ cc

② Converting Between Metric Units of Weight

In science classes we make a distinction between weight and **mass.** Mass is the amount of material in an object. Weight is a measure of the pull of gravity on an object. The farther you are from the center of Earth, the less you weigh. If you were in an astronaut suit floating in outer space you would be weightless. The mass of your body, however, would not change. Throughout this book we will refer to *weight* only. The technical difference between weight and mass is not something we need to pay attention to in everyday life.

In the metric system the basic unit of weight is the gram. A **gram** is the weight of the water in a box that is 1 centimeter on each side. To get an idea of how small a gram is, we note that two small paper clips weigh about 1 gram. A gram is only about 0.035 ounce.

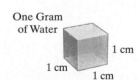

One Gram of Water

1 cm 1 cm 1 cm

One kilogram is 1000 times larger than a gram. A kilogram weighs about 2.2 pounds. Some of the measures of weight in the metric system are shown in the following chart.

> **COMMON METRIC WEIGHT MEASUREMENTS**
>
> 1 metric ton (t) = 1,000,000 grams
> 1 kilogram (kg) = 1000 grams
> 1 gram (g)
> 1 milligram (mg) = 0.001 gram

> Use this prefix chart as a guide when you change one metric prefix to another. Move the decimal point in the same direction and the same number of places.
>
>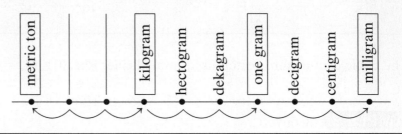

We will focus exclusively on the metric ton, kilogram, gram, and milligram. We convert weight measurements the same way we convert volume and length measurements. The metric ton is also called a **megagram.** *Mega* means "million" so a megagram is 1,000,000 grams.

Mc EXAMPLE 4   Convert.

(a) 2 t = _____ kg   **(b)** 0.42 kg = _____ g

Solution   **(a)** 2 t = 2000 kg   **(b)** 0.42 kg = 420 g

Student Practice 4   Convert.

(a) 3.2 t = _____ kg

(b) 7.08 kg = _____ g

EXAMPLE 5   Convert.

(a) 283 kg = _____ t   **(b)** 7.98 mg = _____ g

Solution

(a) 283 kg = 0.283 t   **(b)** 7.98 mg = 0.00798 g

Student Practice 5   Convert.

(a) 59 kg = _____ t   **(b)** 28.3 mg = _____ g

EXAMPLE 6   If a chemical costs $0.03 per gram, what will it cost per kilogram?

Solution   Since there are 1000 grams in a kilogram, a chemical that costs $0.03 per gram would cost 1000 times as much per kilogram.

$$1000 \times \$0.03 = \$30.00$$

The chemical would cost $30.00 per kilogram.

Student Practice 6   If coffee costs $10.00 per kilogram, what will it cost per gram?

A kilogram is slightly more than 2 pounds. A 3-week-old miniature pinscher weighs about 1 kilogram.

EXAMPLE 7   Select the most reasonable weight for Tammy's Toyota.

(a) 820 t   **(b)** 820 g   **(c)** 820 kg   **(d)** 820 mg

Solution   The most reasonable answer is **(c)**, 820 kg. The other weight values are much too large or much too small. Since a kilogram is slightly more than 2 pounds, we see that this weight, 820 kg, most closely approximates the weight of a car.

Student Practice 7   Select the most reasonable weight for Hank, a starting linebacker for the college football team.

(a) 120 kg   **(b)** 120 g   **(c)** 120 mg

SIDELIGHT: Using Very Small Units
When dealing with very small particles or atomic elements, scientists sometimes use units smaller than a gram.

> **SMALL WEIGHT MEASUREMENTS**
>
> 1 milligram = 0.001 gram
> 1 microgram = 0.000001 gram
> 1 nanogram = 0.000000001 gram
> 1 picogram = 0.000000000001 gram

We could make the following conversions.

$$2.6 \text{ picograms} = 0.0026 \text{ nanogram}$$
$$29.7 \text{ micrograms} = 0.0297 \text{ milligram}$$
$$58 \text{ nanograms} = 58{,}000 \text{ picograms}$$
$$58 \text{ nanograms} = 0.058 \text{ microgram}$$

6.3 Exercises
MyMathLab®

Watch the videos
in MyMathLab

Download the
MyDashBoard App

Verbal and Writing Skills, Exercises 1–6

Write the metric unit that best represents each measurement.

1. one thousand liters

2. one thousandth of a liter

3. one thousandth of a gram

4. one thousand grams

5. one thousandth of a kilogram

6. one thousand milliliters

Perform each conversion.

7. 9 kL = _____ L

8. 5 kL = _____ L

9. 12 L = _____ mL

10. 25 L = _____ mL

11. 18.9 mL = _____ L

12. 31.5 mL = _____ L

13. 752 L = _____ kL

14. 368 L = _____ kL

15. 5.652 kL = _____ mL

16. 14.3 kL = _____ mL

17. 82 mL = _____ cc

18. 18.9 mL = _____ cm³

19. 24,418 mL = _____ kL

20. 8835 mL = _____ kL

21. 74 L = _____ cm³

22. 122 L = _____ cm³

23. 216 g = _____ kg

24. 2940 g = _____ kg

25. 35 mg = _____ g

26. 13 mg = _____ g

27. 6328 mg = _____ g

28. 986 mg = _____ g

29. 2.92 kg = _____ g

30. 14.6 kg = _____ g

31. 2.4 t = _____ kg

32. 9500 kg = _____ t

Fill in the blanks with the correct values.

33. 7 mL = _____ L = _____ kL

34. 23 mL = _____ L = _____ kL

35. 84 cm³ = _____ L = _____ kL

36. 0.584 kL = _____ L = _____ cc

37. 0.033 kg = _____ g = _____ mg

38. 0.098 kg = _____ g = _____ mg

39. 2.58 metric tons = _____ kg = _____ g

40. 7183 g = _____ kg = _____ t

41. *Jar Size* Alice bought a jar of apple juice at the store. Choose the most reasonable measurement for its contents.

    (a) 0.32 kL   **(b)** 0.32 L   **(c)** 0.32 mL

42. *Injection Dosage* A nurse gave an injection of insulin to a diabetic patient. Choose the most reasonable measurement for the dose.

    (a) 4 kL   **(b)** 4 L   **(c)** 4 mL

43. *Dinosaurs* One of the heaviest dinosaurs was the Argentinosaurus. Choose the most reasonable measurement for its weight.

    (a) 100 t   **(b)** 100 kg   **(c)** 100 g

44. *College Textbooks* Robert Tobey bought a new psychology textbook. Choose the most reasonable measurement for its weight.

    (a) 0.49 t   **(b)** 0.49 kg   **(c)** 0.49 g

Find a convenient unit of measure and add.

45. 83 L + 822 mL + 30.1 L

46. 28 L + 115.2 L + 385 mL

47. 20 g + 52 mg + 1.5 kg

48. 73 mg + 124 g + 5 kg

Mixed Practice *Write true or false for each statement.*

49. 1 milliliter = 0.001 liter

50. 1 metric ton = 1000 grams

51. Orange juice can be purchased in milliliter containers at the grocery store.

52. Small amounts of medicine are often measured in liters.

53. A reasonable weight for an adult man is 500,000 grams.

54. A bottle of soda might contain 1000 mL.

55. A nickel weighs about 5 grams.

56. A convenient size for a family purchase is 2 kilograms of ground beef.

Applications

57. *Food Costs* Starbucks sells 453-gram bags of ground coffee for $8.99. Rhonda bought 3.624 kg. How much did she spend on coffee?

58. *Food Cost* Randy's Premier Pizza needs to order tomato sauce from the distributor. The sauce comes in 4000-gram jars for $6.80. If Randy orders 56 kg, how much will he pay for the tomato sauce?

59. *Biogenetic Research* A very rare essence of an almost extinct flower found in the Amazon jungle of South America is extracted by a biogenetic company trying to copy and synthesize it. The company estimates that if the procedure is successful, the product will cost the company $850 per milliliter to produce. How much will it cost the company to produce 0.4 liter of the engineered essence?

60. *Price of Gold* One day in August 2010, the price of gold was $39.56 per gram. At that price, how much would a kilogram of gold cost?

To Think About

Carbon Dioxide Emissions *In countries outside the United States, very heavy items are measured in terms of **metric tons***. *A metric ton is defined as 1000 kilograms (or 2025 pounds).*

Carbon dioxide being released into the air fuels the green-house effect and warms Earth's atmosphere. Consider the bar graph, which displays the billions of metric tons of carbon dioxide emitted by two categories: countries such as the United States, Canada, Japan, Australia, New Zealand, and the countries of Western Europe, considered industrialized countries, and the countries of Asia, the Middle East, Africa, Central America, and South America, considered to be developing countries.

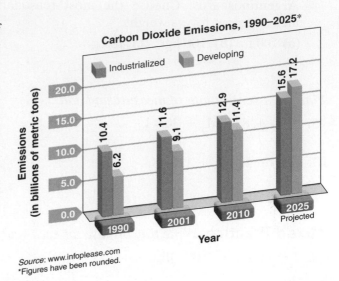

61. By how many metric tons did carbon dioxide emissions by industrialized countries increase from 1990 to 2010?

62. By how many metric tons did carbon dioxide emissions by developing countries increase from 1990 to 2010?

63. How many kilograms of carbon dioxide are expected to be emitted by developing countries in 2025?

64. How many kilograms of carbon dioxide are expected to be emitted by industrialized countries in 2025?

65. If the same *percent of increase* that occurred from 2001 to 2010 also occurs from 2025 to 2034, how many metric tons of carbon dioxide will be emitted by industrialized countries in 2034?

66. If the same *percent of increase* that occurred from 2001 to 2010 also occurs from 2025 to 2034, how many metric tons of carbon dioxide will be emitted by developing countries in 2034?

Cumulative Review

67. **[5.3.2]** 14 out of 70 is what percent?

68. **[5.3.2]** What is 23% of 250?

69. **[5.4.2]** ***Home Theater Costs*** Marilyn bought a new home theater system for $4800. After a 10% discount was taken, a 5% sales tax was added. What did Marilyn pay for the theater system?

70. **[5.5.1]** ***Commission Sales*** A salesperson earns a commission of 8%. She sold furniture worth $8960. How much commission did she earn?

Quick Quiz 6.3

1. Convert 671 grams to kilograms.

2. Convert 8.52 liters to milliliters.

3. Convert 45.62 milligrams to grams.

4. **Concept Check** Explain how you would convert 54 kilograms to milligrams.

How Am I Doing?   Sections 6.1–6.3

How are you doing with your homework assignments in Sections 6.1 to 6.3? Do you feel you have mastered the material so far? Do you understand the concepts you have covered? Before you go further in the textbook, take some time to do each of the following problems.

6.1 *Convert. When necessary, express your answer as a decimal rounded to the nearest hundredth.*

1. 48 feet = ____ yards

2. 24 quarts = ____ gallons

3. 3 miles = ____ yards

4. 4.5 tons = ____ pounds

5. 22 minutes = ____ seconds

6. 48 pints = ____ gallons

7. Isabel is making a fish dinner for her roommates. She purchases 2 pounds 4 ounces of haddock for $6.80 per pound. How much did Isabel spend on the haddock?

6.2 *Perform each conversion.*

8. 6.75 km = ____ m

9. 73.9 m = ____ cm

10. 34 cm = ____ mm

11. 27 mm = ____ m

12. 5296 mm = ____ cm

13. 482 m = ____ km

Convert to meters and add.

14. 1.2 km + 192 m + 984 m

15. 305 cm + 82.5 m + 6150 mm

16. The wire that connects a motor to a control panel in Tokyo is 3.4 meters long. One portion of the wire measuring 78 centimeters has double insulation. Another portion of the wire measuring 128 centimeters has triple insulation. The remaining part of the wire has single insulation. How long is the portion of the wire that has single insulation?

6.3 *Perform each conversion.*

17. 5.66 L = ____ mL

18. 535 g = ____ kg

19. 56.3 kg = ____ t

20. 4.8 kL = ____ L

21. 568 mg = ____ g

22. 8.9 L = ____ cm^3

23. A wholesaler sells peanut butter in canisters of 5000 g for $7.75. The Boston Rescue Mission needs to buy 75 kg to restock its pantry. How much will the mission spend on peanut butter if they buy from this wholesaler?

24. Italy Foods receives a special shipment of olive oil in 5000-gram cans. Italy Foods sells these cans for $32.50. If Ricardo's Restaurant needs 35 kg of olive oil and purchases it from Italy Foods, how much will it cost?

25. An antibacterial spray costs $36 per liter to produce. The Smiths purchased 200 milliliters of the spray. They know the owner of the company that produces the spray and he sold it to them at cost. How much did they pay for 200 milliliters?

26. Last month gold cost $38.40 per gram. Mariah purchased 0.5 kilogram of gold at that price. How much did she pay?

Now turn to page SA-11 for the answer to each of these problems. Each answer also includes a reference to the objective in which the problem is first taught. If you missed any of these problems, you should stop and review the Examples and Student Practice problems in the referenced objective. A little review now will help you master the material in the upcoming sections of the test.

1. _____
2. _____
3. _____
4. _____
5. _____
6. _____
7. _____
8. _____
9. _____
10. _____
11. _____
12. _____
13. _____
14. _____
15. _____
16. _____
17. _____
18. _____
19. _____
20. _____
21. _____
22. _____
23. _____
24. _____
25. _____
26. _____

6.4 Converting Units

Student Learning Objectives

After studying this section, you will be able to:

① Convert units of length, volume, or weight between the metric and American systems.

② Convert between Fahrenheit and Celsius degrees of temperature.

① Converting Units of Length, Volume, or Weight Between the Metric and American Systems

So far we've seen how to convert units when working *within* either the American or the metric system. Many people, however, work in *both* the metric and the American systems. If you study fields such as chemistry, electromechanical technology, business, X-ray technology, nursing, or computers, you will probably need to convert measurements between the two systems.

To convert between American units and metric units, it is helpful to have equivalent values. The most commonly used equivalents are listed in the following table. Most of these are approximate.

Equivalent Measures

	American to Metric	Metric to American
Units of length	1 mile ≈ 1.61 kilometers 1 yard ≈ 0.914 meter 1 foot ≈ 0.305 meter 1 inch = 2.54 centimeters[*]	1 kilometer ≈ 0.62 mile 1 meter ≈ 1.09 yards 1 meter ≈ 3.28 feet 1 centimeter ≈ 0.394 inch
Units of volume	1 gallon ≈ 3.79 liters 1 quart ≈ 0.946 liter	1 liter ≈ 0.264 gallon 1 liter ≈ 1.06 quarts
Units of weight	1 pound ≈ 0.454 kilogram 1 ounce ≈ 28.35 grams	1 kilogram ≈ 2.2 pounds 1 gram ≈ 0.0353 ounce

[*]exact value

Remember that to convert from one unit to another you multiply by a fraction that is equivalent to 1. Create a fraction from the equivalent measures table so that the unit in the denominator cancels the unit you are changing. In this book, if we do a calculation that involves rounding, we will use the ≈ sign. If no rounding is needed, we will use the = sign.

To change 5 miles to kilometers, we look in the table and find that 1 mile ≈ 1.61 kilometers. We will use the unit fraction

$$\frac{1.61 \text{ kilometers}}{1 \text{ mile}}$$

because we want to have miles in the denominator.

$$5 \text{ miles} \times \frac{1.61 \text{ kilometers}}{1 \text{ mile}} = 5 \times 1.61 \text{ kilometers} = 8.05 \text{ kilometers}$$

Thus 5 miles ≈ 8.05 kilometers. Because multiplication is easier than division, it's easier to use a fraction with a 1 in the denominator.

Is this the only way to do the problem? No. To make the previous conversion, we also could have used the relationship that 1 kilometer ≈ 0.62 mile. Again, we want to have miles in the denominator, so we use

$$\frac{1 \text{ kilometer}}{0.62 \text{ mile}}$$

$$5 \text{ miles} \times \frac{1 \text{ kilometers}}{0.62 \text{ mile}} = \frac{5}{0.62} \approx 8.06 \text{ kilometers.}$$

Using this approach, we find that 5 miles ≈ 8.06 kilometers. This is not the same result we obtained before. The discrepancy between the results of the two conversions has occurred because the numbers given in the equivalent measures table are approximations.

Often we have to make conversions in order to make comparisons. For example, is the 6 inches of attic insulation commonly used in the northeastern United States more or less than the 16 centimeters of attic insulation commonly used in Sweden? In order to find out we would have to convert 6 inches to centimeters. What unit fraction would we use? What result would we get?

EXAMPLE 1   Convert 3 feet to meters.

Solution

$$3 \text{ feet} \times \frac{0.305 \text{ meter}}{1 \text{ foot}} = 0.915 \text{ meter}$$

Student Practice 1   Convert 7 feet to meters.

NOTE TO STUDENT: *Fully worked-out solutions to all of the Student Practice problems can be found at the back of the text starting at page SP-1.*

Unit abbreviations are quite common, so we will use them for the remainder of this section. We list them here for your reference.

American Measure (Alphabetical Order)	Standard Abbreviation
feet	ft
gallon	gal
inch	in.
mile	mi
ounce	oz
pound	lb
quart	qt
yard	yd

Metric Measure	Standard Abbreviation
centimeter	cm
gram	g
kilogram	kg
kilometer	km
liter	L
meter	m
millimeter	mm

EXAMPLE 2

(a) Convert 26 m to yd.
(b) Convert 1.9 km to mi.
(c) Convert 14 gal to L.
(d) Convert 2.5 L to qt.

Solution

(a) $26 \text{ m} \times \dfrac{1.09 \text{ yd}}{1 \text{ m}} = 28.34 \text{ yd}$

(b) $1.9 \text{ km} \times \dfrac{0.62 \text{ mi}}{1 \text{ km}} = 1.178 \text{ mi}$

(c) $14 \text{ gal} \times \dfrac{3.79 \text{ L}}{1 \text{ gal}} = 53.06 \text{ L}$

(d) $2.5 \text{ L} \times \dfrac{1.06 \text{ qt}}{1 \text{ L}} = 2.65 \text{ qt}$

Student Practice 2

(a) Convert 17 m to yd.
(b) Convert 29.6 km to mi.
(c) Convert 26 gal to L.
(d) Convert 6.2 L to qt.

Some conversions require more than one step.

EXAMPLE 3   Convert 235 cm to ft. Round to the nearest hundredth of a foot.

Solution   Our first fraction converts centimeters to inches. Our second fraction converts inches to feet.

$$235 \text{ cm} \times \frac{0.394 \text{ in.}}{1 \text{ cm}} \times \frac{1 \text{ ft}}{12 \text{ in.}} = \frac{92.59}{12} \text{ ft}$$

$$\approx 7.72 \text{ ft (rounded to the nearest hundredth)}$$

Student Practice 3   Convert 180 cm to ft.

The same rules can be followed for a rate such as 50 miles per hour.

If the unit to be converted is in the numerator, be sure to express that unit in the denominator of the unit fraction. If the unit to be converted is in the denominator, be sure to express that unit in the numerator of the unit fraction.

EXAMPLE 4   Convert 100 km/hr to mi/hr.

Solution   We need to multiply by a unit fraction. The fraction we multiply by must have kilometers in the denominator.

$$\frac{100 \text{ km}}{\text{hr}} \times \frac{0.62 \text{ mi}}{1 \text{ km}} = 62 \text{ mi/hr}$$

Thus 100 km/hr is approximately equal to 62 mi/hr.

Student Practice 4   Convert 88 km/hr to mi/hr.

Sometimes we need more than one unit fraction to make the conversion of two rates. We will see how this is accomplished in Example 5.

EXAMPLE 5   A rocket carrying a communication satellite is launched from a rocket launch pad. It is traveling at 700 miles per hour. How many feet per second is the rocket traveling? Round to the nearest whole number.

Solution

$$\frac{700 \text{ miles}}{\text{hr}} \times \frac{5280 \text{ ft}}{1 \text{ mile}} \times \frac{1 \text{ hr}}{60 \text{ min}} \times \frac{1 \text{ min}}{60 \text{ sec}}$$

$$= \frac{700 \times 5280 \text{ ft}}{60 \times 60 \text{ sec}} = \frac{3{,}696{,}000 \text{ ft}}{3600 \text{ sec}} \approx 1027 \text{ ft/sec}$$

The missile is traveling at approximately 1027 feet per second.

Student Practice 5   In 2003, a Concorde jet flew from Boston to Paris at a speed of 900 miles per hour. What was the speed of the Concorde jet in feet per second?

▲ SIDELIGHT: From Square Yards to Square Meters

Suppose we consider a rectangle that measures 2 yards wide by 4 yards long. The area would be 2 yards × 4 yards = 8 square yards. How could you change 8 square yards to square meters? Suppose that we look at 1 square yard. Each side is 1 yard long, which is equivalent to 0.914 meter.

$$\text{Area} = 1 \text{ yard} \times 1 \text{ yard} \approx 0.914 \text{ meter} \times 0.914 \text{ meter}$$
$$\text{Area} = 1 \text{ square yard} \approx 0.8354 \text{ square meter}$$

Thus 1 yd² ≈ 0.8354 m². Therefore

$$8 \text{ yd}^2 \times \frac{0.8354 \text{ m}^2}{1 \text{ yd}^2} = 6.6832 \text{ m}^2.$$

8 square yards ≈ 6.6832 square meters.

② Converting Between Fahrenheit and Celsius Degrees of Temperature

In the metric system, temperature is measured on the **Celsius scale.** Water boils at 100° (100°C) and freezes at 0° (0°C) on the Celsius scale. In the **Fahrenheit system,** water boils at 212° (212°F) and freezes at 32° (32°F).

> To convert Celsius to Fahrenheit, we can use the formula
>
> $$F = 1.8 \times C + 32,$$
>
> where C is the number of Celsius degrees and F is the number of Fahrenheit degrees.

Mc **EXAMPLE 6** When the temperature is 35°C, what is the Fahrenheit reading?

Solution

$$\begin{aligned} F &= 1.8 \times C + 32 \\ &= 1.8 \times 35 + 32 \\ &= 63 + 32 \\ &= 95 \end{aligned}$$

The temperature is 95°F.

Student Practice 6 Convert 20°C to Fahrenheit temperature.

> To convert Fahrenheit temperature to Celsius, we can use the formula
>
> $$C = \frac{5 \times F - 160}{9},$$
>
> where F is the number of Fahrenheit degrees and C is the number of Celsius degrees.

Calculator

Converting Temperatures

You can use your calculator to convert temperature readings between Fahrenheit and Celsius. To convert 30°C to Fahrenheit temperature, enter

1.8 $\boxed{\times}$ 30 $\boxed{+}$ 32 $\boxed{=}$

Display

$$\boxed{86}$$

The temperature is 86°F.

To convert 82.4°F to Celsius temperature, enter

5 $\boxed{\times}$ 82.4 $\boxed{-}$ 160

$\boxed{=}$ $\boxed{\div}$ 9 $\boxed{=}$

Display

$$\boxed{28}$$

The temperature is 28°C.

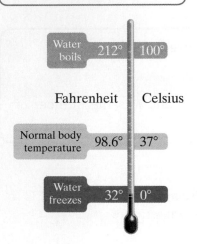

	Fahrenheit	Celsius
Water boils	212°	100°
Normal body temperature	98.6°	37°
Water freezes	32°	0°

EXAMPLE 7 When the temperature is 50°F, what is the Celsius reading?

Solution

$$C = \frac{5 \times F - 160}{9}$$

$$= \frac{5 \times 50 - 160}{9}$$

$$= \frac{250 - 160}{9}$$

$$= \frac{90}{9}$$

$$= 10$$

The temperature is 10°C.

Student Practice 7 When the temperature is 86°F, what is the Celsius reading?

If you spend a long time in a country that uses the Celsius scale, you may find it helpful to refer to this conversion chart. It will make it much more convenient to predict what clothes to wear when you see the weather report in that country.

Celsius	30°	25°	20°	10°	0°	−10°	−20°
Fahrenheit	86°	77°	68°	50°	32°	14°	−4°

Verbal and Writing Skills, Exercises 1–4

1. Which metric measure is approximately the same length as a yard? Which unit is larger?

2. Which metric measure of volume is approximately the same as a quart? Which unit is larger?

3. Which American measure is approximately twice the length of a centimeter?

4. Which metric measure is approximately double a pound?

Perform each conversion. Round to the nearest hundredth when necessary.

5. 8 ft to m	**6.** 11 ft to m	**7.** 9 in. to cm
8. 13 in. to cm	**9.** 32 m to yd	**10.** 115 m to yd
11. 30.8 yd to m	**12.** 42.5 yd to m	**13.** 82 mi to km
14. 68 mi to km	**15.** 9.25 m to yd	**16.** 15.35 m to yd
17. 17.5 cm to in.	**18.** 19.6 cm to in.	**19.** 200 m to ft
20. 500 m to ft	**21.** 5 km to mi	**22.** 16 km to mi
23. 48 gal to L	**24.** 63 gal to L	**25.** 23 qt to L
26. 28 qt to L	**27.** 19 L to gal	**28.** 15 L to gal
29. 4.5 L to qt	**30.** 6.5 L to qt	**31.** 82 kg to lb

32. 45 kg to lb

33. 130 lb to kg

34. 155 lb to kg

35. 26 oz to g

36. 34 oz to g

37. 152 kg to lb

38. 208 kg to lb

Mixed Practice *Perform each conversion. Round to the nearest hundredth if necessary.*

39. 158 g to oz

40. 105 g to oz

41. 35 ft to cm

42. 22 ft to cm

43. 55 km/hr to mi/hr

44. 120 km/hr to mi/hr

45. 400 ft/sec to mi/hr (Round to the nearest whole number.)

46. 300 ft/sec to mi/hr (Round to the nearest whole number.)

47. A wire that is 13 mm wide is how many inches wide?

48. A bolt that is 7 mm wide is how many inches wide?

49. 85°C to Fahrenheit

50. 120°C to Fahrenheit

51. 12°C to Fahrenheit

52. 21°C to Fahrenheit

53. 140°F to Celsius

54. 131°F to Celsius

55. 95°F to Celsius

56. 88°F to Celsius

Applications *Solve. Round to the nearest hundredth when necessary.*

57. Speed Limit The speed limit on a New Zealand highway is 90 km/hr. Molly is driving at 65 mi/hr. Is Molly speeding?

58. Spain Bike Trip John and Sandy Westphal traveled to Spain for a bike tour. The woman leading the tour told the participants they would be biking 75 km the first day, 83 km the second day, and 78 km the third day. How many miles did John and Sandy bike in those three days?

59. Fuel Consumption Pierre had a Jeep imported into France. During a trip from Paris to Lyon, he used 38 liters of gas. The tank, which he filled before starting the trip, holds 15 gallons of gas. How many liters of gas were left in the tank when he arrived?

60. Drinking Water It is recommended that each visitor to Death Valley, California, have 2 gallons of drinking water available. Rachel brought six 1-liter bottles of water. Does she have the recommended amount? Why?

61. Weight Records One of the heaviest females documented in medical records weighed 540 kg when admitted to the hospital in 1993. What would have been her weight in pounds?

62. Weight of a Child The average weight for a 10-year-old boy is 32 kilograms. What is the average weight in pounds?

63. Male Height Record According to the Guinness Book of World Records, the tallest male reached a height of 2.72 m. What would his height be in feet?

64. Female Height Record According to the Guinness Book of World Records, the tallest female reached a height of 2.48 m. What would her height be in feet?

65. Australia Rock Climbing Tourists who visit Ayers Rock in central Australia in the summer begin climbing at four o'clock in the morning, when the temperature is 19° Celsius. They do this because after seven o'clock in the morning, the temperature can reach 45°C and can cause climbers to die of dehydration. What are equivalent Fahrenheit temperatures?

66. Medication A prescription label says that a medication should be stored at room temperature, which is between 15°C and 30°C. What is this temperature range in Fahrenheit?

67. Biology There are 96,550 km of blood vessels in the human body. How many miles of blood vessels is this?

68. Earth Science The distance around Earth at the equator (the circumference) is approximately 40,325 km. How many miles is this?

Round to four decimal places.

▲ **69.** 28 square inches = ? square centimeters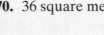

▲ **70.** 36 square meters = ? square yards

To Think About

▲ **71.** *Geometry* Frederick Himlein is planning to carpet his rectangular living room, which measures 8 yards by 4 yards. He has found some carpet in New York City that he likes that costs $28 per square yard. While visiting family friends in Germany, his wife Gertrude found some carpet that costs $30 per square meter. The company has a New York City office and can sell it in America for the same price. Frederick says that the German carpet is much too expensive. How much would it cost to carpet the living room with the American carpet? How much would it cost to carpet the living room with the German carpet? How much difference in cost is there between these two choices? (Round to the nearest dollar.)

▲ **72.** *Geometry* Jacques Bertoude is planning to carpet his rectangular family room, which measures 7 yards by 5 yards. His wife found some carpet in Boston that she likes, and it costs $24 per square yard. While Jacques was visiting his father in Paris, he found some carpet that costs $26 per square meter. The company has a Boston office and can sell it in America for the same price. Jacques told his wife that the carpet in Paris was a better buy. She does not agree. How much would it cost to carpet the family room with the American carpet? How much would it cost to carpet the family room with the French carpet? How much difference in cost is there between these two choices? (Round to the nearest dollar.)

Cumulative Review *Do the operations in the correct order.*

73. **[1.6.2]** $3^4 \times 2 - 5 + 12$

74. **[1.6.2]** $96 + 24 \div 4 \times 3$

Is the equation a proportion?

75. **[4.2.2]** $\dfrac{6}{11} \stackrel{?}{=} \dfrac{18}{31}$

76. **[4.2.2]** $\dfrac{1\frac{1}{2}}{12} \stackrel{?}{=} \dfrac{2}{16}$

Quick Quiz 6.4 *Round to the nearest hundredth when necessary.*

1. Convert 5 ounces to grams.

2. Convert 24 kilometers to miles.

3. Convert 6 liters to quarts.

4. **Concept Check** Explain how you would convert a speed of 65 miles per hour to a speed in kilometers per hour.

6.5 Solving Applied Measurement Problems

① Solving Applied Problems Involving Metric and American Units

Once again, we will be using the Mathematics Blueprint for solving applied problems that we used previously.

Student Learning Objective

After studying this section, you will be able to:

① Solve applied problems involving metric and American units.

▲ **EXAMPLE 1** A triangular support piece holds a solar panel. The sketch shows the dimensions of the triangle. Find the perimeter of this triangle. Express the answer in *feet*.

Solution

Mathematics Blueprint for Problem Solving

Gather the Facts	What Am I Asked to Do?	How Do I Proceed?	Key Points to Remember
The triangle has three sides: $2\frac{1}{4}$ yd, $1\frac{1}{4}$ yd, and $2\frac{3}{4}$ yd.	Find the perimeter.	Add the lengths of the three sides. Then change the answer from yards to feet.	Be sure to change the number of yards to an improper fraction before multiplying by 3 to obtain feet.

1. *Understand the problem.* The perimeter is the sum of the lengths of the sides. Remember to convert the yards to feet.

2. *Solve and state the answer.*

   Add the three sides.

$$2\frac{1}{4} \text{ yards}$$
$$1\frac{1}{4} \text{ yards}$$
$$+\ 2\frac{3}{4} \text{ yards}$$
$$\overline{5\frac{5}{4} \text{ yards} = 6\frac{1}{4} \text{ yards}}$$

Convert $6\frac{1}{4}$ yards to feet using the fact that 1 yard = 3 feet. To make the calculation easier, we will change $6\frac{1}{4}$ to $\frac{25}{4}$.

$$6\frac{1}{4}\text{ yards} \times \frac{3 \text{ feet}}{1 \text{ yard}} = \frac{25}{4}\text{ yards} \times \frac{3 \text{ feet}}{1 \text{ yard}} = \frac{75}{4} \text{ feet} = 18\frac{3}{4} \text{ feet}$$

The perimeter of the triangle is $18\frac{3}{4}$ feet.

3. *Check.* We will check by estimating the answer.

$$2\frac{1}{4} \text{ yd} \approx 2 \text{ yd} \qquad 1\frac{1}{4} \text{ yd} \approx 1 \text{ yd} \qquad 2\frac{3}{4} \text{ yd} \approx 3 \text{ yd}$$

Now we add the three sides, using our estimated values.

$$2 + 1 + 3 = 6 \text{ yards} \qquad 6 \text{ yards} = 18 \text{ feet}$$

Solar panel
Side support
$2\frac{1}{4}$ yards
$1\frac{1}{4}$ yards
$2\frac{3}{4}$ yards
Roof support

Continued on next page

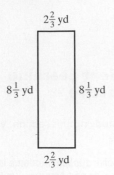

$2\frac{2}{3}$ yd

$8\frac{1}{3}$ yd $8\frac{1}{3}$ yd

$2\frac{2}{3}$ yd

Our estimated answer, 18 feet, is close to our calculated answer, $18\frac{3}{4}$ feet. Thus our answer seems reasonable. ✓

▲ **Student Practice 1** Find the perimeter of the rectangle on the left. Express the answer in *feet*.

EXAMPLE 2 How many 210-*liter* gasoline barrels can be filled from a tank of 5.04 *kiloliters* of gasoline?

Solution

1. *Understand the problem.*

Mathematics Blueprint for Problem Solving

Gather the Facts	What Am I Asked to Do?	How Do I Proceed?	Key Points to Remember
We have 5.04 kiloliters of gasoline. We are going to divide the gasoline into smaller barrels that hold 210 liters each.	Find out how many of these smaller 210-liter barrels can be filled.	We need to get all measurements in the same units. We choose to convert 5.04 kiloliters to liters. Then we divide that result by 210 to find out how many barrels can be filled.	To convert 5.04 kiloliters to liters, we move the decimal point three places to the right.

2. *Solve and state the answer.* First we convert 5.04 kiloliters to liters.

$$5.04 \text{ kiloliters} = 5040 \text{ liters}$$

Now we find out how many barrels can be filled. How many 210-liter barrels will 5040 liters fill? Visualize fitting 210-liter barrels into a big barrel that holds 5040 liters. (Use rectangular barrels.)

We need to divide

$$\frac{5040 \text{ liters}}{210 \text{ liters}} = 24.$$

Thus we can fill 24 of the 210-liter barrels.

3. *Check.* Estimate each value. First 5.04 kiloliters is approximately 5 kiloliters or 5000 liters. 210-liter barrels hold approximately 200 liters. How many times does 200 fit into 5000?

$$\frac{5000}{200} = 25$$

We estimate that 25 barrels can be filled. This is very close to our calculated value of 24 barrels. Thus our answer is reasonable. ✓

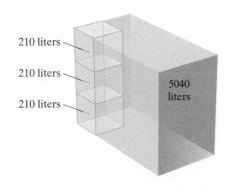

210 liters

210 liters

210 liters

5040 liters

Student Practice 2 A lab assistant must use 18.06 liters of solution to fill 42 jars. How many milliliters of the solution will go into each jar?

Applications *Solve. Round to the nearest hundredth when necessary.*

▲ **1.** Find the perimeter of the triangle. Express your answer in feet.

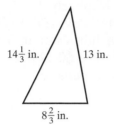

$14\frac{1}{3}$ in. 13 in.

$8\frac{2}{3}$ in.

▲ **2.** Find the perimeter of the triangle. Express your answer in feet.

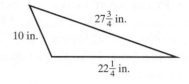

$27\frac{3}{4}$ in.

10 in.

$22\frac{1}{4}$ in.

▲ **3.** *Petting Zoo Fence* At the county fair, a triangular area is being fenced for a petting zoo. One side is 42 yards long and a second side is 65 yards long. If there are 480 feet of fencing available, how much is left for the third side? Express your answer in yards.

▲ **4.** *Farm Fencing* A farmer has 630 feet of fencing to fence in a triangular area for his pigs. One side will be 85 yards and a second side will be 70 yards. How much fencing will the farmer have left for the third side? Express your answer in yards.

▲ **5.** *Doorway Insulation* A rectangular doorway measures 90 centimeters × 200 centimeters. Weatherstripping is applied on the top and the two sides. The weatherstripping costs $6.00 per meter. What did it cost to weatherstrip the door?

▲ **6.** *Window Insulation* A rectangular picture window measures 87 centimeters × 152 centimeters. Window insulation is applied along all four sides. The insulation costs $7.00 per meter. What will it cost to insulate the window?

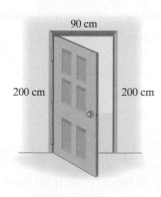

90 cm

200 cm   200 cm

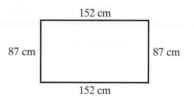

152 cm

87 cm   87 cm

152 cm

7. *Parking Space Dimensions* A stretch of road 1.863 kilometers long has 230 parking spaces of equal length painted in white on the pavement. How many meters long is each parking space?

8. *Leather Horse Equipment* A tack supply company, which makes saddles, fittings, bridles, and other equipment for horses and riders, has a length of braided leather 12.4 meters long. The leather must be cut into 4 equal pieces. How many centimeters long will each piece be?

9. ***Track and Field*** When Gary's father was in high school, he ran the 880 yd run. Gary is on the high school track team and is training for the 800 m race. Which distance is longer? By how much?

10. ***Longest Snake*** The longest snake ever captured was an Asiatic Reticulated Python that measured 33 feet. How many meters is this?

11. ***Gasoline Prices*** Don drove his car from Los Angeles, California, to Tijuana, Mexico. The price of gasoline in California is $3.20 per gallon. In Mexico, the price is $0.89 per liter. Where is gasoline more expensive?

12. ***Food Costs*** Kimberly lives in New Zealand, where bananas sell for $0.80 per kilogram. While visiting her sister in Texas, she noticed that bananas were $0.39 per pound. Where are bananas more expensive?

13. ***Temperature*** While traveling in Scotland, Jenny noticed the high temperature one day was 25°C. What was the high temperature that day in Scotland in the Fahrenheit system? Jenny called her husband in Boston that day and found the high temperature in Boston that day was 86°F. How much hotter was Boston than Scotland that day?

14. ***Temperature*** The temperature in Baghdad, Iraq, today is 39°C. The temperature on this date last year was 95°F. What is the difference in degrees Fahrenheit between the temperature in Baghdad today and the temperature one year ago?

15. ***Cooking*** A Swedish flight attendant is heating passenger meals for the new American airline he is working for. He is used to heating meals at 180°C. His co-worker tells him that the food must be heated at 350°F. What is the difference in temperature in degrees Fahrenheit between the two temperatures? Which temperature is hotter?

16. ***Temperature*** The weather page in an American newspaper listed 106°F as the temperature one day in Cairo, Egypt. This country uses the Celsius scale. How would an Egyptian report the temperature? That same day about 2000 miles north, in Helsinki, Finland, the temperature was 12°C. How much cooler was Helsinki than Cairo on that day?

17. ***Travel in Mexico*** Sharon and James Hanson traveled from Arizona to Acapulco, Mexico. The last day of their trip, they traveled 520 miles and it took them eight hours. The maximum speed limit is 110 kilometers/hour.

    (a) How many kilometers per hour did they average on the last day of the trip?

    (b) Did they break the speed limit?

18. ***Jet Travel*** A small corporate jet travels at 600 kilometers/hour for 1.5 hours. The pilot says the plane will be on time if it travels at 350 miles per hour.

    (a) What is the jet's speed in miles per hour?

    (b) Will it arrive on time?

19. *Concession Stand Sales* The concession stand at the high school sells large 1-pint sodas. On an average night, two sodas are sold each minute. How many gallons of soda are sold per hour?

20. *Leaky Faucet* On May 1, Frank placed an empty bucket under his dripping faucet. After one day, a quart of water had dripped into the bucket. If Frank doesn't fix the faucet until June 1, how many gallons of water will have dripped out?

21. *Tax on Trucks* On January 1, 2011, the state of Hawaii began taxing trucks each year at $0.05 per pound. Sam's empty truck weighs 1.8 ton. What was his annual tax in 2011?

22. *Banana Shipment* 3.4 tons of bananas are off-loaded from a ship that has just arrived from Costa Rica. The port taxes imported fruit at $0.015 per pound. What is the tax on the entire shipment?

23. *Raisin Bran* There are 708 g of cereal in a family-size box of raisin bran. The box contains 12 servings. How many ounces is one serving?

24. *Canned Fruit* A can of peaches contains 16 ounces. Fred discovered that 5 ounces were syrup and the rest was fruit. How many grams of fruit were in the can?

25. *Tree Insecticide* Carlos is mixing a fruit tree spray to retard the damage caused by beetles. He has 16 quarts of spray concentrate available. The old recipe he used in Mexico called for 11 liters of spray concentrate.

(a) How many extra quarts of spray concentrate does he have?

(b) If the spray costs $2.89 per quart, how much will it cost him to prepare the old recipe?

26. *Motor Oil Consumption* Maria bought 18 quarts of motor oil for $2.75 per quart. Her uncle from Mexico is visiting this summer. He said he will use 12 liters of the oil while driving his car this summer.

(a) How many extra quarts of oil did Maria buy? Round to the nearest quart.

(b) How much did this extra oil cost her?

27. *Fuel Efficiency* Jackson's small motorcycle gets 56 kilometers per liter. He drives 392 kilometers to Quebec City, Canada, and gas is $1.09 per liter.

(a) How much does the gas used for the trip cost?

(b) How many miles per gallon does Jackson's motorcycle get?

28. *Fuel Efficiency* Ryan's Volvo station wagon runs on diesel fuel. He gets 25 miles per gallon on the highway.

(a) If he drives 275 kilometers to Mexico City, and diesel fuel costs $4.90 per gallon, how much does the fuel used for the trip cost him? Round to the nearest cent.

(b) How many kilometers per liter does he get?

29. ***Discharge Rate at a Dam*** The flow rate of a safety discharge pipe at a dam in Lowell, Massachusetts, is rated for a maximum of 240,000 gallons per hour. The inspector asked if this flow rate could have handled the floods of 1927. During the floods of 1927 the flow rate at the dam was measured as 440 pints per second. Could the safety discharge pipe safely handle a flow rate of 440 pints per second? Why?

30. ***Water Reservoir*** The old main lines that run water from the Quabbin Reservoir to Boston have some leaks. It is estimated that the lines leak approximately 36,000 gallons per hour. A local newspaper said that the lines leak 80 pints per second. Did the newspaper have the correct information? Why?

Cumulative Review

31. **[4.4.1]** ***Map Scale*** Dori is going to India. While reading his atlas he sees that Bombay is 6 inches from where he will begin his travel. The scale shows that 3 inches represents 7.75 miles on the ground. If he travels by train on a straight track, how far must the train go to take him to his destination?

32. **[4.4.1]** ***Scale Models*** Thompson has a scale model of a famous fishing schooner. Every 2 centimeters on the model represents an actual length of 7.5 yards. The model is 11 centimeters long. How long is the famous fishing schooner?

Quick Quiz 6.5 *Round each answer to the nearest tenth if necessary.*

1. In Switzerland, John and Nancy Tobey rode the mountain train to the top of the Jungfrau. The guide said it would be 28°F at the top of the mountain. The outdoor thermometer actually read 2°C. How close was the guide's prediction to the actual temperature? Indicate your answer in Fahrenheit.

2. A rectangular box measures 4 cm by 8 cm. What is the perimeter of the box in inches?

3. The speed limit on a Mexican road is 100 km/hr. Maria is driving a distance of 45 miles. How long will it take her if she drives at the speed limit? Indicate your answer in minutes.

4. **Concept Check** Suppose you are reading a map that has a scale showing that 3 inches represents 6.5 miles on the ground. Explain how you would find out how many miles there are between two cities that are 5 inches apart on the map.

Did You Know...

That Balancing Your Checkbook Can Help You Keep Track of What You Are Spending?

BALANCING YOUR FINANCES

Understanding the Problem:

One of the first steps in saving money is to determine your current spending habits. The first step in that process is learning to balance your finances.

Terry balanced his checkbook once a month when he received his bank statement. Below is a table that records the deposits Terry made for the month of May. The beginning balance for May was $300.50.

Date	Deposit
May 1	$200.00
May 3	$150.50
May 10	$120.25
May 25	$50.00
May 28	$25.00

Keeping a Record of Checks:

Terry needs to know if he is depositing enough money to cover his monthly expenses. Below is a table that records each check Terry wrote for the month of May.

Date	Check Number	Checks
May 2	102	$238.50
May 6	103	$75.00
May 12	104	$200.00
May 28	105	$28.56
May 30	106	$36.00

Finding the Facts:

Step 1: Terry needs to know how much he deposits into the bank every month.

Task 1: *Determine how much Terry deposited in the bank in May.*

Step 2: Terry needs to know how much he spends each month.

Task 2: *Determine the total amount of the checks Terry wrote in May.*

Task 3: *Based on the given information, will Terry be able to cover all his expenses for the month of May?*

Making a Decision:

Step 3: Terry needs to know if he can continue to spend money at the same rate, or if he needs to cut back on his spending.

Task 4: *Assuming all the checks cleared for May, what would Terry's balance be at the beginning of June?*

Task 5: *If Terry continues these spending habits, what will happen?*

Applying the Situation to Your Life:

Knowing your monthly income and spending habits can help you to save. Balance your checkbook and monitor your spending habits each month, and try to cut out unnecessary expenses. If you are charged ATM fees for making withdrawals from your checking account with an ATM card be sure to subtract those costs from your checkbook balance. If you pay monthly fees to the bank for the cost of your checking account be sure to subtract those costs from your checkbook balance.

Task 6: *How often do you balance your checkbook?*

Task 7: *Do you have any unnecessary expenses that can be cut out?*

Chapter 6 Organizer

Topic and Procedure	Examples	— You Try It
Changing from one American unit to another, *p. 365* 1. Find the equality statement that relates what you want to find and what you know. 2. Form a unit fraction. The denominator will contain the units of the original measurement. 3. Multiply by the unit fraction and simplify.	**(a)** Convert 210 inches to feet. First use 12 inches = 1 foot. Then use the unit fraction $= \dfrac{1 \text{ foot}}{12 \text{ inches}}$ $210 \text{ in.} \times \dfrac{1 \text{ ft}}{12 \text{ in.}} = \dfrac{210}{12} \text{ ft}$ $= 17.5 \text{ ft}$ **(b)** Convert 86 yards to feet. First use 1 yard = 3 feet. Then use the unit fraction $= \dfrac{3 \text{ feet}}{1 \text{ yard}}$ $86 \text{ yd} \times \dfrac{3 \text{ ft}}{1 \text{ yd}} = 86 \times 3 \text{ ft} = 258 \text{ ft}$	1. **(a)** Convert 45 yards to feet. **(b)** Convert 88 ounces to pounds.
Changing from one metric unit to another, *pp. 372, 382, 383* When you change from one prefix to another by moving to the *left* in the prefix guide, move the decimal point to the *left* the same number of places. kilo- = 1000 hecto- = 100 deka- = 10 one unit = 1 deci- = 0.1 centi- = 0.01 milli- = 0.001 When you change from one prefix to another by moving to the *right* in the prefix guide, move the decimal point to the *right* the same number of places.	**(a)** Change 7.2 meters to kilometers. Move three decimal places to the left. $0.007.2$ $7.2 \text{ m} = 0.0072 \text{ km}$ **(b)** Change 196 centimeters to meters. Move two places to the left. $196.$ $196 \text{ cm} = 1.96 \text{ m}$ **(c)** Change 17.3 liters to milliliters. Move three decimal places to the right. 17.300 $17.3 \text{ L} = 17,300 \text{ mL}$	2. **(a)** Change 1500 kilometers to meters. **(b)** Change 25 centimeters to millimeters. **(c)** Change 12,500 milliliters to liters.

Changing from American units to metric units, *p. 390*
 	
 1. From the list of approximate equivalent measures, pick an equality statement that begins with the unit in the original measurement.
 	
    1 mi ≈ 1.61 km
    1 yd ≈ 0.914 m
    1 ft ≈ 0.305 m
    1 in. = 2.54 cm (exact)
    1 gal ≈ 3.79 L
    1 qt ≈ 0.946 L
    1 lb ≈ 0.454 kg
    1 oz ≈ 28.35 g
 	
 2. Multiply by a unit fraction.
	(a) Convert 7 gallons to liters.
	 Use 1 gal ≈ 3.79 L
	
	 $7 \text{ gal} \times \dfrac{3.79 \text{ L}}{1 \text{ gal}} = 26.53 \text{ L}$
	
	 (b) Convert 18 pounds to kilograms.
	 Use 1 lb ≈ 0.454 kg
	
	 $18 \text{ lb} \times \dfrac{0.454 \text{ kg}}{1 \text{ lb}} = 8.172 \text{ kg}$
		3. **(a)** Convert 6.5 yards to meters.
		
		 (b) Convert 8 miles to kilometers.

Changing from metric units to American units, *p. 390*
 	
 1. From the list of approximate equivalent measures, pick an equality statement that begins with the unit in the original measurement and ends with the unit you want.
 	
    1 km ≈ 0.62 mi
    1 m ≈ 3.28 ft
    1 m ≈ 1.09 yd
    1 cm ≈ 0.394 in.
    1 L ≈ 0.264 gal
    1 L ≈ 1.06 qt
    1 kg ≈ 2.2 lb
    1 g ≈ 0.0353 oz
 	
 2. Multiply by a unit fraction.
	(a) Convert 605 grams to ounces.
	 Use 1 g ≈ 0.0353 oz
	
	 $605 \text{ g} \times \dfrac{0.0353 \text{ oz}}{1 \text{ g}} = 21.3565 \text{ oz}$
	
	 (b) Convert 80 km/hr to mi/hr.
	 Use 1 km ≈ 0.62 mi
	
	 $80 \dfrac{\text{km}}{\text{hr}} \times \dfrac{0.62 \text{ mi}}{1 \text{ km}} = 49.6 \text{ mi/hr}$
		4. **(a)** Convert 50 meters to feet.
		
		 (b) Convert 120 km/hr to mi/hr.

Topic and Procedure	Examples	You Try It
Changing from Celsius to Fahrenheit temperature, p. 393 1. To convert Celsius to Fahrenheit, we use the formula $$F = 1.8 \times C + 32.$$ 2. We replace C by the Celsius temperature. 3. We calculate to find the Fahrenheit temperature.	Convert 65°C to Fahrenheit. Use $F = 1.8 \times C + 32$ $$F = 1.8 \times 65 + 32$$ $$F = 117 + 32 = 149$$ 65°C is 149°F.	5. Convert 40°C to Fahrenheit.
Changing from Fahrenheit to Celsius temperature, p. 393 1. To convert Fahrenheit to Celsius, we use the formula $$C = \frac{5 \times F - 160}{9}.$$ 2. We replace F by the Fahrenheit temperature. 3. We calculate to find the Celsius temperature.	Convert 50°F to Celsius. Use $C = \dfrac{5 \times F - 160}{9}$ $$C = \frac{5 \times 50 - 160}{9}$$ $$C = \frac{250 - 160}{9} = \frac{90}{9} = 10$$ 50°F is 10°C.	6. Convert 85°F to Celsius.

Chapter 6 Review Problems

Section 6.1

Convert. When necessary, express your answer as a decimal. Round to the nearest hundredth.

1. 33 ft = _____ yd

2. 5 mi = _____ yd

3. 126 in. = _____ ft

4. 2.5 mi = _____ ft

5. 14 ft = _____ in.

6. 7 tons = _____ lb

7. 8 oz = _____ lb

8. 3.5 lb = _____ oz

9. 15 gal = _____ qt

10. 31 pt = _____ qt

Section 6.2

Convert. Do not round.

11. 56 cm = _____ mm

12. 1763 mm = _____ cm

13. 13.25 m = _____ cm

14. 10,000 m = _____ km

15. 9.2 km = _____ m

16. 2400 cm = _____ m

Change all units to meters and add.

17. 6.2 m + 121 cm + 0.52 m

18. 0.024 km + 1.8 m + 983 cm

Section 6.3

Convert. Do not round.

19. 17 kL = _____ L

20. 8000 L = _____ kL

21. 196 kg = _____ g

22. 95 mg = _____ g

23. 3500 g = _____ kg

24. 15.1 g = _____ mg

25. 765 cc = _____ mL

26. 423 cm^3 = _____ mL

27. 0.256 L = _____ cm^3

Section 6.4

Perform each conversion. Round to the nearest hundredth.

28. 42 kg = _____ lb

29. 9 ft = _____ m

30. 45 mi = _____ km

31. 14 cm = _____ in.

32. 20 lb = _____ kg

33. 50 yd = _____ m

34. 80 km/hr = _____ mi/hr

35. 12°C = _____ F

36. 32°C = _____ F

37. 221°F = _____ C

38. 32°F = _____ C

39. 13 L = _____ gallons

40. 27 quarts = _____ L

Section 6.5

Solve. Round to the nearest hundredth when necessary.

41. Find the perimeter of the rectangle.
   (a) Express your answer in meters.
   (b) Express your answer in kilometers

▲ **42.** Find the perimeter of the triangle.
   (a) Express your answer in feet.
   (b) Express your answer in inches.

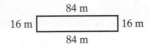

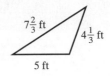

▲ **43.** *Geometry* Allison's living room measures 5 yd by 18 ft. What is the area of her living room in square feet? What is the area in square yards?

44. *Food Cost* The unit price on a box of Rice Krispies was $0.16 per ounce. The net weight was 510 grams. How much did the cereal cost?

45. *Metric Conversion* Lucia is from Mexico, where distances are measured in kilometers. While in San Diego, she rented a car and drove 70 mi/hr on the highway. In Mexico, she never drives faster than 100 km/hr. Was she driving faster than this in San Diego?

46. *Cooking* A German cook wishes to bake a cake at 185° Celsius. The oven is set at 390° Fahrenheit. By how many degrees Fahrenheit is the oven temperature different from what is desired? Is the oven too hot or not hot enough?

47. *Flagpole* A flagpole is 19 meters long. The bottom $\frac{1}{5}$ of it is coated with a special water seal before being placed in the ground. How many centimeters long is the portion that has the water seal?

48. *Horse Racing* In January 2005, a horse named Mr. Light set a record time for the mile, finishing in 91.41 seconds. What was the horse's speed in miles per hour?

49. *Gasoline Cost* When buying gas in Canada, Greg Salzman was told by the attendant that in U.S. currency he was paying $1.05 per liter. How much did the gas cost per gallon?

▲ **50.** *Geometry* The driveway of Sir Arthur Jensen in London is 4 meters wide and 12 meters long. How many square feet of sealer does he need to cover his driveway?

How Am I Doing?   Chapter 6 Test

 **MATH COACH**   **MyMathLab**®   **You Tube**™

After you take this test read through the Math Coach on pages 411 and 412. Math Coach videos are available via MyMathLab and YouTube. Step-by-step test solutions in the Chapter Test Prep Videos are also available via MyMathLab and YouTube. (Search "TobeyBasicCollMath" and click on "Channels.")

Convert. *Express your answer as a decimal rounded to the nearest hundredth when necessary.*

1. 1.6 tons = ____ lb

2. 19 ft = ____ in.

3. 21 gal = ____ qt

4. 36,960 ft = ____ mi

5. 1800 sec = ____ min

6. 3 cups = ____ qt

7. 8 oz = ____ lb

8. 5.5 yd = ____ ft

Perform each conversion. *Do not round.*

9. 9.2 km = ____ m

10. 9.88 cm = ____ m

11. 46 mm = ____ cm

12. 12.7 m = ____ cm

13. 0.936 cm = ____ mm

14. 46 L = ____ kL

15. 28.9 mg = ____ g

16. 983 g = ____ kg

17. 0.92 L = ____ mL

18. 9.42 g = ____ mg

Perform each conversion. *Round to the nearest hundredth when necessary.*

19. 42 mi = ____ km

20. 1.78 yd = ____ m

21. 9 cm = ____ in.

22. 30 km = ____ mi

23. 7.3 kg = ____ lb

24. 3 oz = ____ g

1. _____ ☐
2. _____ ☐
3. _____ ☐
4. _____ ☐
5. _____ ☐
6. _____ ☐
7. _____ ☐
8. _____ ☐
9. _____ ☐
10. _____ ☐
11. _____ ☐
12. _____ ☐
13. _____ ☐
14. _____ ☐
15. _____ ☐
16. _____ ☐
17. _____ ☐
18. _____ ☐
19. _____ ☐
20. _____ ☐
21. _____ ☐
22. _____ ☐
23. _____ ☐
24. _____ ☐

Solve. *Round to the nearest hundredth.*

25. 15 gallons = ____ L

26. 3 L = ____ quarts

▲ **27.** A rectangular picture frame measures 3 m × 7 m.

3 m

7 m   7 m

3 m

(a) What is the perimeter of the picture frame in meters?

(b) What is the perimeter of the picture frame in yards?

28. The temperature is 80°F today. Kristen's computer has a warning not to operate above 35°C.

(a) How many degrees Fahrenheit are there between the two temperatures?

(b) Can she use her computer today?

29. A pump is running at 5.5 quarts per minute. How many gallons per hour is this?

30. The speed limit on a Canadian road is 100 km/hr.

(a) How far can Samuel travel at this speed limit in three hours?

(b) If Samuel has to travel 200 miles, how much farther in miles will he need to go after three hours of driving at 100 km/hr?

31. Rick bought 1 lb 6 oz of bananas, 2 lb 2 oz of grapes, and 1 lb 12 oz of plums. How many pounds of fruit did he buy?

Ⓜ© **32.** The warmest day this year in Acapulco, Mexico, was 40°C. What was the Fahrenheit temperature?

MATH COACH

Mastering the skills you need to do well on the test.

Students often make the same types of errors when they do the Chapter 6 Test. Here are some helpful hints to keep you from making those common errors on test problems.

Convert American Units—Problem 6   3 cups = _____ qt

> **Helpful Hint** Recall that there are four cups in one quart. Make sure to round your answer to the nearest hundredth as requested.

Did you use the correct unit fraction for this problem, $\frac{1 \text{ quart}}{4 \text{ cups}}$?

Yes ____ No ____

If you answered No, stop and consider why you should use this unit fraction.

Did you multiply 3 cups $\times \dfrac{1 \text{ quart}}{4 \text{ cups}}$?

Yes ____ No ____

If you answered No, go back and perform this operation.

Now go back and rework Problem 6 using these suggestions.

Convert Metric Units—Problem 16   983 g = ____ kg

> **Helpful Hint** Remember that a kilogram is much larger than a gram so there will be fewer kilograms than grams when the conversion is made.

Did you remember that when converting from grams (smaller unit) to kilograms (larger unit), you must move the decimal point to the LEFT?

Yes ____ No ____

If you answered No, go back and perform this step.

Did you remember to move the decimal point three places?

Yes ____ No ____

If you answered No, stop and make this adjustment in your work.

Need help? Watch the MATH COACH videos in MyMathLab® or on YouTube™.

411

Convert Between Metric and American Units—Problem 22 $30 \text{ km} = \underline{\quad} \text{ mi}$

> **Helpful Hint** Recall that 1 kilometer is approximately equal to 0.62 mile.

Did you remember to choose a unit fraction with kilometers in the denominator and miles in the numerator?

Yes ____   No ____

If you answered No, stop and consider which unit fraction is best for this problem.

Did you multiply $30 \text{ km} \times \dfrac{0.62 \text{ mi}}{1 \text{ km}}$?

Yes ____   No ____

If you answered No, stop and perform this operation.

Convert Temperatures—Problem 32   The warmest day this year in Acapulco, Mexico was 40°C. What was the Fahrenheit temperature?

> **Helpful Hint** When looking for a Fahrenheit temperature, we use the formula $F = 1.8 \times C + 32$, where F is the number of Fahrenheit degrees and C is the number of Celsius degrees.

Did you choose the correct formula, $F = 1.8 \times C + 32$?

Yes ____   No ____

If you answered No, go back and think about why this is the correct formula for this problem.

Did you substitute for C correctly by writing $F = 1.8 \times 40 + 32$?

Yes ____   No ____

If you answered No, stop and make this correction to your work.

Did you perform the multiplication of 1.8×40 before adding 32?

Yes ____   No ____

If you answered No, go back and follow the correct order of operations

If you answered Problem 32 incorrectly, please rework the problem using these suggestions.

Need more help? Look for section examples marked with $\mathbb{MC}$ to review.

412

Cumulative Test for Chapters 1–6

This test provides a comprehensive review of the key objectives for Chapters 1–6.

1. Add.   12,655
$+\ 8,915$

2. Subtract.   30,075
$-\ 19,328$

3. Multiply.   4256
$\times\ \ \ 534$

4. Divide. $7\overline{)18,669}$

5. Evaluate. $2 \times (6 - 3)^3 \div 9 + 15$

6. Estimate the quotient. $27,551 \div 322$

7. Reduce. $\dfrac{24}{60}$

8. Write as a mixed number. $\dfrac{85}{10}$

9. Multiply. $2\dfrac{1}{3} \times 3\dfrac{3}{4}$

10. Divide. $\dfrac{9}{10} \div \dfrac{3}{5}$

11. Add. $4\dfrac{2}{3} + 5\dfrac{1}{2}$

12. Evaluate. $\left(\dfrac{1}{2}\right)^2 + \dfrac{7}{8} \times \dfrac{1}{3}$

13. Write 0.48 as a fraction.

14. Round to the nearest thousandth.   1.86374

15. Add. $32.15 + 8.5 + 320.42$

16. Divide. $28\overline{)9.8}$

17. Write $\dfrac{13}{20}$ as a decimal.

18. David bought 0.75 of a pound of a specialty cheese. If the cheese costs $9.80 per pound, how much did David pay?

19. Write as a unit rate.   384 points in 12 games

20. Is the statement a proportion? $\dfrac{5.5}{8} \overset{?}{=} \dfrac{22}{32}$

1. _____

2. _____

3. _____

4. _____

5. _____

6. _____

7. _____

8. _____

9. _____

10. _____

11. _____

12. _____

13. _____

14. _____

15. _____

16. _____

17. _____

18. _____

19. _____

20. _____

21. _____

22. _____

23. _____

24. _____

25. _____

26. _____

27. _____

28. _____

29. _____

30. _____

31. _____

32. _____

33. _____

34. _____

21. Solve the proportion. $\dfrac{0.4}{n} = \dfrac{2}{30}$

▲**22.** A piece of wire 6.5 centimeters long weighs 68 grams. What will a 20-centimeter length of the same wire weigh? (Round to the nearest hundredth.)

23. Write as a fraction. 85%   **24.** Write as a percent. $\dfrac{3}{40}$

25. Find 15% of 800.

26. What percent of 74 is 148?

27. 0.5% of what number is 100?

28. Cynthia is purchasing a living room set at a 15% discount. The list price for the set is $1220. How much will she pay for the living room set?

Perform each conversion.

29. 38 qt = _____ gal   **30.** 3.7 kL = _____ L

31. 5 cm = _____ m   **32.** 42 lb = _____ oz

▲**33.** Find the perimeter in meters of this triangle.

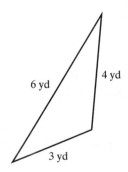

6 yd   4 yd   3 yd

34. Change 15°C to Fahrenheit temperature. Now find the difference between 15°C and 15°F. Which figure represents the higher temperature?

Tourists from all over the world come to see the Great Pyramid of Giza. The construction of this pyramid required a substantial amount of mathematics. You will learn most of these mathematical skills in this chapter.

Geometry

7.1 Angles

① Understanding and Using Angles

Geometry is a branch of mathematics that deals with the properties of and relationships between figures in space. One of the simplest figures is a *line*. A **line** extends indefinitely, but a portion of a line, called a **line segment,** has a beginning and an end. A **ray** is a part of a line that has only one endpoint and goes on forever in one direction. An **angle** is made up of two rays that start at a common endpoint. The two rays are called the **sides** of the angle. The point at which they meet is called the **vertex** of the angle.

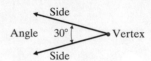

The "amount of opening" of an angle can be measured. Angles are commonly measured in **degrees.** In the preceding sketch the angle measures 30 degrees, or 30°. The symbol ° indicates degrees. If you fix one side of an angle and keep moving the other side, the angle measure will get larger and larger until eventually you have gone around in one complete revolution.

One complete revolution is 360°.

One-half revolution is 180°.

One-fourth revolution is 90°.

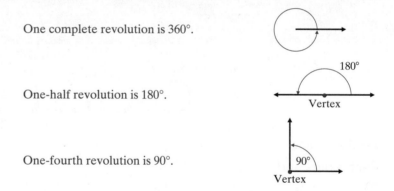

We call two lines **perpendicular** when they meet at an angle of 90°. A 90° angle is called a **right angle.** A 90° angle is often indicated by a small □ at the vertex. Thus when you see ∟ you know that the angle measures 90° and also that the sides are perpendicular to each other. The following three angles are right angles.

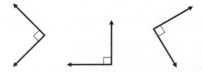

Often, to avoid confusion, angles are labeled with letters. Suppose we consider the angle with a vertex at point *B*. This angle can be called ∠*ABC*

or $\angle CBA$. Notice that when three letters are used, the middle letter is the vertex. This angle can also be called $\angle B$ or $\angle x$.

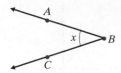

Now consider the following angles. We could label angle y as $\angle DEF$ or angle $\angle FED$. However, we could not label it as $\angle E$ because this would be unclear. If we refer to $\angle E$, people would not know for sure whether we mean $\angle DEF$, $\angle FEG$, or $\angle DEG$. In cases where there might be some confusion, the use of the three-letter label is always preferred.

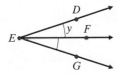

Certain types of angles are commonly encountered. It is important to learn their names. An angle that measures 180° is called a **straight angle.** Angle ABC in the following figure is a straight angle. As we mentioned previously, this is one-half of a revolution.

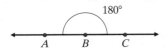

An angle whose measure is between but not including 0° and 90° is called an **acute angle.** $\angle DEF$ and $\angle GHJ$ are both acute angles.

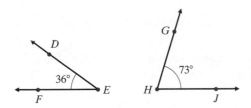

An angle whose measure is between but not including 90° and 180° is called an **obtuse angle.** $\angle ABC$ and $\angle JKL$ are both obtuse angles.

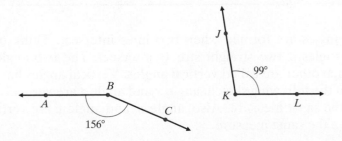

Surveyors make accurate measurements of angles so that reliable maps of land regions and buildings can be made.

EXAMPLE 1 In the following sketch, determine which angles are acute, obtuse, right, or straight angles.

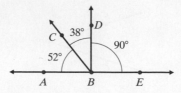

Solution $\angle ABC$ and $\angle CBD$ are acute angles, $\angle CBE$ is an obtuse angle, $\angle ABD$ and $\angle DBE$ are right angles, and $\angle ABE$ is a straight angle.

Student Practice 1 In the following sketch, determine which angles are acute, obtuse, right, or straight angles.

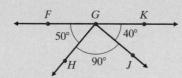

NOTE TO STUDENT: Fully worked-out solutions to all of the Student Practice problems can be found at the back of the text starting at page SP-1.

Two angles whose measures have a sum of 90° are called **complementary angles.** We can therefore say that each angle is the **complement** of the other. Two angles whose measures have a sum of 180° are called **supplementary angles.** In this case we say that each angle is the **supplement** of the other.

EXAMPLE 2 Angle A measures 39°.

(a) Find the complement of angle A.

(b) Find the supplement of angle A.

Solution

(a) Complementary angles have a sum of 90°. So the complement of angle A measures $90° - 39° = 51°$.

(b) Supplementary angles have a sum of 180°. So the supplement of angle A measures $180° - 39° = 141°$.

Student Practice 2 Angle B measures 83°.

(a) Find the complement of angle B.

(b) Find the supplement of angle B.

Four angles are formed when two lines intersect. Think of how you have four angles if two straight streets intersect. The two angles that are opposite each other are called **vertical angles.** Vertical angles have the same measure. In the following sketch, angle x and angle z are vertical angles, and they have the same measure. Also, angle w and angle y are vertical angles, so they have the same measure.

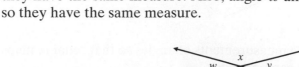

Now suppose we consider two angles that have a common side and a common vertex, such as angle w and angle x. Two angles that share a common side are called **adjacent** angles. Adjacent angles of intersecting lines are supplementary. If we know that the measure of angle x is $120°$, then we also know that the measure of angle w is $60°$.

EXAMPLE 3   In the following sketch, two lines intersect, forming four angles. The measure of angle a is $55°$. Find the measure of all the other angles.

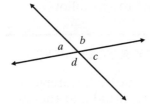

Solution   Since $\angle a$ and $\angle c$ are vertical angles, we know that they have the same measure. Thus we know that $\angle c$ measures $55°$.

Since $\angle a$ and $\angle b$ are adjacent angles of intersecting lines, we know that they are supplementary angles. Thus we know that $\angle b$ measures $180° - 55° = 125°$.

Finally, $\angle b$ and $\angle d$ are vertical angles, so we know that they have the same measure. Thus we know that $\angle d$ measures $125°$.

Student Practice 3   In the following sketch, two lines intersect, forming four angles. The measure of angle y is $133°$. Find the measure of all the other angles.

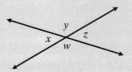

In mathematics there is a common notation for perpendicular lines. If line m is perpendicular to line n, we write $m \perp n$. **Parallel lines** never meet. If line p is parallel to line q, we write $p \parallel q$.

One more situation that is very important in geometry involves lines and angles. A line that intersects two or more lines at different points is called a **transversal**. In the following figure, line m is a transversal that intersects line n and line p. **Alternate interior angles** are two angles that are on opposite sides of the transversal and between the other two lines. In the figure, $\angle c$ and $\angle w$ are alternate interior angles.

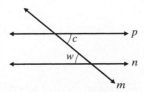

Corresponding angles are two angles that are on the same side of the transversal and are both above (or both below) the other two lines. In the following figure, angle *a* and angle *b* are corresponding angles.

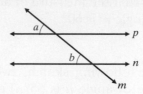

The most important case occurs when the two lines cut by the transversal are parallel. We will state this as follows:

PARALLEL LINES CUT BY A TRANSVERSAL

If two parallel lines are cut by a transversal, then the measures of **corresponding angles are equal** and the measures of **alternate interior angles** are equal.

EXAMPLE 4   In the following figure, $m \parallel n$ and the measure of $\angle a$ is 64°. Find the measures of $\angle b$, $\angle c$, $\angle d$, and $\angle e$.

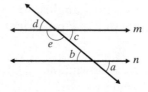

Solution

$\angle a = \angle b = 64°$. $\angle a$ and $\angle b$ are vertical angles.

$\angle b = \angle c = 64°$. $\angle b$ and $\angle c$ are alternate interior angles.

$\angle b = \angle d = 64°$. $\angle b$ and $\angle d$ are corresponding angles.

$\angle e = 180° - 64° = 116°$. $\angle e$ and $\angle d$ are adjacent angles of intersecting lines.

Student Practice 4   In the following figure, $p \parallel q$ and the measure of $\angle x$ is 105°. Find the measures of $\angle w$, $\angle y$, $\angle z$, and $\angle v$.

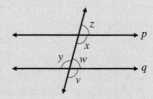

Watch the videos
in MyMathLab

Download the
MyDashBoard App

Verbal and Writing Skills, Exercises 1–8

In your own words, give a definition for each term.

1. acute angle

2. obtuse angle

3. complementary angles

4. supplementary angles

5. vertical angles

6. adjacent angles

7. transversal

8. alternate interior angles

In exercises 9–14, two straight lines intersect at B, as shown in the following sketch.

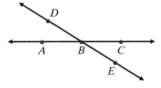

9. Name all the acute angles.

10. Name all the obtuse angles.

11. Name two pairs of angles that have the same measure.

12. Name two pairs of angles that are supplementary.

13. Name two pairs of angles that are complementary, if any exist.

14. Name a pair of vertical angles.

In exercises 15–22, find the measure of each angle, as shown in the following sketch. Assume that angle LOK is a right angle, and angle JOK is a straight angle.

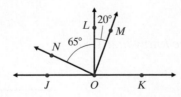

15. ∠LOJ

16. ∠NOL

17. ∠JON

18. ∠NOM

19. ∠JOM

20. ∠MOK

21. ∠NOK

22. ∠KOJ

23. Find the complement of an angle that measures 31°.

24. Find the complement of an angle that measures 86°.

25. Find the supplement of an angle that measures 127°.

26. Find the supplement of an angle that measures 23°.

Find the measure of ∠a.

27.

28.

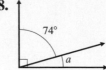

29.

30.

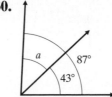

31.

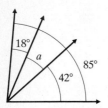

32.

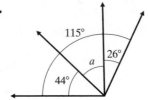

Find the measures of ∠a, ∠b, and ∠c.

33.

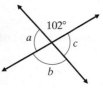

34.

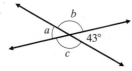

35.

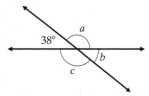

36.

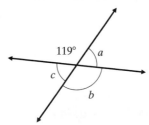

Find the measures of ∠a, ∠b, and ∠c if we know that p ∥ q.

37.

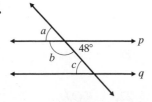

38.

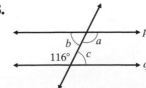

Find the measures of $\angle a$, $\angle b$, $\angle c$, $\angle d$, $\angle e$, $\angle f$, and $\angle g$ if we know that $p \parallel q$.

39.

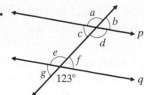

40.

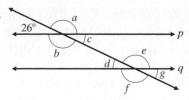

Applications

41. *Leaning Tower of Pisa* The famous Leaning Tower of Pisa used to have an angle of inclination of 84°. During restoration between 1990 and 2001, the tower was moved slightly so the angle of inclination became 86°. Find the angle x, at which the tower deviates from the normal upright position.

42. *Pyramids of Monte Albán* In Mexico the famous pyramids of Monte Albán are visited by thousands of tourists each month. The one most often climbed by tourists is steeper than the pyramids of Egypt and tourists find the climb very challenging. Find the angle x, which indicates the angle of inclination of the pyramid, based on the following sketch.

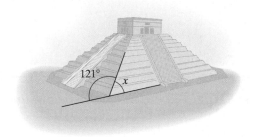

43. *Course of Jet Plane* A jetliner is flying 62° north of east when it leaves the airport in Phoenix. The control tower orders the plane to change course by turning to the right 9°. Describe the new course in terms of how many degrees north of east the plane is flying.

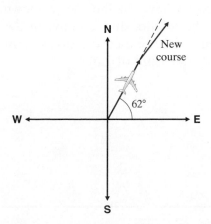

44. *Course of Cruise Ship* A cruise ship is leaving Bermuda and is heading on a course 72° north of west. The captain orders that the ship be turned to the left 9°. Describe the new course in terms of how many degrees north of west the ship is heading.

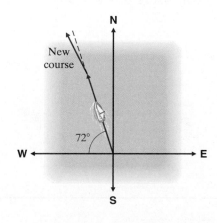

Cumulative Review

45. **[6.4.1]** *Le Tour de France*   The 2010 Tour de France bicycle race was made up of 20 stages for a total of 3642 km. Stages 9–12 were a difficult part of the race, as they took place in a mountainous region of France. The lengths of these stages were 204.5 km, 179 km, 184.5 km, and 210.5 km, respectively. What is the total distance of these four stages? How many miles is this?

46. **[6.4.1]** *Driving in Canada*   While driving in the province of Ontario, Canada, Nathan saw a sign that read TORONTO: 58 KILOMETERS AHEAD. He is driving an American car with an odometer that reads in miles. How many miles farther does he need to drive to reach Toronto? Round to the nearest tenth of a mile.

47. **[5.4.1]** *Jogging Training Program*   Seth has started a jogging program. He is following a training plan that requires him to increase his weekly mileage by no more than 5%. This week he ran 24 miles. According to the training plan, what is the most number of miles he should run next week?

48. **[5.5.2]** *Jogging Training Program*   If Seth increased his weekly jogging mileage from 24 miles per week to 27 miles per week, what would be the percent of increase?

Quick Quiz 7.1   *In the figure below, lines m and n are parallel. The measure of angle a is 124°.*

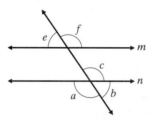

1. Find the measure of ∠c.

2. Find the measure of ∠b.

3. Find the measure of ∠e.

4. **Concept Check**   In the figure shown, explain what the relationship is between ∠e and ∠a. If you know the measure of ∠e, how can you find the measure of ∠a?

7.2 Rectangles and Squares

① Finding the Perimeters of Rectangles and Squares

Geometry has a visual aspect that numbers and abstract ideas do not have. We can take pen in hand and draw a picture of a rectangle that represents a room with certain dimensions. We can easily visualize problems such as "What is the distance around the outside edges of the room (perimeter)?" or "How much carpeting will be needed for the room (area)?"

A rectangle is a four-sided figure like those shown here.

A rectangle has two interesting properties: (1) Any two adjoining sides are perpendicular and (2) the lengths of the opposite sides of a rectangle are equal. By "any two adjoining sides are perpendicular," we mean that any two sides that meet form an angle that measures 90°. We indicate the 90° angle with a small red box □ at each corner. When we say that "the lengths of the opposite sides of a rectangle are equal," we mean that the measure of one side is equal to the measure of the side opposite to it. Thus, we define a **rectangle** as a four-sided figure that has four right angles. If all four sides have the same length, then the rectangle is called a **square.**

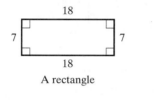

A rectangle

This rectangle is also a square.

A farmer owns some land in the Colorado mountains. It is in the shape of a rectangle. The **perimeter** of a rectangle is the sum of the lengths of all its sides. To find the perimeter of the rectangular field shown in the following figure, we add up the lengths of all the sides of the field.

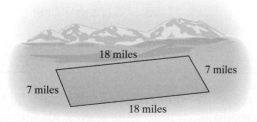

Perimeter = 7 miles + 18 miles + 7 miles + 18 miles
= 50 miles

Thus the perimeter of the field is 50 miles.

We could also use a formula to find the perimeter of a rectangle. In the formula we use letters to represent the measurements of the length and width of the rectangle. Let l represent the length, w represent the width, and P represent the perimeter. Note that the length is the longer side and

Student Learning Objectives

After studying this section, you will be able to:

① Find the perimeters of rectangles and squares.

② Find the perimeters of shapes made up of rectangles and squares.

③ Find the areas of rectangles and squares.

④ Find the areas of shapes made up of rectangles and squares.

the width is the shorter side. Since the perimeter is found by adding up the measurements all around the rectangle, we see that

$$P = w + l + w + l$$
$$= 2l + 2w.$$

When we write $2l$ and $2w$, we mean 2 times l and 2 times w. We can use the formula to find the perimeter of the rectangle.

$$P = 2l + 2w$$
$$= (2)(18 \text{ mi}) + (2)(7 \text{ mi})$$
$$= 36 \text{ mi} + 14 \text{ mi}$$
$$= 50 \text{ mi}$$

Notice that we use parentheses ( ) here to indicate multiplication of 2×18 and 2×7.

Thus the perimeter can be found quickly by using the following formula.

The **perimeter (P) of a rectangle** is twice the length (*l*) plus twice the width (*w*).

$$P = 2l + 2w$$

EXAMPLE 1   A helicopter has a 3-cm by 5.5-cm insulation pad near the control panel that is rectangular. Find the perimeter of the rectangle.

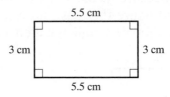

5.5 cm

3 cm   3 cm

5.5 cm

Solution   Length $= l = 5.5$ cm
                           Width $= w = 3$ cm

In the formula for the perimeter of a rectangle, we substitute 5.5 cm for *l* and 3 cm for *w*. Remember, $2l$ means 2 times *l* and $2w$ means 2 times *w*. Thus

$$P = 2l + 2w$$
$$= (2)(5.5 \text{ cm}) + (2)(3 \text{ cm})$$
$$= 11 \text{ cm} + 6 \text{ cm} = 17 \text{ cm}.$$

Student Practice 1 Find the perimeter of the rectangle in the margin.

NOTE TO STUDENT: *Fully worked-out solutions to all of the Student Practice problems can be found at the back of the text starting at page SP-1.*

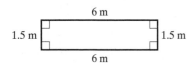

6 m

1.5 m   1.5 m

6 m

A square is a rectangle where all four sides have the same length. Since a rectangle is defined to have four right angles, all squares have four right angles. Some examples of squares are shown below.

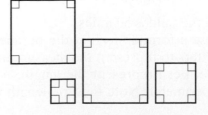

A square, then, is only a special type of rectangle. We can find the perimeter of a square just as we found the perimeter of a rectangle—by adding the measurements of all the sides of the square. Because the lengths of all sides are the same, the formula for the perimeter of a square is very simple. Let *s* represent the length of one side and *P* represent the perimeter. To find the perimeter, we multiply the length of a side by 4.

The **perimeter (P) of a square** is four times the length of a side *(s)*.

$$P = 4s$$

EXAMPLE 2 High Ridge Stables has a new sign at the highway entrance that is in the shape of a square, with each side measuring 8.6 yards. Find the perimeter of the sign.

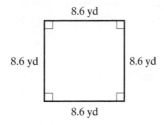

8.6 yd

8.6 yd   8.6 yd

8.6 yd

Solution   Side $= s = 8.6$ yd

$$P = 4s$$
$$= (4)(8.6 \text{ yd})$$
$$= 34.4 \text{ yd}$$

Student Practice 2 Find the perimeter of the square in the margin.

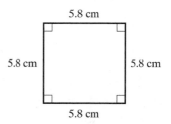

5.8 cm

5.8 cm   5.8 cm

5.8 cm

Since drawing the small red boxes sometimes makes drawings overly complicated, we will assume that all drawings in this chapter that appear to be rectangles and squares do in fact have four 90° angles.

---

② Finding the Perimeters of Shapes Made Up of Rectangles and Squares

Some figures are a combination of rectangles and squares. To find the perimeter of the total figure, look only at the outside edges.

We can apply our knowledge to everyday problems. For example, by knowing how to find the perimeter of a rectangle, we can find out how many feet of picture framing a painting will need or how many feet of weather stripping will be needed to seal a doorway. Consider the following problem.

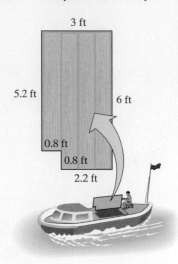

3 ft

5.2 ft

6 ft

0.8 ft

0.8 ft

2.2 ft

EXAMPLE 3 Find the cost of weather stripping needed to seal the edges of the hatch of the boat pictured at left. Weather stripping costs $0.12 per foot.

Solution First we need to find the perimeter of the hatch. The perimeter is the sum of all the edges. We use the sketch at the right.

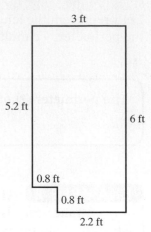

3 ft

5.2 ft

6 ft

0.8 ft

0.8 ft

2.2 ft

$$
\begin{array}{r}
3.0 \text{ ft} \\
6.0 \text{ ft} \\
2.2 \text{ ft} \\
0.8 \text{ ft} \\
0.8 \text{ ft} \\
+\ 5.2 \text{ ft} \\
\hline
18.0 \text{ ft}
\end{array}
$$

The perimeter is 18 ft. Now we calculate the cost.

$$18.0 \ \text{ft} \times \frac{0.12 \text{ dollar}}{\text{ft}} = \$2.16 \text{ for weather stripping materials}$$

Student Practice 3 Find the cost of weather stripping required to seal the edges of the hatch shown below. Weather stripping costs $0.16 per foot.

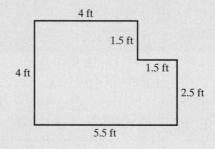

4 ft

1.5 ft

1.5 ft

4 ft

2.5 ft

5.5 ft

③ Finding the Areas of Rectangles and Squares

What do we mean by **area**? Area is the measure of the *surface inside* a geometric figure. For example, for a rectangular room, the area is the amount of floor in that room.

One *square meter* is the measure of a square that is 1 m on each side.

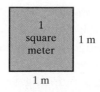

1 square meter

1 m

1 m

We can abbreviate *square meter* as m^2. In fact, all areas are measured in square meters, square feet, square inches, and so on (written as m^2, ft^2, $in.^2$, and so on).

We can calculate the area of a rectangular region if we know its length and its width. To find the area, *multiply* the length by the width.

> The **area (A) of a rectangle** is the length (*l*) times the width (*w*).
>
> $$A = lw$$

EXAMPLE 4 Find the area of the rectangle shown below.

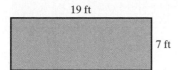

19 ft

7 ft

Solution Our answer must be in square feet because the measures of the length and width are in feet.

$$l = 19 \text{ ft} \qquad w = 7 \text{ ft}$$
$$A = (l)(w) = (19 \text{ ft})(7 \text{ ft}) = 133 \text{ ft}^2$$

The area is 133 square feet.

Student Practice 4 Find the area of the rectangle shown below.

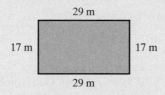

29 m

17 m   17 m

29 m

To find the area of a square, we multiply the length of one side by itself.

> The **area (A) of a square** is the square of the length of one side (*s*).
>
> $$A = s^2$$

EXAMPLE 5 A square measures 9.6 in. on each side. Find its area.

Solution We know our answer will be measured in square inches. We will write this as in.^{2}.

$$
\begin{aligned}
A &= s^2 \\
&= (9.6 \text{ in.})^2 \\
&= (9.6 \text{ in.})(9.6 \text{ in.}) \\
&= 92.16 \text{ in.}^2
\end{aligned}
$$

Student Practice 5 Find the area of a square computer chip that measures 11.8 mm on each side.

④ **Finding the Areas of Shapes Made Up of Rectangles and Squares**

EXAMPLE 6 Consider the shape shown below, which is made up of a rectangle and a square. Find the area of the shaded region.

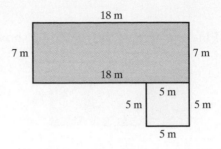

Solution The shaded region is made up of two separate regions. You can think of each separately, and calculate the area of each one. The total area is just the sum of the two separate areas.

$$\text{Area of rectangle} = (7\,\text{m})(18\,\text{m}) = 126\,\text{m}^2$$
$$\text{Area of square} = (5\,\text{m})^2 = 25\,\text{m}^2$$

$$
\begin{array}{ll}
\text{The area of the rectangle} & = 126\,\text{m}^2 \\
+\ \text{The area of the square} & = 25\,\text{m}^2 \\
\hline
\text{The total area is} & = 151\,\text{m}^2
\end{array}
$$

Student Practice 6 Find the area of the shaded region shown in the figure below.

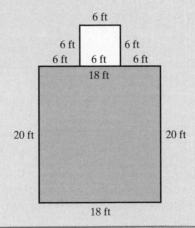

7.2 Exercises
MyMathLab®

Watch the videos
in MyMathLab

Download the
MyDashBoard App

Verbal and Writing Skills, Exercises 1–4

1. A rectangle has two properties: (1) any two adjoining sides are _____ and (2) the lengths of opposite sides are _____.

2. To find the perimeter of a figure, we _____ the lengths of all of the sides.

3. To find the area of a rectangle, we _____ the length by the width.

4. All area is measured in _____ units.

Find the perimeter of the rectangle or square.

5.

5.5 mi

2 mi [   ] 2 mi

5.5 mi

6.

9 cm

1.5 cm [   ] 1.5 cm

9 cm

7.

2.5 ft

9.3 ft

8.

11.3 ft

8.7 ft

9.

4.3 in.

4.3 in. [   ] 4.3 in.

4.3 in.

10.

15.6 ft

15.6 ft [   ] 15.6 ft

15.6 ft

11. Length = 0.84 mm, width = 0.12 mm

12. Length = 9.4 m, width = 4.3 m

13. Length = width = 4.28 km

14. Length = width = 12.35 m

15. Length = 3.2 ft, width = 48 in. (*Hint:* Make the units of length the same.)

16. Length = 10 ft, width = 42 in. (*Hint:* Make the units of length the same.)

Find the perimeter of the square. The length of the side is given.

17. 0.068 mm

18. 0.082 mm

19. $3\frac{1}{2}$ cm

20. $5\frac{3}{4}$ cm

Find the perimeter of each shape made up of rectangles and squares.

21.

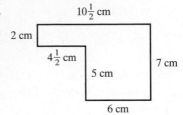

22.

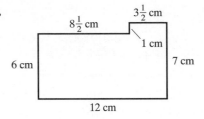

23.

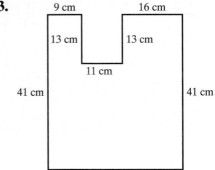

24.

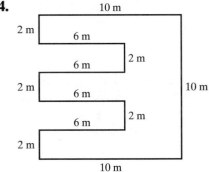

Find the area of the rectangle or square.

25. Length = width = 2.5 ft

26. Length = width = 5.1 m

27. Length = 8 mi, width = 1.5 mi

28. Length = 12.4 mi, width = 8 mi

29. Length = 39 yd, width = 9 ft (*Hint:* Make the units of length the same.)

30. Length = 57 yd, width = 15 ft (*Hint:* Make the units of length the same.)

Mixed Practice

31.

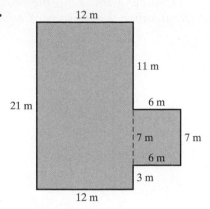

(a) Find the shaded area.

(b) Find the perimeter indicated by the black lines.

32.

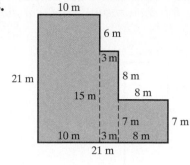

(a) Find the shaded area.

(b) Find the perimeter indicated by the black lines.

Applications *Some of the following exercises will require that you find a perimeter. Others will require that you find an area. Read each problem carefully to determine which you are to find.*

33. ***Dance Floor*** A four-star hotel is building a new dance floor in one of their ballrooms to accommodate wedding receptions and other large gatherings. The dance floor measures 40 ft × 55 ft and will be made of a special three-layered cushioned flooring that costs $15.00 per square foot. How much will the new dance floor cost?

34. ***Lacrosse Team Warm-Up*** The standard size of a boys' lacrosse field is 60 yards wide by 110 yards long. Garfield High School has a field with these dimensions. The lacrosse team is running the perimeter of their field to warm up for a game. If they run around the field 5 times, how many yards will they run? How many yards more would they need to run to make a mile?

35. ***California King Blanket*** Caroline is making a fleece blanket for her bed. Her mattress is a California king size, measuring 72 in. wide by 84 in. long. She wants the blanket to be the same length as the mattress, but the width she wants increased by 12 in.

(a) Find how many square feet the blanket will be.

(b) If Caroline sews a border on all four sides of the blanket, how many feet of border should she buy?

36. ***Scuba Shop Sign*** Sammy's Scuba Shop is installing a new sign measuring 5.4 ft × 8.1 ft. The sign will be framed in purple neon light, which will cost $32.50 per foot. How much will it cost to frame the sign in purple neon light?

37. A farmer has 16 feet of fencing. He constructs a rectangular garden whose sides are whole numbers. He uses all the fence to enclose the garden.

(a) How many possible shapes can the garden have?

(b) What is the area of each possible garden?

(c) Which shape has the largest area?

38. A farmer has 18 feet of fencing. She constructs a rectangular garden whose sides are whole numbers. She uses all the fence to enclose the garden.

(a) How many possible shapes can the garden have?

(b) What is the area of each possible garden?

(c) Which shape has the largest area?

Installation of Carpeting *A family decides to have custom carpeting installed. It will cost $14.50 per square yard. The binding, which runs along the outside edges of the carpet, will cost $1.50 per yard. Find the cost of carpeting and binding for each room. Note that dimensions are given in feet. (Remember, 1 square yard equals 9 square feet.)*

39.

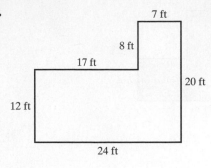

40.

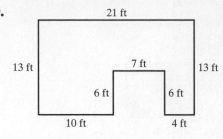

Cumulative Review

41. [3.3.1] Add.   156.8
                        27.2
                     + 39.3

42. [3.3.2] Subtract.   200.57
                          − 193.39

43. [3.4.1] Multiply.   1076
                          × 20.3

44. [3.5.2] Divide. 12.3)‾19.384‾

Quick Quiz 7.2

1. Find the perimeter of a rectangle that measures 2.3 centimeters by 1.5 centimeters.

2. Find the area of a square that measures 11 miles on each side.

3. Find the cost of carpeting a rectangular room that measures 11 feet by 16 feet if the carpeting costs $6 per square foot.

4. **Concept Check** If a rectangle that measures 12 feet by 15 feet is attached to a square that measures 15 feet on a side, explain how you would find the area of the entire region.

7.3 Parallelograms, Trapezoids, and Rhombuses

① Finding the Perimeter and Area of a Parallelogram or a Rhombus

Parallelograms, rhombuses, and trapezoids are figures related to rectangles. Actually, they are in the same "family," the **quadrilaterals** (four-sided figures). For all these figures, the perimeter is the distance around the figure. But there is a different formula for finding the area of each.

A **parallelogram** is a four-sided figure in which both pairs of opposite sides are parallel. The opposite sides of a parallelogram are equal in length.

The following figures are parallelograms. Notice that the adjoining sides need not be perpendicular as in a rectangle.

Student Learning Objectives

After studying this section, you will be able to:

① Find the perimeter and area of a parallelogram or a rhombus.

② Find the perimeter and area of a trapezoid.

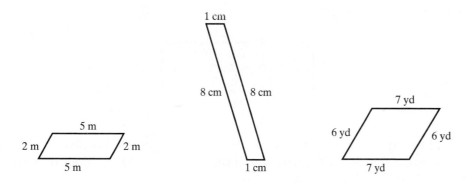

The **perimeter** of a parallelogram is the distance around the parallelogram. It is found by adding the lengths of all the sides of the figure.

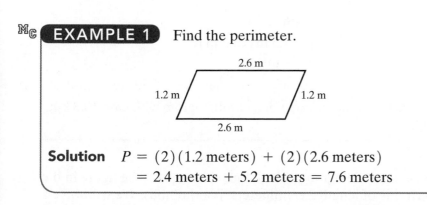

EXAMPLE 1 Find the perimeter.

2.6 m

1.2 m 1.2 m

2.6 m

Solution $P = (2)(1.2 \text{ meters}) + (2)(2.6 \text{ meters})$
$= 2.4 \text{ meters} + 5.2 \text{ meters} = 7.6 \text{ meters}$

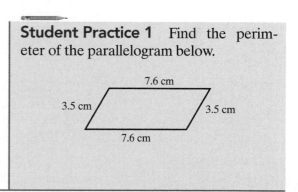

Student Practice 1 Find the perimeter of the parallelogram below.

7.6 cm

3.5 cm 3.5 cm

7.6 cm

To find the **area** of a parallelogram, we multiply the base times the height. Any side of a parallelogram can be considered the **base.** The **height** is the shortest distance between the base and the side opposite the base. The height is a line segment that is perpendicular to the base. When we write the formula for area, we use the lengths of the base (*b*) and the height (*h*).

NOTE TO STUDENT: Fully worked-out solutions to all of the Student Practice problems can be found at the back of the text starting at page SP-1.

height

base

435

> The **area (A) of a parallelogram** is the base *(b)* times the height *(h)*.
>
> $$A = bh$$

Why is the area of a parallelogram equal to the base times the height? What reasoning leads us to that formula? Suppose that we cut off the triangular region on one side of the parallelogram and move it to the other side.

We now have a rectangle.

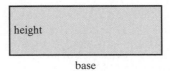

To find the area, we multiply the width by the length. In this case, $A = bh$. Thus finding the area of a parallelogram is like finding the area of a rectangle of length *b* and width *h*: $A = bh$.

EXAMPLE 2 Find the area of a parallelogram with base 7.5 m and height 3.2 m.

Solution
$$A = bh$$
$$= (7.5 \text{ m})(3.2 \text{ m})$$
$$= 24 \text{ m}^2$$

Student Practice 2 Find the area of a parallelogram with base 10.3 km and height 1.5 km.

2 cm

2 cm Rhombus 2 cm

2 cm

A **rhombus** is a parallelogram with all four sides equal. The figure in the margin, with each side of length 2 centimeters, is a rhombus. We will solve a problem involving a rhombus in Example 3.

EXAMPLE 3 A truck is manufactured with an iron brace welded to the truck frame. The brace is shaped like a rhombus. The brace has a base of 9 inches and a height of 5 inches. Find the perimeter and the area of this iron brace.

Solution Since all four sides are equal, we merely multiply

$$P = 4(9 \text{ in.}) = 36 \text{ in.}$$

The perimeter of this brace is 36 inches.

Since the rhombus is a special type of parallelogram, we can use the area formula for a parallelogram. In this case the base is 9 inches and the height is 5 inches.

$$A = bh = (9 \text{ in.})(5 \text{ in.}) = 45 \text{ in.}^2$$

Thus the area of the brace is 45 square inches.

Student Practice 3 An inlaid piece of cherry wood on the front of a hope chest is shaped like a rhombus. This piece has a base of 6 centimeters and a height of 4 centimeters. Find the perimeter and the area of this inlaid piece of cherry wood.

② Finding the Perimeter and Area of a Trapezoid

A **trapezoid** is a four-sided figure with two parallel sides. The parallel sides are called **bases.** The lengths of the bases do not have to be equal. The adjoining sides do not have to be perpendicular.

Sometimes the trapezoid is sitting on a base. Then both bases are horizontal. But be careful. Sometimes the bases are vertical. You can recognize the bases because they are the two parallel sides. This becomes important when you use the formula for finding the area of a trapezoid.

Look at the following trapezoids. See if you can recognize the bases.

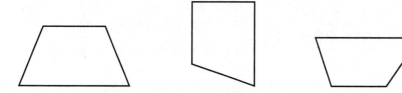

The perimeter of a trapezoid is the sum of the lengths of all of its sides.

EXAMPLE 4 Find the perimeter of the trapezoid on the right.

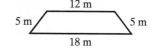

Solution

$$P = 18 \text{ m} + 5 \text{ m} + 12 \text{ m} + 5 \text{ m}$$
$$= 40 \text{ m}$$

Student Practice 4 Find the perimeter of a trapezoid with sides of 7 yd, 15 yd, 21 yd, and 13 yd.

Remember, we often use parentheses as a way to group numbers together. The numbers inside parentheses should be combined first.

$(5)(7 + 2) = (5)(9)$ First we add numbers inside the parentheses.
$\qquad\quad = 45$ Then we multiply.

The formula for the area of a trapezoid uses parentheses in this way.

The **height** of a trapezoid is the distance between the two parallel sides. The area of a trapezoid is one-half the height times the sum of the bases. (This means you add the bases *first.*)

Now this can be written $\dfrac{h}{2} \cdot (b + B)$ or $h\left(\dfrac{b + B}{2}\right)$ or $\dfrac{h(b + B)}{2}$.

Some students like to remember $h\left(\dfrac{b + B}{2}\right)$ because it is the height times the average of the bases.

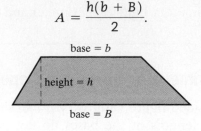

The **area (A) of a trapezoid** with a shorter base (b), a longer base (B), and height (h) is

$$A = \frac{h(b + B)}{2}.$$

base = b

height = h

base = B

EXAMPLE 5   A roadside sign is in the shape of a trapezoid. It has a height of 30 ft, and the bases are 60 ft and 75 ft.

(a) What is the area of the sign?

(b) If 1 gallon of paint covers 200 ft², how many gallons of paint will be needed to paint the sign?

Solution

(a) We use the trapezoid formula with $h = 30, b = 60$, and $B = 75$.

$$A = \frac{h(b + B)}{2}$$

$$= \frac{(30\text{ ft})(60\text{ ft} + 75\text{ ft})}{2}$$

$$= \frac{(30\text{ ft})(135\text{ ft})}{2} = \frac{4050}{2}\text{ft}^2 = 2025\text{ ft}^2$$

(b) Each gallon covers 200 ft², so we multiply the area by the fraction $\dfrac{1\text{ gal}}{200\text{ ft}^2}$. This fraction is equivalent to 1.

$$2025\ \cancel{\text{ft}^2} \times \frac{1\text{ gal}}{200\ \cancel{\text{ft}^2}} = \frac{2025}{200}\text{gal}$$
$$= 10.125\text{ gal}$$

Thus 10.125 gallons of paint would be needed. In real life we would buy 11 gallons of paint.

Student Practice 5   A corner parking lot is shaped like a trapezoid. The trapezoid has a height of 140 yd. The bases measure 180 yd and 130 yd.

(a) Find the area of the parking lot.

(b) If 1 gallon of sealant will cover 100 square yards of the parking lot, how many gallons are needed to cover the entire parking lot?

Some area problems involve two or more separate regions. Remember, areas can be added or subtracted.

EXAMPLE 6 Find the area of the following piece for inlaid wood-work made by a master carpenter. Since this shape is hard to cut, it is made of one trapezoid and one rectangle laid together.

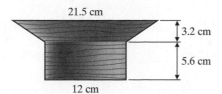

21.5 cm
3.2 cm
5.6 cm
12 cm

Solution We separate the area into two portions and find the area of each portion separately.

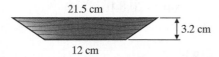

21.5 cm
3.2 cm
12 cm

The area of the trapezoid is

$$A = \frac{h(b + B)}{2}$$

$$= \frac{(3.2 \text{ cm})(12 \text{ cm} + 21.5 \text{ cm})}{2}$$

$$= \frac{(3.2 \text{ cm})(33.5 \text{ cm})}{2}$$

$$= \frac{107.2}{2} \text{ cm}^2$$

$$= 53.6 \text{ cm}^2.$$

The area of the rectangle is

$$A = lw$$

$$= (12 \text{ cm})(5.6 \text{ cm})$$

$$= 67.2 \text{ cm}^2.$$

We now add each area.

$$\begin{array}{r} 67.2 \text{ cm}^2 \\ + \ 53.6 \text{ cm}^2 \\ \hline 120.8 \text{ cm}^2 \end{array}$$

The total area of the piece for inlaid woodwork is 120.8 cm^2.

Student Practice 6 Find the area of the piece for inlaid woodwork shown in the margin. The shape is made of one trapezoid and one rectangle.

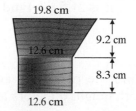

19.8 cm
9.2 cm
12.6 cm
8.3 cm
12.6 cm

Verbal and Writing Skills, Exercises 1–4

1. The perimeter of a parallelogram is found by _____ the lengths of all the sides of the figure.

2. To find the area of a parallelogram, multiply the base times the _____.

3. The height of a parallelogram is a line segment that is _____ to the base.

4. The area of a trapezoid is one-half the height times the _____ of the bases.

Find the perimeter of the parallelogram.

5. One side measures 2.8 m and a second side measures 17.3 m.

6. One side measures 15.2 ft and a second side measures 20.8 ft.

7.

15.6 in.

9.2 in.   9.2 in.

15.6 in.

8.

12.3 in.

2.6 in.   2.6 in.

12.3 in.

Find the area of the parallelogram.

9. The base is 17.6 m and the height is 20.15 m.

10. The base is 12.4 m and the height is 20.5 m.

11. ***Music Theatre Seating*** The preferred seating area at the South Shore Music Theatre is in the shape of a parallelogram. Its base is 28 yd and its height is 21.5 yd. Find the area.

12. ***Courtyard*** A courtyard is shaped like a parallelogram. Its base is 126 yd and its height is 28 yd. Find its area.

13. Find the perimeter and the area of a rhombus with height 6 meters and base 12 meters.

14. Find the perimeter and the area of a rhombus with height 9 yards and base 14 yards.

15. ***Kite*** Walter made his son Daniel a kite that was in the shape of a rhombus. The height of the kite is 1.5 feet. The length of the base of the kite is 2.4 feet. Find the perimeter and the area of the kite.

16. ***State Park*** The lawn in front of Bradley Palmer State Park is constructed in the shape of a rhombus. The height of the lawn region is 17 feet. The length of the base is 25 feet. Find the perimeter and the area of this lawn.

Find the perimeter of the trapezoid.

17.
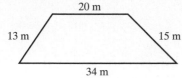
20 m
13 m   15 m
34 m

18.
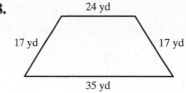
24 yd
17 yd   17 yd
35 yd

19. The two bases are 55 ft and 135 ft. The other two sides are 80.5 ft and 75.5 ft.

20. The two bases are 15 m and 23.5 m. The other two sides are 38.5 m and 45 m.

Find the area of the trapezoid.

21. The height is 12 yd and the bases are 9.6 yd and 10.2 yd.

22. The height is 15 cm and the bases are 18.3 cm and 9.8 cm.

23. *Diving in Key West* An underwater diving area for snorkelers and scuba divers located in Key West, Florida, is designated by buoys and ropes, making the diving section into the shape of a trapezoid on the surface of the water. The trapezoid has a height of 265 meters. The bases are 300 meters and 280 meters. Find the area of the designated diving area.

24. *Swimming Zone* The swimming zone at Lake Calhoun Beach is in the shape of a trapezoid. The shoreline is 40 feet long. The other base is a rope in the water that measures 30 feet long, and the height is 24 feet. What is the area of the swimming zone?

Mixed Practice

(a) Find the area of the entire shape made of trapezoids, parallelograms, squares, and rectangles.
(b) Name the object that is shaded in orange.
(c) Name the object that is shaded in yellow.

25.
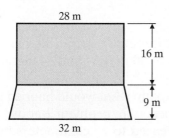
28 m
16 m
9 m
32 m

26.
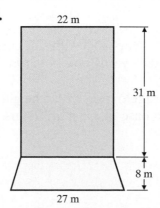
22 m
31 m
8 m
27 m

27.

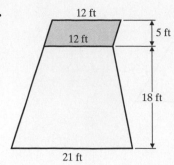

28.

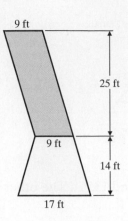

Applications

Carpeting in Conference Center Each of the following shapes represents the lobby of a conference center. The lobby will be carpeted at a cost of $22 per square yard. How much will the carpeting cost?

29.

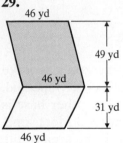

30.

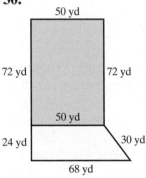

Cumulative Review   *Complete each conversion.*

31. **[6.1.2]** 40 qt = _____ gal

32. **[6.2.2]** 500 cm = _____ m

33. **[6.1.2]** 4 yd = _____ in.

34. **[6.3.2]** 8.2 kg = _____ g

Quick Quiz 7.3

1. Find the perimeter of a trapezoid with sides measuring 9 yards, 15 yards, 9 yards, and 17 yards.

2. Find the area of a trapezoid with a height of 9 meters and bases of 30 meters and 34 meters.

3. Find the area of a parallelogram with a height of 2.5 centimeters and a base of 4.8 centimeters.

4. **Concept Check** Explain what would happen to the area of the trapezoid in problem 2 above if the height was increased to 16 centimeters but the length of each base remained the same.

7.4 Triangles

① Finding the Measures of Angles in a Triangle

A **triangle** is a three-sided figure with three angles. The prefix *tri-* means "three." Some triangles are shown.

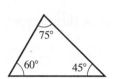 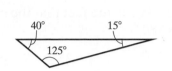

Student Learning Objectives

After studying this section, you will be able to:

① Find the measures of angles in a triangle.

② Find the perimeter and the area of a triangle.

Although all triangles have three sides, not all triangles have the same shape. The shape of a triangle depends on the sizes of the angles and the lengths of the sides.

We will begin our study of triangles by looking at the angles. Although the sizes of the angles in triangles may be different, the sum of the angle measures of any triangle is always 180°.

> The sum of the measures of the angles in a triangle is 180°.

Why is this? Perhaps you are wondering why all the angles of a triangle have measures that add up to 180°.

Remember, a straight angle is 180°.

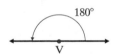

Suppose you take any triangle with ∠*A*, ∠*B*, and ∠*C*. Now cut off each corner of the triangle.

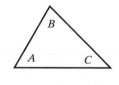

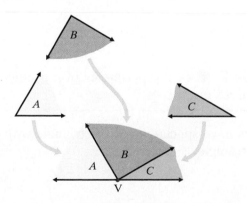

Now move the three angles so that they all have the same vertex and the middle angle shares an adjacent side with each of the other two angles. The three angles form a straight angle. Thus, the sum of the measures in a triangle is 180°.

We can use this fact to find the measure of an unknown angle in a triangle if we know the measures of the other two angles.

443

EXAMPLE 1 In the triangle below, angle A measures 35° and angle B measures 95°. Find the measure of angle C.

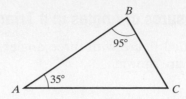

Solution We will use the fact that the sum of the measures of the angles of a triangle is 180°.

$$35 + 95 + x = 180$$
$$130 + x = 180$$

What number x when added to 130 equals 180? Since $130 + 50 = 180$, x must equal 50.

Angle C must measure 50°.

Student Practice 1 In a triangle, angle B measures 125° and angle C measures 15°. What is the measure of angle A?

NOTE TO STUDENT: Fully worked-out solutions to all of the Student Practice problems can be found at the back of the text starting at page SP-1.

② Finding the Perimeter and the Area of a Triangle

Recall that the perimeter of any figure is the sum of the lengths of its sides. Thus the perimeter of a triangle is the sum of the lengths of its three sides.

EXAMPLE 2 Find the perimeter of a triangular sail whose sides are 12 ft, 14 ft, and 17 ft.

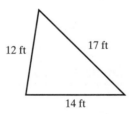

Solution $P = 12 \text{ ft} + 14 \text{ ft} + 17 \text{ ft} = 43 \text{ ft}$

Student Practice 2 Find the perimeter of a triangle whose sides are 10.5 m, 10.5 m, and 8.5 m.

Some triangles have special names. A triangle with two equal sides is called an **isosceles triangle.**

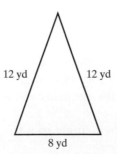

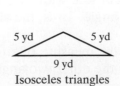

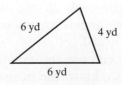

Isosceles triangles

A triangle with three equal sides is called an **equilateral triangle.** All angles in an equilateral triangle are exactly 60°.

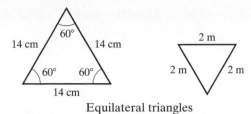

Equilateral triangles

A **scalene triangle** has no two sides of equal lengths and no two angles of equal measure.

A triangle with one 90° angle is called a **right triangle.**

The **height** of any triangle is the distance of a line drawn from a vertex perpendicular to the opposite side or an extension of the opposite side. The height may be one of the sides in a right triangle. The **base** of a triangle is perpendicular to the height.

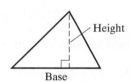
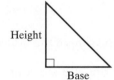
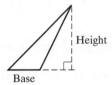

To find the area of a triangle, we need to be able to identify its height and base. The area of any triangle is half of the product of the base times the height of the triangle. The height is measured from the vertex above the base to that base.

The **area (A) of a triangle** is the base (b) times the height (h) divided by 2.

$$A = \frac{bh}{2}$$

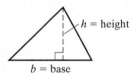

Where does the 2 come from in the formula $A = \frac{bh}{2}$? Why does this formula for the area of a triangle work? Suppose that we construct a triangle with base b and height h.

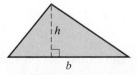

Now let us make an exact copy of the triangle and turn the copy around to the right exactly 180°. Carefully place the two triangles together. We now have a parallelogram of base b and height h. The area of a parallelogram is $A = bh$.

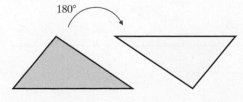
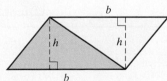

Because the parallelogram has area $A = bh$ and is made up of two triangles of identical shape and area, the area of one of the triangles is the area of the parallelogram divided by 2. Thus the area of a triangle is $A = \dfrac{bh}{2}$.

EXAMPLE 3 Find the area of the triangle.

Solution

$$A = \frac{bh}{2} = \frac{(23 \text{ m})(16 \text{ m})}{2} = \frac{368 \text{ m}^2}{2} = 184 \text{ m}^2$$

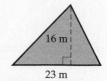

Student Practice 3 Find the area of the triangle in the margin.

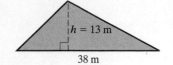

In some geometric shapes, triangles are combined with rectangles, squares, parallelograms, and trapezoids.

EXAMPLE 4 Find the area of the side of the house shown in the margin.

Solution Because the lengths of opposite sides of a rectangle are equal, the triangle has a base of 24 ft. Thus we can calculate its area.

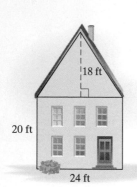

$$A = \frac{bh}{2} = \frac{(24 \text{ ft})(18 \text{ ft})}{2} = \frac{432 \text{ ft}^2}{2} = 216 \text{ ft}^2$$

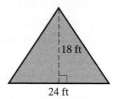

The area of the rectangle is $A = lw = (24 \text{ ft})(20 \text{ ft}) = 480 \text{ ft}^2$.

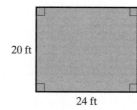

Now we find the sum of the two areas.

$$\begin{array}{r} 216 \text{ ft}^2 \\ + 480 \text{ ft}^2 \\ \hline 696 \text{ ft}^2 \end{array}$$

Thus the area of the side of the house is 696 square feet.

Student Practice 4 Find the area of the figure.

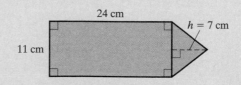

Verbal and Writing Skills, Exercises 1–6

1. A 90° angle is called a _____ angle.

2. The sum of the angle measures of a triangle is _____.

3. Explain in your own words how you would find the measure of an unknown angle in a triangle if you knew the measures of the other two angles.

4. If you were told that a triangle was an isosceles triangle, what could you conclude about the sides of that triangle?

5. If you were told that a triangle was an equilateral triangle, what could you conclude about the sides of the triangle?

6. How do you find the area of a triangle?

Write true *or* false *for each statement.*

7. Two lines that meet at a 90° angle are perpendicular.

8. A right triangle has two angles of 90°.

9. The sum of the angles of a triangle is 180°.

10. The three angles of a scalene triangle all have different measures.

11. An equilateral triangle has one angle greater than 90°.

12. The two equal sides of an isosceles triangle are always longer than the third side.

13. To find the area of a triangle, multiply its base by its height.

14. To find the perimeter of an equilateral triangle, you can multiply the length of one of the sides by 3.

Find the missing angle in the triangle.

15. Two angles are 36° and 74°.

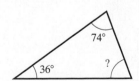

16. Two angles are 23° and 95°.

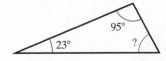

17. Two angles are 44.6° and 52.5°.

18. Two angles are 94.5° and 68.2°.

Find the perimeter of the triangle.

19. A scalene triangle whose sides are 18 m, 45 m, and 55 m

20. A scalene triangle whose sides are 27 m, 44 m, and 23 m

21. An isosceles triangle whose sides are 45.25 in., 35.75 in., and 35.75 in.

22. An isosceles triangle whose sides are 36.2 in., 47.65 in., and 47.65 in.

23. An equilateral triangle whose side measures $3\frac{1}{3}$ mi.

24. An equilateral triangle whose side measures $12\frac{2}{3}$ ft.

Find the area of the triangle.

25.

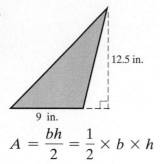

12.5 in.

9 in.

$$A = \frac{bh}{2} = \frac{1}{2} \times b \times h$$

26.

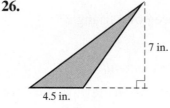

7 in.

4.5 in.

$$A = \frac{bh}{2} = \frac{1}{2} \times b \times h$$

27. The base is 17.5 cm and the height is 9.5 cm.

28. The base is 5.8 cm and the height is 10.4 cm.

29. The base is $3\frac{1}{2}$ yd and the height is $4\frac{1}{3}$ yd.

30. The base is $13\frac{1}{2}$ ft and the height is $6\frac{1}{3}$ ft.

Mixed Practice *Find the area of the shaded region.*

31.

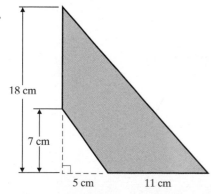

18 cm

7 cm

5 cm   11 cm

32.

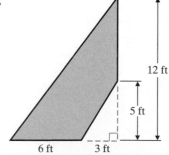

12 ft

5 ft

6 ft   3 ft

33.

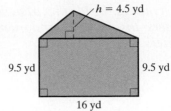

$h = 4.5$ yd

9.5 yd   9.5 yd

16 yd

34.

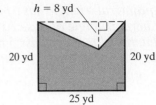

$h = 8$ yd

20 yd   20 yd

25 yd

Applications

Area of Siding on a Building *Find the total area of all four vertical sides of the building.*

35.

12 ft

15 ft

20 ft   30 ft

36.

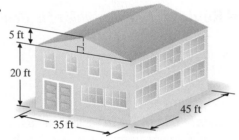

5 ft

20 ft

35 ft   45 ft

Coating on Test Plane Wings *The top surface of the wings of a test plane must be coated with a special lacquer that costs $90 per square meter. Find the cost to coat the shaded wing surface of the plane.*

37.

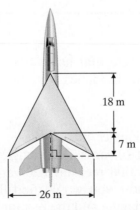

18 m

7 m

26 m

38.

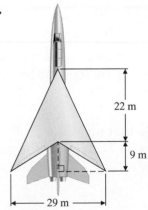

22 m

9 m

29 m

To Think About *An equilateral triangle has a base of 20 meters and a height of h meters. Inside that triangle is constructed a second equilateral triangle of base 10 meters and a height of 0.5h meters. Inside the second triangle is constructed a third equilateral triangle of base 5 meters and a height of 0.25h meters.*

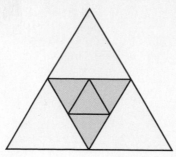

39. What percent of the area of the largest triangle is the area of the smallest triangle?

40. What percent of the perimeter of the largest triangle is the perimeter of the smallest triangle?

Cumulative Review *Find the value of n. Round to the nearest hundredth.*

41. [4.3.2] $\dfrac{5}{n} = \dfrac{7.5}{18}$

42. [4.3.2] $\dfrac{n}{\frac{3}{4}} = \dfrac{7}{\frac{1}{8}}$

43. [4.4.1] *Coastal Cleanup* In 1986, a member of The Ocean Conservancy organized a beach cleanup in Texas where 2800 volunteers collected 124 tons of trash from 122 miles of coastline. Today this is a nationwide event. How many tons of trash would you expect could be collected with 3500 volunteers? How many miles of coastline could be cleaned up? (*Source:* www.oceanconservancy.org)

44. [4.4.1] *Transatlantic Plane Flight* Recently, an airline found that after the transatlantic flight to Frankfurt, 68 people out of 300 passengers kept their in-flight magazines after being encouraged to take the magazines with them to read at their leisure. On a similar flight carrying 425 people, how many in-flight magazines would the airline expect to be taken? Round to the nearest whole number.

Quick Quiz 7.4

1. Find the perimeter of a triangle that has sides measuring 22.8 meters, 21.9 meters, and 36.7 meters.

2. Find the area of a triangle that has a base of 17 inches and a height of 12 inches.

3. A triangle has an angle that measures 75.4° and another that measures 53.7°. What is the measure of the third angle?

4. Concept Check A triangle has a base of 20 yards and a height of 20 yards. The triangle is attached to a rectangle that measures 20 yards by 15 yards. Explain how you would find the combined area of the triangle and the rectangle.

7.5 Square Roots

Student Learning Objectives

After studying this section, you will be able to:

① Evaluate the square root of a perfect square.

② Approximate the square root of a number that is not a perfect square.

① Evaluating the Square Root of a Perfect Square

We know that by using the formula $A = s^2$ we can quickly find the area of a square with a side of 3 in. We simply square 3 in. That is, $A = (3 \text{ in.})(3 \text{ in.}) = 9 \text{ in.}^2$ for an answer. Sometimes we want to ask another kind of question. If a square has an area of 64 in.^{2}, what is the length of its sides?

The answer is 8 in. Why? The skill we need to find a number when we are given the square of that number is called *finding the square root*. The square root of 64 is 8.

If a number is a product of two identical factors, then either factor is called a **square root.**

> The square root of 64 is 8 because $(8)(8) = 64$.
> The square root of 9 is 3 because $(3)(3) = 9$.

The symbol for finding the square root of a number is $\sqrt{}$. To write the square root of 64, we write $\sqrt{64} = 8$. Sometimes we speak of finding the square root of a number as *taking* the square root of the number, or we can say that we will *evaluate* the square root of the number. Thus to take the square root of 9, we write $\sqrt{9} = 3$; to evaluate the square root of 9, we write $\sqrt{9} = 3$.

EXAMPLE 1 Find.

(a) $\sqrt{25}$ **(b)** $\sqrt{121}$

Solution

(a) $\sqrt{25} = 5$ because $(5)(5) = 25$.
(b) $\sqrt{121} = 11$ because $(11)(11) = 121$.

Student Practice 1 Find.

(a) $\sqrt{49}$ **(b)** $\sqrt{169}$

NOTE TO STUDENT: Fully worked-out solutions to all of the Student Practice problems can be found at the back of the text starting at page SP-1.

If square roots are added or subtracted, they must be evaluated *first*, then added or subtracted.

EXAMPLE 2 Find. $\sqrt{25} + \sqrt{36}$

Solution $\sqrt{25} = 5$ because $(5)(5) = 25$.
                       $\sqrt{36} = 6$ because $(6)(6) = 36$.

Thus $\sqrt{25} + \sqrt{36} = 5 + 6 = 11$.

Student Practice 2 Find. $\sqrt{49} - \sqrt{4}$

451

When a whole number is multiplied by itself, the number that is obtained is called a **perfect square.**

$$36 \text{ is a perfect square because } (6)(6) = 36.$$
$$49 \text{ is a perfect square because } (7)(7) = 49.$$

The numbers 20 or 48 are *not* perfect squares. There is no *whole number* that when squared—multiplied by itself—yields 20 or 48. Consider 20. $4^2 = 16$, which is less than 20. $5^2 = 25$, which is more than 20. We realize, then, that the square root of 20 is between 4 and 5 because 20 is between 16 and 25. Since there is no whole number between 4 and 5, no whole number squared equals 20. Since the square root of a perfect square is a whole number, we can say that 20 is *not* a perfect square.

It is helpful to know the first 15 perfect squares. Take a minute to complete the following table.

Number, n	1	2	3	4	5	6	7	8	9	10	11	12	13	14	15
Number Squared, n^2	1	4	9	16										196	225

EXAMPLE 3

(a) Is 81 a perfect square?   **(b)** If so, find $\sqrt{81}$.

Solution

(a) Yes. 81 is a perfect square because $(9)(9) = 81$.
(b) $\sqrt{81} = 9$

Student Practice 3

(a) Is 144 a perfect square?   **(b)** If so, find $\sqrt{144}$.

② Approximating the Square Root of a Number That Is Not a Perfect Square

If a number is not a perfect square, we can only approximate its square root. This can be done by using a square root table such as the one that follows. Except for exact values such as $\sqrt{4} = 2.000$, all values are rounded to the nearest thousandth.

Number, n	Square Root of the Number, $\sqrt{n}$	Number, n	Square Root of the Number, $\sqrt{n}$
1	1.000	8	2.828
2	1.414	9	3.000
3	1.732	10	3.162
4	2.000	11	3.317
5	2.236	12	3.464
6	2.449	13	3.606
7	2.646	14	3.742

A square root table is located on page A-15. It gives you the square root of whole numbers up to 200. Square roots can also be found with any calculator that has a square root key. Usually the key looks like this $\boxed{\sqrt{\ }}$ or this $\boxed{\sqrt{x}}$. To find the square root of 8 on most calculators, enter the number 8 and press $\boxed{\sqrt{\ }}$ or $\boxed{\sqrt{x}}$. You will see displayed 2.8284271. On some calculators, you must enter the square root key first followed by the number. (Your calculator may display fewer or more digits.) Remember, no matter how many digits your calculator displays, when we find $\sqrt{8}$, we have only an **approximation.** It is not an exact answer. To emphasize this we use the $\approx$ notation to mean "is approximately equal to." Thus $\sqrt{8} \approx 2.828$.

EXAMPLE 4   Find approximate values using the square root table or a calculator. Round to the nearest thousandth.

(a) $\sqrt{2}$ **(b)** $\sqrt{12}$ **(c)** $\sqrt{7}$

Solution

(a) $\sqrt{2} \approx 1.414$ **(b)** $\sqrt{12} \approx 3.464$ **(c)** $\sqrt{7} \approx 2.646$

Student Practice 4   Approximate to the nearest thousandth.

(a) $\sqrt{3}$ **(b)** $\sqrt{13}$ **(c)** $\sqrt{5}$

Calculator

 Square Roots

Locate the square root key $\boxed{\sqrt{\ }}$ on your calculator.

1. To find $\sqrt{289}$, enter

$$289 \; \boxed{\sqrt{\ }}$$

The display should read

$$\boxed{17}$$

2. To find $\sqrt{194}$, enter

$$194 \; \boxed{\sqrt{\ }}$$

The display should read

$$\boxed{13.928388}$$

This is just an approximation of the actual square root. We will round the answer to the nearest thousandth.

$$\sqrt{194} \approx 13.928$$

Your calculator may require you to enter the square root key first and then the number.

EXAMPLE 5   Approximate to the nearest thousandth of an inch the length of the side of a square that has an area of 6 in.^{2}.

Solution $\qquad\qquad\qquad \sqrt{6 \text{ in.}^2} \approx 2.449 \text{ in.}$

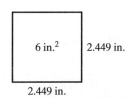

Thus, to the nearest thousandth of an inch, the side measures 2.449 in.

Student Practice 5   Approximate to the nearest thousandth of a meter the length of the side of a square that has an area of 22 m^2.

7.5 Exercises

MyMathLab®

Watch the videos
in MyMathLab

Download the
MyDashBoard App

Verbal and Writing Skills, Exercises 1–6

1. Why is $\sqrt{25} = 5$?

2. $\sqrt{49}$ is read "the _____ of 49."

3. 25 is a perfect square because its square root, 5, is a _____ number.

4. Is 32 a perfect square? Why or why not?

5. How can you approximate the square root of a number that is not a perfect square?

6. How would you find $\sqrt{0.04}$?

Find each square root. Do not use a calculator. Do not refer to a table of square roots.

7. $\sqrt{9}$ **8.** $\sqrt{16}$ **9.** $\sqrt{64}$ **10.** $\sqrt{81}$

11. $\sqrt{144}$ **12.** $\sqrt{196}$ **13.** $\sqrt{0}$ **14.** $\sqrt{225}$

15. $\sqrt{169}$ **16.** $\sqrt{121}$ **17.** $\sqrt{100}$ **18.** $\sqrt{324}$

In exercises 19–28, evaluate the square roots first, then add, subtract, or multiply the results. Do not use a calculator or a square root table.

19. $\sqrt{49} + \sqrt{9}$ **20.** $\sqrt{25} + \sqrt{64}$ **21.** $\sqrt{81} + \sqrt{1}$

22. $\sqrt{0} + \sqrt{121}$ **23.** $\sqrt{225} - \sqrt{144}$ **24.** $\sqrt{169} - \sqrt{64}$

25. $\sqrt{169} - \sqrt{121} + \sqrt{36}$ **26.** $\sqrt{196} + \sqrt{36} - \sqrt{16}$

27. $\sqrt{4} \times \sqrt{121}$ **28.** $\sqrt{225} \times \sqrt{9}$

29. (a) Is 256 a perfect square?   **30. (a)** Is 289 a perfect square?
    (b) If so, find $\sqrt{256}$.   **(b)** If so, find $\sqrt{289}$.

Use a table of square roots or a calculator with a square root key to approximate to the nearest thousandth.

31. $\sqrt{18}$ **32.** $\sqrt{45}$ **33.** $\sqrt{76}$

34. $\sqrt{90}$ **35.** $\sqrt{200}$ **36.** $\sqrt{186}$

Find the length of the side of the square. If the area is not a perfect square, approximate by using a square root table or a calculator with a square root key. Round to the nearest thousandth.

37. A square with area 34 m^2 **38.** A square with area 62 m^2

39. A square with area 136 m^2 **40.** A square with area 200 m^2

Mixed Practice *Evaluate the square roots first. Then combine the results. Use a calculator or square root table when needed. Round to the nearest thousandth.*

41. $\sqrt{36} + \sqrt{20}$ **42.** $\sqrt{40} + \sqrt{100}$ **43.** $\sqrt{198} - \sqrt{49}$ **44.** $\sqrt{145} - \sqrt{64}$

Applications

Basketball Court *High school basketball is played on a standard rectangular court that measures 92 feet in length and 50 feet in width. Some middle schools have smaller basketball courts that measure 80 feet in length and 42 feet in width.*

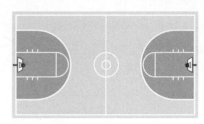

45. The diagonal of a standard high school basketball court measures $\sqrt{10,964}$ feet in length. Find the length of this diagonal to the nearest tenth of a foot.

46. The diagonal of the smaller basketball court found in some middle schools measures $\sqrt{8164}$ feet in length. Find the length of this diagonal to the nearest tenth of a foot.

47. *Baseball* The distance from second base to home plate on a professional baseball field is $\sqrt{16,200}$ ft. Find this distance to the nearest tenth of a foot.

48. *Television Measurements* Television screens are measured diagonally. The screen of the Sony WEGA Rear-Projection SXRD HDTV measures 33 in. high, 50 in. wide and has a diagonal measurement of about $\sqrt{3589}$ in. Find the length of this diagonal to the nearest whole inch.

Using a calculator with a square root key, evaluate and round to the nearest thousandth.

 49. $\sqrt{456} + \sqrt{322}$

50. $\sqrt{578} + \sqrt{984}$

Cumulative Review

51. [7.2.3] *Australia Zoo* The Taronga Zoo in Sydney, Australia, has a viewing tank for its platypuses. The tank is 60 in. high and 80 in. wide. What is the area of the front of the rectangular tank?

52. [6.4.1] *Somersaulting World Record* Ashrita Furman of New York holds over 100 world records, one of which is somersaulting for a distance of 19.68 kilometers. How many miles is this? Round to the nearest tenth.

53. [6.4.1] *30-km Race* North Shore Community College is hosting a 30-km running race. How many miles is the race?

54. [6.4.1] *South American Beetle* The world's largest beetle, Titanus giganteus, is found in South America and can reach a length of 17 cm. How many inches is this?

Quick Quiz 7.5 *Evaluate the following without the use of a calculator or a table of square roots.*

1. $\sqrt{64}$

2. $\sqrt{36} + \sqrt{144}$

3. Find the length of the side of a square with area 196 ft^2.

4. Concept Check Explain how you would find the length of the side of a tiny square with an area of 0.81 cm^2 without using a calculator.

How Am I Doing? Sections 7.1–7.5

How are you doing with your homework assignments in Sections 7.1 to 7.5? Do you feel you have mastered the material so far? Do you understand the concepts you have covered? Before you go further in the textbook, take some time to do each of the following problems.

7.1

1. Find the complement of an angle that is 72°.

2. Find the supplement of an angle that is 63°.

3. Find the measure of angle *a*, angle *b*, and angle *c* in the sketch to the right, which shows two intersecting straight lines.

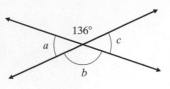

7.2

Find the perimeter of each rectangle or square.

4. Length = 6.5 m, width = 2.5 m

5. Length = width = 3.5 m

Find the area of each square or rectangle.

6. Length = width = 4.8 cm

7. Length = 5.8 yd, width = 3.9 yd

7.3

Find the perimeter.

8. A parallelogram with one side measuring 9.2 yd and another side measuring 3.6 yd.

9. A trapezoid with sides measuring 17 ft, 15 ft, $25\frac{1}{2}$ ft, and $21\frac{1}{2}$ ft.

Find the area.

10. A parallelogram with a base of 27 in. and a height of 13 in.

11. A trapezoid with a height of 9 in. and bases of 16 in. and 22 in.

12.

7 m

9 m

7 m

4 m

10 m

7.4

13. A triangle has two angles measuring 22.5° and 54.5°. Find the measure of the third angle.

14. Find the perimeter of a triangle whose sides measure $9\frac{1}{2}$ in., 4 in., and $6\frac{1}{2}$ in.

15. Find the area of a triangle with a base of 21 m and a height of 10 m.

Applications of 7.1 to 7.4

16. A college entrance has a sign shaped like this figure.

   (a) How many square feet of paint are needed to cover the sign?

   (b) How many feet of trim are needed to cover the edge (perimeter) of the sign?

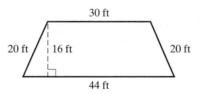

7.5

Evaluate exactly.

17. $\sqrt{64}$

18. $\sqrt{225} + \sqrt{16}$

19. $\sqrt{169}$

20. $\sqrt{256}$

21. Approximate $\sqrt{46}$ using a square root table or a calculator with a square root key. Round to the nearest thousandth.

Now turn to page SA-13 for the answer to each of these problems. Each answer also includes a reference to the objective in which the problem is first taught. If you missed any of these problems, you should stop and review the Examples and Student Practice problems in the referenced objective. A little review now will help you master the material in the upcoming sections of the text.

13. _____

14. _____

15. _____

16. (a) _____

   (b) _____

17. _____

18. _____

19. _____

20. _____

21. _____

7.6 The Pythagorean Theorem

Student Learning Objectives

After studying this section, you will be able to:

① Find the hypotenuse of a right triangle given the length of each leg.

② Find the length of a leg of a right triangle given the lengths of the hypotenuse and the other leg.

③ Solve applied problems using the Pythagorean Theorem.

④ Solve for the missing sides of special right triangles.

① Finding the Hypotenuse of a Right Triangle Given the Length of Each Leg

The Pythagorean Theorem is a mathematical idea formulated long ago. It is as useful today as it was when it was discovered. The Pythagoreans lived in Italy about 2500 years ago. They studied various mathematical properties. They discovered that for any right triangle, the square of the hypotenuse equals the sum of the squares of the two legs of the triangle. This relationship is known as the **Pythagorean Theorem.** The side opposite the right angle is called the **hypotenuse;** the other two sides are called the legs of the right triangle.

The Pythagoreans discovered that this theorem could be used to find the length of the third side of any right triangle if the lengths of two of the sides were known. We still use this theorem today for the very same reason.

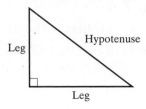

$$(\text{hypotenuse})^2 = (\text{leg})^2 + (\text{leg})^2$$

Note how the Pythagorean Theorem applies to the right triangle shown here.

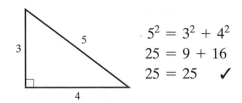

$$5^2 = 3^2 + 4^2$$
$$25 = 9 + 16$$
$$25 = 25 \quad ✓$$

In a right triangle, the hypotenuse is the longest side. It is always opposite the largest angle, the right angle. The legs are the two shorter sides. When we know each leg of a right triangle, we use the following property.

$$\text{Hypotenuse} = \sqrt{(\text{leg})^2 + (\text{leg})^2}$$

EXAMPLE 1 Find the hypotenuse of a right triangle with legs of 5 in. and 12 in.

Solution
$$\begin{aligned}\text{Hypotenuse} &= \sqrt{(5)^2 + (12)^2} \\ &= \sqrt{25 + 144} \qquad \text{Square each value first.} \\ &= \sqrt{169} \qquad \text{Add together the two values.} \\ &= 13 \text{ in.} \qquad \text{Take the square root.}\end{aligned}$$

Student Practice 1 Find the hypotenuse of a right triangle with legs of 8 m and 6 m.

NOTE TO STUDENT: *Fully worked-out solutions to all of the Student Practice problems can be found at the back of the text starting at page SP-1.*

Sometimes we cannot find the hypotenuse exactly. In those cases, we often approximate the square root by using a calculator or a square root table.

Mc **EXAMPLE 2** Find the hypotenuse of a right triangle with legs of 4 m and 5 m. (See figure in margin below.) Round to the nearest thousandth.

Solution

$$\text{Hypotenuse} = \sqrt{(4)^2 + (5)^2}$$
$$= \sqrt{16 + 25} \qquad \text{Square each value first.}$$
$$= \sqrt{41} \text{ m} \qquad \text{Add the two values together.}$$

Using the square root table or a calculator, we have the hypotenuse ≈ 6.403 m.

Student Practice 2 Find the hypotenuse of a right triangle with legs of 3 cm and 7 cm. Round to the nearest thousandth.

② **Finding the Length of a Leg of a Right Triangle Given the Lengths of the Hypotenuse and the Other Leg**

When we know the hypotenuse and one leg of a right triangle, we find the length of the other leg by using the following property.

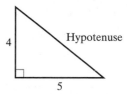

$$\text{Leg} = \sqrt{(\text{hypotenuse})^2 - (\text{leg})^2}$$

EXAMPLE 3 A right triangle has a hypotenuse of 15 cm and a leg of 12 cm. Find the length of the other leg.

Solution
$$\text{Leg} = \sqrt{(15)^2 - (12)^2}$$
$$= \sqrt{225 - 144} \qquad \text{Square each value first.}$$
$$= \sqrt{81} \qquad \text{Subtract.}$$
$$= 9 \text{ cm} \qquad \text{Find the square root.}$$

Student Practice 3 A right triangle has a hypotenuse of 17 m and a leg of 15 m. Find the length of the other leg.

EXAMPLE 4 A sail for a sailboat is in the shape of a right triangle. The right triangle has a hypotenuse of 14 feet and a leg of 8 feet. Find the length of the other leg. Round to the nearest thousandth.

Solution
$$\text{Leg} = \sqrt{(14)^2 - (8)^2}$$
$$= \sqrt{196 - 64} \qquad \text{Square each value first.}$$
$$= \sqrt{132} \text{ feet} \qquad \text{Subtract the two numbers.}$$

Using a calculator or a square root table, we can see that the leg ≈ 11.489 feet.

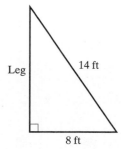

Student Practice 4 A right triangle has a hypotenuse of 10 m and a leg of 5 m. Find the length of the other leg. Round to the nearest thousandth.

③ Solving Applied Problems Using the Pythagorean Theorem

Certain applied problems call for the use of the Pythagorean Theorem in the solution.

EXAMPLE 5   A pilot flies 13 mi east from Pennsville to Salem. She then flies 5 mi south from Salem to Elmer. What is the straight-line distance from Pennsville to Elmer? Round to the nearest tenth of a mile.

Solution

1. *Understand the problem.*
   It might help to draw a picture.

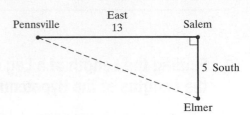

The distance from Pennsville to Elmer is the hypotenuse of the triangle.

2. *Solve and state the answer.*

$$\text{Hypotenuse} = \sqrt{(\text{leg})^2 + (\text{leg})^2}$$
$$= \sqrt{(13)^2 + (5)^2}$$
$$= \sqrt{169 + 25}$$
$$= \sqrt{194}$$
$$\sqrt{194} \approx 13.928$$

Rounded to the nearest tenth, the distance is 13.9 mi.

3. *Check.* Work backward to check. Use the Pythagorean Theorem.

$$13.9^2 \overset{?}{\approx} 13^2 + 5^2 \quad \text{(We use} \approx \text{because 13.9 is an approximate answer.)}$$
$$193.21 \overset{?}{\approx} 169 + 25$$
$$193.21 \approx 194 \quad \checkmark$$

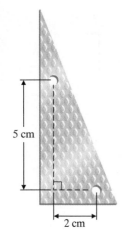

5 cm

2 cm

Student Practice 5   Find the distance to the nearest thousandth between the centers of the holes in the triangular metal plate in the margin.

EXAMPLE 6   A 25-ft ladder is placed against a building at a point 22 ft from the ground. What is the distance of the base of the ladder from the building? Round to the nearest tenth.

Solution

1. *Understand the problem.* Draw a picture.

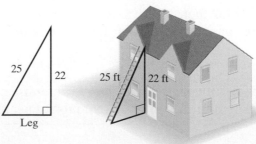

2. Solve and state the answer.

$$\text{Leg} = \sqrt{(\text{hypotenuse})^2 - (\text{leg})^2}$$
$$= \sqrt{(25)^2 - (22)^2}$$
$$= \sqrt{625 - 484}$$
$$= \sqrt{141}$$
$$\sqrt{141} \approx 11.874$$

If we round to the nearest tenth, the base of the ladder is 11.9 ft from the building.

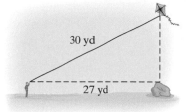

Student Practice 6   A kite is out on 30 yd of string. The kite is directly above a rock. The rock is 27 yd from the boy flying the kite. How far above the rock is the kite? Round to the nearest tenth.

④ Solving for the Missing Sides of Special Right Triangles

If we use the Pythagorean Theorem and some other facts from geometry, we can find a relationship among the sides of two special right triangles. The first special right triangle is one that contains an angle that measures 30° and one that measures 60°. We call this a 30°–60°–90° right triangle.

> In a 30°–60°–90° triangle, the length of the leg opposite the 30° angle is $\frac{1}{2}$ the length of the hypotenuse.

Notice that the hypotenuse of the first triangle is 10 m and the side opposite the 30° angle is exactly $\frac{1}{2}$ of that, or 5 m. The second triangle has a hypotenuse of 15 yd. The side opposite the 30° angle is exactly $\frac{1}{2}$ of that, or 7.5 yd.

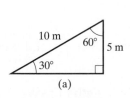
(a)

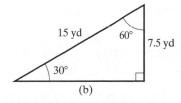
(b)

The second special right triangle is one that contains exactly two angles that each measure 45°. We call this the 45°–45°–90° right triangle.

> In a 45°–45°–90° triangle, the lengths of the sides opposite the 45° angles are equal. The length of the hypotenuse is equal to $\sqrt{2} \times$ the length of either leg.

We will use the decimal approximation $\sqrt{2} \approx 1.414$ with this property.

$$\text{Hypotenuse} = \sqrt{2} \times 7$$
$$\approx 1.414 \times 7$$
$$\approx 9.898 \text{ cm}$$

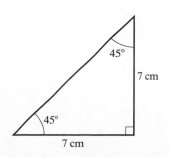

> **EXAMPLE 7** Find the requested sides of each special triangle. Round to the nearest tenth.
>
> **(a)** Find the lengths of sides y and x.   **(b)** Find the length of hypotenuse z.

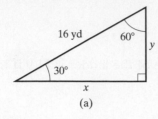

(a)

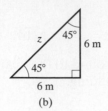

(b)

Solution

(a) In a $30°$–$60°$–$90°$ triangle, the side opposite the $30°$ angle is $\frac{1}{2}$ of the hypotenuse.

$$\frac{1}{2} \times 16 = 8$$

Therefore, $y = 8$ yd.

When we know two sides of a right triangle, we find the third side using the Pythagorean Theorem.

$$\begin{aligned} \text{Leg} &= \sqrt{(\text{hypotenuse})^2 - (\text{leg})^2} \\ &= \sqrt{16^2 - 8^2} = \sqrt{256 - 64} \\ &= \sqrt{192} \approx 13.856 \end{aligned}$$

Thus $x = 13.9$ yd rounded to the nearest tenth.

(b) In a $45°$–$45°$–$90°$ triangle, we have the following.

$$\begin{aligned} \text{Hypotenuse} &= \sqrt{2} \times \text{leg} \\ &\approx 1.414(6) \\ &= 8.484 \end{aligned}$$

Rounded to the nearest tenth, the hypotenuse $= 8.5$ m.

Student Practice 7 Find the requested sides of each special triangle. Round to the nearest tenth.

(a) Find the lengths of sides y and x.   **(b)** Find the length of hypotenuse z.

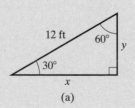

(a)

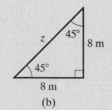

(b)

Verbal and Writing Skills, Exercises 1 and 2

1. Explain in your own words how to obtain the length of the hypotenuse of a right triangle if you know the length of each of the legs of the triangle.

2. Explain in your own words how to obtain the length of one leg of a right triangle if you know the length of the hypotenuse and the length of the other leg.

Find the unknown side of the right triangle. Use a calculator or square root table when necessary and round to the nearest thousandth.

3.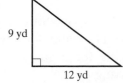

9 yd

12 yd

4.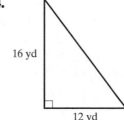

16 yd

12 yd

5.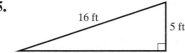

16 ft

5 ft

6.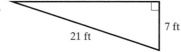

21 ft

7 ft

Find the unknown side of the right triangle using the information given. Round to the nearest thousandth.

7. leg = 11 m, leg = 3 m

8. leg = 9 m, leg = 6 m

9. leg = 10 m, leg = 10 m

10. leg = 7 m, leg = 7 m

11. hypotenuse = 11 ft, leg = 4 ft

12. hypotenuse = 15 yd, leg = 12 yd

13. hypotenuse = 14 yd, leg = 10 yd

14. hypotenuse = 25 ft, leg = 18 ft

15. leg = 12 m, leg = 9 m

16. leg = 6 m, leg = 8 m

17. hypotenuse = 16 ft, leg = 11 ft

18. hypotenuse = 21 m, leg = 6 m

Applications, Exercises 19–24 *Solve. Round to the nearest tenth.*

19. *Loading Ramp* Find the length of the ramp to the back of a truck.

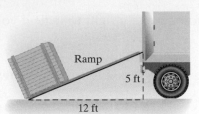

20. *Telephone Pole* Find the length of the guy wire supporting the telephone pole.

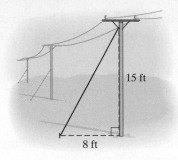

21. *Construction of a Steel Plate* A construction project requires a stainless steel plate with holes drilled as shown. Find the distance between the centers of the holes in this triangular plate.

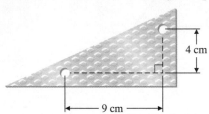

22. *Running Out of Gas* Juan's car runs out of gas in Los Lunas, New Mexico. He walks 4 mi west and then 3 mi south looking for a gas station. How far is he from his starting point?

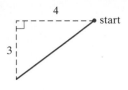

23. *Kite Flying* Barbara is flying her dragon kite on 32 yd of string. The kite is directly above the edge of a pond. The edge of the pond is 30 yd from where the kite is tied to the ground. How far is the kite above the pond?

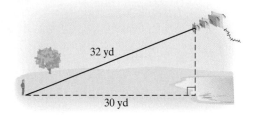

24. *Ladder Distance* A 20-ft ladder is placed against a college classroom building at a point 18 ft above the ground. What is the distance from the base of the ladder to the building?

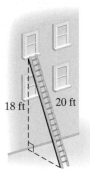

Using your knowledge of special right triangles, find the length of each leg. Round to the nearest tenth.

25.

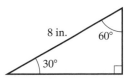

26.

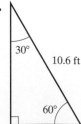

Using your knowledge of special right triangles, find the length of the hypotenuse. Round to the nearest tenth.

27.

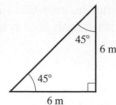

45°
6 m

45°
6 m

28.

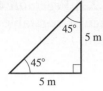

45°
5 m

45°

5 m

29.

45°

18 cm

45°

30.

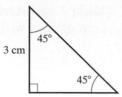

45°

3 cm

45°

To Think About

31. *Flagpole Construction* A carpenter is going to use a wooden flagpole 10 in. in diameter, from which he will shape a rectangular base. The base will be 7 in. wide. The carpenter wishes to make the base as tall as possible, minimizing any waste. How tall will the rectangular base be? (Round to the nearest tenth.)

10

?

7

32. *Shortwave Antenna* A 4-m shortwave antenna is placed on a garage roof that is 2 m above the lower part of the roof. The base of the garage roof is 16 m wide. How long is an antenna support from point *A* to point *B*?

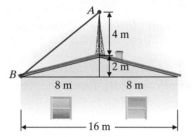

A

4 m

B

2 m

8 m   8 m

16 m

33. *Campus Walkway* Natasha needs to walk from the campus library to her dormitory. The library is 0.4 mi directly east of the student common, and her dormitory is 0.25 mi directly south of the common. There is a straight walkway from her dormitory to the library. How long is this walkway? Round to the nearest hundredth.

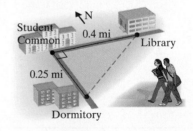

N

Student Common

0.4 mi

Library

0.25 mi

Dormitory

34. *Picture Frame Construction* Nancy makes picture frames from strips of wood, plastic, and metal. She often measures the lengths of the frame's diagonals to be sure the corners are true right angles. If the dimensions of a frame are 7 in. by 9 in., how long would the diagonal measures be? (Round to the nearest hundredth.)

Mixed Practice
Find the approximate value of the unknown side of the right triangle. Round to the nearest thousandth.

35. The two legs are 14 cm and 5 cm.

36. The hypotenuse is 45 yd and one leg is 43 yd.

Cumulative Review

37. **[7.4.2]** *Land Area* Find the area of a triangular piece of land with height 22 m and base 31 m.

38. **[7.2.3]** *Vegetable Garden* Find the area of a rectangular vegetable garden with length 20.5 ft and width 14.5 ft.

39. **[7.2.3]** *Lighthouse Window* Find the area of a square window in a lighthouse that measures 21 inches on each side.

40. **[7.3.1]** *Roof of a Building* Find the area of the parallelogram-shaped roof of a building with a height of 48 yd and a base of 88 yd.

Quick Quiz 7.6

You may use a calculator or a square root table to complete the following problems. Round your answers to the nearest hundredth.

1. One leg of a triangle measures 10 feet. Another leg measures 5 feet. Find the length of the hypotenuse.

2. The hypotenuse of a triangle is 26 centimeters. One leg of the triangle is 24 centimeters. What is the length of the other leg of the triangle?

3. Stephanie is hiking at the Grand Canyon. She walks north for exactly 3 miles. Then she walks east for exactly 8 miles. If you draw a straight line from her finishing point to her starting point, how many miles is she from where she started?

4. **Concept Check** You look up at a plane that is flying at a distance of exactly two miles from your position. The plane is flying at an altitude of exactly 1.5 miles above the land and drops a package of supplies. Explain how you would find the distance from your position to the supplies.

7.7 Circles

① Finding the Area and Circumference of a Circle

Every point on a circle is the same distance from the center of the circle, so the circle looks the same all around. In geometry we study the relationship between the parts of a circle and learn how to calculate the distance around a circle as well as the area of a circle.

A **circle** is a two-dimensional flat figure for which all points are at an equal distance from a given point. This given point is called the **center** of the circle.

Student Learning Objectives

After studying this section, you will be able to:

① **Find the area and circumference of a circle.**

② **Solve area problems containing circles and other geometric shapes.**

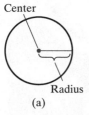

Center

Radius

(a)

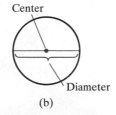

Center

Diameter

(b)

A **radius** is a line segment from the center to a point on the circle.

A **diameter** is a line segment across the circle that passes through the center with end-points on the circle.

We often use the words **radius** and **diameter** to mean the lengths of those segments. Note that the plural of radius is radii. Clearly, then,

$$\text{diameter} = 2 \times \text{radius} \quad \text{or} \quad d = 2r.$$

We could also say that

$$\text{radius} = \text{diameter} \div 2 \quad \text{or} \quad r = \frac{d}{2}.$$

The distance around the circle is called the **circumference.**

There is a special number called **pi,** which we denote by the symbol π. π is the number we get when we divide the circumference of a circle by the diameter $\frac{C}{d} = \pi$. The value of π is approximately 3.14159265359. We can approximate π to any number of digits. For all work in this book we will use the following.

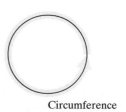

Circumference

> π is approximately 3.14, rounded to the nearest hundredth.

When we approximate π with 3.14 in this section, the answers are *approximate values.*

> We find the **circumference (C) of a circle** by multiplying the length of the diameter (*d*) times π.
>
> $$C = \pi d$$

NOTE TO STUDENT: Fully worked-out solutions to all of the Student Practice problems can be found at the back of the text starting at page SP-1.

EXAMPLE 1   Find the circumference of a quarter if we know the diameter is 2.4 cm. Use $\pi \approx 3.14$. Round to the nearest tenth.

Solution

$$C = \pi d = (3.14)(2.4 \text{ cm})$$
$$= 7.536 \text{ cm} \approx 7.5 \text{ cm (rounded to the nearest tenth)}$$

Student Practice 1   Find the circumference of a circle if its diameter is 9 m. Use $\pi \approx 3.14$. Round to the nearest tenth.

An alternative formula is $C = 2\pi r$. Remember, $d = 2r$. We can use this formula to find the circumference if we are given the length of the radius.

When solving word problems involving circles, be careful. Ask yourself, "Is the radius given, or is the diameter given?" Then do the calculations accordingly.

EXAMPLE 2   A bicycle tire has a diameter of 24 in. How many feet does the bicycle travel if the wheel makes one revolution?

Solution

1. *Understand the problem.* The distance the wheel travels when it makes 1 revolution is the circumference of the tire. Think of the tire unwinding.

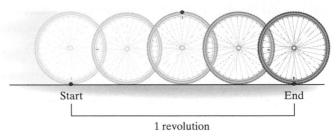

Start   End

1 revolution

We are given the *diameter*. The diameter is given in *inches*. The answer should be in *feet*.

2. *Solve and state the answer.* Since we are given the diameter, we will use $C = \pi d$. We use 3.14 for π.

$$C = \pi d$$
$$= (3.14)(24 \text{ in.}) = 75.36 \text{ in.}$$

We will change 75.36 inches to feet.

$$75.36 \text{ in.} \times \frac{1 \text{ ft}}{12 \text{ in.}} = 6.28 \text{ ft}$$

When the wheel makes 1 revolution, the bicycle travels 6.28 ft.

3. Check. We estimate to check. Since $\pi \approx 3.14$, we will use 3 for π.

$$C \approx (3)(24 \text{ in.}) \times \frac{1 \text{ ft}}{12 \text{ in.}} \approx 6 \text{ ft} \quad \checkmark$$

Student Practice 2 A bicycle tire has a diameter of 30 in. How many feet does the bicycle travel if the wheel makes two revolutions?

The **area (A) of a circle** is the product of π times the radius (*r*) squared.

$$A = \pi r^2$$

EXAMPLE 3

(a) Estimate the area of a circle whose radius is 6 cm.

(b) Find a more exact area of a circle whose radius is 6 cm. Use $\pi \approx 3.14$. Round to the nearest tenth.

Solution

(a) Since π is approximately equal to 3.14, we will use 3 for π to estimate the area.

$$\begin{aligned}
A &= \pi r^2 \\
&\approx (3)(6 \text{ cm})^2 \\
&\approx (3)(6 \text{ cm})(6 \text{ cm}) \\
&\approx (3)(36 \text{ cm}^2) \\
&\approx 108 \text{ cm}^2
\end{aligned}$$

Thus our *estimated* area is 108 cm^2.

(b) Now let's compute a more exact area.

$$\begin{aligned}
A &= \pi r^2 \\
&= (3.14)(6 \text{ cm})^2 \\
&= 3.14(6 \text{ cm})(6 \text{ cm}) \\
&= (3.14)(36 \text{ cm}^2) \quad \text{We } must \text{ square the radius first before} \\
&= 113.04 \text{ cm}^2 \quad \text{multiplying by 3.14.} \\
&\approx 113.0 \text{ cm}^2 \text{ (rounded to the nearest tenth)}
\end{aligned}$$

Our exact answer is close to the value 108 that we found in part (a).

Student Practice 3 Find the area of a circle whose radius is 5 km. Use $\pi \approx 3.14$. Round to the nearest tenth. Estimate to check.

The formula for the area of a circle uses the length of the radius. If we are given the diameter, we can use the property that $r = \dfrac{d}{2}$.

EXAMPLE 4 Lexie Hatfield wants to buy a circular braided rug that is 8 ft in diameter. Find the cost of the rug at $35 a square yard.

Solution

1. Understand the problem. We are given the *diameter in feet*. We will need to find the radius. The cost of the rug is in *square yards*. We will need to change square feet to square yards.

Continued on next page

2. *Solve and state the answer.* Find the radius.

$$r = \frac{d}{2}$$

$$= \frac{8 \text{ ft}}{2}$$

$$= 4 \text{ ft}$$

Use 3.14 for π.

$$A = \pi r^2$$

$$= (3.14)(4 \text{ ft})^2$$

$$= (3.14)(16 \text{ ft}^2)$$

$$= 50.24 \text{ ft}^2$$

Change square feet to square yards. Since 1 yd = 3 ft, $(1 \text{ yd})^2 = (3 \text{ ft})^2$. That is, $1 \text{ yd}^2 = 9 \text{ ft}^2$.

$$50.24 \text{ ft}^2 \times \frac{1 \text{ yd}^2}{9 \text{ ft}^2} \approx 5.58 \text{ yd}^2$$

Find the cost.

$$\frac{\$35}{1 \text{ yd}^2} \times 5.58 \text{ yd}^2 = \$195.30$$

3. *Check.* You may use a calculator to check.

Student Practice 4 Dorrington Little wants to buy a circular pool cover that is 10 ft in diameter. Find the cost of the pool cover at \$12 a square yard.

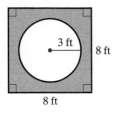

3 ft
8 ft
8 ft

② Solving Area Problems Containing Circles and Other Geometric Shapes

Several applied area problems have a circular region combined with another region.

EXAMPLE 5 Find the area of the shaded region in the margin. Use $\pi \approx 3.14$. Round to the nearest tenth.

Solution We will subtract two areas to find the shaded region.

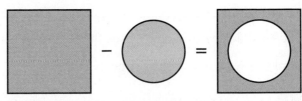

Area of the square − area of the circle = area of the shaded region

$$A = s^2 \qquad\qquad A = \pi r^2$$

$$= (8 \text{ ft})^2 \qquad\qquad = (3.14)(3 \text{ ft})^2$$

$$= 64 \text{ ft}^2 \qquad\qquad = (3.14)(9 \text{ ft}^2)$$

$$\qquad\qquad\qquad = 28.26 \text{ ft}^2$$

Area of the square   area of the circle   area of the shaded region

      64 ft²   –   28.26 ft²   =   35.74 ft²

                                              ≈ 35.7 ft²

                                    (rounded to nearest tenth)

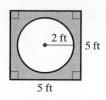

Student Practice 5 Find the area of the shaded region in the margin. Use $\pi \approx 3.14$. Round to the nearest tenth.

Many geometric shapes involve the semicircle. A **semicircle** is one-half of a circle. The area of a semicircle is therefore one-half of the area of a circle.

EXAMPLE 6 Find the area of the shaded region. Use $\pi \approx 3.14$. Round to the nearest tenth.

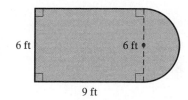

9 ft

Solution First we will find the area of the semicircle with diameter 6 ft.

$$r = \frac{d}{2} = \frac{6 \text{ ft}}{2} = 3 \text{ ft}$$

The radius is 3 ft. The area of a semicircle with radius 3 ft is

$$A_{\text{semicircle}} = \frac{\pi r^2}{2} = \frac{(3.14)(3 \text{ ft})^2}{2} = \frac{(3.14)(9 \text{ ft}^2)}{2}$$

$$= \frac{28.26 \text{ ft}^2}{2} = 14.13 \text{ ft}^2.$$

Now we add the area of the rectangle.

$$A = lw = (9 \text{ ft})(6 \text{ ft}) = 54 \text{ ft}^2$$

$$\begin{array}{ll} 54.00 \text{ ft}^2 & \text{area of rectangle} \\ + \ 14.13 \text{ ft}^2 & \text{area of semicircle} \\ \hline 68.13 \text{ ft}^2 & \text{total area} \end{array}$$

Rounded to the nearest tenth, area = 68.1 ft².

Student Practice 6 Find the area of the shaded region below. Use $\pi \approx 3.14$. Round to the nearest tenth.

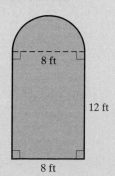

Verbal and Writing Skills, Exercises 1–6

1. The distance around a circle is called the
_____.

2. The radius is a line segment from the _____
to a point on the circle.

3. The diameter is two times the _____ of
the circle.

4. Explain in your own words how to *estimate* the
area of a circle if you are given the diameter.

5. Explain in your own words how to find the circumference of a circle if you are given the radius.

6. Explain in your own words how to find the area
of a semicircle if you are given the radius.

In all exercises, use $\pi \approx 3.14$. Round to the nearest hundredth.

Find the diameter of a circle if the radius has the value given.

7. $r = 29$ in.

8. $r = 18$ in.

9. $r = 8\frac{1}{2}$ mm

10. $r = 14\frac{1}{4}$ yd

Find the radius of a circle if the diameter has the value given.

11. $d = 45$ yd

12. $d = 65$ yd

13. $d = 32.18$ ft

14. $d = 25.48$ cm

Find the circumference of the circle.

15. diameter $= 32$ cm

16. diameter $= 17$ cm

17. radius $= 18.5$ in.

18. radius $= 27.3$ in.

Travel Distance on a Bicycle *A bicycle wheel makes five revolutions. Determine how far the bicycle travels in feet.*

19. The diameter of the wheel is 32 in.

20. The diameter of the wheel is 24 in.

Find the area of each circle.

21. radius = 5 yd

22. radius = 7 yd

23. radius = 8.5 in.

24. radius = 12.5 in.

25. diameter = 32 cm

26. diameter = 52 cm

Water Sprinkler Distribution *A water sprinkler sends water out in a circular pattern. Determine how large an area is watered.*

27. The radius of watering is 12 ft.

28. The radius of watering is 8 ft.

Radio Signal Distribution *A radio station sends out radio waves in all directions from a tower at the center of the circle of broadcast range. Determine how large an area is reached.*

29. The diameter is 90 mi.

30. The diameter is 120 mi.

Find the area of the shaded region.

31.
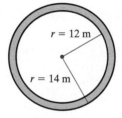

32.
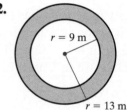

33.
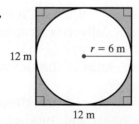

34.
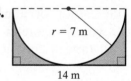

35.
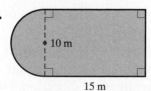

36.
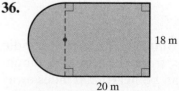

Fertilizing a Playing Field *Find the cost of fertilizing a playing field at $0.20 per square yard for the conditions stated.*

37. The rectangular part of the field is 120 yd long and the diameter of each semicircle is 40 yd.

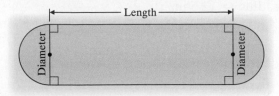

38. The rectangular part of the field is 110 yd long and the diameter of each semicircle is 50 yd.

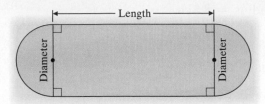

Applications *Use* $\pi \approx 3.14$ *in exercises 39–48. Round to the nearest hundredth.*

39. ***Manhole Cover*** A manhole cover has a diameter of 3 ft. What is the length of the brass grip-strip that encircles the cover, making it easier to manage?

40. ***Ship Porthole*** A porthole window on a freighter ship has a diameter of 2 ft. What is the length of the insulating strip that encircles the window and keeps out wind and moisture?

41. ***Truck Travel*** Jimmy's truck has tires with a radius of 30 inches. How many feet does his truck travel if the wheel makes nine revolutions?

42. ***Car Travel*** Elena's car has tires with a radius of 14 in. How many feet does her car travel if the wheel makes 35 revolutions?

43. ***Car Travel*** Lucy's Volkswagen Jetta has tires with a radius of 7.5 in. How many revolutions do her wheels make in 1 mile? (*Hint:* First determine how many inches are in 1 mile.)

44. ***Car Travel*** Mickey's Hummer has tires with a radius of 8.5 in. How many revolutions do his wheels make in 1 mile? (*Hint:* First determine how many inches are in 1 mile.)

45. ***Basketball Court*** In the center of a new basketball court, a circle with a diameter of 8 ft must be marked with tape before it is painted.
(a) How long will the tape be?
(b) The circle is then painted. How large an area must be painted?

46. ***Food Delivery*** Great Wall Chinese Restaurant will make deliveries within a 1.5-mile radius of the restaurant.
(a) How many square miles is the delivery area?
(b) Jiang has one delivery 1.5 mi straight west of the restaurant, and another 1.5 mi directly east of the restaurant. If he is able to drive around the edge of the delivery area, how many miles will he drive between the two deliveries?

47. ***Mountain Radio Station Broadcast*** A radio station broadcasts from the top of a mountain. The signal can be heard 200 miles away in every direction. How many square miles are in the receiving range of the radio station?

48. ***Distance of Sound Travel*** The sound of an explosion at a fireworks factory could be heard 300 miles away in every direction. How many square miles were in the area where people heard the explosion?

To Think About

49. *Value of Pizza Slice* Noah discovered that a 16-in.-diameter pizza costs $12.00. A 12-in.-diameter pizza costs $8.00. The 12-in.-diameter pizza is cut into six slices. The 16-in.-diameter pizza is cut into eight slices.

(a) What is the cost per slice of the 16-in.-diameter pizza? How many square inches of pizza are in one slice?

(b) What is the cost per slice of the 12-in.-diameter pizza? How many square inches of pizza are in one slice?

(c) If you want more value for your money, which slice of pizza should you buy?

50. *Value of Pizza Slice* Olivia discovered that a 14-in.-diameter pizza costs $10.00. It is cut into eight slices. A 12.5 in. × 12.5 in. square pizza costs $12.00. It is cut into nine slices.

(a) What is the cost of one slice of the 14-in.-diameter pizza? How many square inches of pizza are in one slice?

(b) What is the cost of one slice of the 12.5 in. × 12.5 in. square pizza? How many square inches of pizza are in one slice?

(c) If you want more value for your money, which slice of pizza should you buy?

Cumulative Review

51. [5.3A.2] Find 25% of 120.

52. [5.3A.2] What is 0.5% of 60?

53. [5.3A.2] 10% of what number is 7?

54. [5.3A.2] 19% of what number is 570?

Quick Quiz 7.7 *In the following problems, use* $\pi \approx 3.14$. *Round to the nearest hundredth.*

1. What is the circumference of a circle with a diameter of 9 inches?

2. What is the area of a circle with a radius of 11 meters?

3. An engineer constructs a rectangular steel plate that measures 3.5 centimeters by 3.1 centimeters. In the center of the plate he drills a hole with a radius of 1.5 centimeters. What is the area of the rectangular plate AFTER he drills the hole in it? (*Hint:* Draw a sketch of the circle inside the rectangle.)

4. Concept Check A carpenter makes a semicircle with a radius of 3 feet. Explain how you would find the area of the semicircle.

7.8 Volume

Student Learning Objectives

After studying this section, you will be able to:

① Find the volume of a rectangular solid (box).

② Find the volume of a cylinder.

③ Find the volume of a sphere.

④ Find the volume of a cone.

⑤ Find the volume of a pyramid.

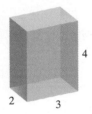

NOTE TO STUDENT: Fully worked-out solutions to all of the Student Practice problems can be found at the back of the text starting at page SP-1.

① Finding the Volume of a Rectangular Solid (Box)

How much grain can that shed hold? How much water is in the polluted lake? How much air is inside a basketball? These are questions of **volume.** In this section we compute the volume of several three-dimensional geometric figures: the rectangular solid (box), cylinder, sphere, cone, and pyramid.

We can start with a box 1 in. × 1 in. × 1 in.

This is a **cube** with a side of 1 inch.

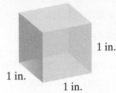

1 in.
1 in.
1 in.

This box has a volume of 1 cubic inch (written 1 in.^{3}). We can use this as a **unit of volume.** Volume is measured in cubic units such as cubic meters (abbreviated m^3) or cubic feet (abbreviated ft^3). When we measure volume, we are measuring the space inside an object.

> The **volume (V) of a rectangular solid** (box) is the length (l) times the width (w) times the height (h).
>
> $$V = lwh$$
>
> h
> w
> l

EXAMPLE 1 Find the volume of a box of width 2 ft, length 3 ft, and height 4 ft.

Solution $V = lwh = (3 \text{ ft})(2 \text{ ft})(4 \text{ ft}) = (6)(4) \text{ ft}^3 = 24 \text{ ft}^3$

Student Practice 1 Find the volume of a box of width 5 m, length 6 m, and height 2 m.

If all sides of the box are equal, the solid is called a cube, and the formula is $V = s^3$ where s is the side of the cube.

② Finding the Volume of a Cylinder

Cylinders are the shape we observe when we see a tin can or a tube.

> The **volume (V) of a cylinder** is the area of its circular base (πr^2) times the height (h).
>
> $$V = \pi r^2 h$$
>
> h
> r

We will continue to use 3.14 as an approximation for π, as we did in Section 7.7, in all volume problems requiring the use of π.

7 in.

$r = 3$ in.

EXAMPLE 2   Find the volume of a cylinder of radius 3 in. and height 7 in. Round to the nearest tenth.

Solution $V = \pi r^2 h = (3.14)(3 \text{ in.})^2(7 \text{ in.})$
$$= (3.14)(9 \text{ in.}^2)(7 \text{ in.})$$
$$= (28.26 \text{ in.}^2)(7 \text{ in.})$$
$$= 197.82 \text{ in.}^3 \approx 197.8 \text{ in.}^3 \text{ rounded to the nearest tenth}$$

Student Practice 2   Find the volume of a cylinder of radius 2 in. and height 5 in. Round to the nearest tenth.

TO THINK ABOUT: **Comparing Volume Formulas** Take a minute to compare the formulas for the volumes of a rectangular solid and a cylinder. Do you see how they are similar? Consider the area of the base of each figure. In each case, what must you multiply the base area by to obtain the volume of the solid?

③ Finding the Volume of a Sphere

Have you ever considered how you would find the volume of the inside of a ball? How many cubic inches of air are inside a basketball? To answer these questions we need a volume formula for a *sphere*.

> The **volume (V) of a sphere** is 4 times π times the radius (r) cubed divided by 3.
>
> $$V = \frac{4\pi r^3}{3}$$ r

EXAMPLE 3   Find the volume of a sphere with radius 3 m. Round to the nearest tenth.

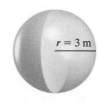

$r = 3$ m

Solution $V = \dfrac{4\pi r^3}{3} = \dfrac{(4)(3.14)(3 \text{ m})^3}{3}$
$$= \frac{(4)(3.14)(27) \text{ m}^3}{3}$$
$$= (12.56)(9) \text{ m}^3 = 113.04 \text{ m}^3$$
$$\approx 113.0 \text{ m}^3 \text{ rounded to the nearest tenth}$$

Student Practice 3   Find the volume of a sphere with radius 6 m. Round to the nearest tenth.

④ Finding the Volume of a Cone

We see the shape of a cone when we look at the sharpened end of a wooden pencil or at an ice cream cone. To find the volume of a cone we use the following formula.

The **volume (V) of a cone** is π times the radius (r) of the base squared times the height (h) divided by 3.

$$V = \frac{\pi r^2 h}{3}$$

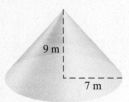

EXAMPLE 4   Find the volume of a cone of radius 7 m and height 9 m. Round to the nearest tenth.

Solution $V = \dfrac{\pi r^2 h}{3}$

$$= \frac{(3.14)(7\ \text{m})^2(9\ \text{m})}{3}$$

$$= \frac{(3.14)(49\ \text{m}^2)(9\ \text{m})}{3}$$

$$= (3.14)(49)(3)\ \text{m}^3$$

$$= (153.86)(3)\ \text{m}^3$$

$$= 461.58\ \text{m}^3$$

$$\approx 461.6\ \text{m}^3 \text{ rounded to the nearest tenth}$$

Student Practice 4   Find the volume of a cone of radius 5 m and height 12 m. Round to the nearest tenth.

⑤ Finding the Volume of a Pyramid

You have seen pictures of the great pyramids of Egypt. These amazing stone structures are over 4000 years old.

The **volume (V) of a pyramid** is obtained by multiplying the area (B) of the base of the pyramid by the height (h) and dividing by 3.

$$V = \frac{Bh}{3}$$

EXAMPLE 5 Find the volume of a pyramid with height 6 m, length of base 7 m, and width of base 5 m.

Solution The base is a rectangle.

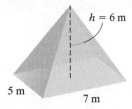

$$\text{Area of base} = (7 \text{ m})(5 \text{ m}) = 35 \text{ m}^2$$

Substituting the area of the base 35 m² and the height of 6 m, we have

$$V = \frac{Bh}{3} = \frac{(35 \text{ m}^2)(6 \text{ m})}{3}$$
$$= (35)(2) \text{ m}^3$$
$$= 70 \text{ m}^3$$

Student Practice 5 Find the volume of a pyramid with the dimensions given.

(a) height 10 m, width of base 6 m, length of base 6 m

(b) height 15 m, width of base 7 m, length of base 8 m

Verbal and Writing Skills, Exercises 1–6

In this section, we have studied six volume formulas. They are $V = lwh$, $V = \pi r^2 h$, $V = \dfrac{4\pi r^3}{3}$, $V = s^3$, $V = \dfrac{\pi r^2 h}{3}$, *and* $V = \dfrac{Bh}{3}$.

For each of the following figures, state (a) the name of the figure and (b) the correct formula for its volume.

1.

2.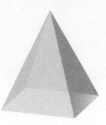

3.

4.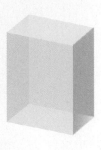

5.

6.

Find the volume. Use $\pi \approx 3.14$. *Round to the nearest tenth unless otherwise directed.*

7. a rectangular solid with width = 12 mm, length = 30 mm, height = 1.5 mm

8. a rectangular solid with width = 3.5 mm, length = 22 mm, height = 10 mm

9. a cylinder with radius 3 m and height 8 m

10. a cylinder with radius 2 m and height 7 m

11. a cylinder with diameter 22 m and height 17 m

12. a cylinder with diameter 30 m and height 9 m

13. a sphere with radius 9 yd

14. a sphere with radius 12 yd

15. a pyramid with a base of 18 ft² and a height of 35 feet

16. a pyramid with a base of 30 ft² and a height of 42 feet

17. a cube with side 0.6 cm (Round to the nearest thousandth.)

18. a cube with side 0.8 cm (Round to the nearest thousandth.)

19. a cone with a radius of 3 yd and a height of 7 yd (Round to the nearest hundredth.)

20. a cone with a radius of 9 yd and a height of 5 yd (Round to the nearest hundredth.)

Exercises 21 and 22 involve hemispheres. A hemisphere is exactly one-half of a sphere.

21. Find the volume of a hemisphere with radius = 7 m.

22. Find the volume of a hemisphere with radius = 6 m.

Mixed Practice   *Find the volume. Use $\pi \approx 3.14$. Round to the nearest tenth.*

23. a cone with a height of 14 cm and a radius of 8 cm

24. a cone with a height of 12 cm and a radius of 9 cm

25. a cone with a height of 12.5 ft and a radius of 7 ft

26. a cone with a height of 14.2 ft and a radius of 9 ft

27. a pyramid with a height of 10 m and a square base of 7 m on a side

28. a pyramid with a height of 7 m and a square base of 3 m on a side

29. a pyramid with a height of 10 m and a rectangular base measuring 8 m by 14 m

30. a pyramid with a height of 5 m and a rectangular base measuring 6 m by 12 m

Applications   *Use $\pi \approx 3.14$ when necessary.*

31. *Garden Mulch* Lexi Salzman has a large rectangular vegetable garden measuring 9 ft by 16 ft. An employee at the local nursery recommended putting down mulch 3 in. thick to prevent weeds from growing. Each bag of mulch covers 3 cubic feet. How many bags should Lexi purchase for her vegetable garden?

32. *Driveway Construction* Caleb Salzman wants to put down a crushed-stone driveway to his summer camp. The driveway is 7 yd wide and 120 yd long. The crushed stone is to be 4 in. thick. How many cubic yards of stone will he need?

Pipe Insulation A collar of Styrofoam is made to insulate a pipe. Find the volume of the unshaded region (which represents the collar). The large radius R is to the outer rim. The small radius r is to the edge of the insulation.

33. $r = 3$ in.
$R = 5$ in.
$h = 20$ in.

34. $r = 4$ in.
$R = 6$ in.
$h = 25$ in.

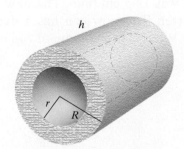

35. *Astronomy* Jupiter has a radius of approximately 45,000 mi. Earth has a radius of approximately 3950 mi. Assuming both planets are spheres, what is the difference in volume between Earth and Jupiter?

36. *Softballs and Golf Balls* A softball has a diameter of about 3.8 in. Most golf balls are about 1.7 in. in diameter. What is the difference in volume between the softball and golf ball? Round your answer to the nearest tenth.

37. *Shipping a Fragile Object* Lora Connelly has a fragile glass box in the shape of a rectangular solid of width 6 in., length 18 in., and height 12 in. It is being shipped in a larger box of width 12 in., length 22 in., and height 16 in. All of the space between the glass box and shipping box will be packed with Styrofoam packing "peanuts." How many cubic inches of the shipping box will be Styrofoam peanuts?

38. *Moving Truck* A U-Haul moving truck has a storage space measuring 6.5 feet high by 7 feet wide by 12 feet long. How many cubic feet of storage space are there? If $\frac{3}{4}$ of the space is full, how many cubic feet of space are being used?

39. *Cone-Shaped Cups* In Nicole's office, there is a 5-gallon (about 1150 cubic inches) water dispenser with cone-shaped paper cups. The cups have a height of 3 inches and a radius is 1.5 inches. How many cubic inches of water can one cup hold? How many cups will the entire dispenser fill?

40. *Radar Nose Cone* The nose cone of a passenger jet is used to receive and send radar. It is made of a special aluminum alloy that costs $4.00 per cm^3. The cone has a radius of 5 cm and a height of 9 cm. What is the cost of the aluminum needed to make this *solid* nose cone?

41. *Root Beer Can* The old Smith Root Beer can was 13.5 cm high. The new can is 1.4 cm shorter. The new can has a diameter of 6.6 cm. What is the volume of the new can?

42. *Swimming Pool* Dan and Connie are installing a new circular in-ground swimming pool. The pool will measure 9 feet deep and 20 feet in diameter. How many cubic feet of earth will need to be removed for the pool to be installed?

43. *Stone Pyramid* Suppose that a stone pyramid has a rectangular base that measures 87 yd by 130 yd. Also suppose that the pyramid has a height of 70 yd. Find the volume.

44. *Stone Pyramid* Suppose the pyramid in exercise 43 is made of solid stone. It is not hollow like the pyramids of Egypt. It is composed of layer after layer of cut stone. The stone weighs 422 lb per cubic yard. How many pounds does the pyramid weigh? How many tons does the pyramid weigh?

Cumulative Review

45. **[2.8.1]** Add. $7\frac{1}{3} + 2\frac{1}{4}$

46. **[2.8.2]** Subtract. $9\frac{1}{8} - 2\frac{3}{4}$

47. **[2.4.3]** Multiply. $2\frac{1}{4} \times 3\frac{3}{4}$

48. **[2.5.3]** Divide. $7\frac{1}{2} \div 4\frac{1}{5}$

49. **[2.8.3]** Evaluate the following expression.

$$\left(\frac{5}{8} - \frac{1}{4}\right)^2 + \frac{7}{32}$$

50. **[2.8.3]** Evaluate the following expression.

$$\left(6\frac{5}{6} + 2\frac{3}{4}\right) \times \frac{2}{3}$$

Quick Quiz 7.8 *Use $\pi \approx 3.14$ in the following problems. Round all answers to the nearest hundredth.*

1. Find the volume of a sphere with a radius of 4 centimeters.

2. Find the volume of a pyramid with a height of 8 yards and a rectangular base of 7 yards on one side and 6 yards on the other side.

3. Find the volume of a cylinder with a radius of 3 meters and a height of 13 meters.

4. **Concept Check** Suppose a new cylinder is formed similar to the cylinder described in problem 3 but the new radius is 4 meters while the height is unchanged. Explain how to determine how much larger the volume of the new cylinder is compared to the original cylinder.

7.9 Similar Geometric Figures

Student Learning Objectives

After studying this section, you will be able to:

① Find the corresponding parts of similar triangles.

② Find the corresponding parts of similar geometric figures.

① Finding the Corresponding Parts of Similar Triangles

In English, "similar" means that the things being compared are, in general, alike. But in mathematics, "similar" means that the things being compared are alike in a special way—they are *alike in shape,* even though they may be different in size. So photographs that are enlarged produce images *similar* to the original; a floor plan of a building is *similar* to the actual building; a model car is *similar* to the actual vehicle.

Two triangles with the same shape but not necessarily the same size are called **similar triangles.** Here are two pairs of similar triangles.

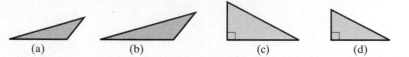

(a) (b) (c) (d)

The two triangles at right are similar. The smallest angle in the first triangle is angle *A.* The smallest angle in the second triangle is angle *D.* Both angles measure 36°. We say that angle *A* and angle *D* are **corresponding angles** in these similar triangles.

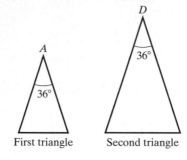

First triangle   Second triangle

> The **corresponding angles** of similar triangles are equal.

The following two triangles are similar. Notice the **corresponding sides.**

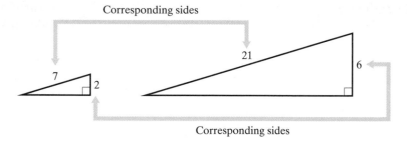

Corresponding sides

Corresponding sides

We see that the ratio of 7 to 21 is the same as the ratio of 2 to 6.

$$\frac{7}{21} = \frac{2}{6} \quad \text{is obviously true since} \quad \frac{1}{3} = \frac{1}{3}.$$

> The corresponding sides of similar triangles have the same ratio.

We can use the fact that corresponding sides of similar triangles have the same ratio to find the lengths of the missing sides of triangles.

EXAMPLE 1 These two triangles are similar. Find the length of side n. Round to the nearest tenth.

Solution The ratio of 12 to 19 is the same as the ratio of 5 to n.

$$\frac{12}{19} = \frac{5}{n}$$

$12n = (5)(19)$ Cross-multiply.

$12n = 95$ Simplify.

$$\frac{12n}{12} = \frac{95}{12}$$ Divide each side by 12.

$n = 7.91\overline{6}$

≈ 7.9 Round to the nearest tenth.

Side n has length 7.9.

Student Practice 1 The two triangles in the margin are similar. Find the length of side n. Round to the nearest tenth.

Similar triangles are not always oriented the same way. You may find it helpful to rotate one of the triangles so that the similarity is more apparent.

EXAMPLE 2 These two triangles are similar. Name the sides that correspond.

NOTE TO STUDENT: Fully worked-out solutions to all of the Student Practice problems can be found at the back of the text starting at page SP-1.

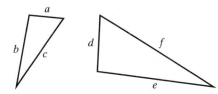

Solution First we turn the second triangle so that the shortest side is on the top, the intermediate side is to the left, and the longest side is on the right.

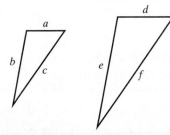

Now the shortest side of each triangle is on the top, the longest side of each triangle is on the right, and so on. We can see that

  a corresponds to d.
  b corresponds to e.
  c corresponds to f.

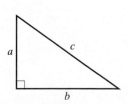

Student Practice 2 The two triangles in the margin are similar. Name the sides that correspond.

> The perimeters of similar triangles have the same ratio as the corresponding sides.

Similar triangles can be used to find distances or lengths that are difficult to measure.

EXAMPLE 3   A flagpole casts a shadow of 36 ft. At the same time, a nearby tree that is 3 ft tall has a shadow of 5 ft. How tall is the flagpole?

Solution

1. **Understand the problem.** The shadows cast by the sun shining on vertical objects at the same time of day form similar triangles. We draw a picture.

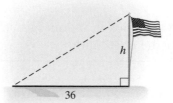

2. **Solve and state the answer.** Let h = the height of the flagpole. Thus we can say h is to 3 as 36 is to 5.

$$\frac{h}{3} = \frac{36}{5}$$
$$5h = (3)(36)$$
$$5h = 108$$
$$\frac{5h}{5} = \frac{108}{5}$$
$$h = 21.6$$

The flagpole is about 21.6 feet tall.

3. **Check.** The check is up to you.

Student Practice 3   What is the height (h) of the side wall of the building in the margin if the two triangles are similar?

② Finding the Corresponding Parts of Similar Geometric Figures

Geometric figures such as rectangles, trapezoids, and circles can also be similar figures.

> The corresponding sides of similar geometric figures have the same ratio.

EXAMPLE 4   The two rectangles shown here are similar because the corresponding sides of the two rectangles have the same ratio. Find the width of the larger rectangle.

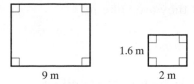

Solution   Let w = the width of the larger rectangle.

$$\frac{w}{1.6} = \frac{9}{2}$$
$$2w = (1.6)(9)$$
$$2w = 14.4$$
$$\frac{2w}{2} = \frac{14.4}{2}$$
$$w = 7.2$$

The width of the larger rectangle is 7.2 meters.

Student Practice 4   The two rectangles in the margin are similar. Find the width of the larger rectangle.

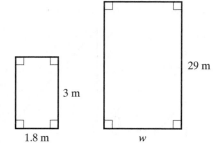

The perimeters of similar figures—whatever the figures—have the same ratio as their corresponding sides. Circles are a special case. All circles are similar. The circumferences of two circles have the same ratio as their radii.

TO THINK ABOUT: Comparing Two Areas How would you find the relationship between the areas of two similar geometric figures? Consider the following two similar rectangles.

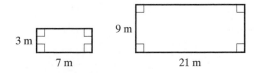

The area of the smaller rectangle is $(3 \text{ m})(7 \text{ m}) = 21 \text{ m}^2$. The area of the larger rectangle is $(9 \text{ m})(21 \text{ m}) = 189 \text{ m}^2$. How could you have predicted this result?

The ratio of small width to large width is $\frac{3}{9} = \frac{1}{3}$. The small rectangle has sides that are $\frac{1}{3}$ as large as the large rectangle. The ratio of the area of the small rectangle to the area of the large rectangle is $\frac{21}{189} = \frac{1}{9}$. Note that $\left(\frac{1}{3}\right)^2 = \frac{1}{9}$.

Thus we can develop the following principle: The areas of two similar figures are in the same ratio as the square of the ratio of two corresponding sides.

Verbal and Writing Skills, Exercises 1–4

1. Similar figures may be different in _____ but they are alike in _____.

2. The corresponding sides of similar triangles have the same _____.

3. The perimeters of similar figures have the same ratio as their corresponding _____.

4. You are given the lengths of the sides of a large triangle and the length of a corresponding side of a smaller, similar triangle. Explain in your own words how to find the perimeter of the smaller triangle.

For each pair of similar triangles, find the missing side n. Round to the nearest tenth when necessary.

5.

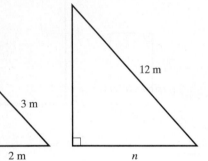

6.

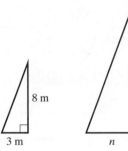

7.

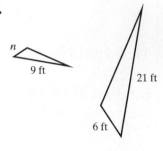

8.

9.

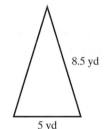

10.

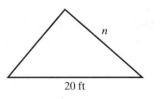

Each pair of triangles is similar. Determine the three pairs of corresponding sides in each case.

11.

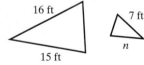

12.

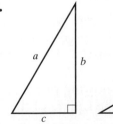

Applications

13. ***Sculptor*** A sculptor is designing her new triangular masterpiece. In her scale drawing, the shortest side of the triangular piece to be made measures 8 cm. The longest side of the drawing measures 25 cm. The longest side of the actual triangular piece to be sculpted must be 10.5 m long. How long will the shortest side of this piece be? Round to the nearest tenth.

14. ***Landscape Architect*** The zoo has hired a landscape architect to design the triangular lobby of the children's petting zoo. In his scale drawing, the longest side of the lobby is 9 cm. The shortest side of the lobby is 5 cm. The longest side of the actual lobby will be 30 m. How long will the shortest side of the actual lobby be? Round to the nearest tenth.

15. ***Photography*** Matt took a photograph of the entire group at the class reunion. He brought it to a professional photography studio and asked that the 4-in.-by-6-in. photo be enlarged to a poster that is 39 in. long. What is the smaller dimension (width) of the poster?

16. ***Family Room Addition*** Greg and Marcia are adding a new family room onto their home. On the blueprints, the room measures 2 in. wide by 3 in. long. The actual room is similar in shape with a width of 11 feet. What will be the length of the family room?

17. ***Blueprints of New House*** On the blueprints of their new home, Ben and Heather notice the bathtub measures 2 in. by 5 in. They know that the actual bathtub will be 5.5 ft long. How wide will the tub be?

18. ***Theater Company Props*** A theater company's prop designer sends drawings of the props to the person in charge of construction. An upcoming play will require a large brick wall to stretch across the stage floor. In the designer's drawing, the wall measures $\frac{1}{4}$ ft high by $\frac{3}{4}$ ft long. If the length of the stage is 36 ft, how tall will the wall be?

Length of Shadows In exercises 19 and 20, a flagpole casts a shadow. At the same time, a nearby tree casts a shadow. Use the sketch to find the height n of each flagpole.

19.

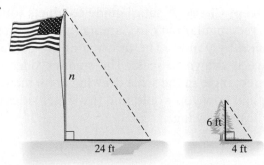

20.

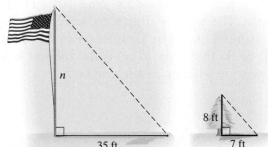

21. ***Length of Shadows*** Leigh is standing outside the shopping mall. She is 5.5 feet tall and her shadow measures 18 feet long. The exterior wall of the department store casts a shadow of 144 feet. How tall is the wall?

22. ***Length of Shadows*** Thomas is rock climbing in Utah. He is 6 feet tall and his shadow measures 8 feet long. The rock he wants to climb casts a shadow of 610 feet. How tall is the rock he is about to climb?

Each pair of figures is similar. Find the missing side. Round to the nearest tenth when necessary.

23.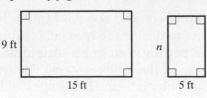

24.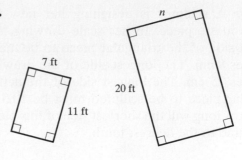

25.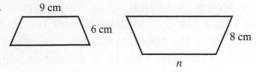

26.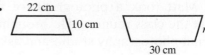

Cumulative Review *Calculate. Use the correct order of operations.*

27. **[1.6.2]** $2 \times 3^2 + 4 - 2 \times 5$

28. **[1.6.2]** $100 \div (8 - 3)^2 \times 2^3$

29. **[2.8.3]** $\dfrac{4}{5} \times \dfrac{5}{3} - \dfrac{1}{3}$

30. **[2.8.3]** $\dfrac{8}{5} \div 3 - \dfrac{1}{3}$

Quick Quiz 7.9 *Round all answers to the nearest hundredth.*

1. Noah is 6 feet tall. At 3 P.M. his shadow is 7 feet long. He is climbing a mountain cliff near Boulder, Colorado. The top of the cliff casts a shadow in the valley of 100 feet. How tall is the cliff? (*Hint:* Assume they are similar triangles.)

2. Two triangles are similar. The larger triangle has one side of 14 meters and one side of 15 meters. The longest side of the larger triangle is not known. The shortest side of the smallest triangle is 3 meters long. What is the length of the next-largest side?

3. Olivia has an architect's plan for her new house. The plan is drawn to scale. The plan measures 7 inches by 12 inches. Her new house will be 36 feet long. (This is the longest side.) How wide will her new house be?

4. **Concept Check** A safari guide conducts tours in a rectangular park in Zambia that measures 5 miles by 9 miles. The drawing in the guide's office is drawn to scale. The smaller sides of the rectangle in the scale drawing are 6 inches. Explain how you would find the larger sides in the scale drawing.

7.10 Solving Applied Problems Involving Geometry

Student Learning Objective

After studying this section, you will be able to:

① Solve applied problems involving geometric shapes.

① Solving Applied Problems Involving Geometric Shapes

We can solve many real-life problems with the geometric knowledge we now have. Our everyday world is filled with objects that are geometric in shape, so we can use our knowledge of geometry to find length, area, or volume. How far is the automobile trip? How much framing, edging, or fencing is required? How much paint, siding, or roofing is required? How much can we store? All of these questions can be answered with geometry.

If it is helpful to you, use the Mathematics Blueprint for Problem Solving to organize a plan to solve these applied problems.

EXAMPLE 1 A professional painter can paint 90 ft^2 of wall space in 20 minutes. How long will it take the painter to paint four walls with the following dimensions: 14 ft × 8 ft, 12 ft × 8 ft, 10 ft × 7 ft, and 8 ft × 7 ft?

Solution

1. **Understand the problem.**

Mathematics Blueprint for Problem Solving

Gather the Facts	What Am I Asked to Do?	How Do I Proceed?	Key Points to Remember
Painter paints four walls: 14 ft × 8 ft 12 ft × 8 ft 10 ft × 7 ft 8 ft × 7 ft Painter can paint 90 ft^2 in 20 minutes.	Find out how long it will take the painter to paint the four walls.	**(a)** Find the total area to be painted. **(b)** Then find how long it will take to paint the total area.	The area of each rectangular wall is length × width. To get the time, we set up a proportion.

2. **Solve and state the answer.**
   (a) Find the total area of the four walls.
      Each wall is a rectangle. The first one is 8 ft wide and 14 ft long.

$$A = lw$$
$$= (14\,\text{ft})(8\,\text{ft}) = 112\,\text{ft}^2$$

We find the areas of the other three walls.

$$(12\,\text{ft})(8\,\text{ft}) = 96\,\text{ft}^2 \quad (10\,\text{ft})(7\,\text{ft}) = 70\,\text{ft}^2 \quad (8\,\text{ft})(7\,\text{ft}) = 56\,\text{ft}^2$$

The total area is obtained by adding.

$$\begin{array}{r} 112\,\text{ft}^2 \\ 96\,\text{ft}^2 \\ 70\,\text{ft}^2 \\ +\ \ 56\,\text{ft}^2 \\ \hline 334\,\text{ft}^2 \end{array}$$

Continued on next page

(b) Determine how long it will take to paint the four walls.

Now we set up a proportion. If 90 ft^2 can be done in 20 minutes, then 334 ft^2 can be done in t minutes.

$$\frac{90 \text{ ft}^2}{20 \text{ minutes}} = \frac{334 \text{ ft}^2}{t \text{ minutes}}$$

$$\frac{90}{20} = \frac{334}{t}$$

$$90(t) = 334(20)$$

$$90t = 6680$$

$$\frac{90t}{90} = \frac{6680}{90}$$

$$t \approx 74 \qquad \text{We round our answer to the nearest minute.}$$

Thus the work can be done in approximately 74 minutes.

3. Check. Estimate the answer.

$$14 \times 8 \approx 10 \times 8 = 80 \text{ ft}^2$$
$$12 \times 8 \approx 10 \times 8 = 80 \text{ ft}^2$$
$$10 \times 7 = 70 \text{ ft}^2$$
$$8 \times 7 = 56 \text{ ft}^2$$

If we estimate the sum of the number of square feet, we will have

$$80 + 80 + 70 + 56 = 286 \text{ ft}^2.$$

Now since 60 minutes = 1 hour, we know that if you can paint 90 ft^2 in 20 minutes, you can paint 270 ft^2 in one hour. Our estimate of 286 ft^2 is slightly more than 270 ft^2, so we would expect that the answer would be slightly more than one hour.

Thus our calculated value of 74 minutes (1 hour 14 minutes) seems reasonable. ✓

Student Practice 1 Mike rented an electric floor sander. It will sand 80 ft^2 of hardwood floor in 15 minutes. He needs to sand the floors in three rooms. The floor dimensions are 24 ft $\times$ 13 ft, 12 ft $\times$ 9 ft, and 16 ft $\times$ 3 ft. How long will it take him to sand the floors in all three rooms?

NOTE TO STUDENT: Fully worked-out solutions to all of the Student Practice problems can be found at the back of the text starting at page SP-1.

EXAMPLE 2 Carlos and Rosetta want to put vinyl siding on the front of their home in West Chicago. The house dimensions are shown in the figure at right. The door dimensions are 6 ft $\times$ 3 ft. The windows measure 2 ft $\times$ 4 ft.

(a) Excluding windows and doors, how many square feet of siding will be needed?

(b) If the siding costs $2.25 per square foot, how much will it cost to side the front of the house?

Solution

1. *Understand the problem.*

Mathematics Blueprint for Problem Solving

Gather the Facts	What Am I Asked to Do?	How Do I Proceed?	Key Points to Remember
House measures 19 ft × 25 ft. Windows measure 2 ft × 4 ft. Door measures 6 ft × 3 ft. Siding costs $2.25 per square foot.	Find the cost to put siding on the front of the house.	**(a)** Find the area of the entire front of the house by multiplying 19 ft by 25 ft. Find the area of one window by multiplying 2 ft by 4 ft. Find the area of the door by multiplying 6 ft by 3 ft. **(b)** Multiply desired area by $2.25.	**(a)** To obtain the desired area we must subtract the area of nine windows and one door from the area of the entire front. **(b)** We must multiply the resulting area by the cost of the siding per foot.

2. *Solve and state the answer.*

(a) Find the area of the front of the house.

We will find the area of the large rectangle representing the front of the house. Then we will subtract the area of the windows and the door.

$$\text{Area of each window} = (2\text{ ft})(4\text{ ft}) = 8\text{ ft}^2$$
$$\text{Area of 9 windows} = (9)(8\text{ ft}^2) = 72\text{ ft}^2$$
$$\text{Area of 1 door} = (6\text{ ft})(3\text{ ft}) = 18\text{ ft}^2$$
$$\text{Area of 9 windows} + 1\text{ door} = 90\text{ ft}^2$$
$$\text{Area of large rectangle} = (19\text{ ft})(25\text{ ft}) = 475\text{ ft}^2$$

$$\begin{array}{ll} \text{Total area of front of house} & 475\text{ ft}^2 \\ -\ \text{Area of 9 windows and 1 door} & -\ \ 90\text{ ft}^2 \\ \hline = \text{Total area to be covered} & 385\text{ ft}^2 \end{array}$$

We see that 385 ft^2 of siding will be needed.

(b) Find the cost of the siding.

$$\text{Cost} = 385\ \cancel{\text{ft}^2} \times \frac{\$2.25}{1\ \cancel{\text{ft}^2}} = \$866.25$$

The cost to put up siding on the front of the house will be $866.25.

3. *Check.* We leave the check up to you.

Student Practice 2   A sketch of the roof of a commercial building is shown in the margin.

(a) What is the area of the roof?

(b) How much would it cost to install new roofing on the roof area shown if the roofing costs $2.75 per square yard?

    (*Hint*: 9 square feet = 1 square yard.)

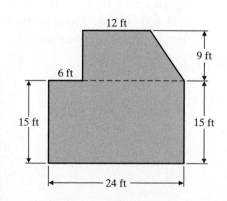

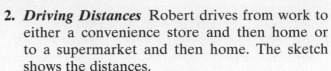

Watch the videos
in MyMathLab

Download the
MyDashBoard App

Applications *Round to the nearest tenth unless otherwise directed.*

1. ***Driving Distances*** Monica drives to work each day from Bethel to Bridgeton. The following sketch shows the two possible routes.

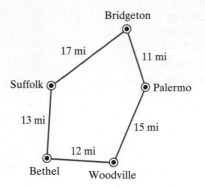

(a) How many miles is the trip if she drives through Suffolk? What is her average speed if this trip takes her 0.6 hour?

(b) How many miles is the trip if she drives through Woodville and Palermo? What is her average speed if this trip takes her 0.7 hour?

(c) Over which route does she travel at a more rapid rate?

2. ***Driving Distances*** Robert drives from work to either a convenience store and then home or to a supermarket and then home. The sketch shows the distances.

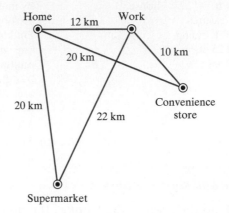

(a) How far does he travel if he goes from work to the supermarket and then home? How fast does he travel if the trip takes 0.6 hour?

(b) How far does he travel if he goes from work to the convenience store and then home? How fast does he travel if the trip takes 0.5 hour?

(c) Over which route does he travel at a more rapid rate?

3. ***Hanging Wallpaper*** A professional wallpaper hanger can wallpaper 120 ft^2 in 35 minutes. She will be papering four walls in a house. They measure 7 ft × 10 ft, 7 ft × 14 ft, 6 ft × 10 ft, and 6 ft × 8 ft. How many minutes will it take her to paper all four walls?

4. ***Painting an Apartment*** Dave and Linda McCormick are painting the walls of their apartment in Seattle. When they worked together at Linda's mother's house, they were able to paint 80 ft^2 in 25 minutes. The living room they wish to paint has one wall that measures 16 feet by 7 feet, one wall that measures 14 feet by 7 feet, and two walls that measure 12 feet by 7 feet. How long will it take Dave and Linda together to paint the living room of their apartment? Round to the nearest minute.

5. *Painting Exterior of House* The Crawfords need to paint the outside of their house. The front and back each measure 55 ft by 24 ft, and each side measures 32 ft by 24 ft. There are sixteen windows and two doors, measuring 4 ft by 2 ft and 7 ft by 3 ft, respectively. What is the total area to be painted? If one gallon of paint covers 350 square feet, how many gallons of paint do they need? Assume only whole gallons can be purchased.

6. *Tiling a Kitchen* A new restaurant is ordering tile for a large wall in the kitchen. The wall measures 19 feet long by 8 feet high, and will be tiled the entire length, but only to three-fourths of the wall height. The tile costs $4 per square foot. How much will the tile cost?

7. *Carpeting a Recreation Room* The floor area of the recreation room at Yvonne's house is shown in the following drawing. How much will it cost to carpet the room if the carpet costs $15 per square yard?

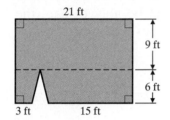

8. *Aluminum Siding on a Barn* The front view of a barn is shown in the following diagram. The cost of aluminum siding is $18 per square yard. How much will it cost to put siding on the front of the barn?

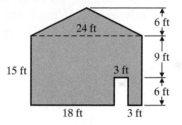

9. *Gold Filling for a Tooth* A dentist places a gold filling in the shape of a cylinder with a hemispherical top in a patient's tooth. The radius r of the filling is 1 mm. The height of the cylinder is 2 mm. Find the volume of the filling. If dental gold costs $95 per cubic millimeter, how much did the gold for the filling cost?

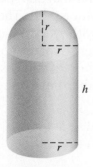

10. *City Sewer System* Find the volume of a concrete connector for the city sewer system. A diagram of the connector is shown. It is shaped like a box with a hole of diameter 2 m. If it is formed using concrete that costs $1.20 per cubic meter, how much will the necessary concrete cost?

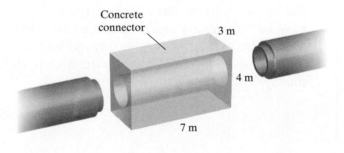

11. *Satellite Orbit* The Landsat satellite orbits Earth in an almost circular pattern. Assume that the radius of orbit (distance from center of Earth to the satellite) is 6500 km.

    (a) How many kilometers long is one orbit of the satellite? (That is, what is the circumference of the orbit path?)

    (b) If the satellite goes through one orbit around Earth in two hours, what is its speed in kilometers per hour?

13. *Valentine's Day Candy* For Valentine's Day, a company makes decorative cylinder-shaped canisters and fills them with cinnamon candies. Each canister is 10 in. high and has a radius of 2 in. They want to make 400 canisters and need to determine how much candy to buy. What is the total number of cubic inches that will be filled with candy?

12. *City Park* The North City Park is constructed in a shape that includes the region outside one-fourth of a circle. It is shaded in this sketch.

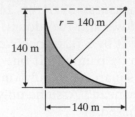

    (a) What is the perimeter of the park?

    (b) How much will it cost to place a fence around the park at $15 per meter?

14. *Cargo Box of Ford F-150* Manuel has two large gardens, each measuring 15 feet by 11 feet. He wants to put wood chips 2 inches deep in each garden. The cargo box of his Ford F-150, when he uses the cover, has a volume of 55.5 cubic feet. Will Manuel be able to haul the load of wood chips in his covered cargo box?

Cumulative Review *Divide.*

15. **[1.5.3]** $16\overline{)2048}$

16. **[1.5.3]** $42\overline{)12,936}$

17. **[3.5.2]** $1.3\overline{)0.325}$

18. **[3.5.2]** $0.52\overline{)2.5324}$

Quick Quiz 7.10 *Use $\pi \approx 3.14$. Round all answers to the nearest hundredth.*

1. Consider a slightly different field from the one shown in the figure for test questions 34 and 35 on page 509 of the text. This new field is 140 yards long and 50 yards wide. The rounded ends of the new field have radii of 25 yards. What is the area of the new athletic field?

2. Michael wants to paint the side of his house. The side is a rectangle 20 feet high and 34 feet long. There are 7 windows on that side of the house. Each window measures 3 feet by 2 feet. Assuming he does not want to paint over the windows, how many square feet of area will he need to cover with paint?

3. Camp Cherith has built a new rope-obstacle course for the campers. The field that contains the course is triangular in shape. The field has a base of 200 meters and a height of 140 meters. The entire field needs to be sprayed with weed killer that costs $0.05 per square meter. How much will the weed killer cost?

4. **Concept Check** Suppose there is a second field at Camp Cherith, described in problem 3. The new field has a base of 300 meters and the same height as the other field. The new field needs to be sprayed with weed killer that costs $0.20 per square meter. Explain how you would find out how much more it will cost to spray the new field than the other field.

Did You Know...

That You Can Save Thousands of Dollars in Fuel Costs with a Hybrid Vehicle Compared to a Less-Efficient Vehicle?

AFFORDABLE TRANSPORTATION

Understanding the Problem:

Michele wants to reduce her auto expenses. Her current vehicle is a 2003 SUV that gets only 16 miles per gallon. Different options for a replacement vehicle are:

- **Option A:** A small SUV that gets 20 miles per gallon (MPG) and sells for $21,540.
- **Option B:** A hybrid model of the same SUV, which gets 32 MPG and sells for $28,150.
- **Option C:** The most fuel-efficient hybrid car, which gets 60 MPG and sells for $23,770.

She wants to determine the cost and potential savings she would enjoy from the three different types of cars.

Making a Plan:

We need to calculate the cost to own the different vehicles.

Step 1: Michele drives about 15,000 miles a year. She wants to know what fuel costs would be for the three options.

Task 1: Determine the number of gallons of gasoline that each car (Options A, B, and C) would use if they are driven 15,000 miles in one year.

Task 2: If the average price of gasoline is $3.60 per gallon (Feb 2011), how much would it cost to fuel each car for 15,000 miles?

Analyzing the Options:

Step 2: The SUV in Option B would cost Michele $6610 more to purchase than the SUV in Option A. Compare the different costs of ownership.

Task 3: How much gas money would Michele save in one year with Option B compared to Option A?

Task 4: How many years would it take for Michele to save $6610 in gas money with Option B compared to Option A?

Task 5: How much gas money would Michele save after five years with Option B compared to Option A? After 10 years?

Step 3: The car in Option C would cost Michele $2230 more to purchase than the SUV in Option A. Compare the different costs of ownership.

Task 6: How much gas money would Michele save in one year with Option C compared to Option A?

Task 7: How many years would it take for Michele to save $2230 in gas money with Option C compared to Option A?

Task 8: How much gas money would Michele save after five years with Option C compared to Option A? After 10 years?

Making a Decision:

Step 4: Even though Michele enjoyed the comfort of her SUV, she could not ignore the potential savings in gas money she saw with Option C. She decided it was time to make a smart financial decision instead of one based on comfort. She expects to have her new car for at least five years. At that point, Option C will have saved her $9000 in gas money!

Applying the Situation to Your Life:

The number of miles you drive per year as well as the price you pay for a gallon of gas is going to affect this analysis when applied to your own situation. The more miles you drive, the greater the savings from driving a fuel-efficient vehicle. Also, the higher the price of gas, the more you will save by switching.

Chapter 7 Organizer

Topic and Procedure	Examples	✏️ You Try It
Perimeter of a rectangle, p. 425 $$P = 2l + 2w$$	Find the perimeter of a rectangle with width = 3 m and length = 8 m. $$P = (2)(8\text{ m}) + (2)(3\text{ m})$$ $$= 16\text{ m} + 6\text{ m} = 22\text{ m}$$	1. Find the perimeter of a rectangle with width = 5 ft and length = 12 ft.
Perimeter of a square, p. 427 $$P = 4s$$	Find the perimeter of a square with side $s = 6$ m. $$P = (4)(6\text{ m}) = 24\text{ m}$$	2. Find the perimeter of a square with side $s = 8.5$ in.
Area of a rectangle, p. 429 $$A = lw$$	Find the area of a rectangle with width = 2 m and length = 7 m. $$A = (7\text{ m})(2\text{ m}) = 14\text{ m}^2$$	3. Find the area of a rectangle with width = 3 m and length = 10 m.
Area of a square, p. 429 $$A = s^2$$	Find the area of a square with a side of 4 m. $$A = s^2 = (4\text{ m})^2 = 16\text{ m}^2$$	4. Find the area of a square with a side of 9 ft.
Perimeter of parallelograms, trapezoids, and triangles, pp. 435, 437, 444 Add up the lengths of all sides.	Find the perimeter of a triangle with sides of 3 m, 6 m, and 4 m. $$3\text{ m} + 6\text{ m} + 4\text{ m} = 13\text{ m}$$	5. Find the perimeter of a triangle with sides of 7 in., 21 in., and 15 in.
Area of a parallelogram, p. 435 $A = bh$ b = length of base h = height 	Find the area of a parallelogram with a base of 12 m and a height of 7 m. $$A = bh = (12\text{ m})(7\text{ m}) = 84\text{ m}^2$$	6. Find the area of a parallelogram with a base of 13 cm and a height of 6 cm.
Area of a trapezoid, p. 437 $$A = \frac{h(b + B)}{2}$$ b = length of shorter base B = length of longer base h = height 	Find the area of a trapezoid whose height is 12 m and whose bases are 17 m and 25 m. $$A = \frac{(12\text{ m})(17\text{ m} + 25\text{ m})}{2} = \frac{(12\text{ m})(42\text{ m})}{2}$$ $$= \frac{504\text{ m}^2}{2} = 252\text{ m}^2$$	7. Find the area of a trapezoid whose height is 8 ft and whose bases are 3.5 ft and 12.5 ft.
The sum of the measures of the three interior angles of a triangle is 180°, p. 443 In a triangle, to find one missing angle if two are given: **1.** Add up the two known angles. **2.** Subtract the sum from 180°.	Find the missing angle if two known angles in a triangle are 60° and 70°. $$60° + 70° = 130°$$ $$\begin{array}{r} 180° \\ -\ 130° \\ \hline 50° \end{array}$$ The missing angle is 50°.	8. Find the missing angle if two known angles in a triangle are 43° and 58°.
Area of a triangle, p. 444 $$A = \frac{bh}{2}$$ b = base h = height 	Find the area of a triangle whose base is 1.5 m and whose height is 3 m. $$A = \frac{bh}{2} = \frac{(1.5\text{ m})(3\text{ m})}{2} = \frac{4.5\text{ m}^2}{2} = 2.25\text{ m}^2$$	9. Find the area of a triangle whose base is 10 cm and whose height is 6.5 cm.
Evaluating square roots of numbers that are perfect squares, p. 451 If a number is a product of two identical factors, then either factor is called a square root.	**(a)** $\sqrt{0} = 0$ because $(0)(0) = 0$ **(b)** $\sqrt{4} = 2$ because $(2)(2) = 4$ **(c)** $\sqrt{100} = 10$ because $(10)(10) = 100$ **(d)** $\sqrt{169} = 13$ because $(13)(13) = 169$	10. Evaluate each square root. **(a)** $\sqrt{1}$ **(b)** $\sqrt{9}$ **(c)** $\sqrt{144}$ **(d)** $\sqrt{225}$

Topic and Procedure	Examples	✏ You Try It
Approximating the square root of a number that is not a perfect square, p. 452 **1.** If a calculator with a square root key is available, enter the number and then press the $\boxed{\sqrt{x}}$ or $\boxed{\sqrt{}}$ key. The approximate value will be displayed. **2.** If using a square root table, find the number n, then look for the square root of that number. The approximate value will be rounded to the nearest thousandth.	**1.** Find on a calculator. **(a)** $\sqrt{13}$ **(b)** $\sqrt{182}$ Round to the nearest thousandth. **(a)** 13 $\boxed{\sqrt{x}}$ 3.60555128 rounds to 3.606. **(b)** 182 $\boxed{\sqrt{x}}$ 13.49073756 rounds to 13.491. **2.** Find from a square root table. **(a)** $\sqrt{31}$ **(b)** $\sqrt{33}$ **(c)** $\sqrt{34}$ To the nearest thousandth, the approximate values are as follows. **(a)** $\sqrt{31} \approx 5.568$ **(b)** $\sqrt{33} \approx 5.745$ **(c)** $\sqrt{34} \approx 5.831$	**11.** Use a calculator or square root table to approximate the square roots to the nearest thousandth. **(a)** $\sqrt{18}$ **(b)** $\sqrt{140}$ **(c)** $\sqrt{39}$

Number, n	Square Root of That Number, $\sqrt{n}$
31	5.568
32	5.657
33	5.745
34	5.831

Topic and Procedure	Examples	You Try It
Finding the hypotenuse of a right triangle when given the length of each leg, p. 458 $\text{Hypotenuse} = \sqrt{(\text{leg})^2 + (\text{leg})^2}$ 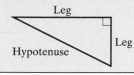	Find the hypotenuse of a triangle with legs of 9 m and 12 m. $\begin{aligned} \text{hypotenuse} &= \sqrt{(12)^2 + (9)^2} \\ &= \sqrt{144 + 81} = \sqrt{225} \\ &= 15 \text{ m} \end{aligned}$	**12.** Find the hypotenuse of a triangle with legs of 7 in. and 11 in. Round to the nearest thousandth.
Finding the leg of a right triangle when given the length of the other leg and the hypotenuse, p. 459 $\text{Leg} = \sqrt{(\text{hypotenuse})^2 - (\text{leg})^2}$	Find the leg of a right triangle. The hypotenuse is 14 in. and the other leg is 12 in. Round to nearest thousandth. $\text{leg} = \sqrt{(14)^2 - (12)^2} = \sqrt{196 - 144} = \sqrt{52}$ Using a calculator or a square root table, the leg ≈ 7.211 in.	**13.** Find the leg of a right triangle. The hypotenuse is 20 ft and the other leg is 12 ft.
Solving applied problems involving the Pythagorean Theorem, p. 460 **1.** Read the problem carefully. **2.** Draw a sketch. **3.** Label the two sides that are given. **4.** If the hypotenuse is unknown, use $\text{hypotenuse} = \sqrt{(\text{leg})^2 + (\text{leg})^2}.$ **5.** If one leg is unknown, use $\text{leg} = \sqrt{(\text{hypotenuse})^2 - (\text{leg})^2}.$	A boat travels 5 mi south and then 3 mi east. How far is it from the starting point? Round to the nearest tenth. $\begin{aligned} \text{hypotenuse} &= \sqrt{(5)^2 + (3)^2} \\ &= \sqrt{25 + 9} = \sqrt{34} \end{aligned}$ Using a calculator or a square root table, the distance is approximately 5.8 mi. 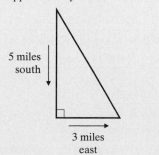	**14.** A ship travels 8 mi north and then 3 mi west. How far is the ship from the starting point? Round to the nearest tenth.
The special 30°–60°–90° right triangle, p. 461 The length of the leg opposite the 30° angle is $\frac{1}{2} \times$ the length of the hypotenuse.	Find y. $y = \dfrac{1}{2}(26\text{ m}) = 13\text{ m}$ 	**15.** Find y. 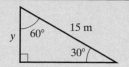

Topic and Procedure	Examples	✏️ You Try It
The special 45°–45°–90° right triangle, p. 461 The sides opposite the 45° angles are equal. The hypotenuse is $\sqrt{2} \times$ the length of either leg.	Find z. $z = \sqrt{2}(13\text{ m}) \approx (1.414)(13\text{ m}) = 18.382\text{ m}$ 	16. Find z.
Radius and diameter of a circle, p. 467 r = radius d = diameter $r = \dfrac{d}{2}$ $d = 2r$	**(a)** What is the radius of a circle with diameter 50 in.? $r = \dfrac{50\text{ in.}}{2} = 25\text{ in.}$ **(b)** What is the diameter of a circle with radius 16 in.? $d = (2)(16\text{ in.}) = 32\text{ in.}$	17. **(a)** A circle has a diameter of 28 ft. What is the radius? **(b)** A circle has a radius of 5.5 in. What is the diameter?
Pi, p. 467 π is a decimal that goes on forever. It can be approximated by as many decimal places as needed. $\pi = \dfrac{\text{circumference of a circle}}{\text{diameter of same circle}}$	Use $\pi \approx 3.14$ for all calculations. Unless otherwise directed, round your final answer to the nearest tenth when any calculation involves π.	
Circumference of a circle, p. 467 $C = \pi d$	Find the circumference of a circle with diameter 12 ft. $C = \pi d = (3.14)(12\text{ ft}) = 37.68\text{ ft}$ $\approx 37.7\text{ ft}$ (rounded to the nearest tenth)	18. Find the circumference of a circle with diameter 6 m.
Area of a circle, p. 469 $A = \pi r^2$ **1.** Square the radius first. **2.** Then multiply the result by 3.14.	Find the area of a circle with radius 7 ft. $A = \pi r^2 = (3.14)(7\text{ ft})^2 = (3.14)(49\text{ ft}^2)$ $= 153.86\text{ ft}^2$ $\approx 153.9\text{ ft}^2$ (rounded to the nearest tenth)	19. Find the area of a circle with radius 9 ft.
Volume of a rectangular solid (box), p. 476 $V = lwh$	Find the volume of a box whose dimensions are 5 m by 8 m by 2 m. $V = (8\text{ m})(5\text{ m})(2\text{ m}) = (40)(2)\text{ m}^3 = 80\text{ m}^3$	20. Find the volume of a box whose dimensions are 4 ft by 7 ft by 10 ft.
Volume of a cylinder, p. 476 r = radius $\quad h$ = height $\quad V = \pi r^2 h$ **1.** Square the radius first. **2.** Then multiply the result by 3.14 and by the height. 	Find the volume of a cylinder with radius 7 m and height 3 m. $V = \pi r^2 h = (3.14)(7\text{ m})^2(3\text{ m})$ $= (3.14)(49)(3)\text{ m}^3$ $= (153.86)(3)\text{ m}^3 = 461.58\text{ m}^3$ $\approx 461.6\text{ m}^3$ (rounded to the nearest tenth)	21. Find the volume of a cylinder with radius 3 in. and height 12 in.
Volume of a sphere, p. 477 $V = \dfrac{4\pi r^3}{3}$ r = radius 	Find the volume of a sphere of radius 3 m. $V = \dfrac{4\pi r^3}{3} = \dfrac{(4)(3.14)(3\text{ m})^3}{3}$ $= \dfrac{(4)(3.14)\overset{9}{\cancel{(27)}}\text{ m}^3}{\underset{1}{\cancel{3}}}$ $= (12.56)(9)\text{ m}^3 = 113.04\text{ m}^3$ $\approx 113.0\text{ m}^3$ (rounded to the nearest tenth)	22. Find the volume of a sphere of radius 5 ft.
Volume of a cone, p. 477 $V = \dfrac{\pi r^2 h}{3}$ r = radius h = height 	Find the volume of a cone of height 9 m and radius 7 m. $V = \dfrac{\pi r^2 h}{3} = \dfrac{(3.14)(7\text{ m})^2(9\text{ m})}{3}$ $= \dfrac{(3.14)(7^2)\overset{3}{\cancel{(9)}}\text{ m}^3}{\underset{1}{\cancel{3}}} = (3.14)(49)(3)\text{ m}^3$ $= (153.86)(3)\text{ m}^3 = 461.58\text{ m}^3$ $\approx 461.6\text{ m}^3$ (rounded to the nearest tenth)	23. Find the volume of a cone of height 14 m and radius 2 m.

Topic and Procedure	Examples	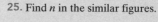 You Try It
Volume of a pyramid, p. 478 $V = \dfrac{Bh}{3}$ B = area of the base h = height 1. Find the area of the base. 2. Multiply this area by the height and divide the result by 3.	Find the volume of a pyramid whose height is 6 m and whose rectangular base is 10 m by 12 m. **1.** $B = (12\text{ m})(10\text{ m}) = 120\text{ m}^2$ **2.** $V = \dfrac{(120)(6)\text{ m}^3}{3} = (120)(2)\text{ m}^3 = 240\text{ m}^3$	**24.** Find the volume of a pyramid whose height is 10 ft and whose rectangular base is 4.5 ft by 8 ft.
Similar figures, corresponding sides, p. 484 The corresponding sides of similar figures have the same ratio.	Find n in the following similar figures. $\dfrac{n}{4} = \dfrac{9}{3}$ $3n = 36$ $n = 12\text{ m}$ n 9 m 4 m 3 m	**25.** Find n in the similar figures. 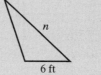 n 8 ft 6 ft 4 ft
Similar figures, corresponding perimeters, p. 486 The perimeters of similar figures have the same ratio as the corresponding sides. For reasons of space, the procedure for the areas of similar figures is not given here but may be found in the text (see p. 487).	The following two figures are similar. Find the perimeter of the larger figure. $\dfrac{6}{12} = \dfrac{29}{p}$ $6p = (12)(29)$ $6p = 348$ $\dfrac{6p}{6} = \dfrac{348}{6}$ $p = 58$ The perimeter of the larger figure is 58 m. 12 m 6 m 5 m 7 m 3 m 8 m	**26.** The two figures are similar. Find the perimeter of the larger figure. 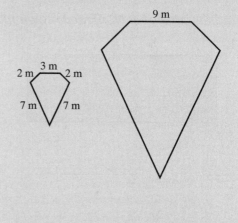 9 m 3 m 2 m 2 m 7 m 7 m

Chapter 7 Review Problems

Round to the nearest tenth. Use $\pi \approx 3.14$ in all calculations requiring the use of π.

Section 7.1

1. Find the complement of an angle of $76°$.

2. Find the supplement of an angle of $76°$.

3. Find the measures of $\angle a$, $\angle b$, and $\angle c$ in the following sketch.

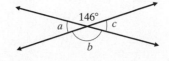
146°
a c
b

4. Find $\angle s$, $\angle t$, $\angle u$, $\angle w$, $\angle x$, $\angle y$, and $\angle z$ in the following sketch if we know that line p is parallel to line q.

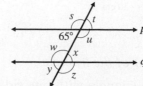
s t
65° u p
w x
y z q

Section 7.2

Find the perimeter of the square or rectangle.

5. length = 9.5 m, width = 2.3 m

6. length = width = 12.7 yd

Find the area of the square or rectangle.

7. length = 5.9 cm, width = 2.8 cm

8. length = width = 7.2 in.

Find the perimeter of each object made up of rectangles and squares.

9.

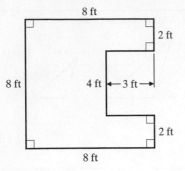

10.

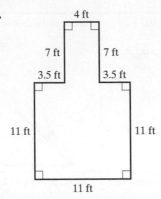

Find the area of each shaded region made up of rectangles and squares.

11.

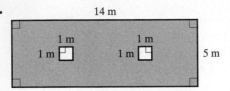

12.

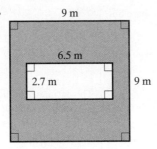

Section 7.3

Find the perimeter of the parallelogram or trapezoid.

13. Two sides of the parallelogram are 38.5 m and 14 m.

14. The sides of the trapezoid are 5 mi, 22 mi, 5 mi, and 30 mi.

Find the area of the parallelogram or trapezoid.

15. A parallelogram has a base of 70 ft and a height of 50 ft.

16. A trapezoid has a height of 18 yd and bases of 21 yd and 19 yd.

Find the total area of each region made up of parallelograms, trapezoids, and rectangles.

17.

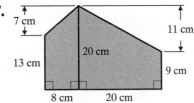

18.

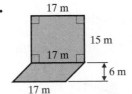

Section 7.4

Find the perimeter of the triangle.

19. An isosceles triangle with one side 18 ft and the other two sides 21 ft.

20. An equilateral triangle with each side 15.5 ft.

Find the measure of the third angle in the triangle.

21. Two known angles are 28° and 45°.

22. A right triangle with one angle of 35°.

Find the area of the triangle.

23. base = 8.5 m, height = 12.3 m

24. base = 12.5 m, height = 9.5 m

Find the total area of each region made up of triangles and rectangles.

25.

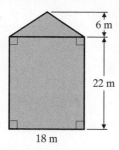

26.

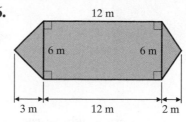

Section 7.5

Evaluate exactly.

27. $\sqrt{81}$

28. $\sqrt{64}$

29. $\sqrt{121}$

30. $\sqrt{144} + \sqrt{16}$

31. $\sqrt{100} - \sqrt{36} + \sqrt{196}$

Approximate using a square root table or a calculator with a square root key. Round to the nearest thousandth.

32. $\sqrt{62}$

33. $\sqrt{165}$

34. $\sqrt{180}$

Section 7.6

Find the unknown side. If the answer cannot be obtained exactly, use a square root table or a calculator with a square root key. Round to the nearest hundredth when necessary.

35.

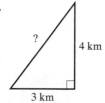

36.

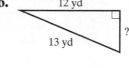

37.

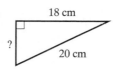

38.

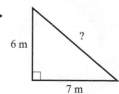

Round to the nearest tenth.

39. ***Construction of a Metal Plate*** Find the distance between the centers of the holes of the metal plate shown.

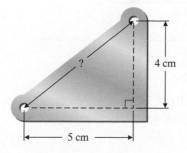

40. ***Wheelchair Ramp*** A builder constructed a wheelchair ramp with the following dimensions. Find the length of the ramp.

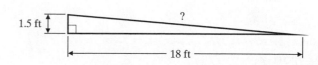

41. *Shed Construction* A shed is built with the following dimensions. Find the distance from the peak of the roof to the horizontal support brace.

42. *Replacing a Door* Find the width of a replacement door if it is 6 ft tall and the diagonal measures 7 ft.

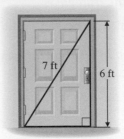

Section 7.7

43. What is the diameter of a circle whose radius is 53 cm?

44. What is the radius of a circle whose diameter is 126 cm?

45. Find the circumference of a circle with diameter 20 m.

46. Find the circumference of a circle with radius 9 in.

Find the area of each circle.

47. radius = 9 m

48. diameter = 8.6 ft

Find the area of each shaded region made up of circles, semicircles, rectangles, trapezoids, and parallelograms. Round your answer to the nearest tenth.

49.

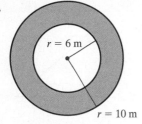

50.

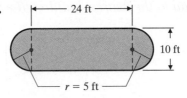

51.

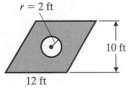

52.

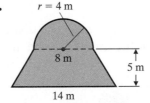

Section 7.8

In exercises 53–57, find the volume. Round answers to the nearest tenth.

53. *U-Haul Truck* U-Haul advertises a moving truck with a storage area measuring 20.8 ft by 7.5 ft by 8.1 ft. Find the volume of the storage area.

54. *Soccer Ball* Find the volume of a soccer ball with radius 4.5 inches.

55. *Garbage Can* Find the volume of a garbage can that is 3 ft high and has a radius of 1.5 ft.

56. *Sculpture* Find the volume of a sculpture in the shape of a pyramid that is 15 m high and whose square base measures 7 m by 7 m.

57. Chemical Pollution A chemical has polluted a volume of ground in a cone shape. The depth of the cone is 30 yd. The radius of the cone is 17 yd. Find the volume of polluted ground.

Section 7.9

Find n in each set of similar triangles.

58.

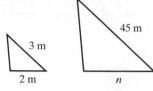

59.

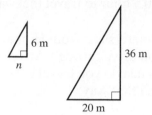

Determine the perimeter of the unlabeled figure.

60.

61.

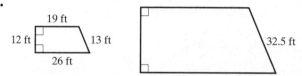

62. Banner Construction Anastasio is in charge of decorations for the "International Cars of the Future" show. He has designed a rectangular banner that will hang in front of the first-prize-winning car, so that all he has to do is push a button and the banner will fly up into the ceiling space to reveal the car. The model of the banner used 12 square yards of fabric. The dimensions of the actual banner will be $3\frac{1}{2}$ times the dimensions of the model. How much fabric will the finished banner need?

Section 7.10

63. Chemistry Lab Tank A conical tank holds acid in a chemistry lab. The tank has a radius of 9 in. and a height of 24 in. How many cubic inches does the tank hold? The acid weighs 16 g per cubic inch. What is the weight of the acid if the tank is full?

64. Carpeting a Recreation Room The Wilsons are carpeting a recreation room with the dimensions shown. Carpeting costs $8 per square yard. How much will the carpeting cost?

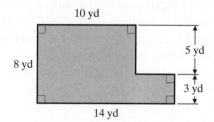

65. Gas Tank Storage The Suburban Gas Company has a spherical gas tank. The diameter of the tank is 90 meters. Find the volume of the spherical tank.

66. Hot Water Tank Charlie and Ginny have a cylindrical hot water tank that is 5 feet high and has a diameter of 18 inches. How many cubic feet does the tank hold?

67. Hot Water Tank The tank in exercise 66 is filled with water. One cubic foot of water is about 7.5 gallons. How many gallons does the tank hold?

68. Lawn of a High School The front lawn at Central High School in Hanover is in the shape of a trapezoid. The bases of the trapezoid are 45 feet and 50 feet. The height of the trapezoid is 35 feet. What is the area of the front lawn?

69. Driving Distances

   (a) In the following diagram, how many kilometers is it from Homeville to Seaview if you drive through Ipswich? How fast do you travel if it takes 0.5 hour to travel that way?

   (b) How many kilometers is it from Homeville to Seaview if you drive through Acton and Westville? How fast do you travel if it takes 0.8 hour to travel that way?

   (c) Over which route do you travel at a more rapid rate?

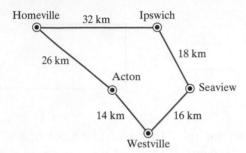

70. Silo Capacity A silo has a cylindrical shape with a hemispherical dome. It has dimensions as shown in the following figure.

   (a) What is its volume in cubic feet?

   (b) If 1 cubic foot ≈ 0.8 bushel, how many bushels of grain will it hold?

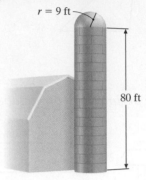

71. Farm Production During 2005, U.S. farms produced an estimated 2.757 billion bushels of soybeans. (*Source:* U.S. Department of Agriculture) Each bushel takes up 1.244 cubic feet of storage. How many cubic feet of storage were needed for the 2005 soybean crop?

72. Soybean Storage If all of the soybeans in exercise 71 were stored in a huge rectangular storage bin that is 10,000 feet wide and 20,000 feet long, how many feet high would the storage bin need to be?

73. Pet Aquarium The largest aquarium at PetsMart measures 2.25 ft by 4 ft by 2 ft. Water weighs about 62 pounds per cubic foot. How many pounds of water does the aquarium hold? There are about 8.6 pounds in one gallon. How many gallons of water does this aquarium hold? Round to the nearest whole gallon.

74. Pet Aquarium It is recommended that 1.5 inches of gravel be placed in the bottom of the aquarium in exercise 73. How many cubic inches of gravel are needed? Assume the base of the aquarium is 4 ft by 2 ft.

75. Pony Rides At the county fair a pony is tied to a 30-ft rope. The pony gives children rides by walking in a circle 5 times with the rope pulled taut. How many feet does the pony walk for each ride?

76. Art Exhibit Hall Carol Kirk manages an art exhibit hall in Elmer. The floor of the hall consists of a large rectangle with a semicircle at each end. The rectangular part measures 18 yards by 25 yards. Each semicircle has a diameter of 18 yards. What is the perimeter of the floor?

77. Rope Lighting Barry Tice has decided to install a special rope lighting along the perimeter of the art exhibit hall in exercise 76. They can order the lighting in 150-foot spools. How many spools do they need to order?

78. Flying a Kite A kite is flying exactly 30 feet above the edge of a pond. The person flying the kite is using exactly 33 feet of string. Assuming that the string is so tight that it forms a straight line, how far is the person standing from the edge of the pond? Round to the nearest tenth.

After you take this test read through the Math Coach on pages 510–511. Math Coach videos are available via MyMathLab and YouTube. Step-by-step test solutions in the Chapter Test Prep Videos are also available via MyMathLab and YouTube. (Search "TobeyBasicCollMath" and click on "Channels.")

1. In the following figure, lines m and n are parallel, and the measure of angle a is 52°. Find the measure of angle b, angle c, and angle e.

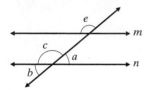

Find the perimeter.

2. a rectangle that measures 9 yd × 11 yd

3. a square with side 6.3 ft

MC 4. a parallelogram with sides measuring 6.5 m and 3.5 m

5. a trapezoid with sides measuring 22 m, 13 m, 32 m, and 13 m

6. a triangle with sides measuring 58.6 m, 32.9 m, and 45.5 m

Find the area. Round to the nearest tenth.

7. a rectangle that measures 10 yd × 18 yd

8. a square 10.2 m on a side

9. a parallelogram with a height of 6 m and a base of 13 m

MC 10. a trapezoid with a height of 9 m and bases of 7 m and 25 m

11. a triangle with a base of 4 cm and a height of 6 cm

Evaluate exactly.

12. $\sqrt{144}$

13. $\sqrt{169}$

14. Find the complement of an angle that measures 63°.

15. Find the supplement of an angle that measures 107°.

16. A triangle has an angle that measures 12.5° and another that measures 83.5°. What is the measure of the third angle?

1. _____ ☐

2. _____ ☐

3. _____ ☐

4. _____ ☐

5. _____ ☐

6. _____ ☐

7. _____ ☐

8. _____ ☐

9. _____ ☐

10. _____ ☐

11. _____ ☐

12. _____ ☐

13. _____ ☐

14. _____ ☐

15. _____ ☐

16. _____ ☐

Approximate using a square root table or a calculator with a square root key. Round to the nearest thousandth.

17. $\sqrt{54}$ **18.** $\sqrt{135}$

In exercises 19 and 20, find the unknown side. Use a calculator or a square root table to approximate square roots to the nearest thousandth.

Mc 19.

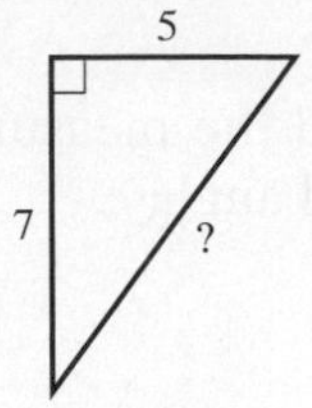

20.

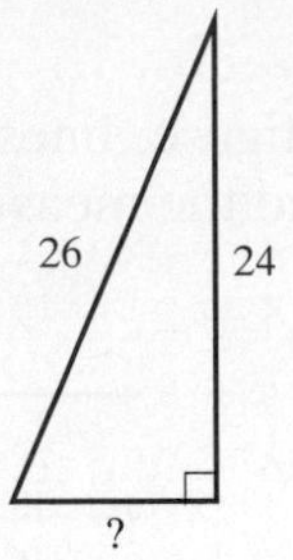

19

In exercises 21–24, round to the nearest hundredth.

21. Find the distance between the centers of the holes drilled in a rectangular metal plate with the dimensions labeled in the following sketch.

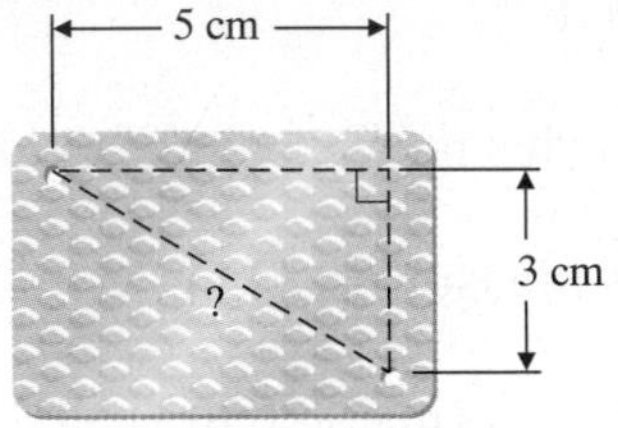

22. A 15-ft-tall ladder is placed so that it reaches 12 ft up on the wall of a house. How far is the base of the ladder from the wall of the house?

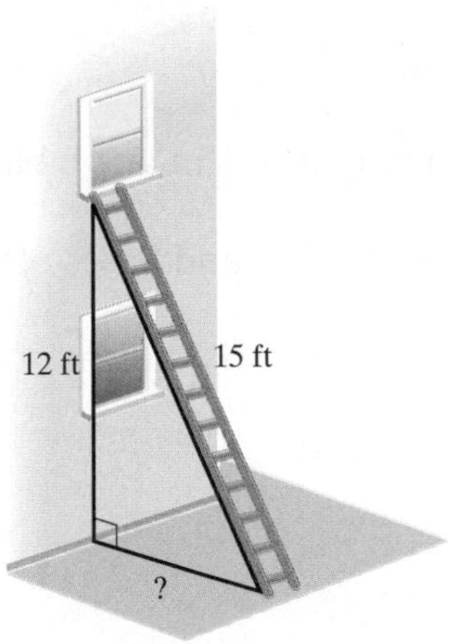

22

23. Find the circumference of a circle with diameter 18 ft.

Mc 24. Find the area of a circle with diameter 12 ft.

Find the shaded area of each region made up of circles, semicircles, rectangles, squares, trapezoids, and parallelograms. Round to the nearest tenth.

25.

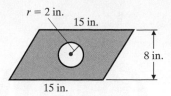

r = 2 in.
15 in.
8 in.
15 in.

26.

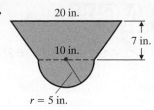

20 in.
7 in.
10 in.
r = 5 in.

Find the volume. Round to the nearest tenth when necessary.

27. a rectangular box measuring 3.5 m by 20 m by 10 m

28. a cone with height 12 m and radius 8 m

29. a sphere of radius 3 m

30. a cylinder of height 2 ft and radius 9 ft

31. a pyramid of height 14 m and whose rectangular base measures 4 m by 3 m

Each pair of triangles is similar. Find the missing side n.

32.

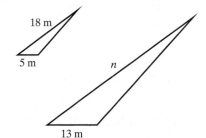

18 m
5 m
n
13 m

33.

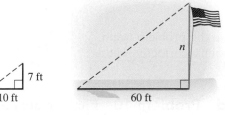

7 ft
10 ft
n
60 ft

Solve. An athletic field has the dimensions shown in the figure below. Assume you are considering only the darker green shaded area. Use $\pi \approx 3.14$.

34. What is the area of the athletic field?

35. How much will it cost to fertilize it at $0.40 per square yard?

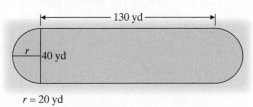

130 yd
r
40 yd
r = 20 yd

25. ☐

26. ☐

27. ☐

28. ☐

29. ☐

30. ☐

31. ☐

32. ☐

33. ☐

34. ☐

35. ☐

Total Correct: ☐

MATH COACH

Mastering the skills you need to do well on the test.

Students often make the same types of errors when they do the Chapter 7 Test. Here are some helpful hints to keep you from making those common errors on test problems.

Find the Perimeter of a Parallelogram—Problem 4   Find the perimeter of a parallelogram with sides measuring 6.5 m and 3.5 m.

> **Helpful Hint** Remember that a parallelogram is a four-sided figure with opposite sides that are equal in length. The perimeter is the distance around the figure.

Did you realize that two sides of the parallelogram each measure 6.5 m, while the other two sides each measure 3.5 m?

Yes ____ No ____

If you answered No, go back and make this correction in your work.

Did you write the problem as either 6.5 + 6.5 + 3.5 + 3.5 OR 2(6.5) + 2(3.5)?

Yes ____ No ____

If you answered No, stop and rewrite the problem again. Be careful to align the decimal points correctly when adding or multiplying.

Did you include the correct units, m, with your answer?

Yes ____ No ____

If you answered No, then make sure to include these units in your final answer.

If you answered Problem 4 incorrectly, please go back and rework this problem using these suggestions.

Find the Area of a Trapezoid—Problem 10   Find the area of a trapezoid with a height of 9 m and bases of 7 m and 25 m.

> **Helpful Hint** You can use the formula for the area of a trapezoid
> $$A = \frac{h(b + B)}{2},$$
> where A equals the area, h is the height, b is the shorter base, and B is the longer base.

When you substituted the height and each base into the formula, did you obtain the equation $A = \frac{9(7 + 25)}{2}$?

Yes ____ No ____

If you answered No, stop and make the necessary corrections to your work.

Did you add the numbers within the parentheses first before multiplying by 9 and then dividing the result by 2?

Yes ____ No ____

If you answered No, go back and follow the correct order of operations.

Did you include the correct units, m^2, with your answer?

Yes ____ No ____

If you answered No, then make sure to include these units in your final answer.

Need help? Watch the MATH COACH videos in MyMathLab® or on You Tube™.

510

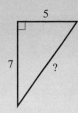

> **Helpful Hint** When the measures of two legs of a right triangle are known, you can find the hypotenuse by using the formula
>
> $$\text{Hypotenuse} = \sqrt{(\text{leg})^2 + (\text{leg})^2}$$

Did you substitute correctly to write the expression $\sqrt{(5)^2 + (7)^2}$ OR $\sqrt{(7)^2 + (5)^2}$?

Yes ____ No ____

If you answered No, stop and look at the problem again to make sure you understand which sides are known and which side is unknown.

Did your problem simplify to $\sqrt{74}$ before approximating the square root?

Yes ____ No ____

If you answered No, go back and perform the necessary calculations again.

Find the Area of a Circle—Problem 24   Find the area of a circle with diameter 12 ft.

> **Helpful Hint** Try to memorize the formula for the area of a circle, $A = \pi r^2$, where A is the area of a circle, r is the circle's radius, and π is pi. Remember that we typically use the value 3.14 as an approximation for π.

Did you remember to divide the diameter by 2 to find the radius?

Yes ____ No ____

If you answered No, go back and complete this step.

Did you remember to square the radius first to obtain the expression 3.14×36?

Yes ____ No ____

If you answered No, stop and make the necessary corrections to your work. Be sure to round the result to the nearest hundredth and add the correct units to your final answer.

Your final answer should have $A \approx$ written to its left. Note that this is the most accurate way to write your answer because an approximation was used for π.

Now go back and rework the problem using these suggestions.

Need more help? Look for section examples marked with $\mathbb{M}_{\mathbb{C}}$ to review.

Have you ever attended a professional women's basketball game? If so, you were probably amazed at the level of athletic ability of the team players. The Women's National Basketball Association (WNBA) was formed in 1996 as the women's counterpart to the NBA. The 12 teams play a regular season each year that starts in June and ends in September. Elaborate statistics are maintained for the WNBA just as they are for the NBA. Many of the statistical measurements that are used are studied in this chapter.

Statistics

8.1 Circle Graphs

① Reading a Circle Graph with Numerical Values

Statistics is that branch of mathematics that collects and studies data. Once the data is collected, it must be organized so that the information is easily readable. We use **graphs** to give a visual representation of the data that is easy to read. Graphs appeal to the eye. Their visual nature allows them to communicate information about the complicated relationships among statistical data. For this reason, newspapers often use graphs to help their readers quickly grasp information.

Circle graphs are especially helpful for showing the relationship of parts to a whole. The entire circle represents 100%; the pie-shaped pieces represent the subcategories. The following circle graph divides the 10,000 students at Westline College into five categories. We will use this graph for Examples 1–4.

Distribution of Students at Westline College

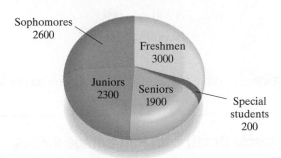

EXAMPLE 1 What is the largest category of students?

Solution The largest pie-shaped section of the circle is labeled "Freshmen." Thus the largest category is freshmen students.

Student Practice 1 What is the smallest category of students?

NOTE TO STUDENT: Fully worked-out solutions to all of the Student Practice problems can be found at the back of the text starting at page SP-1.

EXAMPLE 2

(a) How many students are sophomores or juniors?

(b) What percent of the students are sophomores or juniors?

Solution

(a) There are 2600 sophomores and 2300 juniors. If we add these two numbers, we have $2600 + 2300 = 4900$. Thus we see that there are 4900 students who are sophomores or juniors.

(b) 4900 out of 10,000 are sophomores or juniors.

$$\frac{4900}{10,000} = 0.49 = 49\%$$

Student Practice 2

(a) How many students are freshmen or special students?

(b) What percent of the students are freshmen or special students?

EXAMPLE 3   What is the ratio of freshmen to seniors?

Solution   Number of freshmen $\longrightarrow$ 3000   Thus $\dfrac{3000}{1900} = \dfrac{30}{19}$.
Number of seniors $\longrightarrow$ 1900

The ratio of freshmen to seniors is $\dfrac{30}{19}$.

Student Practice 3   What is the ratio of freshmen to sophomores?

EXAMPLE 4   What is the ratio of seniors to the total number of students?

Solution   There are 1900 seniors. If we only consider the data in the circle graph, we find the total of all the students by adding the number of students in each section of the graph. There are 10,000 students. The ratio of seniors to the total number of students is

$$\frac{1900}{10,000} = \frac{19}{100}.$$

Student Practice 4   What is the ratio of freshmen to the total number of students?

② Reading a Circle Graph with Percentage Values

Together, the Great Lakes form the largest body of fresh water in the world. The total area of these five lakes is about 94,680 mi^2. The percentage of this total area taken up by each of the Great Lakes is shown in the circle graph below.

Percentage of Area Occupied by Each of the Great Lakes

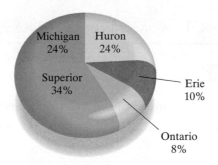

Source: U.S. Department of the Interior

EXAMPLE 5   What percent of the total area is occupied by Lake Erie or Lake Ontario?

Solution   If we add 10% for Lake Erie and 8% for Lake Ontario, we get

$$10\% + 8\% = 18\%.$$

Thus 18% of the area is occupied by Lake Erie or Lake Ontario.

Student Practice 5   What percent of the total area is occupied by Lake Superior or Lake Michigan?

EXAMPLE 6   How many of the total 94,680 mi² are occupied by Lake Michigan? Round to the nearest whole number.

Solution   Remember that we multiply the percent times the base to obtain the amount. Here 24% of 94,680 is what is occupied by Lake Michigan.

$$(0.24)(94,680) = n$$
$$22,723.2 = n$$

Rounded to the nearest whole number, 22,723 mi² are occupied by Lake Michigan.

Student Practice 6   How many of the total 94,680 mi² are occupied by Lake Superior? Round to the nearest whole number.

Sometimes a circle graph is used to investigate the distribution of one part of a larger group. For example, it is projected that in the year 2018, there will be 19,882,000 students enrolled in U.S. colleges and universities. The circle graph shows how the students will be distributed by age.

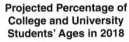

Projected Percentage of College and University Students' Ages in 2018

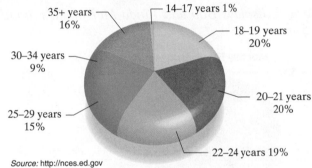

35+ years 16%
14–17 years 1%
18–19 years 20%
30–34 years 9%
20–21 years 20%
25–29 years 15%
22–24 years 19%

Source: http://nces.ed.gov

EXAMPLE 7

(a) What percent of the students represented in this circle graph are projected to be between 20 and 24 years old?

(b) Of the projected 19,882,000 college and university students in 2018, how many will be between 30 and 34 years old?

Solution

(a) We add 20% to 19% to obtain 39%. Thus 39% of the students represented by this graph are projected to be between 20 and 24 years old.

(b) We take 9% of the 19,882,000 students expected to be enrolled in colleges and universities. Thus we have $(0.09)(19,882,000) = 1,789,380$. Approximately 1,789,380 college and university students will be between 30 and 34 years old in 2018.

Student Practice 7

(a) What percent of the students represented in this circle graph are projected to be 21 years or younger?

(b) How many college and university students are projected to be 35 years or older in 2018?

Verbal and Writing Skills *Suppose that you must create a circle graph for 4000 students who attend Springfield Community College.*

1. If 25% of the students live within 5 miles of the college, how would you determine how many students live within 5 miles of the college?

2. If 45% of the students live more than 8 miles from the college, how would you determine how many students live more than 8 miles from the college?

3. How would you construct a pie slice that describes the students who live within 5 miles of the college?

4. You plan to create a circle graph with three slices: one for those who live within 5 miles of the college, one for those who live between 5 and 8 miles from the college, and one for those who live more than 8 miles from the college. Explain how you would find how many students live between 5 and 8 miles from the college.

Applications

Monthly Budget *The following circle graph displays Bob and Linda McDonald's monthly $2700 family budget. Use the circle graph to answer exercises 5–14.*

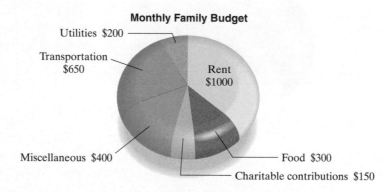

Monthly Family Budget

Utilities $200
Transportation $650
Rent $1000
Miscellaneous $400
Food $300
Charitable contributions $150

5. What category takes the largest amount of the budget?

6. Which two categories take the least amounts of the budget?

7. How much money is allotted each month for utilities?

8. How much money is allotted each month for transportation (this includes car payments, insurance, and gas)?

9. How much money in total is allotted each month for transportation or charitable contributions?

10. How much money is allotted for food or rent?

11. What is the ratio of money spent for transportation to money spent on utilities?

12. What is the ratio of money spent on rent to money spent on miscellaneous items?

13. What is the ratio of money spent on rent to the total amount of the monthly budget?

14. What is the ratio of money spent on food to the total amount of the monthly budget?

Age Distribution *It is projected that in the year 2020, there will be approximately 336 million people in the United States. The following circle graph shows the age distribution of these people. Use the circle graph to answer exercises 15–24.*

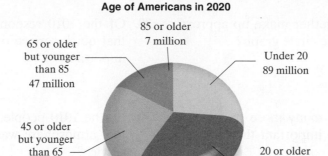

Age of Americans in 2020

85 or older
7 million

65 or older
but younger
than 85
47 million

Under 20
89 million

45 or older
but younger
than 65
84 million

20 or older
but younger
than 45
109 million

Source: www.census.gov

15. What age group will have the smallest number of people?

16. What age group will have the largest number of people?

17. How many people will be 65 years old or older but younger than 85?

18. How many people will be 20 years old or older but younger than 45?

19. How many people will be younger than 65?

20. How many people will be 45 or older?

21. What will be the ratio of the number of people younger than 45 to the number of people 45 years old or older?

22. What will be the ratio of the number of people 20 or older to the number of people under 20?

23. What will be the ratio of the number of people under 20 to the total population?

24. What will be the ratio of the number of people 85 or older to the total population?

Restaurant Preferences *In a survey, 1010 people were asked which aspect of dining out was most important to them. The results are shown in the circle graph below. Use the graph to answer exercises 25–30. Round all answers to the nearest whole number.*

What Consumers Want Most in a Restaurant

Reasonable prices
22%

Great
food
56%

Atmosphere
11%

Quick service
8%

Don't know/None of these
3%

Source: National Restaurant Association

25. What percent of respondents felt atmosphere or quick service was most important?

26. What percent of respondents did not feel that great food was most important?

27. Which two categories together make up approximately three-fourths of the circle graph?

28. Of the 1010 respondents, how many responded that quick service or reasonable prices was most important?

29. Of the 1010 people, how many more people felt that great food was more important than reasonable prices?

30. Of the 1010 people, how many more people felt that atmosphere was more important than quick service?

Vehicle Production In 2009, the top seven countries for vehicle production were those displayed in the following circle graph. A total of 42,000,000 vehicles were manufactured that year. The approximate percentages of the total, rounded to the nearest tenth, are given for each country. Use the graph to answer exercises 31–38.

Vehicle Production in 2009

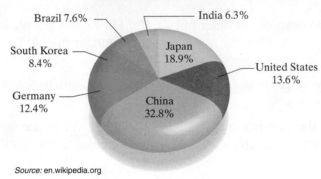

Brazil 7.6% India 6.3%

South Korea 8.4% Japan 18.9%

Germany 12.4% United States 13.6%

China 32.8%

Source: en.wikipedia.org

31. Approximately how many vehicles were produced in China?

32. Approximately how many vehicles were produced in the United States?

33. What percentage of the vehicles were produced in Asia (China, Japan, and South Korea)?

34. What percentage of the vehicles were produced in non-Asian countries (Brazil, Germany, United States, India)?

35. What percentage of the vehicles were *not* manufactured in China or Japan?

36. What percentage of the vehicles were *not* manufactured in the Western hemisphere (United States, Brazil)?

37. How many more vehicles were produced in China than in Japan and the United States combined?

38. How many more vehicles were produced in Japan than in India and Brazil combined?

Cumulative Review

▲ **39.** **[7.4.2]** *Geometry* Find the area of a right triangle with base 12 ft and height 20 ft.

▲ **40.** **[7.3.1]** *Geometry* Find the area of a parallelogram with base of 17 in. and height 12 in.

▲ **41.** **[7.10.1]** *Paint Coverage* How many gallons of paint will it take to cover the four sides of a barn with two sides that measure 7 yd by 12 yd and two sides that measure 7 yd by 20 yd? Assume that a gallon of paint covers 28 square yards.

▲ **42.** **[7.10.1]** *Reflector Construction* A circular reflector has a radius of 8 cm. How many grams of silver will it take to cover the reflector if each gram will cover 64 cm^2? Assume that the reflector is covered on one side only. (Use $\pi \approx 3.14$.) Round to the nearest whole number.

Quick Quiz 8.1 *Recently a group of car dealers estimated the number of new cars and SUVs sold in New England in 2010 to be 850,000 vehicles. They estimated the distribution by category as shown in the circle graph below. Use this graph to answer the following questions.*

1. What percent of the vehicles sold were two-door coupes or four-door sedans?

2. How many of the vehicles sold were SUVs or minivans?

3. How many of the vehicles sold were *not* station wagons?

4. **Concept Check** Explain how you would find the total number of the vehicles sold that were station wagons, four-door sedans, or minivans.

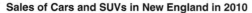

Sales of Cars and SUVs in New England in 2010

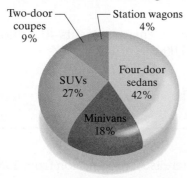

Two-door coupes 9%
Station wagons 4%
SUVs 27%
Four-door sedans 42%
Minivans 18%

8.2 Bar Graphs and Line Graphs

Student Learning Objectives

After studying this section, you will be able to:

1. Read and interpret a bar graph.

2. Read and interpret a double-bar graph.

3. Read and interpret a line graph.

4. Read and interpret a comparison line graph.

NOTE TO STUDENT: Fully worked-out solutions to all of the Student Practice problems can be found at the back of the text starting at page SP-1.

① Reading and Interpreting a Bar Graph

Bar graphs are helpful for seeing changes over a period of time. Bar graphs or line graphs are especially helpful when the same type of data is repeatedly studied. The following bar graph shows the approximate population of California from 1950 to 2010.

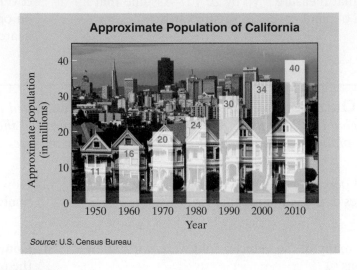

Approximate Population of California

Source: U.S. Census Bureau

EXAMPLE 1 What was the approximate population of California in 2010?

Solution The bar for 2010 rises to 40. This represents 40 million; thus the approximate population was 40,000,000.

Student Practice 1 What was the approximate population of California in 1980?

EXAMPLE 2 What was the increase in population from 1980 to 1990?

Solution The bar for 1980 rises to 24. Thus the approximate population was 24,000,000. The bar for 1990 rises to 30. Thus the approximate population was 30,000,000. To find the increase in population from 1980 to 1990, we subtract.

$$30,000,000 - 24,000,000 = 6,000,000$$

The answer is 6 million or 6,000,000.

Student Practice 2 What was the increase in population from 1950 to 1970?

② Reading and Interpreting a Double-Bar Graph

Double-bar graphs are useful for making comparisons. For example, when a company is analyzing its sales, it may want to compare different years or different quarters. The following double-bar graph illustrates the sales of new cars at a Ford dealership for two different years, 2009 and 2010. The sales are recorded for each quarter of the year.

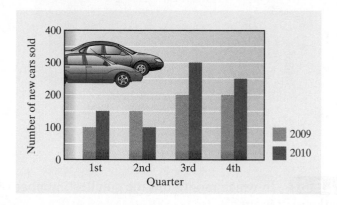

EXAMPLE 3 How many cars were sold in the second quarter of 2009?

Solution The bar rises to 150 for the second quarter of 2009. Therefore, 150 cars were sold.

Student Practice 3 How many cars were sold in the fourth quarter of 2010?

EXAMPLE 4 How many more cars were sold in the third quarter of 2010 than in the third quarter of 2009?

Solution From the double-bar graph, we see that 300 cars were sold in the third quarter of 2010 and that 200 cars were sold in the third quarter of 2009.

$$
\begin{array}{r}
300 \\
-\ 200 \\
\hline
100
\end{array}
$$

Thus, 100 more cars were sold.

Student Practice 4 How many fewer cars were sold in the second quarter of 2010 than in the second quarter of 2009?

③ Reading and Interpreting a Line Graph

A **line graph** is useful for showing trends over a period of time. In a line graph only a few points are actually plotted from measured values. The points are then connected by straight lines to show a trend. The intervening values between points may not lie exactly on the line. The following line graph shows the number of customers per month coming into a restaurant in a vacation community.

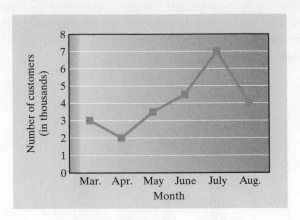

EXAMPLE 5 In which month did the smallest number of customers come into the restaurant?

Solution The lowest point on the graph occurs for the month of April. Thus the fewest number of customers came in April.

Student Practice 5 In which month did the greatest number of customers come into the restaurant?

EXAMPLE 6

(a) Approximately how many customers came into the restaurant during the month of June?

(b) From May to June, did the number of customers increase or decrease?

Solution

(a) Notice that the dot is halfway between 4 and 5. This represents a value halfway between 4000 and 5000 customers. Thus we would estimate that 4500 customers came during the month of June.

(b) From May to June the line goes up, so the number of customers increased.

Student Practice 6

(a) Approximately how many customers came into the restaurant during the month of May?

(b) From March to April, did the number of customers increase or decrease?

EXAMPLE 7 Between what two months was the increase in the number of customers the largest?

Solution The line from June to July goes upward at the steepest angle. This represents the largest increase. (You can check this by reading the numbers from the left axis.) Thus the greatest increase in customers was between June and July.

Student Practice 7 Between what two months did the biggest decrease occur?

④ Reading and Interpreting a Comparison Line Graph

Two or more sets of data can be compared by using a **comparison line graph.** A comparison line graph shows two or more line graphs together. A different style for each line distinguishes them. Note that using a blue line and a red line in the following graph makes it easy to read.

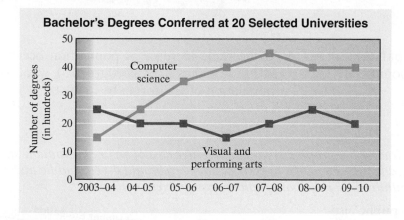

EXAMPLE 8 How many bachelor's degrees in computer science were awarded in the academic year 2009–2010?

Solution Because the dot corresponding to 2009–2010 is at 40 and the scale is in hundreds, we have $40 \times 100 = 4000$. Thus 4000 degrees were awarded in computer science in 2009–2010.

Student Practice 8 How many bachelor's degrees in visual and performing arts were awarded in the academic year 2009–2010?

EXAMPLE 9 In what academic year were more degrees awarded in the visual and performing arts than in computer science?

Solution The only year when more bachelor's degrees were awarded in the visual and performing arts was the academic year 2003–2004.

Student Practice 9 What was the first academic year in which more degrees were awarded in computer science than in the visual and performing arts?

8.2 Exercises

MyMathLab®

Watch the videos
in MyMathLab

Download the
MyDashBoard App

Applications

California Population The following bar graph shows the approximate population of California from 1960 to 2020. Use the graph to answer exercises 1–6.

1. What was the approximate population in 1960?

2. What is the population projected to be in 2020?

3. What was the approximate population in 1980?

4. What was the approximate population in 2000?

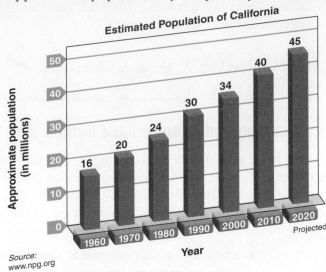

5. Between what years did the population of California increase by the largest amount?

6. If the same increase in population occurs from 2020 to 2040 as from 2000 to 2020, what will the population of California be in 2040?

Cost of Higher Education The following double-bar graph displays the average cost of an undergraduate student's tuition, fees, and room and board for the academic years 2003 to 2007 at both 2-year and 4-year public institutions. Figures have been rounded to the nearest hundred. Use the bar graph to answer exercises 7–18.

7. What was the average cost at a 2-year public institution in 2003–04?

8. What was the average cost at a 4-year public institution in 2006–07?

9. How much higher was the average cost at a 4-year institution than at a 2-year institution in 2007–08?

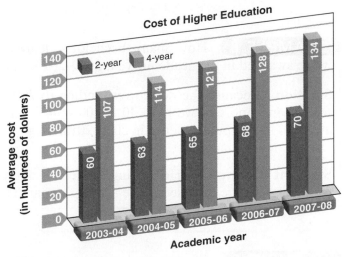

10. How much less was the average cost at a 2-year institution than at a 4-year institution in 2005–06?

11. In which academic year was there the smallest difference between the average costs at a 2-year institution and a 4-year institution?

12. In which academic year was there the greatest difference between the average costs at 2-year and 4-year institutions?

13. By how much did the average cost at a 2-year institution increase from 2006–07 to 2007–08?

14. By how much did the average cost at a 4-year institution increase from 2005–06 to 2006–07?

15. James decided to attend a 2-year college for two years, and then transfer to a 4-year college for two years. If he started college in the fall of 2003 and paid the average cost for each of the four years, how much did he spend on his four years of education?

16. Monica attended a 2-year community college for two years starting in the fall of 2006, and then transferred to a 4-year school. Assuming she paid average costs, how much did she save during the first two years by choosing to start her education at a 2-year college?

17. What was the percent increase in the average cost at a public 2-year institution from 2003–04 to 2007–08?

18. What was the percent increase in the average cost at a public 4-year institution from 2003–04 to 2007–08?

Baseball Players' Salaries *The following line graph shows how the average major league baseball player's salary has increased over a 12-year period. Notice the average salary is given for odd-numbered years only. Also note that the left scale does not start at zero. Use the graph to answer exercises 19–24.*

19. What was the average baseball player's salary in 1997?

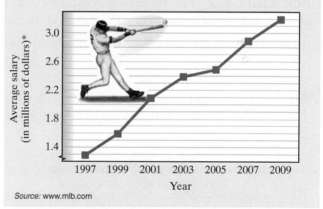

Source: www.mlb.com

*Figures have been rounded.

20. What was the average baseball player's salary in 2009?

21. Which 2-year period had the smallest increase?

22. Which 2-year period had the largest increase?

23. Compare the average salary in 1999 to the average salary in 2001. By how much did the average salary increase from 1999 to 2001?

24. If the average salary continues to increase every two years by the same amount that it increased from 2007 to 2009, what will the average salary be in 2017?

Springfield Rainfall *The following comparison line graph indicates the rainfall for the last six months of two different years in Springfield. Use the graph to answer exercises 25–30.*

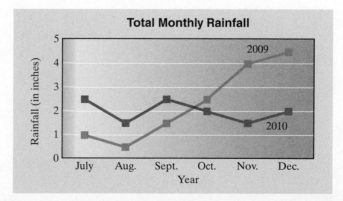

25. In September 2010, how many inches of rain were recorded?

26. In October 2009 how many inches of rain were recorded?

27. During what months was the rainfall of 2010 less than the rainfall of 2009?

28. During what months was the rainfall of 2010 greater than the rainfall of 2009?

29. How many more inches of rain fell in November 2009 than in October 2009?

30. How many more inches of rain fell in September 2009 than in August 2009?

Cumulative Review *Do each calculation in the proper order.*

31. **[1.6.2]** $(5 + 6)^2 - 18 \div 9 \times 3$

32. **[2.8.3]** $\dfrac{1}{5} + \left(\dfrac{1}{5} - \dfrac{1}{6} \right) \times \dfrac{2}{3}$

33. **[5.4.1]** For the 2018–2019 academic year, it is projected that 1,821,000 bachelor's degrees will be awarded in the United States. Of these, 59.2% will be awarded to women. How many bachelor's degrees will be awarded to women in 2018–2019? (*Source:* www.nces.ed.gov)

34. **[5.4.1]** The Appalachian Trail is 2175 miles long. This trail accounts for about 31.25% of the total miles of national scenic trails in the United States. How many total national scenic trail miles are there in the United States? Round to the nearest whole mile. (*Source: Time Almanac 2010*)

Quick Quiz 8.2 *The following comparison line graph shows the number of houses and the number of condominiums built in the years 1990 to 2010 in Essex County. Use the graph to answer the problems below.*

1. How many condominiums were built in the year 2005?

2. How many more homes were built in 1990 compared to the number of condominiums?

3. During what year was the number of homes built closest to the number of condominiums built?

4. **Concept Check** If the increase in the number of condominiums from 2000 to 2010 continues at the same rate until 2020, explain how you would find the number of condominiums that would be constructed in 2020.

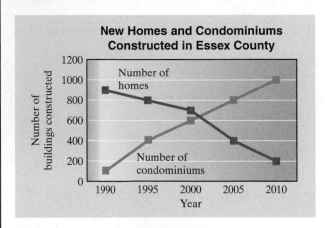

How Am I Doing?　Sections 8.1–8.2

How are you doing with your homework assignments in Sections 8.1 and 8.2? Do you feel you have mastered the material so far? Do you understand the concepts you have covered? Before you go further in the textbook, take some time to do each of the following problems.

8.1

In 2009, the top six most-visited U.S. national parks had about 27,000,000 visitors. The circle graph displays the approximate percentages of the total that visited each park.

Most-Visited U.S. National Parks in 2009

Source: www.infoplease.com

1. What percentage of the visitors went to Yosemite National Park?

2. To which park did the greatest number of visitors go?

3. What percent of the visitors went to Olympic or Yellowstone National Parks?

4. How many of the 27,000,000 visitors went to Grand Canyon National Park?

5. How many of the total visitors went to Yosemite or Yellowstone National Parks?

8.2

The following double-bar graph indicates the number of new housing starts in Springfield during each quarter of 2009 and 2010.

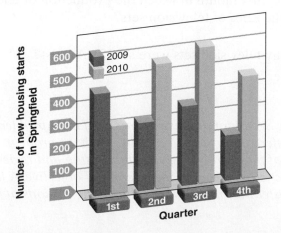

6. How many housing starts were there in Springfield in the first quarter of 2009?

1. _____

2. _____

3. _____

4. _____

5. _____

6. _____

7. How many housing starts were there in Springfield in the second quarter of 2010?

8. When were the smallest number of housing starts in Springfield?

9. When were the greatest number of housing starts in Springfield?

10. How many more housing starts were there in the third quarter of 2010 than in the third quarter of 2009?

11. How many fewer housing starts were there in the first quarter of 2010 than in the first quarter of 2009?

The line graph indicates sales and production of color television sets by a major manufacturer during the specified months.

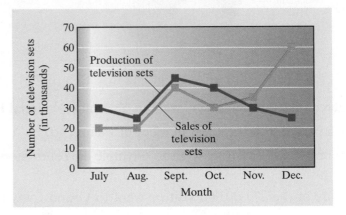

12. During what months was the production of television sets the lowest?

13. During what month were the sales of television sets the highest?

14. What was the first month in which the production of television sets was lower than the sales of television sets?

15. (a) How many television sets were sold in August?
    (b) November?

Now turn to page SA-14 for the answer to each of these problems. Each answer also includes a reference to the objective in which the problem is first taught. If you missed any of these problems, you should stop and review the Examples and Student Practice problems in the referenced objective. A little review now will help you master the material in the upcoming sections of the text.

8.3 Histograms

① Understanding and Interpreting a Histogram

Student Learning Objectives

After studying this section, you will be able to:

① Understand and interpret a histogram.

② Construct a histogram from raw data.

In business and in higher education you are often asked to take data and organize them in some way. This section shows you the technique for making a *histogram*—a type of bar graph.

Suppose that a mathematics professor announced the results of a class test. The 40 students in the class scored between 50 and 99 on the test. The results are displayed in the following chart.

Scores on the Test	Class Frequency (Number of Students)
50–59	4
60–69	6
70–79	16
80–89	8
90–99	6

The results in the table can be organized in a special type of bar graph known as a **histogram.** In a histogram the width of each bar is the same. The width represents the range of scores on the test. This is called a **class interval.** The height of each bar gives the class frequency for each class interval. The **class frequency** is the number of times a score occurs in a particular class interval. Be sure to notice that the bars touch each other. This is a main difference between a bar graph and a histogram. Use the histogram below to do Examples 1 and 2.

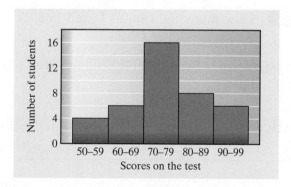

EXAMPLE 1 How many students scored a B on the test if the professor considers a test score of 80–89 a B?

Solution Since the 80–89 bar rises to a height of 8, eight students scored a B on the test.

Student Practice 1 How many students scored a D on the test if the professor considers a test score of 60–69 a D?

NOTE TO STUDENT: Fully worked-out solutions to all of the Student Practice problems can be found at the back of the text starting at page SP-1.

EXAMPLE 2 How many students scored less than 80 on the test?

Solution From the histogram, we see that there are three different bar heights to be included. Four tests were scored 50–59, six tests were scored 60–69, and 16 tests were scored 70–79. When we combine $4 + 6 + 16 = 26$, we can see that 26 students scored less than 80 on the test.

Student Practice 2 How many students scored greater than 69 on the test?

The following histogram tells us about the length of life of 110 new light bulbs tested at a research center. The number of hours the bulbs lasted is indicated on the horizontal scale. The number of bulbs lasting that long is indicated on the vertical scale. Use this histogram for Examples 3 and 4.

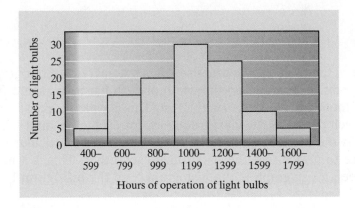

Hours of operation of light bulbs

EXAMPLE 3 How many light bulbs lasted between 1400 and 1599 hours?

Solution The bar with a range of 1400–1599 hours rises to 10. Thus 10 light bulbs lasted that long.

Student Practice 3 How many light bulbs lasted between 800 and 999 hours?

EXAMPLE 4 How many light bulbs lasted less than 1000 hours?

Solution We see that there are three different bar heights to be included. Five bulbs lasted 400–599 hours, 15 bulbs lasted 600–799 hours, and 20 bulbs lasted 800–999 hours. We add $5 + 15 + 20 = 40$. Thus 40 light bulbs lasted less than 1000 hours.

Student Practice 4 How many light bulbs lasted more than 1199 hours?

② Constructing a Histogram from Raw Data

To construct a histogram, we start with *raw data,* data that have not yet been organized or interpreted. We perform the following steps.

---

1. Select class intervals of equal width for the data.

2. Make a table with class intervals and a *tally* (count) of how many numbers occur in each interval. Add up the tally to find the class frequency for each class interval.

3. Draw the histogram.

    First we will practice making the table. Later we will use the table to draw the histogram.

EXAMPLE 5 Each of the following numbers represents the number of kilowatt-hours of electricity used in 15 homes during a one-month period. Create a set of class intervals for this data and then determine the frequency for each class interval.

770	520	850	900	1100
1200	1150	730	680	900
1160	590	670	1230	980

Solution

1. We select class intervals of equal width for the data. We choose intervals of 200. We might have chosen smaller or larger intervals, but we choose 200 because it gives us a convenient number of intervals to work with, as we will see.

2. We make a table. We write down the class intervals, then count (tally) how many numbers occur within each interval. Then we write the total. This is the class frequency.

Kilowatt-Hours Used *(Class Interval)*	*Tally*	*Frequency*
500–699	||||	4
700–899	|||	3
900–1099	|||	3
1100–1299	|||||	5

Student Practice 5 Each of the following numbers represents the weight in pounds of a new car.

2250	1760	2000	2100	1900
1640	1820	2300	2210	2390
2150	1930	2060	2350	1890

Continued on next page

Complete the following table to determine the frequency for each class interval for the preceding data.

Weight in Pounds (Class Interval)	Tally	Frequency
1600–1799		
1800–1999		
2000–2199		
2200–2399		

One of the purposes of a histogram is to give you a visual sense of how the data is distributed. For example, if you look at the raw data of Example 5 you may be left with the sense that the home used very different amounts of electricity during the month. For most of us, looking at the raw data does not help us understand the situation. However, once we construct a histogram such as the one in Example 6, we are able to see patterns and trends of electricity use.

EXAMPLE 6 Draw a histogram from the table in Example 5.

Solution

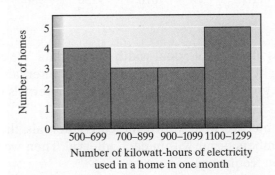

Student Practice 6 Draw a histogram using the data in Student Practice problem 5.

Note: Usually it is desirable for the class intervals to be of equal size. However, sometimes data is collected in such a way that this is not possible. We will see this situation in Example 7. Here government data was collected with unequal class intervals.

EXAMPLE 7 Draw a histogram for the following table of recent data showing the number of people in the United States, in each of five age categories.

Age Category	Number of People in the United States
19 or younger	84,151,000
20–34	63,597,000
35–54	85,982,000
55–64	36,275,000
65 or older	40,229,000

Source: U.S. Census Bureau

Solution

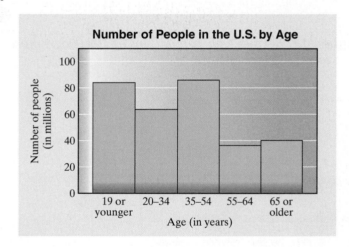

Student Practice 7 Based on the preceding histogram, between what two age categories is there the greatest difference in population in the United States?

8.3 Exercises

MyMathLab®

Watch the videos
in MyMathLab

Download the
MyDashBoard App

Verbal and Writing Skills

1. Describe two main differences between a bar graph and a histogram.

2. Suppose you had test data 22, 24, 33, 44, 55, 66, 38, 48, and 60. If you were going to have a histogram with three class intervals, explain how you would pick the intervals.

3. Explain in your own words what is meant by class frequency.

4. Jason made a histogram and it had the following class intervals: 300–400, 400–500, 500–600, 600–700. Explain why that is not a good choice of class intervals.

Applications

City Population *Recent data showing the number of U.S. cities with populations of 100,000 or more is depicted in the following histogram. Use the histogram to answer exercises 5–12.*

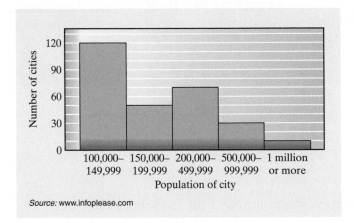

Source: www.infoplease.com

5. How many U.S. cities have a population of 100,000–149,999?

6. How many U.S. cities have a population of 200,000–499,999?

7. How many U.S. cities have a population of 1 million or more?

8. How many U.S. cities have a population of 150,000–199,999?

9. How many U.S. cities have a population of 500,000 or more?

10. How many U.S. cities have a population of 100,000 or more?

11. How many U.S. cities have between 100,000 and 199,999 people?

12. How many U.S. cities have between 150,000 and 999,999 people?

Book Sales *A large company composed of three bookstores studied its yearly report to find out its customers' spending habits. The company sold a total of 70,000 books. The following histogram indicates the number of books sold in certain price ranges. Use the histogram to answer exercises 13–22.*

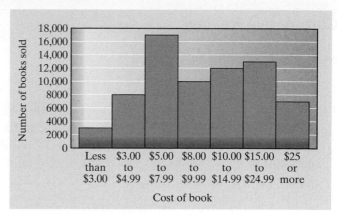

13. How many books priced at $3.00 to $4.99 were sold?

14. How many books priced at $25.00 or more were sold?

15. Which price category of books did the bookstore sell the most of?

16. What price category of books did the bookstore sell the least of?

17. How many books priced at less than $8.00 were sold?

18. How many books priced at more than $9.99 were sold?

19. How many books priced between $5.00 and $24.99 were sold?

20. How many books priced between $3.00 and $9.99 were sold?

21. What percent of the 70,000 books sold were over $14.99? Round to the nearest tenth.

22. What percent of the 70,000 books sold were under $8.00?

Boston Temperature *The numbers in the following chart are the daily high temperatures in degrees Fahrenheit in Boston during February. In exercises 23–30, determine the frequencies of the class intervals for this data.*

23°	26°	30°	18°	42°	17°	19°
51°	42°	38°	36°	12°	18°	14°
20°	24°	26°	30°	18°	17°	16°
35°	38°	40°	33°	19°	22°	26°

	Temperature (Class Interval)	Tally	Frequency		Temperature (Class Interval)	Tally	Frequency
23.	12°–16°	_____	_____	**24.**	17°–21°	_____	_____
25.	22°–26°	_____	_____	**26.**	27°–31°	_____	_____
27.	32°–36°	_____	_____	**28.**	37°–41°	_____	_____
29.	42°–46°	_____	_____	**30.**	47°–51°	_____	_____

31. Construct a histogram using the table prepared in exercises 23–30.

32. How many days in February was the temperature in Boston greater than 36°?

33. How many days in February was the temperature in Boston less than 27°?

Cumulative Review

34. [4.3.2] Solve for m. $\dfrac{182}{m} = \dfrac{25}{19}$

35. [4.3.2] Solve for n. $\dfrac{n}{18} = \dfrac{3.5}{9}$

36. [4.4.1] *Gas Mileage* Warren discovered that he could drive 375 miles on 7.5 gallons of gas in his new Honda Civic Hybrid. The tank of this hybrid car holds 12.3 gallons of gas. How many miles can he drive on a full tank?

37. [4.4.1] *Snowfall on Mount Washington* Tim and Judy Newitt worked as scientists on Mount Washington last year. They found that every 23 in. of snow corresponded to 2 in. of water. During the month of January they measured 150 in. of snow at the mountain weather observatory. How many inches of water does this correspond to? Round to the nearest tenth of an inch.

Quick Quiz 8.3 *The number of times per month that people visit the YMCA gym in Springfield is displayed on the histogram below. Use this histogram to answer questions 1–4.*

1. How many people visit the gym 9–12 times per month?

2. How many people visit the gym more than four times per month?

3. How many more people visit the gym 9–12 times per month compared to those who visit only 1–4 times per month?

4. **Concept Check** During a promotion month last summer, nonmembers were allowed to visit the gym for free. The number of people who visited the gym 1–4 times a month tripled. The number of people who visited the gym 5–8 times a month doubled. Explain how you would find how many people visited the gym between one and eight times per month.

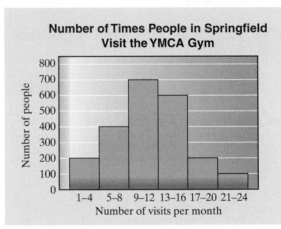

8.4 Mean, Median, and Mode

① Finding the Mean of a Set of Numbers

We often want to know the "middle value" of a group of numbers. In this section we learn that, in statistics, there is more than one way of describing this middle value: there is the *mean* of the group of numbers, there is the *median* of the group of numbers, and in most cases there is a *mode* of the group of numbers. In some situations it's more helpful to look at the mean, in others it's more helpful to look at the median, and in yet others the mode. We'll learn to tell which situations lend themselves to one or the other.

The **mean** of a set of values is the sum of the values divided by the number of values. The mean is often called the **average.**

The mean value is often rounded to a certain decimal-place accuracy.

Student Learning Objectives

After studying this section, you will be able to:

① Find the mean of a set of numbers.

② Find the median of a set of numbers.

③ Find the mode of a set of numbers.

EXAMPLE 1 Carl recorded the miles per gallon achieved by his car for the last two months. His results were as follows:

Week	1	2	3	4	5	6	7	8
Miles per Gallon	26	24	28	29	27	25	24	23

What is the mean miles-per-gallon figure for the last eight weeks? Round to the nearest whole number.

Solution

$$\text{Sum of values} \longrightarrow \quad \frac{26 + 24 + 28 + 29 + 27 + 25 + 24 + 23}{8} \longleftarrow \text{Number of values}$$

$$= \frac{206}{8} = 25.75 \approx 26 \text{ rounded to the nearest whole number}$$

The mean miles-per-gallon figure is 26.

Student Practice 1 Bob and Wally kept records of their phone bills for the last six months. Their bills were $39.20, $43.50, $81.90, $34.20, $51.70, and $48.10. Find the mean monthly bill. Round to the nearest cent.

NOTE TO STUDENT: Fully worked-out solutions to all of the Student Practice problems can be found at the back of the text starting at page SP-1.

② Finding the Median of a Set of Numbers

If a set of numbers is arranged in order from smallest to largest, the **median** is that value that has the same number of values above it as below it.

If the numbers are not arranged in order, then the first step in finding the median is to put the numbers in order.

EXAMPLE 2   Find the median value of the following costs for microwave ovens: $100, $60, $120, $200, $190, $120, $320, $290, $180.

Solution   We must arrange the numbers in order from smallest to largest (or largest to smallest).

$$\underbrace{\$60, \$100, \$120, \$120}_{\substack{\text{four} \\ \text{numbers}}} \quad \underset{\substack{\uparrow \\ \text{middle} \\ \text{number}}}{\$180} \quad \underbrace{\$190, \$200, \$290, \$320}_{\substack{\text{four} \\ \text{numbers}}}$$

Thus $180 is the median cost.

Student Practice 2   Find the median value of the following weekly salaries: $320, $150, $400, $600, $290, $150, $450.

If a list of numbers contains an even number of items, then of course there is no one middle number. In this situation we obtain the median by taking the average of the two middle numbers.

EXAMPLE 3   Find the median of the following numbers: 26, 31, 39, 33, 13, 16, 18, 38.

Solution   First we place the numbers in order from smallest to largest.

$$\underbrace{13, 16, 18}_{\substack{\text{three} \\ \text{numbers}}} \quad \underset{\substack{\uparrow \\ \text{two middle} \\ \text{numbers}}}{26, 31} \quad \underbrace{33, 38, 39}_{\substack{\text{three} \\ \text{numbers}}}$$

The average (mean) of 26 and 31 is

$$\frac{26 + 31}{2} = \frac{57}{2} = 28.5.$$

Thus the median value is 28.5.

Student Practice 3   Find the median value of the following numbers: 126, 105, 88, 100, 90, 118.

SIDELIGHT   When would someone want to use the mean, and when would someone want to use the median? Which is more helpful?

The mean, or average, is used more frequently. It is most helpful when the data are distributed fairly evenly, that is, when no one value is "much larger" or "much smaller" than the rest.

For example, suppose a company had employees with annual salaries of $9000, $11,000, $14,000, $15,000, $17,000, and $20,000. All the salaries fall within a fairly limited range. The mean salary

$$\frac{9000 + 11,000 + 14,000 + 15,000 + 17,000 + 20,000}{6} \approx \$14,333.33$$

gives us a reasonable idea of the typical salary.

However, suppose the company had six employees with salaries of $9000, $11,000, $14,000, $15,000, $17,000, and $90,000. Talking about the mean salary, which is $26,000, is deceptive. No one earns a salary very close to the mean salary. The typical worker in that company does not earn around $26,000. In this case, the median value is more appropriate. Here the median is $14,500. See exercises 43 and 44 in Exercises 8.4 for more on this.

③ Finding the Mode of a Set of Numbers

Another value that is sometimes used to describe a set of data is the mode. The **mode** of a set of data is the number or numbers that occur most often.

EXAMPLE 4   The following numbers are the weights of automobiles measured in pounds:

$$2345, 2567, 2785, 2967, 3105, 3105, 3245, 3546.$$

Find the mode of these weights.

Solution   The value 3105 occurs twice, whereas each of the other values occurs just once. Thus the mode is 3105 pounds.

Student Practice 4   The following numbers are the heights in inches of 10 male students in Basic Mathematics: 64, 66, 67, 69, 70, 71, 71, 73, 75, 76. Find the mode of these heights.

A set of numbers may have more than one mode.

EXAMPLE 5   The following numbers are finish times for 12 high school students who ran a distance of one mile. The finish times are measured in seconds.

$$290, 272, 268, 260, 290, 272, 330, 355, 368, 290, 370, 272$$

Find the mode of these finish times.

Solution   First we need to arrange the numbers in order from smallest to largest and include all repeats.

$$260, 268, 272, 272, 272, 290, 290, 290, 330, 355, 368, 370$$

Now we can see that the value 272 occurs three times, as does the value 290. Thus the modes for these finish times are 272 seconds and 290 seconds.

Student Practice 5   The following numbers are distances in miles that 16 students traveled to take classes at Massasoit Community College each day.

$$2, 5, 8, 3, 12, 15, 28, 8, 3, 14, 16, 31, 33, 27, 3, 28$$

Find the mode of these distances.

A set of numbers may have **no mode** at all. For example, the set of numbers 50, 60, 70, 80, 90 has no mode because each number occurs just once. The set of numbers 33, 33, 44, 44, 55, 55 has no mode because each number occurs twice. If all numbers occur the same number of times, there is no mode.

MyMathLab®

Watch the videos
in MyMathLab

Download the
MyDashBoard App

Verbal and Writing Skills

1. Explain the difference between a median and a mean.

2. Explain why some sets of numbers have one mode, others two modes, and others no modes.

Applications *In exercises 3–12, find the mean. Round to the nearest tenth unless otherwise directed.*

3. *Coffeehouse Customers* The numbers of customers who were served at Grinders Coffeehouse between 8:00 A.M. and 9:00 A.M. in the past seven days were as follows: 30, 29, 28, 35, 34, 37, 31.

4. *Pizza Delivery* The numbers of pizzas delivered by Papa John's over the last 7 days were as follows: 28, 17, 18, 21, 24, 30, 30.

5. *Average Rainfall* The average amount of rain in Jackson, Mississippi, for the first six months of the year is recorded as follows. Find the mean number of inches for these months.

Jan	Feb	Mar	Apr	May	June
5.3 in.	4.7 in.	5.8 in.	5.6 in.	5.1 in.	3.3 in.

Source: www.countrystudies.us

6. *Average Temperature* The average high temperature in Montgomery, Alabama, for July through December is recorded as follows. Find the mean high temperature for these months.

July	Aug	Sep	Oct	Nov	Dec
91°F	90°F	87°F	78°F	68°F	60°F

Source: www.countrystudies.us

7. *Baseball* The captain of the college baseball team achieved the following results:

	Game 1	Game 2	Game 3	Game 4	Game 5
Hits	0	2	3	2	2
Times at Bat	5	4	6	5	4

Find his batting average by dividing his total number of hits by the total times at bat. Round to the nearest thousandth if necessary.

8. *Bowling* The captain of the college bowling team had the following results after practice:

	Practice 1	Practice 2	Practice 3	Practice 4
Score (Pins)	541	561	840	422
Number of Games	3	3	4	2

Find her bowling average by dividing the total number of pins scored by the total number of games.

9. *Population of Guam* The population on the island of Guam has increased significantly over the last 30 years. It is expected to continue to increase in 2020. Find an approximate value for the mean population for this 40-year period from the following population chart.

1980 Population	1990 Population	2000 Population	2010 Population	2020* Population
107,000	134,000	152,000	182,000	204,000

*estimated
Sources: U.S. Census Bureau, www.brittanica.com

10. *Population of Virgin Islands* The population on the U.S. Virgin Islands increased from 1980 to 2000 but since then has stayed relatively unchanged. A slight decrease is projected for 2020. Find the approximate value for the mean population from 1980 to 2020 by using all the values in the following population chart.

1980 Population	1990 Population	2000 Population	2010 Population	2020* Population
98,000	104,000	109,000	110,000	109,000

*estimated
Source: U.S. Census Bureau

11. *Gas Used on a Trip* Frank and Wally traveled to the West Coast during the summer. The number of miles they drove and the number of gallons of gas they used are recorded in the following chart.

	Day 1	Day 2	Day 3	Day 4
Miles Driven	276	350	391	336
Gallons of Gas	12	14	17	14

Find the average miles per gallon achieved by the car on the trip by dividing the total number of miles driven by the total number of gallons used.

12. *Gas Used on a Trip* Cindy and Andrea traveled to Boston this fall. The number of miles they drove and the number of gallons of gas they used are recorded in the following chart.

	Day 1	Day 2	Day 3	Day 4
Miles Driven	260	375	408	416
Gallons of Gas	10	15	17	16

Find the average miles per gallon achieved by the car on the trip by dividing the total number of miles driven by the total number of gallons used.

In exercises 13–24, find the median value.

13. 126, 232, 180, 195, 229

14. 548, 554, 560, 539, 512

15. 12.4, 11.6, 11.9, 12.1, 12.5, 11.9

16. 8.2, 8.1, 7.8, 8.8, 7.9, 7.5

17. *Annual Salary* The annual salaries of six staff members at a local college are $28,500, $32,700, $42,000, $38,250, $40,750, and $35,800.

18. *Family Income* The annual incomes of six families are $24,000, $60,000, $32,000, $18,000, $29,000, and $35,000.

19. *Waiter Workload* The number of tables Carl waited on at his job the past 7 days were 12, 10, 21, 25, 31, 18, 28.

20. *Swimming Times* The number of minutes Ashley spent swimming the past seven days were 35, 60, 45, 40, 50, 80, and 45.

21. *Phone Bills* The phone bills for Dr. Price's cellular phone over the last seven months were as follows: $109, $207, $420, $218, $97, $330, and $185.

22. *Compact Disc Prices* The prices of the same compact disc sold at several different music stores or by mail order were as follows: $15.99, $11.99, $5.99, $12.99, $14.99, $9.99, $13.99, $7.99, and $10.99.

23. *Grade Point Averages* The grade point averages (GPA) for eight students were 1.8, 1.9, 3.1, 3.7, 2.0, 3.1, 2.0, and 2.4.

24. *Food Purchases* The numbers of pounds of smoked turkey breast purchased at a deli by the last eight customers were 1.2, 2.0, 1.7, 2.5, 2.4, 1.6, 1.5, and 2.3.

25. *Largest U.S. Businesses* The table below shows the revenue of the four largest businesses in the United States for the year 2005. Determine the median revenue based on this data.

Company	Chevron	Exxon Mobil	General Motors	Wal-Mart Stores
Revenue, in millions of dollars	$189,481	$339,938	$192,604	$315,654

Source: www.infoplease.com

26. *NBA Leaders* At the end of the 2009–2010 National Basketball Association regular season, the top four scoring leaders with their season point totals were as shown below. The number of points per game is also given. Determine the mean number of total points and the mean number of points per game based on this data.

Player	Kevin Durant (Oklahoma City Thunder)	LeBron James (Cleveland Cavaliers)	Dwyane Wade (Miami Heat)	Dirk Nowitzki (Dallas Mavericks)
Total points	2472	2258	2045	2027
Points per game	30.1	29.7	26.6	25.0

Source: www.nba.com

Find the mean. Round to the nearest cent when necessary.

27. *Business Owners Salary* The salaries of eight small business owners in Big Rapids are $30,000, $74,500, $47,890, $89,000, $57,645, $78,090, $110,370, and $65,800.

28. *Price of Laptop Computer* The prices of nine laptop computers with a Core 2 Duo chip are $5679, $6902, $1530, $2738, $2999, $4105, $3655, $5980, and $4430.

In exercises 29 and 30, find the median.

29. 2576, 8764, 3700, 5000, 7200, 4700, 9365, 1987

30. 15.276, 21.375, 18.90, 29.2, 14.77, 19.02

31. *Holiday Shopping* It took Jenny 5 days to complete her holiday shopping. The amounts she spent during these five days were $120.50, $66.74, $80.95, $210.52, and $45.00. Find the mean and the median.

32. *Property Taxes* The amounts the Dayton family has paid in property taxes the past five years are $1381, $1405, $1405, $1520, $1592. Find the mean and the median.

In exercises 33–38, find the mode.

33. 60, 65, 68, 60, 72, 59, 80

34. 86, 84, 82, 87, 84, 88, 90

35. 121, 150, 116, 150, 121, 181, 117, 123

36. 144, 143, 140, 141, 149, 144, 141, 150

37. *Bicycle Prices* The last seven bicycles sold at the Skol Bike shop cost $249, $649, $269, $259, $269, $249, and $269.

38. *HDTV Television Prices* The last seven HDTV televisions sold at the local Circuit City cost $550, $380, $550, $780, $1250, $640 and $550.

Mixed Practice

39. *Life Expectancy* In 2010, the countries with the highest life expectancy were Japan: 82.6 years; Hong Kong: 82.2 years; Iceland: 81.8 years; Switzerland: 81.7 years; Australia: 81.2 years; Spain: 80.9 years; Sweden: 80.9 years. Find the mean, the median, and the mode. (*Source:* www.infoplease.com)

40. *Commuter Passengers* The numbers of passengers taking the Rockport to Boston train during the last seven days were 568, 388, 588, 688, 750, 900, and 388. Find the mean, the median, and the mode.

41. *Salary of Employees* A local travel office has 10 employees. Their monthly salaries are $1500, $1700, $1650, $1300, $1440, $1580, $1820, $1380, $2900, and $6300.

  (a) Find the mean.

  (b) Find the median.

  (c) Find the mode.

  (d) Which of these numbers best represents what the typical person earns? Why?

42. *Track Running Times* A college track star in California ran the 100-meter event in eight track meets. Her times were 11.7 seconds, 11.6 seconds, 12.0 seconds, 12.1 seconds, 11.9 seconds, 18 seconds, 11.5 seconds, and 12.4 seconds.

  (a) Find the mean.

  (b) Find the median.

  (c) Find the mode.

  (d) Which of these numbers represents her typical running time? Why?

43. *Number of Phone Calls* Sally made a record of the number of phone calls she received each night this last week.

Day of the Week	Sun	Mon	Tues	Wed	Thurs	Fri	Sat
Number of Phone Calls	23	3	2	3	7	10	11

  (a) Find the mean. Round your answer to the nearest tenth.

  (b) Find the median.

  (c) Find the mode.

  (d) Which of these three measures best represents the number of phone calls Sally receives on a typical night? Why?

44. *Number of Overnight Business Trips* David has to travel as a representative for his computer software company. He made a record of the number of nights he had to spend away from home on business travel during the first seven months of the year.

Month of the Year	Jan	Feb	Mar	April	May	June	July
Number of Nights Spent Away from Home on Business	3	7	8	6	9	28	3

  (a) Find the mean. Round to the nearest tenth.

  (b) Find the median.

  (c) Find the mode.

  (d) Which of these three measures best represents the number of nights that David has to spend away from home on business? Why?

Cumulative Review *Round to the nearest tenth. Use $\pi \approx 3.14$.*

▲ **45.** **[7.10.1]** *Geometry* A triangular piece of insulation is located under the dash of a Ford Explorer. It has a base of 7 inches and a height of 5.5 inches. What is the area of this piece of insulation?

▲ **46.** **[7.10.1]** *Geometry* The Canaan Family Farm has two fields that are irrigated with a rotating sprinkler system. This system waters a circular area. The system is designed to deliver 2 gallons per hour for each square foot of the field. If each circular area has a radius of 40 feet, how many gallons per hour are needed to water these fields?

▲ **47.** **[7.10.1]** *Sign Costs* Collette Daniels has made a huge advertising sign in the shape of a rhombus. The sign has a base of 5 feet and a height of 4 feet. The sign is made out of aluminum that costs $16.50 per square foot. How much did it cost for the aluminum used to make the sign?

▲ **48.** **[7.10.1]** *Coffee Cost* A medium coffee at Coffee Time costs $1.98 and comes in a paper cup with a height of 7 in. and a radius of 1.5 in. What is the cost of the coffee per cubic inch? Round to the nearest cent.

Quick Quiz 8.4 *Dr. Tobey asked his 4:00 P.M. Basic College Mathematics students how many times a year they order a pizza. Here are the responses:*

$$1, 3, 8, 35, 16, 8, 5, 17, 24, 15$$

1. Find the median number of times students order a pizza.

2. Find the mean number of times students order a pizza.

3. Find the mode of the number of times students order a pizza.

4. **Concept Check** Professor Blair wants to conduct a survey of her students to determine how many times a year they order Chinese food. Would you select the mean, the median, or the mode for this survey? Explain your reasoning.

Did You Know...

That You Can Save Money by Turning Down Your Thermostat This Winter?

ADJUST THE THERMOSTAT

Understanding the Problem:
Mark heats his home with oil. He is interested in how much he needs to budget for the heating season and how changing the settings on his thermostat will affect that amount.

Making a Plan:
Mark needs to calculate the cost of his current oil usage and potential savings.

Step 1: During the winter of 2009 Mark paid $2.95 per gallon for home heating oil. The price has since increased to $3.55 per gallon during the winter of 2011.

Task 1: *The oil company will deliver only 100 gallons of heating oil or more. Determine the cost of 100 gallons in December 2009 and November 2011.*

Task 2: *Mark knows he will need a 100-gallon heating oil delivery every month during the winter (November and December 2011 and January, February, March 2012). What can Mark expect to pay in home heating oil costs for the upcoming five months of winter?*

Making a Decision:
Step 2: Mark knows that for every 1° change in his thermostat setting, he can see a 2% change in his utility bills.

Task 3: *Mark typically keeps his thermostat at 72°. How much will he save on his heating costs if he turns the thermostat down to 68° for the entire winter?*

Task 4: *How low would Mark have to set his thermostat to save one month's worth of heating costs ($355)?*

Applying the Situation to Your Life:
If you lower your thermostat a few degrees, you too can realize savings on your heating bill. If you are not comfortable with the temperature set lower, you can still turn the heat down when you are away at work or school, as well as when you are sleeping. A programmable thermostat that will adjust the temperature depending on the time of day would make it easy to do that.

Task 5: *At what temperature do you keep your thermostat?*

Task 6: *Calculate your savings if you lower your thermostat during the winter.*

Chapter 8 Organizer

Topic and Procedure	Examples	✏️ You Try It
Circle graphs, p. 513 The following circle graph describes the ages of the 200 men and women of the Grover City police force. **Age Distribution of Grover City Police Force** 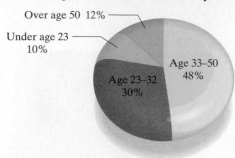 Over age 50 12% Under age 23 10% Age 33–50 48% Age 23–32 30%	**(a)** What percent of the police force is between 23 and 32 years old? 30% **(b)** How many men and women in the police force are over 50 years old? 12% of 200 = (0.12)(200) = 24 people	**1. (a)** What percent of the police force is over age 50? **(b)** How many men and women in the police force are between 33 and 50 years old?
Bar graphs and double-bar graphs, pp. 520–521 The following double-bar graph illustrates the sales of LCD HDTV sets by a major store chain for 2009 and 2010 in three regions of the country. 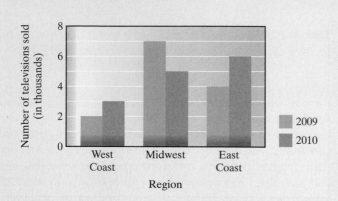	**(a)** How many LCD HDTV sets were sold by the chain on the East Coast in 2010? 6000 sets **(b)** How many *more* LCD HDTV sets were sold in 2010 than in 2009 on the West Coast? 3000 sets were sold in 2010; 2000 sets were sold in 2009. $\begin{array}{r} 3000 \\ -\ 2000 \\ \hline 1000 \end{array}$ sets more in 2010	**2. (a)** How many LCD HDTV sets were sold by the chain in the Midwest in 2010? **(b)** How many *more* LCD HDTV sets were sold in the Midwest than on the West Coast in 2009?
Line graphs and comparison line graphs, pp. 521–523 The following line graph indicates the number of visitors to Wetlands State Park during a four-month period in 2009 and 2010. 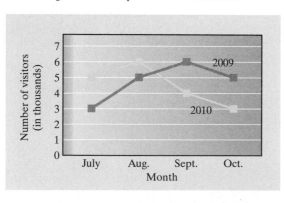	**(a)** How many visitors came to the park in July 2009? 3000 visitors **(b)** In what months were there more visitors in 2009 than in 2010? September and October **(c)** The sharpest *decrease* in attendance took place between what two months? Between August 2010 and September 2010	**3. (a)** How many visitors came to the park in August 2010? **(b)** In what months were there more visitors in 2010 than 2009? **(c)** The sharpest *increase* in attendance took place between what two months?

Topic and Procedure	Examples	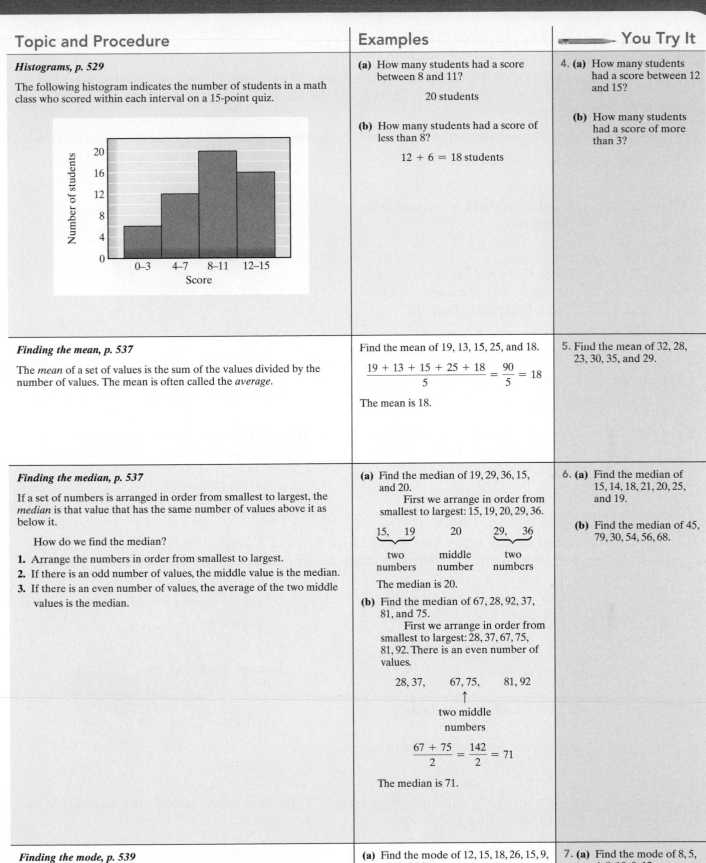 You Try It
Histograms, p. 529 The following histogram indicates the number of students in a math class who scored within each interval on a 15-point quiz.	**(a)** How many students had a score between 8 and 11? 20 students **(b)** How many students had a score of less than 8? 12 + 6 = 18 students	**4. (a)** How many students had a score between 12 and 15? **(b)** How many students had a score of more than 3?
Finding the mean, p. 537 The *mean* of a set of values is the sum of the values divided by the number of values. The mean is often called the *average*.	Find the mean of 19, 13, 15, 25, and 18. $\dfrac{19 + 13 + 15 + 25 + 18}{5} = \dfrac{90}{5} = 18$ The mean is 18.	**5.** Find the mean of 32, 28, 23, 30, 35, and 29.
Finding the median, p. 537 If a set of numbers is arranged in order from smallest to largest, the *median* is that value that has the same number of values above it as below it. How do we find the median? **1.** Arrange the numbers in order from smallest to largest. **2.** If there is an odd number of values, the middle value is the median. **3.** If there is an even number of values, the average of the two middle values is the median.	**(a)** Find the median of 19, 29, 36, 15, and 20. First we arrange in order from smallest to largest: 15, 19, 20, 29, 36. 15, 19 20 29, 36 two numbers middle number two numbers The median is 20. **(b)** Find the median of 67, 28, 92, 37, 81, and 75. First we arrange in order from smallest to largest: 28, 37, 67, 75, 81, 92. There is an even number of values. 28, 37, 67, 75, 81, 92 two middle numbers $\dfrac{67 + 75}{2} = \dfrac{142}{2} = 71$ The median is 71.	**6. (a)** Find the median of 15, 14, 18, 21, 20, 25, and 19. **(b)** Find the median of 45, 79, 30, 54, 56, 68.
Finding the mode, p. 539 The *mode* of a set of values is the value that occurs most often. A set of values may have more than one mode or no mode.	**(a)** Find the mode of 12, 15, 18, 26, 15, 9, 12, and 27. The modes are 12 and 15. **(b)** Find the mode of 4, 8, 15, 21, and 23. There is no mode.	**7. (a)** Find the mode of 8, 5, 4, 8, 10, 9, 12. **(b)** Find the mode of 9, 15, 11, 12, 9, 16, 10, 12, 14.

Chapter 8 Review Problems

Computer Manufacturers *A student found that there were a total of 140 personal computers owned by students in the dormitory. The following circle graph displays the distribution of manufacturers of these computers. Use the graph to answer exercises 1–6.*

Distribution of Computers in a Dormitory by Manufacturer

1. How many personal computers were manufactured by Lenovo?

2. How many personal computers were manufactured by Dell or Compaq?

3. What is the ratio of the number of computers manufactured by Lenovo to the number of computers manufactured by Hewlett-Packard?

4. What is the ratio of the number of computers manufactured by Dell to the number of computers manufactured by Apple?

5. What percent of the 140 computers were manufactured by Compaq? Round to the nearest tenth.

6. What percent of the computers were manufactured by Apple? Round to the nearest tenth.

College Majors *Bradford College offers majors in 6 areas: business, science, social science, language arts, education, and art. The distribution by category is displayed in the circle graph below. Use the graph to answer exercises 7–12.*

7. What percent of the students are majoring in business or social science?

8. What percent of the students are majoring in an area other than business?

9. Which area has the least number of students?

10. Which two areas together make up one-fifth of the graph?

11. If Bradford College has 8000 students, how many of them are majoring in language arts?

12. How many more students are majoring in business than education?

Majors at Bradford College

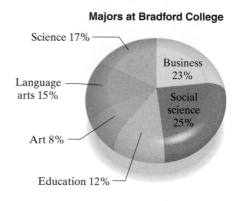

Health Expenditures *The following double-bar graph shows the increase of certain health expenditures in the United States from 1995 to 2010. Use the graph to answer exercises 13–18.*

13. How much was spent on prescription drugs in 2000?

14. How much was spent on physician/clinical services in 2010?

15. What was the increase in the amount spent on physician/clinical services from 2005 to 2010?

16. What was the increase in the amount spent on prescription drugs from 2000 to 2005?

17. In which five-year period is there the greatest difference in the amount spent on prescription drugs?

18. What is the ratio of the amount spent on prescription drugs to that spent on physician/clinical services in 2010?

Corn and Soybean Production *The following double-bar graph shows the number of billion bushels (1 bushel is about 9 gallons), rounded to the nearest tenth, of corn and soybeans produced in the United States from 2002 to 2010. Use the bar graph to answer exercises 19–26.*

19. How many bushels of corn were produced in 2008?

20. How many bushels of soybeans were produced in 2006?

21. Between which two years was there the greatest change in soybean production?

22. How many total bushels of corn and soybeans were produced in 2010?

23. In what year was the difference between the production of corn and soybeans the greatest?

24. In what year was the difference between the production of corn and soybeans the smallest?

25. For the years 2002 to 2010, what was the average U.S. corn production?

26. If the same increase in corn production occurs from 2010 to 2014 as from 2006 to 2010, how many bushels of corn will be produced in 2014?

Graduating Students *The following line graph shows the numbers of graduates of Norfolk College during the last six years. Use the graph to answer exercises 27–32.*

27. How many Norfolk College students graduated in 2009?

28. How many more Norfolk College students graduated in 2010 than in 2005?

29. How many fewer Norfolk College students graduated in 2009 than in 2008?

30. Between what two years was there the smallest increase in the number of graduating students?

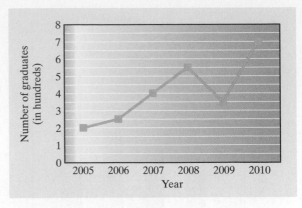

31. Find the average number of graduating students for all six years.

32. What was the percentage of increase in the number of graduating students from 2007 to 2008?

Ice Cream Cone Sales *The following comparison line graph shows the number of ice cream cones purchased at the Junction Ice Cream Stand during a five-month period in 2009 and 2010. Use this graph to answer exercises 33–38.*

33. How many ice cream cones were purchased in July 2010?

34. How many ice cream cones were purchased in August 2009?

35. How many more ice cream cones were purchased in May 2009 than in May 2010?

36. How many more ice cream cones were purchased in August 2010 than in August 2009?

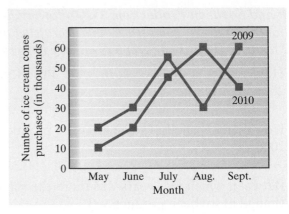

37. How many total ice cream cones were purchased between May and August of 2009?

38. At the location of the ice cream stand, July 2009 was warm and sunny, whereas August 2009 was cold and rainy. Describe how the weather might have played a role in the trend shown on the graph from July to August 2009.

Master's Degrees *The following comparison line graph shows the approximate number of master's degrees awarded in the United States for physical sciences (astronomy, chemistry, and physics) and mathematics and statistics for selected years. Notice the vertical axis does not start at zero. We have zoomed in on the graph in order to see more details. Use the graph to answer exercises 39–46.*

39. How many physical science master's degrees were awarded in 2000?

40. How many more mathematics and statistics master's degrees than physical science master's degrees were awarded in 2008?

41. In which years were more physical science than mathematics and statistics master's degrees awarded?

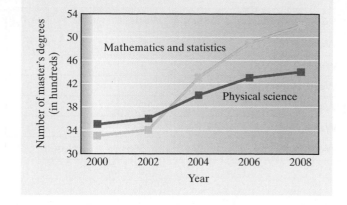

42. In which years were more mathematics and statistics than physical science master's degrees awarded?

43. Between which years is there the largest increase in the number of mathematics and statistics master's degrees awarded?

44. Between which years is there the largest increase in the number of physical science master's degrees?

45. Where on the graph is there the most dramatic change in any two-year period?

46. If the same change in the number of mathematics and statistics master's degrees occurs from 2008 to 2012 as from 2004 to 2008, how many master's degrees in mathematics and science will be awarded in 2012?

Women's Shoe Sales *The following histogram shows the numbers of pairs of women's shoes sold at the grand opening of a new shoe store. Use the histogram to answer exercises 47–52.*

47. How many pairs sold were size 8–8.5?

48. How many pairs sold were size 10 or higher?

49. How many pairs sold were between size 7 and size 9.5?

50. Before the grand opening, the store had 200 pairs of women's shoes. What percent of these were sold during the grand opening?

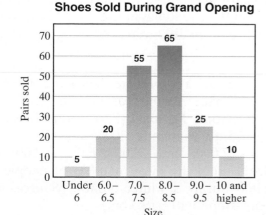

51. How many more pairs of size 8–8.5 were sold than size 6–6.5?

52. What is the ratio of pairs sold of size 9–9.5 to total pairs sold?

Color Television Manufacturing *During the last 28 days, a major manufacturer produced 400 new HDTV television sets each day. The manufacturer recorded the number of defective television sets produced each day. The results are shown in the following chart. In exercises 53–57, determine frequencies for the class intervals for this data.*

	Mon.	Tues.	Wed.	Thurs.	Fri.
Week 1	3	5	8	2	0
Week 2	13	6	3	4	1
Week 3	0	2	16	5	7
Week 4	12	10	17	5	4
Week 5	1	7	8	12	13
Week 6	14	0	3	closed	

	Number of Defective Televisions Produced (Class Interval)	Tally	Frequency
53.	0–3	_____	____
54.	4–7	_____	____
55.	8–11	_____	____
56.	12–15	_____	____
57.	16–19	_____	____

58. Construct a histogram using the table prepared in exercises 53–57.

59. Based on the data of exercises 53–57, how often were between 0 and 7 defective television sets identified in the production?

Find the mean.

60. *Temperatures in Los Angeles* The maximum temperature readings in Los Angeles for the last seven days in July were: 86°, 83°, 88°, 95°, 97°, 100°, and 81°.

61. *Gas Heating Expenses* The LeBlanc family's gas bills for January through June were $145, $162, $95, $67, $43, and $26. Round to the nearest cent.

62. *Visitors at Yellowstone National Park* The approximate number of people who visited Yellowstone National Park from November of 2005 to March of 2006 was Nov., 12,000; Dec., 17,000; Jan., 24,000; Feb., 29,000; Mar., 19,000. Find the mean number of people who visited during a winter month. (*Source:* www.yellowstone-natl-park.com)

63. *Rental Car Employees* The numbers of employees throughout the nation employed annually by Freedom Rent-A-Car for the last six years were 882, 913, 1017, 1592, 1778, and 1936.

Find the median.

64. *Cost of Trucks* The costs of eight trucks purchased by the highway department: $28,500, $29,300, $21,690, $35,000, $37,000, $43,600, $45,300, $38,600.

65. *Cost of Houses* The costs of 10 houses recently purchased in Stillwater: $98,000, $150,000, $120,000, $139,000, $170,000, $156,000, $135,000, $144,000, $154,000, $126,000.

In exercises 66 and 67, find the median and the mode.

66. *San Diego Zoo* The ages of the last 16 people who have passed through the entrance of the San Diego Zoo: 28, 30, 15, 54, 77, 79, 10, 8, 43, 38, 28, 31, 4, 7, 34, 35.

67. *Pizza Deliveries* The daily numbers of deliveries made by the Northfield House of Pizza: 21, 16, 15, 3, 19, 24, 13, 18, 9, 31, 36, 25, 28, 14, 15, 26.

68. *Test Scores* The scores on eight tests taken by Wong Yin in calculus last semester were 96, 98, 88, 100, 31, 89, 94, and 98. Which is a better measure of his usual score, the *mean* or the *median*? Why?

69. *Sales of Cars* The ten salespeople at People's Dodge sold the following numbers of cars last month: 13, 16, 8, 4, 5, 19, 15, 18, 39, 12. Which is a better measure of the usual sales of these salespersons, the *mean* or the *median*? Why?

70. Barbara made a record of the number of hours she uses the computer each day in her apartment.

Day of the Week	Sun.	Mon.	Tues.	Wed.	Thurs.	Fri.	Sat.
Number of Hours Spent on the Computer	2	3	2	4	7	12	5

(a) Find the mean. Round your answer to the nearest tenth if necessary.

(b) Find the median.

(c) Find the mode.

(d) Which of these three measures best represents the number of hours she uses the computer on a typical day? Why?

How Am I Doing? Chapter 8 Test

After you take this test read through the Math Coach on pages 557–558. Math Coach videos are available via MyMathLab and YouTube. Step-by-step test solutions in the Chapter Test Prep Videos are also available via MyMathLab and YouTube. (Search "TobeyBasicCollMath" and click on "Channels.")

A state highway safety commission recently reported the results of inspecting 300,000 automobiles. The following circle graph depicts the percent of automobiles that passed and the percent that had one or more safety violations. Use this graph to answer questions 1–5.

1. What percent of the automobiles passed inspection?

2. What percent of the automobiles had two safety violations?

3. What percent of the automobiles had more than two safety violations?

4. If 300,000 automobiles were inspected, how many of them had one safety violation?

5. If 300,000 automobiles were inspected, how many of them had two violations or three violations?

The following double-bar graph shows an estimate of how the average college tuition and fees costs per year have increased at public and private colleges since 1991. Use the graph to answer questions 6–11.

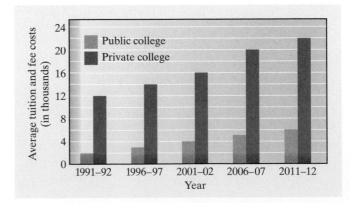

1. _____

2. _____

3. _____

4. _____

5. _____

6. What was the average cost per year at a private college in 1991–1992?

7. What was the average cost per year at a public college in 1996–1997?

8. How much did the average cost per year at private colleges increase from 1991–1992 to 2011–2012?

9. How much did the average cost per year at public colleges increase from 1996–1997 to 2011–2012?

10. How much more did a year at private college cost than a year at public college in 1996–1997?

Mꞅ **11.** How much more did a year at private college cost than a year at public college in 2011–2012?

A research study by 10 midwestern universities produced the following line graph. Use the graph to answer questions 12–16.

12. Approximately how many more years is a 45-year-old American man expected to live if he smokes?

13. Approximately how many more years is a 55-year-old American man expected to live if he does not smoke?

Mꞅ **14.** According to this graph, approximately how much longer is a 25-year-old nonsmoker expected to live than a 25-year-old smoker?

15. According to this graph, at what age is the difference between the life expectancy of a smoker and a nonsmoker the greatest?

6.	☐
7.	☐
8.	☐
9.	☐
10.	☐
11.	☐
12.	☐
13.	☐
14.	☐
15.	☐

16. According to this graph, at what age is the difference between the life expectancy of a smoker and a nonsmoker the smallest?

The following histogram was prepared by a consumer research group. Use the histogram to answer questions 17–20.

17. How many color television sets lasted 6–8 years?

18. How many color television sets lasted 3–5 years?

19. How many color television sets lasted more than 11 years?

ᴹᴄ **20.** How many color television sets lasted 9–14 years?

A chemistry student had the following scores on ten quizzes in her chemistry class: 10, 16, 15, 12, 18, 17, 14, 10, 13, 20.

21. Find the mean quiz score.

ᴹᴄ **22.** Find the median quiz score.

23. Find the mode quiz score.

24. Which of the values—mean, median, or mode—best represents her typical quiz score?

16. _____ ☐

17. _____ ☐

18. _____ ☐

19. _____ ☐

20. _____ ☐

21. _____ ☐

22. _____ ☐

23. _____ ☐

24. _____ ☐

Total Correct: ☐

MATH COACH

Mastering the skills you need to do well on the test.

Students often make the same types of errors when they do the Chapter 8 Test. Here are some helpful hints to keep you from making these common errors on test problems.

Interpreting a Double Bar Graph—Problem 11

Please refer to the double bar graph on page 554 in order to answer this question.

How much more did a year at private college cost than a year at public college in 2011–2012?

> **Helpful Hint** If every bar height is not labeled, then you must examine the distance between two labeled values and determine what the bar height represents.

Did you determine that the bar height for public college costs in 2011–2012 was 6?

Yes ____ No ____

If you answered No, look carefully at the first bar for 2011–2012. Use a sheet of paper to align horizontally so that you can accurately read the scale. Do you see that it aligns to the line that appears halfway between 4 and 8?

Did you determine that the bar height for private college costs in 2011–2012 was 22?

Yes ____ No ____

If you answered No, look carefully at the second bar for 2011–2012. Use a sheet of paper to align horizontally so that you can accurately read the scale. Do you see that it aligns to the line that appears halfway between 20 and 24?

Now go back and rework the problem using these suggestions. Remember to write your answer in thousands of dollars.

Interpreting a Comparison Line Graph—Problem 14

Please refer to the comparison line graph on page 555 in order to answer this question.

According to this graph, approximately how much longer is a 25-year-old nonsmoker expected to live than a 25-year-old smoker?

> **Helpful Hint** If every tic mark on the vertical scale is not labeled, then you must examine the distance between two labeled values closest to the desired value and determine the value of each tic mark between those two labeled values.

Did you determine that the life expectancy for nonsmokers at age 25 is 48?

Yes ____ No ____

If you answered No, look carefully at the first point of the top line on the graph. Use a sheet of paper to align horizontally. Notice that this first point appears in between the labeled values 40 and 50. There are four tic marks between 40 and 50. They represent 42, 44, 46, and 48. The first point aligns with the tic mark for 48.

Did you determine that the life expectancy for smokers at age 25 is 36?

Yes ____ No ____

If you answered No, look carefully at the first point of the lower line on the graph. Use a sheet of paper to align horizontally. Notice that this first point appears in between the labeled values 30 and 40. There are four tic marks between 30 and 40. They represent 32, 34, 36, and 38. The first point aligns with the tic mark for 36.

Now go back and rework the problem using these suggestions. Remember to include the unit "years" with your answer.

Need help? Watch the **MATH COACH** videos in MyMathLab® or on You Tube™.

557

Interpreting a Histogram—Problem 20

Please refer to the histogram on page 556 in order to answer this question.

How many color television sets lasted 9–14 years?

Did you realize that the number of color television sets that lasted 9–14 years includes the bar for 9–11 years added to the bar for 12–14 years?

Yes _____ No _____

If you answered No, reread the question and examine the graph more closely.

Did you realize that the value for the 9–11 years bar is 45 and the value for the 12–14 years bar is 15?

Yes _____ No _____

If you answered No, examine the graph and interpret the scale more carefully.

If you answered Problem 20 incorrectly, go back and rework the problem using these suggestions.

Finding the Median—Problem 22

A chemistry student had the following scores on ten quizzes in her chemistry class: 10, 16, 15, 12, 18, 17, 14, 10, 13, 20. Find the median quiz score.

Did you arrange the quiz scores from smallest to largest to obtain 10, 10, 12, 13, 14, 15, 16, 17, 18, 20?

Yes _____ No _____

If you answered No, go back and perform this step. Remember that any duplicate numbers should be placed next to each other in the list.

Did you identify the two middle values as 14 and 15?

Yes _____ No _____

If you answered No, stop and count to see which values are in fifth and sixth place within the ordered list of ten numbers.

Did you calculate the average of the two middle values?

Yes _____ No _____

If you answered No, go back and perform this step.

Need more help? Look for section examples marked with $\mathbb{M_C}$ to review.

558

Keeping track of positive and negative balances becomes a very important task in the life of anyone who has a bank or investment account. In this chapter, you will become more proficient in the use of positive and negative numbers. If you master this material, it will save you money and help you avoid costly mistakes in your financial decisions.

Signed Numbers

9.1 Adding Signed Numbers

① Adding Two Signed Numbers with the Same Sign

In Chapters 1–8 we worked with whole numbers, fractions, and decimals. In this chapter we enlarge the set of numbers we work with to include numbers that are less than zero. Many real-life situations require using numbers that are less than zero. A debt that is owed, a financial loss, temperatures that fall below zero, and elevations that are below sea level can be expressed only in numbers that are less than zero, or negative numbers.

The following is a graph of the financial reports of four small airlines for the year. It shows **positive numbers**—those numbers that rise above zero—and **negative numbers**—those numbers that fall below zero. The positive numbers represent money gained. The negative numbers represent money lost.

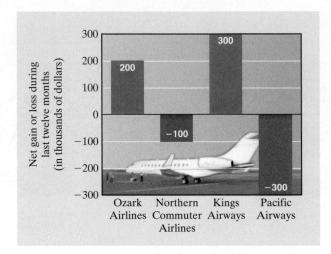

A value of −100,000 is shown for Northern Commuter Airlines. This means that Northern Commuter Airlines lost $100,000 during the year. A value of −300,000 is recorded for Pacific Airways. What does this mean?

Another way to picture positive and negative numbers is on a number line. A **number line** is a line on which each number is associated with a point. The numbers may be positive or negative, whole numbers, fractions, or decimals. Positive numbers are to the right of zero on the number line. Negative numbers are to the left of zero on the number line. Zero is neither positive nor negative.

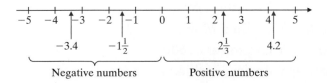

Positive numbers can be written with a plus sign—for example, +2—but this is not usually done. Positive 2 is usually written as 2. It is understood that the *sign* of the number is positive although it is not written. Negative numbers must always have the negative sign so that we know they are negative numbers. Negative 2 is written as −2. The *sign* of the number is negative. The set of positive numbers, negative numbers, and zero is called the set of **signed numbers.**

Order Signed numbers are named in **order** on the number line. Smaller numbers are to the left. Larger numbers are to the right. For any two numbers on the number line, the number on the left is less than the number on the right.

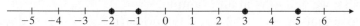

We use the symbol $<$ to mean "is less than." Thus the mathematical sentence $-2 < -1$ means "-2 is less than -1." We use the symbol $>$ to mean "is greater than." Thus the mathematical sentence $5 > 3$ means "5 is greater than 3."

EXAMPLE 1   In each case, replace the ? with $<$ or $>$.

(a) $-8 \: ? \: -4$ (b) $7 \: ? \: 1$ (c) $-2 \: ? \: 0$
(d) $-6 \: ? \: 3$ (e) $2 \: ? \: -5$

Solution
(a) Since -8 lies to the left of -4, we know that $-8 < -4$.
(b) Since 7 lies to the right of 1, we know that $7 > 1$.
(c) Since -2 lies to the left of 0, we know that $-2 < 0$.
(d) Since -6 lies to the left of 3, we know that $-6 < 3$.
(e) Since 2 lies to the right of -5, we know that $2 > -5$.

Student Practice 1 In each case, replace the ? with $<$ or $>$.

(a) $4 \: ? \: 2$ (b) $-5 \: ? \: -3$ (c) $0 \: ? \: -6$
(d) $-2 \: ? \: 1$ (e) $5 \: ? \: -7$

NOTE TO STUDENT: *Fully worked-out solutions to all of the Student Practice problems can be found at the back of the text starting at page SP-1.*

Absolute Value Sometimes we are only interested in the distance a number is from zero. For example, the distance from 0 to $+3$ is 3. The distance from 0 to -3 is also 3. Notice that distance is always a positive number, regardless of which direction we travel on the number line. This distance is called the *absolute value*.

> The **absolute value** of a number is the distance between that number and zero on the number line.

The symbol for absolute value is $| \ |$. When we write $|5|$, we are looking for the distance from 0 to 5 on the number line. Thus $|5| = 5$. This is read, "The absolute value of 5 is 5." $|-5|$ is the distance from 0 to -5 on the number line. Thus $|-5| = 5$. This is read, "The absolute value of -5 is 5."

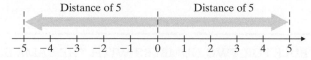

Other examples of absolute value are shown next.

$$|6| = 6 \qquad |-3| = 3$$
$$|7.2| = 7.2 \qquad \left|-\frac{1}{5}\right| = \frac{1}{5}$$
$$|0| = 0 \qquad |-26| = 26$$

When we find the absolute value of any nonzero number, we always get a positive value. We use the concept of absolute value to develop rules for adding signed numbers. We begin by looking at addition of numbers with the same sign. Although you are already familiar with the addition of positive numbers, we will look at an example.

Suppose that we earn $52 one day and earn $38 the next day. To learn what our two-day total is, we add the positive numbers. We earn

$$\$52 + \$38 = +\$90.$$

We can show this sum on a number line by drawing an arrow that starts at 0 and points 52 units to the right (because 52 is positive). At the end of this arrow we draw a second arrow that points 38 units to the right. Note that the second arrow ends at the sum, 90.

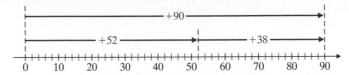

The money is coming in, and the plus sign records a gain. Notice that we added the numbers and that the sign of the sum is the same as the sign of the addends.

Now let's consider an example of addition of two negative numbers.

Suppose that we consider money spent as negative dollars. If we spend $52 one day (−$52) and we spend $38 the next day (−$38), we must add two negative numbers. What is our financial position?

$$-\$52 + (-\$38) = -\$90$$

We have spent $90. The negative sign tells us the direction of the money: out. Notice that we added the numbers and that the sign of the sum is the same as the sign of the addends.

We can illustrate this sum on a number line by drawing a line that starts at 0 and points 52 units to the left (because −52 is negative). At the end of this arrow we add a second arrow that points 38 units to the left. This second arrow ends at the sum, −90.

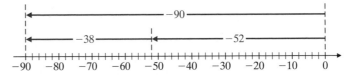

These examples suggest the addition rule for two numbers with the same sign.

ADDITION RULE FOR TWO NUMBERS WITH THE SAME SIGN

To add two numbers with the same sign:

1. Add the absolute values of the numbers.

2. Use the common sign in the answer.

EXAMPLE 2 Add.   **(a)** $7 + 5$ **(b)** $-3.2 + (-5.6)$

Solution

(a)
$$\begin{array}{r} 7 \\ +\ 5 \\ \hline 12 \end{array}$$
We add the absolute value of the numbers 7 and 5. The positive sign, although not written, is common to both numbers. The answer is a positive 12. (The + sign is not written.)

(b)
$$\begin{array}{r} -3.2 \\ +\ -5.6 \\ \hline -8.8 \end{array}$$
We add the absolute value of the numbers 3.2 and 5.6.

We use a negative sign in our answer because we added two negative numbers.

Student Practice 2 Add. **(a)** $9 + 14$ **(b)** $-4.5 + (-1.9)$

These rules can be applied to fractions as well.

EXAMPLE 3 Add. **(a)** $\dfrac{5}{18} + \dfrac{1}{3}$ **(b)** $-\dfrac{1}{7} + \left(-\dfrac{3}{5}\right)$

Student Practice 3 Add.

(a) $\dfrac{5}{12} + \dfrac{1}{4}$

(b) $-\dfrac{1}{6} + \left(-\dfrac{2}{7}\right)$

Solution

(a) The LCD = 18. The first fraction already has the LCD.

$$\frac{5}{18} = \frac{5}{18}$$
$$+\ \frac{1}{3} \times \frac{6}{6} = +\ \frac{6}{18}$$
$$\frac{11}{18}$$

We add two positive numbers, so the answer is positive.

(b) The LCD = 35.

$$\frac{1}{7} \times \frac{5}{5} = \frac{5}{35}$$

Because $\dfrac{1}{7} = \dfrac{5}{35}$ it follows that $-\dfrac{1}{7} = -\dfrac{5}{35}$.

$$\frac{3}{5} \times \frac{7}{7} = \frac{21}{35}$$

Because $\dfrac{3}{5} = \dfrac{21}{35}$ it follows that $-\dfrac{3}{5} = -\dfrac{21}{35}$. Thus

$$\begin{array}{l} -\dfrac{1}{7} \\ +\ -\dfrac{3}{5} \end{array} \text{ is equivalent to } \begin{array}{l} -\dfrac{5}{35} \\ +\ -\dfrac{21}{35} \\ \hline -\dfrac{26}{35} \end{array}$$

We add two negative numbers, so the answer is negative.

It can also be written by adding $\left(\dfrac{-5}{35}\right) + \left(\dfrac{-21}{35}\right)$ to get $\dfrac{-26}{35}$. The negative sign can be placed in the numerator or in front of the fraction bar.

It is interesting to see how often negative numbers appear in statements about the federal budget. Observe the data in the following bar graph.

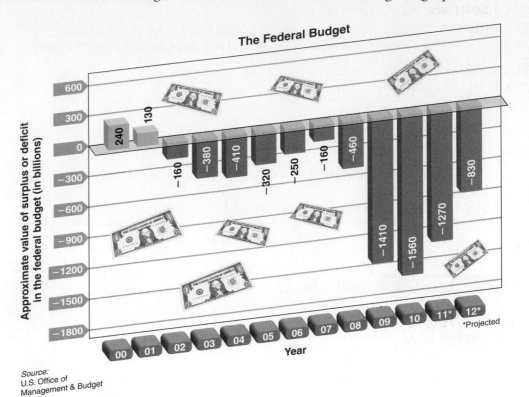

Source:
U.S. Office of
Management & Budget

EXAMPLE 4 Find the total value of surplus or deficit for the two years 2005 and 2006.

Solution We add $(-\$320 \text{ billion}) + (-\$250 \text{ billion})$ to obtain $-\$570$ billion.

The total deficit for these two years is $570,000,000,000.

Student Practice 4 Find the total value of surplus or deficit for the two years 2002 and 2011.

② Adding Two Signed Numbers with Different Signs

Let's look at some real-life situations involving addition of signed numbers with different signs. Suppose that we earn $52 one day and we spend $38 the next day. If we combine (add) the two transactions, it would look like

$$\$52 + (-\$38) = +\$14.$$

On a number line we draw an arrow that starts at zero and points 52 units to the right. From the end of this arrow we draw an arrow that points 38 units to the left. (Remember, the arrow points to the left for a negative number.)

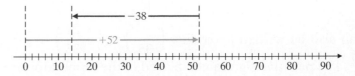

This is a situation with which we are familiar. What we actually do is subtract. That is, we take the difference between $52 and $38. Notice that the sign of the larger number is positive and that the sign of the answer is also positive.

Let's look at another situation. Suppose we spend $52, and we earn $38. The situation would look like this.

$$(-\$52) + \$38 = -\$14$$

On a number line we draw an arrow that starts at zero and points 52 units to the left. Again the arrow must point to the left to represent a negative number.

From the end of this arrow we draw an arrow that points 38 units to the right.

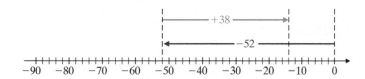

On our number line we end up at -14.

In our real-life situation we end up owing $14, which is represented by a negative number. To find the sum, we actually find the difference between $52 and $38. Notice that if we do not account for sign, the larger number is 52. The sign of that number is negative and the sign of the answer is also negative. This suggests the addition rule for two numbers with different signs.

ADDITION RULE FOR TWO NUMBERS WITH DIFFERENT SIGNS

To add two numbers with different signs:

1. Find the difference between the larger absolute value and the smaller absolute value.

2. Use the sign of the number with the larger absolute value.

EXAMPLE 5 Add.

(a) $8 + (-10)$ **(b)** $-16.6 + 12.3$ **(c)** $\dfrac{3}{4} + \left(-\dfrac{2}{3}\right)$

Solution

(a)

$$\begin{array}{r} 8 \\ + \ -10 \\ \hline -2 \end{array}$$

$\longleftarrow$ The signs are different, so we find the difference: $10 - 8 = 2$.

The sign of the number with the larger absolute value is negative, so the answer is negative.

Continued on next page

(b) -16.6
$+ 12.3$
-4.3 ⟵—— The signs are different, so we find the difference.

The sign of the number with the larger absolute value is negative, so the answer is negative.

(c) $\dfrac{3}{4} + \left(-\dfrac{2}{3}\right) = \dfrac{9}{12} + \left(-\dfrac{8}{12}\right) = \dfrac{9 + (-8)}{12} = \dfrac{1}{12}$

The signs are different, so we find the difference. The sign of the number with the larger absolute value is positive, so the answer is positive.

Student Practice 5   Add.

(a) $7 + (-12)$ **(b)** $-20.8 + 15.2$ **(c)** $\dfrac{5}{6} + \left(-\dfrac{3}{4}\right)$

Notice that in **(b)** of Example 5, the number with the larger absolute value is on top. This makes the numbers easier to subtract. Because addition is commutative, we could have written **(a)** as

$$-10$$
$$+ 8.$$

This makes the computation easier. If you are adding two numbers with different signs, place the number with the larger absolute value on top so that you can find the difference easily. As noted, the commutative property of addition holds for signed numbers.

COMMUTATIVE PROPERTY OF ADDITION

For any signed numbers a and b,

$$a + b = b + a.$$

EXAMPLE 6   Last night the temperature dropped to $-14°F$. From that low, today the temperature rose $34°F$. What was the high temperature today?

Solution   We want to add $-14°F$ and $34°F$. Because addition is commutative, it does not matter whether we add $-14 + 34$ or $34 + (-14)$.

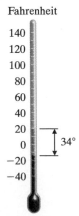

$34°F$
$+ -14°F$ The 34 is larger than 14. The difference between 34
$20°F$ ⟵—— and 14 is 20. The number with the larger absolute
value is positive, so the answer is positive.

Student Practice 6   Last night the temperature dropped to $-19°F$. From that low, today the temperature rose $28°F$. What was the high temperature today?

③ Adding Three or More Signed Numbers

We can add three or more numbers using these rules. Since addition is associative, we may group the numbers to be added in reverse order. That is, it does not matter which two numbers are added first.

EXAMPLE 7 Add. $24 + (-16) + (-10)$

Solution We can go from left to right and start with 24, or we can start with -16.

Step 1	$\begin{array}{r} 24 \\ +\ -16 \\ \hline 8 \end{array}$	or	**Step 1**	$\begin{array}{r} -16 \\ +\ -10 \\ \hline -26 \end{array}$
Step 2	$\begin{array}{r} 8 \\ +\ -10 \\ \hline -2 \end{array}$	or	**Step 2**	$\begin{array}{r} -26 \\ +\ 24 \\ \hline -2 \end{array}$

Student Practice 7 Add. $36 + (-21) + (-18)$

ASSOCIATIVE PROPERTY OF ADDITION

For any three signed numbers a, b, and c,

$$(a + b) + c = a + (b + c).$$

If there are many numbers to add, it may be easier to add the positive numbers and the negative numbers separately and then combine the results.

EXAMPLE 8 The results of a new company's operations over five months are listed in the following table. What is the company's overall profit or loss over the five-month period?

Net Operations Profit/Loss Statement in Dollars

Month	Profit	Loss
January	30,000	
February		−50,000
March		−10,000
April	20,000	
May	15,000	

Continued on next page

Solution   First we will add separately the positive numbers and the negative numbers.

$$
\begin{array}{rr}
30{,}000 & \\
20{,}000 & -50{,}000 \\
+\ 15{,}000 & +\ -10{,}000 \\
\hline
65{,}000 & -60{,}000
\end{array}
$$

Now we add the positive number 65,000 and the negative number −60,000.

$$
\begin{array}{r}
65{,}000 \\
+\ -60{,}000 \\
\hline
5\ 000
\end{array}
$$

The company had an overall profit of $5000 for the five-month period.

Student Practice 8   The results of the next five months of operations for the same company are listed in the following table. What is the overall profit or loss over this five-month period?

Net Operations Profit/Loss Statement in Dollars

Month	Profit	Loss
June		−20,000
July	30,000	
August	40,000	
September		−5000
October		−35,000

Verbal and Writing Skills, Exercises 1 and 2

1. Explain in your own words how to add two signed numbers if the signs are the same.

2. Explain in your own words how to add two signed numbers if one number is positive and one number is negative.

In each case, replace the ? with < or >.

3. $-9 \; ? \; 2$ **4.** $-8 \; ? \; 6$ **5.** $-3 \; ? \; -5$ **6.** $-7 \; ? \; -14$

7. $5 \; ? \; -2$ **8.** $6 \; ? \; -3$ **9.** $-12 \; ? \; -10$ **10.** $-15 \; ? \; -13$

Simplify each absolute value expression.

11. $|7|$ **12.** $|6|$ **13.** $|-16|$ **14.** $|-18|$

Add each pair of signed numbers that have the same sign.

15. $-6 + (-11)$ **16.** $-5 + (-13)$ **17.** $-4.9 + (-2.1)$ **18.** $-8.3 + (-3.7)$

19. $8.9 + 7.6$ **20.** $12.5 + 7.8$ **21.** $\dfrac{1}{5} + \dfrac{2}{7}$ **22.** $\dfrac{5}{6} + \dfrac{1}{4}$

23. $-2\dfrac{1}{2} + \left(-\dfrac{1}{2}\right)$ **24.** $-5\dfrac{1}{4} + \left(-\dfrac{3}{4}\right)$

Add each pair of signed numbers that have different signs.

25. $14 + (-5)$ **26.** $15 + (-6)$ **27.** $-17 + 12$ **28.** $-21 + 15$

29. $-36 + 58$ **30.** $-42 + 57$ **31.** $-9.3 + 6.05$ **32.** $-7.2 + 4.04$

33. $\dfrac{1}{12} + \left(-\dfrac{3}{4}\right)$ **34.** $\dfrac{7}{20} + \left(-\dfrac{19}{20}\right)$

Mixed Practice *Add.*

35. $\dfrac{7}{9} + \left(-\dfrac{2}{9}\right)$ **36.** $\dfrac{5}{12} + \left(-\dfrac{7}{12}\right)$ **37.** $-18 + (-4)$ **38.** $-34 + (-2)$

39. $1.48 + (-2.2)$ **40.** $3.72 + (-4.1)$ **41.** $-125 + (-238)$ **42.** $-514 + (-176)$

43. $13 + (-9)$ **44.** $-18 + 7$ **45.** $-3\frac{3}{4} + \left(-1\frac{7}{10}\right)$ **46.** $4\frac{7}{8} + \left(-1\frac{3}{16}\right)$

47. $-7.56 + 13.8$ **48.** $-6.89 + 15.9$ **49.** $-5 + \left(-\frac{1}{2}\right)$ **50.** $\frac{11}{12} + (-5)$

51. $-20.5 + 18.1 + (-12.3)$ **52.** $-10.8 + (-14.3) + 12.7$

53. $11 + (-9) + (-10) + 8$ **54.** $(-13) + 8 + (-12) + 17$

55. $-7 + 6 + (-2) + 5 + (-3) + (-5)$ **56.** $(-5) + 12 + (-7) + (-3) + 4 + 7$

57. $\left(-\frac{1}{5}\right) + \left(-\frac{2}{3}\right) + \frac{4}{25}$ **58.** $\left(-\frac{1}{7}\right) + \left(-\frac{5}{21}\right) + \frac{3}{14}$

Applications

Profit and Loss Statements *Use signed numbers to represent the total profit or loss for a company after the following reports.*

59. A $43,000 loss in February followed by a $51,000 loss in March.

60. An $18,000 loss in September followed by a $22,000 loss in October.

61. A $9500 profit in November followed by a $17,000 loss in December.

62. An $11,000 loss in July followed by an $8400 profit in August.

63. An $18,500 loss in January, a $12,300 profit in February, and a $15,000 profit in March.

64. A $4500 loss in March, a $7800 loss in April, and a $14,000 profit in May.

Solve.

65. ***Temperature Change*** One night in Juneau, Alaska, the temperature was $-5°$F. By the next morning, the temperature had dropped $13°$F. What was the morning temperature?

66. ***Temperature Change*** One morning in Winnipeg, Manitoba, the temperature was $-9°$F. By that afternoon, the temperature had risen $15°$F. What was the afternoon temperature?

67. ***Temperature Change*** This morning, the temperature was $-5°$F. This evening, the temperature rose $4°$F. What was the new temperature?

68. ***Temperature Change*** Last night the temperature was $-9°$F. The temperature dropped $16°$F this afternoon. What was the afternoon temperature?

69. *Stock Market* The following list of signed numbers is the daily loss or gain of one share of Starbucks Corporation stock for the week of September 27–October 1, 2010: +0.12, −0.23, −0.38, −0.38, +0.05. What was the net loss or gain for the week?

70. *Stock Market* The following list of signed numbers is the daily loss or gain of one share of Hershey Foods Corporation stock for the week of September 27–October 1, 2010: −0.61, +0.26, +0.07, −0.20, −0.07. What was the net loss or gain for the week?

71. *Football* In three plays of a football game, the quarterback threw passes that lost 8 yards, gained 13 yards, and lost 6 yards. What was the total gain or loss of the three plays?

72. *Football* In three plays of a football game, the quarterback threw passes that gained 20 yards, lost 13 yards, and gained 5 yards. What was the total gain or loss of the three plays?

To Think About

73. *Checking Accounts* Bob examined his checking account register. He thought his balance was $89.50. However, he forgot to subtract an ATM withdrawal of $50.00. Since the ATM that he used was at a different bank, he was also charged $2.50 for using the ATM. What was the actual balance in his checking account?

74. *Checking Accounts* Nancy examined her checking account register. She thought her balance was $97.40. However, she forgot to subtract a check of $95.00 that she had made out the previous week. She also forgot to subtract the monthly $4.50 fee charged by her bank for having a checking account. What was the actual balance in her checking account?

Cumulative Review

▲ **75. [7.8.3] *Geometry*** Use $V = \dfrac{4\pi r^3}{3}$ to find the volume of a sphere of radius 6 feet. Use $\pi \approx 3.14$ and round to the nearest tenth.

▲ **76. [7.8.5] *Geometry*** Use $V = \dfrac{Bh}{3}$ to find the volume of a pyramid whose rectangular base measures 9 meters by 7 meters and whose height is 10 meters.

77. [8.4.2] Find the median value. 62, 59, 60, 57, 60, 61

78. [8.4.1] Find the mean. $36, $42, $39, $39, $41, $43

Quick Quiz 9.1 *Add the following.*

1. $16 + (-3) + 5 + (-12)$

2. $5.9 + (-7.4)$

3. $-4\dfrac{2}{3} + 1\dfrac{1}{3}$

4. Concept Check In calculating an addition problem such as $4 + (-12) + 23 + (-15)$ some students add from left to right. Other students first add the positive numbers, then add the negative numbers, and then add the two results. Explain which method you prefer and why.

9.2 Subtracting Signed Numbers

① Subtracting One Signed Number from Another

We begin our discussion by defining the word **opposite**. The opposite of a positive number is a negative number with the same absolute value. For example, the opposite of 7 is -7.

The opposite of a negative number is a positive number with the same absolute value. For example, the opposite of -9 is 9. If a number is the opposite of another number, these two numbers are at an equal distance from zero on the number line.

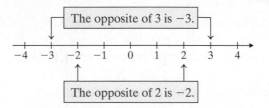

The sum of a number and its opposite is zero.

$$-3 + 3 = 0 \qquad 2 + (-2) = 0$$

We will use the concept of opposite to develop a way to subtract signed numbers.

Let's think about how a checking account works. Suppose that you deposit $25 and the bank adds a service charge of $5 for a new checkbook. Your account looks like this:

$$\$25 + (-\$5) = \$20$$

Suppose instead that you deposit $25 and the bank adds no charge. The next day, you write a check for $5. The result of these two transactions is

$$\$25 - \$5 = \$20$$

Note that your account has the same amount of money ($20) in both cases.

We see that adding a negative 5 to 25 is the same as subtracting a positive 5 from 25. That is, $25 + (-5) = 20$ and $25 - 5 = 20$.

Subtracting is equivalent to adding the opposite.

Subtracting	*Adding the Opposite*
$25 - 5 = 20$	$25 + (-5) = 20$
$19 - 6 = 13$	$19 + (-6) = 13$
$7 - 3 = 4$	$7 + (-3) = 4$
$15 - 5 = 10$	$15 + (-5) = 10$

We define a rule for the subtraction of signed numbers.

SUBTRACTION OF SIGNED NUMBERS

To subtract signed numbers, add the opposite of the second number to the first number.

   Thus, to do a subtraction problem, we first change it to an equivalent addition problem in which the first number does not change but the second number is replaced by its opposite. Then we follow the rules for *addition* of signed numbers.

EXAMPLE 1

Solution   Subtract. $-8 - (-2)$

$$-8 - (-2)$$

$$-8 + 2 \longleftarrow$$ Write the opposite of -2, which is 2.

Change subtraction to addition.

Now we use the rules of addition for two numbers with opposite signs.

$$-8 + 2 = -6$$

Student Practice 1 Subtract. $-10 - (-5)$

NOTE TO STUDENT: Fully worked-out solutions to all of the Student Practice problems can be found at the back of the text starting at page SP-1.

EXAMPLE 2   Subtract.   **(a)** $7 - 8$ **(b)** $-12 - 16$

Solution

(a) $7 - 8$

$$7 + (-8) \longleftarrow$$ Write the opposite of 8, which is -8.

Change subtraction to addition.

   Now we use the rules of addition for two numbers with opposite signs.

$$7 + (-8) = -1$$

(b) $-12 - 16$

$$-12 + (-16) \longleftarrow$$ Write the opposite of 16, which is -16.

Change subtraction to addition.

   Now we follow the rules of addition of two numbers with the same sign.

$$-12 + (-16) = -28$$

Student Practice 2 Subtract.   **(a)** $5 - 12$ **(b)** $-11 - 17$

   Sometimes the numbers we subtract are fractions or decimals.

Mc **EXAMPLE 3**   Subtract.

(a) $5.6 - (-8.1)$

(b) $-\dfrac{6}{11} - \left(-\dfrac{1}{22}\right)$

Solution

(a) Change the subtraction to adding the opposite. Then add.

$$5.6 - (-8.1) = 5.6 + 8.1$$
$$= 13.7$$

(b) Change the subtraction to adding the opposite. Then add.

$$-\dfrac{6}{11} - \left(-\dfrac{1}{22}\right) = -\dfrac{6}{11} + \dfrac{1}{22}$$

$$= -\dfrac{6}{11} \times \dfrac{2}{2} + \dfrac{1}{22} \quad \begin{array}{l}\text{We see that the LCD} = 22.\\ \text{We change } \frac{6}{11} \text{ to a fraction}\\ \text{with a denominator of 22.}\end{array}$$

$$= -\dfrac{12}{22} + \dfrac{1}{22} \quad \text{Add.}$$

$$= -\dfrac{11}{22} = -\dfrac{1}{2}$$

Student Practice 3   Subtract.

(a) $3.6 - (-9.5)$

(b) $-\dfrac{5}{8} - \left(-\dfrac{5}{24}\right)$

Remember that in performing subtraction of two signed numbers:

1. The first number does not change.
2. The subtraction sign is changed to addition.
3. We write the opposite of the second number.
4. We find the result of this addition problem.

Calculator

 Negative
Numbers

To enter a negative number on most scientific calculators, find the key marked $\boxed{+/-}$. To enter the number -3, press the key 3 and then the key $\boxed{+/-}$. The display should read

$$\boxed{-3}$$

To find $(-32) + (-46)$, enter

$$32\,\boxed{+/-}\,\boxed{+}\,46\,\boxed{+/-}$$

$$\boxed{=}$$

The display should read

$$\boxed{-78}$$

Try the following.

(a) $-756 + 184$
(b) $92 + (-51)$
(c) $-618 - (-824)$
(d) $-36 + (-10) - (-15)$

Note: The $\boxed{+/-}$ key changes the sign of a number from $+$ to $-$ or from $-$ to $+$.

Think of each subtraction problem as a problem of adding the opposite.

If you see $7 - 10$, think $7 + (-10)$.
If you see $-3 - 19$, think $-3 + (-19)$.
If you see $6 - (-3)$, think $6 + (+3)$.

EXAMPLE 4   Subtract.

(a) $6 - (+3)$

(b) $-\dfrac{1}{2} - \left(-\dfrac{1}{3}\right)$

(c) $2.7 - (-5.2)$

Solution

(a) $6 - (+3) = 6 + (-3) = 3$

(b) $-\dfrac{1}{2} - \left(-\dfrac{1}{3}\right) = -\dfrac{1}{2} + \dfrac{1}{3} = -\dfrac{3}{6} + \dfrac{2}{6} = -\dfrac{1}{6}$

(c) $2.7 - (-5.2) = 2.7 + 5.2 = 7.9$

Student Practice 4   Subtract.

(a) $20 - (-5)$

(b) $-\dfrac{1}{5} - \left(-\dfrac{1}{2}\right)$

(c) $3.6 - (-5.5)$

② Solving Problems Involving Both Addition and Subtraction of Signed Numbers

EXAMPLE 5 Perform the following set of operations, working from left to right. $-8 - (-3) + (-5)$

Solution $-8 - (-3) + (-5) = -8 + 3 + (-5)$ First we change
$$= -5 + (-5) = -10$$ subtracting a -3
to adding a 3.

Student Practice 5 Perform the following set of operations, working from left to right.

$$-5 - (-9) + (-14)$$

③ Solving Simple Applied Problems That Involve the Subtraction of Signed Numbers

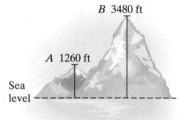

When we want to find the difference in altitude between two mountains, we subtract. We subtract the lower altitude from the higher altitude. Look at the illustration at the right. The difference in altitude between A and B is 3480 feet − 1260 feet = 2220 feet.

Land that is below sea level is considered to have a negative altitude. The Dead Sea is 1312 feet below sea level. Look at the following illustration. The difference in altitude between C and D is 2590 feet − (−1312 feet) = 3902 feet.

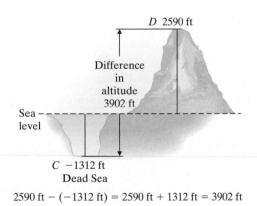

$$2590 \text{ ft} - (-1312 \text{ ft}) = 2590 \text{ ft} + 1312 \text{ ft} = 3902 \text{ ft}$$

EXAMPLE 6 Find the difference in temperature between 38°F during the day in Anchorage, Alaska, and −26°F at night.

Solution We subtract the lower temperature from the higher temperature.

$$38 - (-26) = 38 + 26 = 64$$

The difference is 64°F.

Student Practice 6 Find the difference in temperature between 31°F during the day in Fairbanks, Alaska, and −37°F at night.

Subtract the signed numbers by adding the opposite of the second number to the first number.

1. $-9 - (-3)$ **2.** $-7 - (-5)$ **3.** $-12 - (-7)$ **4.** $-10 - (-2)$ **5.** $3 - 9$

6. $5 - 12$ **7.** $-14 - 3$ **8.** $-6 - 18$ **9.** $-16 - (-25)$ **10.** $-12 - (-20)$

11. $46 - (-39)$ **12.** $53 - (-28)$ **13.** $12 - 30$ **14.** $10 - 14$ **15.** $-12 - (-15)$

16. $-17 - (-30)$ **17.** $150 - 210$ **18.** $500 - 150$ **19.** $300 - (-256)$ **20.** $420 - (-300)$

21. $-2.5 - 4.2$ **22.** $-4.1 - 3.9$ **23.** $6.2 - 14.9$ **24.** $8.5 - 19.2$

Mixed Practice *Subtract the signed numbers by adding the opposite of the second number to the first number.*

25. $-10.9 - (-2.3)$ **26.** $-6.8 - (-2.9)$ **27.** $20.23 - (-12.71)$ **28.** $18.24 - (-11.35)$

29. $\dfrac{1}{4} - \left(-\dfrac{3}{4}\right)$ **30.** $\dfrac{5}{7} - \left(-\dfrac{6}{7}\right)$ **31.** $-\dfrac{5}{6} - \dfrac{1}{3}$ **32.** $-\dfrac{3}{5} - \dfrac{1}{10}$

33. $-2\dfrac{3}{10} - \left(-3\dfrac{5}{6}\right)$ **34.** $-7\dfrac{5}{8} - \left(-12\dfrac{2}{3}\right)$ **35.** $\dfrac{2}{9} - \dfrac{5}{7}$ **36.** $\dfrac{3}{5} - \dfrac{11}{12}$

Perform each set of operations, working from left to right.

37. $2 - (-8) + 5$ **38.** $5 - (-5) + 7$ **39.** $-5 - 6 - (-11)$

40. $-8 - 5 - (-17)$ **41.** $21 - (-15) - (-10)$ **42.** $29 - (-10) - (-16)$

43. $-16 - (-6) - 12$ **44.** $-15 - (-1) - 19$ **45.** $9 - 3 - 2 - 6$

46. $12 - 5 - 4 - 8$ **47.** $-2.4 - 7.1 + 1.3 - (-2.8)$ **48.** $-3.8 + 1.6 - 5.7 - (-2.2)$

Applications *Use your knowledge of signed numbers to answer exercises 49–54.*

49. *Altitude Change* The highest point in California is Mt. Whitney at 14,494 feet. The lowest point is −282 feet in Death Valley. How far above Death Valley is Mt. Whitney?

50. *Altitude Change* The highest point in Africa is Mount Kilimanjaro at 19,340 ft. The lowest point is −502 ft at Lake 'Asal. How far above Lake 'Asal is Mount Kilimanjaro?

51. *Temperature Change* Find the difference in temperature in Alta, Utah, between 23°F during the day and −19°F at night.

52. *Temperature Change* Find the difference in temperature in Fairbanks, Alaska, between 27°F during the day and −33°F at night.

53. *Temperature Change* In Thule, Greenland, yesterday, the temperature was −29°F. Today the temperature rose 16°F. What is the new temperature?

54. *Height Change* Find the difference in height between the top of a hill 642 feet high and a crack caused by an earthquake 57 feet below sea level.

Profit and Loss Statements *A company's profit and loss statement in dollars for five months is shown in the following table.*

Month	Profit	Loss
January	18,700	
February		−34,700
March		−6300
April	43,600	
May		−12,400

55. What is the change in the profit/loss status of the company from the first of January to the end of February?

56. What is the change in the profit/loss status of the company from the first of February to the end of March?

57. What is the change in the profit/loss status of the company from the first of March to the end of May?

58. What is the change in the profit/loss status of the company from the first of January to the end of May?

59. In January 2000 (before the New York Stock Exchange began trading shares in decimal price increments in the year 2001), the value of one share of a certain stock was $15\frac{1}{2}$. During the next three days, the value fell $1\frac{1}{2}$, rose $2\frac{3}{4}$, and fell $3\frac{1}{4}$. What was the value of one share at the end of the three days?

60. In April 2000 (before the New York Stock Exchange began trading shares in decimal price increments in the year 2001), the value of one share of a certain stock was $32\frac{1}{4}$. During the next three days, the value rose $3\frac{1}{4}$, fell $3\frac{3}{4}$, and rose $2\frac{3}{4}$. What was the value of one share at the end of these three days?

Cumulative Review *In exercises 61–62, perform the operations in the proper order.*

61. [1.6.2] $20 \times 2 \div 10 + 4 - 3$

62. [1.6.2] $2 + 3 \times (5 + 7) \div 9$

Quick Quiz 9.2 *Subtract the signed numbers.*

1. $\dfrac{3}{7} - \left(-\dfrac{9}{14}\right)$

2. $-8.7 - (-3.2)$

3. $-34 - 48$

4. Concept Check Explain how you would perform the indicated operations to evaluate $8 - (-13) + (-5)$.

9.3 Multiplying and Dividing Signed Numbers

Student Learning Objectives

After studying this section, you will be able to:

① Multiply and divide two signed numbers.

② Multiply three or more signed numbers.

NOTE TO STUDENT: Fully worked-out solutions to all of the Student Practice problems can be found at the back of the text starting at page SP-1.

① Multiplying and dividing two signed numbers

Recall the different ways we can indicate multiplication.

$$3 \times 5 \qquad 3 \cdot 5 \qquad (3)(5) \qquad 3(5)$$

It is common to use parentheses to mean multiplication.

> **EXAMPLE 1** Evaluate. **(a)** $(7)(8)$ **(b)** $3(12)$
>
> **Solution**
>
> **(a)** $(7)(8) = 56$ **(b)** $3(12) = 36$

Student Practice 1 Evaluate.

(a) $(6)(9)$ **(b)** $7(12)$

Now suppose we multiply a positive number times a negative number. What will happen? Let us look for a pattern.

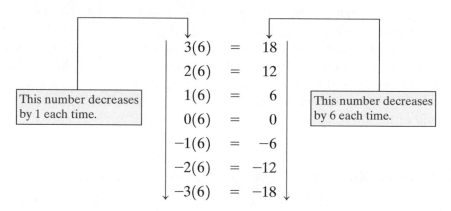

This number decreases by 1 each time.

$$3(6) = 18$$
$$2(6) = 12$$
$$1(6) = 6$$
$$0(6) = 0$$
$$-1(6) = -6$$
$$-2(6) = -12$$
$$-3(6) = -18$$

This number decreases by 6 each time.

Our pattern suggests that when we multiply a positive number times a negative number, we get a negative number. Thus we will state the following rule.

> **MULTIPLICATION RULE FOR TWO NUMBERS WITH DIFFERENT SIGNS**
>
> To multiply two numbers with different signs, multiply the absolute values. The result is negative.

> **EXAMPLE 2** Multiply. **(a)** $2(-8)$ **(b)** $(-3)(25)$
>
> **Solution** In each case, we are multiplying two signed numbers with different signs. We will always get a negative number for an answer.
>
> **(a)** $2(-8) = -16$ **(b)** $(-3)(25) = -75$

Student Practice 2 Multiply.

(a) $(-8)(5)$ **(b)** $3(-60)$

A similar rule applies to division of two numbers when the signs are not the same.

DIVISION RULE FOR TWO NUMBERS WITH DIFFERENT SIGNS

To divide two numbers with different signs, divide the absolute values. The result is negative.

EXAMPLE 3 Divide. **(a)** $-20 \div 5$ **(b)** $36 \div (-18)$

Solution

(a) $-20 \div 5 = -4$ **(b)** $36 \div (-18) = -2$

Student Practice 3 Divide.

(a) $-50 \div 25$ **(b)** $49 \div (-7)$

What happens if we multiply $(-2)(-6)$? What sign will we obtain? Let us once again look for a pattern.

$$
\begin{array}{rcr}
3(-6) & = & -18 \\
2(-6) & = & -12 \\
1(-6) & = & -6 \\
0(-6) & = & 0 \\
-1(-6) & = & 6 \\
-2(-6) & = & 12 \\
-3(-6) & = & 18
\end{array}
$$

This number decreases by 1 each time.

This number increases by 6 each time.

Our pattern suggests that when we multiply a negative number by a negative number, we get a positive number. A similar pattern occurs for division. Thus we are ready to state the following rule.

MULTIPLICATION AND DIVISION RULE FOR TWO NUMBERS WITH THE SAME SIGN

To multiply or divide two numbers with the same sign, multiply or divide the absolute values. The sign of the result is positive.

EXAMPLE 4 Multiply. **(a)** $-5(-6)$ **(b)** $\left(-\dfrac{1}{2}\right)\left(-\dfrac{3}{5}\right)$

Solution   In each case, we are multiplying two numbers with the same sign. We will always obtain a positive number.

(a) $-5(-6) = 30$ **(b)** $\left(-\dfrac{1}{2}\right)\left(-\dfrac{3}{5}\right) = \dfrac{3}{10}$

Student Practice 4 Multiply.

(a) $-10(-6)$ **(b)** $\left(-\dfrac{1}{3}\right)\left(-\dfrac{2}{7}\right)$

Because division is related to multiplication, we find, just as in multiplication, that whenever we divide two numbers with the same sign, the result is a positive number.

EXAMPLE 5   Divide.   **(a)** $(-50) \div (-2)$ **(b)** $(-9.9) \div (-3)$

Solution

(a) $(-50) \div (-2) = 25$ **(b)** $(-9.9) \div (-3) = 3.3$

Student Practice 5   Divide.

(a) $-78 \div (-2)$ **(b)** $(-1.2) \div (-0.5)$

---

② Multiplying Three or More Signed Numbers

When multiplying more than two numbers, multiply any two numbers first, then multiply the result by another number. Continue until each factor has been used.

EXAMPLE 6   Multiply. $5(-2)(-3)$

Solution

$$5(-2)(-3) = -10(-3) \quad \text{First multiply } 5(-2) = -10.$$
$$= 30 \quad \text{Then multiply } -10(-3) = 30.$$

Student Practice 6   Multiply. $(-6)(3)(-4)$

EXAMPLE 7   Chemists have determined that a phosphate ion has an electrical charge of -3. If 10 phosphate ions are removed from a substance, what is the change in the charge of the remaining substance?

Solution   Removing 10 ions from a substance can be represented by the number -10. We can use multiplication to determine the result of removing 10 ions each having an electrical charge of -3.

$$(-3)(-10) = 30$$

Thus the change in charge would be $+30$.

Student Practice 7   An oxide ion has an electrical charge of -2. If 6 oxide ions are added to a substance, what is the change in the charge of the new substance?

EXAMPLE 8 Travis Tobey went outside his house to measure the temperature at 4:00 P.M. for seven days in October in Copper Center, Alaska. His temperature readings in degrees Fahrenheit were $-11°$, $-8°$, $-15°$, $-3°$, $5°$, $12°$, and $-1°$. Find the average temperature for this seven-day period.

Solution To find the average, we take the sum of the seven days of temperature readings and divide by seven.

$$\frac{-11 + (-8) + (-15) + (-3) + 5 + 12 + (-1)}{7} = \frac{-21}{7} = -3$$

The average temperature was $-3°F$.

Student Practice 8 In March Tony Pitkin measured the morning temperature at his farm in North Dakota at 7:00 A.M. each day. For a six-day period the temperatures in degrees Fahrenheit were $17°$, $19°$, $2°$, $-4°$, $-3°$, and $-13°$. What was the average temperature at his farm over this six-day period?

MyMathLab®

Watch the videos
in MyMathLab

Download the
MyDashBoard App

Verbal and Writing Skills, Exercises 1 and 2

1. In your own words, state the rule for multiplying two signed numbers if the signs are the same.

2. In your own words, state the rule for multiplying two signed numbers if the signs are different.

Multiply.

3. $(12)(3)$

4. $(15)(5)$

5. $(-20)(-3)$

6. $(-30)(-6)$

7. $(-20)(8)$

8. $(-15)(12)$

9. $(3)(-22)$

10. $(4)(-34)$

11. $(2.5)(-0.6)$

12. $(8.5)(-0.3)$

13. $(-12.5)(-2.25)$

14. $(-7.35)(-10.5)$

15. $\left(-\dfrac{2}{5}\right)\left(\dfrac{3}{7}\right)$

16. $\left(-\dfrac{5}{12}\right)\left(-\dfrac{2}{3}\right)$

17. $\left(-\dfrac{6}{5}\right)\left(-\dfrac{5}{2}\right)$

18. $\left(-\dfrac{9}{4}\right)\left(-\dfrac{10}{27}\right)$

Divide.

19. $-64 \div 8$

20. $-63 \div 7$

21. $\dfrac{48}{-6}$

22. $\dfrac{52}{-13}$

23. $\dfrac{-150}{-25}$

24. $\dfrac{-180}{-45}$

25. $-25 \div (-5)$

26. $-36 \div (-4)$

27. $-\dfrac{4}{9} \div \left(-\dfrac{16}{27}\right)$

28. $-\dfrac{3}{20} \div \left(-\dfrac{6}{5}\right)$

29. $\dfrac{-\dfrac{4}{5}}{-\dfrac{7}{10}}$

30. $\dfrac{-\dfrac{28}{17}}{-\dfrac{14}{3}}$

31. $50.28 \div (-6)$

32. $30.45 \div (-5)$

33. $\dfrac{45.6}{-8}$

34. $\dfrac{58.1}{-7}$

35. $\dfrac{-21,000}{-700}$

36. $\dfrac{-450,000}{-9000}$

Mixed Practice *Multiply or divide.*

37. $5(-9)$

38. $6(-11)$

39. $(-12)(-4)$

40. $(-10)(-7)$

41. $\dfrac{15}{-3}$

42. $\dfrac{36}{-2}$

43. $-30 \div (-3)$

44. $-50 \div (-5)$

45. $(-1.4)(2)$

46. $(-1.6)(3)$

47. $0.028 \div (-1.4)$

48. $0.069 \div (-2.3)$

49. $\left(-\dfrac{3}{5}\right)\left(-\dfrac{5}{7}\right)$

50. $\left(-\dfrac{2}{11}\right)\left(-\dfrac{11}{9}\right)$

51. $\dfrac{12}{5} \div \left(-\dfrac{3}{10}\right)$

52. $\dfrac{11}{4} \div \left(-\dfrac{33}{6}\right)$

Multiply.

53. $10(-5)(-3)$ **54.** $5(-4)(-6)$ **55.** $(-6)(7)(-2)$ **56.** $(-4)(3)(-8)$

57. $2(-8)(3)\left(-\dfrac{1}{3}\right)$ **58.** $7(-2)(-5)\left(\dfrac{1}{7}\right)$ **59.** $(-20)(6)(-30)(-5)$ **60.** $(-50)(-20)(3)(-6)$

61. $8(-3)(-5)(0)(-2)$ **62.** $9(-6)(-4)(-3)(0)$

Perform each set of operations. Simplify your answer. Work from left to right.

63. $\left(-\dfrac{2}{3}\right)\left(-\dfrac{3}{4}\right)\left(-\dfrac{5}{6}\right)$ **64.** $\left(-\dfrac{2}{3}\right) \div \left(-\dfrac{2}{3}\right)\left(\dfrac{3}{5}\right)$

Applications

65. *Stock Market* Paul owns 70 shares of a stock whose value went up \$2.60 per share. He also owns 120 shares of a stock whose value went down \$0.90 per share. How much did Paul gain or lose?

66. *Stock Market* Catalina owns 120 shares of a stock whose value went up \$0.80 per share. She also owns 85 shares of a stock whose value went down \$2.20 per share. How much did Catalina gain or lose?

67. *Temperature Records* In Missoula, Montana, Bob recorded the following temperature readings in degrees Fahrenheit at 7 A.M. each morning for eight days in a row: $-12°$, $-14°$, $-3°$, $5°$, $8°$, $-1°$, $-10°$, and $-23°$. What was the average temperature?

68. *Temperature Records* In Rhinelander, Wisconsin, Nancy recorded the following temperature readings in degrees Fahrenheit at 9 A.M. each morning for seven days in a row: $-8°$, $-5°$, $-18°$, $-22°$, $-6°$, $3°$, and $7°$. What was the average temperature?

69. *Underwater Photography* An underwater photographer made seven sets of underwater shots. He started at the surface, dropped down ten feet, and shot some pictures. Then he dropped ten more feet and shot more pictures. He continued this pattern until he had dropped seven times. How far beneath the surface was he at that point?

70. *Company Losses* A pharmaceutical company reported losses of \$1.5 million for five quarters in a row. What was the amount of total losses after five quarters?

Electrical Charges Ions are atoms or groups of atoms with positive or negative electrical charges. The charges of some ions are given in the following box.

aluminum +3	chloride −1	magnesium +2
oxide −2	phosphate −3	silver +1

In exercises 71–74, find the total charge.

71. 11 aluminum ions

72. 8 oxide ions

73. 6 magnesium ions and 4 chloride ions (*Hint:* Multiply first, and then add.)

74. 6 phosphate ions and 15 silver ions (*Hint:* Multiply first, and then add.)

75. Eight chloride ions are removed from a substance. What is the change in the charge of the remaining substance?

76. Six oxide ions are removed from a substance. What is the change in the charge of the remaining substance?

Golf *A round of golf is nine holes. Pine Hills Golf Course is a par 3 course. This means that the expected number of strokes it takes to complete each hole is 3. The following table indicates points above or below par for each hole. For example, if someone shoots a 5 on one of the holes (that is, takes 5 strokes to complete the hole), this is called a double bogey and is 2 points over par.*

Birdie -1	Bogey $+1$
Eagle -2	Double Bogey $+2$

77. During a round of golf with his friend, Louis got two birdies, one eagle, and two double bogeys. He made par on the rest of the holes. How many points above or below par for the entire nine-hole course is this? (*Hint:* Multiply first, and then add.)

78. Judy shot three birdies, two bogeys, and one double bogey. On the other three holes, she made par. How many points above or below par for the entire nine-hole course is this? (*Hint:* Multiply first, and then add.)

Cumulative Review

▲ **79.** [7.3.1] *Geometry* Find the area of a parallelogram with height of 6 inches and base of 15 inches.

▲ **80.** [7.3.2] *Geometry* Find the area of a trapezoid with height of 12 meters and bases of 18 meters and 26 meters.

Quick Quiz 9.3

Multiply.

1. $(-5)(-9)$

2. $(-2)(3)(-4)(-3)$

Divide.

3. $-156 \div (-4)$

4. Concept Check A student was doing the problem $(-4) + (-8) = -12$ and comparing it to the problem $(-4)(-8) = +32$. The student commented, "In the first problem two negative numbers give you a negative answer. In the second problem two negative numbers give you a positive answer. This is confusing!" Explain how the student could keep from being confused with these two types of problems.

How Am I Doing? Sections 9.1–9.3

How are you doing with your homework assignments in Sections 9.1 to 9.3? Do you feel you have mastered the material so far? Do you understand the concepts you have covered? Before you go further in the textbook, take some time to do each of the following problems.

9.1
Add.

1. $-7 + (-12)$

2. $-23 + 19$

3. $7.6 + (-3.1)$

4. $8 + (-5) + 6 + (-9)$

5. $\dfrac{8}{9} + \left(-\dfrac{2}{3}\right)$

6. $-\dfrac{5}{6} + \left(-\dfrac{1}{3}\right)$

7. $-2.8 + (-4.2)$

8. $-3.7 + 5.4$

9.2
Subtract.

9. $13 - 21$

10. $-26 - 15$

11. $\dfrac{5}{17} - \left(-\dfrac{9}{17}\right)$

12. $-19 - (-7)$

13. $-12.5 - 3.8$

14. $3.5 - 9.1$

15. $21 - (-21)$

16. $\dfrac{2}{3} - \left(-\dfrac{3}{5}\right)$

9.3
Multiply or divide.

17. $(-3)(-8)$

18. $-48 \div (-12)$

19. $-72 \div 9$

20. $(5)(-4)(2)(-1)\left(-\dfrac{1}{4}\right)$

21. $\dfrac{72}{-3}$

22. $\dfrac{-\dfrac{2}{3}}{-\dfrac{11}{12}}$

23. $(-8)(-2)(-4)$

24. $120 \div (-12)$

Mixed Practice

Perform the indicated operations. Simplify your answer.

25. $18 - (-6)$

26. $-7(-3)$

27. $-15 \div 10$

28. $1.6 + (-1.8) + (-3.4)$

29. $2.9 - 3.5$

30. $-\dfrac{1}{3} + \left(-\dfrac{2}{5}\right)$

31. $\left(-\dfrac{7}{10}\right)\left(-\dfrac{2}{7}\right)$

32. $5\dfrac{1}{2} \div \left(-\dfrac{1}{2}\right)$

33. For six days in February, the temperature in Fargo, North Dakota, was $2°F$, $-6°F$, $-10°F$, $-8°F$, $-3°F$, and $4°F$. What was the average temperature for these six days?

Now turn to page SA-16 for the answer to each of these problems. Each answer also includes a reference to the objective in which the problem is first taught. If you missed any of these problems, you should stop and review the Examples and Student Practice problems in the referenced objective. A little review now will help you master the material in the upcoming sections of the text.

1.
2.
3.
4.
5.
6.
7.
8.
9.
10.
11.
12.
13.
14.
15.
16.
17.
18.
19.
20.
21.
22.
23.
24.
25.
26.
27.
28.
29.
30.
31.
32.
33.

9.4 Order of Operations with Signed Numbers

Student Learning Objective

After studying this section, you will be able to:

① Calculate with signed numbers using more than one operation.

① Calculating with Signed Numbers Using More Than One Operation

The order of operations we discussed for whole numbers applies to signed numbers as well. The rules that we will use in this chapter are listed in the following box.

ORDER OF OPERATIONS FOR SIGNED NUMBERS

With grouping symbols:

Do first   1. Perform operations inside the parentheses.

       2. Simplify any expressions with exponents, and find any square roots.

       3. Multiply or divide from left to right.

Do last   4. Add or subtract from left to right.

EXAMPLE 1 Perform the indicated operations in the proper order.

$$-6 \div (-2)(5)$$

Solution Multiplication and division are of equal priority. So we work from left to right and divide first.

$$\underbrace{-6 \div (-2)}(5)$$
$$= \quad 3 \quad (5) = 15$$

Student Practice 1 Perform the indicated operations in the proper order.

$$20 \div (-5)(-3)$$

NOTE TO STUDENT: *Fully worked-out solutions to all of the Student Practice problems can be found at the back of the text starting at page SP-1.*

EXAMPLE 2 Perform the indicated operations in the proper order.

(a) $7 + 6(-2)$ **(b)** $9 \div 3 - 16 \div (-2)$

Solution

(a) Multiplication and division must be done first. We begin with $6(-2)$.

$$7 + \underbrace{6(-2)}$$
$$= 7 + (-12) = -5$$

(b) There is no multiplication, but there is division, and we do that first.

$$\underbrace{9 \div 3} - \underbrace{16 \div (-2)}$$
$$= \quad 3 \quad - \quad (-8) \quad = 3 + 8 \qquad \text{Transform subtraction to adding}$$
$$= 11 \qquad\qquad\qquad \text{the opposite.}$$

Student Practice 2 Perform the indicated operations in the proper order.

(a) $25 \div (-5) + 16 \div (-8)$ **(b)** $9 + 20 \div (-4)$

If a fraction has operations written in the numerator, in the denominator, or both, these operations must be done first. Then the fraction may be simplified or the division carried out.

EXAMPLE 3 Perform the indicated operations in the proper order.

$$\frac{7(-2) + 4}{8 \div (-2)(5)}$$

Solution

$$\frac{7(-2) + 4}{8 \div (-2)(5)} = \frac{-14 + 4}{(-4)(5)}$$

We perform the multiplication and division first in the numerator and the denominator, respectively.

$$= \frac{-10}{-20}$$

Simplify the fraction.

$$= \frac{1}{2}$$

Note that the answer is positive since a negative number divided by a negative number gives a positive result.

Student Practice 3 Perform the indicated operations in the proper order.

$$\frac{9(-3) - 5}{2(-4) \div (-2)}$$

Some problems involve both parentheses and exponents and will require additional steps.

EXAMPLE 4 Perform the indicated operations in the proper order.

$$4(6 - 9) + (-2)^3 + 3(-5)$$

Solution

$4(-3) + (-2)^3 + 3(-5)$ First we combine the numbers inside the parentheses.

$= 4(-3) + (-8) + 3(-5)$ Next we simplify the expression with an exponent. We obtain $(-2)^3 = (-2)(-2)(-2) = -8$.

$= -12 + (-8) + (-15)$ Next we perform each multiplication.

$= -35$ Now we add three numbers.

Student Practice 4 Perform the indicated operations in the proper order.

$$-2(-12 + 15) + (-3)^4 + 2(-6)$$

Be sure to use extra caution with problems involving fractions or decimals. It is easy to make an error in finding a common denominator or in placing a decimal point.

EXAMPLE 5 Perform the indicated operations in the proper order.

$$\left(\frac{1}{2}\right)^3 + 2\left(\frac{3}{4} - \frac{3}{8}\right) \div \left(-\frac{3}{5}\right)$$

Solution

$$\left(\frac{1}{2}\right)^3 + 2\left(\frac{6}{8} - \frac{3}{8}\right) \div \left(-\frac{3}{5}\right) \quad \text{First find the LCD and write } \frac{3}{4} = \frac{6}{8}.$$

$$= \left(\frac{1}{2}\right)^3 + 2\left(\frac{3}{8}\right) \div \left(-\frac{3}{5}\right) \quad \text{Next we combine the two fractions inside the parentheses.}$$

$$= \frac{1}{8} + 2\left(\frac{3}{8}\right) \div \left(-\frac{3}{5}\right) \quad \text{Next we simplify } \left(\frac{1}{2}\right)^3 = \left(\frac{1}{2}\right)\left(\frac{1}{2}\right)\left(\frac{1}{2}\right) = \frac{1}{8}.$$

$$= \frac{1}{8} + \frac{3}{4} \div \left(-\frac{3}{5}\right) \quad \text{Next multiply: } \left(\frac{2}{1}\right)\left(\frac{3}{8}\right) = \frac{3}{4}.$$

$$= \frac{1}{8} + \frac{3}{4} \times \left(-\frac{5}{3}\right) \quad \text{To divide two fractions, invert the second fraction and multiply.}$$

$$= \frac{1}{8} + \left(-\frac{5}{4}\right) \quad \text{Multiply } \left(\frac{3}{4}\right)\left(-\frac{5}{3}\right) = -\frac{5}{4}.$$

$$= \frac{1}{8} + \left(-\frac{10}{8}\right) \quad \text{Change } -\frac{5}{4} \text{ to the equivalent } -\frac{10}{8}.$$

$$= -\frac{9}{8} \text{ or } -1\frac{1}{8} \quad \text{Add the two fractions.}$$

Student Practice 5 Perform the indicated operations in the proper order.

$$\left(\frac{1}{5}\right)^2 + 4\left(\frac{1}{5} - \frac{3}{10}\right) \div \frac{2}{3}$$

9.4 Exercises MyMathLab®

Watch the videos in MyMathLab   Download the MyDashBoard App

Perform the indicated operations in the proper order.

1. $-8 \div (-4)(3)$

2. $-20 \div (-5)(6)$

3. $50 \div (-25)(4)$

4. $70 \div (-10)(2)$

5. $16 + 32 \div (-4)$

6. $15 - (-18) \div 3$

7. $24 \div (-3) + 16 \div (-4)$

8. $(-42) \div (-6) + 24 \div (-4)$

9. $3(-4) + 5(-2) - (-3)$

10. $6(-3) + 8(-1) - (-2)$

11. $-4(1.5 - 2.3)$

12. $-10(2.6 - 1.9)$

13. $5 - 30 \div 3$

14. $8 - 70 \div 5$

15. $36 \div 12(-2)$

16. $9(-6) + 6$

17. $3(-4) + 6(-2) - 3$

18. $-6(7) + 8(-3) + 5$

19. $11(-6) - 3(12)$

20. $(12)(-5) - 10(6)$

21. $16 - 4(8) + 18 \div (-9)$

22. $20 - 3(-2) + (-20) \div (-5)$

In exercises 23–32, simplify the numerator and denominator first, using the proper order of operations. Then reduce the fraction if possible.

23. $\dfrac{8 + 6 - 12}{3 - 6 + 5}$

24. $\dfrac{11 - 3 - 2}{-9 - 5 + 8}$

25. $\dfrac{6(-2) + 4}{6 - 3 - 5}$

26. $\dfrac{12 - 4 + 10}{5(-3) + 9}$

27. $\dfrac{2(8) \div 4 - 5}{-35 \div (-7)}$

28. $\dfrac{-25 \div 5 + (-3)(-4) - 7}{8 - (18 \div 3)}$

29. $\dfrac{24 \div (-3) - (6 - 2)}{-5(4) + 8}$

30. $\dfrac{6(-3) \div (-2) + 5}{32 \div (-8)}$

31. $\dfrac{12 \div 3 + (-2)(2)}{9 - 9 \div (-3)}$

32. $\dfrac{8(-4) - 4}{(-42) \div (-6) + (12 - 10)}$

Perform the operations in the proper order.

33. $3(2 - 6) + 4^2$

34. $-3(9 - 2) + 7^2$

35. $12 \div (-6) + (7 - 2)^3$

36. $-20 \div 2 + (6 - 3)^4$

37. $\left(-1\dfrac{1}{2}\right)(-4) - 3\dfrac{1}{4} \div \dfrac{1}{4}$

38. $(6)\left(-2\dfrac{1}{2}\right) + 2\dfrac{1}{3} \div \dfrac{1}{3}$

39. $\left(\dfrac{3}{5} - \dfrac{2}{5}\right)^2 + \left(\dfrac{3}{2}\right)\left(-\dfrac{1}{5}\right)$

40. $\left(\dfrac{5}{8} - \dfrac{1}{8}\right)^2 - \left(-\dfrac{5}{6}\right)\left(\dfrac{8}{5}\right)$

41. $(1.2)^2 - 3.6(-1.5)$

42. $(0.7)^2 - 2.8(-5.5)$

Applications

Temperature Averages *The following chart gives the average monthly high and low temperatures in degrees Fahrenheit for the city of Noril'sk, one of Russia's northernmost cities.*

	Jan.	Feb.	Mar.	Apr.	May	June	July	Aug.	Sept.	Oct.	Nov.	Dec.
High	−13	−11	0	13	29	50	64	59	44	21	−1	−9
Low	−27	−21	−14	−3	17	38	49	46	35	12	−10	−18

Use the chart to solve Exercises 43–48. Round to the nearest tenth of a degree when necessary.

43. What is the average low temperature in Noril'sk during December, January, and February?

44. What is the average high temperature in Noril'sk during December, January, and February?

45. What is the average low temperature in Noril'sk from January through June?

46. What is the average high temperature in Noril'sk from October through March?

47. What is the average yearly high temperature in Noril'sk?

48. What is the average yearly low temperature in Noril'sk?

Cumulative Review

49. [6.2.2] ***Metric Measure*** A telephone wire that is 3840 meters long is how long measured in kilometers?

50. [6.3.2] ***Metric Measure*** A jar with 36.8 grams of protein contains how many milligrams of protein?

Quick Quiz 9.4 *Perform the indicated operations in the proper order. Simplify all answers.*

1. $5(-4) - 6(2 - 5)^2$

2. $-3.2 - 8.5 - (-3.0) + 2(0.4)$

3. $\dfrac{5 + 26 \div (-2)}{-2(3) + 4(-3)}$

4. Concept Check Explain in what order to perform the operations in the expression $4(3 - 9) + 5^2$.

9.5 Scientific Notation

① Changing Numbers in Standard Notation to Scientific Notation

Scientists who frequently work with very large or very small measurements use a certain way to write numbers, called *scientific notation*. In our usual way of writing a number, which we call "standard notation" or "ordinary form," we would express the distance to the nearest star, Proxima Centauri, as 24,800,000,000,000 miles. In scientific notation, we more conveniently write this as

$$2.48 \times 10^{13} \text{ miles.}$$

For a very small number, like two millionths, the standard notation is

$$0.000002.$$

The same quantity in scientific notation is

$$2 \times 10^{-6}.$$

Notice that each number in scientific notation has two parts: (1) a number that is 1 or greater but less than 10, which is multiplied by (2) a power of 10. That power is either a whole number or the negative of a whole number. In this section we learn how to go back and forth between standard and scientific notation.

> A positive number is in **scientific notation** if it is in the form $a \times 10^n$, where a is a number greater than (or equal to) 1 and less than 10, and n is an integer.

We begin our investigation of scientific notation by looking at large numbers. We recall from Section 1.6 that

$$10 = 10^1$$
$$100 = 10^2$$
$$1000 = 10^3.$$

The number of zeros tells us the number for the exponent. To write a number in scientific notation, we want the first number to be greater than or equal to 1 and less than 10, and the second number to be a power of 10.

Let's see how this can be done. Consider the following.

$$6700 = \underbrace{6.7}_{\substack{\text{greater than 1} \\ \text{and less than 10}}} \times 1000 = 6.7 \times \underset{\substack{\uparrow \\ \text{a power} \\ \text{of 10}}}{10^3}$$

Let us look at two more cases.

$$530 = 5.3 \times 100 \qquad\qquad 156{,}000 = 1.56 \times 100{,}000$$
$$= \underbrace{5.3 \times 10^2} \qquad\qquad\qquad = \underbrace{1.56 \times 10^5}$$

These numbers are in scientific notation.

Now that we have seen some examples of numbers in scientific notation, let us think through the steps in the next example.

Student Learning Objectives

After studying this section, you will be able to:

① Change numbers in standard notation to scientific notation.

② Change numbers in scientific notation to standard notation.

③ Add and subtract numbers in scientific notation.

It is important to remember that all numbers *greater than or equal to* 10 always have a positive exponent when expressed in scientific notation.

EXAMPLE 1   Write in scientific notation.

(a) 9826

(b) 163,457

Solution

(a)

What power?

$$9826. \ = \ 9.826 \times 10$$

starting position of decimal point

ending position of decimal point

The decimal point moved 3 places to the left. We therefore use 3 for the power of 10.

$$9826 = 9.826 \times 10^3$$

(b)

What power?

$$163,457. \ = \ 1.63457 \times 10$$

starting position of decimal point

ending position of decimal point

The decimal point moved 5 places to the left. We therefore use 5 for the power of 10.

$$163,457 = 1.63457 \times 10^5$$

Student Practice 1   Write in scientific notation.

(a) 3729

(b) 506,936

NOTE TO STUDENT: *Fully worked-out solutions to all of the Student Practice problems can be found at the back of the text starting at page SP-1.*

Calculator

 Standard to Scientific

Many calculators have a setting that will display numbers in scientific notation. Consult your manual on how to do this. To convert 154.32 into scientific notation, first change your setting to display in scientific notation. Often this is done by pressing $\boxed{\text{SCI}}$ or $\boxed{\text{2nd}}$ $\boxed{\text{SCI}}$. Often SCI is then displayed on the calculator.

Then enter:

$$154.32 \ \boxed{=}$$

Display:

$$\boxed{1.5432 \quad 02}$$

1.5432   02 means 1.5432 × 10².

Note that your calculator display may show the power of 10 in a different manner. Be sure to change your setting back to the regular display when you are done.

When changing to scientific notation, all zeros to the *right* of the final non-zero digit may be eliminated. This does not change the value of your answer.

Now we will look at numbers that are *less than* 1. In the introduction to this section we saw that 0.000002 can be written as 2×10^{-6}. How is it possible to have a negative exponent? Let's take another look at the powers of 10.

$$10^3 = 1000$$
$$10^2 = 100$$
$$10^1 = 10$$
$$10^0 = 1 \qquad \text{Recall that any number to the zero power is 1.}$$

Now, following this pattern, what would you expect the next number on the left of the equals sign to be? What would be the next number on the right of the equals sign? Each number on the right is one-tenth the number above it. We continue the pattern.

$$10^{-1} = 0.1$$
$$10^{-2} = 0.01$$
$$10^{-3} = 0.001$$

Thus you can see how it is possible to have negative exponents. We use negative exponents to write numbers that are less than 1 in scientific notation.

Let's look at 0.76. This number is less than 1. Recall that in our definition for scientific notation the first number must be greater than or equal to 1 and less than 10. To change 0.76 to such a number, we will have to move the decimal point.

$$0.76 = 7.6 \times 10^{-1}$$

The decimal point moves 1 place to the right. We therefore put a -1 for the power of 10. We use a negative exponent because the original number is less than 1. Let's look at two more cases.

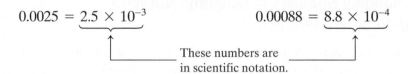

$$0.0025 = 2.5 \times 10^{-3} \qquad\qquad 0.00088 = 8.8 \times 10^{-4}$$

These numbers are
in scientific notation.

When we start with a positive number that is less than 1 and write it in scientific notation, we will get a result with 10 to a negative power. Think carefully through the steps of the following example.

EXAMPLE 2   Write in scientific notation.

(a) 0.036   **(b)** 0.72

Solution

(a) We change the given number to a number greater than or equal to 1 and less than 10. Thus we change 0.036 to 3.6.

What power?

$$0.036 = 3.6 \times 10$$

starting position
of decimal point

ending position
of decimal point

The decimal point moved 2 places to the right. Since the original number is less than 1, we use -2 for the power of 10.

$$0.036 = 3.6 \times 10^{-2}$$

(b)

What power?

$$0.72 = 7.2 \times 10$$

starting position
of decimal point

ending position
of decimal point

The decimal point moved 1 place to the right. Because the original number is less than 1, we use -1 for the power of 10.

$$0.72 = 7.2 \times 10^{-1}$$

Student Practice 2   Write in scientific notation.

(a) 0.076   **(b)** 0.982

It is very helpful when writing numbers in scientific notation to remember these two concepts.

1. A number that is larger than or equal to 10 and written in scientific notation will always have a positive exponent as the power of 10.

2. A positive number that is smaller than 1 and written in scientific notation will always have a negative exponent as the power of 10.

② Changing Numbers in Scientific Notation to Standard Notation

Often we are given a number in scientific notation and want to write it in standard notation—that is, we want to write it in what we consider "ordinary form." To do this, we reverse the process we've just used. If the number in scientific notation has a positive power of 10, we know the number is greater than or equal to 10. Therefore, we move the decimal point to the right to convert to standard notation.

EXAMPLE 3 Write in standard notation.

$$5.8671 \times 10^4$$

Solution

$5.8671 \times 10^4 = 58,671$ Move the decimal point four places to the right.

Student Practice 3 Write in standard notation.

$$6.543 \times 10^3$$

In some cases, we will need to add zeros as we move the decimal point to the right.

EXAMPLE 4 Write in standard notation.

(a) 9.8×10^5 **(b)** 3×10^3

Solution

(a) $9.8 \times 10^5 = 980,000$ Move the decimal point five places to the right. Add four zeros.

(b) $3 \times 10^3 = 3000$ Move the decimal point three places to the right. Add three zeros.

Student Practice 4 Write in standard notation.

(a) 4.3×10^5 **(b)** 6×10^4

If the number in scientific notation has a negative power of 10, we know the number is less than 1. Therefore, we move the decimal point to the left to convert to standard notation. In most cases, we need to add zeros as we move the decimal point to the left.

$\mathbb{M}_{\mathbb{C}}$ **EXAMPLE 5**   Write in standard notation.

(a) 2.48×10^{-3} **(b)** 1.2×10^{-4}

Solution

(a) $2.48 \times 10^{-3} = 0.00248$ Move the decimal point three places to the left. Add two zeros between the decimal point and 2.

(b) $1.2 \times 10^{-4} = 0.00012$ Move the decimal point four places to the left. Add three zeros between the decimal point and 1.

Student Practice 5   Write in standard notation.

(a) 7.72×10^{-3}

(b) 2.6×10^{-5}

③ Adding and Subtracting Numbers in Scientific Notation

Scientists calculate with numbers in scientific notation when they study galaxies (the huge) and when they study microbes (the tiny). To add or subtract numbers in scientific notation, the numbers must have the same power of 10.

> Numbers in scientific notation may be added or subtracted if they have the same power of 10. We add or subtract the decimal part and leave the power of 10 unchanged.

EXAMPLE 6   Add. 5.89×10^{-20} meters $+ 3.04 \times 10^{-20}$ meters

Solution
$$
\begin{array}{r}
5.89 \times 10^{-20} \text{ meters} \\
+ \ 3.04 \times 10^{-20} \text{ meters} \\
\hline
8.93 \times 10^{-20} \text{ meters}
\end{array}
$$

Student Practice 6   Add.

$$6.85 \times 10^{22} \text{ kilograms} + 2.09 \times 10^{22} \text{ kilograms}$$

EXAMPLE 7   Add. $7.2 \times 10^6 + 5.2 \times 10^5$

Solution   Note that these two numbers have different powers of ten, so we cannot add them as written. We will change one number from scientific notation to another form.

Now $5.2 \times 10^5 = 520{,}000$ can be rewritten as 0.52×10^6. Now we can add.

$$
\begin{array}{r}
7.2 \ \times 10^6 \\
+ \ 0.52 \times 10^6 \\
\hline
7.72 \times 10^6
\end{array}
$$

Student Practice 7   Subtract. $4.36 \times 10^5 - 3.1 \times 10^4$

9.5 Exercises

MyMathLab®

Watch the videos
in MyMathLab

Download the
MyDashBoard App

Verbal and Writing Skills, Exercises 1–4

1. Why does scientific notation use exponents with base 10?

2. Explain how to write 0.787 in scientific notation.

3. What are the two parts of a number in scientific notation?

4. If $a \times 10^6$ is a number written in scientific notation, what are the possible values of a?

Write in scientific notation.

5. 120

6. 340

7. 1900

8. 5200

9. 26,300

10. 78,100

11. 288,000

12. 554,000

13. 10,000

14. 100,000

15. 12,000,000

16. 28,000,000

17. 0.0931

18. 0.0242

19. 0.00279

20. 0.00613

21. 0.82

22. 0.17

23. 0.00054

24. 0.00092

25. 0.00000531

26. 0.00000198

27. 0.000008

28. 0.00000007

Write in standard notation.

29. 5.36×10^4

30. 2.19×10^4

31. 5.334×10^3

32. 7.0235×10^5

33. 4.6×10^{12}

34. 3.8×10^{11}

35. 6.2×10^{-2}

36. 3.5×10^{-2}

37. 8.99×10^{-3}

38. 4.05×10^{-1}

39. 9×10^{11}

40. 2×10^{12}

41. 3.862×10^{-8}

42. 8.139×10^{-9}

Mixed Practice

43. Write in scientific notation. 35,689

44. Write in scientific notation. 76,371

45. Write in standard notation. 3.3×10^{-4}

46. Write in standard notation. 5.7×10^{-2}

47. Write in scientific notation. 0.00278

48. Write in scientific notation. 0.0000134

49. Write in standard notation. 1.88×10^6

50. Write in standard notation. 3.49×10^6

Applications *Write in scientific notation.*

51. *Light Speed* In 1 year, light will travel 5,878,000,000,000 miles.

52. *Oceans* The world's oceans hold approximately 326,000,000,000,000,000,000 gallons of water.

▲ **53.** *Red Blood Cells* The average volume of a red blood cell is 0.000000000000092 liter.

▲ **54.** *Hydrogen Atom* The radius of a hydrogen atom is 0.000000005 meter.

Write in standard notation.

55. *Nanosecond* An electrical signal can travel one foot in one nanosecond, which is 10^{-9} second.

56. *Femtosecond* A common measurement used in laser technology is the femtosecond, which is 10^{-15} second.

▲ **57.** *Human Blood* The diameter of a red corpuscle of human blood is about 7.5×10^{-5} centimeter.

▲ **58.** *Mercury Drill* Recently a tiny hole with an approximate diameter of 3.16×10^{-10} meter was produced by Doctors Heckl and Maddocks using a chemical method involving a mercury drill.

59. *Volcano Eruption* During the eruption of the Taupo volcano in New Zealand in 130 AD, approximately 1.4×10^{10} tons of pumice were carried up in the air.

60. *National Debt* On October 5, 2010, the United States' national debt was approximately 1.3549×10^{13} dollars.

Add.

61. 3.38×10^7 dollars $+ 5.63 \times 10^7$ dollars

62. 8.17×10^9 atoms $+ 2.76 \times 10^9$ atoms

63. *Planets* The mass of Earth is 5.87×10^{21} tons, and the mass of Venus is 4.81×10^{21} tons. What is the total mass of the two planets?

64. *Planets* The masses of Mars, Pluto, and Mercury are 6.34×10^{20} tons, 1.76×10^{20} tons, and 3.21×10^{20} tons, respectively. What is the total mass of the three planets?

Subtract.

65. 4×10^8 feet $- 3.76 \times 10^7$ feet

66. 9×10^{10} meters $- 1.26 \times 10^9$ meters

▲ **67.** *Geography* The two largest continents are Asia and Africa with areas of 1.76×10^7 square miles and 1.16×10^7 square miles, respectively. How much larger is Asia than Africa?

▲ **68.** *Geography* The total ocean area and land area of Earth is approximately 1.39×10^8 square miles and 5.747×10^7 square miles, respectively. What is the total area of Earth covered by land or ocean?

To Think About

69. *Planets* In 2010, astronomers discovered the first credible example of a planet outside our solar system that could sustain life. The planet's name is Gliese 581g, and is 20 light-years from Earth. A light-year is approximately 5.88 trillion miles. Find in scientific notation how many miles Gliese 581g is from Earth.

70. *Planets* Kepler-7b is a planet outside our solar system and was discovered by astronomers in 2010. Kepler-7b is 2300 light-years from Earth. A light-year is approximately 5.88 trillion miles. Find in scientific notation how many miles Kepler-7b is from Earth.

Cumulative Review *Calculate.*

71. **[3.4.1]** 12.5×0.21

72. **[3.5.2]** $0.53 \overline{)0.13674}$

73. **[7.10.1]** *Radar Construction* A radar mounting plate is constructed in the shape of a parallelogram. Two sides of the parallelogram are 9 feet and 13 feet. A rubber piece of safety bumper must be placed around all sides of the parallelogram. It costs $4 per foot. How much will it cost to place the safety bumper around the plate?

74. **[7.9.1]** *Mountain Climbing* Michael is mountain climbing in the Rocky Mountains. He is 70 inches tall. At 2 P.M. his shadow measures 95 inches. He wants to climb to the top of a cliff that casts a shadow of 800 feet at 2 P.M. How tall is the cliff? Round to the nearest whole number.

Quick Quiz 9.5 *Write in scientific notation.*

1. 0.000345

2. 568,300

Write in standard notation.

3. 8.34×10^{-6}

4. **Concept Check** Explain how to place the zeros and commas in the correct locations to write the number 5.398×10^8 in standard notation.

Did You Know...
That You Can Save Money by Purchasing Store Brand Products?

FOOD AND RICE PRICES

Understanding the Problem:
Many large stores and chains have their own brands of products. Store brand products often cost much less than the equivalent name brand products. For example, a one-pound bag of rice from a national name brand can cost $2.50. A one-pound bag of the same rice from a store brand might cost only $1.00.

Making a Plan:
Lucy and her family enjoy rice as part of their dinner four times a week. Lucy has a family of four that consumes approximately 2/3 cup of rice with each meal. Lucy wants to calculate the cost of feeding her family and see if they can save money.

Step 1: Lucy needs to know how much rice her family consumes in a week.

Task 1: How many cups of rice does Lucy's family eat in a week?

Task 2: If each cup of raw rice weighs approximately 21 ounces, what is the weight in ounces of the rice Lucy's family eats in a week?

Task 3: There are 16 ounces in a pound. Find the number of pounds of raw rice Lucy's family eats each week.

Step 2: Lucy notices that her supermarket has its own brand of rice that is much less expensive than the brand she normally buys. She usually buys name brand rice for $2.66 per pound. The store brand is only 88 cents per pound.

Task 4: Find out how much Lucy spends per week on the name brand rice.

Task 5: Find out how much Lucy could save per week by buying the same amount of rice from the store brand.

Task 6: Find the percent savings that Lucy gets by buying the store brand rice instead of the name brand.

Finding a Solution:
Step 3: Lucy realizes that she can have this kind of savings each week on all the food she buys by choosing the store brand products over the name brand.

Task 7: If Lucy typically spends $162 per week on groceries by buying the name brand products, how much could she save in a year by purchasing all store brand products?

Task 8: Lucy finds that her family still prefers some of the name brand products. She continues to buy some of the name brand products and some of the store brands. She finds that her weekly grocery bill is reduced to around $130 per week. If this continues, how much will she save in the course of a year?

Applying the Situation to Your Life:
You should calculate how much you spend on groceries every week. Then see how you can reduce that by purchasing store brands. You may not change to store brands for all your shopping, but for every case where you do, you can save money.

Chapter 9 Organizer

Topic and Procedure	Examples	✏ You Try It
Absolute value, p. 561 The absolute value of a number is the distance between the number and zero on a number line.	**(a)** $\lvert -6 \rvert = 6$ **(b)** $\lvert 3 \rvert = 3$ **(c)** $\lvert 0 \rvert = 0$	1. Find each absolute value. **(a)** $\lvert -10 \rvert$ **(b)** $\lvert 18 \rvert$ **(c)** $\lvert -2.5 \rvert$
Adding signed numbers with the same sign, p. 562 To add two numbers with the same sign: **1.** Add the absolute values of the numbers. **2.** Use the common sign in the answer.	**(a)** $12 + 5 = 17$ **(b)** $-6 + (-8) = -14$ **(c)** $-5.2 + (-3.5) = -8.7$ **(d)** $-\dfrac{1}{7} + \left(-\dfrac{3}{7}\right) = -\dfrac{4}{7}$	2. Add. **(a)** $13 + 4$ **(b)** $-12 + (-5)$ **(c)** $-3.2 + (-3.5)$ **(d)** $-\dfrac{3}{8} + \left(-\dfrac{2}{8}\right)$
Adding signed numbers with different signs, p. 564 To add two numbers with different signs: **1.** Subtract the absolute values of the numbers. **2.** Use the sign of the number with the larger absolute value.	**(a)** $14 + (-8) = 6$ **(b)** $-4 + 8 = 4$ **(c)** $-3.2 + 7.1 = 3.9$ **(d)** $\dfrac{5}{13} + \left(-\dfrac{8}{13}\right) = -\dfrac{3}{13}$	3. Add. **(a)** $9 + (-8)$ **(b)** $-22 + 5$ **(c)** $-9.5 + 4.7$ **(d)** $\dfrac{1}{12} + \left(-\dfrac{5}{12}\right)$
Subtracting signed numbers, p. 572 To subtract signed numbers, add the opposite of the second number to the first number.	**(a)** $-9 - (-3) = -9 + 3 = -6$ **(b)** $5 - (-7) = 5 + 7 = 12$ **(c)** $8 - 12 = 8 + (-12) = -4$ **(d)** $-4 - 13 = -4 + (-13) = -17$ **(e)** $-\dfrac{1}{12} - \left(-\dfrac{5}{12}\right) = -\dfrac{1}{12} + \dfrac{5}{12} = \dfrac{4}{12} = \dfrac{1}{3}$	4. Subtract. **(a)** $8.1 - (-1.3)$ **(b)** $-10 - 3$ **(c)** $-34 - (-16)$ **(d)** $1.2 - 1.9$ **(e)** $-\dfrac{2}{5} - \left(-\dfrac{3}{5}\right)$
Multiplying or dividing signed numbers with different signs, pp. 578–579 To multiply or divide two numbers with different signs, multiply or divide the absolute values. The result is negative.	**(a)** $7(-3) = -21$ **(b)** $(-6)(4) = -24$ **(c)** $(-36) \div 2 = -18$ **(d)** $\dfrac{41.6}{-8} = -5.2$	5. Multiply or divide. **(a)** $(-5)(9)$ **(b)** $(-0.5)(24)$ **(c)** $63 \div (-3)$ **(d)** $\dfrac{-17}{3.4}$
Multiplying or dividing signed numbers with the same sign, p. 579 To multiply or divide two numbers with the same sign, multiply or divide the absolute values. The sign of the result is positive.	**(a)** $(0.5)(0.3) = 0.15$ **(b)** $(-6)(-2) = 12$ **(c)** $\dfrac{-20}{-2} = 10$ **(d)** $\dfrac{-\frac{1}{3}}{-\frac{1}{7}} = \left(-\dfrac{1}{3}\right)\left(-\dfrac{7}{1}\right) = \dfrac{7}{3}$	6. Multiply or divide. **(a)** $\left(\dfrac{1}{3}\right)(9)$ **(b)** $(-7)(-8)$ **(c)** $\dfrac{-30}{-15}$ **(d)** $\dfrac{-\frac{1}{2}}{-\frac{1}{4}}$

Order of operations for signed numbers, p. 586
 	
With grouping symbols:
Do first   **1.** Perform operations inside the parentheses.
     2. Simplify any expressions with exponents, and find any square roots.
     3. Multiply or divide from left to right.
Do last   **4.** Add or subtract from left to right.
	Perform the operations in the proper order.
	$$-2(12 - 8) + (-3)^3 + 4(-6)$$
	$$= -2(4) + (-3)^3 + 4(-6)$$
	$$= -2(4) + (-27) + 4(-6)$$
	$$= -8 + (-27) + (-24)$$
	$$= -59$$
		7. Perform the operations in the proper order. $(-2)^2 + 3(-5) + 5(9 - 12)$

Topic and Procedure	Examples	✏️ You Try It
Simplifying fractions with combined operations in numerator and denominator, p. 587 1. Perform the operations in the numerator. 2. Perform the operations in the denominator. 3. Simplify the fraction.	Perform the operations in the proper order. $$\frac{7(-4) - (-2)}{8 - (-5)}$$ The numerator is $7(-4) - (-2) = -28 - (-2) = -26.$ The denominator is $8 - (-5) = 8 + 5 = 13.$ Thus the fraction becomes $\dfrac{-26}{13} = -2.$	8. Perform the operations in the proper order. $\dfrac{-2(5 - 2) + 8}{3(-4)}$
Writing a number in scientific notation, p. 591 1. Move the decimal point to the position immediately to the right of the first nonzero digit. 2. Count the number of decimal places you moved the decimal point. 3. Multiply the result by a power of 10 equal to the number of places moved. If the number you started with is larger than or equal to 10, use a positive exponent. If the number you started with is less than 1, use a negative exponent.	**(a)** $178 = 1.78 \times 10^2$ **(b)** $25{,}000{,}000 = 2.5 \times 10^7$ **(c)** $0.006 = 6 \times 10^{-3}$ **(d)** $0.00001732 = 1.732 \times 10^{-5}$	9. Write in scientific notation. **(a)** 3124 **(b)** 0.000588 **(c)** 180,000,000 **(d)** 0.000009
Changing from scientific notation to standard notation, p. 594 1. If the exponent of 10 is *positive*, move the decimal point to the *right* as many places as the exponent shows. Insert extra zeros as necessary. 2. If the exponent of 10 is *negative*, move the decimal point to the *left* as many places as the exponent shows. *Note:* Remember that numbers in scientific notation that have positive exponents are always greater than or equal to 10. Numbers in scientific notation that have negative exponents are always less than 1.	**(a)** $8 \times 10^6 = 8{,}000{,}000$ **(b)** $1.23 \times 10^4 = 12{,}300$ **(c)** $7 \times 10^{-2} = 0.07$ **(d)** $8.45 \times 10^{-5} = 0.0000845$	10. Write in standard notation. **(a)** 4×10^5 **(b)** 4×10^{-3} **(c)** 2.526×10^6 **(d)** 8.13×10^{-2}

Chapter 9 Review Problems

Section 9.1

Add.

1. $-20 + 5$

2. $-18 + 4$

3. $-3.6 + (-5.2)$

4. $10.4 + (-7.8)$

5. $-\dfrac{1}{5} + \left(-\dfrac{1}{3}\right)$

6. $\dfrac{9}{10} + \left(-\dfrac{5}{2}\right)$

7. $20 + (-14)$

8. $12 + (-7) + (-8) + 3$

Section 9.2

Subtract.

9. $25 - 36$

10. $12 - 40$

11. $14.5 - (-6)$

12. $-11.4 - 5.8$

13. $-5.2 - 7.1$

14. $-\dfrac{2}{5} - \left(-\dfrac{1}{3}\right)$

Perform the indicated operations from left to right.

15. $5 - (-2) - (-6)$

16. $-15 - (-3) + 9$

17. $9 - 8 - 6 - 4$

Section 9.3

Multiply or divide.

18. $\left(-\dfrac{2}{7}\right)\left(-\dfrac{1}{5}\right)$

19. $(5.2)(-1.5)$

20. $-60 \div (-20)$

21. $-18 \div (-3)$

22. $\dfrac{70}{-14}$

23. $\dfrac{-13.2}{-2.2}$

24. $\dfrac{-\dfrac{3}{4}}{\dfrac{1}{6}}$

25. $\dfrac{-\dfrac{1}{3}}{-\dfrac{7}{9}}$

26. $3(-5)(-2)$

27. $(-2)(3)(-6)(-1)$

Section 9.4

Perform the indicated operations in the proper order.

28. $10 + 40 \div (-4)$

29. $2(-6) + 3(-4) - (-13)$

30. $36 \div (-12) + 50 \div (-25)$

31. $50 \div 25(-4)$

32. $-3.5 \div (-5) - 1.2$

33. $2.5(-2) + 3.8$

In exercises 34–36, simplify the numerator and denominator first, using the proper order of operations. Then reduce the fraction if possible.

34. $\dfrac{8 - 17 + 1}{6 - 10}$

35. $\dfrac{9 - 3 + 4(-3)}{2 - (-6)}$

36. $\dfrac{20 \div (-5) - (-6)}{(2)(-2)(-5)}$

Perform the operations in the proper order.

37. $2(7 - 11)^2 - 4^3$

38. $-50 \div (-10) + (5 - 3)^4$

39. $\left(\dfrac{2}{3}\right)^2 - \dfrac{3}{8}\left(\dfrac{8}{5}\right)$

40. $(1.2)^2 + (2.8)(-0.5)$

41. $1.4(4.7 - 4.9) - 12.8 \div (-0.2)$

Section 9.5

Write in scientific notation.

42. 4160

43. 3,700,000

44. 200,000

45. 0.007

46. 0.0000218

47. 0.00000763

Write in standard notation.

48. 1.89×10^4

49. 3.76×10^3

50. 3.14×10^5

51. 7.52×10^{-2}

52. 6.61×10^{-3}

53. 9×10^{-7}

Add or subtract. Express your answer in scientific notation.

54. $2.42 \times 10^7 + 5.76 \times 10^7$

55. $6.11 \times 10^{10} + 3.87 \times 10^{10}$

56. $3.42 \times 10^{14} - 1.98 \times 10^{14}$

57. $1.76 \times 10^{26} - 1.08 \times 10^{26}$

58. *Astronomy* The distance from Earth to the sun is approximately 93,000,000 miles. How many feet is that? Write your answer in scientific notation.

59. *Alpha Centauri* Alpha Centauri is the closest star system to our own solar system, at approximately 280,000 astronomical units away. One astronomical unit is equal to the distance from Earth to the sun. Using the information given in exercise 58, calculate how many miles Alpha Centauri is from our solar system. Write your answer in scientific notation.

60. *Atomic Particles* The mass of a proton is about 1.67 yg (yoctograms), and the mass of an electron is about 0.00091 yg. If *yocto-* means 0.000000000000000000000001, write the mass of a proton and that of an electron in grams in scientific notation.

▲ **61.** *Saturn's Rings* The rings of Saturn are approximately 2.5×10^8 meters in diameter. Write this number in standard form.

Mixed Practice

62. *Moon* The average distance to the moon is 384.4 Mm (megameters). If a megameter is equal to 10^6 meters, write out the number in meters.

63. *Football* In three plays of a football game, the quarterback threw passes that lost 5 yards, gained 6 yards, and lost 7 yards. What was the total gain or loss of the three plays?

64. *Small Plane* Fred is 6 feet tall and is standing at the lowest point of Death Valley, California, which is 282 feet below sea level. A small plane flies directly over Fred at an altitude of 2400 feet above sea level. Find the distance from the top of Fred's head to the plane.

65. *Checking Account* Max has overdrawn his checking account by $18. His bank charged him $20 for an overdraft fee. He quickly deposited $40. What is his current balance?

66. *Temperature Statistics* In Duluth, Minnesota, the high temperature in degrees Fahrenheit for five days during January was $-16°$, $-18°$, $-5°$, $3°$, and $-12°$. What was the average high temperature for these five days?

67. *Golf* Frank played golf with his friend Samuel. They played nine holes of golf on a special practice course. The expected number of strokes for each hole is 3. A birdie is 1 below par. An eagle is 2 below par. A bogey is 1 above par. A double bogey is 2 above par. Frank played on par for 1 hole and got two birdies, one eagle, four bogeys, and one double bogey on the rest of the course. How many points above or below par was Frank on this nine-hole course?

How Am I Doing?   Chapter 9 Test

 MATH COACH   MyMathLab®   You Tube™

After you take this test read through the Math Coach on pages 606–607. Math Coach videos are available via MyMathLab and YouTube. Step-by-step test solutions in the Chapter Test Prep Videos are also available via MyMathLab and YouTube. (Search "TobeyBasicCollMath" and click on "Channels.")

Add.

1. $-26 + 15$

2. $-31 + (-12)$

3. $12.8 + (-8.9)$

4. $-3 + (-6) + 7 + (-4)$

5. $-5\dfrac{3}{4} + 2\dfrac{1}{4}$

MC 6. $-\dfrac{1}{4} + \left(-\dfrac{5}{8}\right)$

Subtract.

7. $-32 - 6$

8. $23 - 18$

9. $\dfrac{4}{5} - \left(-\dfrac{1}{3}\right)$

10. $-50 - (-7)$

MC 11. $-2.5 - (-6.5)$

12. $-8.5 - 2.8$

13. $\dfrac{1}{12} - \left(-\dfrac{5}{6}\right)$

14. $-15 - (-15)$

Multiply or divide.

15. $(-20)(-6)$

16. $27 \div \left(-\dfrac{3}{4}\right)$

17. $-40 \div (-4)$

18. $(-9)(-1)(-2)(4)\left(\dfrac{1}{4}\right)$

19. $\dfrac{-39}{-13}$

20. $\dfrac{-\dfrac{3}{5}}{\dfrac{6}{7}}$

21. $(-12)(0.5)(-3)$

22. $96 \div (-3)$

1. ☐
2. ☐
3. ☐
4. ☐
5. ☐
6. ☐
7. ☐
8. ☐
9. ☐
10. ☐
11. ☐
12. ☐
13. ☐
14. ☐
15. ☐
16. ☐
17. ☐
18. ☐
19. ☐
20. ☐
21. ☐
22. ☐

Perform the indicated operations in the proper order.

23. $7 - 2(-5)$

24. $-2.5 - 1.2 \div (-0.4)$

25. $18 \div (-3) + 24 \div (-12)$

$\mathbb{M}\mathbb{C}$ **26.** $-6(-3) - 4(3 - 7)^2$

27. $1.3 - 9.5 - (-2.5) + 3(-0.5)$

28. $-48 \div (-6) - 7(-2)^2$

29. $\dfrac{3 + 8 - 5}{(-4)(6) + (-6)(3)}$

30. $\dfrac{5 + 28 \div (-4)}{7 - (-5)}$

Write in scientific notation.

31. 80,540

32. 0.000007

Write in standard notation.

$\mathbb{M}\mathbb{C}$ **33.** 9.36×10^{-5}

34. 7.2×10^4

Solve.

35. In Chicago the high temperatures in degrees Fahrenheit for five days during February were $-14°$, $-8°$, $-5°$, $7°$, and $-11°$. What was the average high temperature for these five days?

▲ **36.** A rectangular computer chip is 5.8×10^{-5} meter wide and 7.8×10^{-5} meter long. Find the perimeter of the chip and express your answer in scientific notation.

37. The lowest recorded temperature in Antarctica is $-128.6°$F in Vostock II on July 21, 1983. The highest recorded temperature in Antarctica is $58.3°$F in Hope Bay on January 5, 1974. What is the difference between these temperatures?

23. _____ ☐

24. _____ ☐

25. _____ ☐

26. _____ ☐

27. _____ ☐

28. _____ ☐

29. _____ ☐

30. _____ ☐

31. _____ ☐

32. _____ ☐

33. _____ ☐

34. _____ ☐

35. _____ ☐

36. _____ ☐

37. _____ ☐

Total Correct: ☐

MATH COACH

Mastering the skills you need to do well on the test.

Students often make the same types of errors when they do the Chapter 9 Test. Here are some helpful hints to keep you from making these common errors on test problems.

Adding Signed Numbers—Problem 6   Add. $-\dfrac{1}{4} + \left(-\dfrac{5}{8}\right)$

> **Helpful Hint** Be sure to find the least common denominator (LCD) for the two fractions before adding.

Did you identify the LCD as 8?

Yes [ ]   No [ ]

If you answered No, go back and make this change.

Did you change $-\dfrac{1}{4}$ to $-\dfrac{2}{8}$ before adding the fractions?

Yes [ ]   No [ ]

If you answered No, stop and make this conversion.

Is your final answer negative?

Yes [ ]   No [ ]

If you answered No, review the sign rules and perform the addition again.

If you answered Problem 6 incorrectly, please go back and rework the problem using these suggestions.

Subtracting Signed Numbers—Problem 11   Subtract. $-2.5 - (-6.5)$

> **Helpful Hint** Remember to rewrite the problem as adding the first number plus the opposite of the second number.

Did you rewrite the problem correctly as $-2.5 + (6.5)$?

Yes [ ]   No [ ]

If you answered No, stop and make this correction to your work.

Did you realize that your answer would be a positive number?

Yes [ ]   No [ ]

If you answered No, review the sign rules and do the problem again.

Need help? Watch the MATH COACH videos in MyMathLab® or on You Tube™.

606

Helpful Hint First, combine numbers inside the parentheses. Next, raise numbers to a power. Then perform multiplication from left to right. Finally, perform subtraction from left to right.

Did you first combine $3 - 7$ inside the parentheses to obtain -4?

Yes _____ No _____

If you answered No, please review the Helpful Hint above and try that step again.

Did you next calculate $(-4)^2 = 16$?

Yes _____ No _____

If you answered No, please review the Helpful Hint above and try that step again.

Remember to follow the order of operations and use the Helpful Hint for the last two steps in the problem.

Changing Numbers from Scientific Notation to Standard Notation— Problem 33 $\quad 9.36 \times 10^{-5}$

Helpful Hint When a number in scientific notation has a negative power of 10, then the number is less than 1. We move the decimal point to the left and insert zeros as needed.

Did you move the decimal point five places to the LEFT?

Yes _____ No _____

If you answered No, please review the Helpful Hint above and try that step again.

Did you remember to insert four zeros to the left of the digit 9?

Yes _____ No _____

If you answered No, remember that for numbers smaller than one, there must be a zero for every place to the right of the decimal point until you have a nonzero digit.

Now go back and rework the problem using these suggestions.

Need more help? Look for section examples marked with $\mathbb{MC}$ to review.

An adequate supply of clean drinking water each day. It is something that we in America take for granted. However, there are severe water shortages in 80 countries of the world. Children are the most vulnerable, for they are the most susceptible to waterborne diseases. They are the greatest beneficiaries of having new wells drilled that allow all the people of a community to have clean water. Scientists, engineers, and construction workers use the algebra of this chapter to construct new wells in rural communities that previously did not have clean drinking water.

Introduction to Algebra

10.1 Variables and Like Terms

① Recognizing the Variables in an Equation or a Formula

In algebra we reason and solve problems by means of symbols. A **variable** is a symbol, usually a letter of the alphabet, that stands for a number. We can use the variable even though we may not know what number the variable stands for. We can find that number by following a logical order of steps. These are the rules of algebra.

We begin by taking a closer look at variables. In the formula for the area of a circle, the equation $A = \pi r^2$ contains two variables. r represents the value of the radius. A represents the value of the area. π is a known value. We often use the decimal approximation 3.14 for π.

Student Learning Objectives

After studying this section, you will be able to:

① Recognize the variables in an equation or a formula.

② Combine like terms containing a variable.

▲ **EXAMPLE 1**   Name the variables in each equation.

(a) $A = lw$

(b) $V = \dfrac{4\pi r^3}{3}$

Solution

(a) $A = lw$ The variables are A, l, and w.

(b) $V = \dfrac{4\pi r^3}{3}$ The variables are V and r.

▲ **Student Practice 1**   Name the variables.

(a) $A = \dfrac{bh}{2}$

(b) $V = lwh$

NOTE TO STUDENT: Fully worked-out solutions to all of the Student Practice problems can be found at the back of the text starting at page SP-1.

We have seen various ways to indicate multiplication. For example, three times n can be written as $3 \times n$. In algebra we usually do not write the multiplication symbol. We can simply write $3n$ to mean three times n. A number just to the left of a variable indicates that the number is multiplied by the variable. Thus $4ab$ means 4 times a times b.

A number just to the left of a set of parentheses also means multiplication. Thus $5(w)$ means $5 \times w$ or $5w$, and $3(n + 8)$ means $3 \times (n + 8)$. So a product can be written with or without a multiplication sign.

▲ **EXAMPLE 2**   Write the formula without a multiplication sign.

(a) $V = \dfrac{B \times h}{3}$

(b) $A = \dfrac{h \times (B + b)}{2}$

Solution

(a) $V = \dfrac{Bh}{3}$

(b) $A = \dfrac{h(B + b)}{2}$

▲ **Student Practice 2**   Write the formula without a multiplication sign.

(a) $P = 2 \times w + 2 \times l$

(b) $A = \pi \times r^2$

② Combining Like Terms Containing a Variable

Recall that when we work with measurements we combine like quantities. A carpenter, for example, might perform the following calculations.

$$20 \text{ m} - 3 \text{ m} = 17 \text{ m}$$
$$7 \text{ in.} + 9 \text{ in.} = 16 \text{ in.}$$

We cannot combine quantities that are not the same. We cannot add 5 yd + 7 gal. We cannot subtract 12 lb − 3 in.

Similarly, when using variables, we can add or subtract only when the same variable is used. For example, we can add $4a + 5a = 9a$, but we cannot add $4a + 5b$.

A **term** is a number, a variable, or a product of a number and one or more variables separated from other terms in an expression by a + sign or a − sign. In the expression $2x + 4y + (-1)$ there are three terms: $2x$, $4y$, and -1. **Like terms** have identical variables and identical exponents, so in the expression $3x + 4y + (-2x)$, the two terms $3x$ and $(-2x)$ are called *like terms*. The terms $2x^2y$ and $5xy$ are not like terms, since the exponents for the variable x are not the same. To combine like terms, you combine the numbers, called the **numerical coefficients,** that are directly in front of the terms by using the rules for adding signed numbers. Then you use this new number as the coefficient of the variable.

EXAMPLE 3 Combine like terms. $5x + 7x$

Solution We add $5 + 7 = 12$. Thus $5x + 7x = 12x$.

Student Practice 3 Combine like terms. $9x + 2x$

When we combine signed numbers, we try to combine them mentally. Thus to combine $7 - 9$, we think $7 + (-9)$ and we write -2. In a similar way, to combine $7x - 9x$, we think $\boxed{7x + (-9x)}$ and we write $-2x$. Your instructor may ask you to write out this "think" step as part of your work. In the following example we show this extra step inside a $\boxed{}$ box. You should determine from your instructor whether he or she feels it is necessary for you to show this step.

EXAMPLE 4 Combine like terms.

(a) $3x + 7x - 15x$ **(b)** $9x - 12x - 11x$

Solution

(a) $3x + 7x - 15x = \boxed{3x + 7x + (-15x)} = 10x + (-15x) = -5x$

(b) $9x - 12x - 11x = \boxed{9x + (-12x) + (-11x)} = -3x + (-11x)$
$$= -14x$$

Student Practice 4 Combine like terms.

(a) $8x - 22x + 5x$ **(b)** $19x - 7x - 12x$

A variable without a numerical coefficient is understood to have a coefficient of 1.

$7x + y$ means $7x + 1y$. $\qquad$ $5a - b$ means $5a - 1b$.

EXAMPLE 5   Combine like terms.

(a) $3x - 8x + x$

(b) $12x - x - 20.5x$

Solution

(a) $3x - 8x + x = 3x - 8x + 1x = -5x + 1x = -4x$

(b) $12x - x - 20.5x = 12x - 1x - 20.5x = 11.0x - 20.5x = -9.5x$

Student Practice 5   Combine like terms.

(a) $9x - 12x + x$

(b) $5.6x - 8x - x$

Numbers cannot be combined with variable terms.

EXAMPLE 6   Combine like terms.

$$7.8 - 2.3x + 9.6x - 10.8$$

Solution   In each case, we combine the numbers separately and the variable terms separately. It may help to use the commutative and associative properties first.

$$7.8 - 2.3x + 9.6x - 10.8$$
$$= 7.8 - 10.8 - 2.3x + 9.6x$$
$$= -3 + 7.3x$$

Student Practice 6   Combine like terms.

$$17.5 - 6.3x - 8.2x + 10.5$$

There may be more than one variable in a problem. Keep in mind, however, that only like terms may be combined.

Mc **EXAMPLE 7**   Combine like terms.

(a) $5x + 2y + 8 - 6x + 3y - 4$ **(b)** $\dfrac{3}{4}x - 12 + \dfrac{1}{6}x + \dfrac{2}{3}$

Solution   For convenience, we will rearrange the problem to place like terms next to each other. This is an optional step; you do not need to do this.

(a) $5x - 6x + 2y + 3y + 8 - 4 = -1x + 5y + 4$
$$= -x + 5y + 4$$

(b) $\dfrac{3}{4}x + \dfrac{1}{6}x - 12 + \dfrac{2}{3} = \dfrac{9}{12}x + \dfrac{2}{12}x - \dfrac{36}{3} + \dfrac{2}{3}$
$$= \dfrac{11}{12}x - \dfrac{34}{3}$$

The order of the terms in an answer is not important in this type of problem. The answer to part (a) could have been $5y + 4 - x$ or $4 + 5y - x$. Often we give the answer with the letters in alphabetical order.

Student Practice 7   Combine like terms.

(a) $2w + 3z - 12 - 5w - z - 16$

(b) $\dfrac{3}{5}x + 5 - \dfrac{7}{15}x - \dfrac{1}{3}$

Verbal and Writing Skills, Exercises 1–4

1. In your own words, write a definition for the word *variable*.

2. In your own words, define *like terms*.

3. Why is it that you cannot combine like terms with a problem such as $3x^2y + 5xy^2$?

4. Why is it that you cannot combine like terms with a problem such as $7xy + 9x$?

Name the variables in each equation.

5. $G = 5xy$

6. $S = 3\pi r^3$

7. $p = \dfrac{4ab}{3}$

8. $p = \dfrac{7ab}{4}$

Write each equation without multiplication signs.

9. $r = 3 \times m + 5 \times n$

▲ **10.** $P = 2 \times w + 2 \times l$

11. $H = 2 \times a - 3 \times b$

12. $A = \dfrac{a \times b + a \times c}{3}$

Combine like terms.

13. $-16x + 26x$

14. $-12x + 40x$

15. $2x - 8x + 5x$

16. $8x - 12x + x$

17. $-\dfrac{1}{2}x + \dfrac{3}{4}x + \dfrac{1}{12}x$

18. $\dfrac{2}{5}x - \dfrac{2}{3}x + \dfrac{7}{15}x$

19. $8x - x + 10 - 6$

20. $x + 15x - 12 + 8$

21. $1.3x + 10 - 2.4x - 3.6$

22. $4.2x + 5 - 1.7x - 5.5$

23. $16x + 9y - 11 + 21x$

24. $20x - 8y - 5 - 10x$

Mixed Practice   *Combine like terms.*

25. $\left(3\dfrac{1}{2}\right)x - 32 - \left(1\dfrac{1}{6}\right)x - 18$

26. $19 - \left(4\dfrac{1}{4}\right)x + \left(2\dfrac{3}{8}\right)x - 8$

27. $7a - c + 6b - 3c - 10a$

28. $10b + 3a - 8b - 5c + a$

29. $\frac{1}{2}x + \frac{1}{7}y - \frac{3}{4}x + \frac{5}{21}y$

30. $\frac{1}{4}x + \frac{1}{3}y - \frac{7}{12}x - \frac{1}{2}y$

31. $7.3x + 1.7x + 4 - 6.4x - 5.6x - 10$

32. $3.1x + 2.9x - 8 - 12.8x - 3.2x + 3$

33. $-7.6n + 1.2 + 11.2m - 3.5n - 8.1m$

34. $4.5n - 5.9m + 3.9 - 7.2n + 9m$

To Think About

▲ **35.** **(a)** Find the perimeter of the triangle.

(b) If each side of the triangle is doubled, what is the new perimeter?

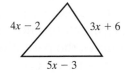

▲ **36.** **(a)** Find the perimeter of this four-sided figure.

(b) If each side is doubled, what is the new perimeter?

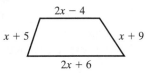

Cumulative Review *Solve for n.*

37. **[4.3.1]** $3 \times n = 36$

38. **[4.3.1]** $8 \times n = 64$

39. **[4.3.2]** $\frac{n}{6} = \frac{12}{15}$

40. **[4.3.2]** $\frac{n}{9} = \frac{36}{40}$

41. **[5.3.2]** Find 10% of 80.

42. **[5.3.2]** Find 50% of 80.

Quick Quiz 10.1 *Combine like terms.*

1. $6a - 5b - 3a - 9b$

2. $\frac{1}{3}x - \frac{4}{5}y - \frac{3}{4}x + \frac{3}{25}y$

3. $-12x + 22y - 34 - 6x - 7y - 13$

4. **Concept Check** Explain how you would combine like terms in the following expression without making any sign errors.

$-8.2x - 3.4y + 6.7z - 3.1x + 5.6y - 9.8z$

Student Learning Objectives

After studying this section, you will be able to:

1. Remove parentheses using the distributive property.

2. Simplify expressions by removing parentheses and combining like terms.

① Removing Parentheses Using the Distributive Property

What do we mean by the word *property* in mathematics? A **property** is an essential characteristic. A property of addition is an essential characteristic of addition. In this section we learn about the distributive property and how to use this property to simplify expressions.

Sometimes we encounter expressions like $4(x + 3)$. We'd like to be able to simplify the expression so that no parentheses appear. Notice that this expression contains two operations, multiplication and addition. We will use the distributive property to distribute the 4 to both terms in the addition statement. That is,

$$4(x + 3) \quad \text{is equal to} \quad 4(x) + 4(3).$$

The numerical coefficient 4 can be "distributed over" the expression $x + 3$ by multiplying the 4 by each of the terms in the parentheses. The expression $4(x) + 4(3)$ can be simplified to $4x + 12$, which has no parentheses. Thus we can use the distributive property to remove the parentheses.

Using variables, we can write the distributive property two ways.

DISTRIBUTIVE PROPERTIES OF MULTIPLICATION OVER ADDITION

If a, b, and c are signed numbers, then

$$a(b + c) = ab + ac \quad \text{and} \quad (b + c)a = ba + ca.$$

A numerical example shows that the distributive property works.

$$7(4 + 6) = 7(4) + 7(6)$$
$$7(10) = 28 + 42$$
$$70 = 70$$

When the number is to the left of the parentheses, we use

$$a(b + c) = ab + ac.$$

There is also a distributive property of multiplication over subtraction: $a(b - c) = ab - ac$.

EXAMPLE 1 Simplify.

(a) $8(x + 5)$ **(b)** $-3(x + 3y)$ **(c)** $6(3a - 7b)$

Solution

(a) $8(x + 5) = 8x + 8(5) = 8x + 40$
(b) $-3(x + 3y) = -3x + (-3)(3y) = -3x + (-9y) = -3x - 9y$
(c) $6(3a - 7b) = 6(3a) - 6(7b) = 18a - 42b$

NOTE TO STUDENT: Fully worked-out solutions to all of the Student Practice problems can be found at the back of the text starting at page SP-1.

Student Practice 1 Simplify.

(a) $7(x + 5)$ **(b)** $-4(x + 2y)$ **(c)** $5(6a - 2b)$

Sometimes the number is to the right of the parentheses, so we use

$$(b + c)a = ba + ca.$$

EXAMPLE 2   Simplify. $(5x + y)(2)$

Solution
$$(5x + y)(2) = (5x)(2) + (y)(2)$$
$$= 10x + 2y$$

Notice that we write our final answer with the numerical coefficient to the left of the variable. We would not leave $(y)(2)$ as an answer but would write $2y$.

Student Practice 2 Simplify. $(x + 3y)(8)$

Sometimes the distributive property is used with three terms within the parentheses. Also, when parentheses are used inside parentheses, the outside ( ) are changed to bracket [ ] notation.

EXAMPLE 3   Simplify.

(a) $-5(x + 2y - 8)$ **(b)** $(1.5x + 3y + 7)(2)$

Solution

(a) $-5(x + 2y - 8) = -5[x + 2y + (-8)]$
$$= -5(x) + (-5)(2y) + (-5)(-8)$$
$$= -5x + (-10y) + 40$$
$$= -5x - 10y + 40$$

(b) $(1.5x + 3y + 7)(2) = (1.5x)(2) + (3y)(2) + (7)(2)$
$$= 3x + 6y + 14$$

In this example every step is shown in detail. You may find that you do not need to write so many steps.

Student Practice 3 Simplify.

(a) $-5(x + 4y + 5)$ **(b)** $(2.2x + 5.5y + 6)(3)$

The parentheses may contain four terms. The coefficients may be decimals or fractions.

EXAMPLE 4   Simplify. $\frac{2}{3}\left(x + \frac{1}{2}y - \frac{1}{4}z + \frac{1}{5}\right)$

Solution
$$\frac{2}{3}\left(x + \frac{1}{2}y - \frac{1}{4}z + \frac{1}{5}\right)$$
$$= \frac{2}{3}(x) + \frac{2}{3}\left(\frac{1}{2}y\right) + \frac{2}{3}\left(-\frac{1}{4}z\right) + \frac{2}{3}\left(\frac{1}{5}\right)$$
$$= \frac{2}{3}x + \frac{1}{3}y - \frac{1}{6}z + \frac{2}{15}$$

Student Practice 4   Simplify.
$$\frac{3}{2}\left(\frac{1}{2}x - \frac{1}{3}y + 4z - \frac{1}{2}\right)$$

② Simplifying Expressions by Removing Parentheses and Combining Like Terms

After removing parentheses we may have a chance to combine like terms. The direction "simplify" means remove parentheses, combine like terms, and leave the answer in the most basic form.

EXAMPLE 5 Simplify. $2(x + 3y) + 3(4x + 2y)$

Solution

$$2(x + 3y) + 3(4x + 2y) = 2x + 6y + 12x + 6y \quad \text{Use the distributive property.}$$
$$= 14x + 12y \quad \text{Combine like terms.}$$

Student Practice 5 Simplify. $3(2x + 4y) + 2(5x + y)$

EXAMPLE 6 Simplify. $2(x - 3y) - 5(2x + 6)$

Solution

$$2(x - 3y) - 5(2x + 6) = 2x - 6y - 10x - 30 \quad \text{Use the distributive property.}$$
$$= -8x - 6y - 30 \quad \text{Combine like terms.}$$

Notice that in the final step, only the x terms could be combined. There are no other like terms.

Student Practice 6 Simplify.

$$-4(x - 5) + 3(-3y + 2x)$$

Verbal and Writing Skills, Exercises 1–4

1. A _____ is a symbol, usually a letter of the alphabet, that stands for a number.

2. What is the variable in the expression $5x + 9$?

3. Identify the like terms in the expression $3x + 2y - 1 + x - 3y$.

4. Explain the distributive property in your own words. Give an example.

Simplify.

5. $9(3x - 2)$

6. $8(4x - 5)$

7. $-2(x + y)$

8. $-5(x + y)$

9. $-6(-2.4x + 5y)$

10. $-4(3.5x - 8y)$

11. $(-3x + 7y)(-10)$

12. $(-2x + 8y)(-12)$

13. $(6a - 5b)(8)$

14. $(4a - 13b)(7)$

15. $(-8y - 7z)(-3)$

16. $(-4y - 9z)(-7)$

17. $4(p + 9q - 10)$

18. $6(2p - 7q + 11)$

19. $3\left(\dfrac{1}{5}x + \dfrac{2}{3}y - \dfrac{1}{4}\right)$

20. $6\left(\dfrac{1}{3}x + \dfrac{3}{4}y - \dfrac{1}{6}\right)$

21. $-15(-2a - 3.2b + 4.5)$

22. $-14(-5a + 1.4b - 2.5)$

23. $(8a + 12b - 9c - 5)(4)$

24. $(-7a + 11b - 10c - 9)(7)$

25. $-2(1.3x - 8.5y - 5z + 12)$

26. $-5(1.8x - 0.1y - 7z - 2)$

27. $\dfrac{1}{2}\left(2x - 3y + 4z - \dfrac{1}{2}\right)$

28. $\dfrac{1}{5}(-20s - 10t + 30)$

29. $-\dfrac{1}{3}(9s - 30t - 63)$

30. $-\dfrac{1}{6}(-18s + 6t - 24)$

Applications

▲ **31.** *Geometry* The perimeter of a rectangle is $P = 2(l + w)$. Write this formula without parentheses and without multiplication signs.

▲ **32.** *Geometry* The surface area of a rectangular solid is $S = 2(lw + lh + wh)$. Write this formula without parentheses and without multiplication signs.

▲ **33.** *Geometry* The area of a trapezoid is $A = \dfrac{h(B + b)}{2}$. Write this formula without parentheses and without multiplication signs.

▲ **34.** *Geometry* The surface area of a cylinder is $S = 2\pi r(h + r)$. Write this formula without parentheses and without multiplication signs.

Simplify. Be sure to combine like terms.

35. $4(5x - 1) + 7(x - 5)$

36. $8(4x + 3) + 2(x - 15)$

37. $10(4a + 5b) - 8(6a + 2)$

38. $11(2a - 3b) - 2(a + 5)$

39. $1.5(x + 2.2y) + 3(2.2x + 1.6y)$

40. $2.4(x + 3.5y) + 2(1.4x + 1.9y)$

41. $2(3b + c - 2a) - 5(a - 2c + 5b)$

42. $3(-4a + c + 4b) - 4(2c + b - 6a)$

To Think About

▲ **43.** Illustrate the distributive property by using the area of two rectangles.

▲ **44.** Show that multiplication is distributive over subtraction by using the area of two rectangles.

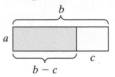

Cumulative Review

▲ **45.** **[7.2.1]** *Geometry* Find the perimeter of a rectangular door with a length of 7.5 ft and a width of 4 ft.

▲ **46.** **[7.4.2]** *Geometry* Find the area of a triangle with base 8.5 in. and height 15 in.

Quick Quiz 10.2 *Simplify.*

1. $3\left(\dfrac{5}{6}x - \dfrac{7}{12}y\right)$

2. $-3.5(2x - 3y + z - 4)$

3. $2(-3x + 7y) - 5(2x - 9y)$

4. **Concept Check** Explain the steps that are needed to simplify $-3(2x + 5y) + 4(5x - 1)$.

10.3 Solving Equations Using the Addition Property

① Solving Equations Using the Addition Property

One of the most important skills in algebra is that of **solving an equation.** Starting from an equation with a variable whose value is unknown, we transform the equation into a simpler, equivalent equation by performing a logical step. In the following sections we'll learn some of the logical steps for solving an equation successfully.

We begin with the concept of an equation. An **equation** is a mathematical statement that says that two expressions are equal. The statement $x + 5 = 13$ means that some number (x) plus 5 equals 13. Experience with the basic addition facts tells us that $x = 8$ since $8 + 5 = 13$. Therefore, 8 is the solution to the equation. The **solution** of an equation is that number which makes the equation true. What if we did not know that $8 + 5 = 13$? How could we find the value of x? Let's look at the first logical step in solving this equation.

ADDITION PROPERTY OF EQUATIONS

You may add the same number to each side of an equation to obtain an equivalent equation.

Suppose we want to make the equation $x + 5 = 13$ into a simpler equation, such as $x = $ some number. What can we add to each side of the equation so that $x + 5$ becomes simply x? We can add the opposite of $+5$. Let's see what happens when we do this addition.

$$x + 5 = 13$$
$$x + 5 + (-5) = 13 + (-5) \quad \text{Remember to add } -5 \text{ to both sides of the equation.}$$
$$x + 0 = 8$$
$$x = 8$$

We found the solution to the equation.

Notice that in the second step, the number 5 was removed from the left side of the equation. We know that we may add a number to both sides of the equation. But what number should we choose to add? We always add the opposite of the number we want to remove from one side of the equation.

EXAMPLE 1 Solve. $x - 9 = 3$

Solution We want to isolate the variable x.

$$x - 9 = 3 \quad \text{Think: "Add the opposite of } -9 \text{ to both sides of the equation."}$$
$$x - 9 + 9 = 3 + 9$$
$$x + 0 = 12$$
$$x = 12$$

Student Practice 1 Solve. $x + 7 = -8$

NOTE TO STUDENT: Fully worked-out solutions to all of the Student Practice problems can be found at the back of the text starting at page SP-1.

When we solve an equation we want to isolate the variable. We do this by performing the same operation on both sides of the equation.

Equations can contain integers, decimals, or fractions.

EXAMPLE 2 Solve. Check your solution to (a).

(a) $x + 1.5 = 4$

(b) $\dfrac{3}{8} = x - \dfrac{3}{4}$

Solution We use the addition property to solve these equations since each equation involves only addition or subtraction. In each case we want to isolate the variable.

(a)

$$x + 1.5 = 4 \quad \text{Think: "Add the opposite of 1.5 to both sides}$$
$$\text{of the equation."}$$
$$x + 1.5 + (-1.5) = 4 + (-1.5)$$
$$x = 2.5$$

Check.

$$x + 1.5 = 4 \quad \text{Substitute 2.5 for } x \text{ in the original equation.}$$
$$2.5 + 1.5 \overset{?}{=} 4$$
$$4 = 4 \checkmark$$

(b)

$$\dfrac{3}{8} = x - \dfrac{3}{4} \qquad \text{Note that here the variable is on the right-hand side.}$$

$$\dfrac{3}{8} + \dfrac{3}{4} = x - \dfrac{3}{4} + \dfrac{3}{4} \qquad \text{Think: "Add the opposite of } -\dfrac{3}{4} \text{ to both sides of the}$$
$$\text{equation."}$$

$$\dfrac{3}{8} + \dfrac{6}{8} = x + 0 \qquad \text{We need to change } \dfrac{3}{4} \text{ to } \dfrac{6}{8}.$$

$$\dfrac{9}{8} \text{ or } 1\dfrac{1}{8} = x$$

Student Practice 2 Solve. Check your solution to (a).

(a) $y - 3.2 = 9$

(b) $\dfrac{2}{3} = x + \dfrac{1}{6}$

EXAMPLE 3 Solve $2x + 7 = x + 9$.

Solution We want to remove the $+7$ on the left-hand side of the equation.

$$2x + 7 = x + 9$$
$$2x + 7 + (-7) = x + 9 + (-7) \quad \text{Add } -7 \text{ to both sides of the equation.}$$
$$2x = x + 2 \qquad \text{Now we need to remove the } x \text{ on the right-}$$
$$\text{hand side of the equation.}$$
$$2x + (-x) = x + (-x) + 2$$
$$x = 2 \qquad \text{Add } -x \text{ to both sides of the equation.}$$

Student Practice 3 Solve $3x - 5 = 2x + 1$. Check your solution.

MyMathLab®

Watch the videos
in MyMathLab

Download the
MyDashBoard App

Verbal and Writing Skills, Exercises 1–4

1. An _____ is a mathematical statement that says that two expressions are equal.

2. The _____ of an equation is that number which makes the equation true.

3. To use the addition property, we add to both sides of the equation the _____ of the number we want to remove from one side of the equation.

4. To use the addition property to solve the equation $x - 8 = 9$, we add _____ to both sides of the equation.

Solve for the variable.

5. $y - 12 = 20$

6. $y - 14 = 27$

7. $x + 6 = 15$

8. $x + 8 = 12$

9. $x + 16 = -2$

10. $x + 15 = -5$

11. $14 + x = -11$

12. $10 + y = -9$

13. $-12 + x = 7$

14. $-20 + y = 10$

15. $5.2 = x - 4.6$

16. $2.8 = y - 7.2$

17. $y + 8.2 = -3.4$

18. $x + 7.5 = -9.3$

19. $x - 25.2 = -12$

20. $y - 29.8 = -15$

21. $\dfrac{4}{5} = x + \dfrac{2}{5}$

22. $\dfrac{5}{8} = x + \dfrac{3}{8}$

23. $x - \dfrac{3}{5} = \dfrac{2}{5}$

24. $y - \dfrac{2}{7} = \dfrac{6}{7}$

25. $x + \dfrac{2}{3} = -\dfrac{5}{6}$

26. $y + \dfrac{1}{2} = -\dfrac{3}{4}$

27. $\dfrac{1}{4} + y = 2\dfrac{3}{8}$

28. $\dfrac{1}{5} + x = 3\dfrac{7}{10}$

Solve for the variable. You may need to use the addition property twice.

29. $3x - 5 = 2x + 9$

30. $5x + 1 = 4x - 3$

31. $5x + 12 = 4x - 1$

32. $8x - 3 = 7x - 12$

33. $7x - 9 = 6x - 7$

34. $12x + 1 = 11x + 9$

35. $18x + 28 = 17x + 19$

36. $14x - 8 = 13x - 10$

Mixed Practice *Solve for the variable.*

37. $y - \dfrac{1}{2} = 6$

38. $x + 1.2 = -3.8$

39. $5 = z + 13$

40. $-15 = -6 + x$

41. $-5.9 + y = -4.7$

42. $z + \dfrac{2}{3} = \dfrac{7}{12}$

43. $2x - 1 = x + 5$

44. $3x - 8 = 2x - 15$

45. $3.6x - 8 = 2.6x + 4$

46. $5.4y + 3 = 4.4y - 1$

47. $6x - 12 = 7x - 5$

48. $9x + 6 = 10x - 9$

To Think About

49. In the equation $x + 9 = 12$, we add -9 to both sides of the equation to solve. What would you do to solve the equation $3x = 12$? Why?

50. In the equation $x - 7 = -12$, we add $+7$ to both sides of the equation to solve. What would you do to solve the equation $4x = -12$? Why?

Cumulative Review

51. **[10.1.2]** Combine like terms.
$5x - y + 3 - 2x + 4y$

52. **[10.2.2]** Simplify. $7(2x + 3y) - 3(5x - 1)$

53. **[8.4.1]** Find the mean. $85, $78, $92, $83, $72

54. **[8.4.2]** Find the median of the following test scores. 84, 90, 98, 86, 93, 80, 85

Quick Quiz 10.3 *Solve for the variable.*

1. $x - 8.4 = -10.6$

2. $8x - 15 = 7x + 20$

3. $5 + 6x = 5x - 5$

4. **Concept Check** Explain the steps that are needed to solve the following equation: $-7x + 5 = -8x - 13$.

10.4 Solving Equations Using the Division or Multiplication Property

① Solving Equations Using the Division Property

Recall that when we solve an equation using the addition property, we transform the equation to a simpler one where $x =$ some number. We use the same idea to solve the equation $3n = 75$. Think: "What can we do to the left side of the equation so that n stands alone?" If we divide the left side of the equation by 3, we will obtain $1 \cdot n$. In this way n stands alone, since $\frac{3}{3} \cdot n = 1 \cdot n = n$. Remember, however, whatever you do to one side of the equation, you must do to the other side of the equation. Our goal once again is to isolate the variable.

$$3n = 75$$
$$\frac{3n}{3} = \frac{75}{3}$$
$$1 \cdot n = 25$$
$$n = 25$$

This is another important procedure used to solve equations.

Student Learning Objectives

After studying this section, you will be able to:

① Solve equations using the division property.

② Solve equations using the multiplication property.

> **DIVISION PROPERTY OF EQUATIONS**
>
> You may divide each side of an equation by the same nonzero number to obtain an equivalent equation.

EXAMPLE 1 Solve for n. $6n = 72$

Solution

$6n = 72$	The variable n is multiplied by 6.
$\dfrac{6n}{6} = \dfrac{72}{6}$	Divide each side by 6.
$1 \cdot n = 12$	We have $72 \div 6 = 12$.
$n = 12$	Since $1 \cdot n = n$

Student Practice 1 Solve for n. $8n = 104$

Sometimes the coefficient of the variable is a negative number. Therefore, in solving problems of this type, we need to divide each side of the equation by that negative number.

EXAMPLE 2 Solve for the variable. $-3n = 20$

Solution

$-3n = 20$	The coefficient of n is -3.
$\dfrac{-3n}{-3} = \dfrac{20}{-3}$	Divide each side of the equation by -3.
$n = -\dfrac{20}{3}$	Watch your signs!

Student Practice 2 Solve for the variable.

$$-7n = 30$$

NOTE TO STUDENT: *Fully worked-out solutions to all of the Student Practice problems can be found at the back of the text starting at page SP-1.*

Sometimes the coefficient of the variable is a decimal. In solving problems of this type, we need to divide each side of the equation by that decimal number.

EXAMPLE 3 Solve for the variable. $2.5y = 20$

Solution

$$2.5y = 20 \qquad \text{The coefficient of } y \text{ is } 2.5.$$

$$\frac{2.5y}{2.5} = \frac{20}{2.5} \qquad \text{Divide each side of the equation by } 2.5.$$

$$\begin{array}{r} 8 \\ 2.5_\wedge\overline{)20.0_\wedge} \end{array}$$

$$y = 8$$

To check, substitute 8 for y in the original equation.

$$2.5(8) \overset{?}{=} 20$$
$$20 = 20 \quad \checkmark$$

It is always best to check the solution to equations involving decimals or fractions.

Student Practice 3 Solve for the variable. Check your solution.

$$3.2x = 16$$

② Solving Equations Using the Multiplication Property

Sometimes the coefficient of the variable is a fraction, as in the equation $\frac{3}{4}x = 6$. Think: "What can we do to the left side of the equation so that x will stand alone?" Recall that when you multiply a fraction by its reciprocal, the product is 1. That is, $\frac{4}{3} \cdot \frac{3}{4} = 1$. We will use this idea to solve the equation. But remember, whatever you do to one side of the equation, you must do to the other side of the equation.

$$\frac{3}{4}x = 6$$

$$\frac{4}{3} \cdot \frac{3}{4}x = 6 \cdot \frac{4}{3}$$

$$1x = \frac{\overset{2}{\cancel{6}}}{1} \cdot \frac{4}{\cancel{3}}$$

$$x = 8$$

> **MULTIPLICATION PROPERTY OF EQUATIONS**
> You may multiply each side of an equation by the same nonzero number to obtain an equivalent equation.

EXAMPLE 4 Solve for the variable and check your solution for (a).

(a) $\dfrac{5}{8}y = 1\dfrac{1}{4}$

(b) $1\dfrac{1}{2}z = 3$

Solution

(a) $\dfrac{5}{8}y = 1\dfrac{1}{4}$

$\dfrac{5}{8}y = \dfrac{5}{4}$ Change the mixed number to a fraction. It will be easier to work with.

$\dfrac{8}{5} \cdot \dfrac{5}{8}y = \dfrac{5}{4} \cdot \dfrac{8}{5}$ Multiply both sides of the equation by $\dfrac{8}{5}$ because $\dfrac{8}{5} \cdot \dfrac{5}{8} = 1$.

$1 \cdot y = 2$

$y = 2$

Check.

$\dfrac{5}{8}y = 1\dfrac{1}{4}$

$\dfrac{5}{8}(2) \overset{?}{=} 1\dfrac{1}{4}$ Substitute 2 for y in the original equation.

$\dfrac{10}{8} \overset{?}{=} 1\dfrac{1}{4}$

$1\dfrac{2}{8} \overset{?}{=} 1\dfrac{1}{4}$

$1\dfrac{1}{4} = 1\dfrac{1}{4}$ ✓

(b) $1\dfrac{1}{2}z = 3$

$\dfrac{3}{2}z = 3$ Change the mixed number to a fraction.

$\dfrac{2}{3} \cdot \dfrac{3}{2}z = 3 \cdot \dfrac{2}{3}$ Multiply both sides of the equation by $\dfrac{2}{3}$. Why?

$z = 2$

It is always a good idea to check the solution to an equation involving fractions. We leave the check for this solution up to you.

Student Practice 4 Solve for the variable and check your solution.

(a) $\dfrac{1}{6}y = 2\dfrac{2}{3}$

(b) $3\dfrac{1}{5}z = 4$

Hint: Remember to write all mixed numbers as improper fractions before solving linear equations. An alternate method that may also be used is to convert the mixed number to decimal form. Thus equations like $4\frac{1}{8}x = 12$ can first be written as $4.125x = 12$.

Verbal and Writing Skills, Exercises 1–4

1. How is an equation similar to a balance scale?

2. The division property states that we may divide each side of an equation by _____ to obtain an equivalent equation.

3. To change $\frac{3}{4}x = 5$ to a simpler equation, multiply both sides of the equation by _____.

4. Given the equation $1\frac{3}{5}y = 2$, we multiply both sides of the equation by $\frac{5}{8}$ to solve for y. Why?

Solve for the variable.

5. $4x = 36$

6. $8x = 56$

7. $7y = -28$

8. $5y = -45$

9. $-9x = 16$

10. $-7y = 22$

11. $-12x = -144$

12. $-11y = -121$

13. $-64 = -4m$

14. $-88 = -8m$

15. $0.6x = 6$

16. $0.5y = 50$

17. $17.5 = 2.5t$

18. $21.6 = 2.4n$

19. $-0.5x = 6.75$

20. $-0.8y = 10.32$

Solve for the variable.

21. $\frac{5}{8}x = 5$

22. $\frac{3}{4}x = 3$

23. $\frac{2}{5}y = 4$

24. $\frac{7}{8}y = 14$

25. $\frac{3}{5}n = \frac{3}{4}$

26. $\frac{2}{3}z = \frac{1}{3}$

27. $-\frac{2}{9}x = \frac{4}{5}$

28. $-\frac{4}{5}x = \frac{8}{3}$

29. $\frac{1}{2}x = -2\frac{1}{4}$

30. $\frac{3}{4}y = -3\frac{3}{8}$

31. $\left(-3\frac{1}{3}\right)z = -20$

32. $\left(-4\frac{1}{2}\right)z = -45$

Mixed Practice *Solve for the variable.*

33. $-60 = -10x$

34. $-75 = -15x$

35. $\dfrac{2}{3}x = -6$

36. $\dfrac{4}{5}x = -8$

37. $1.5x = 0.045$

38. $1.6x = 0.064$

39. $12 = -\dfrac{3}{5}x$

40. $20 = -\dfrac{5}{6}x$

Cumulative Review

41. **[10.1.2]** Combine like terms.
$6 - 3x + 5y + 7x - 12y$

42. **[10.2.2]** Simplify.
$-2(3a - 5b + c) + 5(-a + 2b - 5c)$

43. **[5.5.2]** *Soybean Prices in United States* In 2007, the average price for a bushel of soybeans in the United States was $10.10. By 2009, the price per bushel had fallen to $9.50. Find the percentage of decrease in the price per bushel of soybeans from 2007 to 2009. (*Source:* www.hpj. com) Round your answer to the nearest tenth of a percent.

▲ **44.** **[5.4.1]** *Geography* Greenland, the world's largest island, has a total area of 2,166,086 square kilometers. Of these, 1,755,637 square kilometers are ice-covered. What percent of Greenland is covered in ice? What percent is not ice-covered? Round your answers to the nearest tenth of a percent.

Quick Quiz 10.4 *Solve for the variable.*

1. $-4x = 15$

2. $-3.5 = -0.5x$

3. $-\dfrac{3}{4}x = \dfrac{9}{2}$

4. Concept Check Explain the steps you would take to solve the equation $12 - 5 = -14x$.

10.5 Solving Equations Using Two Properties

Student Learning Objectives

After studying this section, you will be able to:

① Use two properties to solve an equation.

② Solve equations where the variable is on both sides of the equals sign.

③ Solve equations with parentheses.

① Using Two Properties to Solve an Equation

To solve an equation, we take logical steps to change the equation to a simpler equivalent equation. The simpler equivalent equation is $x =$ some number. To do this, we use the addition property, the division property, or the multiplication property. In this section you will use more than one property to solve complex equations. Each time you use a property, you take a step toward solving the equation. At each step you try to isolate the variable. That is, you try to get the variable to stand alone.

EXAMPLE 1 Solve $3x + 18 = 27$. Check your solution.

Solution We want only x terms on the left and only numbers on the right. We begin by removing 18 from the left side of the equation.

$$3x + 18 + (-18) = 27 + (-18)$$ Add the opposite of 18 to both sides of the equation so that $3x$ stands alone.

$$3x = 9$$

$$\frac{3x}{3} = \frac{9}{3}$$ Divide both sides of the equation by 3 so that x stands alone.

$$x = 3$$ The solution to the equation is $x = 3$.

Check.

$$3(3) + 18 \overset{?}{=} 27$$ Substitute 3 for x in the original equation.

$$9 + 18 \overset{?}{=} 27$$

$$27 = 27 \ ✓$$

Student Practice 1 Solve $5x + 13 = 33$. Check your solution.

NOTE TO STUDENT: *Fully worked-out solutions to all of the Student Practice problems can be found at the back of the text starting at page SP-1.*

Note the variable is on the right-hand side in the next example.

EXAMPLE 2 Solve $-41 = 9x - 5$. Check your solution.

Solution We begin by removing -5 from the right-hand side of the equals sign.

$$-41 + 5 = 9x - 5 + 5$$ Add the opposite of -5 to both sides of the equation.

$$-36 = 9x$$

$$\frac{-36}{9} = \frac{9x}{9}$$ Divide both sides of the equation by 9.

$$-4 = x$$

The check is left up to you.

Student Practice 2 Solve $-50 = 7x - 8$. Check your solution.

② Solving Equations Where the Variable Is on Both Sides of the Equals Sign

Sometimes variables appear on both sides of the equation. When this occurs, we need to isolate the variable. This requires us to add a variable term to each side of the equation.

EXAMPLE 3   Solve. $8x = 5x - 21$

Solution   We want to remove the $5x$ from the right-hand side of the equation so that all of the variables are on one side of the equation and all of the numbers are on the other side of the equation.

$$8x + (-5x) = 5x + (-5x) - 21 \quad \text{Add the opposite of } 5x \text{ to both sides of the}$$
$$3x = -21 \qquad\qquad \text{equation.}$$
$$\frac{3x}{3} = \frac{-21}{3} \qquad\qquad \text{Divide both sides of the equation by 3.}$$
$$x = -7$$

The check is left up to you.

Student Practice 3   Solve. $4x = -8x + 42$

Suppose there is a variable term and a numerical term on each side of an equation. We want to collect all the variables on one side of the equation and all the numerical terms on the other side of the equation. To do this, we will have to use the addition property twice.

EXAMPLE 4   Solve. $2x + 9 = 5x - 3$

Solution   We begin by collecting the numerical terms on the right-hand side of the equation.

$$2x + 9 + (-9) = 5x - 3 + (-9) \quad \text{Add } -9 \text{ to both sides of the equation.}$$
$$2x = 5x - 12$$

Now we want to collect all the variable terms on the left-hand side of the equation.

$$2x + (-5x) = 5x + (-5x) - 12 \quad \text{Add } -5x \text{ to both sides of the equation.}$$
$$-3x = -12$$
$$\frac{-3x}{-3} = \frac{-12}{-3} \qquad\qquad \text{Divide both sides of the equation by } -3.$$
$$x = 4$$

Check: $2(4) + 9 \overset{?}{=} 5(4) - 3 \qquad$ Substitute 4 for x in the original equation.
$$8 + 9 \overset{?}{=} 20 - 3$$
$$17 = 17 \qquad ✓$$

Student Practice 4   Solve. $4x - 7 = 9x + 13$

If there are like terms on one side of the equation, these should be combined first. Then proceed as before.

$\mathbb{M}_\mathbb{C}$ **EXAMPLE 5** Solve for the variable.
$-5 + 2y + 8 = 7y + 23$

Solution We begin by combining the like terms on the left-hand side of the equation.

$$-5 + 2y + 8 = 7y + 23$$
$$2y + 3 = 7y + 23 \qquad \text{Combine } -5 + 8.$$
$$2y + (-7y) + 3 = 7y + (-7y) + 23 \qquad \text{Remove the variable on the right-hand side of the equation.}$$
$$-5y + 3 = 23$$
$$-5y + 3 + (-3) = 23 + (-3) \qquad \text{Remove the 3 on the left-hand side of the equation.}$$
$$-5y = 20$$
$$\frac{-5y}{-5} = \frac{20}{-5} \qquad \text{Divide both sides of the equation by } -5.$$
$$y = -4$$

Check:

$$-5 + 2(-4) + 8 \stackrel{?}{=} 7(-4) + 23$$
$$-5 + (-8) + 8 \stackrel{?}{=} -28 + 23$$
$$-5 = -5 \checkmark$$

Student Practice 5 Solve for the variable.

$$4x - 23 = 3x + 7 - 2x$$

It is wise to check your answer when solving this type of linear equation. The chance of making a simple error with signs is quite high. Checking gives you a chance to detect this type of error.

③ **Solving Equations with Parentheses**

> **PROCEDURE TO SOLVE EQUATIONS**
>
> 1. Remove any parentheses by using the distributive property.
> 2. Combine like terms on each side of the equation.
> 3. Add the appropriate value to both sides of the equation to get all numbers on one side.
> 4. Add the appropriate term to both sides of the equation to get all variable terms on the other side.
> 5. Divide both sides of the equation by the numerical coefficient of the variable term.
> 6. Check by substituting the solution back into the original equation.

You have probably noticed that steps 3 and 4 are interchangeable. You can do step 3 and then step 4, or you can do step 4 and then step 3.

If a problem contains one or more sets of parentheses, remove them using the distributive property. Then combine like terms on each side of the equation. Then solve.

EXAMPLE 6 Isolate the variable on the right-hand side. Then solve for x.

$$7x - 3(x - 4) = 9(x + 2)$$

Solution

$7x - 3x + 12 = 9x + 18$	Remove parentheses by using the distributive property.
$4x + 12 = 9x + 18$	Add like terms.
$4x + 12 + (-18) = 9x + 18 + (-18)$	Add -18 to each side.
$4x + (-6) = 9x$	Simplify.
$4x + (-4x) - 6 = 9x + (-4x)$	Add $-4x$ to each side. This isolates the variable on the right-hand side.
$-6 = 5x$	Simplify.
$\dfrac{-6}{5} = \dfrac{5x}{5}$	Divide each side of the equation by 5.
$-\dfrac{6}{5} = x$	We obtain a solution that is a fraction.

Student Practice 6 Isolate the variable on the right-hand side. Then solve for x. $8(x - 3) + 5x = 15(x - 2)$

Verbal and Writing Skills, Exercises 1 and 2

1. Explain how you would decide what to add to each side of the equation as you begin to solve $-5x - 6 = 29$ for x.

2. Explain how you would decide what to add to each side of the equation as you begin to solve $-11x = 4x - 45$ for x.

Check to see whether the given answer is a solution to the equation.

3. Is $x = 3$ a solution to $2x + 5 = 7 - 4x$?

4. Is $x = 6$ a solution to $5 - 3x = -4x + 1$?

5. Is $x = \dfrac{1}{2}$ a solution to $8x - 2 = 10 - 16x$?

6. Is $x = \dfrac{1}{3}$ a solution to $12x - 7 = 3 - 18x$?

Solve.

7. $15x - 10 = 35$

8. $12x - 30 = 6$

9. $6x - 9 = -12$

10. $8x - 7 = -13$

11. $-9x = 3x - 10$

12. $-3x = 7x + 14$

13. $14x - 10 = 18$

14. $12x + 15 = -9$

15. $0.26 = 2x - 0.34$

16. $0.78 = 3x - 0.12$

17. $\dfrac{2}{3}x - 5 = 17$

18. $\dfrac{3}{5}x - 2 = -14$

19. $18 - 2x = 4x + 6$

20. $3x + 4 = 7x - 12$

21. $9 - 8x = 3 - 2x$

22. $8 + x = 3x - 6$

23. $5z + 6 = 3z - 2$

24. $4x + 10 = 6x - 4$

25. $1.2 + 0.3x = 0.6x - 2.1$

26. $1.2 + 0.5y = -0.8 - 0.3y$

27. $0.2x + 0.6 = -0.8 - 1.2x$

28. $0.4x + 0.5 = -1.9 - 0.8x$

29. $-10 + 6y + 2 = 3y - 26$

30. $9 - 4x + 1 = 4x + 42$

31. $-y + 7 = 14 + 2y - 6$

32. $-x - 2 = -13 + 3x + 8$

33. $-30 - 12y + 18 = -24y + 13 + 7y$

34. $15 - 18y - 21 = 15y - 22 - 29y$

35. $3(2x - 5) - 5x = 1$

36. $4(2x - 1) - 7x = 9$

37. $5(y - 2) = 2(2y + 3) - 16$

38. $13 + 7(2y - 1) = 5(y + 6)$

39. $8x + 4(4 - x) = 2x - 18$

40. $10x - 3(x - 4) = 9x - 8$

41. $5x + 9 = \dfrac{1}{3}(3x - 6)$

42. $6x + 5 = \dfrac{1}{4}(8x - 4)$

43. $-2x - 5(x + 1) = -3(2x + 5)$

44. $-3x - 2(x + 1) = -4(x - 1)$

Cumulative Review *Use $\pi \approx 3.14$. Round to the nearest tenth.*

45. **[7.8.3]** *Geometry* A topographic globe has a radius of 46 centimeters. Find the volume of this sphere.

▲ **46.** **[7.10.1]** *Geometry* Find the area of the shaded region in the given figure.

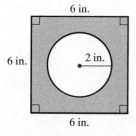

6 in.

6 in.   2 in.

6 in.

6 in.

Quick Quiz 10.5 *Solve for the variable.*

1. $0.7 - 0.3x = 0.9x - 4.1$

2. $-6 - 2x + 9 = 12 + 3x + 11$

3. $-4(x + 6) + 9 = 12 + 3(x + 5)$

4. **Concept Check** Explain the steps you would take to solve the following equation: $7x - 3(x - 6) = 2(x - 3) + 8$.

How Am I Doing? Sections 10.1–10.5

How are you doing with your homework assignments in Sections 10.1 to 10.5? Do you feel you have mastered the material so far? Do you understand the concepts you have covered? Before you go further in the textbook, take some time to do each of the following problems.

10.1

Combine like terms.

1. $23x - 40x$

2. $-8y + 12y - 3y$

3. $6a - 5b - 9a + 7b$

4. $8y - 9 + x + 3y - 7x - 1$

5. $7x - 14 + 5y + 8 - 7y + 9x$

6. $4a - 7b + 3c - 5b$

10.2

Simplify.

7. $6(7x - 3y)$

8. $-4\left(\dfrac{1}{2}a - \dfrac{1}{4}b + 3\right)$

9. $-2(1.5a + 3b - 6c - 5)$

10. $4(x + 3y) - 2(5x - y)$

11. $(9x + 4y)(-2)$

12. $(7x - 3y)(-3)$

10.3

Solve for the variable.

13. $5 + x = 42$

14. $x + 2.5 = 6$

15. $y + \dfrac{4}{5} = -\dfrac{3}{10}$

16. $-12 = -20 + x$

10.4

Solve for the variable.

17. $-9y = -72$

18. $2.7y = 27$

19. $\dfrac{3}{5}x = \dfrac{9}{10}$

20. $84 = -7x$

10.5

Solve for the variable.

21. $-7 + 6m = 25$

22. $12 - 5a = -8a + 15$

23. $5(x - 1) = 7 - 3(x - 4)$

24. $3x + 7 = 5(5 - x)$

25. $5x - 18 = 2(x + 3)$

26. $8x - 5(x + 2) = -3(x - 5)$

27. $12 + 4y - 7 = 6y - 9$

28. $0.3x + 0.4 = 0.7x - 1.2$

Now turn to page SA-18 for the answer to each of these problems. Each answer also includes a reference to the objective in which the problem is first taught. If you missed any of these problems, you should stop and review the Examples and Student Practice problems in the referenced objective. A little review now will help you master the material in the upcoming sections of the text.

1.

2.

3.

4.

5.

6.

7.

8.

9.

10.

11.

12.

13.

14.

15.

16.

17.

18.

19.

20.

21.

22.

23.

24.

25.

26.

27.

28.

10.6 Translating English to Algebra

① Translating English into Mathematical Equations Using Two Given Variables

In the preceding section you learned how to solve equations. We can use equations to solve applied problems, but before we can do that, we need to know how to write an equation that will represent the situation in a word problem.

In this section we practice *translating* to help you to write your own mathematical equations. That is, we translate English expressions into algebraic expressions. In the next section we'll apply this translation skill to a variety of word problems.

The following chart presents the mathematical symbols generally used in translating English phrases into equations.

Student Learning Objectives

After studying this section, you will be able to:

① Translate English into mathematical equations using two given variables.

② Write algebraic expressions for several quantities using one given variable.

The English Phrase:	Is Usually Represented by the Symbol:
greater than increased by more than added to sum of	+
less than decreased by smaller than fewer than shorter than difference of	−
multiplied by of product of times	×
double	2 ×
triple	3 ×
divided by ratio of quotient of	÷
is was has costs equals represents amounts to	=

An English sentence describing the relationship between two or more quantities can often be translated into a short equation using variables. For example, if we say in English "Bob's salary is $1000 greater than Fred's salary," we can express the mathematical relationship by the equation

$$b = 1000 + f$$

where b represents Bob's salary and f represents Fred's salary.

EXAMPLE 1 Translate the English sentence into an equation using variables. Use r to represent Roberto's weight and j to represent Juan's weight.

> Roberto's weight is 42 pounds more than Juan's weight.

Solution Roberto's weight   is   42 pounds   more than   Juan's weight.
  ↓   ↓   ↓   ↓   ↓
  r $=$ 42 $+$ j

The equation $r = 42 + j$ could also be written as

$$r = j + 42.$$

Both are correct translations because addition is commutative.

Student Practice 1 Translate the English sentence into an equation using variables. Use t to represent Tom's height and a to represent Abdul's height. *Tom's height is 7 inches more than Abdul's height.*

NOTE TO STUDENT: Fully worked-out solutions to all of the Student Practice problems can be found at the back of the text starting at page SP-1.

When translating the phrase "less than" or "fewer than," be sure that the number that appears before the phrase is the value that is subtracted. Words like "costs," "weighs," or "has the value of" are translated into an equals sign ($=$).

EXAMPLE 2 Translate the English sentence into an equation using variables. Use c to represent the cost of the chair in dollars and s to represent the cost of the sofa in dollars.

> The chair costs $200 less than the sofa.

Solution   The chair costs $200 *less than* the sofa.
  ↓   ↓   ↓   ↓
  c $=$ s $-$ 200

Note the order of the equation. We subtract 200 from s, so we have $s - 200$. It would be incorrect to write $200 - s$. Do you see why?

Student Practice 2 Translate the English sentence into an equation using variables. Use n to represent the number of students in the noon class and m to represent the number of students in the morning class. *The noon class has 24 fewer students than the morning class.*

In a similar way we can translate the phrases "more than" or "greater than," but since addition is commutative, we find it easier to write the mathematical symbols in the same order as the words in the English sentence.

EXAMPLE 3 Translate the English sentence into an equation using variables. Use *f* for the cost of a 14-foot truck and *e* for the cost of an 11-foot truck. *The daily cost of a 14-foot truck is 20 dollars more than the daily cost of an 11-foot truck.*

Solution The 14-foot truck cost is 20 more than the 11-foot truck cost.

$$f = 20 + e$$

Student Practice 3 Translate the English sentence into an equation using variables. Use *t* to represent the number of boxes carried on Thursday and *f* to represent the number of boxes carried on Friday. *On Thursday Adrianne carried five more boxes into the dorm than she did on Friday.*

EXAMPLE 4 Translate the following English sentence into an equation using the variables indicated. Use *l* for the length and *w* for the width of the rectangle. *The length of a rectangle is 3 feet shorter than double the width.*

Solution The length of the rectangle is compared to the width. Therefore, we begin with the width. We have

$$w = \text{the width of the rectangle.}$$

Now the length of the rectangle is 3 feet shorter than double the width. Double the width is $2w$. If it is 3 feet shorter than double the width, we will have to take away 3 from the $2w$. Therefore,

$$2w - 3 = \text{the length of the rectangle.}$$

So we have

$$l = 2w - 3.$$

Student Practice 4 Translate the following English sentence into an equation using the variables indicated. Use *l* for the length and *w* for the width of the rectangle. *The length of a rectangle is 7 feet longer than double the width.*

② Writing Algebraic Expressions for Several Quantities Using One Given Variable

In each of the examples so far we have used two *different variables*. Now we'll learn how to write algebraic expressions for several quantities using the *same variable*. In the next section we'll use this skill to write and solve equations.

A mathematical expression that contains a variable is often called an **algebraic expression.**

EXAMPLE 5 Write algebraic expressions for Bob's salary and Fred's salary. Use the letter *b*. *Fred's salary is $150 more than Bob's salary.*

Solution

$$\text{Let } b = \text{Bob's salary.}$$

$$\text{So therefore } \underbrace{b + 150}_{\text{\$150 more than Bob's salary}} = \text{Fred's salary.}$$

Notice that Fred's salary is described in terms of Bob's salary. Thus it is logical to let Bob's salary be *b* and then to express Fred's salary as $150 more than Bob's.

Student Practice 5 Write algebraic expressions for Sally's trip and Melinda's trip. Use the letter *s*. *Melinda's trip is 380 miles longer than Sally's trip.*

EXAMPLE 6 Write algebraic expressions for the size of each of two angles of a triangle. Use the letter *A*. *Angle B of the triangle is 34° less than angle A.*

Solution

$$\text{Let } A = \text{the number of degrees in angle } A.$$

$$\text{So therefore } \underbrace{A - 34}_{34° \text{ less than angle } A.} = \text{the number of degrees in angle } B.$$

Student Practice 6 Write algebraic expressions for the height of each of two buildings. Use the letter *m*. *Larson Center is 126 feet shorter than McCormick Hall.*

Often in algebra when we write expressions for one or two unknown quantities, we use the letter *x*.

EXAMPLE 7 Write algebraic expressions for the length of each of three sides of a triangle. Use the letter *x*. *The second side is 4 inches longer than the first. The third side is 7 inches shorter than triple the length of the first side.*

Solution Since the other two sides are described in terms of the first side, we start by writing an expression for the first side.

$$\text{Let } x = \text{the length of the first side.}$$

$$\text{So therefore } x + 4 = \text{the length of the second side.}$$

$$\text{So therefore } 3x - 7 = \text{the length of the third side.}$$

Student Practice 7 Write an algebraic expression for the length of each of three sides of a triangle. Use the letter *x*. *The second side is double the length of the first side. The third side is 6 inches longer than the first side.*

Verbal and Writing Skills *Translate the English sentence into an equation using the variables indicated.*

1. ***Weight Comparison*** Harry weighs 34 pounds more than Rita. Use h for Harry's weight and r for Rita's weight.

2. ***Cereal Boxes*** The large cereal box contains 7 ounces more than the small box of cereal. Use l for the number of ounces in the large cereal box and s for the number of ounces in the small cereal box.

3. ***Jewelry*** The bracelet costs $107 less than the necklace. Use b for the cost of the bracelet and n for the cost of the necklace.

4. ***Education*** There were 42 fewer students taking algebra in the fall semester than the spring semester. Use s to represent the number of students registered in spring and f to represent the number of students registered in fall.

5. ***Temperature Comparisons*** The temperature in New Delhi, India, was 14°F more than the temperature in Athens, Greece. Use n for the temperature in New Delhi and a for the temperature in Athens.

6. ***Temperature Comparisons*** The temperature in Quebec City was 21°F less than the temperature in Rome. Use q for the temperature in Quebec City and r for the temperature in Rome.

▲ 7. ***Geometry*** The length of a rectangle is 7 meters longer than double the width. Use l for the length and w for the width of the rectangle.

▲ 8. ***Geometry*** The length of a rectangle is 8 meters shorter than double the width. Use l for the length and w for the width of the rectangle.

▲ 9. ***Geometry*** The length of a rectangle is 2 meters shorter than triple the width. Use l for the length and w for the width of the rectangle.

▲ 10. ***Geometry*** The length of a parallelogram is 10 ft shorter than double the width. Use l for the length and w for the width of the parallelogram.

11. ***Football*** During a college football game, Miami scored 10 points more than triple the number of points scored by Temple. Use m to represent the number of points Miami scored and t to represent the number of points Temple scored.

12. ***Football*** During a college football game, Texas scored 2 points more than double the number of points scored by Iowa State. Use T to represent the number of points Texas scored and I to represent the number of points Iowa State scored.

13. ***Hours Worked*** The combined number of hours Tina and Louise work at Leo's Pizzeria is 32 hours. Use t for the number of hours Tina works and l for the number of hours Louise works.

14. ***High School Play*** The attendance at the high school play was greater on Saturday night than on Friday night. The difference in attendance between the two nights was 138. Use s for the attendance on Saturday and f for the attendance on Friday.

15. **Hourly Wage** The product of your hourly wage and the amount of time worked is $500. Let h = the hourly wage and t = the number of hours worked.

16. **Education** The ratio of men to women at Central College is 5 to 3. Let m = the number of men and w = the number of women.

Write algebraic expressions for each quantity using the given variable.

17. **Airfare** The airfare from Chicago to San Diego was $135 more than the airfare from Chicago to Phoenix. Use the letter p.

18. **Electronics Cost** The cost of a Microsoft Zune 4 MP3 player was $75 more than the cost of an Apple iPod shuffle MP3 player. Use the letter c.

▲19. **Geometry** Angle A of the triangle is 46° less than angle B. Use the letter b.

▲20. **Geometry** The top of the box is 38 centimeters shorter than the side of the box. Use the letter s.

21. **Tallest Buildings** The Burj Khalifa in the United Arab Emirates is 1267 ft taller than the Willis Tower in Chicago. Use the letter w.

22. **Largest Lakes** The two largest lakes in the world are the Caspian Sea and Lake Superior. The maximum depth of the Caspian Sea is 1771 ft more than the maximum depth of Lake Superior. Use the letter d.

23. **Reading Books** During the summer, Nina read twice as many books as Aaron. Molly read five more books than Aaron. Use the letter a.

24. **Tip Salary** Sam made $12 more in tips than Lisa one Friday night. Brenda made $6 less than Lisa. Use the letter l.

▲25. **Geometry** The length of a box is 5 inches longer than its height. The width is triple the height. Use the letter h.

▲26. **Geometry** The height of a box is 7 inches longer than the width. The length is 1 inch shorter than double the width. Use the letter w.

▲27. **Triangles** The measure of the second angle of a triangle is double the measure of the first. The measure of the third angle is 14° smaller than the measure of the first. Use the letter x.

▲28. **Triangles** The measure of the second angle of a triangle is triple the measure of the first. The measure of the third angle is 36° larger than the measure of the first. Use the letter x.

Cumulative Review *Perform the indicated operations in the proper order.*

29. **[9.4.1]** $-6 - (-7)(2)$

30. **[9.4.1]** $5 - 5 + 8 - (-4) + 2 - 15$

31. **[10.5.3]** Solve for x. $-2(3x + 5) + 12 = 8$

32. **[10.5.2]** Solve for y. $3y - 4 - 5y = 10 - y$

33. **[5.4.1]** *Basketball* The following are statistics for the top 3-point shooters in the NBA during the 2009–2010 season. Fill in the blanks to complete the table. Round to nearest whole number in columns A and B. Round to three decimal places in column C.

Player	Team	A 3-Point Attempts	B 3-Point Shots Made	C 3-Point %
Kyle Korver	Utah Jazz	110	59	
Mike Miller	Washington Wizards		82	0.480
Daniel Gibson	Cleveland Cavaliers	149		0.477

Quick Quiz 10.6

1. Translate the English sentence into an equation using the variables indicated. Use t to represent the truck's MPG (miles per gallon). Use c to represent the car's MPG. *Charlie's truck gets 12 miles per gallon less than his car gets.*

Write an algebraic expression for each quantity using the given variable.

2. The length of a rectangle is 3 inches longer than double the width. Use the letter w.

3. The number of SUVs in the college parking lot is half the number of compact cars. The number of trucks in the college parking lot is 35 more than the number of compact cars. Use the variable c.

4. **Concept Check** In Dr. Tobey's Basic Mathematics class, 12 more students have part-time jobs than full-time jobs. The students wanted to describe this relationship with algebraic expressions. One student said let p = the number of students with part-time jobs and let $p - 12$ = the number of students with full-time jobs. Another student said let f = the number of students with full-time jobs and let $f + 12$ = the number of students with part-time jobs. Which student is right? Are both right? Explain your answer.

10.7 Solving Applied Problems

Student Learning Objectives

After studying this section, you will be able to:

1. Solve problems involving comparisons.

2. Solve problems involving geometric formulas.

3. Solve problems involving rates and percents.

① Solving Problems Involving Comparisons

To solve the following problem, we use the three steps for problem solving with which you are familiar, plus another step: *Write an equation.*

EXAMPLE 1 A 12-foot piece of wood is cut into two pieces. The longer piece is 3.5 feet longer than the shorter piece. What is the length of each piece?

Solution You may find it helpful to use the Mathematics Blueprint for Problem Solving to organize the data and make a plan for solving.

Mathematics Blueprint for Problem Solving

Gather the Facts	What Am I Asked to Do?	How Do I Proceed?	Key Points to Remember
The piece of wood is 12 feet long. It is cut into two pieces. One piece is 3.5 feet longer than the other.	Find the length of each piece.	Let x = length of shorter piece and use x to write an expression for the longer piece.	Make an equation by adding the length of both pieces to get 12 feet.

1. **Understand the problem.** Draw a diagram.

   Shorter piece   Longer piece

   |←———— 12 ft ————→|

   Since the longer piece is described in terms of the shorter piece, we let the variable represent the shorter piece. Let x = the length of the shorter piece. The longer piece is 3.5 feet longer than the shorter piece.
   So therefore $x + 3.5$ = the length of the longer piece. The sum of the lengths of the two pieces is 12 feet. We write an equation.

2. **Write an equation.**

$$x + (x + 3.5) = 12$$

3. **Solve and state the answer.**

$$x + x + 3.5 = 12$$
$$2x + 3.5 = 12 \qquad \text{Combine like terms.}$$
$$2x + 3.5 + (-3.5) = 12 + (-3.5) \qquad \text{Add } -3.5 \text{ to each side.}$$
$$2x = 8.5 \qquad \text{Combine like terms.}$$
$$\frac{2x}{2} = \frac{8.5}{2} \qquad \text{Divide each side by 2.}$$
$$x = 4.25$$

The shorter piece is 4.25 feet long.

$$x + 3.5 = \text{length of the longer piece}$$
$$4.25 + 3.5 = 7.75$$

The longer piece is 7.75 feet long.

4. *Check.* We verify solutions to word problems by making sure that all the calculated values satisfy the original conditions. Do the lengths of the two pieces add up to 12 feet?

$$4.25 + 7.75 \stackrel{?}{=} 12$$
$$12 = 12 \quad \checkmark$$

Is one piece 3.5 feet longer than the other?

$$7.75 \stackrel{?}{=} 3.5 + 4.25$$
$$7.75 = 7.75 \quad \checkmark$$

Student Practice 1 An 18-foot board is cut into two pieces. The longer piece is 4.5 feet longer than the shorter piece. What is the length of each piece?

NOTE TO STUDENT: Fully worked-out solutions to all of the Student Practice problems can be found at the back of the text starting at page SP-1.

Sometimes three items are compared. Let a variable represent the quantity to which the other two quantities are compared. Then write an expression for the other two quantities.

EXAMPLE 2 Professor Jones is teaching 332 students in three sections of general psychology this semester. His noon class has 23 more students than his 8:00 A.M. class. His 2:00 P.M. class has 36 fewer students than his 8:00 A.M. class. How many students are in each class?

Solution

1. *Understand the problem.* Each class enrollment is described in terms of the enrollment in the 8:00 A.M. class.

Let x = the number of students in the 8:00 A.M. class.

The noon class has 23 more students than the 8:00 A.M. class.

So therefore $x + 23$ = the number of students in the noon class.

The 2:00 P.M. class has 36 fewer students than the 8:00 A.M. class.

Therefore $x - 36$ = the number of students in the 2:00 P.M. class.

The total enrollment for the three sections is 332.
You can draw a diagram.

2. *Write an equation.*

$$x + (x + 23) + (x - 36) = 332$$

Continued on next page

3. *Solve and state the answer.*

$$x + x + 23 + x - 36 = 332$$

$$3x - 13 = 332 \qquad \text{Combine like terms.}$$

$$3x + (-13) + 13 = 332 + 13 \qquad \text{Add 13 to each side.}$$

$$3x = 345 \qquad \text{Simplify.}$$

$$\frac{3x}{3} = \frac{345}{3} \qquad \text{Divide each side by 3.}$$

$$x = 115 \qquad \text{8:00 A.M. class}$$

$$x + 23 = 115 + 23 = 138 \qquad \text{noon class}$$

$$x - 36 = 115 - 36 = 79 \qquad \text{2:00 P.M. class}$$

Thus there are 115 students in the 8:00 A.M. class, 138 students in the noon class, and 79 students in the 2:00 P.M. class.

4. *Check.* Do the numbers of students in the classes total 332?

$$115 + 138 + 79 \overset{?}{=} 332$$

$$332 = 332 \quad \checkmark$$

Does the noon class have 23 more students than the 8:00 A.M. class?

$$138 \overset{?}{=} 23 + 115$$

$$138 = 138 \quad \checkmark$$

Does the 2:00 P.M. class have 36 fewer students than the 8:00 A.M. class?

$$79 \overset{?}{=} 115 - 36$$

$$79 = 79 \quad \checkmark$$

Student Practice 2 The city airport had a total of 349 departures on Monday, Tuesday, and Wednesday. There were 29 more departures on Tuesday than on Monday. There were 16 fewer departures on Wednesday than on Monday. How many departures occurred on each day?

② Solving Problems Involving Geometric Formulas

The following applied problems concern the geometric properties of two-dimensional figures. The problems involve perimeter or the measure of the angles in a triangle.

Recall that when we double something, we are multiplying by 2. That is, if something is x units, then double that value is $2x$. Triple that value is $3x$.

▲ **EXAMPLE 3** A farmer wishes to fence in a rectangular field with 804 feet of fence. The length is to be 3 feet longer than *double the width.* How long and how wide is the field?

Solution

1. *Understand the problem.* The perimeter of a rectangle is given by $P = 2w + 2l$.

Let w = the width.

The length is 3 feet longer than double the width.

$$\text{Length} = 3 + 2w$$

Therefore $2w + 3$ = the length.

You may wish to draw a diagram and label the figures with the given facts.

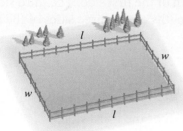
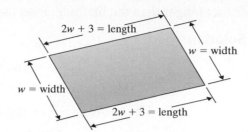

2. *Write an equation.* Substitute the given facts into the perimeter formula.

$$2w + 2l = P$$
$$2w + 2(2w + 3) = 804$$

3. *Solve and state the answer.*

$2w + 2(2w + 3) = 804$	
$2w + 4w + 6 = 804$	Use the distributive property.
$6w + 6 = 804$	Combine like terms.
$6w + 6 + (-6) = 804 + (-6)$	Add −6 to each side.
$6w = 798$	Simplify.
$\dfrac{6w}{6} = \dfrac{798}{6}$	Divide each side by 6.
$w = 133$	

The width is 133 feet.
The length $= 2w + 3$. When $w = 133$, we have

$$2(133) + 3 = 266 + 3 = 269.$$

Thus the length is 269 feet.

4. *Check.* Is the length 3 feet longer than double the width?

$$269 \stackrel{?}{=} 3 + (2)(133)$$
$$269 \stackrel{?}{=} 3 + 266$$
$$269 = 269 \ \checkmark$$

Is the perimeter 804 feet?

$$2(133) + 2(269) \stackrel{?}{=} 804$$
$$266 + 538 \stackrel{?}{=} 804$$
$$804 = 804 \ \checkmark$$

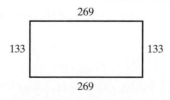

▲ **Student Practice 3**   What are the length and width of a rectangular field that has a perimeter of 772 feet and a length that is 8 feet longer than double the width?

▲ **EXAMPLE 4**   The perimeter of a triangular rug section is 21 feet. The length of the second side is double the length of the first side. The third side is 3 feet longer than the first side. Find the lengths of the three sides of the rug.

Solution

$$\text{Let } x = \text{the length of the first side.}$$

$$\text{Therefore } 2x = \text{the length of the second side.}$$

$$\text{Therefore } x + 3 = \text{the length of the third side.}$$

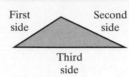
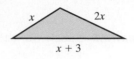

The distance around the three sides totals 21 feet.
Thus

$$x + 2x + (x + 3) = 21 \qquad \text{Use the perimeter formula.}$$
$$4x + 3 = 21 \qquad \text{Combine like terms.}$$
$$4x + 3 + (-3) = 21 + (-3) \qquad \text{Add } -3 \text{ to each side.}$$
$$4x = 18 \qquad \text{Simplify.}$$
$$\frac{4x}{4} = \frac{18}{4} \qquad \text{Divide each side by 4.}$$
$$x = 4.5$$

The first side is 4.5 feet long.

$$2x = 2(4.5) = 9 \text{ feet}$$

The second side is 9 feet long.

$$x + 3 = 4.5 + 3 = 7.5 \text{ feet}$$

The third side is 7.5 feet long.

Check.
Do the three sides add up to a perimeter of 21 feet?

$$4.5 + 9 + 7.5 \stackrel{?}{=} 21$$
$$21 = 21 \quad \checkmark$$

Is the length of the second side double the length of the first side?

$$9 \stackrel{?}{=} 2(4.5)$$
$$9 = 9 \quad \checkmark$$

Is the third side 3 feet longer than the first side?

$$7.5 \stackrel{?}{=} 3 + 4.5$$
$$7.5 = 7.5 \quad \checkmark$$

▲ **Student Practice 4**   The perimeter of a triangle is 36 meters. The second side is double the first side. The third side is 10 meters longer than the first side. Find the length of each side. Check your solutions.

▲ EXAMPLE 5   A triangle has three angles, *A*, *B*, and *C*. The measure of angle *C* is triple the measure of angle *B*. The measure of angle *A* is 105° larger than the measure of angle *B*. Find the measure of each angle. Check your answer.

Solution

$$\text{Let } x = \text{ the number of degrees in angle } B.$$
$$\text{Therefore } 3x = \text{ the number of degrees in angle } C.$$
$$\text{Therefore } x + 105 = \text{ the number of degrees in angle } A.$$

The sum of the interior angles of a triangle is 180°. Thus we can write the following.

$$x + 3x + (x + 105) = 180$$
$$5x + 105 = 180$$
$$5x + 105 + (-105) = 180 + (-105)$$
$$5x = 75$$
$$\frac{5x}{5} = \frac{75}{5}$$
$$x = 15$$

Angle *B* measures 15°.

$$3x = (3)(15) = 45$$

Angle *C* measures 45°.

$$x + 105 = 15 + 105 = 120$$

Angle *A* measures 120°.

Check.
Do the angles total 180°?

$$15 + 45 + 120 \overset{?}{=} 180$$
$$180 = 180 \quad ✓$$

Is angle *C* triple angle *B*?

$$45 \overset{?}{=} (3)(15)$$
$$45 = 45 \quad ✓$$

Is angle *A* 105° larger than angle *B*?

$$120 \overset{?}{=} 105 + 15$$
$$120 = 120 \quad ✓$$

▲ **Student Practice 5**   The measure of angle *C* of a triangle is triple the measure of angle *A*. The measure of angle *B* is 30° less than the measure of angle *A*. Find the measure of each angle.

③ Solving Problems Involving Rates and Percents

You can use equations to solve problems that involve rates and percents. Recall that the commission a salesperson earns is based on the total sales made. For example, a saleswoman earns $40 if she gets a 4% commission and she sells $1000 worth of products. That is, 4% of $1000 = $40. Sometimes a salesperson earns a base salary. The commission will then be added to the base salary to determine the total salary. You can find the total salary if you know the amount of sales. How would you find the amount of sales if the salary were known? We will use an equation.

EXAMPLE 6 This month's salary for an appliance saleswoman was $3000. This includes her base monthly salary of $1800 plus a 5% commission on total sales. Find her total sales for the month.

Solution

| total salary of $3000 | = | base salary of $1800 | + | 5% commission on total sales |

Let s = the amount of total sales.
Then $0.05s$ = the amount of commission earned from the sales.

$$3000 = 1800 + 0.05s$$
$$1200 = 0.05s$$
$$\frac{1200}{0.05} = \frac{0.05s}{0.05}$$
$$24{,}000 = s$$

She sold $24,000 worth of appliances.

Check.
Does 5% of $24,000 added to $1800 yield a salary of $3000?

$$0.05(24{,}000) + 1800 \stackrel{?}{=} 3000$$
$$1200 + 1800 \stackrel{?}{=} 3000$$
$$3000 = 3000 \quad \checkmark$$

Student Practice 6 A salesperson at a boat dealership earns $1000 a month plus a 3% commission on the total sales of the boats he sells. Last month he earned $3250. What was the total sales of the boats he sold?

Applications *Solve using an equation. Show what you let the variable equal.*

1. ***Carpentry*** A 16-foot board is cut into two pieces. The longer piece is 5.5 feet longer than the shorter piece. What is the length of each piece?

2. ***Carpentry*** A 20-foot board is cut into two pieces. The longer piece is 4.5 feet longer than the shorter piece. What is the length of each piece?

3. ***Rugby*** During a rugby game, Japan scored 22 points less than France. A total of 80 points were scored. How many points did each team score?

4. ***Cross-Country*** In a cross-country race, St. Mark's scored 27 points less than Thayer. A total of 63 points were scored. How many points were scored by each team?

5. ***Car Wash*** The Business Club's thrice-yearly car wash serviced 398 cars this year. A total of 84 more cars participated in May than in November. A total of 43 fewer cars were washed in July than in November. How many cars were washed during each month?

6. ***Scrabble*** In the game Scrabble, wooden tiles with letters on them are placed on a board to spell words. There are three times as many A tiles as there are G tiles. The number of O tiles is two more than twice the number of G tiles. The total number of A, O, and G tiles is 20. How many tiles of each are there?

Solve using an equation. Show what you let the variable equal. Check your answers.

7. ***Furniture*** A 12-foot solid cherry wood tabletop is cut into two pieces to allow for an insert later on. Of the two original pieces, the shorter piece is 4.7 feet shorter than the longer piece. What is the length of each piece?

8. ***Painting Supplies*** Stella Tobey has created a huge painting 18 feet long. Her goal is to cut the canvas and have two pieces of the same painting. The longer piece of canvas will be 6.5 feet longer than the shorter piece. What will the length of each piece be?

▲ 9. ***Game Board*** The playing board of a new game has a perimeter of 76 inches. It was designed so that the length is 4 inches shorter than double the width. What are the dimensions of the playing board?

▲ 10. ***New Room*** Marcus and Joannie are having a new family room added on to their home. They have designed the room to have a length 6 feet less than double the width. If the perimeter of the room is 78 feet, find the dimensions of the family room.

▲ 11. ***Triangular Flag*** An unusual triangular wall flag at the United Nations has a perimeter of 199 millimeters. The second side is 20 millimeters longer than the first side. The third side is 4 millimeters shorter than the first side. Find the length of each side.

▲ **12.** *Texas Oil Field* There is a triangular piece of land adjoining an oil field in Texas, with a perimeter of 271 meters. The length of the second side is double the first side. The length of the third side is 15 meters longer than the first side. Find the length of each side.

▲ **13.** *Puzzle* Nathaniel's geometric puzzle has a triangular puzzle piece with a perimeter of 44 centimeters. The length of the second side is double the first side. The length of the third side is 12 centimeters longer than the first side. Find the length of each side.

▲ **14.** *Triangular Pennant* Josiah has an unusual triangular pennant with a perimeter of 63 inches. The length of the first side is twice the length of the second side. The third side is 3 inches longer than twice the second side. Find the length of each side.

▲ **15.** *Geometry* A triangle has three angles, *A*, *B*, and *C*. Angle *B* is triple angle *A*. Angle *C* is 40° larger than angle *A*. Find the measure of each angle.

▲ **16.** *Geometry* A triangle has three angles, *G*, *H*, and *I*. Angle *H* is triple angle *I*. Angle *G* is 15° less than angle *I*. Find the measure of each angle.

17. *Sales Commission* A saleswoman at a car dealership earns $1200 per month plus a 5% commission on her total sales. If she earned $5000 last month, what was the amount of her sales?

18. *Sales Commission* A salesman at a jewelry store earns $1000 per month plus a 6% commission on his total sales. If he earned $2200 last month, what was the amount of his sales?

19. *Real Estate* A real estate agent charges $100 to place a rental listing plus 12% of the yearly rent. An apartment in Central City was rented by the agent for one year. She charged the landowner $820. How much did the apartment cost to rent for one year?

20. *Real Estate* A real estate agent charges $50 to place a rental listing plus 9% of the yearly rent. An apartment in the town of West Longmeadow was rented by the agent for one year. He charged the landowner $482. How much did the apartment cost to rent for one year?

21. *Mural* A community in South Florida has decided to paint a mural. Adults and children each have one section, but since there are more children interested in participating than adults, the children are awarded the larger piece of wall. If the wall is 32 feet long and the child section is 6.2 feet longer than the adult section, what is the length of each section of wall?

22. *Children's Theater* The new play at the children's theater attracted 321 people on opening night. There were 67 more children in attendance than adults. How many children attended? How many adults attended?

To Think About

23. ***Organ Transplants*** In 2009, the combined number of heart, liver, and pancreas transplants performed in the United States was 8910. The number of heart transplants was 316 more than five times the number of pancreas transplants. The number of liver transplants was 123 less than seventeen times the number of pancreas transplants. How many transplants of each organ were performed? (*Source:* www.optn.transplant.hrsa.org)

24. ***Vacation Days*** The average number of paid vacation days and holidays per year for the top nine countries is given in the table below. The numbers for Austria and Portugal are missing, but the values are the same for both countries. The average number of vacation days for all nine countries is 31.6. (*Source:* www.thesocietypages.org) Find the values for Austria and Portugal. Round to the nearest whole number.

Finland	39 days
Austria	?
Portugal	?
Spain	34 days
Italy	33 days

France	31 days
Norway	27 days
Sweden	25 days
Germany	25 days

25. ***Workers*** The table below gives the number of U.S. workers, in thousands, for selected occupations in 2009. The numbers are missing for customer service representatives and registered nurses, but the values are the same for both occupations. The average number of workers for all eight occupations is 2323 thousand. (*Source:* www.bls.gov) Find the values for customer service representatives and registered nurses. Round to the nearest whole number. (Remember the table gives the number of workers in thousands.)

Retail salespersons	4198
Cashiers	3396
Food prep workers/servers	2540
Waiters/waitresses	2293

Customer service representatives	?
Registered nurses	?
Security guards	929
Accountants/auditors	978

Cumulative Review

26. **[5.3.2]** What percent of 20 is 12?

27. **[5.3.2]** 38% of what number is 190?

28. **[4.3.2]** Solve the proportion. $\dfrac{x}{12} = \dfrac{10}{15}$

29. **[6.1.2]** How many ounces are in 5 pounds?

Quick Quiz 10.7 *Solve using an equation.*

1. Melinda earns $125 less per week than Barbara. The combined income of these two people is $437 per week. How much per week does each person earn?

2. At Middlesex Community College twice as many students work part time as full time. The number of students who do not work at all is 1200 less than the number who work part time. There are 6000 students at the college this semester. How many work full time? How many work part time? How many do not work while attending college?

3. A rectangular field has a perimeter of 176 yards. The length is 7 yards longer than double the width. Find the dimensions of the rectangle.

4. **Concept Check** The first angle of a triangle is twice as large as the second angle. The third angle is 10 degrees less than the second angle. Explain how you would write an expression for each of the three angles. How would you set up an equation to find the measures of the angles? Explain how you would solve the equation.

Did You Know...

That Having a Budget Can Help You Control How You Spend and Save Your Money?

TIME TO BUDGET

Understanding the Problem:

One way to improve your financial situation is to learn to manage the money you have with a budget. A budget can maximize your efforts to ensure you have enough money to cover your fixed expenses, as well as your variable expenses.

Michael is a teacher. One of Michael's goals is to go back to school to earn his master's degree in education. He knows that this will not only help him further his career, but also provide better financial stability in the long run. His net monthly income is currently $2500. So, Michael knows he'll need to put himself on a budget for a period of time in order to save money to go back to school.

Making a Plan:

Michael needs to budget his expenses to help control his spending and to save for college. He wants to see how long it will take to save up the needed money.

Step 1: Research shows that consumer credit counseling services recommend allocating the following percentages for each category of the monthly budget:

Housing	25%
Transportation	10%
Savings	5%
Utilities	5%
Debt Payments	20%
Food	15%
Misc.	20%

Task 1: If Michael follows this plan, how much will he put away in savings by the end of one year?

Task 2: If Michael follows this plan, how much will he spend on food by the end of one year?

Step 2: After investigating several schools in his area, Michael chooses to attend a state college that offers a one-year program for the degree he wishes to pursue. Michael will need $3000 for tuition and fees the first semester, and $450 for textbooks.

Task 3: What is the total cost for tuition, fees, and textbooks for one semester?

Finding a Solution:

Step 3: By controlling his spending Michael is able to save for his master's degree.

Task 4: How many months will Michael have to save to pay for one year (two semesters) at this college?

Task 5: If Michael can increase his savings to 10% of his monthly budget by making cuts in other areas, how long would he have to save to pay for one year (two semesters)?

Step 4: Once Michael earns his advanced degree, his salary will increase on the following schedule:

• Year 1: an additional $4200
• Year 2: an additional $4800
• Year 3: an additional $5200
• Year 4: an additional $5500
• Year 5: an additional $5800

Task 6: If Michael goes back to his original budget how much will he be saving per month with his new income after five years?

Task 7: How much will he have available per month for misc. spending?

Applying the Situation to Your Life:

Having a budget can help you control your spending.

Task 8: Do you have a budget?

Task 9: How would you adjust Michael's budget to fit your needs?

Task 10: You may not be planning to get a master's degree but probably you are thinking of saving up for some important purchase. Are you thinking of buying a new car that is more fuel-efficient? Are you planning a special trip? How could you adjust Michael's budget to help you with your goal?

Chapter 10 Organizer

Topic and Procedure	Examples	✏ You Try It
Combining like terms, p. 610 If the terms are like terms, combine the numerical coefficients directly in front of the variables.	Combine like terms. **(a)** $7x - 8x + 2x = -1x + 2x = x$ **(b)** $3a - 2b - 6a - 5b = -3a - 7b$ **(c)** $a - 2b + 3 - 5a = -4a - 2b + 3$	1. Combine like terms. **(a)** $10a - a + 3a$ **(b)** $8a + 10b - 5a - 12b$ **(c)** $18x - 6 - 7y + x$
The distributive properties, p. 614 $a(b + c) = ab + ac$ and $(b + c)a = ba + ca$	Simplify. **(a)** $5(x - 4y) = 5x - 20y$ **(b)** $3(a + 2b - 6) = 3a + 6b - 18$ **(c)** $(-2x + y)(7) = -14x + 7y$	2. Simplify. **(a)** $3(a - 6b)$ **(b)** $-4(2x - 3y - 1)$ **(c)** $(5a - b)8$
Problems involving parentheses and like terms, p. 616 1. Remove the parentheses using the distributive property. 2. Combine like terms.	Simplify. $$2(4x - y) - 3(-2x + y) = 8x - 2y + 6x - 3y$$ $$= 14x - 5y$$	3. Simplify. $-3(4x - 5y) - (x + 3y)$
Solving equations using the addition property, p. 619 1. Add the appropriate value to both sides of the equation so that the variable is on one side and a number is on the other side of the equals sign. 2. Check by substituting your answer back into the original equation.	Solve for x. $x - 2.5 = 7$ $$x - 2.5 + 2.5 = 7 + 2.5$$ $$x + 0 = 9.5$$ $$x = 9.5$$ Check. $9.5 - 2.5 \overset{?}{=} 7$ $$7 = 7 \checkmark$$	4. Solve for x. Check your solution. $x + 4.2 = 9$
Solving equations using the division property, p. 623 1. Divide both sides of the equation by the numerical coefficient of the variable. 2. Check by substituting your answer back into the original equation.	Solve for x. $-12x = 60$ $$\frac{-12x}{-12} = \frac{60}{-12}$$ $$x = -5$$ Check. $(-12)(-5) \overset{?}{=} 60$ $$60 = 60 \checkmark$$	5. Solve for x. Check your solution. $-15x = 90$
Solving equations using the multiplication property, p. 624 1. Multiply both sides of the equation by the reciprocal of the numerical coefficient of the variable. 2. Check by substituting your answer back into the original equation.	Solve for x. $\dfrac{3}{4}x = \dfrac{5}{8}$ $$\frac{4}{3} \cdot \frac{3}{4}x = \frac{5}{8} \cdot \frac{4}{3}$$ $$x = \frac{5}{6}$$ Check. $\left(\dfrac{3}{4}\right)\left(\dfrac{5}{6}\right) \overset{?}{=} \dfrac{5}{8}$ $$\frac{5}{8} = \frac{5}{8} \checkmark$$	6. Solve for x. Check your solution. $-\dfrac{2}{3}x = \dfrac{5}{9}$
Solving equations using more than one property, p. 628 1. Remove any parentheses by using the distributive property. 2. Combine like terms on each side of the equation. 3. Add the appropriate value to both sides of the equation to get all numbers on one side. 4. Add the appropriate term to both sides of the equation to get all variable terms on the other side. 5. Divide both sides of the equation by the numerical coefficient of the variable term. 6. Check by substituting back into the original equation.	Solve for x. $$5x - 2(6x - 1) = 3(1 + 2x) + 12$$ $$5x - 12x + 2 = 3 + 6x + 12$$ $$-7x + 2 = 15 + 6x$$ $$-7x + 2 + (-2) = 15 + (-2) + 6x$$ $$-7x = 13 + 6x$$ $$-7x + (-6x) = 13 + 6x + (-6x)$$ $$-13x = 13$$ $$x = -1$$ Check. $$5(-1) - 2[6(-1) - 1] \overset{?}{=} 3[1 + 2(-1)] + 12$$ $$-5 - 2(-6 - 1) \overset{?}{=} 3(1 - 2) + 12$$ $$-5 - 2(-7) \overset{?}{=} 3(-1) + 12$$ $$9 = 9 \checkmark$$	7. Solve for x. Check your solution. $7(2x - 5) - (8x + 1) = -5(x - 2) - 13$

Topic and Procedure	Examples	You Try It
Translating an English sentence into an equation, p. 635 When translating English into an equation: replace "greater than" by + and "less than" by −. See complete table on page 635.	Translate a comparison in English into an equation using two given variables. Use t to represent Thursday's temperature and w to represent Wednesday's temperature. The temperature Thursday was 12 degrees higher than the temperature on Wednesday. 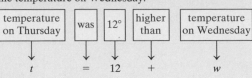 $$t \quad = \quad 12 \quad + \quad w$$	8. Translate the English sentence into an equation. Use a to represent Ana's age and f to represent her father's age. *Ana's age is 32 less than her father's age.*
Writing algebraic expressions for several quantities, p. 637 1. Use a variable to describe the quantity that other quantities are described in terms of. 2. Write an expression in terms of that variable for each of the other quantities.	Write algebraic expressions for the size of each angle of a triangle. The second angle of a triangle is 7° less than the first angle. The third angle of a triangle is double the first angle. Use the letter x. Since two angles are described in terms of the first angle, we let the variable x represent that angle. Let x = the number of degrees in the first angle. Let $x - 7$ = the number of degrees in the second angle. Let $2x$ = the number of degrees in the third angle.	9. Write algebraic expressions for the size of each angle. Use the letter x. *The first angle of a triangle is 12° more than the second angle. The third angle is 18° less than the second angle.*

Solving applied problems using equations, p. 642
 	
1. *Understand the problem.*
  (a) Draw a sketch.
  (b) Choose a variable.
  (c) Represent other variables in terms of the first variable.
2. *Write an equation.*
3. *Solve the equation and state the answer.*
4. *Check.*
	The perimeter of a field is 128 meters. The length of this rectangular field is 4 meters less than triple the width. Find the dimensions of the field.
	
	1. *Understand the problem.*
	  Let w = the width of the rectangle in meters.
	  Therefore $3w - 4$ = the length of the rectangle in meters.
	2. *Write an equation.*
	  Perimeter = 2(width) + 2(length)
	    $128 = 2(w) + 2(3w - 4)$
	3. *Solve and state the answer.*
	  $128 = 2w + 6w - 8$
	  $128 = 8w - 8$
	   $17 = w$ The width is 17 meters.
	  $3w - 4 = 3(17) - 4 = 47$
	  The length is 47 meters.
	4. *Check.* Is the perimeter 128 meters?
	  Does $17 + 47 + 17 + 47 = 128$? Yes.
	  Is the length 4 less than triple the width?
	  Is $47 = 3(17) - 4$? $47 = 47$ ✓
		10. Swenson Hall houses 260 college students on three floors. There are twice as many students on the second floor as on the first floor. There are 32 more students on the third floor as on the first floor. Find the number of students on each floor.

Chapter 10 Review Problems

Section 10.1
Combine like terms.

1. $-8a + 6 - 5a - 3$

2. $\dfrac{1}{3}x + \dfrac{1}{3} + \dfrac{5}{9} + \dfrac{1}{2}x$

3. $5x + 2y - 7x - 9y$

4. $3x - 7y + 8x + 2y$

5. $5x - 9y - 12 - 6x - 3y + 18$

6. $8a - 11b + 15 - b + 5a - 19$

Section 10.2

Simplify.

7. $-3(5x + y)$

8. $-4(2x + 3y)$

9. $2(x - 3y + 4)$

10. $5(6a - 8b + 5)$

11. $-12\left(\dfrac{3}{4}a - \dfrac{1}{6}b - 1\right)$

12. $5(1.2x + 3y - 5.5)$

Simplify.

13. $2(x + 3y) - 4(x - 2y)$

14. $2(5x - y) - 3(x + 2y)$

15. $-2(a + b) - 3(2a + 8)$

16. $-4(a - 2b) + 3(5 - a)$

Section 10.3

Solve for the variable.

17. $x - 3 = 9$

18. $x + 8.3 = 20$

19. $-8 = x - 12$

20. $2.4 = x - 5$

21. $3.1 + x = -9$

22. $x + \dfrac{1}{2} = 3\dfrac{3}{4}$

23. $y + \dfrac{5}{8} = -\dfrac{1}{8}$

24. $2x + 20 = 25 + x$

25. $7y + 12 = 8y + 3$

Section 10.4

Solve for the variable.

26. $8x = -20$

27. $-12y = 60$

28. $1.5x = 9$

29. $-1.4y = -12.6$

30. $\dfrac{3}{4}x = 6$

31. $\dfrac{2}{9}x = \dfrac{5}{18}$

Section 10.5

Solve for the variable.

32. $5x - 3 = 27$

33. $8x - 5 = 19$

34. $10 - x = -3x - 6$

35. $9x - 3x + 18 = 36$

36. $4 + 3x - 8 = 12 + 5x + 4$

37. $-2(3x + 5) = 4x + 8 - x$

38. $2(3x - 4) = 7 - 2x + 5x$

39. $5 - (y + 7) = 10 + 3(y - 4)$

Section 10.6

Translate the English sentence into an equation using the variables indicated.

40. *Vehicle Weight* The weight of a truck is 3000 pounds more than the weight of a car. Use w for the weight of the truck and c for the weight of the car.

41. *Education* Professor Garrison's evening psychology class has 12 more students than the afternoon class. Use e for the number of students in the evening class and a for the number of students in the afternoon class.

▲ **42. Geometry** The number of degrees in angle *A* is triple the number of degrees in angle *B*. Use *A* for the number of degrees in angle *A* and *B* for the number of degrees in angle *B*.

▲ **43. Geometry** The length of a rectangle is 3 inches shorter than double the width of the rectangle. Use *w* for the width of the rectangle in inches and *l* for the length of the rectangle in inches.

Write an algebraic expression for each quantity using the given variable.

44. Salary Comparison Michael's salary is $2050 more than Roberto's salary. Use the letter *r*.

▲ **45. Geometry** The length of the second side of a triangle is double the length of the first side of the triangle. Use the letter *x*.

46. Summer Employment During the summer, Carmen worked 12 days more than double the number of days Dennis worked. Use the letter *d*.

47. Book Sale The number of fiction books sold at the library's annual sale was 225 more than the number of nonfiction books sold. Use the letter *n*.

Section 10.7

Solve using an equation. Show what you let the variable equal.

48. Plumbing A 60-ft length of pipe is divided into two pieces. One piece is 6.5 ft longer than the other. Find the length of each piece.

49. Salary Comparison Two clerks work in a store. The new employee earns $28 less per week than the experienced employee. Together they earn $412 per week. What is the weekly salary of each person?

50. Fast-Food Restaurant A local fast-food restaurant had twice as many customers in March as in February. It had 3000 more customers in April than in February. Over the three months, 45,200 customers came to the restaurant. How many came each month?

51. Trip Distance During three days of travel, Anthony drove from Augusta, Maine, to Baltimore, Maryland, a distance of 670 miles. He drove twice as many miles Saturday than on Friday. He drove 30 miles more on Sunday than on Friday. How many miles did he drive each day?

▲ **52. Geometry** A rectangle has a perimeter of 72 in. The length is 3 in. less than double the width. Find the dimensions of the rectangle.

▲ **53. Geometry** A triangle has three angles, *X, Y,* and *Z*. Angle *Y* is double the measure of angle *Z*. Angle *X* is 12 degrees smaller than angle *Z*. Find the measure of each angle.

▲ **54. Football** A regulation NFL football field is in the shape of a rectangle. The width of the field is 67 yards shorter than the length. The perimeter of the field is 346 yards. Find the width and length of the field.

55. Trip Distance Ellen and Laurie drove from San Francisco, California, to Seattle, Washington, a distance of 810 miles. It took two days to make the trip. They drove 106 more miles on Sunday than on Saturday. How many miles did they drive each day?

56. Education During the second week of July, the North Lake Community College admissions office received 156 more applications than it did during the first week of July. During the third week of July, it received 142 fewer applications than it did during the first week of July. During these three weeks it received 800 applications. How many were received each week?

57. Sales Commission Megan receives an 8% commission on the furniture that she sells. Last month her total salary was $3050. Her base salary for the month was $1500. What was the cost of the furniture she sold last month?

How Am I Doing? Chapter 10 Test

 MATH COACH MyMathLab® You Tube™

After you take this test read through the Math Coach on pages 659–660. Math Coach videos are available via MyMathLab and YouTube. Step-by-step test solutions in the Chapter Test Prep Videos are also available via MyMathLab and YouTube. (Search "TobeyBasicCollMath" and click on "Channels.")

Combine like terms.

1. $5a - 11a$

MC 2. $\dfrac{1}{3}x + \dfrac{5}{8}y - \dfrac{1}{5}x + \dfrac{1}{2}y$

3. $\dfrac{1}{4}a - \dfrac{2}{3}b + \dfrac{3}{8}a$

4. $6a - 5b - 5a - 3b$

5. $7x - 8y + 2z - 9z + 8y$

6. $x + 5y - 6 - 5x - 7y + 11$

Simplify.

7. $5(12x - 5y)$

8. $4\left(\dfrac{1}{2}x - \dfrac{5}{6}y\right)$

9. $-1.5(3a - 2b + c - 8)$

MC 10. $2(-3a + 2b) - 5(a - 2b)$

Solve for the variable.

11. $-5 - 3x = 19$

12. $x - 3.45 = -9.8$

13. $-5x + 9 = -4x - 6$

MC 14. $8x - 2 - x = 3x - 9 - 10x$

15. $0.5x + 0.6 = 0.2x - 0.9$

16. $-\dfrac{5}{6}x = \dfrac{7}{12}$

1.	☐
2.	☐
3.	☐
4.	☐
5.	☐
6.	☐
7.	☐
8.	☐
9.	☐
10.	☐
11.	☐
12.	☐
13.	☐
14.	☐
15.	☐
16.	☐

Translate the English sentence into an equation using the variables indicated.

17. The second floor of Trabor Laboratory has 15 more classrooms than the first floor. Use s to represent the number of classrooms on the second floor and f to represent the number of classrooms on the first floor.

18. The north field yields 15,000 fewer bushels of wheat than the south field. Use n to represent the number of bushels of wheat in the north field and s to represent the number of bushels of wheat in the south field.

Write an algebraic expression for each quantity using the given variable.

▲ **19.** The first angle of a triangle is half the second angle. The third angle of the triangle is twice the second angle. Use the variable s.

Mc
▲ **20.** The length of a rectangle is 5 inches shorter than double the width. Use the letter w.

Solve using an equation.

21. The number of acres of land in the old Smithfield farm is three times the number of acres of land in the Prentice farm. Together the two farms have 348 acres. How many acres of land are there on each farm?

22. Sam earns $1500 less per year than Marcia does. The combined income of the two people is $46,500 per year. How much does each person earn?

23. During the fall semester, 183 students registered for Introduction to Biology. The morning class has 24 fewer students than the afternoon class. The evening class has 12 more students than the afternoon class. How many students registered for each class?

▲ **24.** A rectangular field has a perimeter of 118 feet. The width is 8 feet longer than half the length. Find the dimensions of the rectangle.

MATH COACH

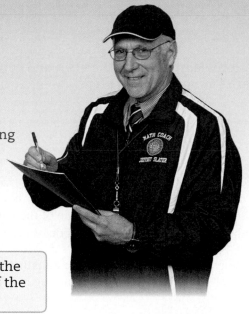

Mastering the skills you need to do well on the test.

Students often make the same types of errors when they do the Chapter 10 Test. Here are some helpful hints to keep you from making these common errors on test problems.

Combining Like Terms with Variables and Fractional Coefficients—Problem 2 $\frac{1}{3}x + \frac{5}{8}y - \frac{1}{5}x + \frac{1}{2}y$

> **Helpful Hint** First, identify the like terms. Then, find the LCD for the coefficients of the x terms and find the LCD for the coefficients of the y terms.

Did you correctly identify $\frac{1}{3}x$ and $-\frac{1}{5}x$ as like terms and $\frac{5}{8}y$ and $\frac{1}{2}y$ as like terms?

Yes ____ No ____

If you answered No, stop and review the definition of like terms.

Was 15 your LCD for the coefficients of the x terms?

Yes ____ No ____

If you answered No, go back and review how to find the LCD of two fractions.

Did you transform $\frac{1}{3}$ to $\frac{5}{15}$ as the coefficient of the first term? Did you transform $-\frac{1}{5}$ to $-\frac{3}{15}$ as the coefficient of the third term?

Yes ____ No ____

If you answered No, stop and carefully review how to write equivalent fractions using the LCD. Now follow the same procedure to find the LCD for the y terms and rewrite those fractions before combining the like terms.

If you answered Problem 2 incorrectly, go back and rework the problem using these suggestions.

Simplifying Expressions Containing Parentheses—Problem 10   Simplify. $2(-3a + 2b) - 5(a - 2b)$

> **Helpful Hint** First, use the distributive property to remove parentheses before doing any other operations. Double-check your work and be careful to avoid sign errors.

After removing the parentheses, did you obtain $-6a + 4b - 5a + 10b$?

Yes ____ No ____

If you answered No, stop and redo the distributive property. Remember to use extra care when multiplying by -5 as you remove the second set of parentheses.

Was $-11a$ your answer when adding $-6a$ and $-5a$?

Yes ____ No ____

If you answered No, go back and perform the addition again. Be careful of your signs. Recall that adding two negative numbers always results in a negative number.

Now go back and rework the problem using these suggestions.

Need help? Watch the MATH COACH videos in MyMathLab® or on YouTube™.

659

Solving Equations with Variables on Both Sides—Problem 14   Solve for the variable.

$8x - 2 - x = 3x - 9 - 10x$

> **Helpful Hint** Be sure to collect and combine like terms on each side of the equation before performing any other steps.

Did you combine like terms on each side to obtain
$7x - 2 = -7x - 9$?

Yes _____   No _____

If you answered No, go back and identify like terms. Then combine the x terms on the left side of the equation and combine the x terms on the right side of the equation.

Did you add $7x$ to each side of the equation and then add 2 to each side of the equation, resulting in $14x = -7$?

Yes _____   No _____

If you answered No, stop and perform these steps.

Consider what the last step would be to solve the equation for x. Now go back and complete this problem again.

Writing an Algebraic Expression for Each Quantity—Problem 20   The length of a rectangle is 5 inches shorter than double the width. Use the letter w.

> **Helpful Hint** Since length is compared to width, let w = the width.

Did you notice that "double the width" could be written as $2w$?

Yes _____   No _____

If you answered No, reread the problem and write this expression down first.

Did you realize that "5 inches shorter than double the width" translates to $2w - 5$?

Yes _____   No _____

If you answered No, go back and perform this step in the translation.

Remember to write an expression for length and an expression for width in your final answer.

If you answered Problem 20 incorrectly, these go back and rework the problem using these suggestions.

Need more help? Look for section examples marked with ^{M}C to review.

Practice Final Examination

This examination is based on Chapters 1–10 of the book. There are 10 questions covering the content of each chapter.

Chapter 1

1. Write in words. 82,367

2. Add. 13,428
  + 16,905

3. Add. 19
  23
  16
  45
  + 70

4. Subtract. 89,071
  − 54,968

Multiply.

5. 78
  × 54

6. 2035
  × 107

In questions 7 and 8, divide. (Be sure to indicate the remainder if one exists.)

7. $7\overline{)1106}$

8. $26\overline{)15,756}$

9. Evaluate. Perform operations in the proper order. $3^4 + 20 \div 4 \times 2 + 5^2$

10. Melinda traveled 512 miles in her car. The car used 16 gallons of gas on the entire trip. How many miles per gallon did the car achieve?

Chapter 2

11. Reduce the fraction. $\dfrac{14}{30}$

12. Change to an improper fraction. $3\dfrac{9}{11}$

13. Add. $\dfrac{1}{10} + \dfrac{3}{4} + \dfrac{4}{5}$

14. Add. $2\dfrac{1}{3} + 3\dfrac{3}{5}$

15. Subtract. $4\dfrac{5}{7} - 2\dfrac{1}{2}$

16. Multiply. $1\dfrac{1}{4} \times 3\dfrac{1}{5}$

17. Divide. $\dfrac{7}{9} \div \dfrac{5}{18}$

18. Divide. $\dfrac{5\frac{1}{2}}{3\frac{1}{4}}$

19. Lucinda jogged $1\frac{1}{2}$ miles on Monday, $3\frac{1}{4}$ miles on Tuesday, and $2\frac{1}{10}$ miles on Wednesday. How many miles in all did she jog over the three-day period?

20. A butcher has $11\frac{2}{3}$ pounds of steak. She wishes to place them in several equal-size packages. Each package will hold $2\frac{1}{3}$ pounds of steak. How many packages can be made?

1. _____

2. _____

3. _____

4. _____

5. _____

6. _____

7. _____

8. _____

9. _____

10. _____

11. _____

12. _____

13. _____

14. _____

15. _____

16. _____

17. _____

18. _____

19. _____

20. _____

21. Express as a decimal. $\dfrac{719}{1000}$

22. Write in reduced fractional notation. 0.86

23. Fill in the blank with $<$, $=$, or $>$. 0.315 _____ 0.309

24. Round to the nearest hundredth. 506.3782

25. Add.
    9.6
    3.82
    1.05
    $+\ 7.3$

26. Subtract.
    3.61
    $-\ 2.853$

27. Multiply.
    1.23
    $\times\ 0.4$

28. Divide. $0.24\overline{)0.8856}$

29. Write as a decimal. $\dfrac{13}{16}$

30. Evaluate by performing operations in proper order.
    $0.7 + (0.2)^3 - 0.08(0.03)$

Chapter 4

31. Write a rate in simplest form to compare 7000 students to 215 faculty.

32. Is this a proportion? $\dfrac{12}{15} = \dfrac{17}{21}$

Solve the proportion. Round to the nearest tenth when necessary.

33. $\dfrac{5}{9} = \dfrac{n}{17}$

34. $\dfrac{3}{n} = \dfrac{7}{18}$

35. $\dfrac{n}{12} = \dfrac{5}{4}$

36. $\dfrac{n}{7} = \dfrac{36}{28}$

Solve using a proportion. Round to the nearest hundredth when necessary.

37. Bob earned $2000 for painting three houses. How much would he earn for painting five houses?

38. Two cities that are actually 200 miles apart appear 6 inches apart on a map. Two other cities are 325 miles apart. How far apart will they appear on the same map?

39. Roberta earned $115 last week on her part-time job. She had $12 withheld for federal income tax. Last year she earned $6000 on her part-time job. Assuming the same rate, how much was withheld for federal income tax last year?

40. Malaga's recipe feeds 18 people and calls for 1.2 pounds of butter. If she wants to feed 24 people, how many pounds of butter does she need?

21. _____
22. _____
23. _____
24. _____
25. _____
26. _____
27. _____
28. _____
29. _____
30. _____
31. _____
32. _____
33. _____
34. _____
35. _____
36. _____
37. _____
38. _____
39. _____
40. _____

Chapter 5

Round to the nearest hundredth when necessary in problems 41–44.

41. Write as a percent. 0.0063

42. Change $\frac{17}{80}$ to a percent.

43. Write as a decimal. 164%

44. What percent of 300 is 52?

Round to the nearest tenth when necessary in problems 45–50.

45. Find 6.3% of 4800.

46. 145 is 58% of what number?

47. 126% of 3400 is what number?

48. Pauline bought a new car. She got an 8% discount. The car listed for $18,800. How much did she pay for the car?

49. A total of 1260 freshmen were admitted to Central College. This is 28% of the student body. How big is the student body?

50. There are 11.28 centimeters of water in the rain gauge this week. Last week the rain gauge held 8.40 centimeters of water. What is the percent of increase from last week to this week?

Chapter 6

Convert. Express your answers as decimals.

51. 17 quarts = _____ gallons

52. 3.25 tons = _____ pounds

53. 16 feet = _____ inches

54. 5.6 kilometers = _____ meters

55. 69.8 grams = _____ kilogram

56. 2.48 milliliters = _____ liter

57. 12 miles = _____ kilometers

58. 9.62 centimeters = _____ meter

59. 3 miles = _____ feet

60. Two metal sheets are 0.623 centimeter and 0.74 centimeter thick, respectively. An insulating foil is 0.0428 millimeter thick. When all three layers are placed tightly together, what is the total thickness? Express your answer in centimeters.

Chapter 7

Round to the nearest hundredth when necessary. Use $\pi \approx 3.14$ when necessary.

▲ **61.** Find the perimeter of a rectangle that is 6 meters long and 1.2 meters wide.

▲ **62.** Find the perimeter of a trapezoid with sides of 82 centimeters, 13 centimeters, 98 centimeters, and 13 centimeters.

41. _____

42. _____

43. _____

44. _____

45. _____

46. _____

47. _____

48. _____

49. _____

50. _____

51. _____

52. _____

53. _____

54. _____

55. _____

56. _____

57. _____

58. _____

59. _____

60. _____

61. _____

62. _____

▲ **63.** Find the area of a triangle with base 6 feet and height 1.8 feet.

▲ **64.** Find the area of a trapezoid with bases of 12 meters and 8 meters and a height of 7.5 meters.

▲ **65.** Find the area of a circle with radius 6 meters.

▲ **66.** Find the circumference of a circle with diameter 18 meters.

▲ **67.** Find the volume of a cone with a radius of 4 centimeters and a height of 10 centimeters.

▲ **68.** Find the volume of a rectangular pyramid with a base of 12 feet by 19 feet and a height of 2.7 feet.

▲ **69.** Find the area of this object, consisting of a square and a triangle.

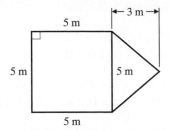

▲ **70.** In the following pair of similar triangles, find n.

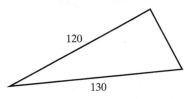

Chapter 8

The following double-bar graph indicates the quarterly profits for Westar Corporation in 2009 and 2010. Use this graph to answer questions 71 and 72.

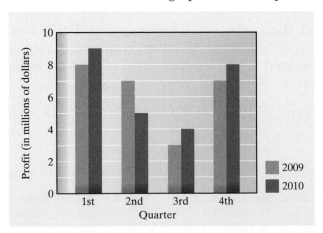

63. _____

64. _____

65. _____

66. _____

67. _____

68. _____

69. _____

70. _____

71. What were the profits in the fourth quarter of 2010?

72. How much greater were the profits in the first quarter of 2010 than the profits in the first quarter of 2009?

The following line graph depicts the average annual temperature at West Valley for the years 1970, 1980, 1990, 2000, and 2010.

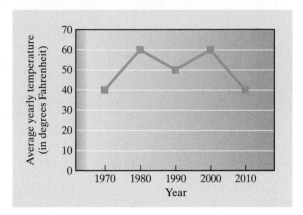

73. What was the average temperature in 1990?

74. In what 10-year period did the average temperature show the greatest decline?

The following histogram shows the number of students in each age category at Center City College.

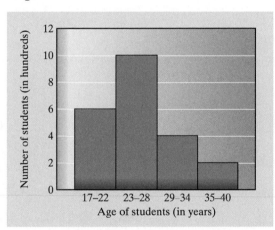

75. How many students are between 17 and 22 years old?

76. How many students are between 23 and 34 years old?

77. Find the *mean* and the *median* of the following. 8, 12, 16, 17, 20, 22. Round to the nearest hundredth.

78. Evaluate exactly. $\sqrt{49} + \sqrt{81}$

79. Approximate to the nearest thousandth using a calculator or a square root table. $\sqrt{123}$

71. _____

72. _____

73. _____

74. _____

75. _____

76. _____

77. _____

78. _____

79. _____

80. Find the unknown side of the right triangle.

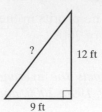

Chapter 9

Add.

81. $-8 + (-2) + (-3)$

82. $-\dfrac{1}{4} + \dfrac{3}{8}$

Subtract.

83. $9 - 12$

84. $-20 - (-3)$

85. Multiply. $(2)(-3)(4)(-1)$

86. Divide. $-\dfrac{2}{3} \div \dfrac{1}{4}$

Perform the indicated operations in the proper order.

87. $(-16) \div (-2) + (-4)$

88. $12 - 3(-5)$

89. $7 - (-3) + 12 \div (-6)$

90. $\dfrac{(-3)(-1) + (-4)(2)}{(0)(6) + (-5)(2)}$

Chapter 10

Combine like terms.

91. $5x - 3y - 8x - 4y$

92. $5 + 2a - 8b - 12 - 6a - 9b$

Simplify.

93. $-2(x - 3y - 5)$

94. $-2(4x + 2) - 3(x + 3y)$

Solve for the variable.

95. $5 - 4x = -3$

96. $5 - 2(x - 3) = 15$

97. $7 - 2x = 10 + 4x$

98. $-3(x + 4) = 2(x - 5)$

Solve using an equation.

99. There are 12 more students taking history than math. There are twice as many students taking psychology as there are students taking math. There are 452 students enrolled in these three subjects. Assume each student is taking only one of these three courses. How many are taking history? How many are taking math?

▲ **100.** A rectangle has a perimeter of 106 meters. The length is 5 meters longer than double the width. Find the length and width of the rectangle.

Appendix A Consumer Finance Applications

A.1 Balancing a Checking Account

① Calculating a Checkbook Balance

If you have a checking account, you should keep records of the checks written, ATM withdrawals, deposits, and other transactions on a check register. To find the amount of money in a checking account you subtract debits and add credits to the balance in the account. Debits are checks written, withdrawals made, or any other amount charged to a checking account. Credits include deposits made, as well as any other money credited to the account.

Student Learning Objectives

After studying this section, you will be able to:

① Calculate a checkbook balance.

② Balance a checkbook.

EXAMPLE 1   Jesse Holm had a balance of $1254.32 in his checking account before writing five checks and making a deposit. On 9/2 Jesse wrote check #243 to the Manor Apartments for $575, check #244 to the Electric Company for $23.41, and check #245 to the Gas Company for $15.67. Then on 9/3, he wrote check #246 to Jack's Market for $125.57, check #247 to Clothing Mart for $35.85, and made a $634.51 deposit. Record the checks and deposit in Jesse's check register and then find Jesse's ending balance.

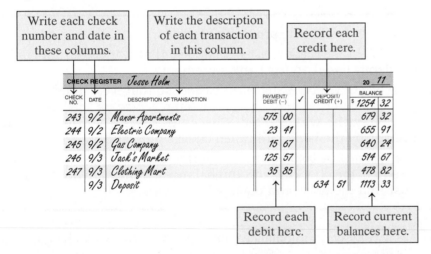

Solution   To find the ending balance, we subtract each check written and add the deposit to the current balance. Then we record these amounts in the check register.

$$\begin{array}{cccccc}
1254.32 & 679.32 & 655.91 & 640.24 & 514.67 & 478.82 \\
-\ 575.00 & -\ 23.41 & -\ 15.67 & -\ 125.57 & -\ 35.85 & +\ 634.51 \\
\hline
679.32 & 655.91 & 640.24 & 514.67 & 478.82 & 1113.33
\end{array}$$

Jesse's balance is 1113.33.

Student Practice 1   My Chung Nguyen had a balance of $1434.52 in her checking account before writing three checks and making a deposit. On 3/1 My Chung wrote check #144 to the Leland Mortgage Company for $908 and check #145 to the Phone Company for $33.21. Then on 3/2 she wrote

NOTE TO STUDENT: Fully worked-out solutions to all of the Student Practice problems can be found at the back of the text starting at page SP-1.

Continued on next page

check #146 to Sam's Food Market for $102.37 and made a $524.41 deposit. Record the checks and deposit in My Chung's check register, and then find My Chung's ending balance.

CHECK REGISTER *My Chung Nguyen*							20 *11*
CHECK NO.	DATE	DESCRIPTION OF TRANSACTION	PAYMENT/ DEBIT (−)	✓	DEPOSIT/ CREDIT (+)	BALANCE $	

② Balancing a Checkbook

The bank provides customers with bank statements each month. This statement lists the checks the bank paid, ATM withdrawals, deposits made, and all other debits and credits made to a checking account. It is very important to verify that these bank records match ours. We must make sure that we deducted all debits and added all credits in our check register. This is called **balancing a checkbook.** Balancing our checkbook allows us to make sure that the balance we think we have in our checkbook is correct. If our checkbook does not balance, we must look for any mistakes.

> To balance a checkbook, proceed as follows.
>
> 1. First, *adjust the check register balance* so that it includes all credits and debits listed on the bank statement.
>
> 2. Then, *adjust the bank statement balance* so that it includes all credits and debits that may not have been received by the bank when the statement was printed. Checks that were written, but not received by the bank, are called **checks outstanding.**
>
> 3. Finally, *compare both balances* to verify that they are equal. If they are equal, the checking account balances. If they are not equal, we must find the error and make adjustments.

There are several ways to balance a checkbook. Most banks include a form you can fill out to assist you in this process.

EXAMPLE 2 Balance Jesse's checkbook using his check register and bank statement.

CHECK REGISTER *Jesse Holm*							20 *11*
CHECK NO.	DATE	DESCRIPTION OF TRANSACTION	PAYMENT/ DEBIT (−)	✓	DEPOSIT/ CREDIT (+)	BALANCE $ *1254* *32*	
243	9/2	Manor Apartments	575 00	✓		679 32	
244	9/2	Electric Company	23 41	✓		655 91	
245	9/2	Gas Company	15 67	✓		640 24	
246	9/3	Jack's Market	125 57	✓		514 67	
247	9/3	Clothing Mart	35 85			478 82	
	9/3	Deposit		✓	634 51	1113 33	
248	9/12	College Bookstore	168 96	✓		944 37	
	9/18	ATM	100 00	✓		844 37	
249	9/25	Telephone Company	43 29	✓		801 08	
250	9/30	Sports Emporium	40 00			761 08	
	10/1	Deposit			530 90	1291 98	

Bank Statement: JESSE HOLM   9/1/2011 to 9/30/2011

Beginning Balance $1254.32
Ending Balance $831.68

Checks cleared by the bank

#243	$575.00	#245	$15.67	#248	$168.96
#244	$23.41	#246*	$125.57	#249	$43.29

*Indicates that the next check in the sequence is outstanding (hasn't cleared).

Deposits

9/3 $634.51

Other withdrawals

9/18 ATM $100.00   Service charge $5.25

Solution Follow steps 1–6 on the checking account balancing form below.

CHECKING RECONCILEMENT		This form is provided to assist you in balancing your checking account.

List checks outstanding* not charged to your checking account		
CHECK NO.	AMOUNT	
247	35	85
250	40	00
TOTAL	75	85

*and ATM withdrawals

	Period ending	9/30 ,20 11
1. Check Register Balance	$	1291.98
Subtract any charges listed on the bank statement which you have not previously deducted from your balance. —	$	5.25
Adjusted Check Register Balance	$	1286.73
2. **Enter** the ending balance shown on the bank statement.	$	831.68
3. **Enter** deposits made later than the ending date on the bank statement. +	$	530.90
+	$	
+	$	
TOTAL (Step 2 plus Step 3)	$	1362.58
4. In your check register, **check off** all the checks paid. In the area provided to the left, **list** numbers and amounts of all outstanding checks and ATM withdrawals.		
5. **Subtract** the total amount in Step 4. —	$	75.85
6. This adjusted bank balance should equal the adjusted Check Register Balance from Step 1.	$	1286.73

The balances in steps **1** and **6** are equal, so Jesse's checkbook is balanced.

Student Practice 2　Balance Anthony's checkbook using his check register, his bank statement, and the given checking account balacing form.

CHECK REGISTER Anthony Maida							20 11
CHECK NO.	DATE	DESCRIPTION OF TRANSACTION	PAYMENT/ DEBIT (−)	✓	DEPOSIT/ CREDIT (+)	BALANCE	
						$ 1823	00
211	7/2	Apple Apartments	985 00			838	00
	7/9	ATM	101 50			736	50
212	7/10	Leland Groceries	98 87			637	63
213	7/21	The Gas Company	45 56			592	07
	7/21	Deposit			687 10	1279	17
214	7/28	Cellular for Less	59 98			1219	19
215	7/28	The Electric Company	89 75			1129	44
216	7/28	Sports World	129 99			999	45
217	7/28	Leland Groceries	205 99			793	46
	7/30	ATM	141 50			651	96
	8/1	Deposit			398 50	1050	46

Bank Statement: ANTHONY MAIDA　　7/1/2011 to 8/1/2011

Beginning Balance　$1823.00
Ending Balance　　$934.95

Checks cleared by the bank

#211	$985.00	#213	$45.56	#216	$129.99
#212	$98.87	#214*	$59.98		

*Indicates that the next check in the sequence is outstanding (hasn't cleared).

Deposits

7/21　$687.10

Other withdrawals

7/9 ATM	$101.50	Service charge	$3.50
7/30 ATM	$141.50	Check purchase	$9.25

CHECKING RECONCILEMENT		This form is provided to assist you in balancing your checking account.

List checks outstanding* not charged to your checking account	
CHECK NO.	AMOUNT
TOTAL	

*and ATM withdrawals

	Period ending	,20
1. Check Register Balance	$	
Subtract any charges listed on the bank statement which you have not previously deducted from your balance. —	$	
Adjusted Check Register Balance	$	
2. **Enter** the ending balance shown on the bank statement.	$	
3. **Enter** deposits made later than the ending date on the bank statement. +	$	
+	$	
+	$	
TOTAL (Step 2 plus Step 3)	$	
4. In your check register, **check off** all the checks paid. In the area provided to the left, **list** numbers and amounts of all outstanding checks and ATM withdrawals.		
5. **Subtract** the total amount in Step 4. —	$	
6. This adjusted bank balance should equal the adjusted Check Register Balance from Step 1.	$	

Note: To calculate the service charges, we add 3.50 + 9.25 = 12.75.

1. Shin Karasuda had a balance of $532 in his checking account before writing four checks and making a deposit. On 6/1 he wrote check #122 to the Mini Market for $124.95 and check #123 to Better Be Dry Cleaners for $41.50. Then on 6/9 he made a $384.10 deposit, wrote check #124 to Macy's Department Store for $72.98, and check #125 to Costco for $121.55. Record the checks and deposit in Shin's check register, and then find the ending balance.

CHECK REGISTER	Shin Karasuda					20 11
CHECK NO.	DATE	DESCRIPTION OF TRANSACTION	PAYMENT/ DEBIT (−)	✓	DEPOSIT/ CREDIT (+)	BALANCE $

2. Mary Beth O'Brian had a balance of $493 in her checking account before writing four checks and making a deposit. On 9/4 she wrote check #311 to Ben's Garage for $213.45 and #312 to Food Mart for $132.50. Then on 9/5 she made a $387.50 check deposit, wrote check #313 to the Shoe Pavilion for $69.98, and check #314 to the Electric Company for $92.45. Record the checks and deposit in Mary Beth's check register, and then find the ending balance.

CHECK REGISTER	Mary Beth O'Brian					20 11
CHECK NO.	DATE	DESCRIPTION OF TRANSACTION	PAYMENT/ DEBIT (−)	✓	DEPOSIT/ CREDIT (+)	BALANCE $

3. The Harbor Beauty Salon had a balance of $2498.90 in its business checking account on 3/4. The manager made a deposit on the same day for $786 and wrote check #734 to the Beauty Supply Factory for $980. Then on 3/9 he wrote check #735 to the Water Department for $131.85 and check #736 to the Electric Company for $251.50. On 3/19 he made a $2614.10 deposit and wrote two payroll checks: #737 to Ranik Ghandi for $873 and #738 to Eduardo Gomez for $750. Record the checks and deposits in the Harbor Beauty Salon's check register, and then find the ending balance.

CHECK REGISTER	Harbor Beauty Salon					20 11
CHECK NO.	DATE	DESCRIPTION OF TRANSACTION	PAYMENT/ DEBIT (−)	✓	DEPOSIT/ CREDIT (+)	BALANCE $

4. Joanna's Coffee Shop had a balance of $1108.50 in its business checking account on 1/7. The owner made a deposit on the same day for $963 and wrote check #527 to the Restaurant Supply Company for $492. Then on 1/11 she wrote check #528 to the Gas Company for $122.45 and check #529 to the Electric Company for $321.20. Then on 1/12 she made a $1518.20 deposit and wrote two payroll checks: #530 to Sara O'Conner for $579, and #531 to Nlegan Raskin for $466. Record the checks and deposits in Joanna's Coffee Shop's check register, and then find the ending balance.

CHECK REGISTER	Joanna's Coffee Shop					20 11
CHECK NO.	DATE	DESCRIPTION OF TRANSACTION	PAYMENT/ DEBIT (−)	✓	DEPOSIT/ CREDIT (+)	BALANCE $

5. On 3/3 Justin Larkin had $321.94 in his checking account before he withdrew $101.50 at the ATM. On 3/7 he made a $601.90 deposit and wrote checks to pay the following bills: check #114 to the Third Street Apartments for $550, check #115 to the Cable Company for $59.50, check #116 to the Electric Company for $43.50, and check #117 to the Gas Company for $15.90. Does Justin have enough money left in his checking account to pay $99 for his car insurance?

6. On 5/2 Leon Jones had $423.54 in his checking account when he withdrew $51.50 at the ATM. On 5/9 he made a $601.80 deposit and wrote checks to pay the following bills: check #334 to Fasco Car Finance for $150.25, check #335 to A-1 Car Insurance for $89.20, check #336 to the Telephone Company for $33.40, and check #337 to the Apple Apartments for $615. Does Leon have enough money left in his checking account to pay $59 for his electric bill?

7. Balance the monthly statement for the Carson Maid Service on the checking account balancing form shown below.

CHECK REGISTER	*Justin Larkin*				20 *11*
CHECK NO.	DATE	DESCRIPTION OF TRANSACTION	PAYMENT/ DEBIT (−)	✓ DEPOSIT/ CREDIT (+)	BALANCE $

CHECK REGISTER	*Leon Jones*				20 *11*
CHECK NO.	DATE	DESCRIPTION OF TRANSACTION	PAYMENT/ DEBIT (−)	✓ DEPOSIT/ CREDIT (+)	BALANCE $

Bank Statement: CARSON MAID SERVICE   4/1/2011 to 4/30/2011

Beginning Balance $1721.50
Ending Balance $1564.82

Checks cleared by the bank

#102	$422.33	#104	$320.00	#108	$450.10
#103	$510.50	#105*	$320.00		

*Indicates that the next check in the sequence is outstanding (hasn't cleared).

Deposits

4/11 $1890.00

Other withdrawals

Service charge $4.50
Check purchase $19.25

CHECK REGISTER		*Carson Maid Service*				20 *11*
CHECK NO.	DATE	DESCRIPTION OF TRANSACTION	PAYMENT/ DEBIT (−)	✓	DEPOSIT/ CREDIT (+)	BALANCE $ 1721 50
102	4/3	A&R Cleaning Supplies	422 33			1299 17
103	4/9	Allison De Julio	510 50			788 67
104	4/9	Mai Vu	320 00			468 67
105	4/9	Jon Veldez	320 00			148 67
	4/11	Deposit			1890 00	2038 67
106	4/20	Mobil Gas Company	355 35			1683 32
107	4/25	Peterson Property Mangt. warehouse rent	525 00			1158 32
108	4/29	Jack's Garage	450 10			708 22
	5/2	Deposit			540 00	1248 22

CHECKING RECONCILEMENT   This form is provided to assist you in balancing your checking account.

List checks outstanding* not charged to your checking account

CHECK NO.	AMOUNT
TOTAL	

*and ATM withdrawals

Period ending ,20

1. Check Register Balance $
  Subtract any charges listed on the bank statement which you have not previously deducted from your balance. − $
  Adjusted Check Register Balance $

2. Enter the ending balance shown on the bank statement. $

3. Enter deposits made later than the ending date on the bank statement. + $
  + $
  + $

TOTAL (Step 2 plus Step 3) $

4. In your check register, **check off** all the checks paid. In the area provided to the left, **list** numbers and amounts of all outstanding checks and ATM withdrawals.

5. Subtract the total amount in Step 4. − $

6. This adjusted bank balance should equal the adjusted Check Register Balance from Step 1. $

8. Balance the monthly statement for The Flower Shop on the checking account balancing form shown below.

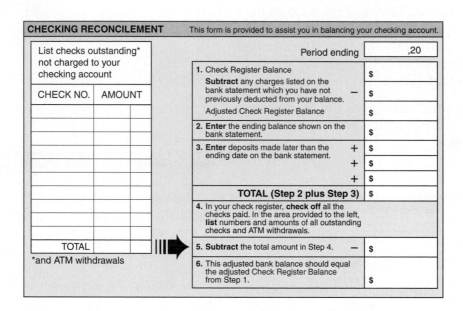

CHECK REGISTER	*The Flower Shop*						20 _11_	
CHECK NO.	DATE	DESCRIPTION OF TRANSACTION	PAYMENT/ DEBIT (−)		✓	DEPOSIT/ CREDIT (+)	BALANCE $ 3459	40
502	9/7	Whole Sale Flower Company	733	67			2725	73
503	9/7	Alexsandra Kruse	580	20			2145	53
504	9/7	Jose Sanchez	430	50			1715	03
505	9/7	Kamir Kosedag	601	90			1113	13
	9/11	Deposit				2654 00	3767	13
506	9/21	L&S Pottery	466	84			3300	29
507	9/24	Peterson Property Mangt. (warehouse rent)	985	00			2315	29
508	9/28	Barton Electric Company	525	60			1789	69
	10/2	Deposit				540 00	2329	69

Bank Statement: THE FLOWER SHOP   9/1/2011 to 9/30/2011

Beginning Balance $3459.40
Ending Balance $3220.28

Checks cleared by the bank

#502 $733.67   #504 $430.50   #508 $525.60
#503 $580.20   #505* $601.90
*Indicates that the next check in the sequence is outstanding (hasn't cleared).

Deposits

9/11 $2654.00

Other withdrawals

Service charge $4.75
Check purchase $16.50

CHECKING RECONCILEMENT   This form is provided to assist you in balancing your checking account.

List checks outstanding* not charged to your checking account

CHECK NO.	AMOUNT	
TOTAL		

*and ATM withdrawals

Period ending _____ ,20

1. Check Register Balance $ _____
   Subtract any charges listed on the bank statement which you have not previously deducted from your balance.   − $ _____
   Adjusted Check Register Balance $ _____

2. **Enter** the ending balance shown on the bank statement. $ _____

3. **Enter** deposits made later than the ending date on the bank statement.   + $ _____
   + $ _____
   + $ _____

   TOTAL (Step 2 plus Step 3) $ _____

4. In your check register, **check off** all the checks paid. In the area provided to the left, **list** numbers and amounts of all outstanding checks and ATM withdrawals.

5. **Subtract** the total amount in Step 4.   − $ _____

6. This adjusted bank balance should equal the adjusted Check Register Balance from Step 1. $ _____

9. On 2/1 Jeremy Sirk had a balance of $672.10 in his checking account. On the same day he deposited $735 of his paycheck into his checking account and wrote check #233 to Stanton Sporting Goods for $92.99 and check #234 to the Garden Apartments for $680. Then on 2/20 he wrote check #235 to the Gas Company for $31.85, check #236 to the Cable Company for $51.50, check #237 to Ralph's Market for $173.98, and also made an ATM withdrawal for $101.50. Then on 3/1 he made an $814.10 deposit, wrote check #238 to State Farm Insurance for $98, and made another ATM withdrawal for $41.50. Record the checks and deposits in Jeremy's check register, and then find the ending balance.

CHECK REGISTER	*Jeremy Sirk*						20 _11_	
CHECK NO.	DATE	DESCRIPTION OF TRANSACTION	PAYMENT/ DEBIT (−)		✓	DEPOSIT/ CREDIT (+)	BALANCE $	

10. On 8/1 Shannon Mending had a balance of $525.90 in her checking account. Later that same day she deposited $588.23 into her checking account and wrote check #333 to Verizon Telephone Company for $33.20 and check #334 to Discount Car Insurance for $332.50. Then on 8/8 she wrote check #335 to the Walden Market for $21.35, made an ATM withdrawal for $81.50, and wrote check #336 to the Cable Company for $41.50. Then on 9/1 she made a $904.10 deposit, wrote check #337 to Next Day Dry Cleaners for $33.50, and check #338 to Marty's Dress Shop for $87.99. Record the checks and deposits in Shannon's check register, and then find the ending balance.

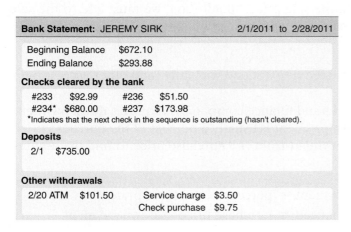

11. Refer to exercise 9 and use the bank statement and checking account balancing form below to balance Jeremy's checking account.

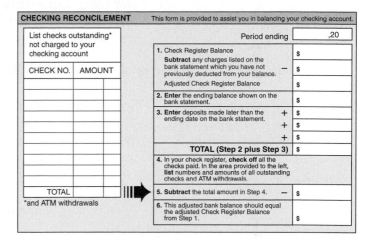

12. Refer to exercise 10 and use the bank statement and checking account balancing form below to balance Shannon's checking account.

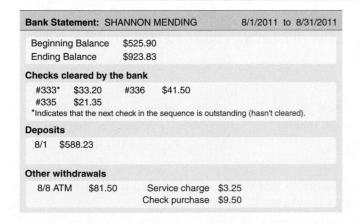

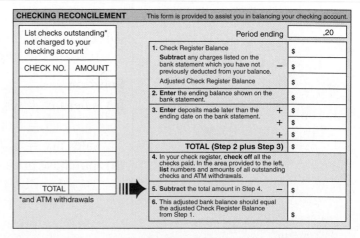

A.2 Determining the Best Deal When Purchasing a Vehicle

Student Learning Objectives

After studying this section, you will be able to:

1. Find the true purchase price of a vehicle.

2. Find the total cost of a vehicle to determine the best deal.

When we buy a car there are several facts to consider in order to determine which car has the best price. The sale price offered by a car dealer or seller is only one factor we must consider—others include the interest rate on the loan, sales tax, license fee, and sale promotions such as cash rebates or 0% interest. We must also consider the cost of options we choose such as extended warranties, sun roof, tinted glass, etc. In this section we will see how to calculate the total purchase price and determine the best deal when buying a vehicle.

1 Finding the True Purchase Price of a Vehicle

In most states there is a sales tax and a license or title fee that must be paid on all vehicles purchased. To find the true purchase price for a vehicle, we *add* these extra costs to the sale price. In addition, we must add to the sale price the cost of any extended warranties and extra options or accessories we buy.

purchase price = sale price + sales tax + license fee + extended warranty (and other accessories)

Sometimes lenders (banks and finance companies) require a **down payment.** The amount of the down payment is usually a percent of the purchase price. We subtract the down payment from the purchase price to find the amount we must finance.

down payment = percent × purchase price
amount financed = purchase price − down payment

EXAMPLE 1 Daniel bought a truck that was on sale for $28,999 in a city that has a 6% sales tax and a 2% license fee.

(a) Find the sales tax and license fee Daniel paid.

(b) Daniel also bought an extended warranty for $1550. Find the purchase price of the truck.

Solution

(a) sales tax = 6% of sale price   license fee = 2% of sale price
           = 0.06 × 28,999   = 0.02 × 28,999
   sales tax = $1739.94   license fee = $579.98

(b) purchase price
     = sale price + sales tax + license fee + extended warranty
     = 28,999 + 1739.94 + 579.98 + 1550
   purchase price = $32,868.92

Student Practice 1 Huy Nguyen bought a van that was on sale for $24,999 in a city that has a 7% sales tax and a 2% license fee.

(a) Find the sales tax and license fee Huy paid.

(b) Huy also bought an extended warranty for $1275. Find the purchase price of the van.

EXAMPLE 2 The purchase price of a car Jerome plans to buy is $19,999. In order to qualify for the loan on the car, Jerome must make a down payment of 20% of the purchase price.

(a) Find the down payment.

(b) Find the amount financed.

Solution

(a) down payment = percent × purchase price

down payment = 20% × 19,999

= 0.20 × 19,999

down payment = $3999.80

(b) amount financed = purchase price − down payment

= 19,999 − 3999.80

amount financed = $15,999.20

Student Practice 2 The purchase price of a Jeep Cheryl plans to buy is $32,499. In order to qualify for the loan on the Jeep, Cheryl must make a down payment of 15% of the purchase price.

(a) Find the down payment.   **(b)** Find the amount financed.

② Finding the Total Cost of a Vehicle to Determine the Best Deal

When we borrow money to buy a car, we often pay interest on the loan and we must consider this extra cost when we calculate the **total cost** of the vehicle. If we know the amount of the car payment and the number of months it will take to pay off the loan, we can find the total payments on the car (the amount we borrowed plus interest) by multiplying the monthly payment amount times the number of months of the loan. Then we must add the down payment to that amount.

total cost = (monthly payment × number of months in loan) + down payment

EXAMPLE 3   Marvin went to two dealerships to find the best deal on the truck he plans to purchase. From which dealership should Marvin buy the truck so that the *total cost* of the truck is the least expensive?

Dealership 1	Dealership 2
• Purchase price: $39,999	• Purchase price: $36,499
• Financing option: 2% financing with $5000 down payment	• Financing option: 5% financing with no down payment
• Monthly payments: $743.73 per month for 48 months	• Monthly payments: $638.73 per month for 60 months

Solution   First, we find the total cost of the truck at Dealership 1.

total cost = (monthly payment × number of months in loan) + down payment

= (743.73 × 48)   + 5000   We multiply, then add.

= 40,699.04

The total cost of the truck at Dealership 1 is $40,699.04.

Next, we find the total cost of the truck at Dealership 2.

total cost = (monthly payment × number of months in loan) + down payment

= (638.73 × 60)   + 0   There is no down payment.

= 38,323.80

The total cost of the truck at Dealership 2 is $38,323.80.

We see that the best deal on the truck Marvin plans to buy is at Dealership 2.

Student Practice 3   Phoebe went to two dealerships to find the best deal on the minivan she plans to purchase. From which dealership should Phoebe buy the minivan so that the *total cost* of the minivan is the least expensive?

Dealership 1	Dealership 2
• Purchase price: $22,999	• Purchase price: $23,999
• Financing option: 4% financing with no down payment	• Financing option: 2% financing with $5500 down payment
• Monthly payments: $398.65 per month for 60 months	• Monthly payments: $393.10 per month for 48 months

1. A college student buys a car and pays a 6% sales tax on the $21,599 sale price. How much sales tax did the student pay?

2. A high school teacher buys a minivan and pays a 5% sales tax on the $26,800 sale price. How much sales tax did the teacher pay?

3. Frances is planning to buy a four-door sedan that is on sale for $18,999. She must pay a 2% license fee. Find the license fee.

4. Mai Vu saw an ad for a short-bed truck that is on sale for $17,599. If she buys the truck she must pay a 2% license fee. Find the license fee.

5. John must make a 10% down payment on the purchase price of the $42,450 sports car he is planning to buy. Find the down payment.

6. Kamir must make a 15% down payment on the purchase price of the $31,500 extended cab truck he is planning to buy. Find the down payment.

7. Tabatha bought a minivan that was on sale for $24,899 in a city that has a 5% sales tax and a 2% license fee.
   (a) Find the sales tax and license fee Tabatha paid.
   (b) Tabatha also bought an extended warranty for $1100. Find the purchase price of the minivan.

8. Dante bought a truck that was on sale for $32,499 in a city that has a 7% sales tax and a 2% license fee.
   (a) Find the sales tax and license fee Dante paid.
   (b) Dante also bought an extended warranty for $1600. Find the purchase price of the truck.

9. Jeremiah bought a sports car that was on sale for $44,799 in a city that has a 7% sales tax and a 2% license fee.
   (a) Find the sales tax and license fee Jeremiah paid.
   (b) Jeremiah also bought an extended warranty for $2100. Find the purchase price of the sports car.

10. Dawn bought a four-door sedan that was on sale for $31,899 in a city that has a 6% sales tax and a 2% license fee.
    (a) Find the sales tax and license fee Dawn paid.
    (b) Dawn also bought an extended warranty for $1300. Find the purchase price of the sedan.

11. The purchase price of an SUV that a soccer coach plans to buy is $49,999. In order to qualify for the loan on the SUV, the coach must make a down payment of 15% of the purchase price.
    (a) Find the down payment.
    (b) Find the amount financed.

12. The purchase price of a flat-bed truck a contractor plans to buy is $39,999. In order to qualify for the loan on the truck, the contractor must make a down payment of 10% of the purchase price.
    (a) Find the down payment.
    (b) Find the amount financed.

13. Tammy went to two dealerships to find the best deal on the truck she plans to purchase. From which dealership should Tammy buy the truck so that the *total cost* of the truck is the least expensive?

Dealership 1	Dealership 2
• Purchase price: $35,999	• Purchase price: $32,499
• Financing option: 2% financing with $5000 down payment	• Financing option: 5% financing with no down payment
• Monthly payments: $658.73 per month for 48 months	• Monthly payments: $568.73 per month for 60 months

14. John went to two dealerships to find the best deal on a luxury SUV for his company to purchase. From which dealership should John buy the SUV so that the *total cost* of the SUV is the least expensive?

Dealership 1	Dealership 2
• Purchase price: $49,999	• Purchase price: $48,499
• Financing option: 2% financing with $8000 down payment	• Financing option: 6% financing with no down payment
• Monthly payments: $892.48 per month for 48 months	• Monthly payments: $856.82 per month for 60 months

To Think About

Natasha went to three car dealerships to check the prices of the same Ford two-door coupe. The city where the dealerships are located has a sales tax of 5% and a license fee of 2%. All dealerships offer extended warranties that are 3 years/70,000 miles. Use the following information gathered by Natasha to answer exercises 15 and 16.

Dealership 1—Ford Coupe	Dealership 2—Ford Coupe	Dealership 3—Ford Coupe
• $24,999 plus $2000 rebate	• $23,799; dealer pays sales tax	• $23,999
• extended warranty $1350	• extended warranty $1450	• free extended warranty

15. (a) Which dealership offers the least expensive *purchase price?* State this amount.

(b) Each dealership offers a *different interest rate* on a 60-month loan without a down payment, resulting in the following monthly payments:

Dealership 1	Dealership 2	Dealership 3
$480.65/month	$485.46/month	$496.44/month

From which dealership should Natasha buy the Ford coupe so that the *total cost* of the car is the least expensive? State this amount.

(c) Compare the results of parts **(a)** and **(b)**. What conclusion can you make?

16. (a) Which dealership offers the most expensive *purchase price?* State this amount.

(b) Each dealership offers a *different interest rate* on a 48-month loan without a down payment, resulting in the following monthly payments:

Dealership 1	Dealership 2	Dealership 3
$589.27/month	$594.07/month	$603.07/month

From which dealership should Natasha buy the Ford coupe so that the *total cost* of the car is the most expensive? State this amount.

(c) Compare the results of parts **(a)** and **(b)**. What conclusion can you make?

Appendix B Tables

Table of Basic Addition Facts

+	0	1	2	3	4	5	6	7	8	9
0	0	1	2	3	4	5	6	7	8	9
1	1	2	3	4	5	6	7	8	9	10
2	2	3	4	5	6	7	8	9	10	11
3	3	4	5	6	7	8	9	10	11	12
4	4	5	6	7	8	9	10	11	12	13
5	5	6	7	8	9	10	11	12	13	14
6	6	7	8	9	10	11	12	13	14	15
7	7	8	9	10	11	12	13	14	15	16
8	8	9	10	11	12	13	14	15	16	17
9	9	10	11	12	13	14	15	16	17	18

Table of Basic Multiplication Facts

×	0	1	2	3	4	5	6	7	8	9	10	11	12
0	0	0	0	0	0	0	0	0	0	0	0	0	0
1	0	1	2	3	4	5	6	7	8	9	10	11	12
2	0	2	4	6	8	10	12	14	16	18	20	22	24
3	0	3	6	9	12	15	18	21	24	27	30	33	36
4	0	4	8	12	16	20	24	28	32	36	40	44	48
5	0	5	10	15	20	25	30	35	40	45	50	55	60
6	0	6	12	18	24	30	36	42	48	54	60	66	72
7	0	7	14	21	28	35	42	49	56	63	70	77	84
8	0	8	16	24	32	40	48	56	64	72	80	88	96
9	0	9	18	27	36	45	54	63	72	81	90	99	108
10	0	10	20	30	40	50	60	70	80	90	100	110	120
11	0	11	22	33	44	55	66	77	88	99	110	121	132
12	0	12	24	36	48	60	72	84	96	108	120	132	144

Table of Prime Factors

Number	Prime Factors	Number	Prime Factors	Number	Prime Factors	Number	Prime Factors
2	prime	52	$2^2 \times 13$	102	$2 \times 3 \times 17$	152	$2^3 \times 19$
3	prime	53	prime	103	prime	153	$3^2 \times 17$
4	2^2	54	2×3^3	104	$2^3 \times 13$	154	$2 \times 7 \times 11$
5	prime	55	5×11	105	$3 \times 5 \times 7$	155	5×31
6	2×3	56	$2^3 \times 7$	106	2×53	156	$2^2 \times 3 \times 13$
7	prime	57	3×19	107	prime	157	prime
8	2^3	58	2×29	108	$2^2 \times 3^3$	158	2×79
9	3^2	59	prime	109	prime	159	3×53
10	2×5	60	$2^2 \times 3 \times 5$	110	$2 \times 5 \times 11$	160	$2^5 \times 5$
11	prime	61	prime	111	3×37	161	7×23
12	$2^2 \times 3$	62	2×31	112	$2^4 \times 7$	162	2×3^4
13	prime	63	$3^2 \times 7$	113	prime	163	prime
14	2×7	64	2^6	114	$2 \times 3 \times 19$	164	$2^2 \times 41$
15	3×5	65	5×13	115	5×23	165	$3 \times 5 \times 11$
16	2^4	66	$2 \times 3 \times 11$	116	$2^2 \times 29$	166	2×83
17	prime	67	prime	117	$3^2 \times 13$	167	prime
18	2×3^2	68	$2^2 \times 17$	118	2×59	168	$2^3 \times 3 \times 7$
19	prime	69	3×23	119	7×17	169	13^2
20	$2^2 \times 5$	70	$2 \times 5 \times 7$	120	$2^3 \times 3 \times 5$	170	$2 \times 5 \times 17$
21	3×7	71	prime	121	11^2	171	$3^2 \times 19$
22	2×11	72	$2^3 \times 3^2$	122	2×61	172	$2^2 \times 43$
23	prime	73	prime	123	3×41	173	prime
24	$2^3 \times 3$	74	2×37	124	$2^2 \times 31$	174	$2 \times 3 \times 29$
25	5^2	75	3×5^2	125	5^3	175	$5^2 \times 7$
26	2×13	76	$2^2 \times 19$	126	$2 \times 3^2 \times 7$	176	$2^4 \times 11$
27	3^3	77	7×11	127	prime	177	3×59
28	$2^2 \times 7$	78	$2 \times 3 \times 13$	128	2^7	178	2×89
29	prime	79	prime	129	3×43	179	prime
30	$2 \times 3 \times 5$	80	$2^4 \times 5$	130	$2 \times 5 \times 13$	180	$2^2 \times 3^2 \times 5$
31	prime	81	3^4	131	prime	181	prime
32	2^5	82	2×41	132	$2^2 \times 3 \times 11$	182	$2 \times 7 \times 13$
33	3×11	83	prime	133	7×19	183	3×61
34	2×17	84	$2^2 \times 3 \times 7$	134	2×67	184	$2^3 \times 23$
35	5×7	85	5×17	135	$3^3 \times 5$	185	5×37
36	$2^2 \times 3^2$	86	2×43	136	$2^3 \times 17$	186	$2 \times 3 \times 31$
37	prime	87	3×29	137	prime	187	11×17
38	2×19	88	$2^3 \times 11$	138	$2 \times 3 \times 23$	188	$2^2 \times 47$
39	3×13	89	prime	139	prime	189	$3^3 \times 7$
40	$2^3 \times 5$	90	$2 \times 3^2 \times 5$	140	$2^2 \times 5 \times 7$	190	$2 \times 5 \times 19$
41	prime	91	7×13	141	3×47	191	prime
42	$2 \times 3 \times 7$	92	$2^2 \times 23$	142	2×71	192	$2^6 \times 3$
43	prime	93	3×31	143	11×13	193	prime
44	$2^2 \times 11$	94	2×47	144	$2^4 \times 3^2$	194	2×97
45	$3^2 \times 5$	95	5×19	145	5×29	195	$3 \times 5 \times 13$
46	2×23	96	$2^5 \times 3$	146	2×73	196	$2^2 \times 7^2$
47	prime	97	prime	147	3×7^2	197	prime
48	$2^4 \times 3$	98	2×7^2	148	$2^2 \times 37$	198	$2 \times 3^2 \times 11$
49	7^2	99	$3^2 \times 11$	149	prime	199	prime
50	2×5^2	100	$2^2 \times 5^2$	150	$2 \times 3 \times 5^2$	200	$2^3 \times 5^2$
51	3×17	101	prime	151	prime		

Table of Square Roots

Square Root Values Are Rounded to the Nearest Thousandth Unless the Answer Ends in .000

n	$\sqrt{n}$	n	$\sqrt{n}$	n	$\sqrt{n}$	n	$\sqrt{n}$	n	$\sqrt{n}$
1	1.000	41	6.403	81	9.000	121	11.000	161	12.689
2	1.414	42	6.481	82	9.055	122	11.045	162	12.728
3	1.732	43	6.557	83	9.110	123	11.091	163	12.767
4	2.000	44	6.633	84	9.165	124	11.136	164	12.806
5	2.236	45	6.708	85	9.220	125	11.180	165	12.845
6	2.449	46	6.782	86	9.274	126	11.225	166	12.884
7	2.646	47	6.856	87	9.327	127	11.269	167	12.923
8	2.828	48	6.928	88	9.381	128	11.314	168	12.961
9	3.000	49	7.000	89	9.434	129	11.358	169	13.000
10	3.162	50	7.071	90	9.487	130	11.402	170	13.038
11	3.317	51	7.141	91	9.539	131	11.446	171	13.077
12	3.464	52	7.211	92	9.592	132	11.489	172	13.115
13	3.606	53	7.280	93	9.644	133	11.533	173	13.153
14	3.742	54	7.348	94	9.695	134	11.576	174	13.191
15	3.873	55	7.416	95	9.747	135	11.619	175	13.229
16	4.000	56	7.483	96	9.798	136	11.662	176	13.266
17	4.123	57	7.550	97	9.849	137	11.705	177	13.304
18	4.243	58	7.616	98	9.899	138	11.747	178	13.342
19	4.359	59	7.681	99	9.950	139	11.790	179	13.379
20	4.472	60	7.746	100	10.000	140	11.832	180	13.416
21	4.583	61	7.810	101	10.050	141	11.874	181	13.454
22	4.690	62	7.874	102	10.100	142	11.916	182	13.491
23	4.796	63	7.937	103	10.149	143	11.958	183	13.528
24	4.899	64	8.000	104	10.198	144	12.000	184	13.565
25	5.000	65	8.062	105	10.247	145	12.042	185	13.601
26	5.099	66	8.124	106	10.296	146	12.083	186	13.638
27	5.196	67	8.185	107	10.344	147	12.124	187	13.675
28	5.292	68	8.246	108	10.392	148	12.166	188	13.711
29	5.385	69	8.307	109	10.440	149	12.207	189	13.748
30	5.477	70	8.367	110	10.488	150	12.247	190	13.784
31	5.568	71	8.426	111	10.536	151	12.288	191	13.820
32	5.657	72	8.485	112	10.583	152	12.329	192	13.856
33	5.745	73	8.544	113	10.630	153	12.369	193	13.892
34	5.831	74	8.602	114	10.677	154	12.410	194	13.928
35	5.916	75	8.660	115	10.724	155	12.450	195	13.964
36	6.000	76	8.718	116	10.770	156	12.490	196	14.000
37	6.083	77	8.775	117	10.817	157	12.530	197	14.036
38	6.164	78	8.832	118	10.863	158	12.570	198	14.071
39	6.245	79	8.888	119	10.909	159	12.610	199	14.107
40	6.325	80	8.944	120	10.954	160	12.649	200	14.142

Solutions to Student Practice

Chapter 1   1.1 Student Practice

1. (a) $3182 = 3000 + 100 + 80 + 2$
   (b) $520{,}890 = 500{,}000 + 20{,}000 + 800 + 90$
   (c) $709{,}680{,}059 = 700{,}000{,}000 + 9{,}000{,}000 + 600{,}000$
                                  $+ 80{,}000 + 50 + 9$

2. (a) 492   (b) 80,427

3. (a) 7   (b) 9   (c) 4000
   (d) 900,000 for the first 9; 9 for the last 9

4. two hundred sixty-seven million, three hundred fifty-eight thousand, nine hundred eighty-one

5. (a) two thousand, seven hundred thirty-six
   (b) nine hundred eighty thousand, three hundred six
   (c) twelve million, twenty-one

6. The estimated world population on July 16, 2010, was six billion, eight hundred fifty-six million, three hundred forty thousand, one hundred eighty-nine.

7. (a) 803   (b) 30,229

8. (a) 13,000   (b) 88,000   (c) 10,000

1.2 Student Practice

1. (a) $6+5=11$ (b) $9+4=13$ (c) $3+0=3$

2.
```
7
6     7 + 6 = 13
5         13 + 5 = 18
8             18 + 8 = 26
+2                26 + 2 = 28
28
```

3.
```
1
7    10
2
9    10
+3
22
```

4.
```
  8246
+ 1702
  9948
```

5.
```
  ¹56
+ 36
  92
```

6.
```
 ²¹789
   63
+ 297
 1149
```

7. (a)
```
 ¹¹¹127
  9876
+  342
 10,345
```
(b) Check by adding in opposite order.
```
 ¹¹¹342
  9876
+  127
 10,345
```
same

8.
```
 ¹²¹²18,316
  24,789
+ 22,965
  66,070 total women
```

9.
```
   1000
   2000
   1000
+  2000
   6000 ft
```

1.3 Student Practice

1. (a) $9-6=3$ (b) $12-5=7$ (c) $17-8=9$ (d) $14-0=14$ (e) $18-9=9$

2.
```
  7695
- 3481
  4214
```

3.
```
  ²3̸ ¹⁴4̸
-  1 6
   1 8
```

4.
```
  6 ⁸9̸ ¹³3̸
- 4 2 6
  2 6 7
```

5.
```
 ⁸9̸ ¹⁰0̸ ⁶7̸ ¹⁰0̸
- 5 8 8 6
  3 1 8 4
```

6. (a)
```
  8964
-  985
  7979
```
(b)
```
  50,000
- 32,508
  17,492
```

7. Subtraction
```
  9763
- 5732
  4031
```
[IT CHECKS] Checking by addition
```
  5732
+ 4031
  9763
```

8. (a)
```
  284,000
-  96,327
  187,673
```
[IT CHECKS] Checking by addition
```
   96,327
+ 187,673
  284,000
```
(b)
```
  8,526,024
- 6,397,518
  2,128,506
```
[IT CHECKS] Checking by addition
```
  6,397,518
+ 2,128,506
  8,526,024
```

9. (a) $17 = 12 + x$
$17 - 12 = x$
$5 = x$
5 vessels left in the afternoon.
(b) $22 = 10 + x$
$22 - 10 = x$
$12 = x$
12 hikers were still on the mountain.

10. (a)
```
  42,206,743
- 28,634,896
  13,571,847
```
(b)
```
  14,227,799
- 11,198,655
   3,029,144
```

11. (a) From the bar graph:
2010 sales   114
2009 sales   − 78
Sales increase   36
(b) From the bar graph:
Springfield   91
Riverside   − 78
        13 more homes
(c) 2010 sales   271
2009 sales   − 240
        31
2011 sales   284
2010 sales   − 271
        13
Therefore, the greatest increase in sales occurred between 2009 and 2010.

1.4 Student Practice

1. (a) $\begin{array}{r} 8 \\ \times\ 8 \\ \hline 64 \end{array}$ **(b)** $\begin{array}{r} 7 \\ \times\ 6 \\ \hline 42 \end{array}$ **(c)** $\begin{array}{r} 5 \\ \times\ 8 \\ \hline 40 \end{array}$ **(d)** $\begin{array}{r} 9 \\ \times\ 7 \\ \hline 63 \end{array}$ **(e)** $\begin{array}{r} 9 \\ \times\ 9 \\ \hline 81 \end{array}$

2. $\begin{array}{r} 3021 \\ \times\ 3 \\ \hline 9063 \end{array}$

3. $\begin{array}{r} \overset{2}{4}3 \\ \times\ 8 \\ \hline 344 \end{array}$

4. $\begin{array}{r} \overset{5\,6}{5}79 \\ \times\ 7 \\ \hline 4053 \end{array}$

5. (a) $1267 \times 10 = 12{,}670$ (one zero)
   (b) $1267 \times 1000 = 1{,}267{,}000$ (three zeros)
   (c) $1267 \times 10{,}000 = 12{,}670{,}000$ (four zeros)
   (d) $1267 \times 1{,}000{,}000 = 1{,}267{,}000{,}000$ (six zeros)

6. (a) $9 \times 60{,}000 = 9 \times 6 \times 10{,}000 = 54 \times 10{,}000 = 540{,}000$
   (b) $15 \times 400 = 15 \times 4 \times 100 = 60 \times 100 = 6000$
   (c) $270 \times 800 = 27 \times 8 \times 10 \times 100 = 216 \times 1000 = 216{,}000$

7. $\begin{array}{r} 323 \\ \times\ 32 \\ \hline 646 \\ 9690 \\ \hline 10{,}336 \end{array}$

8. $\begin{array}{r} 385 \\ \times\ 69 \\ \hline 3465 \\ 23100 \\ \hline 26{,}565 \end{array}$

9. $\begin{array}{r} 34 \\ \times\ 20 \\ \hline 0 \\ 680 \\ \hline 680 \end{array}$

10. $\begin{array}{r} 130 \\ \times\ 50 \\ \hline 0 \\ 6500 \\ \hline 6500 \end{array}$

11. $\begin{array}{r} 923 \\ \times\ 675 \\ \hline 4615 \\ 6461 \\ 5538 \\ \hline 623{,}025 \end{array}$

12. $25 \times 4 \times 17 = (25 \times 4) \times 17 = 100 \times 17 = 1700$

13. $8 \times 4 \times 3 \times 25 = 8 \times 3 \times 4 \times 25$
   $= 8 \times 3 \times (4 \times 25)$
   $= 24 \times 100$
   $= 2400$

14. $\begin{array}{r} 17348 \\ \times\ 378 \\ \hline 138\ 784 \\ 1\ 214\ 36 \\ 5\ 204\ 4 \\ \hline 6{,}557{,}544 \end{array}$
   The total sales of cars was $6,557,544.

15. Area = 5 yards × 7 yards = 35 square yards

1.5 Student Practice

1. (a) $4)\overline{36}$ with quotient 9 **(b)** $5)\overline{25}$ with quotient 5 **(c)** $9)\overline{72}$ with quotient 8 **(d)** $6)\overline{30}$ with quotient 5

2. (a) $\dfrac{7}{1} = 7$ **(b)** $\dfrac{9}{9} = 1$ **(c)** $\dfrac{0}{5} = 0$ **(d)** $\dfrac{12}{0}$ cannot be done

3. $\begin{array}{r} 7\ R\ 3 \\ 6)\overline{45} \\ 42 \\ \hline 3 \end{array}$ **Check** $\begin{array}{r} 6 \\ \times\ 7 \\ \hline 42 \\ +\ 3 \\ \hline 45 \end{array}$

4. $\begin{array}{r} 21\ R\ 3 \\ 6)\overline{129} \\ 12 \\ \hline 9 \\ 6 \\ \hline 3 \end{array}$ **Check** $\begin{array}{r} 21 \\ \times\ 6 \\ \hline 126 \\ +\ 3 \\ \hline 129 \end{array}$

5. $\begin{array}{r} 529\ R\ 5 \\ 8)\overline{4237} \\ 40 \\ \hline 23 \\ 16 \\ \hline 77 \\ 72 \\ \hline 5 \end{array}$

6. $\begin{array}{r} 7\ R\ 19 \\ 32)\overline{243} \\ 224 \\ \hline 19 \end{array}$

7. $\begin{array}{r} 1278\ R\ 9 \\ 33)\overline{42183} \\ 33 \\ \hline 91 \\ 66 \\ \hline 258 \\ 231 \\ \hline 273 \\ 264 \\ \hline 9 \end{array}$

8. $\begin{array}{r} 25\ R\ 27 \\ 128)\overline{3227} \\ 256 \\ \hline 667 \\ 640 \\ \hline 27 \end{array}$

9. $\begin{array}{r} 16{,}852 \\ 7)\overline{117{,}964} \end{array}$ **Check** $\begin{array}{r} 16{,}852 \\ \times\ 7 \\ \hline 117{,}964 \end{array}$
   The cost of one car is $16,852.

10. $\begin{array}{r} 367 \\ 14)\overline{5138} \end{array}$ The average speed was 367 mph.

1.6 Student Practice

1. (a) $12 \times 12 \times 12 \times 12 = 12^4$
   (b) $2 \times 2 \times 2 \times 2 \times 2 \times 2 = 2^6$

2. (a) $12^2 = 12 \times 12 = 144$
   (b) $6^3 = 6 \times 6 \times 6 = 216$
   (c) $2^6 = 2 \times 2 \times 2 \times 2 \times 2 \times 2 = 64$
   (d) $1^{10} = 1 \times 1 \times 1 \times 1 \times 1 \times 1 \times 1 \times 1 \times 1 \times 1 = 1$

3. (a) $7^3 + 8^2 = (7)(7)(7) + (8)(8) = 343 + 64 = 407$
(b) $9^2 + 6^0 = (9)(9) + 1 = 81 + 1 = 82$
(c) $5^4 + 5 = (5)(5)(5)(5) + 5 = 625 + 5 = 630$

4. $7 + 4^3 \times 3 = 7 + 64 \times 3$ Exponents
$ = 7 + 192$ Multiply
$ = 199$ Add

5. $37 - 20 \div 5 + 2 - 3 \times 4$
$= 37 - 4 + 2 - 3 \times 4$ Divide
$= 37 - 4 + 2 - 12$ Multiply
$= 33 + 2 - 12$ Subtract
$= 35 - 12$ Add
$= 23$ Subtract

6. $4^3 - 2 + 3^2$
$= 4 \times 4 \times 4 - 2 + 3 \times 3$ Evaluate exponents.
$= 64 - 2 + 9$ $4^3 = 64$ and $3^2 = 9$.
$= 62 + 9$ Subtract
$= 71$ Add

7. $(17 + 7) \div 6 \times 2 + 7 \times 3 - 4$
$= 24 \div 6 \times 2 + 7 \times 3 - 4$ Combine inside parentheses
$= 4 \times 2 + 7 \times 3 - 4$ Divide
$= 8 + 7 \times 3 - 4$ Multiply
$= 8 + 21 - 4$ Multiply
$= 29 - 4$ Add
$= 25$ Subtract

8. $5^2 - 6 \div 2 + 3^4 + 7 \times (12 - 10)$
$= 5^2 - 6 \div 2 + 3^4 + 7 \times 2$ Combine inside parentheses
$= 25 - 6 \div 2 + 81 + 7 \times 2$ Exponents
$= 25 - 3 + 81 + 7 \times 2$ Divide
$= 25 - 3 + 81 + 14$ Multiply
$= 22 + 81 + 14$ Subtract
$= 103 + 14$ Add
$= 117$ Add

1.7 Student Practice

1. $6\,5\,,5\,2\,8$ Locate the thousands round-off place.

$6\,5\,,(5)\,2\,8$ The first digit to the right is 5 or more. We will increase the thousands digit by 1.

$6\,6\,,0\,0\,0$ All digits to the right of the thousands place are replaced by zero.

2. $1\,7(2),9\,6\,3 = 170,000$ to the nearest ten thousand.

1.7 Student Practice (continued)

3. (a) $5\,3\,,2\,8\,2 = 53,280$ to the nearest ten. The digit to the right of the tens place was less than 5.

(b) $1\,6\,4\,,4\,8\,5 = 164,000$ to the nearest thousand. The digit to the right of the thousands place was less than 5.

(c) $1\,,3\,6\,5\,,2\,7\,3 = 1,400,000$ to the nearest hundred thousand. The digit to the right of the hundred thousands place was greater than 5.

4. (a) $9\,3\,5\,,6\,8\,2 = 936,000$ to the nearest thousand. The digit to the right of the thousands place is greater than 5.

(b) $9\,3\,5\,,6\,8\,2 = 900,000$ to the nearest hundred thousand. The digit to the right of the hundred thousands place is less than 5.

(c) $9\,3\,5\,,6\,8\,2 = 1,000,000$ to the nearest million. The digit to the right of the millions place is greater than 5.

5. $9,460,000,000,000,000$ meters $= 9,500,000,000,000,000$ meters to the nearest hundred trillion meters.

6.
Actual Sum	Estimated Sum
3456	3000
9876	10000
5421	5000
+ 1278	+ 1000
20,031	19,000 Close to the actual sum

7.
697	700
35	40
+ 19	+ 20
	760

We estimate that the total cost is $760. (The exact answer is $751, so we can see that our answer is quite close.)

8. Estimate: $10,000 + 10,000 + 20,000 + 60,000 = 100,000$
This is significantly different from 81,358, so we would suspect that an error has been made. In fact, Ming did make an error. The exact sum is actually 101,358!

9. Estimate: $40,000,000 - 20,000,000 = 20,000,000$
We estimate that 20,000,000 more people lived in California than in Texas.

10. Estimate: $9000 \times 7000 = 63,000,000$
We estimate the product to be 63,000,000.

11. $\frac{2000}{40)\overline{80,000}}$ Our estimate is 2000.

12. $\frac{33,333 \text{ R } 20}{60)\overline{2,000,000}}$ Our estimate is $33,333 for one truck.

1.8 Student Practice

Student Practice 1

1. Understand the problem.

Mathematics Blueprint for Problem Solving

Gather the Facts	What Am I Asked to Do?	How Do I Proceed?	Key Points to Remember
The deductions are $135, $28, $13, and $34.	Find out the total amount of deductions.	I must add the four deductions to obtain the total.	Watch out! Gross pay of $1352 is not needed to solve the problem.

2. Solve and state the answer:
$135 + 28 + 13 + 34 = 210$
The total amount taken out of Diane's paycheck is $210.

3. *Check.* Estimate to see if the answer is reasonable.

Student Practice 2

1. Understand the problem.

Mathematics Blueprint for Problem Solving

Gather the Facts	What Am I Asked to Do?	How Do I Proceed?	Key Points to Remember
Obama had 69,456,897 votes. McCain had 59,934,814 votes.	Find out by how many votes Obama beat McCain.	I must subtract the amounts.	Obama is a Democrat and McCain is a Republican.

2. Solve and state the answer:

$$\begin{array}{r} 69,456,897 \\ -\ 59,934,814 \\ \hline 9,522,083 \end{array}$$

Obama beat McCain by 9,522,083 votes.

3. *Check.* Estimate to see if the answer is reasonable.

Student Practice 3

1. Understand the problem.

Mathematics Blueprint for Problem Solving

Gather the Facts	What Am I Asked to Do?	How Do I Proceed?	Key Points to Remember
1 gallon is 1024 fluid drams.	Find out how many fluid drams in 9 gallons.	I need to multiply 1024 by 9.	I must use fluid drams as the unit of measure in my answer.

2. Solve and state the answer:

$$\begin{array}{r} 1024 \\ \times\ \ \ 9 \\ \hline 9216 \end{array}$$

There are 9216 fluid drams in 9 gallons.

3. *Check.* Estimate to see if the answer is reasonable.

Student Practice 4

1. Understand the problem.

Mathematics Blueprint for Problem Solving

Gather the Facts	What Am I Asked to Do?	How Do I Proceed?	Key Points to Remember
Donna bought 45 shares of stock. She paid $1620 for them.	Find out the cost per share of stock.	I need to divide 1620 by 45.	Use dollars as the unit of measure in the answer.

2. Solve and state the answer:

$$\begin{array}{r} 36 \\ 45\overline{)1620} \\ \underline{135} \\ 270 \\ \underline{270} \\ 0 \end{array}$$

Donna paid $36 per share for the stock.

3. *Check.* Estimate to see if the answer is reasonable.

Student Practice 5

1. Understand the problem. We will make an imaginary bill of sale.

2. Solve and state the answer. We do the calculation and enter the results in the bill of sale.

Customer: Anderson Dining Commons

Quantity	Item	Cost per Item	Amount for This Item
50	Tables	$200	$10,000 (50 × $200 = $10,000)
180	Chairs	$ 40	$ 7200 (180 × $40 = $7200)
6	Moving Carts	$ 65	$390 (6 × $65 = $390)
		Total	$17,590 (sum of the three amounts)

The total cost of purchase was $17,590.

3. *Check.* Estimate to see if the answer is reasonable.

Student Practice 6

1. Understand the problem.

Mathematics Blueprint for Problem Solving

Gather the Facts	What Am I Asked to Do?	How Do I Proceed?	Key Points to Remember
Old balance: $498 New deposits: $607 $163 Interest: $36 Withdrawals: $19 $158 $582 $74	Find her new balance after the transactions.	**(a)** Add the new deposits and interest to the old balance. **(b)** Add the withdrawals. **(c)** Subtract the results of steps (a) and (b).	Deposits and interest are added and withdrawals are subtracted from savings accounts.

2. Solve and state the answer:

$$\textbf{(a)} \quad \begin{array}{r} 498 \\ 607 \\ 163 \\ +\ 36 \\ \hline 1304 \end{array} \qquad \textbf{(b)} \quad \begin{array}{r} 19 \\ 158 \\ 582 \\ +\ 74 \\ \hline 833 \end{array} \qquad \textbf{(c)} \quad \begin{array}{r} 1304 \\ -\ 833 \\ \hline 471 \end{array}$$

Her balance this month is $471.

3. *Check.* Estimate to see if the answer is reasonable.

Student Practice 7

1. Understand the problem.

Mathematics Blueprint for Problem Solving

Gather the Facts	What Am I Asked to Do?	How Do I Proceed?	Key Points to Remember
Odometer reading at end of trip: 51,118 miles Odometer reading at start of trip: 50,698 miles Used on trip: 12 gallons of gas	Find the number of miles per gallon that the car obtained on the trip.	**(a)** Subtract the two odometer readings. **(b)** Divide that number by 12.	The gas tank was full at the beginning of the trip. 12 gallons fills the tank at the end of the trip.

2. Solve and state the answer:

$$\begin{array}{r} 51{,}118 \\ -\ 50{,}698 \\ \hline 420 \end{array}$$ odometer at end of trip
odometer at start of trip
miles traveled on trip

$$\frac{420 \text{ miles}}{12 \text{ gallons of gas used}} = 12\overline{)420} \quad \begin{array}{r} 35 \\ \underline{36} \\ 60 \\ \underline{60} \\ 0 \end{array}$$

Deidre obtained 35 miles per gallon on the trip.

3. *Check.* Estimate to see if the answer is reasonable.

Chapter 2   2.1 Student Practice

1. (a) Four parts out of twelve are shaded. The fraction is $\frac{4}{12}$.

   (b) Three parts out of six are shaded. The fraction is $\frac{3}{6}$.

   (c) Two parts out of three are shaded. The fraction is $\frac{2}{3}$.

2. (a) Shade $\frac{4}{5}$ of the object.

   (b) Shade $\frac{3}{7}$ of the group.

3. (a) $\frac{9}{17}$ represents 9 players out of 17.

   (b) The total class is $382 + 351 = 733$.

   The fractional part that is men is $\frac{382}{733}$.

   (c) $\frac{7}{8}$ of a yard of material.

4. Total number of defective items: $1 + 2 = 3$. Total number of items: $7 + 9 = 16$. A fraction that represents the portion of the items that were defective is $\frac{3}{16}$.

2.2 Student Practice

1. (a) $18 = 2 \times 9$
   $= 2 \times 3 \times 3$
   $= 2 \times 3^2$

   (b) $72 = 8 \times 9$
   $= 2 \times 2 \times 2 \times 3 \times 3$
   $= 2^3 \times 3^2$

   (c) $400 = 10 \times 40$
   $= 5 \times 2 \times 5 \times 8$
   $= 5 \times 2 \times 5 \times 2 \times 2 \times 2$
   $= 2^4 \times 5^2$

2. (a) $\frac{30}{42} = \frac{30 \div 6}{42 \div 6} = \frac{5}{7}$

   (b) $\frac{60}{132} = \frac{60 \div 12}{132 \div 12} = \frac{5}{11}$

3. (a) $\frac{120}{135} = \frac{2 \times 2 \times 2 \times \cancel{3} \times \cancel{5}}{3 \times 3 \times \cancel{3} \times \cancel{5}} = \frac{8}{9}$

   (b) $\frac{715}{880} = \frac{\cancel{5} \times \cancel{11} \times 13}{2 \times 2 \times 2 \times 2 \times \cancel{5} \times \cancel{11}} = \frac{13}{16}$

4. (a) $\frac{84}{108} \overset{?}{=} \frac{7}{9}$ (b) $\frac{3}{7} \overset{?}{=} \frac{79}{182}$

   $84 \times 9 \overset{?}{=} 108 \times 7$ $3 \times 182 \overset{?}{=} 7 \times 79$

   $756 = 756$ Yes $546 \neq 553$ No

2.3 Student Practice

1. (a) $4\frac{3}{7} = \frac{4 \times 7 + 3}{7} = \frac{28 + 3}{7} = \frac{31}{7}$

   (b) $6\frac{2}{3} = \frac{6 \times 3 + 2}{3} = \frac{18 + 2}{3} = \frac{20}{3}$

   (c) $19\frac{4}{7} = \frac{19 \times 7 + 4}{7} = \frac{133 + 4}{7} = \frac{137}{7}$

2. (a) $4\overline{)17}$ so $\frac{17}{4} = 4\frac{1}{4}$
   $\underline{16}$
   1

   (b) $5\overline{)36}$ so $\frac{36}{5} = 7\frac{1}{5}$
   $\underline{35}$
   1

   (c) $27\overline{)116}$ so $\frac{116}{27} = 4\frac{8}{27}$
   $\underline{108}$
   8

   (d) $13\overline{)91}$ so $\frac{91}{13} = 7$
   $\underline{91}$
   0

3. $\frac{51}{15} = \frac{\overset{1}{\cancel{3}} \times 17}{\cancel{3} \times 5} = \frac{17}{5}$

4. $\frac{16}{80} = \frac{1}{5}$ so $3\frac{16}{80} = 3\frac{1}{5}$.

5. $\frac{1001}{572} = 1\frac{429}{572}$

   Now the fraction $\frac{429}{572} = \frac{3 \times \overset{1}{\cancel{11}} \times \overset{1}{\cancel{13}}}{2 \times 2 \times \cancel{11} \times \cancel{13}} = \frac{3}{4}$.

   Thus $\frac{1001}{572} = 1\frac{429}{572} = 1\frac{3}{4}$.

2.4 Student Practice

1. (a) $\frac{6}{7} \times \frac{3}{13} = \frac{6 \times 3}{7 \times 13} = \frac{18}{91}$

   (b) $\frac{1}{5} \times \frac{11}{12} = \frac{1 \times 11}{5 \times 12} = \frac{11}{60}$

2. $\frac{55}{72} \times \frac{16}{33} = \frac{5 \cdot 11}{2 \cdot 2 \cdot 2 \cdot 3 \cdot 3} \times \frac{2 \cdot 2 \cdot 2 \cdot 2}{3 \cdot 11}$
   $= \frac{\cancel{2} \cdot \cancel{2} \cdot \cancel{2} \cdot 2 \cdot 5 \cdot \cancel{11}}{\cancel{2} \cdot \cancel{2} \cdot \cancel{2} \cdot 3 \cdot 3 \cdot 3 \cdot \cancel{11}}$
   $= \frac{10}{27}$

3. (a) $7 \times \frac{5}{13} = \frac{7}{1} \times \frac{5}{13} = \frac{35}{13}$ or $2\frac{9}{13}$

   (b) $\frac{13}{4} \times 8 = \frac{13}{\cancel{4}} \times \frac{\overset{2}{\cancel{8}}}{1} = \frac{26}{1} = 26$

4. $\frac{3}{\cancel{8}} \times \overset{12,300}{\cancel{98,400}} = \frac{3}{1} \times 12,300 = 36,900$

   There are 36,900 square feet in the wetland area.

5. (a) $2\frac{1}{6} \times \frac{4}{7} = \frac{13}{\cancel{6}} \times \frac{\overset{2}{\cancel{4}}}{7} = \frac{26}{21}$ or $1\frac{5}{21}$

   (b) $10\frac{2}{3} \times 13\frac{1}{2} = \frac{\overset{16}{\cancel{32}}}{\cancel{3}} \times \frac{\overset{9}{\cancel{27}}}{\cancel{2}} = \frac{144}{1} = 144$

   (c) $\frac{3}{5} \times 1\frac{1}{3} \times \frac{5}{8} = \frac{\overset{1}{\cancel{3}}}{\cancel{5}} \times \frac{\overset{1}{\cancel{4}}}{\cancel{3}} \times \frac{\overset{1}{\cancel{5}}}{\cancel{8}} = \frac{1}{2}$

   (d) $3\frac{1}{5} \times 2\frac{1}{2} = \frac{\overset{8}{\cancel{16}}}{\cancel{5}} \times \frac{\overset{1}{\cancel{5}}}{\cancel{2}} = \frac{8}{1} = 8$

6. Area $= 1\frac{1}{5} \times 4\frac{5}{6} = \frac{\overset{1}{\cancel{6}}}{5} \times \frac{29}{\cancel{6}} = \frac{29}{5} = 5\frac{4}{5}$

   The area is $5\frac{4}{5}$ square meters.

7. Since $8 \cdot 10 = 80$ and $9 \cdot 9 = 81$,

   we know that $\frac{8}{9} \cdot \frac{10}{9} = \frac{80}{81}$.

   Therefore $x = \frac{10}{9}$.

2.5 Student Practice

1. (a) $\frac{7}{13} \div \frac{3}{4} = \frac{7}{13} \times \frac{4}{3} = \frac{28}{39}$

   (b) $\frac{16}{35} \div \frac{24}{25} = \frac{\overset{2}{\cancel{16}}}{\cancel{35}} \times \frac{\overset{5}{\cancel{25}}}{\cancel{24}} = \frac{10}{21}$

2. (a) $\dfrac{3}{17} \div 6 = \dfrac{3}{17} \div \dfrac{6}{1} = \dfrac{3}{17} \times \dfrac{1}{\cancel{6}_2} = \dfrac{1}{34}$

   (b) $14 \div \dfrac{7}{15} = \dfrac{14}{1} \div \dfrac{7}{15} = \dfrac{\cancel{14}^2}{1} \times \dfrac{15}{\cancel{7}_1} = 30$

3. (a) $1 \div \dfrac{11}{13} = \dfrac{1}{1} \times \dfrac{13}{11} = \dfrac{13}{11}$ or $1\dfrac{2}{11}$

   (b) $\dfrac{14}{17} \div 1 = \dfrac{14}{17} \times \dfrac{1}{1} = \dfrac{14}{17}$

   (c) $\dfrac{3}{11} \div 0$ Division by zero is undefined.

   (d) $0 \div \dfrac{9}{16} = \dfrac{0}{1} \times \dfrac{16}{9} = \dfrac{0}{9} = 0$

4. (a) $1\dfrac{1}{5} \div \dfrac{7}{10} = \dfrac{6}{5} \div \dfrac{7}{10} = \dfrac{6}{\cancel{5}_1} \times \dfrac{\cancel{10}^2}{7} = \dfrac{12}{7}$ or $1\dfrac{5}{7}$

   (b) $2\dfrac{1}{4} \div 1\dfrac{7}{8} = \dfrac{9}{4} \div \dfrac{15}{8} = \dfrac{\cancel{9}^3}{\cancel{4}_1} \times \dfrac{\cancel{8}^2}{\cancel{15}_5} = \dfrac{6}{5}$ or $1\dfrac{1}{5}$

5. (a) $\dfrac{5\dfrac{2}{3}}{7} = 5\dfrac{2}{3} \div 7 = \dfrac{17}{3} \times \dfrac{1}{7} = \dfrac{17}{21}$

   (b) $\dfrac{1\dfrac{2}{5}}{2\dfrac{1}{3}} = 1\dfrac{2}{5} \div 2\dfrac{1}{3} = \dfrac{7}{5} \div \dfrac{7}{3} = \dfrac{\cancel{7}}{5} \times \dfrac{3}{\cancel{7}_1} = \dfrac{3}{5}$

6. $x \div \dfrac{3}{2} = \dfrac{22}{36}$

   $x \cdot \dfrac{2}{3} = \dfrac{22}{36}$

   $\dfrac{11}{12} \cdot \dfrac{2}{3} = \dfrac{22}{36}$ Thus $x = \dfrac{11}{12}$.

7. $19\dfrac{1}{4} \div 14 = \dfrac{\cancel{77}^{11}}{4} \times \dfrac{1}{\cancel{14}_2} = \dfrac{11}{8}$ or $1\dfrac{3}{8}$

   Each piece will be $1\dfrac{3}{8}$ feet long.

2.6 Student Practice

1. The multiples of 14 are 14, 28, 42, 56, 70, 84, . . .
   The multiples of 21 are 21, 42, 63, 84, 105, 126, . . .
   42 is the least common multiple of 14 and 21.
2. The multiples of 10 are 10, 20, 30, 40 . . .
   The multiples of 15 are 15, 30, 45 . . .
   30 is the least common multiple of 10 and 15.
3. 54 is a multiple of 6. We know that $6 \times 9 = 54$.
   The least common multiple of 6 and 54 is 54.
4. (a) The LCD of $\dfrac{3}{4}$ and $\dfrac{11}{12}$ is 12.
      12 can be divided by 4 and 12.
   (b) The LCD of $\dfrac{1}{7}$ and $\dfrac{8}{35}$ is 35.
      35 can be divided by 7 and 35.
5. The LCD of $\dfrac{3}{7}$ and $\dfrac{5}{6}$ is 42.
   42 can be divided by 7 and 6.
6. (a) $14 = 2 \times 7$
      $10 = 2 \times 5$
      LCD $= 2 \times 5 \times 7 = 70$
   (b) $15 = 3 \times 5$
      $50 = 2 \times 5 \times 5$
      LCD $= 2 \times 3 \times 5 \times 5 = 150$
   (c) $16 = 2 \times 2 \times 2 \times 2$
      $12 = 2 \times 2 \times 3$
      LCD $= 2 \times 2 \times 2 \times 2 \times 3 = 48$

7. $49 = 7 \times 7$
   $21 = 7 \times 3$
   $7 = 7 \times 1$
   LCD $= 7 \times 7 \times 3 = 147$
8. (a) $\dfrac{3}{5} = \dfrac{3}{5} \times \dfrac{8}{8} = \dfrac{24}{40}$ (c) $\dfrac{2}{7} = \dfrac{2}{7} \times \dfrac{4}{4} = \dfrac{8}{28}$

   (b) $\dfrac{7}{11} = \dfrac{7}{11} \times \dfrac{4}{4} = \dfrac{28}{44}$ $\dfrac{3}{4} = \dfrac{3}{4} \times \dfrac{7}{7} = \dfrac{21}{28}$

9. (a) $20 = 2 \times 2 \times 5$
      $15 = 3 \times 5$
      LCD $= 2 \times 2 \times 3 \times 5 = 60$
   (b) $\dfrac{3}{20} = \dfrac{3}{20} \times \dfrac{3}{3} = \dfrac{9}{60}$ $\dfrac{11}{15} = \dfrac{11}{15} \times \dfrac{4}{4} = \dfrac{44}{60}$

10. (a) $64 = 2 \times 2 \times 2 \times 2 \times 2 \times 2$
      $80 = 2 \times 2 \times 2 \times 2 \times 5$
      LCD $= 2 \times 2 \times 2 \times 2 \times 2 \times 2 \times 5 = 320$
   (b) $\dfrac{5}{64} = \dfrac{5}{64} \times \dfrac{5}{5} = \dfrac{25}{320}$

      $\dfrac{3}{80} = \dfrac{3}{80} \times \dfrac{4}{4} = \dfrac{12}{320}$

2.7 Student Practice

1. $\dfrac{3}{17} + \dfrac{12}{17} = \dfrac{15}{17}$

2. (a) $\dfrac{1}{12} + \dfrac{5}{12} = \dfrac{6}{12} = \dfrac{1}{2}$

   (b) $\dfrac{13}{15} + \dfrac{7}{15} = \dfrac{20}{15} = \dfrac{4}{3}$ or $1\dfrac{1}{3}$

3. (a) $\dfrac{5}{19} - \dfrac{2}{19} = \dfrac{3}{19}$ (b) $\dfrac{21}{25} - \dfrac{6}{25} = \dfrac{15}{25} = \dfrac{3}{5}$

4. LCD $= 15$ $\dfrac{2}{15} \qquad = \qquad \dfrac{2}{15}$

   $+ \dfrac{1}{5} \times \dfrac{3}{3} = \qquad + \dfrac{3}{15}$
   $\rule{5cm}{0.4pt}$
   $\qquad\qquad\qquad \dfrac{5}{15} = \dfrac{1}{3}$

5. LCD $= 48$ $\dfrac{5}{12} \times \dfrac{4}{4} = \dfrac{20}{48}$ $\dfrac{5}{16} \times \dfrac{3}{3} = \dfrac{15}{48}$

   $\dfrac{5}{12} + \dfrac{5}{16} = \dfrac{20}{48} + \dfrac{15}{48} = \dfrac{35}{48}$

6. LCD $= 48$

   $\dfrac{3}{16} \times \dfrac{3}{3} = \dfrac{9}{48}$ $\dfrac{1}{8} \times \dfrac{6}{6} = \dfrac{6}{48}$ $\dfrac{1}{12} \times \dfrac{4}{4} - \dfrac{4}{48}$

   $\dfrac{3}{16} + \dfrac{1}{8} + \dfrac{1}{12} = \dfrac{9}{48} + \dfrac{6}{48} + \dfrac{4}{48} = \dfrac{19}{48}$

7. LCD $= 96$ $\dfrac{9}{48} \times \dfrac{2}{2} = \dfrac{18}{96}$ $\dfrac{5}{32} \times \dfrac{3}{3} = \dfrac{15}{96}$

   $\dfrac{9}{48} - \dfrac{5}{32} = \dfrac{18}{96} - \dfrac{15}{96} = \dfrac{3}{96} = \dfrac{1}{32}$

8. LCD $= 20$ $\dfrac{9}{10} \times \dfrac{2}{2} = \dfrac{18}{20}$ $\dfrac{1}{4} \times \dfrac{5}{5} = \dfrac{5}{20}$

   $\dfrac{9}{10} - \dfrac{1}{4} = \dfrac{18}{20} - \dfrac{5}{20} = \dfrac{13}{20}$

   There is $\dfrac{13}{20}$ gallon left.

9. The LCD of $\dfrac{3}{10}$ and $\dfrac{23}{25}$ is 50.

   $\dfrac{3}{10} \times \dfrac{5}{5} = \dfrac{15}{50}$ Now rewriting: $x + \dfrac{15}{50} = \dfrac{46}{50}$

   $\dfrac{23}{25} \times \dfrac{2}{2} = \dfrac{46}{50}$ $\dfrac{31}{50} + \dfrac{15}{50} = \dfrac{46}{50}$

   So, $x = \dfrac{31}{50}$

10. $\dfrac{15}{16} + \dfrac{3}{40}$

$\dfrac{15}{16} \times \dfrac{40}{40} = \dfrac{600}{640}$ $\qquad$ $\dfrac{3}{40} \times \dfrac{16}{16} = \dfrac{48}{640}$

Thus $\dfrac{15}{16} + \dfrac{3}{40} = \dfrac{600}{640} + \dfrac{48}{640} = \dfrac{648}{640} = \dfrac{81}{80}$ or $1\dfrac{1}{80}$

2.8 Student Practice

1. $\quad 5\dfrac{1}{12}$

$\dfrac{+ \ 9\dfrac{5}{12}}{14\dfrac{6}{12} = 14\dfrac{1}{2}}$

2. The LCD is 20.

$\dfrac{1}{4} \times \dfrac{5}{5} = \dfrac{5}{20}$ $\qquad$ $\dfrac{2}{5} \times \dfrac{4}{4} = \dfrac{8}{20}$

$6\dfrac{1}{4} = \quad 6\dfrac{5}{20}$

$\dfrac{+ \ 2\dfrac{2}{5} = \ + \ 2\dfrac{8}{20}}{8\dfrac{13}{20}}$

3. LCD = 12 $\quad 7\boxed{\dfrac{1}{4} \times \dfrac{3}{3}} = \quad 7\dfrac{3}{12}$

$\dfrac{+3\boxed{\dfrac{5}{6} \times \dfrac{2}{2}} = \ + \ 3\dfrac{10}{12}}{10\dfrac{13}{12} = 10 + 1\dfrac{1}{12} = 11\dfrac{1}{12}}$

4. LCD = 12 $\quad 12\dfrac{5}{6} = \quad 12\dfrac{10}{12}$

$\dfrac{-7\dfrac{5}{12} = \ - \ 7\dfrac{5}{12}}{5\dfrac{5}{12}}$

5. (a) LCD = 24 $\quad 9\boxed{\dfrac{1}{8} \times \dfrac{3}{3}} = \quad 9\dfrac{3}{24} = \quad 8\dfrac{27}{24}$

$\qquad \qquad \qquad - \ 3\boxed{\dfrac{2}{3} \times \dfrac{8}{8}} = \ - \ 3\dfrac{16}{24} = \ - \ 3\dfrac{16}{14}$

$\qquad \qquad \qquad \qquad \qquad \qquad \qquad \qquad \qquad 5\dfrac{11}{24}$

Borrow 1 from 9:

$9\dfrac{3}{24} = 8 + 1\dfrac{3}{24} = 8\dfrac{27}{24}$

(b) $\quad 18 \quad = \quad 17\dfrac{18}{18}$

$\dfrac{- \ 6\dfrac{7}{18} = \ - \ 6\dfrac{7}{18}}{11\dfrac{11}{8}}$

6. $\quad 6\dfrac{1}{4} = \quad 6\dfrac{3}{12} = \quad 5\dfrac{15}{12}$

$\dfrac{- \ 4\dfrac{2}{3} = \ - \ 4\dfrac{8}{12} = \ - \ 4\dfrac{8}{12}}{1\dfrac{7}{12}}$

They had $1\dfrac{7}{12}$ gallons left over.

7. $\dfrac{3}{5} - \dfrac{1}{15} \times \dfrac{10}{13}$

$= \dfrac{3}{5} - \dfrac{2}{39} \qquad$ LCD = $5 \cdot 39 = 195$

$= \dfrac{117}{195} - \dfrac{10}{195}$

$= \dfrac{107}{195}$

8. $\dfrac{1}{7} \times \dfrac{5}{6} + \dfrac{5}{3} \div \dfrac{7}{6} = \dfrac{1}{7} \times \dfrac{5}{6} + \dfrac{5}{3} \times \dfrac{6}{7}$

$= \dfrac{5}{42} + \dfrac{10}{7} \qquad$ LCD = 42

$= \dfrac{5}{42} + \dfrac{60}{42}$

$= \dfrac{65}{42}$ or $1\dfrac{23}{42}$

2.9 Student Practice

Student Practice 1

1. Understand the problem.

Mathematics Blueprint for Problem Solving

Gather the Facts	What Am I Asked to Do?	How Do I Proceed?	Key Points to Remember
Gas amounts: $18\dfrac{7}{10}$ gal $15\dfrac{2}{5}$ gal $14\dfrac{1}{2}$ gal	Find out how many gallons of gas she bought altogether.	Add the three amounts.	When adding mixed numbers, the LCD is needed for the fractions.

2. Solve and state the answer:

$$\text{LCD} = 10 \quad 18\frac{7}{10} = \quad 18\frac{7}{10}$$

$$15\frac{2}{5} = \quad 15\frac{4}{10}$$

$$14\frac{1}{2} = \quad + 14\frac{5}{10}$$

$$47\frac{16}{10} = 48\frac{6}{10}$$

$$= 48\frac{3}{5}$$

The total is $48\frac{3}{5}$ gallons.

3. *Check.* Estimate to see if the answer is reasonable.

Student Practice 2

1. Understand the problem.

Mathematics Blueprint for Problem Solving

Gather the Facts	What Am I Asked to Do?	How Do I Proceed?	Key Points to Remember
Poster: $12\frac{1}{4}$ in. Top border: $1\frac{3}{8}$ in. Bottom border: 2 in.	Find the length of the inside portion of the poster.	**(a)** Add the two border lengths. **(b)** Subtract this total from the poster length.	When adding mixed numbers, the LCD is needed for the fractions.

2. Solve and state the answer:

(a)
$$1\frac{3}{8}$$
$$+ 2$$
$$3\frac{3}{8}$$

(b)
$$12\frac{1}{4} = \quad 12\frac{2}{8} = \quad 11\frac{10}{8}$$
$$- 3\frac{3}{8} = \quad - 3\frac{3}{8} = \quad - 3\frac{3}{8}$$
$$8\frac{7}{8}$$

The length of the inside portion is $8\frac{7}{8}$ inches.

3. *Check.* Estimate to see if the answer is reasonable or work backward to check.

Student Practice 3

1. Understand the problem.

Mathematics Blueprint for Problem Solving

Gather the Facts	What Am I Asked to Do?	How Do I Proceed?	Key Points to Remember
Regular tent uses $8\frac{1}{4}$ yards. Large tent uses $1\frac{1}{2}$ times the regular. She makes 6 regular and 16 large tents.	Find out how many yards of cloth will be needed to make the tents.	Find the amount used for regular tents, and the amount used for large tents. Then add the two.	Large tents use $1\frac{1}{2}$ times the amount of a regular tent.

2. Solve and state the answer:

We multiply $6 \times 8\frac{1}{4}$ for regular tents and $16 \times 1\frac{1}{2} \times 8\frac{1}{4}$ for

large tents. Then add total yardage.

Regular tents: $6 \times 8\frac{1}{4} = \overset{3}{\cancel{6}} \times \frac{33}{\underset{2}{\cancel{4}}} = \frac{99}{2} = 49\frac{1}{2}$

Large tents: $16 \times 1\frac{1}{2} \times 8\frac{1}{4} = \overset{\overset{2}{\cancel{8}}}{\cancel{16}} \times \frac{3}{\underset{1}{\cancel{2}}} \times \frac{33}{\underset{1}{\cancel{4}}} = \frac{198}{1} = 198$

Total yardage for all tents is $198 + 49\frac{1}{2} = 247\frac{1}{2}$ yards.

3. *Check.* Estimate to see if the answer is reasonable.

Student Practice 4

1. Understand the problem.

Mathematics Blueprint for Problem Solving

Gather the Facts	What Am I Asked to Do?	How Do I Proceed?	Key Points to Remember
He purchases 12-foot boards. Each shelf is $2\frac{3}{4}$ ft. He needs four shelves for each bookcase and he is making two bookcases.	**(a)** Find out how many boards he needs to buy. **(b)** Find out how many feet of shelving are actually needed. **(c)** Find out how many feet will be left over.	Find out how many $2\frac{3}{4}$-ft shelves he can get from one board. Then see how many boards he needs to make all eight shelves.	There will be three answers to this problem. Don't forget to calculate the leftover wood.

2. Solve and state the answer:

We want to know how many $2\frac{3}{4}$-ft shelves are in a 12-ft board.

$$12 \div 2\frac{3}{4} = \frac{12}{1} \div \frac{11}{4} = \frac{12}{1} \times \frac{4}{11} = \frac{48}{11} = 4\frac{4}{11}$$

He will get 4 shelves from each board with some left over.

(a) For two bookcases, he needs eight shelves. He gets four shelves out of each board. $8 \div 4 = 2$. He will need two 12-ft boards.

(b) He needs 8 shelves at $2\frac{3}{4}$ feet.

$$8 \times 2\frac{3}{4} = 8 \times \frac{11}{4} = 22$$

He actually needs 22 feet of shelving.

(c)
 24 feet of shelving bought
$-$ 22 feet of shelving used
 2 feet of shelving left over.

3. *Check.* Work backward to check the answer.

Student Practice 5

1. Understand the problem.

Mathematics Blueprint for Problem Solving

Gather the Facts	What Am I Asked to Do?	How Do I Proceed?	Key Points to Remember
Distance is $199\frac{3}{4}$ miles. He uses $8\frac{1}{2}$ gallons of gas.	Find out how many miles per gallon he gets.	Divide the distance by the number of gallons.	Change mixed numbers to improper fractions before dividing.

2. Solve and state the answer: $199\dfrac{3}{4} \div 8\dfrac{1}{2} = \dfrac{799}{4} \div \dfrac{17}{2}$

$$= \frac{\overset{47}{\cancel{799}}}{\underset{2}{\cancel{4}}} \times \frac{\overset{1}{\cancel{2}}}{\underset{1}{\cancel{17}}}$$

$$= \frac{47}{2} = 23\frac{1}{2}$$

He gets $23\frac{1}{2}$ miles per gallon.

3. *Check.* Estimate to see if the answer is reasonable.

Chapter 3 3.1 Student Practice

1. (a) 0.073 seventy-three thousandths
  (b) 4.68   four and sixty-eight hundredths
  (c) 0.0017 seventeen ten-thousandths
  (d) 561.78 five hundred sixty-one and seventy-eight hundredths

2. seven thousand, eight hundred sixty-three and $\frac{4}{100}$ dollars

3. (a) $\frac{9}{10} = 0.9$ **(b)** $\frac{136}{1000} = 0.136$

  (c) $2\frac{56}{100} = 2.56$ **(d)** $34\frac{86}{1000} = 34.086$

4. (a) $0.37 = \frac{37}{100}$ **(b)** $182.3 = 182\frac{3}{10}$

  (c) $0.7131 = \frac{7131}{10,000}$ **(d)** $42.019 = 42\frac{19}{1000}$

5. (a) $8.5 = 8\frac{5}{10} = 8\frac{1}{2}$ **(b)** $0.58 = \frac{58}{100} = \frac{29}{50}$

  (c) $36.25 = 36\frac{25}{100} = 36\frac{1}{4}$ **(d)** $106.013 = 106\frac{13}{1000}$

6. $\frac{2}{1,000,000,000} = \frac{1}{500,000,000}$

The concentration of PCBs is $\frac{1}{500,000,000}$.

3.2 Student Practice

1. Since $4 < 5$, therefore $5.74 \; < \; 5.75$.

$$5.7\overset{\longleftarrow}{4} \qquad 5.7\overset{\longleftarrow}{5}$$

2. $0.894 > 0.890$, so $0.894 > 0.89$
3. 2.45, 2.543, 2.46, 2.54, 2.5
It is helpful to add extra zeros and to place the decimals that begin with 2.4 in a group and the decimals that begin with 2.5 in the other.
    2.450, 2.460, 2.543, 2.540, 2.500
In order, we have from smallest to largest
    2.450, 2.460, 2.500, 2.540, 2.543.
It is OK to leave the extra terminal zeros in the answer.
4. 723.88
723.9 Since the digit to right of tenths is greater than 5, we round up.
5. (a) 12.92 6 47
12.926 Since the digit to right of thousandths is less than 5, we drop the digits 4 and 7.
  (b) 0.00 7 892
0.008 Since the digit to right of thousandths is greater than 5, we round up.
6. 15,699.953
15,700.0 Since the digit to right of tenths is five, we round up.

7.

		Rounded to Nearest Dollar
Medical bills	$375.50	$376
Taxes	$971.39	$971
Retirement	$980.49	$980
Charity	$817.65	$818

3.3 Student Practice

1. (a)
```
  1
  9.8
  3.6
+ 5.4
 18.8
```
(b)
```
  1 1 1
300.72
163.75
+291.08
755.55
```
(c)
```
  2
  8.9000
 37.0560
  0.0023
+945.0000
990.9583
```

2.
```
  1   1
93,521.8
+ 1634.8
95,156.6
```
The odometer reading was 95,156.6 miles.

3.
```
  3 1 2 2
$ 80.95
 133.91
 256.47
  53.08
+381.32
$905.73
```

4. (a)
```
   7 18
  38.8
- 26.9
  11.9
```
(b)
```
      9 12
 1 10 2 14  8 10
 2 0 3 4 . 9 0 8
-1 9 8 6 . 3 2 5
   4 8 . 5 8 3
```

5. (a)
```
      9  9
    8 10 10 10
 1 9 . 0 0 0
-1 2 . 5 7 9
  6 . 4 2 1
```
(b)
```
         17 12    9
  1 7 2   10 17
 2 8 3 . 0 7 6
 -  9 6 . 3 8 0
 1 8 6 . 6 9 6
```

6.
```
       10 15 9
    6  0  5 10   11
 8 7, 1 6 0 . 1
-8 2, 3 7 0 . 9
   4 7 8 9 . 2
```
He had driven 4789.2 miles.

7.
```
   4 13
 1 5 . 3
- 1 0 . 8
   4 . 5
```
$x = 4.5$

3.4 Student Practice

1.
```
 0.09    2 decimal places
×  0.6    1 decimal place
 0.054    3 decimal places in product
```

2. (a)
```
    0.47    2 decimal places
 × 0.28    2 decimal places
   376
    94
 0.1316    4 decimal places in product
```
  (b)
```
    0.436    3 decimal places
 × 18.39    2 decimal places
   3924
   1308
  3 488
  4 36
 8.01804    5 decimal places in product
```

3.
```
   0.4264    4 decimal places
 ×    38    0 decimal places
  3 4112
 12 792
 16.2032    4 decimal places in product
```

4. Area = length × width
```
  1.26
× 2.3
  378
 252
 2.898
```
The area is 2.898 square millimeters.

5. (a) $0.0561 \times 10 = 0.561$ Decimal point moved one place to the right.
  (b) $1462.37 \times 100 = 146{,}237.$ Decimal point moved two places to the right.
6. (a) $0.26 \times 1000 = 260.$ Decimal point moved three places to the right. One extra zero needed.
  (b) $5862.89 \times 10{,}000 = 58{,}628{,}900.$ Decimal point moved four places to the right. Two extra zeros needed.

7. $7.684 \times 10^4 = 76,840.$ Decimal point moved four places to the right. One extra zero needed.

8. $156.2 \times 1000 = 156,200$
156.2 kilometers is equal to 156,200 meters.

3.5 Student Practice

1. (a)
```
    0.258
7)1.806
   1 4
    40
    35
    56
    56
     0
```
(b)
```
      0.0058
16)0.0928
     80
    128
    128
      0
```

2.
```
    0.517    The answer rounded to the nearest hundredth is 0.52.
46)23.820
   230
    82
    46
   360
   322
    38
```

3.
```
     186.25
19)3538.75
   19
   163
   152
   118
   114
    47
    38
    95
    95
     0
```
He pays $186.25 per month.

4. (a)
```
        1 .12
0.09∧)0.10∧08
      9
      1 0
        9
       18
       18
        0
```
(b)
```
           46.
0.037∧)1.702∧
       1 48
        222
        222
          0
```

5. (a)
```
         0 .023
1.8∧)0.0∧414
     36
     54
     54
      0
```
(b)
```
            2310.
0.0036∧)8.3160∧
        7 2
        1 11
        1 08
          36
          36
           0
```

6. (a)
```
          13 7 .26   The answer rounded to the nearest tenth
3.8∧)521.6∧00   is 137.3.
     38
     141
     114
     27 6
     26 6
      1 0 0
        7 6
        2 40
        2 28
          12
```

(b)
```
          0 .0211    The answer rounded to the nearest thousandth
8.05∧)0.17∧0000      is 0.021.
      16 10
        900
        805
        950
        805
        145
```

7.
```
          1 5.94
28.5∧)454.4∧00    The truck got approximately 15.9 miles per
      285          gallon.
      169 4
      142 5
       26 9 0
       25 6 5
        1 2 50
        1 1 40
          1 10
```

8.
```
          5.8
0.12∧)0.69∧6    n is 5.8.
      60
       9 6
       9 6
        0
```

9. Find the sum of levels for the years 1980, 1990, and 2000.
```
     25.9
     20.1
   + 16.3
     62.3
```
Then divide by three to obtain the average.
```
      20.7666
3)62.3000
  6
  2
  0
  23
  21
   20
   18
    20
    18
     20
     18
      2
```
The three-year average, rounded to the nearest thousandth, is 20.767 million tons. The five-year average was found to be 20.98 in Example 9. Find the difference between the averages.
```
   20.980    The three-year average differs
 − 20.767    from the five-year average by
   0.213     0.213 million tons.
```

3.6 Student Practice

1. (a)
```
     0.3125
16)5.0000
   4 8
   20
   16
   40
   32
   80
   80
    0
```
$\dfrac{5}{16} = 0.3125$

(b)
```
      0.1375
80)11.0000
   8 0
   3 00
   2 40
     600
     560
     400
     400
       0
```
$\dfrac{11}{80} = 0.1375$

2. (a)

$$
\begin{array}{r}
0.6363 \\
11\overline{)7.0000} \\
\underline{6\,6} \\
40 \\
\underline{33} \\
70 \\
\underline{66} \\
40 \\
\underline{33} \\
7
\end{array}
$$

$$\frac{7}{11} = 0.\overline{63}$$

(b)

$$
\begin{array}{r}
0.533 \\
15\overline{)8.000} \\
\underline{7\,5} \\
50 \\
\underline{45} \\
50 \\
\underline{45} \\
5
\end{array}
$$

$$\frac{8}{15} = 0.5\overline{3}$$

(c)

$$
\begin{array}{r}
0.29545 \\
44\overline{)13.000000} \\
\underline{8\,8} \\
4\,20 \\
\underline{3\,96} \\
240 \\
\underline{220} \\
200 \\
\underline{176} \\
240 \\
\underline{220} \\
20
\end{array}
$$

$$\frac{13}{44} = 0.29\overline{54}$$

3. (a) $2\dfrac{11}{18} = 2 + \dfrac{11}{18}$

$$0.611 = 0.6\overline{1}$$

$$
\begin{array}{r}
18\overline{)11.000} \\
\underline{10\,8} \\
20 \\
\underline{18} \\
20 \\
\underline{18} \\
2
\end{array}
$$

$$2\frac{11}{18} = 2.6\overline{1}$$

(b)

$$
\begin{array}{r}
1.03703 \\
27\overline{)28.00000} \\
\underline{27} \\
1\,00 \\
\underline{81} \\
190 \\
\underline{189} \\
100 \\
\underline{81} \\
19
\end{array}
$$

$$\frac{28}{27} = 1.0\overline{37}$$

4.

$$
\begin{array}{r}
0.7916 \\
24\overline{)19.0000} \\
\underline{16\,8} \\
2\,20 \\
\underline{2\,16} \\
40 \\
\underline{24} \\
160 \\
\underline{144} \\
16
\end{array}
$$

The answer rounded to the nearest thousandth is 0.792.

5. Divide to find the decimal equivalent of $\dfrac{5}{8}$.

$$
\begin{array}{r}
0.625 \\
8\overline{)5.000} \\
\underline{4\,8} \\
20 \\
\underline{16} \\
40 \\
\underline{40} \\
0
\end{array}
$$

In the hundredths place $2 < 3$, so we know

$$0.6\underline{2}5 < 0.6\underline{3}0.$$

Therefore, $\dfrac{5}{8} < 0.63$.

6. $0.3 \times 0.5 + (0.4)^3 - 0.036 = 0.3 \times 0.5 + 0.064 - 0.036$
$= 0.15 + 0.064 - 0.036$
$= 0.214 - 0.036$
$= 0.178$

7. $6.56 \div (2 - 0.36) + (8.5 - 8.3)^2$
$= 6.56 \div (1.64) + (0.2)^2$ Parentheses
$= 6.56 \div 1.64 + 0.04$ Exponent
$= 4 + 0.04$ Divide
$= 4.04$ Add

3.7 Student Practice

Student Practice 1

(a) $385.98 + 875.34 \approx 400 + 900 = 1300$
(b) $0.0932 - 0.0579 \approx 0.09 - 0.06 = 0.03$
(c) $5876.34 \times 0.087 \approx$

$$
\begin{array}{r}
6000 \\
\times\ 0.09 \\
\hline
540.00
\end{array}
$$

(d)

$$46{,}873 \div 8.456 \approx
\begin{array}{r}
6\,250 \\
8\overline{)50{,}000} \\
\underline{48} \\
2\,0 \\
\underline{1\,6} \\
40 \\
\underline{40} \\
0
\end{array}
$$

Student Practice 2

1. Understand the problem.

Mathematics Blueprint for Problem Solving

Gather the Facts	What Am I Asked to Do?	How Do I Proceed?	Key Points to Remember
She worked 51 hours. She gets paid $12.36 per hour for 40 hours. She gets paid time and a half for 11 hours.	Find the amount Melinda earned working 51 hours last week.	Add the earnings of 40 hours at $12.36 per hour to the earnings of 11 hours at overtime pay.	Overtime pay is time and a half, which is $1.5 \times \$12.36$.

2. Solve and state the answer:
   (a) Calculate regular earnings for 40 hours.

$$
\begin{array}{r}
\$12.36 \\
\times\ 40 \\
\hline
\$494.40
\end{array}
$$

   (b) Calculate overtime pay rate.

$$
\begin{array}{r}
\$12.36 \\
\times\ \ 1.5 \\
\hline
6\,180 \\
12\,36 \\
\hline
\$18.540
\end{array}
$$

(c) Calculate overtime earnings for 11 hours.

$$
\begin{array}{r}
\$18.54 \\
\times \qquad 11 \\
\hline
18\,54 \\
185\,4 \\
\hline
\$203.94
\end{array}
$$

(d) Add the two amounts.

$$
\begin{array}{rl}
\overset{1}{\$494}.40 & \text{Regular earnings} \\
+\;\;203.94 & \text{Overtime earnings} \\
\hline
\$698.34 & \text{Total earnings}
\end{array}
$$

Melinda earned $698.34 last week.

3. Check. Regular pay: $40 \times \$12 = \$480$ $\$480$
         Overtime pay: $2 \times \$12 = \$24$ $+\;\;200$
                 $10 \times \$20 = \$200$ $\$680$ The answer is reasonable.

Student Practice 3

1. Understand the problem.

Mathematics Blueprint for Problem Solving

Gather the Facts	What Am I Asked to Do?	How Do I Proceed?	Key Points to Remember
The total amount of steak is 17.4 pounds. Each package contains 1.45 pounds. Prime steak costs $7.60 per pound.	**(a)** Find out how many packages of steak the butcher will have. **(b)** Find the cost of each package.	**(a)** Divide the total, 17.4, by the amount in each package, 1.45, to find the number of packages. **(b)** Multiply the cost of one pound, $7.60, by the amount in one package, 1.45 pounds.	There will be two answers to this problem.

2. Solve and state the answer.

(a)
$$
\begin{array}{r}
12. \\
1.45 \overline{)17.40}\!\wedge \\
14\,5\!\wedge \\
\hline
2\,90 \\
2\,90 \\
\hline
0
\end{array}
$$
The butcher will have 12 packages of steak.

(b)
$$
\begin{array}{r}
\$7.60 \\
\times\;\;1.45 \\
\hline
3800 \\
3040 \\
760 \\
\hline
\$11.0200
\end{array}
$$
Each package will cost $11.02.

3. Check.

(a)
$$
\begin{array}{r}
1.45 \\
\times\;\;12 \\
\hline
290 \\
145 \\
\hline
17.40
\end{array}
$$

(b) $\$8 \times 1.5 = \$12$

The answers are reasonable.

Chapter 4   4.1 Student Practice

1. (a) $\dfrac{36}{40} = \dfrac{9}{10}$ **(b)** $\dfrac{18}{15} = \dfrac{6}{5}$ **(c)** $\dfrac{220}{270} = \dfrac{22}{27}$

2. (a) $\dfrac{200}{450} = \dfrac{4}{9}$

    (b) The total number of students surveyed is
$200 + 450 + 300 + 150 + 100 = 1200$. $\dfrac{300}{1200} = \dfrac{1}{4}$

3. $\dfrac{44 \text{ dollars}}{900 \text{ tons}} = \dfrac{11 \text{ dollars}}{225 \text{ tons}}$

4. $\dfrac{212 \text{ miles}}{4 \text{ hours}} = \dfrac{53 \text{ miles}}{1 \text{ hour}}$ or 53 miles/hour

5.
$$
\begin{array}{rl}
\text{selling price} & \$170.40 \\
-\;\text{purchase price} & -\;129.60 \\
\hline
\text{profit} & \$\;\;40.80
\end{array}
$$
She made a profit of $40.80 on 120 batteries.
$$
\begin{array}{r}
0.34 \\
120 \overline{)40.80} \\
360 \\
\hline
480 \\
480 \\
\hline
0
\end{array}
$$
Her profit was $0.34 per battery.

6. (a) $\dfrac{\$2.04}{12 \text{ ounces}} = \$0.17/\text{ounce}$ $\dfrac{\$2.80}{20 \text{ ounces}} = \$0.14/\text{ounce}$

    (b) Fred saves $0.03/ounce by buying the larger size.

4.2 Student Practice

1. 6 is to 8 as 9 is to 12.
$$\dfrac{6}{8} = \dfrac{9}{12}$$

2. $\dfrac{2 \text{ hours}}{72 \text{ miles}} = \dfrac{3 \text{ hours}}{108 \text{ miles}}$

3. (a) $\dfrac{10}{18} \overset{?}{=} \dfrac{25}{45}$

$$18 \times 25 = 450$$

$\dfrac{10}{18} \Large\times \normalsize \dfrac{25}{45}$ The cross products are equal.

$$10 \times 45 = 450$$

Thus $\dfrac{10}{18} = \dfrac{25}{45}$. This is a proportion.

(b) $\dfrac{42}{100} \overset{?}{=} \dfrac{22}{55}$

$$100 \times 22 = 2200$$
$$\uparrow$$

$\dfrac{42}{100} \bowtie \dfrac{22}{55}$ The cross products are not equal.

$$\downarrow$$
$$42 \times 55 = 2310$$

Thus $\dfrac{42}{100} \neq \dfrac{22}{55}$. This is not a proportion.

4. (a) $\dfrac{2.4}{3} \overset{?}{=} \dfrac{12}{15}$

$$3 \times 12 = 36$$
$$\uparrow$$

$\dfrac{2.4}{3} \bowtie \dfrac{12}{15}$ The cross products are equal.

$$\downarrow$$
$$2.4 \times 15 = 36$$

Thus $\dfrac{2.4}{3} = \dfrac{12}{15}$. This is a proportion.

(b) $\dfrac{2\frac{1}{3}}{6} \overset{?}{=} \dfrac{14}{38}$

$$2\frac{1}{3} \times 38 = \frac{7}{3} \times \frac{38}{1} = \frac{266}{3} = 88\frac{2}{3}$$

$\dfrac{2\frac{1}{3}}{6} \bowtie \dfrac{14}{38}$

$$6 \times 14 = 84$$

The cross products are not equal.

$$2\frac{1}{3} \times 38 = 88\frac{2}{3}$$

Thus $\dfrac{2\frac{1}{3}}{6} \neq \dfrac{14}{38}$. This is not a proportion.

5. (a) $\dfrac{1260}{7} \overset{?}{=} \dfrac{3530}{20}$

$$7 \times 3530 = 24{,}710$$
$$\uparrow$$

$\dfrac{1260}{7} \bowtie \dfrac{3530}{20}$ The cross products are not equal.

$$\downarrow$$
$$1260 \times 20 = 25{,}200$$

The rates are not equal. This is not a proportion.

(b) $\dfrac{2}{11} \overset{?}{=} \dfrac{16}{88}$

$$11 \times 16 = 176$$
$$\uparrow$$

$\dfrac{2}{11} \bowtie \dfrac{16}{88}$ The cross products are equal.

$$\downarrow$$
$$2 \times 88 = 176$$

The rates are equal. This is a proportion.

4.3 Student Practice

1. (a) $5 \times n = 45$

$$\frac{5 \times n}{5} = \frac{45}{5}$$
$$n = 9$$

(b) $7 \times n = 84$

$$\frac{7 \times n}{7} = \frac{84}{7}$$
$$n = 12$$

2. (a) $108 = 9 \times n$

$$\frac{108}{9} = \frac{9 \times n}{9}$$
$$12 = n$$

(b) $210 = 14 \times n$

$$\frac{210}{14} = \frac{14 \times n}{14}$$
$$15 = n$$

3. (a) $15 \times n = 63$

$$\frac{15 \times n}{15} = \frac{63}{15}$$
$$n = 4.2$$

$$\begin{array}{r} 4.2 \\ 15\overline{)63.0} \\ \underline{60} \\ 3\,0 \\ \underline{3\,0} \\ 0 \end{array}$$

(b) $39.2 = 5.6 \times n$

$$\frac{39.2}{5.6} = \frac{5.6 \times n}{5.6}$$
$$7 = n$$

$$\begin{array}{r} 7. \\ 5.6_\wedge\overline{)39.2_\wedge} \\ \underline{39\,2} \\ 0 \end{array}$$

4. $\dfrac{24}{n} = \dfrac{3}{7}$

$$24 \times 7 = n \times 3$$
$$168 = n \times 3$$
$$\frac{168}{3} = \frac{n \times 3}{3}$$
$$56 = n$$

5. $\dfrac{176}{4} = \dfrac{286}{n}$

$$176 \times n = 286 \times 4$$
$$176 \times n = 1144$$
$$\frac{176 \times n}{176} = \frac{1144}{176}$$
$$n = 6.5$$

$$\begin{array}{r} 6.5 \\ 176\overline{)1144.0} \\ \underline{1056} \\ 88\,0 \\ \underline{88\,0} \\ 0 \end{array}$$

6. $\dfrac{n}{30} = \dfrac{\frac{2}{3}}{4}$

$$4 \times n = 30 \times \frac{2}{3}$$
$$4 \times n = 20$$
$$\frac{4 \times n}{4} = \frac{20}{4}$$
$$n = 5$$

7. $\dfrac{n \text{ tablespoons}}{24 \text{ gallons}} = \dfrac{2.5 \text{ tablespoons}}{3 \text{ gallons}}$

$$3 \times n = 24 \times 2.5$$
$$3 \times n = 60$$
$$\frac{3 \times n}{3} = \frac{60}{3}$$
$$n = 20$$

20 tablespoons should be used.

8. $264 \times 2 = 3.5 \times n$

$$528 = 3.5 \times n$$
$$\frac{528}{3.5} = \frac{3.5 \times n}{3.5}$$
$$150.9 \approx n$$

The answer to the nearest tenth is $n = 150.9$.

4.4 Student Practice

1. $\dfrac{27 \text{ defective engines}}{243 \text{ engines produced}} = \dfrac{n \text{ defective engines}}{4131 \text{ engines produced}}$

$$27 \times 4131 = 243 \times n$$
$$111{,}537 = 243 \times n$$
$$\frac{111{,}537}{243} = \frac{243 \times n}{243}$$
$$459 = n$$

Thus we estimate that 459 engines are defective.

2. $\dfrac{9 \text{ gallons of gas}}{234 \text{ miles traveled}} = \dfrac{n \text{ gallons of gas}}{312 \text{ miles traveled}}$

$$9 \times 312 = 234 \times n$$
$$2808 = 234 \times n$$
$$\dfrac{2808}{234} = \dfrac{234 \times n}{234}$$
$$12 = n$$

She will need 12 gallons of gas.

3. $\dfrac{80 \text{ revolutions per minute}}{16 \text{ miles per hour}} = \dfrac{90 \text{ revolutions per minute}}{n \text{ miles per hour}}$

$$80 \times n = 16 \times 90$$
$$80 \times n = 1440$$
$$\dfrac{80 \times n}{80} = \dfrac{1440}{80}$$
$$n = 18$$

Alicia will be riding 18 miles per hour.

4. $\dfrac{4050 \text{ walk in}}{729 \text{ purchase}} = \dfrac{5500 \text{ walk in}}{n \text{ purchase}}$

$$4050 \times n = 729 \times 5500$$
$$4050 \times n = 4{,}009{,}500$$
$$\dfrac{4050 \times n}{4050} = \dfrac{4{,}009{,}500}{4050}$$
$$n = 990$$

Tom will expect 990 people to make a purchase in his store.

5. $\dfrac{50 \text{ bears tagged in 1st sample}}{n \text{ bears in forest}} = \dfrac{4 \text{ bears tagged in 2nd sample}}{50 \text{ bears caught in 2nd sample}}$

$$50 \times 50 = n \times 4$$
$$2500 = n \times 4$$
$$\dfrac{2500}{4} = \dfrac{n \times 4}{4}$$
$$625 = n$$

We estimate that there are 625 bears in the forest.

Chapter 5 5.1 Student Practice

1. (a) $\dfrac{51}{100} = 51\%$ **(b)** $\dfrac{68}{100} = 68\%$

   (c) $\dfrac{7}{100} = 7\%$ **(d)** $\dfrac{26}{100} = 26\%$

2. (a) $\dfrac{238}{100} = 238\%$ **(b)** $\dfrac{121}{100} = 121\%$

3. (a) $\dfrac{0.5}{100} = 0.5\%$ **(b)** $\dfrac{0.06}{100} = 0.06\%$

   (c) $\dfrac{0.003}{100} = 0.003\%$

4. (a) $47\% = \dfrac{47}{100} = 0.47$ **(b)** $2\% = \dfrac{2}{100} = 0.02$

5. (a) $80.6\% = 0.806$ **(b)** $2.5\% = 0.025$
   (c) $0.29\% = 0.0029$ **(d)** $231\% = 2.31$

6. (a) $0.78 = 78\%$ **(b)** $0.02 = 2\%$
   (c) $5.07 = 507\%$ **(d)** $0.029 = 2.9\%$
   (e) $0.006 = 0.6\%$

5.2 Student Practice

1. (a) $71\% = \dfrac{71}{100}$ **(b)** $25\% = \dfrac{25}{100} = \dfrac{1}{4}$

   (c) $8\% = \dfrac{8}{100} = \dfrac{2}{25}$

2. (a) $8.4\% = 0.084 = \dfrac{84}{1000} = \dfrac{21}{250}$

   (b) $28.5\% = 0.285 = \dfrac{285}{1000} = \dfrac{57}{200}$

3. (a) $170\% = 1.70 = 1\dfrac{7}{10}$ **(b)** $288\% = 2.88 = 2\dfrac{88}{100} = 2\dfrac{22}{25}$

4. $7\dfrac{5}{8}\% = 7\dfrac{5}{8} \div 100$

$$= \dfrac{61}{8} \times \dfrac{1}{100}$$
$$= \dfrac{61}{800}$$

5. $12\dfrac{5}{6}\% = 12\dfrac{5}{6} \div 100$

$$= 12\dfrac{5}{6} \times \dfrac{1}{100}$$
$$= \dfrac{77}{6} \times \dfrac{1}{100}$$
$$= \dfrac{77}{600}$$

6. $\dfrac{5}{8}$ $\begin{array}{r} 0.625 \\ 8\overline{)5.000} \end{array}$ 62.5%

7. (a) $\dfrac{21}{25} = 0.84 = 84\%$ **(b)** $\dfrac{7}{16} = 0.4375 = 43.75\%$

8. (a) $\dfrac{7}{9} = 0.77777\overline{7} \approx 0.7778 = 77.78\%$

   (b) $\dfrac{19}{30} = 0.63333\overline{3} \approx 0.6333 = 63.33\%$

9. $\dfrac{7}{12}$ If we divide

$$\begin{array}{r} 0.58 \\ 12\overline{)7.00} \\ \underline{6\ 0} \\ 1\ 00 \\ \underline{96} \\ 4 \end{array}$$

Thus $\dfrac{7}{12} = 0.58\dfrac{4}{12} = 58\dfrac{1}{3}\%$.

10.

Fraction	Decimal	Percent
$\dfrac{23}{99}$	0.2323	23.23%
$\dfrac{129}{250}$	0.516	51.6%
$\dfrac{97}{250}$	0.388	$38\dfrac{4}{5}\%$

5.3A Student Practice

1. What is 26% of 35?

$\;\;\downarrow\;\;\;\downarrow\;\;\;\downarrow\;\;\;\downarrow\;\;\;\downarrow$

$\;\;n\;\;\;=\;26\%\;\times\;35$

2. Find 0.08% of 350.

$\;\;\downarrow\;\;\;\;\;\downarrow\;\;\;\;\downarrow\;\;\;\downarrow$

$\;\;n = 0.08\% \times 350$

3. (a) 58%   of   what   is   400?

$\;\;\;\;\;\downarrow\;\;\;\;\;\downarrow\;\;\;\;\downarrow\;\;\;\;\downarrow\;\;\;\;\;\downarrow$

$\;\;\;\;58\%\;\;\times\;\;\;n\;\;\;=\;\;\;400$

   (b) 9.1   is   135%   of   what?

$\;\;\;\;\downarrow\;\;\;\downarrow\;\;\;\;\downarrow\;\;\;\downarrow\;\;\;\;\downarrow$

$\;\;\;\;9.1\;\;=\;\;135\%\;\;\times\;\;\;n$

4. What percent of 250 is 36?

$\;\;\;\;\underbrace{\qquad\qquad}\;\;\;\downarrow\downarrow\downarrow\;\;\downarrow$

$\;\;\;\;\;\;\;\;\downarrow$

$\;\;\;\;\;\;\;n\;\;\;\;\;\times 250 = 36$

5. (a) 50 is what percent of 20?

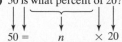

$$50 = n \times 20$$

(b) What percent of 2000 is 4.5?

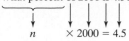

$$n \times 2000 = 4.5$$

6. What is 82% of 350?

$$n = 82\% \times 350$$
$$n = 0.82(350)$$
$$n = 287$$

7. $n = 230\% \times 400$
$$n = (2.30)(400)$$
$$n = 920$$

8. The problem asks: What is 8% of $350?

$$n = 8\% \times \$350$$
$$n = \$28$$

The tax was $28.

9. $32 = 0.4\% \times n$
$$32 = 0.004n$$
$$\frac{32}{0.004} = \frac{0.004n}{0.004}$$
$$8000 = n$$

10. The problem asks: 30% of what is 6?

$$30\% \times n = 6$$
$$0.30n = 6$$
$$\frac{0.30n}{0.30} = \frac{6}{0.30}$$
$$n = 20$$

There are 20 people on the team.

11. What percent of 9000 is 4.5?

$$n \times 9000 = 4.5$$
$$9000n = 4.5$$
$$\frac{9000n}{9000} = \frac{4.5}{9000}$$
$$n = 0.0005$$
$$n = 0.05\%$$

12. $198 = n \times 33$
$$\frac{198}{33} = \frac{33n}{33}$$
$$6 = n$$

Now express n as a percent: 600%

13. The problem asks: 5 is what percent of 16?

$$5 = n \times 16$$
$$\frac{5}{16} = \frac{16n}{16}$$
$$0.3125 = n$$

Now express n as a percent rounded to the nearest tenth: 31.3%

5.3B Student Practice

1. (a) Find 83% of 460.
   p is 83.
(b) 18% of what number is 90?
   p is 18.
(c) What percent of 64 is 8?
   The percent is unknown. Use the variable p.

2. (a) 30% of 52 is 15.6
   $b = 52, a = 15.6$
(b) 170 is 85% of what? Base $= b, a = 170$

3. (a) What is 18% of 240?
   Percent $p = 18$
   Base $b = 240$

Amount is unknown; use the variable a.

(b) What percent of 64 is 4?
   Percent is unknown; use the variable p.
   Base $b = 64$
   Amount $a = 4$

4. Find 340% of 70.
   Percent $p = 340$
   Base $b = 70$
   Amount is unknown; use amount $= a$.

$$\frac{a}{b} = \frac{p}{100} \quad \text{becomes} \quad \frac{a}{70} = \frac{340}{100}$$
$$\frac{a}{70} = \frac{17}{5}$$
$$5a = (70)(17)$$
$$5a = 1190$$
$$\frac{5a}{5} = \frac{1190}{5}$$
$$a = 238$$

Thus 340% of 70 is 238.

5. 68% of what is 476?
   Percent $p = 68$
   Base is unknown; use base $= b$.
   Amount $a = 476$

$$\frac{a}{b} = \frac{p}{100} \quad \text{becomes} \quad \frac{476}{b} = \frac{68}{100}$$
$$\frac{476}{b} = \frac{17}{25}$$
$$(476)(25) = 17b$$
$$11{,}900 = 17b$$
$$\frac{11{,}900}{17} = \frac{17b}{17}$$
$$700 = b$$

Thus 68% of 700 is 476.

6. 216 is 0.3% of what?
   Percent $p = 0.3$
   Base is unknown; use base $= b$.
   Amount $a = 216$

$$\frac{a}{b} = \frac{p}{100} \quad \text{becomes} \quad \frac{216}{b} = \frac{0.3}{100}$$
$$(216)(100) = 0.3b$$
$$21{,}600 = 0.3b$$
$$\frac{21{,}600}{0.3} = \frac{0.3b}{0.3}$$
$$72{,}000 = b$$

Thus $72,000 was exchanged.

7. What percent of 3500 is 105?
   Percent is unknown; use percent $= p$.
   Base $b = 3500$
   Amount $a = 105$

$$\frac{a}{b} = \frac{p}{100} \quad \text{becomes} \quad \frac{105}{3500} = \frac{p}{100}$$
$$\frac{3}{100} = \frac{p}{100}$$
$$300 = 100p$$
$$\frac{300}{100} = \frac{100p}{100}$$
$$3 = p$$

Thus 3% of 3500 is 105.

5.4 Student Practice

1. Method A Let n = number of people with reserved airline tickets.

12% of $n = 4800$

$0.12 \times n = 4800$

$\dfrac{0.12 \times n}{0.12} = \dfrac{4800}{0.12}$

$n = 40,000$

Method B The percent $p = 12$. Use b for the unknown base.
The amount $a = 4800$.

$\dfrac{a}{b} = \dfrac{p}{100}$ becomes $\dfrac{4800}{b} = \dfrac{12}{100}$.

$(4800)(100) = 12b$

$480,000 = 12b$

$\dfrac{480,000}{12} = b$

$40,000 = b$

40,000 people held airline tickets that month.

2. Method A The problem asks: What is 8% of $\$62.30$?

$n = 0.08 \times 62.30$

$n = 4.984$

Method B The percent $p = 8$. The base $b = 62.30$. Use a for the unknown amount.

$\dfrac{a}{b} = \dfrac{p}{100}$ becomes $\dfrac{a}{62.30} = \dfrac{8}{100}$.

$\dfrac{a}{62.3} = \dfrac{2}{25}$

$25a = (2)(62.3)$

$25a = 124.60$

$\dfrac{25a}{25} = \dfrac{124.60}{25}$

$a = 4.984$

The tax is $\$4.98$.

3. Method A The problem asks: 105 is what percent of 130?

$105 = n \times 130$

$\dfrac{105}{130} = n$

$0.8077 \approx n$

Method B

Use p for the unknown percent. The base $b = 130$. The amount $a = 105$.

$\dfrac{a}{b} = \dfrac{p}{100}$ becomes $\dfrac{105}{130} = \dfrac{p}{100}$.

$\dfrac{21}{26} = \dfrac{p}{100}$

$(21)(100) = 26p$

$2100 = 26p$

$\dfrac{2100}{26} = \dfrac{26p}{26}$

$80.769230\ldots = p$

Thus 80.8% of the flights were on time.

Chapter 6 6.1 Student Practice

1. (a) 3 **(b)** 5280 **(c)** 60 **(d)** 7 **(e)** 16 **(f)** 2 **(g)** 4

2. $15,840 \text{ feet} \times \dfrac{1 \text{ mile}}{5280 \text{ feet}} = \dfrac{15,840}{5280} \text{ miles} = 3 \text{ miles}$

3. (a) $18.93 \text{ miles} \times \dfrac{5280 \text{ feet}}{1 \text{ mile}} = 99,950.4 \text{ feet}$

(b) $16\dfrac{1}{2} \text{ inches} \times \dfrac{1 \text{ yard}}{36 \text{ inches}} = \dfrac{33}{2} \times \dfrac{1}{36} \text{ yard}$

$= \dfrac{\overset{11}{\cancel{33}}}{2} \times \dfrac{1}{\underset{12}{\cancel{36}}} \text{ yard} = \dfrac{11}{24} \text{ yard}$

4. $760.5 \text{ pounds} \times \dfrac{16 \text{ ounces}}{1 \text{ pound}} = 760.5 \times 16 \text{ ounces} = 12,168 \text{ ounces}$

4. 100% Cost of meal + tip of $15\% = \$46.00$

Let n = Cost of meal

100% of $n + 15\%$ of $n = \$46.00$

115% of $n = 46.00$

$1.15 \times n = 46.00$

$\dfrac{1.15 \times n}{1.15} = \dfrac{46.00}{1.15}$

$n = 40.00$

They can spend $\$40.00$ on the meal itself.

5. (a) 7% of $\$13,600$ is the discount.

$0.07 \times 13,600 = $ the discount

$\$952$ is the discount.

(b) $\$13,600$ list price

$\quad\underline{-\quad 952}\quad$ discount

$\$12,648$ Amount Betty paid for the car.

5.5 Student Practice

1. Commission = commission rate $\times$ value of sales

Commission $= 6\% \times \$156,000$

$= 0.06 \times 156,000$

$= 9360$

His commission is $\$9360$.

2. $\quad 15,000$

$\underline{-10,500}$

$\quad 4500 \quad$ the amount of decrease

Percent of decrease $= \dfrac{\text{amount of decrease}}{\text{original amount}} = \dfrac{4500}{15,000}$

$= 0.30 = 30\%$

The percent of decrease is 30%.

3. $I = P \times R \times T$

$P = \$5600 \qquad R = 12\% \qquad T = 1 \text{ year}$

$I = 5600 \times 12\% \times 1$

$= 5600 \times 0.12$

$= 672$

The interest is $\$672$.

4. (a) $I = P \times R \times T$

$= 1800 \times 0.11 \times 4$

$= 198 \times 4$

$= 792$

The interest for four years is $\$792$.

(b) $I = P \times R \times T$

$= 1800 \times 0.11 \times \dfrac{1}{2}$

$= 198 \times \dfrac{1}{2}$

$= 99$

The interest for six months is $\$99$.

5. $19 \text{ pints} \times \dfrac{1 \text{ quart}}{2 \text{ pints}} = \dfrac{19}{2} \text{ quarts} = 9.5 \text{ quarts}$

6. Step 1: $26 \text{ yards} \times \dfrac{3 \text{ feet}}{1 \text{ yard}} = 26 \times 3 \text{ feet} = 78 \text{ feet}$

Step 2: $78 \text{ feet} + 2 \text{ feet} = 80 \text{ feet}$

The path is 80 feet long.

7. Step 1: $1\dfrac{3}{4} \text{ days} \times \dfrac{24 \text{ hours}}{1 \text{ day}} = \dfrac{7}{4} \times \dfrac{24}{1} \text{ hours} = 42 \text{ hours parked}$

Step 2: $42 \text{ hours} \times \dfrac{1.50 \text{ dollars}}{1 \text{ hour}} = 63 \text{ dollars}$

She paid $\$63$.

6.2 Student Practice

1. (a) deka- means ten   (b) milli- means thousandth
2. (a) 4 meters = 4.00ᴧ centimeters = 400 cm

   (b) 30 centimeters = 30.0ᴧ millimeters = 300 mm
3. (a) 3 millimeters = 0ᴧ003. meter = 0.003 meter

   (b) 47 centimeters = 0ᴧ00047. kilometer = 0.00047 kilometer
4. The car length would logically be choice (b) 3.8 meters. (A meter is close to a yard and 3.8 yards seems reasonable.)
5. (a) 375 cm = 3ᴧ75 m = 3.75 m

   (b) 46 m = 46.000ᴧ mm = 46,000 mm
6. (a) 389 mm = 0.0389 dam (four places to left)

   (b) 0.48 hm = 4800 cm (four places to right)
7. 782 cm =   7.82 m
      2 m =   2.00 m
    537 m =   537.00 m
               546.82 m

6.3 Student Practice

1. (a) 5 L = 5.000ᴧ mL = 5000 mL

   (b) 84 kL = 84.000ᴧ L = 84,000 L

   (c) 0.732 L = 0.732ᴧ mL = 732 mL
2. (a) 15.8 mL = 0.0158 L

   (b) 12,340 mL = 12.34 L

   (c) 86.3 L = 0.0863 kL
3. (a) 396 mL = 396 cm³

   (because 1 milliliter = 1 cubic centimeter)

   (b) 0.096 L = 96 cm³ = 96 cc
4. (a) 3.2 t = 3200 kg

   (b) 7.08 kg = 7080 g
5. (a) 59 kg = 0.059 t

   (b) 28.3 mg = 0.0283 g
6. A gram is $\frac{1}{1000}$ of a kilogram. If the coffee costs $10.00 per kilogram, then 1 gram would cost $\frac{1}{1000}$ of $10.

   $\frac{1}{1000} \times \$10 = \frac{\$10.00}{1000} = \$0.01$

   The coffee costs $0.01 per gram.
7. (a) 120 kg (A kilogram is slightly more than 2 pounds.)

6.4 Student Practice

1. $7 \text{ feet} \times \dfrac{0.305 \text{ meter}}{1 \text{ foot}} = 2.135 \text{ meters}$

2. (a) $17 \text{ m} \times \dfrac{1.09 \text{ yd}}{1 \text{ m}} = 18.53 \text{ yd}$

   (b) $29.6 \text{ km} \times \dfrac{0.62 \text{ mi}}{1 \text{ km}} = 18.352 \text{ mi}$

6.4 Student Practice (continued)

(c) $26 \text{ gal} \times \dfrac{3.79 \text{ L}}{1 \text{ gal}} = 98.54 \text{ L}$

(d) $6.2 \text{ L} \times \dfrac{1.06 \text{ qt}}{1 \text{ L}} = 6.572 \text{ qt}$

3. $180 \text{ cm} \times \dfrac{0.394 \text{ in.}}{1 \text{ cm}} \times \dfrac{1 \text{ ft}}{12 \text{ in.}} = 5.91 \text{ ft}$

4. $\dfrac{88 \text{ km}}{\text{hr}} \times \dfrac{0.62 \text{ mi}}{1 \text{ km}} = 54.56 \text{ mi/hr}$

5. $\dfrac{900 \text{ miles}}{\text{hr}} \times \dfrac{5280 \text{ ft}}{1 \text{ mile}} \times \dfrac{1 \text{ hr}}{60 \text{ min}} \times \dfrac{1 \text{ min}}{60 \text{ sec}}$

   $= \dfrac{900 \times 5280 \text{ ft}}{60 \times 60 \text{ sec}} = \dfrac{4,752,000 \text{ ft}}{3600 \text{ sec}}$

   $= 1320 \text{ ft/sec}$

   The jet is traveling at 1320 feet per second.
6. $F = 1.8 \times C + 32$

   $= 1.8 \times 20 + 32$

   $= 36 + 32$

   $= 68$

   The temperature is 68°F.
7. $C = \dfrac{5 \times F - 160}{9}$

   $= \dfrac{5 \times 86 - 160}{9}$

   $= \dfrac{430 - 160}{9}$

   $= \dfrac{270}{9}$

   $= 30$

   The temperature is 30°C.

6.5 Student Practice 1

Step 1: $2\frac{2}{3}$ yd

$8\frac{1}{3}$ yd

$2\frac{2}{3}$ yd

$+ 8\frac{1}{3}$ yd
——————
22 yd

Step 2: $22 \text{ yd} \times \dfrac{3 \text{ ft}}{1 \text{ yd}} = 66 \text{ ft}$

The perimeter is 66 ft.

6.5 Student Practice 2

1. *Understand the problem.*

Mathematics Blueprint for Problem Solving

Gather the Facts	What Am I Asked to Do?	How Do I Proceed?	Key Points to Remember
He must use 18.06 liters of solution. He has 42 jars to fill.	Find out how many milliliters of solution will go into each jar.	We need to convert 18.06 liters to milliliters, and then divide that result by 42.	To convert 18.06 liters to milliliters, we move the decimal point three places to the right.

2. *Solve and state the answer:*

   18.06 L = 18,060 mL

   $\dfrac{18,060 \text{ mL}}{42 \text{ jars}} = 430 \text{ mL/jar}$

   Thus, 430 mL of solution will go into each jar.

3. *Check.* 18.06 L is approximately 18 L, or 18,000 mL. There are approximately 40 jars.

   $\dfrac{18,000}{40} = 450$

   The answer is reasonable.

Chapter 7   7.1 Student Practice

1. $\angle FGH$ and $\angle KGJ$ are acute angles, $\angle HGK$ and $\angle FGJ$ are obtuse angles, $\angle HGJ$ is a right angle, and $\angle FGK$ is a straight angle.
2. (a) The complement of angle B measures $90° - 83° = 7°$.
   (b) The supplement of angle B measures $180° - 83° = 97°$.
3. $\angle y$ and $\angle w$ are vertical angles and so have the same measure. Thus $\angle w = 133°$. $\angle y$ and $\angle z$ are adjacent angles, so we know they are supplementary. Thus $\angle z$ measures $180° - 133° = 47°$. Finally, $\angle x$ and $\angle z$ are vertical angles, so we know they have the same measure. Thus $\angle x$ measures $47°$.
4. $\angle z = 180° - 105° = 75°$ ($\angle x$ and $\angle z$ are adjacent angles).
   $\angle y = \angle x = 105°$ ($\angle x$ and $\angle y$ are alternate interior angles).
   $\angle v = \angle x = 105°$ ($\angle v$ and $\angle x$ are corresponding angles).
   $\angle w = 180° - 105° = 75°$ ($\angle w$ and $\angle v$ are adjacent angles).

7.2 Student Practice

1. $P = 2l + 2w$
   $= (2)(6\text{ m}) + 2(1.5\text{ m})$
   $= 12\text{ m} + 3\text{ m} = 15\text{ m}$
2. $P = 4s$
   $= (4)(5.8\text{ cm}) = 23.2\text{ cm}$
3. $P = 4 + 4 + 5.5 + 2.5 + 1.5 + 1.5 = 19\text{ ft}$
   $\text{Cost} = 19\text{ ft} \times \dfrac{0.16\text{ dollar}}{1\text{ ft}} = \$3.04$
4. $A = lw = (29\text{ m})(17\text{ m}) = 493\text{ m}^2$
5. $A = s^2$
   $= (11.8\text{ mm})^2$
   $= (11.8\text{ mm})(11.8\text{ mm})$
   $= 139.24\text{ mm}^2$
6. Area of rectangle $= (18\text{ ft})(20\text{ ft}) = 360\text{ ft}^2$
   Area of square $= (6\text{ ft})^2 = 36\text{ ft}^2$
   Total area $= 396\text{ ft}^2$

7.3 Student Practice

1. $P = (2)(7.6\text{ cm}) + (2)(3.5\text{ cm})$
   $= 15.2\text{ cm} + 7.0\text{ cm} = 22.2\text{ cm}$
2. $A = bh$
   $= (10.3\text{ km})(1.5\text{ km})$
   $= 15.45\text{ km}^2$
3. $P = 4(6\text{ cm}) = 24\text{ cm}$
   $A = bh$
   $= (4\text{ cm})(6\text{ cm}) = 24\text{ cm}^2$
4. $P = 7\text{ yd} + 15\text{ yd} + 21\text{ yd} + 13\text{ yd} = 56\text{ yd}$
5. (a) $A = \dfrac{h(b + B)}{2} = \dfrac{(140\text{ yd})(130\text{ yd} + 180\text{ yd})}{2} = 21{,}700\text{ yd}^2$
   (b) $21{,}700\text{ yd}^2 \times \dfrac{1\text{ gallon}}{100\text{ yd}^2} = 217\text{ gallons}$
   Thus 217 gallons of sealant are needed.
6. The area of the trapezoid is
   $A = \dfrac{(9.2\text{ cm})(12.6\text{ cm} + 19.8\text{ cm})}{2}$
   $= \dfrac{(9.2\text{ cm})(32.4\text{ cm})}{2} = \dfrac{298.08\text{ cm}^2}{2}$
   $= 149.04\text{ cm}^2$.
   The area of the rectangle is
   $A = (8.3\text{ cm})(12.6\text{ cm}) = 104.58\text{ cm}^2$
   Total area $= 149.04\text{ cm}^2 + 104.58\text{ cm}^2 = 253.62\text{ cm}^2$

7.4 Student Practice

1. The sum of the measures of the angles in a triangle is $180°$. The two given angles total $125° + 15° = 140°$. Thus $180° - 140° = 40°$. Angle A must measure $40°$.

2. $P = 10.5\text{ m} + 10.5\text{ m} + 8.5\text{ m} = 29.5\text{ m}$
3. $A = \dfrac{bh}{2} = \dfrac{(38\text{ m})(13\text{ m})}{2} = \dfrac{494\text{ m}^2}{2} = 247\text{ m}^2$
4. Area of rectangle $= (11\text{ cm})(24\text{ cm}) = 264\text{ cm}^2$
   Area of triangle $= \dfrac{(11\text{ cm})(7\text{ cm})}{2} = \dfrac{77\text{ cm}^2}{2} = 38.5\text{ cm}^2$
   Total area $= 264\text{ cm}^2 + 38.5\text{ cm}^2 = 302.5\text{ cm}^2$

7.5 Student Practice

1. (a) $\sqrt{49} = 7$ because $(7)(7) = 49$.
   (b) $\sqrt{169} = 13$ because $(13)(13) = 169$.
2. $\sqrt{49} = 7$ because $(7)(7) = 49$.
   $\sqrt{4} = 2$ because $(2)(2) = 4$.
   Thus $\sqrt{49} - \sqrt{4} = 7 - 2 = 5$.
3. (a) Yes. 144 is a perfect square because $(12)(12) = 144$.
   (b) $\sqrt{144} = 12$
4. (a) $\sqrt{3} \approx 1.732$ (b) $\sqrt{13} \approx 3.606$ (c) $\sqrt{5} \approx 2.236$
5. $\sqrt{22\text{ m}^2} \approx 4.690\text{ m}$
   Thus, to the nearest thousandth of a meter, the side measures 4.690 m.

7.6 Student Practice

1. Hypotenuse $= \sqrt{(8)^2 + (6)^2}$
   $= \sqrt{64 + 36}$ Square each value first.
   $= \sqrt{100}$ Add together the two values.
   $= 10\text{ m}$ Take the square root.
2. Hypotenuse $= \sqrt{(3)^2 + (7)^2}$
   $= \sqrt{9 + 49}$ Square each value first.
   $= \sqrt{58}\text{ cm}$ Add the two values together.
   Using a square root table or a calculator, we have the hypotenuse ≈ 7.616 cm.
3. Leg $= \sqrt{(17)^2 - (15)^2}$
   $= \sqrt{289 - 225}$ Square each value first.
   $= \sqrt{64}$ Subtract.
   $= 8\text{ m}$ Find the square root.
4. Leg $= \sqrt{(10)^2 - (5)^2}$
   $= \sqrt{100 - 25}$ Square each value first.
   $= \sqrt{75}\text{ m}$ Subtract the two numbers.
   Using a calculator or a square root table, we find that the leg ≈ 8.660 m.
5. 1. **Understand the problem.**
   We are given a picture.
   The distance between the centers of the holes is the hypotenuse of the triangle.
   2. **Solve and state the answer.**
   Hypotenuse $= \sqrt{(\text{leg})^2 + (\text{leg})^2}$
   $= \sqrt{(2)^2 + (5)^2}$
   $= \sqrt{4 + 25}$
   $= \sqrt{29}$
   $\sqrt{29} \approx 5.385$
   Rounded to the nearest thousandth, the distance is 5.385 cm.
   3. **Check.**
   Work backward to check. Use the Pythagorean Theorem.
   $5.385^2 \overset{?}{\approx} 2^2 + 5^2$ (We use $\approx$ because 5.385 is an approximate answer.)
   $28.998225 \overset{?}{\approx} 4 + 25$
   $28.998225 \approx 29$ ✓

6. **1. *Understand the problem.***
   We are given a picture.
   2. *Solve and state the answer.*
   $$\text{Leg} = \sqrt{(\text{hypotenuse})^2 - (\text{leg})^2}$$
   $$= \sqrt{(30)^2 - (27)^2}$$
   $$= \sqrt{900 - 729}$$
   $$= \sqrt{171}$$
   $$\sqrt{171} \approx 13.1$$
   If we round to the nearest tenth, the kite is 13.1 yd above the rock.

7. **(a)** In a 30°–60°–90° triangle, the side opposite the 30° angle is $\frac{1}{2}$ of the hypotenuse.
   $$\frac{1}{2} \times 12 = 6$$
   Therefore, $y = 6$ ft.
   When we know two sides of a right triangle, we find the third side using the Pythagorean Theorem.
   $$\text{Leg} = \sqrt{(\text{hypotenuse})^2 - (\text{leg})^2}$$
   $$= \sqrt{(12)^2 - (6)^2} = \sqrt{144 - 36}$$
   $$= \sqrt{108} \approx 10.4$$
   $x = 10.4$ ft rounded to the nearest tenth.
   (b) In a 45°–45°–90° triangle, we have the following:
   $$\text{Hypotenuse} = \sqrt{2} \times \text{leg}$$
   $$\approx 1.414(8)$$
   $$= 11.312 \text{ m}$$
   Rounded to the nearest tenth, the hypotenuse = 11.3 m.

7.7 Student Practice

1. $C = \pi d$
   $$= (3.14)(9 \text{ m})$$
   $$= 28.26 \text{ m}$$
   $C = 28.3$ m rounded to the nearest tenth.

2. $C = \pi d$
   $$= (3.14)(30 \text{ in.})$$
   $$= 94.2 \text{ in.}$$
   Change 94.2 in. to ft.
   $$94.2 \text{ in.} \times \frac{1 \text{ ft}}{12 \text{ in.}} = 7.85 \text{ ft}$$
   When the wheel makes 2 revolutions, the bicycle travels $7.85 \times 2 = 15.7$ ft.

3. $A = \pi r^2$
   $$= (3.14)(5 \text{ km})^2$$
   $$= (3.14)(25 \text{ km}^2)$$
   $$= 78.5 \text{ km}^2$$

4. $r = \dfrac{d}{2} = \dfrac{10 \text{ ft}}{2} = 5 \text{ ft}$
   $A = \pi r^2$
   $$= (3.14)(5 \text{ ft})^2$$
   $$= 78.5 \text{ ft}^2$$
   Change 78.5 ft^2 to yd^2.
   $$78.5 \text{ ft}^2 \times \frac{1 \text{ yd}^2}{9 \text{ ft}^2} \approx 8.7222 \text{ yd}^2$$
   Find the cost: $\dfrac{\$12}{1 \text{ yd}^2} \times 8.7222 \text{ yd}^2 \approx \$104.67$.
   The cost of the pool cover is $104.67.

5. Area of square − area of circle = shaded area
   $A = s^2$
   $$= (5 \text{ ft})^2$$
   $$= 25 \text{ ft}^2$$
   $A = \pi r^2$
   $$= (3.14)(2 \text{ ft})^2$$
   $$= (3.14)(4 \text{ ft}^2)$$
   $$= 12.56 \text{ ft}^2$$
   $25 \text{ ft}^2 - 12.56 \text{ ft}^2 = 12.44 \text{ ft}^2$
   The area is 12.4 ft^2 rounded to the nearest tenth.

6. $r = \dfrac{d}{2} = \dfrac{8 \text{ ft}}{2} = 4 \text{ ft}$
   $$A_{\text{semicircle}} = \frac{\pi r^2}{2}$$
   $$= \frac{(3.14)(4 \text{ ft})^2}{2}$$
   $$= 25.12 \text{ ft}^2$$
   $$A_{\text{rectangle}} = lw = (12 \text{ ft})(8 \text{ ft}) = 96 \text{ ft}^2.$$
   $$\begin{array}{r} 25.12 \text{ ft}^2 \\ + \ 96.00 \text{ ft}^2 \\ \hline 121.12 \text{ ft}^2 \end{array}$$
   The total area is approximately 121.1 ft^2.

7.8 Student Practice

1. $V = lwh$
   $$= (6 \text{ m})(5 \text{ m})(2 \text{ m})$$
   $$= (30)(2) \text{ m}^3$$
   $$= 60 \text{ m}^3$$

2. $V = \pi r^2 h$
   $$= (3.14)(2 \text{ in.})^2(5 \text{ in.})$$
   $$= (3.14)(4 \text{ in.}^2)(5 \text{ in.})$$
   $$= 62.8 \text{ in.}^3$$

3. $V = \dfrac{4\pi r^3}{3} = \dfrac{(4)(3.14)(6 \text{ m})^3}{3} = \dfrac{(4)(3.14)(216) \text{ m}^3}{3}$
   $$= (12.56)(72) \text{ m}^3 = 904.32 \text{ m}^3$$
   The volume is 904.3 m^3 rounded to nearest tenth.

4. $V = \dfrac{\pi r^2 h}{3}$
   $$= \frac{(3.14)(5 \text{ m})^2(12 \text{ m})}{3}$$
   $$= 314.0 \text{ m}^3$$

5. $V = \dfrac{Bh}{3}$
   (a) $B = (6 \text{ m})(6 \text{ m}) = 36 \text{ m}^2$
   $$V = \frac{(36 \text{ m}^2)(10 \text{ m})}{3} = \frac{360 \text{ m}^3}{3} = 120 \text{ m}^3$$
   (b) $B = (7 \text{ m})(8 \text{ m}) = 56 \text{ m}^2$
   $$V = \frac{(56 \text{ m}^2)(15 \text{ m})}{3} = \frac{840 \text{ m}^3}{3} = 280 \text{ m}^3$$

7.9 Student Practice

1. $\dfrac{11}{27} = \dfrac{15}{n}$
   $$11n = (27)(15)$$
   $$11n = 405$$
   $$\frac{11n}{11} = \frac{405}{11}$$
   $$n = 36.\overline{81}$$
   $n = 36.8$ meters, to the nearest tenth.

2. a corresponds to p, b corresponds to m, c corresponds to n

3. $\dfrac{h}{5} = \dfrac{20}{2}$
   $$2h = 100$$
   $$h = 50$$
   The side wall is 50 feet tall.

4. $\dfrac{3}{29} = \dfrac{1.8}{w}$
   $$3w = (1.8)(29)$$
   $$3w = 52.2$$
   $$\frac{3w}{3} = \frac{52.2}{3}$$
   $$w = 17.4 \quad \text{The width is 17.4 meters.}$$

7.10 Student Practice
Student Practice 1

1. Understand the problem.

Mathematics Blueprint for Problem Solving

Gather the Facts	What Am I Asked to Do?	How Do I Proceed?	Key Points to Remember
Mike needs to sand three rooms: 24 ft × 13 ft 12 ft × 9 ft 16 ft × 3 ft He can sand 80 ft² in 15 min.	Find out how long it will take him to sand all three rooms.	(a) Find the total area to be sanded. (b) Then find out how long it will take him to sand the total area.	Area = length × width To get the total time, set up a proportion.

2. Solve and state the answer:

$24 \times 13 = 312$ ft² room 1

$12 \times 9 = 108$ ft² room 2

$16 \times 3 = 48$ ft² room 3

Total area $= 468$ ft²

$$\frac{80 \text{ ft}^2}{15 \text{ min}} = \frac{468 \text{ ft}^2}{t \text{ min}}$$

$$\frac{80}{15} = \frac{468}{t}$$

$$80t = (15)(468)$$

$$80t = 7020$$

$$\frac{80t}{80} = \frac{7020}{80}$$

$$t = 87.75$$

It will take Mike 87.75 minutes to sand the rooms.

3. Check. Estimate to see if the answer is reasonable.

Student Practice 2

1. Understand the problem.

Mathematics Blueprint for Problem Solving

Gather the facts	What Am I Asked to Do?	How Do I Proceed?	Key Points to Remember
The trapezoid has a height of 9 ft. The bases are 18 ft and 12 ft. The rectangular portion measures 24 ft × 15 ft. Roofing costs $2.75 per square yard.	(a) Find the area of the roof. (b) Find the cost to install new roofing.	(a) Find the area of the entire roof. Change square feet to square yards. (b) Multiply by $2.75.	9 square feet = 1 square yard

2. Solve and state the answer:

(a) Area of trapezoid $= \dfrac{1}{2}h(b + B)$

$$= \frac{1}{2}(9 \text{ ft})(12 \text{ ft} + 18 \text{ ft})$$

$$= 135 \text{ ft}^2$$

Area of rectangle $= lw$

$$= (15 \text{ ft})(24 \text{ ft})$$

$$= 360 \text{ ft}^2$$

Total area $= 135$ ft² $+ 360$ ft² $= 495$ ft²

Change square feet to square yards.

$$495 \text{ ft}^2 \times \frac{1 \text{ yd}^2}{9 \text{ ft}^2} = 55 \text{ yd}^2$$

The area of the roof is 55 yd².

(b) Cost $= 55 \text{ yd}^2 \times \dfrac{\$2.75}{1 \text{ yd}^2} = \$151.25$

The cost to install new roofing would be $151.25.

3. Check. Estimate to see if the answers seem reasonable.

Chapter 8   8.1 Student Practice

1. The smallest category of students is special students.
2. (a) 3000 freshmen + 200 special students = 3200
   There are 3200 students who are either freshmen or special students.
   (b) 3200 out of 10,000 are freshmen or special students.
   $$\frac{3200}{10,000} = 0.32 = 32\%$$
3. There are 3000 freshmen and 2600 sophomores. The ratio of freshmen to sophomores is $\frac{3000}{2600} = \frac{15}{13}$.
4. The ratio of freshmen to the total number of students is $\frac{3000}{10,000} = \frac{3}{10}$.
5. The percent of the total area occupied by either Lake Superior or Lake Michigan is: 34% + 24% = 58%.
6. Lake Superior has 34% of the area. 34% of 94,680 mi² is $(0.34)(94,680 \text{ mi}^2) \approx 32,191 \text{ mi}^2$.
7. (a) 20% + 20% + 1% = 41%
   41% of the students represented by the graph will be 21 years or younger.
   (b) Take 16% of 19,882,000
   $(0.16)(19,882,000) = 3,181,120$
   Approximately 3,181,120 college and university students will be 35 years or older in 2018.

8.2 Student Practice

1. The bar rises to 24. The approximate population was 24 million or 24,000,000.
2. $20 - 11 = 9$. The population increased by 9 million or 9,000,000.
3. The bar rises to 250. The number of new cars sold in the fourth quarter of 2010 was 250.
4. $150 - 100 = 50$. Thus, 50 fewer cars were sold.
5. The greatest number of customers came in July, since the highest point of the graph occurs for July.
6. (a) For May, the dot is halfway between 3 and 4, so approximately 3500 customers came during the month of May.
   (b) From March to April, the line goes down, so the number of customers decreased.
7. The line from July to August goes downward at the steepest angle. Thus the greatest decrease occurs between July and August.
8. Because the dot corresponding to 2009–2010 is at 20 and the scale is in hundreds, we have $20 \times 100 = 2000$. Thus, 2000 degrees in visual and performing arts were awarded.
9. The dot for computer science is above the visual and performing arts dot first in 2004–2005. Thus the first academic year with more degrees in computer science was 2004–2005.

8.3 Student Practice

1. The 60–69 bar rises to a height of 6. Thus six students would have a D grade.
2. From the histogram, 16 tests were 70–79, 8 tests were 80–89, and 6 tests were 90–99. When we combine $16 + 8 + 6 = 30$, we can see that 30 students scored greater than 69 on the test.

3. The 800–999 bar rises to a height of 20. Thus 20 light bulbs lasted between 800 and 999 hours.
4. From the histogram, 25 bulbs lasted 1200–1399 hours, 10 lasted 1400–1599 hours, and 5 lasted 1600–1799 hours. When we combine $25 + 10 + 5 = 40$, we can see that 40 light bulbs lasted more than 1199 hours.

5.

Weight in Pounds (Class Interval)	Tally	Frequency
1600–1799	||	2
1800–1999	||||	4
2000–2199	||||	4
2200–2399	⫿⫿⫿	5

6.

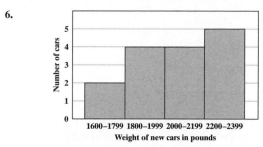

7. The greatest difference occurs between the 35–54 age category and the 55–64 category.

8.4 Student Practice

1. $\frac{\$39.20 + \$43.50 + \$81.90 + \$34.20 + \$51.70 + \$48.10}{6} \approx \$49.77$
   The mean monthly phone bill is $49.77.

2. $150, $150, $290 → $320 ← $400, $450, $600
   three numbers   middle number   three numbers
   Thus, $320 is the median salary.

3. 88, 90   100, 105   118, 126
   two numbers   two middle numbers   two numbers
   $\frac{100 + 105}{2} = \frac{205}{2} = 102.5$
   The median is 102.5.
4. The value 71 occurs twice. The mode is 71 inches.
5. Arrange the values in order from smallest to largest.
   2, 3, 3, 3, 5, 8, 8, 12, 14, 15, 16, 27, 28, 28, 31, 33
   The values 3, 8, and 28 repeat. Since the value 3 occurs three times, while 8 and 28 both occur only twice, the mode is 3.

Chapter 9   9.1 Student Practice

1. (a) 4 lies to the right of 2, so $4 > 2$.
   (b) -5 lies to the left of -3, so $-5 < -3$.
   (c) 0 lies to the right of -6, so $0 > -6$.
   (d) -2 lies to the left of 1, so $-2 < 1$.
   (e) 5 lies to the right of -7, so $5 > -7$.

2. (a)   9
        $+ 14$
         23

   (b) -4.5
        $+ -1.9$
         -6.4

3. (a)

$$\frac{5}{12} = \frac{5}{12}$$
$$+ \frac{1}{4} \times \frac{3}{3} = + \frac{3}{12}$$
$$\overline{}$$
$$\frac{8}{12} = \frac{2}{3}$$

(b) The LCD $= 42$.

$$\frac{1}{6} \times \frac{7}{7} = \frac{7}{42}$$

Because $\frac{1}{6} = \frac{7}{42}$ it follows that $-\frac{1}{6} = -\frac{7}{42}$.

$$\frac{2}{7} \times \frac{6}{6} = \frac{12}{42}$$

Because $\frac{2}{7} = \frac{12}{42}$ it follows that $-\frac{2}{7} = -\frac{12}{42}$.

Thus

$$-\frac{1}{6} \qquad\qquad -\frac{7}{42}$$
$$+ -\frac{2}{7} \text{ is equivalent to } + -\frac{12}{42}$$
$$\overline{} \qquad\qquad \overline{}$$
$$-\frac{19}{42}$$

4. Add $(-\$160 \text{ billion}) + (-\$1270 \text{ billion})$ to obtain $-\$1430$ billion. The total deficit for these two years is $\$1,430,000,000,000$.

5. (a)

$$\begin{array}{r} 7 \\ + -12 \\ \hline -5 \end{array}$$

(b)

$$\begin{array}{r} -20.8 \\ + 15.2 \\ \hline -5.6 \end{array}$$

(c) $\dfrac{5}{6} + \left(-\dfrac{3}{4}\right) = \dfrac{10}{12} + \left(-\dfrac{9}{12}\right) = \dfrac{10 + (-9)}{12} = \dfrac{1}{12}$

6.

$$\begin{array}{r} 28°F \\ + -19°F \\ \hline 9°F \end{array}$$

7.

$$\begin{array}{r} 36 \\ + -21 \\ \hline 15 \end{array} \text{ Then we add } \begin{array}{r} 15 \\ + -18 \\ \hline -3 \end{array}$$

Alternatively,

$$\begin{array}{r} -21 \\ + -18 \\ \hline -39 \end{array} \qquad \begin{array}{r} -39 \\ + 36 \\ \hline -3 \end{array}$$

8.

$$\begin{array}{r} \$30{,}000 \\ + \$40{,}000 \\ \hline \$70{,}000 \end{array} \quad \begin{array}{r} -\$20{,}000 \\ -\$5000 \\ \hline + -\$35{,}000 \end{array} \quad \begin{array}{r} \$70{,}000 \\ + -\$60{,}000 \\ \hline \$10{,}000 \end{array}$$
$$-\$60{,}000$$

The company had an overall profit of $\$10,000$ in the five-month period.

9.2 Student Practice

1. $-10 - (-5) = -10 + 5 = -5$

2. (a) $5 - 12 = 5 + (-12) = -7$

(b) $-11 - 17 = -11 + (-17) = -28$

3. (a) $3.6 - (-9.5) = 3.6 + 9.5 = 13.1$

(b) $-\dfrac{5}{8} - \left(-\dfrac{5}{24}\right) = -\dfrac{5}{8} + \dfrac{5}{24}$

$$= -\dfrac{5}{8} \times \dfrac{3}{3} + \dfrac{5}{24}$$

$$= -\dfrac{15}{24} + \dfrac{5}{24}$$

$$= -\dfrac{10}{24} = -\dfrac{5}{12}$$

4. (a) $20 - (-5) = 20 + 5 = 25$

(b) $-\dfrac{1}{5} - \left(-\dfrac{1}{2}\right) = -\dfrac{1}{5} + \dfrac{1}{2} = -\dfrac{2}{10} + \dfrac{5}{10} = \dfrac{3}{10}$

(c) $3.6 - (-5.5) = 3.6 + 5.5 = 9.1$

5. $-5 - (-9) + (-14) = -5 + 9 + (-14) = 4 + (-14) = -10$

6. $31 - (-37) = 31 + 37 = 68$ The difference is $68°F$.

9.3 Student Practice

1. (a) $(6)(9) = 54$ **(b)** $(7)(12) = 84$

2. (a) $(-8)(5) = -40$ **(b)** $3(-60) = -180$

3. (a) $-50 \div 25 = -2$ **(b)** $49 \div (-7) = -7$

4. (a) $-10(-6) = 60$ **(b)** $\left(-\dfrac{1}{3}\right)\left(-\dfrac{2}{7}\right) = \dfrac{2}{21}$

5. (a) $-78 \div (-2) = 39$ **(b)** $(-1.2) \div (-0.5) = 2.4$

6. $(-6)(3)(-4) = (-18)(-4) = 72$

7. $(-2)(6) = -12$

Thus the change in charge would be -12.

8. $\dfrac{17 + 19 + 2 + (-4) + (-3) + (-13)}{6} = \dfrac{18}{6} = 3$

The average temperature was $3°F$.

9.4 Student Practice

1. $20 \div (-5) \, (-3)$

$$= (-4) \, (-3)$$

$$= 12$$

2. (a) $25 \div (-5) \quad + \quad 16 \div (-8)$ **(b)** $9 + 20 \div (-4)$

$$= (-5) \quad + \quad (-2) \qquad\qquad = 9 + (-5)$$

$$= -7 \qquad\qquad\qquad\qquad = 4$$

3. $\dfrac{9(-3) - 5}{2(-4) \div (-2)} = \dfrac{-27 - 5}{-8 \div (-2)} = \dfrac{-32}{4} = -8$

4. $-2(-12 + 15) + (-3)^4 + 2(-6)$

$$= -2(3) + (-3)^4 + 2(-6)$$

$$= -2(3) + 81 + 2(-6)$$

$$= -6 + 81 + (-12)$$

$$= 63$$

5. $\left(\dfrac{1}{5}\right)^2 + 4\left(\dfrac{1}{5} - \dfrac{3}{10}\right) \div \dfrac{2}{3}$

$$= \left(\dfrac{1}{5}\right)^2 + 4\left(\dfrac{2}{10} - \dfrac{3}{10}\right) \div \dfrac{2}{3}$$

$$= \left(\dfrac{1}{5}\right)^2 + 4\left(-\dfrac{1}{10}\right) \div \dfrac{2}{3}$$

$$= \dfrac{1}{25} + 4\left(-\dfrac{1}{10}\right) \div \dfrac{2}{3}$$

$$= \dfrac{1}{25} + \left(-\dfrac{2}{5}\right) \div \dfrac{2}{3}$$

$$= \dfrac{1}{25} + \left(-\dfrac{2}{5}\right) \times \dfrac{3}{2}$$

$$= \dfrac{1}{25} + \left(-\dfrac{3}{5}\right)$$

$$= \dfrac{1}{25} + \left(-\dfrac{15}{25}\right)$$

$$= -\dfrac{14}{25}$$

9.5 Student Practice

1. (a) Move the decimal point three places to the left.
$3729 = 3.729 \times 10^3$

(b) Move the decimal point five places to the left.
$506{,}936 = 5.06936 \times 10^5$

2. (a) Move the decimal point two places to the right.
$0.076 = 7.6 \times 10^{-2}$

(b) Move the decimal point one place to the right.
$0.982 = 9.82 \times 10^{-1}$

3. Move the decimal point three places to the right.
$6.543 \times 10^3 = 6543$

4. (a) Move the decimal point five places to the right. Add four zeros.
$4.3 \times 10^5 = 430{,}000$

(b) Move the decimal point four places to the right. Add four zeros.
$6 \times 10^4 = 60{,}000$

5. **(a)** Move the decimal point three places to the left. Add two zeros.
$7.72 \times 10^{-3} = 0.00772$
(b) Move the decimal point five places to the left. Add four zeros.
$2.6 \times 10^{-5} = 0.000026$

6. 6.85×10^{22} kilograms
$\underline{+\ 2.09 \times 10^{22} \text{ kilograms}}$
8.94×10^{22} kilograms

7. $3.1 \times 10^4 = 31,000$
But $31,000 = 0.31 \times 10^5$
4.36×10^5
$\underline{-\ 0.31 \times 10^5}$
4.05×10^5

Chapter 10   10.1 Student Practice

1. **(a)** The variables are A, b, and h.
(b) The variables are V, l, w, and h.
2. **(a)** $P = 2w + 2l$
(b) $A = \pi r^2$
3. We add $9 + 2 = 11$, therefore $9x + 2x = 11x$.
4. **(a)** $8x - 22x + 5x = 8x + (-22x) + 5x = -14x + 5x = -9x$
(b) $19x - 7x - 12x = 19x + (-7x) + (-12x)$
$= 12x + (-12x) = 0x = 0$
5. **(a)** $9x - 12x + x = 9x + (-12x) + 1x = -3x + 1x = -2x$
(b) $5.6x - 8x - x = 5.6x - 8x - 1x = -2.4x - 1x = -3.4x$
6. $17.5 - 6.3x - 8.2x + 10.5$
$= 17.5 + 10.5 - 6.3x - 8.2x = 28 - 14.5x$
7. **(a)** $2w + 3z - 12 - 5w - z - 16$
$= 2w - 5w + 3z - 1z - 12 - 16$
$= -3w + 2z - 28$
(b) $\frac{3}{5}x - \frac{7}{15}x + 5 - \frac{1}{3} = \frac{9}{15}x - \frac{7}{15}x + \frac{15}{3} - \frac{1}{3} = \frac{2}{15}x + \frac{14}{3}$

10.2 Student Practice

1. **(a)** $7(x + 5) = 7(x) + 7(5) = 7x + 35$
(b) $-4(x + 2y) = -4(x) + (-4)(2y) = -4x - 8y$
(c) $5(6a - 2b) = 5(6a) - 5(2b) = 30a - 10b$
2. $(x + 3y)(8) = (x)(8) + (3y)(8) = 8x + 24y$
3. **(a)** $-5(x + 4y + 5) = -5(1x) + (-5)(4y) + (-5)(5)$
$= -5x - 20y - 25$
(b) $(2.2x + 5.5y + 6)(3) = (2.2x)(3) + (5.5y)(3) + (6)(3)$
$= 6.6x + 16.5y + 18$
4. $\frac{3}{2}\left(\frac{1}{2}x - \frac{1}{3}y + 4z - \frac{1}{2}\right)$
$= \frac{3}{2}\left(\frac{1}{2}x\right) + \frac{3}{2}\left(-\frac{1}{3}y\right) + \frac{3}{2}(4z) + \frac{3}{2}\left(-\frac{1}{2}\right)$
$= \frac{3}{4}x - \frac{1}{2}y + 6z - \frac{3}{4}$
5. $3(2x + 4y) + 2(5x + y) = 6x + 12y + 10x + 2y = 16x + 14y$
6. $-4(x - 5) + 3(-3y + 2x)$
$= -4x + 20 - 9y + 6x$
$= 2x - 9y + 20$

10.3 Student Practice

1. $x + 7 = -8$
$x + 7 + (-7) = -8 + (-7)$
$x + 0 = -15$
$x = -15$
2. **(a)** $y - 3.2 = 9$ **Check.**
$y - 3.2 + 3.2 = 9.0 + 3.2$ $y - 3.2 = 9$
$y = 12.2$ $12.2 - 3.2 \overset{?}{=} 9$
$9 = 9$ ✓
(b) $\frac{2}{3} = x + \frac{1}{6}$ **Check.**
$\frac{4}{6} + \left(-\frac{1}{6}\right) = x + \frac{1}{6} + \left(-\frac{1}{6}\right)$ $\frac{2}{3} = x + \frac{1}{6}$
$\frac{3}{6} = x$ $\frac{2}{3} \overset{?}{=} \frac{1}{2} + \frac{1}{6}$
$\frac{1}{2} = x$ $\frac{2}{3} \overset{?}{=} \frac{3}{6} + \frac{1}{6}$
$\frac{2}{3} = \frac{4}{6}$ ✓

3. $3x - 5 = 2x + 1$ **Check.**
$3x - 5 + 5 = 2x + 1 + 5$ $3x - 5 = 2x + 1$
$3x = 2x + 6$ $3(6) - 5 \overset{?}{=} 2(6) + 1$
$3x + (-2x) = 2x + (-2x) + 6$ $18 - 5 \overset{?}{=} 12 + 1$
$x = 6$ $13 = 13$ ✓

10.4 Student Practice

1. $8n = 104$
$\frac{8n}{8} = \frac{104}{8}$
$n = 13$
2. $-7n = 30$
$\frac{-7n}{-7} = \frac{30}{-7}$
$n = -\frac{30}{7}$
3. $3.2x = 16$ **Check.**
$\frac{3.2x}{3.2} = \frac{16}{3.2}$ $3.2(5) \overset{?}{=} 16$
$x = 5$ $16 = 16$ ✓
4. **(a)** $\frac{1}{6}y = 2\frac{2}{3}$ **Check.**
$\frac{1}{6}y = \frac{8}{3}$ $\frac{1}{6}y = 2\frac{2}{3}$
$\frac{6}{1} \cdot \frac{1}{6}y = \frac{8}{3} \cdot \frac{6}{1}$ $\frac{1}{6} \cdot 16 \overset{?}{=} 2\frac{2}{3}$
$y = 16$ $\frac{16}{6} \overset{?}{=} \frac{8}{3}$
$\frac{8}{3} = \frac{8}{3}$ ✓
(b) $3\frac{1}{5}z = 4$ **Check.**
$\frac{16}{5}z = 4$ $3\frac{1}{5}z = 4$
$\frac{5}{16} \cdot \frac{16}{5}z = 4 \cdot \frac{5}{16}$ $\frac{16}{5} \cdot \frac{5}{4} \overset{?}{=} 4$
$z = \frac{5}{4} \text{ or } 1\frac{1}{4}$ $4 = 4$ ✓

10.5 Student Practice

1. $5x + 13 = 33$ **Check.**
$5x + 13 + (-13) = 33 + (-13)$ $5(4) + 13 \overset{?}{=} 33$
$5x = 20$ $20 + 13 \overset{?}{=} 33$
$\frac{5x}{5} = \frac{20}{5}$ $33 = 33$ ✓
$x = 4$
2. $-50 = 7x - 8$ **Check.**
$-50 + 8 = 7x - 8 + 8$ $-50 \overset{?}{=} 7(-6) - 8$
$-42 = 7x$ $-50 \overset{?}{=} -42 - 8$
$\frac{-42}{7} = \frac{7x}{7}$ $-50 = -50$ ✓
$-6 = x$

3.
$$4x = -8x + 42$$
$$4x + 8x = -8x + 8x + 42$$
$$12x = 42$$
$$\frac{12x}{12} = \frac{42}{12}$$
$$x = \frac{7}{2} \text{ or } 3\frac{1}{2}$$

4.
$$4x - 7 = 9x + 13$$
$$4x - 7 + 7 = 9x + 13 + 7$$
$$4x = 9x + 20$$
$$4x + (-9x) = 9x + (-9x) + 20$$
$$-5x = 20$$
$$\frac{-5x}{-5} = \frac{20}{-5}$$
$$x = -4$$

5.
$$4x - 23 = 3x + 7 - 2x$$
$$4x - 23 = x + 7$$
$$4x + (-1x) - 23 = (-1x) + x + 7$$
$$3x - 23 = 7$$
$$3x - 23 + 23 = 7 + 23$$
$$3x = 30$$
$$\frac{3x}{3} = \frac{30}{3}$$
$$x = 10$$

6.
$$8(x - 3) + 5x = 15(x - 2)$$
$$8x - 24 + 5x = 15x - 30$$
$$13x - 24 = 15x - 30$$
$$13x - 24 + 30 = 15x - 30 + 30$$
$$13x + 6 = 15x$$
$$13x + (-13x) + 6 = 15x + (-13x)$$
$$6 = 2x$$
$$\frac{6}{2} = \frac{2x}{2}$$
$$3 = x$$

10.6 Student Practice

1. Tom's height   is   7 inches   more than   Abdul's height
   ↓   ↓   ↓   ↓   ↓
   t =   7   + a

2. The noon class   has   24 fewer students than   the morning class
   ↓   ↓   ↓   ↓
   n = m −   24

3. On Thursday she carried 5 more than on Friday. $t = 5 + f$

4. Double the width is $2w$. $l = 2w + 7$

5. Let s = length in miles of Sally's trip
   $s + 380$ = length in miles of Melinda's trip

   380 miles longer than Sally's trip.

6. Let m = height in feet of McCormick Hall
   $m - 126$ = height in feet of Larson Center

   126 feet shorter than McCormick Hall.

7. Let x = length in inches of the first side of the triangle
   $2x$ = length in inches of the second side of the triangle
   $x + 6$ = length in inches of the third side of the triangle

10.7 Student Practice

1. Let x = length in feet of shorter piece of board
   $x + 4.5$ = length in feet of longer piece of board
$$x + (x + 4.5) = 18$$
$$x + x + 4.5 = 18$$
$$2x + 4.5 = 18$$
$$2x + 4.5 + (-4.5) = 18 + (-4.5)$$
$$2x = 13.5$$
$$\frac{2x}{2} = \frac{13.5}{2}$$
$$x = 6.75$$

The shorter piece is 6.75 feet long.
$$x + 4.5 = 6.75 + 4.5 = 11.25$$
The longer piece is 11.25 feet long.
Check.
$$6.75 + 11.25 \overset{?}{=} 18$$
$$18 = 18 \checkmark$$

$$11.25 \overset{?}{=} 6.75 + 4.5$$
$$11.25 = 11.25 \checkmark$$

2. Let x = the number of departures on Monday
   $x + 29$ = the number of departures on Tuesday
   $x - 16$ = the number of departures on Wednesday
$$x + (x + 29) + (x - 16) = 349$$
$$x + x + 29 + x - 16 = 349$$
$$3x + 13 = 349$$
$$3x + 13 + (-13) = 349 + (-13)$$
$$3x = 336$$
$$\frac{3x}{3} = \frac{336}{3}$$
$$x = 112$$
$$x + 29 = 112 + 29 = 141$$
$$x - 16 = 112 - 16 = 96$$
There were 112 departures on Monday, 141 departures on Tuesday, and 96 departures on Wednesday.
Check.
$$112 + 141 + 96 \overset{?}{=} 349$$
$$349 = 349 \checkmark$$
$$141 \overset{?}{=} 112 + 29$$
$$141 = 141 \checkmark$$

$$96 \overset{?}{=} 112 - 16$$
$$96 = 96 \checkmark$$

3. Let w = the width of the field measured in feet
   $2w + 8$ = the length of the field measured in feet
   2(width) + 2(length) = perimeter
$$2(w) + 2(2w + 8) = 772$$
$$2w + 4w + 16 = 772$$
$$6w + 16 = 772$$
$$6w + 16 + (-16) = 772 + (-16)$$
$$6w = 756$$
$$\frac{6w}{6} = \frac{756}{6}$$
$$w = 126$$
The width of the field is 126 feet.
$$2w + 8 = 2(126) + 8 = 252 + 8 = 260$$
The length of the field is 260 feet.
Check.
$$260 \overset{?}{=} 2(126) + 8$$
$$260 \overset{?}{=} 252 + 8$$
$$260 = 260 \checkmark$$
$$2(126) + 2(260) \overset{?}{=} 772$$
$$252 + 520 \overset{?}{=} 772$$
$$772 = 772 \checkmark$$

4. Let x = the length in meters of the first side of the triangle
   $2x$ = the length in meters of the second side of the triangle
   $x + 10$ = the length in meters of the third side of the triangle
$$x + 2x + (x + 10) = 36$$
$$4x + 10 = 36$$
$$4x + 10 + (-10) = 36 + (-10)$$
$$4x = 26$$
$$\frac{4x}{4} = \frac{26}{4}$$
$$x = 6.5$$
The first side of the triangle is 6.5 meters long.
$$2x = 2(6.5) = 13$$
The second side of the triangle is 13 meters long.
$$x + 10 = 6.5 + 10 = 16.5$$
The third side of the triangle is 16.5 meters long.

Check.

$6.5 + 13 + 16.5 \stackrel{?}{=} 36$

$36 = 36$ ✓

$13 \stackrel{?}{=} 2(6.5)$

$13 = 13$ ✓

$16.5 \stackrel{?}{=} 10 + 6.5$

$16.5 = 16.5$ ✓

5. Let x = the measure of angle A in degrees

$3x$ = the measure of angle C in degrees

$x - 30$ = the measure of angle B in degrees

$x + 3x + (x - 30) = 180$

$5x - 30 = 180$

$5x - 30 + 30 = 180 + 30$

$5x = 210$

$\dfrac{5x}{5} = \dfrac{210}{5}$

$x = 42$

Angle A measures $42°$.

$3x = 3(42) = 126$

Angle C measures $126°$.

$x - 30 = 42 - 30 = 12$

Angle B measures $12°$.

Check.

$42 + 126 + 12 \stackrel{?}{=} 180$

$180 = 180$ ✓

$126 \stackrel{?}{=} 3(42)$

$126 = 126$ ✓

$12 \stackrel{?}{=} 42 - 30$

$12 = 12$ ✓

6. Let s = the total amount of sales of the boats in dollars

$0.03s$ = the amount of the commission earned on sales of s dollars

$3250 = 1000 + 0.03s$

$2250 = 0.03s$

$\dfrac{2250}{0.03} = \dfrac{0.03s}{0.03}$

$75,000 = s$

Therefore he sold $75,000 worth of boats for the month.

Check.

$0.03(75,000) + 1000 \stackrel{?}{=} 3250$

$2250 + 1000 \stackrel{?}{=} 3250$

$3250 = 3250$ ✓

Appendix A.1 Balancing a Checking Account

Student Practice

1.

CHECK REGISTER	*My Chung Ngayen*					20 *11*	
CHECK NO.	DATE	DESCRIPTION OF TRANSACTION	PAYMENT/ DEBIT (−)	✓	DEPOSIT/ CREDIT (+)	BALANCE $ *1434*	*52*
144	3/1	*Leland Mortgage Company*	908 00			526	52
145	3/1	*Phone Company*	33 21			493	31
146	3/2	*Sam's Food Market*	102 37			390	94
	3/2	*Deposit*			524 41	915	35

To find the ending balance we subtract each check written and add the deposit to the current balance. Then we record these amounts in the check register.

$$
\begin{array}{cccc}
1434.52 & 526.52 & 493.31 & 390.94 \\
-\ 908.00 & -\ 33.21 & -\ 102.37 & +\ 524.41 \\
\hline
526.52 & 493.31 & 390.94 & 915.35 \\
\end{array}
$$

My Chung's balance is $915.35

2.

CHECKING RECONCILEMENT	This form is provided to assist you in balancing your checking account.

List checks outstanding* not charged to your checking account			Period ending 8/1 ,20 11		
CHECK NO.	AMOUNT		1. Check Register Balance	$	1050.46
215	89 75		Subtract any charges listed on the bank statement which you have not previously deducted from your balance.	− $	12.75
217	205 99		Adjusted Check Register Balance	$	1037.71
			2. Enter the ending balance shown on the bank statement.	$	934.95
			3. Enter deposits made later than the ending date on the bank statement.	+ $	398.50
				+ $	
				+ $	
			TOTAL (Step 2 plus Step 3)	$	1333.45
			4. In your check register, check off all the checks paid. In the area provided to the left, list numbers and amounts of all outstanding checks and ATM withdrawals.		
TOTAL	295 74		5. Subtract the total amount in Step 4.	− $	295.74
* and ATM withdrawals			6. This adjusted bank balance should equal the adjusted Check Register Balance from Step 1.	$	1037.71

The balances in steps **1** and **6** are equal, so Anthony's checkbook is balanced.

Appendix A.2 Determining the Best Deal When Purchasing a Vehicle

Student Practice

1. (a) sales tax = 7% of sale price

= $0.07 \times 24,999$

sales tax = $1749.93

license fee = 2% of sale price

= $0.02 \times 24,999$

license fee = $499.98

(b) purchase price = sale price + sales tax + license fee + extended warranty

= $24,999 + 1749.93 + 499.98 + 1275$

purchase price = $28,523.91

2. (a) down payment = percent × purchase price.

down payment = 15% × 32,499

= $0.15 \times 32,499$

down payment = $4874.85

(b) amount financed = purchase price − down payment

= $32,499 - 4874.85$

amount financed = $27,624.15

3. First, we find the total cost of the minivan at Dealer 1.

total cost = (monthly payment × number of months in loan) + down payment

= $(398.65 \times 60) + 0$ There is no down payment.

= 23,919.00

The total cost of the minivan at Dealership 1 is $23,919.00.

Next, we find the cost of the minivan at Dealer 2.

total cost = (monthly payment × number of months in loan) + down payment

= $(393.10 \times 48) + 5500$ We multiply, then add.

= 24,368.80

The total cost of the minivan at Dealership 2 is $24,368.80.

We see that the best deal on the minivan Phoebe plans to buy is at Dealership 1.

Answers to Selected Exercises

Chapter 1

1.1 Exercises **1.** 6000 + 700 + 30 + 1 **3.** 100,000 + 8000 + 200 + 70 + 6
5. 20,000,000 + 3,000,000 + 700,000 + 60,000 + 1000 + 300 + 40 + 5 **7.** 100,000,000 + 3,000,000 + 200,000 + 60,000 + 700 + 60 + 8
9. 671 **11.** 9863 **13.** 40,885 **15.** 706,200 **17. (a)** 7 **(b)** 30,000 **19. (a)** 2 **(b)** 200,000 **21.** one hundred forty-two
23. nine thousand, three hundred four **25.** thirty-six thousand, one hundred eighteen **27.** one hundred five thousand, two hundred sixty-one
29. fourteen million, two hundred three thousand, three hundred twenty-six **31.** four billion, three hundred two million, one hundred fifty-six thousand,
two hundred **33.** 1561 **35.** 33,809 **37.** 100,079,826 **39.** one thousand, nine hundred sixty-five **41.** 9 million or 9,000,000
43. 42 million or 42,000,000 **45.** 946,000 flights **47.** 54,558,000 passengers **49. (a)** 5 **(b)** 2 **51. (a)** 2 **(b)** 1 **53.** 613,001,033,208,003
55. three quintillion, six hundred eighty-two quadrillion, nine hundred sixty-eight trillion, nine billion, nine hundred thirty-one million, nine hundred
sixty thousand, seven hundred forty-seven **57.** You would obtain 2 E 20. This is 200,000,000,000,000,000,000 in standard notation.

Quick Quiz 1.1 *See Examples noted with Ex.* **1.** 70,000 + 3000 + 900 + 50 + 2 (Ex. 1) **2.** eight million, nine hundred thirty-two thousand,
four hundred seventy-five (Ex. 4) **3.** 964,257 (Ex. 7) **4.** See Student Solutions Manual

1.2 Exercises **1. (a)** You can change the order of the addends without changing the sum. **(b)** You can group the addends in any way without
changing the sum.

3.

+	3	5	4	8	0	6	7	2	9	1
2	5	7	6	10	2	8	9	4	11	3
7	10	12	11	15	7	13	14	9	16	8
5	8	10	9	13	5	11	12	7	14	6
3	6	8	7	11	3	9	10	5	12	4
0	3	5	4	8	0	6	7	2	9	1
4	7	9	8	12	4	10	11	6	13	5
1	4	6	5	9	1	7	8	3	10	2
8	11	13	12	16	8	14	15	10	17	9
6	9	11	10	14	6	12	13	8	15	7
9	12	14	13	17	9	15	16	11	18	10

5. 23 **7.** 26 **9.** 57 **11.** 99 **13.** 4125 **15.** 9994 **17.** 13,861 **19.** 117,240
21. 121 **23.** 1143 **25.** 10,130 **27.** 11,579,426 **29.** 1,135,280,240 **31.** 2,303,820
33. 300 **35.** 335 **37.** $723 **39.** $5549 **41.** 468 feet **43.** 968,106 square miles
45. 16,934,720 yards **47. (a)** 1134 students **(b)** 1392 students **49.** 202 miles
51. 434 feet **53. (a)** $9553 **(b)** $7319 **(c)** $13,047 **55.** 1161 **57.** Answers
may vary. A sample is: You could not group the addends in groups that sum to 10s to make
column addition easier. **58.** seventy-six million, two hundred eight thousand, nine
hundred forty-one. **59.** one hundred twenty-one million, three hundred seventy-four
60. 8,724,396 **61.** 9,051,719 **62.** 28,387,018

Quick Quiz 1.2 *See Examples noted with Ex.* **1.** 212 (Ex. 6)
2. 1615 (Ex. 6) **3.** 1,004,811 (Ex. 7) **4.** See Student Solutions Manual

1.3 Exercises **1.** In subtraction the minuend minus the subtrahend equals the difference. To check the problem we add the subtrahend
and the difference to see if we get the minuend. If we do, the answer is correct. **3.** We know that 1683 + 1592 = 32?5. Therefore if we add
8 tens and 9 tens we get 17 tens, which is 1 hundred and 7 tens. Thus the ? should be replaced by 7. **5.** 5 **7.** 6 **9.** 16 **11.** 9
13. 7 **15.** 6 **17.** 3 **19.** 9 **21.** 21

26
+ 21
47

23. 12

73
+ 12
85

25. 343

36
+ 343
379

27. 321

548
+ 321
869

29. 4203

596
+ 4203
4799

31. 143,235

12,600
+ 143,235
155,835

33. 553,101

433,201
+ 553,101
986,302

35.

19
+ 110
129

Correct

37.

3215
+ 5781
8996

Incorrect
Correct answer: 5381

39.

5020
+ 1020
6040

Incorrect
Correct answer: 1010

41.

33,846
+ 13,023
46,869

Incorrect
Correct answer: 14,023

43. 46 **45.** 92 **47.** 384 **49.** 718 **51.** 10,715 **53.** 34,092 **55.** 7447 **57.** 908,930 **59.** $x = 5$ **61.** $x = 8$ **63.** $x = 27$
65. 422,310 votes **67.** 6,237,414 **69.** $762 **71.** 1,614,533 people **73.** 2,108,790 people **75.** 320,317 people **77.** 93,154 people
79. 93 homes **81.** 13 homes **83.** between 2010 and 2011 **85.** Willow Creek and Harvey **87.** It is true if a and b represent the same
number, for example, if $a = 10$ and $b = 10$. **89.** $550 **91.** 8,466,084 **92.** two hundred ninety-six thousand, three hundred eight **93.** 218
94. 1,174,750

Quick Quiz 1.3 *See Examples noted with Ex.* **1.** 4454 (Ex. 5) **2.** 222,933 (Ex. 8) **3.** 5,638,122 (Ex. 8)
4. See Student Solutions Manual

1.4 Exercises **1. (a)** You can change the order of the factors without changing the product. **(b)** You can group the factors in any way
without changing the product.

3.

×	6	2	3	8	0	5	7	9	12	4
5	30	10	15	40	0	25	35	45	60	20
7	42	14	21	56	0	35	49	63	84	28
1	6	2	3	8	0	5	7	9	12	4
0	0	0	0	0	0	0	0	0	0	0
6	36	12	18	48	0	30	42	54	72	24
2	12	4	6	16	0	10	14	18	24	8
3	18	6	9	24	0	15	21	27	36	12
8	48	16	24	64	0	40	56	72	96	32
4	24	8	12	32	0	20	28	36	48	16
9	54	18	27	72	0	45	63	81	108	36

5. 96   **7.** 70   **9.** 522   **11.** 693   **13.** 1932   **15.** 18,306   **17.** 36,609
19. 31,308   **21.** 100,208   **23.** 3,101,409   **25.** 1560   **27.** 2,715,800   **29.** 482,000
31. 372,560,000   **33.** 8460   **35.** 88,400   **37.** 56,000,000   **39.** 6168   **41.** 7884
43. 5696   **45.** 15,175   **47.** 20,672   **49.** 69,312   **51.** 148,567   **53.** 823,823
55. 1,881,810   **57.** 143,812   **59.** 217,980   **61.** 2,653,296   **63.** 720,000
65. 10,000   **67.** 90,600   **69.** 70   **71.** 308   **73.** 13,596   **75.** 1600   **77.** $x = 0$
79. 384 square feet   **81.** 195 square feet   **83.** $1200   **85.** $3192   **87.** 612 miles
89. $5040   **91.** $33,920,000,000   **93.** 198   **95.** 62   **97.** $x = 8$ **99.** $x = 9$
101. No, it would not always be true. In our number system $62 = 60 + 2$. But in Roman numerals IV $\neq$ I + V. The digit system in Roman numerals involves subtraction. Thus (XII) × (IV) $\neq$ (XII × I) + (XII × V).   **103.** 6756   **104.** 1249   **105.** $805
106. $160   **107.** 3251 people   **108.** $745,200,000,000

Quick Quiz 1.4 *See Examples noted with Ex.* **1.** 174,930 (Ex. 4)   **2.** 5056 (Ex. 7)
3. 207,306 (Ex. 11)   **4.** See Student Solutions Manual

1.5 Exercises
1. (a) When you divide a nonzero number by itself, the result is one.   **(b)** When you divide a number by 1, the result is that number.
(c) When you divide zero by a nonzero number, the result is zero.   **(d)** You cannot divide a number by zero. Division by zero is undefined.
3. 7   **5.** 3   **7.** 5   **9.** 4   **11.** 3   **13.** 6   **15.** 9   **17.** 9   **19.** 6   **21.** 9   **23.** 0   **25.** undefined   **27.** 0   **29.** 1   **31.** 4 R 5
33. 9 R 4   **35.** 25 R 3   **37.** 21 R 7   **39.** 32   **41.** 37   **43.** 322 R 1   **45.** 127 R 1   **47.** 869   **49.** 1238 R 2   **51.** 2056 R 2
53. 2562 R 3   **55.** 30 R 5   **57.** 5 R 7   **59.** 8   **61.** 418 R 8   **63.** 48 R 12   **65.** 845   **67.** 210 R 8   **69.** 14 R 2   **71.** 4 R 4
73. 125   **75.** $x = 37$ **77.** 61,693 runs per day   **79.** $288   **81.** $21,053   **83.** $185   **85.** 165 sandwiches   **87. (a)** 41,808 km   **(b)** 8192 km
89. a and b must represent the same number. For example, if $a = 12$, then $b = 12$.   **91.** 5400   **92.** 1,038,490   **93.** 406,195   **94.** 66,844

Quick Quiz 1.5 *See Examples noted with Ex.* **1.** 467 (Ex. 5)   **2.** 3287 R 3 (Ex. 5)   **3.** 328 (Ex. 7)   **4.** See Student Solutions Manual

How Am I Doing?   Sections 1.1–1.5
1. seventy-eight million, three hundred ten thousand, four hundred thirty-six. (obj. 1.1.3)
2. 30,000 + 8000 + 200 + 40 + 7 (obj. 1.1.1)   **3.** 5,064,122 (obj. 1.1.2)   **4.** 17,487,000 students (obj. 1.1.4)   **5.** 20,080,000 students (obj. 1.1.4)
6. 244 (obj. 1.2.4)   **7.** 50,570 (obj. 1.2.4)   **8.** 1,351,461 (obj. 1.2.4)   **9.** 3993 (obj. 1.3.3)   **10.** 76,311 (obj. 1.3.3)   **11.** 1,981,652 (obj. 1.3.3)
12. 108 (obj. 1.4.1)   **13.** 100,000 (obj. 1.4.3)   **14.** 18,606 (obj. 1.4.2)   **15.** 3740 (obj. 1.4.4)   **16.** 331,420 (obj. 1.4.4)   **17.** 10,605 (obj. 1.5.2)
18. 7376 R 1 (obj. 1.5.2)   **19.** 26 R 8 (obj. 1.5.3)   **20.** 139 (obj. 1.5.3)

1.6 Exercises
1. 5^3 means $5 \times 5 \times 5$. $5^3 = 125$.   **3.** base
5. To ensure consistency we
   1. perform operations inside parentheses
   2. simplify any expressions with exponents
   3. multiply or divide from left to right
   4. add or subtract from left to right
7. 6^4 **9.** 4^7 **11.** 9^4 **13.** 7^1 **15.** 16   **17.** 64   **19.** 36   **21.** 10,000   **23.** 1   **25.** 64   **27.** 243   **29.** 225   **31.** 343   **33.** 256
35. 1   **37.** 625   **39.** 1,000,000   **41.** 169   **43.** 9   **45.** 64   **47.** 10   **49.** 108   **51.** 520   **53.** $56 - 20 = 36$ **55.** $27 - 5 = 22$
57. $48 \div 8 + 4 = 6 + 4 = 10$ **59.** $3 \times 36 - 50 = 108 - 50 = 58$ **61.** $100 + 3 \times 5 = 100 + 15 = 115$ **63.** $20 \div 20 = 1$
65. $950 \div 5 = 190$ **67.** $56 - 18 = 38$ **69.** $9 + 16 \div 4 = 9 + 4 = 13$ **71.** $42 - 4 \div 4 = 42 - 1 = 41$ **73.** $100 - 9 \times 4 = 100 - 36 = 64$
75. $25 + 4 + 27 = 56$ **77.** $8 \times 3 \times 1 \div 2 = 24 \div 2 = 12$ **79.** $144 - 0 = 144$ **81.** $16 \times 6 \div 3 = 96 \div 3 = 32$
83. $60 - 40 + 10 = 20 + 10 = 30$ **85.** $3 + 9 \times 6 + 4 = 3 + 54 + 4 = 61$ **87.** $32 \div 2 \times 16 = 16 \times 16 = 256$
89. $9 \times 6 \div 9 + 4 \times 3 = 6 + 12 = 18$ **91.** $1 + 125 + 9 = 135$ **93.** $1200 - 8(3) \div 6 = 1200 - 4 = 1196$ **95.** $120 \div 40 - 1 = 3 - 1 = 2$
97. $4 + 10 - 1 = 13$ **99.** $7 \times 3 - (2)^3 + 3^0 = 21 - 8 + 1 = 14$ **101.** 86,164 seconds   **103. (a)** 3   **(b)** 2,000,000   **104.** 200,765,909
105. two hundred sixty-one million, seven hundred sixty-three thousand, two   **106.** 1460 feet of fencing will be needed. 120,000 square feet of grass must be planted

Quick Quiz 1.6 *See Examples noted with Ex.* **1.** 12^5 (Ex. 1)   **2.** 1296 (Ex. 2)   **3.** 91 (Ex. 8)   **4.** See Student Solutions Manual

1.7 Exercises
1. Locate the rounding place. If the digit to the right of the rounding place is 5 or greater than 5, round up. If the digit to the right of the rounding place is less than 5, round down.   **3.** 80   **5.** 70   **7.** 170   **9.** 7440   **11.** 2960   **13.** 200   **15.** 2800
17. 7700   **19.** 8000   **21.** 1000   **23.** 28,000   **25.** 800,000   **27.** 15,000,000 stars   **29. (a)** 373,500,000   **(b)** 400,000,000
31. (a) 3,700,000 square miles; 9,600,000 square kilometers   **(b)** 3,710,000 square miles; 9,600,000 square kilometers

33.
```
   800
   300
 + 200
  1300
```
35.
```
   40
   70
  100
 + 20
  230
```
37.
```
 200,000
  50,000
 + 9,000
 259,000
```
39.
```
 300,000
 - 70,000
 230,000
```
41.
```
 800,000
 - 80,000
 720,000
```
43.
```
 30,000,000
 - 20,000,000
 10,000,000
```
45.
```
   50
 × 60
 3000
```
47.
```
 1000
 ×  8
 8000
```
49.
```
   600,000
 ×    300
 180,000,000
```
51. $40\overline{)6000}$ **53.** $40\overline{)400,000}$ **55.** $800\overline{)4,000,000}$

57. Incorrect
```
  400
  500
  900
+ 200
 2000
```
(correct: 150 ... 5000)

59. Incorrect
```
 100,000
  50,000
+ 40,000
 190,000
```
(correct: 150 ... 10,000)

61.
$$\begin{array}{r} 300{,}000 \\ -\ 90{,}000 \\ \hline 210{,}000 \end{array}$$ Correct
63.
$$\begin{array}{r} 80{,}000{,}000 \\ -\ 50{,}000{,}000 \\ \hline 30{,}000{,}000 \end{array}$$ Incorrect
65.
$$\begin{array}{r} 400 \\ \times\ 30 \\ \hline 12{,}000 \end{array}$$ Incorrect
67.
$$\begin{array}{r} 6000 \\ \times\ 70 \\ \hline 420{,}000 \end{array}$$ Correct
69. $40\overline{)80{,}000}$ Correct
71. $400\overline{)200{,}000}$ Correct

73. 400 square feet   **75.** 12,000,000 people   **77.** 30,000 pizzas   **79.** 20,600,000 passengers   **81.** $590{,}000 - 270{,}000 = 320{,}000$ square miles
83. (a) 400,000 hours   **(b)** 20,000 days   **85.** 83   **86.** 27   **87.** 28   **88.** 66   **89.** 367,763   **90.** 87

Quick Quiz 1.7 *See Examples noted with Ex.* **1.** 92,400 (Ex. 1)   **2.** 2,340,000 (Ex. 4)   **3.** 2,400,000,000 (Ex. 10)
4. See Student Solutions Manual

1.8 Exercises **1.** $8800   **3.** 1560 bagels   **5.** 7¢ per ounce   **7.** $64   **9.** 90,837,000 people   **11.** $20,382   **13.** 25,231; 466
15. 800,000 people   **17.** $192   **19.** $1360   **21.** $16,405   **23.** 25 miles per gallon   **25.** There are 54 oak trees, 108 maple trees,
and 756 pine trees. In total there are 936 trees.   **27.** 1401 stations   **29.** 1151 stations   **31.** 57,000,000 households   **33.** 112,000,000 households
35. 343   **36.** 21   **37.** 4788   **38.** 258   **39.** 802   **40.** 23,285   **41.** 526,196,000   **42.** 3,400,603,025

Quick Quiz 1.8 *See Examples noted with Ex.* **1.** $269 (Ex. 4)   **2.** $858 (Ex. 6)   **3.** $126 (Ex. 5)   **4.** See Student Solutions Manual

Use Math to Save Money **1.** $100, $300, $2000, $8000, $8000, $8000, $12,000   **2.** $3 \times \$25 + \$50 + \$200 + 2 \times \$20 = \$365$
3. $100 loan, $300 loan, $2000 car loan   **4.** $40   **5.** $2000 - \$400 = \$1600$; $1600/$240 is 7 months if we round to the nearest whole number.
6. Answers will vary.

You Try It **1.** hundred thousands   **2.** $100{,}000 + 30{,}000 + 2000 + 200 + 50 + 9$ **3.** fifty-eight million, eight hundred seventy-two
thousand, one hundred fifty   **4.** 945   **5.** 4198   **6.** 88,844   **7.** 1920   **8.** 628   **9. (a)** 9^5 **(b)** 2401   **10.** 50   **11. (a)** 339,000
(b) 700,000   **12.** 180,000,000

Chapter 1 Review Problems **1.** eight hundred ninety-two   **2.** one hundred nine thousand, two hundred seventy-six
3. $4000 + 300 + 60 + 4$ **4.** $40{,}000{,}000 + 2{,}000{,}000 + 100{,}000 + 60{,}000 + 6000 + 30 + 7$ **5.** 5302   **6.** 1,328,828   **7.** 115   **8.** 400
9. 150   **10.** 400   **11.** 1007   **12.** 60,100   **13.** 14,703   **14.** 17   **15.** 27   **16.** 171   **17.** 6155   **18.** 80,722   **19.** 6,236,011
20. 5,332,991   **21.** 144   **22.** 0   **23.** 800   **24.** 62,100   **25.** 84,312,000   **26.** 780,000   **27.** 24,444   **28.** 50,056   **29.** 1856
30. 1752   **31.** 25,524   **32.** 87,822   **33.** 268,513   **34.** 543,510   **35.** 255,068   **36.** 111,370   **37.** 7,200,000   **38.** 2,000,000,000
39. 2   **40.** 5   **41.** 0   **42.** 7   **43.** 9   **44.** 7   **45.** undefined   **46.** 4   **47.** 125   **48.** 207   **49.** 2504   **50.** 3064
51. 36,958   **52.** 36,921   **53.** 15,046 R 3   **54.** 7 R 21   **55.** 31 R 15   **56.** 60 R 22   **57.** 195   **58.** 54   **59.** 19   **60.** 21^3
61. 8^5 **62.** 10^6 **63.** 64   **64.** 81   **65.** 125   **66.** 49   **67.** 81   **68.** 216   **69.** 8   **70.** 22   **71.** 22   **72.** 17   **73.** 26
74. 86   **75.** 3360   **76.** 5900   **77.** 42,640   **78.** 12,000   **79.** 23,000   **80.** 202,000   **81.** 4,600,000   **82.** 10,000,000
83.
$$\begin{array}{r} 20{,}000 \\ 8000 \\ +\ 40{,}000 \\ \hline 68{,}000 \end{array}$$
84.
$$\begin{array}{r} 4000 \\ 30{,}000 \\ -\ 20{,}000 \\ \hline 10{,}000 \end{array}$$
85.
$$\begin{array}{r} 3{,}000{,}000 \\ \times\ 900 \\ \hline 2{,}700{,}000{,}000 \end{array}$$
86. $20\overline{)80{,}000}$ **87.** 175 words   **88.** 7020 people   **89.** 10,301 feet   **90.** $3348
91. $74   **92.** $278   **93.** 25 miles per gallon   **94.** $2031   **95.** 41,700,000 tons   **96.** 21,400,000 tons from 1990 to 1995   **97.** 105,500,000 tons
98. 2284   **99.** 7867   **100.** 11,088   **101.** 129   **102.** 29   **103.** $747   **104. (a)** 330 square feet   **(b)** 74 feet

How Am I Doing?   Chapter 1 Test **1.** forty-four million, seven thousand, six hundred thirty-five (obj. 1.1.3)
2. $20{,}000 + 6000 + 800 + 50 + 9$ (obj. 1.1.1)   **3.** 3,581,076 (obj. 1.1.2)   **4.** 831 (obj. 1.2.4)   **5.** 1491 (obj. 1.2.4)   **6.** 318,977 (obj. 1.2.4)
7. 8067 (obj. 1.3.3)   **8.** 172,858 (obj. 1.3.3)   **9.** 5,225,768 (obj. 1.3.3)   **10.** 378 (obj. 1.4.1)   **11.** 4320 (obj. 1.4.4)   **12.** 192,992 (obj. 1.4.4)
13. 129,437 (obj. 1.4.2)   **14.** 3014 R 1 (obj. 1.5.2)   **15.** 2358 (obj. 1.5.2)   **16.** 352 (obj. 1.5.3)   **17.** 14^3 (obj. 1.6.1)   **18.** 64 (obj. 1.6.1)
19. 23 (obj. 1.6.2)   **20.** 50 (obj. 1.6.2)   **21.** 79 (obj. 1.6.2)   **22.** 94,800 (obj. 1.7.1)   **23.** 6,460,000 (obj. 1.7.1)   **24.** 5,300,000 (obj. 1.7.1)
25. 150,000,000,000 (obj. 1.7.2)   **26.** 16,000 (obj. 1.7.2)   **27.** $2148 (obj. 1.8.1)   **28.** 467 feet (obj. 1.8.1)   **29.** $127 (obj. 1.8.2)
30. $292 (obj. 1.8.2)   **31.** 748,000 square feet (obj. 1.8.1)   **32.** 46 feet (obj. 1.8.2)

Chapter 2   **2.1 Exercises** **1.** fraction   **3.** denominator   **5.** N: 3; D: 5   **7.** N: 7; D: 8   **9.** N: 1; D: 17   **11.** $\dfrac{1}{3}$
13. $\dfrac{7}{9}$ **15.** $\dfrac{3}{4}$ **17.** $\dfrac{3}{7}$ **19.** $\dfrac{2}{5}$ **21.** $\dfrac{7}{10}$ **23.** $\dfrac{5}{8}$ **25.** $\dfrac{4}{7}$ **27.** $\dfrac{7}{8}$ **29.** $\dfrac{9}{15}$ **31.**
33.   **35.**   **37.** $\dfrac{51}{95}$ **39.** $\dfrac{329}{950}$ **41.** $\dfrac{89}{211}$ **43.** $\dfrac{9}{26}$
45. $\dfrac{21}{44}$ **47. (a)** $\dfrac{90}{195}$ **(b)** $\dfrac{22}{195}$ **49.** The amount of money each of six business owners gets if the business has a profit of $0.
51. 241   **52.** 13,216   **53.** 146,188   **54.** 1258 R 4

Quick Quiz 2.1 *See Examples noted with Ex.* **1.** $\dfrac{4}{7}$ (Ex. 1)   **2.** $\dfrac{204}{371}$ (Ex. 3)   **3.** $\dfrac{13}{33}$ (Ex. 4)   **4.** See Student Solutions Manual

2.2 Exercises **1.** 11, 19, 41, 5   **3.** composite number   **5.** $56 = 2 \times 2 \times 2 \times 7$ or $2^3 \times 7$ **7.** 3×5 **9.** 5×7 **11.** 7^2 **13.** 2^4
15. 5×11 **17.** $3^2 \times 7$ **19.** $2^2 \times 3 \times 7$ **21.** 2×3^3 **23.** $2^3 \times 3 \times 5$ **25.** $2^3 \times 23$ **27.** prime   **29.** 3×19 **31.** prime
33. 2×31 **35.** prime   **37.** prime   **39.** 11×11 or 11^2 **41.** 5×29 **43.** $\dfrac{18 \div 9}{27 \div 9} = \dfrac{2}{3}$ **45.** $\dfrac{36 \div 12}{48 \div 12} = \dfrac{3}{4}$ **47.** $\dfrac{54 \div 6}{84 \div 6} = \dfrac{9}{14}$

49. $\dfrac{260 \div 10}{290 \div 10} = \dfrac{26}{29}$ **51.** $\dfrac{5 \times 1}{5 \times 6} = \dfrac{1}{6}$ **53.** $\dfrac{2 \times 3 \times 11}{2 \times 2 \times 2 \times 11} = \dfrac{3}{4}$ **55.** $\dfrac{2 \times 3 \times 5}{3 \times 3 \times 5} = \dfrac{2}{3}$ **57.** $\dfrac{2 \times 2 \times 3 \times 5}{3 \times 5 \times 5} = \dfrac{4}{5}$ **59.** $\dfrac{6 \times 8}{6 \times 11} = \dfrac{8}{11}$

61. $\dfrac{9 \times 7}{9 \times 12} = \dfrac{7}{12}$ **63.** $\dfrac{11 \times 8}{11 \times 11} = \dfrac{8}{11}$ **65.** $\dfrac{40 \times 3}{40 \times 5} = \dfrac{3}{5}$ **67.** $\dfrac{11 \times 20}{13 \times 20} = \dfrac{11}{13}$ **69.** $4 \times 28 \overset{?}{=} 16 \times 7$ **71.** no **73.** no **75.** yes
$112 = 112$
yes

77. yes **79.** $\dfrac{3}{4}$ **81.** $\dfrac{1}{8}$ failed; $\dfrac{7}{8}$ passed **83.** $\dfrac{5}{7}$ **85.** $\dfrac{17}{45}$ **87.** $\dfrac{8}{45}$ **89.** 164,050 **90.** 1296 **91.** 960,000 **92.** \$899,303,864

Quick Quiz 2.2 *See Examples noted with Ex.* **1.** $\dfrac{5}{7}$ (Ex. 2) **2.** $\dfrac{1}{6}$ (Ex. 3) **3.** $\dfrac{8}{21}$ (Ex. 3) **4.** See Student Solutions Manual

2.3 Exercises 1. (a) Multiply the whole number by the denominator of the fraction. **(b)** Add the numerator of the fraction to the product formed in step (a). **(c)** Write the sum found in step (b) over the denominator of the fraction. **3.** $\dfrac{7}{3}$ **5.** $\dfrac{17}{7}$ **7.** $\dfrac{83}{9}$ **9.** $\dfrac{32}{3}$

11. $\dfrac{58}{5}$ **13.** $\dfrac{51}{7}$ **15.** $\dfrac{121}{6}$ **17.** $\dfrac{131}{12}$ **19.** $\dfrac{79}{10}$ **21.** $\dfrac{201}{25}$ **23.** $\dfrac{65}{12}$ **25.** $\dfrac{494}{3}$ **27.** $\dfrac{131}{15}$ **29.** $\dfrac{199}{30}$ **31.** $1\dfrac{1}{3}$ **33.** $2\dfrac{3}{4}$

35. $2\dfrac{1}{7}$ **37.** $3\dfrac{3}{8}$ **39.** 25 **41.** $9\dfrac{5}{9}$ **43.** $23\dfrac{1}{3}$ **45.** $6\dfrac{1}{4}$ **47.** $5\dfrac{7}{10}$ **49.** $17\dfrac{1}{2}$ **51.** 13 **53.** 14 **55.** 6 **57.** $7\dfrac{1}{5}$

59. $5\dfrac{1}{2}$ **61.** $4\dfrac{1}{6}$ **63.** $15\dfrac{1}{4}$ **65.** 4 **67.** $\dfrac{12}{5}$ **69.** $\dfrac{15}{4}$ **71.** $2\dfrac{88}{126} = 2\dfrac{44}{63}$ **73.** $2\dfrac{20}{280} = 2\dfrac{1}{14}$ **75.** $1\dfrac{212}{296} = 1\dfrac{53}{74}$

77. $\dfrac{1082}{3}$ yards **79.** $50\dfrac{1}{3}$ acres **81.** $141\dfrac{3}{8}$ pounds **83.** No, 101 is prime and is not a factor of 5687. **85.** 260,247 **86.** 2,000,000,000

87. 300 **88.** $\dfrac{29}{78}$

Quick Quiz 2.3 *See Examples noted with Ex.* **1.** $\dfrac{59}{13}$ (Ex. 1) **2.** $7\dfrac{5}{12}$ (Ex. 2) **3.** 3 (Ex. 3) **4.** See Student Solutions Manual

2.4 Exercises 1. $\dfrac{21}{55}$ **3.** $\dfrac{15}{52}$ **5.** 1 **7.** $\dfrac{1}{16}$ **9.** $\dfrac{12}{55}$ **11.** $\dfrac{21}{8}$ or $2\dfrac{5}{8}$ **13.** $\dfrac{24}{7}$ or $3\dfrac{3}{7}$ **15.** $\dfrac{10}{3}$ or $3\dfrac{1}{3}$ **17.** $\dfrac{1}{6}$ **19.** 3

21. $\dfrac{1}{2}$ **23.** 31 **25.** 0 **27.** $3\dfrac{7}{8}$ **29.** $\dfrac{55}{12}$ or $4\dfrac{7}{12}$ **31.** $\dfrac{69}{50}$ or $1\dfrac{19}{50}$ **33.** 35 **35.** $\dfrac{8}{5}$ or $1\dfrac{3}{5}$ **37.** $\dfrac{7}{9}$ **39.** $\dfrac{38}{3}$ or $12\dfrac{2}{3}$

41. $x = \dfrac{7}{9}$ **43.** $x = \dfrac{8}{9}$ **45.** $37\dfrac{11}{12}$ square miles **47.** 1560 miles **49.** 1629 grams **51.** 5332 students **53.** 377 companies

55. $1\dfrac{8}{9}$ miles **57.** The step of dividing the numerator and denominator by the same number allows us to work with smaller numbers when we do the multiplication. Also, this allows us to avoid the step of having to simplify the fraction in the final answer. **59.** 529 cars

60. 368 calls **61.** $\dfrac{37}{78}$ **62.** $\dfrac{27}{32}$

Quick Quiz 2.4 *See Examples noted with Ex.* **1.** 10 (Ex. 3) **2.** $\dfrac{44}{65}$ (Ex. 1) **3.** $\dfrac{143}{12}$ or $11\dfrac{11}{12}$ (Ex. 5) **4.** See Student Solutions Manual

2.5 Exercises 1. Think of a simple problem like $3 \div \frac{1}{2}$. One way to think of it is how many $\frac{1}{2}$'s can be placed in 3? For example, how many $\frac{1}{2}$-pound rocks could be put in a bag that holds 3 pounds of rocks? The answer is 6. If we inverted the first fraction by mistake, we would have $\frac{1}{3} \times \frac{1}{2} = \frac{1}{6}$. We know this is wrong since there are obviously several $\frac{1}{2}$-pound rocks in a bag that holds 3 pounds of rocks. The answer $\frac{1}{6}$ would make no sense. **3.** $\dfrac{7}{12}$ **5.** $\dfrac{9}{2}$ or $4\dfrac{1}{2}$ **7.** $\dfrac{1}{9}$ **9.** $\dfrac{25}{9}$ or $2\dfrac{7}{9}$ **11.** 1 **13.** $\dfrac{9}{49}$ **15.** $\dfrac{4}{5}$ **17.** $\dfrac{3}{44}$ **19.** $\dfrac{27}{7}$ or $3\dfrac{6}{7}$ **21.** 0 **23.** undefined

25. 10 **27.** $\dfrac{7}{32}$ **29.** $\dfrac{3}{4}$ **31.** $\dfrac{13}{9}$ or $1\dfrac{4}{9}$ **33.** 2 **35.** 5000 **37.** $\dfrac{1}{250}$ **39.** $\dfrac{7}{40}$ **41.** 16 **43.** $\dfrac{7}{18}$ **45.** 2 **47.** 4

49. $\dfrac{68}{27}$ or $2\dfrac{14}{27}$ **51.** 1 **53.** $\dfrac{91}{75}$ or $1\dfrac{16}{75}$ **55.** $\dfrac{5}{12}$ **57.** $\dfrac{30}{19}$ or $1\dfrac{11}{19}$ **59.** 0 **61.** $\dfrac{7}{44}$ **63.** 12 **65.** $x = \dfrac{7}{5}$ **67.** $x = \dfrac{3}{10}$

69. $2\dfrac{1}{4}$ gallons **71.** $37\dfrac{1}{2}$ miles per hour **73.** 58 students **75.** 100 large Styrofoam cups **77.** It took six drill attempts.

79. We estimate by dividing $15 \div 5$, which is 3. The exact value is $2\dfrac{26}{31}$, which is very close. Our answer is off by only $\dfrac{5}{31}$. **81.** thirty-nine million, five hundred seventy-six thousand, three hundred four **82.** $500,000 + 9000 + 200 + 70$ **83.** 1099 **84.** 87,595,631

Quick Quiz 2.5 *See Examples noted with Ex.* **1.** $\dfrac{3}{4}$ (Ex. 1) **2.** $\dfrac{76}{29}$ or $2\dfrac{18}{29}$ (Ex. 4) **3.** $\dfrac{31}{16}$ or $1\dfrac{15}{16}$ (Ex. 2) **4.** See Student Solutions Manual

How Am I Doing? Sections 2.1–2.5 1. $\dfrac{3}{8}$ (obj. 2.1.1) **2.** $\dfrac{8}{69}$ (obj. 2.1.3) **3.** $\dfrac{5}{112}$ (obj. 2.1.3) **4.** $\dfrac{1}{7}$ (obj. 2.2.2)

5. $\dfrac{1}{3}$ (obj. 2.2.2) **6.** $\dfrac{1}{7}$ (obj. 2.2.2) **7.** $\dfrac{7}{8}$ (obj. 2.2.2) **8.** $\dfrac{4}{11}$ (obj. 2.2.2) **9.** $\dfrac{11}{3}$ (obj. 2.3.1) **10.** $\dfrac{46}{3}$ (obj. 2.3.1) **11.** $20\dfrac{1}{4}$ (obj. 2.3.2)

12. $5\dfrac{4}{5}$ (obj. 2.3.2) **13.** $2\dfrac{2}{17}$ (obj. 2.3.2) **14.** $\dfrac{5}{44}$ (obj. 2.4.1) **15.** $\dfrac{2}{3}$ (obj. 2.4.1) **16.** $\dfrac{160}{9}$ or $17\dfrac{7}{9}$ (obj. 2.4.3) **17.** 1 (obj. 2.5.1)

18. $\dfrac{1}{2}$ (obj. 2.5.1) **19.** $\dfrac{69}{13}$ or $5\dfrac{4}{13}$ (obj. 2.5.3) **20.** 21 (obj. 2.5.2)

How Am I Doing?   Test on Sections 2.1–2.5

1. $\frac{33}{40}$ **2.** $\frac{85}{113}$ **3.** $\frac{1}{2}$ **4.** $\frac{5}{7}$ **5.** $\frac{4}{11}$ **6.** $\frac{25}{31}$ **7.** $\frac{5}{14}$ **8.** $\frac{7}{3}$ or $2\frac{1}{3}$

9. $\frac{38}{3}$ **10.** $\frac{33}{8}$ **11.** $6\frac{3}{7}$ **12.** $8\frac{1}{3}$ **13.** $\frac{21}{88}$ **14.** $\frac{7}{4}$ or $1\frac{3}{4}$ **15.** 15   **16.** $\frac{33}{2}$ or $16\frac{1}{2}$ **17.** $\frac{161}{12}$ or $13\frac{5}{12}$ **18.** 80   **19.** $\frac{16}{21}$

20. $\frac{16}{3}$ or $5\frac{1}{3}$ **21.** 7   **22.** $\frac{12}{5}$ or $2\frac{2}{5}$ **23.** $\frac{63}{8}$ or $7\frac{7}{8}$ **24.** 14   **25.** $\frac{16}{3}$ or $5\frac{1}{3}$ **26.** $\frac{23}{8}$ or $2\frac{7}{8}$ **27.** $\frac{13}{16}$ **28.** $\frac{1}{3}$ **29.** $\frac{14}{23}$

30. $\frac{13}{15}$ **31.** $45\frac{15}{16}$ square feet   **32.** 4 cups   **33.** $46\frac{7}{8}$ miles   **34.** 16 full packages; $\frac{3}{8}$ lb left over   **35.** 51 computers

36. 24,600 gallons   **37.** 8 feet   **38.** 6 tents; 7 yards left over   **39.** 41 days

2.6 Exercises

1. 24   **3.** 100   **5.** 60   **7.** 30   **9.** 147   **11.** 10   **13.** 28   **15.** 35   **17.** 18   **19.** 60   **21.** 32   **23.** 90
25. 80   **27.** 105   **29.** 120   **31.** 6   **33.** 12   **35.** 132   **37.** 84   **39.** 120   **41.** 3   **43.** 35   **45.** 20   **47.** 40   **49.** 96

51. 63   **53.** $\frac{21}{36}$ and $\frac{20}{36}$ **55.** $\frac{25}{80}$ and $\frac{68}{80}$ **57.** $\frac{18}{20}$ and $\frac{19}{20}$ **59.** LCD = 35; $\frac{14}{35}$ and $\frac{9}{35}$ **61.** LCD = 24; $\frac{5}{24}$ and $\frac{9}{24}$

63. LCD = 30; $\frac{16}{30}$ and $\frac{5}{30}$ **65.** LCD = 60; $\frac{16}{60}$ and $\frac{25}{60}$ **67.** LCD = 36; $\frac{10}{36}, \frac{11}{36}, \frac{21}{36}$ **69.** LCD = 56; $\frac{3}{56}, \frac{49}{56}, \frac{40}{56}$ **71.** LCD = 63; $\frac{5}{63}, \frac{12}{63}, \frac{56}{63}$

73. (a) LCD = 16   **(b)** $\frac{3}{16}, \frac{12}{16}, \frac{6}{16}$ **75.** 25   **76.** $\frac{19}{6}$ or $3\frac{1}{6}$ **77.** 22

Quick Quiz 2.6 *See Examples noted with Ex.*   **1.** 42 (Ex. 4)   **2.** 140 (Ex. 7)   **3.** $\frac{21}{78}$ (Ex. 8)   **4.** See Student Solutions Manual

2.7 Exercises

1. $\frac{7}{9}$ **3.** $\frac{11}{9}$ or $1\frac{2}{9}$ **5.** $\frac{2}{5}$ **7.** $\frac{17}{44}$ **9.** $\frac{5}{6}$ **11.** $\frac{9}{20}$ **13.** $\frac{7}{8}$ **15.** $\frac{23}{20}$ or $1\frac{3}{20}$ **17.** $\frac{37}{100}$ **19.** $\frac{7}{15}$

21. $\frac{31}{24}$ or $1\frac{7}{24}$ **23.** $\frac{27}{40}$ **25.** $\frac{19}{18}$ or $1\frac{1}{18}$ **27.** 0   **29.** $\frac{5}{12}$ **31.** $\frac{11}{60}$ **33.** $\frac{1}{4}$ **35.** $\frac{2}{3}$ **37.** $\frac{1}{36}$ **39.** 0   **41.** $\frac{5}{12}$ **43.** 1

45. $\frac{11}{30}$ **47.** $\frac{22}{15}$ or $1\frac{7}{15}$ **49.** $x = \frac{3}{14}$ **51.** $x = \frac{5}{33}$ **53.** $x = \frac{17}{30}$ **55.** $\frac{11}{12}$ cup   **57.** $\frac{17}{12}$ or $1\frac{5}{12}$ pounds of nuts; $\frac{7}{8}$ pound of dried fruit

59. $\frac{19}{60}$ of the research paper   **61.** 16 chocolates   **63.** $\frac{7}{40}$ of the membership   **64.** $\frac{3}{17}$ **65.** $\frac{3}{23}$ **66.** $8\frac{13}{14}$ **67.** $\frac{101}{7}$ **68.** $2\frac{8}{9}$ **69.** 7

Quick Quiz 2.7 *See Examples noted with Ex.*   **1.** $\frac{19}{16}$ or $1\frac{3}{16}$ (Ex. 4)   **2.** $\frac{32}{21}$ or $1\frac{11}{21}$ (Ex. 6)   **3.** $\frac{19}{45}$ (Ex. 7)   **4.** See Student Solutions Manual

2.8 Exercises

1. $9\frac{3}{4}$ **3.** $4\frac{1}{7}$ **5.** $17\frac{1}{2}$ **7.** 13   **9.** $\frac{4}{7}$ **11.** $2\frac{1}{16}$ **13.** $9\frac{4}{9}$ **15.** 0   **17.** $4\frac{14}{15}$ **19.** $14\frac{4}{7}$ **21.** $7\frac{2}{5}$

23. $10\frac{3}{10}$ **25.** $41\frac{4}{5}$ **27.** $8\frac{5}{12}$ **29.** $8\frac{1}{6}$ **31.** $73\frac{37}{40}$ **33.** $5\frac{1}{2}$ **35.** $\frac{2}{3}$ **37.** $4\frac{41}{60}$ **39.** $8\frac{8}{15}$ **41.** $102\frac{5}{8}$ **43.** $14\frac{1}{24}$

45. $43\frac{1}{8}$ miles   **47.** $6\frac{9}{10}$ miles   **49.** $2\frac{3}{4}$ inches   **51. (a)** $3\frac{11}{12}$ pounds   **(b)** $4\frac{1}{12}$ pounds   **53.** $\frac{2607}{40}$ or $65\frac{7}{40}$ **55.** We estimate by adding

35 + 24 to obtain 59. The exact answer is $59\frac{7}{12}$. Our estimate is very close. We are off by only $\frac{7}{12}$.   **57.** $\frac{2}{3}$ **59.** 1   **61.** $\frac{3}{2}$ or $1\frac{1}{2}$ **63.** $\frac{3}{5}$

65. $\frac{9}{25}$ **67.** $\frac{1}{4}$ **69.** $\frac{1}{9}$ **71.** $\frac{6}{5}$ or $1\frac{1}{5}$ **73.** 480,000   **74.** 8,529,300

Quick Quiz 2.8 *See Examples noted with Ex.*   **1.** $9\frac{7}{40}$ (Ex. 3)   **2.** $1\frac{43}{60}$ (Ex. 5)   **3.** $\frac{41}{55}$ (Ex. 7)   **4.** See Student Solutions Manual

2.9 Exercises

1. $23\frac{13}{30}$ inches   **3.** 389 gorillas   **5.** $1\frac{9}{16}$ inches   **7.** $9\frac{19}{20}$ miles   **9.** $12\frac{3}{4}$ cups; 26 boxes   **11.** $275\frac{5}{8}$ gallons

13. $106\frac{7}{8}$ nautical miles   **15.** $451 per week   **17. (a)** 55 bracelets   **(b)** $1\frac{2}{5}$ feet   **19. (a)** $14\frac{1}{8}$ ounces of bread   **(b)** $\frac{5}{8}$ ounce

21. (a) $30\frac{1}{2}$ knots   **(b)** 7 hours   **23. (a)** 5485 bushels   **(b)** $11,998\frac{7}{16}$ cubic feet   **(c)** $9598\frac{3}{4}$ bushels   **25.** $\frac{1}{4}$ **26.** $\frac{7}{10}$ **27.** 140   **28.** $\frac{3}{4}$

Quick Quiz 2.9 *See Examples noted with Ex.*   **1.** 168 square feet (Ex. 3)   **2.** 16 packets (Ex. 5)   **3.** $5\frac{3}{8}$ miles (Ex. 1)   **4.** See Student Solutions Manual

Use Math to Save Money

1. $180   **2.** $2160   **3.** Yes   **4.** Yes, there would be $260 left over for the celebration dinner.
5. If the cost of the television is $1000, then the sale price would be only $750. Thus, $510 would be left over for the birthday dinner.   **6.** $150
7. $1800   **8.** Answers will vary.   **9.** Answers will vary.   **10.** Answers will vary.

You Try It

1. $\frac{9}{14}$ **2.** $\frac{17}{23}$ **3.** $2^2 \times 3 \times 5$ **4.** $\frac{3}{10}$ **5.** $\frac{32}{3}$ **6.** $9\frac{1}{3}$ **7. (a)** $\frac{4}{45}$ **(b)** $\frac{5}{7}$ **8.** 11   **9.** $\frac{5}{6}$ **10.** $\frac{24}{7}$ or $3\frac{3}{7}$

11. 120   **12.** $\frac{24}{54}$ **13. (a)** $\frac{8}{15}$ **(b)** $\frac{1}{11}$ **14.** $\frac{11}{6}$ or $1\frac{5}{6}$ **15.** $12\frac{1}{6}$ **16.** $6\frac{9}{20}$ **17.** $\frac{7}{2}$ or $3\frac{1}{2}$

Chapter 2 Review Problems

1. $\frac{3}{8}$ 2. $\frac{5}{12}$ 3. answers will vary 4. answers will vary 5. $\frac{9}{80}$ 6. $\frac{87}{100}$

7. 2×3^3 8. $2^3 \times 3 \times 5$ 9. $2^3 \times 3 \times 7$ 10. prime 11. $2 \times 3 \times 13$ 12. prime 13. $\frac{2}{7}$ 14. $\frac{1}{4}$ 15. $\frac{3}{8}$

16. $\frac{7}{8}$ 17. $\frac{35}{8}$ 18. $\frac{63}{4}$ 19. $\frac{33}{5}$ 20. $5\frac{5}{8}$ 21. $4\frac{16}{21}$ 22. $7\frac{4}{7}$ 23. $3\frac{3}{11}$ 24. $\frac{117}{8}$ 25. $4\frac{1}{8}$ 26. $\frac{20}{77}$ 27. $\frac{7}{15}$ 28. 0

29. $\frac{4}{63}$ 30. $\frac{82}{5}$ or $16\frac{2}{5}$ 31. 16 32. $\$677\frac{1}{4}$ 33. $\frac{261}{2}$ or $130\frac{1}{2}$ square feet 34. $\frac{15}{14}$ or $1\frac{1}{14}$ 35. 1500 36. $\frac{1}{2}$ 37. 8 38. 0

39. $\frac{46}{33}$ or $1\frac{13}{33}$ 40. 12 rolls 41. $\frac{560}{3}$ or $186\frac{2}{3}$ calories 42. 98 43. 100 44. 90 45. $\frac{24}{56}$ 46. $\frac{33}{72}$ 47. $\frac{80}{150}$ 48. $\frac{2}{7}$

49. $\frac{13}{12}$ or $1\frac{1}{12}$ 50. $\frac{11}{40}$ 51. $\frac{23}{70}$ 52. $\frac{44}{45}$ 53. $\frac{61}{75}$ 54. $5\frac{1}{4}$ 55. $8\frac{2}{3}$ 56. $\frac{49}{8}$ or $6\frac{1}{8}$ 57. $\frac{279}{80}$ or $3\frac{39}{80}$ 58. $\frac{9}{10}$ 59. $\frac{3}{10}$

60. $8\frac{29}{40}$ miles 61. $283\frac{1}{12}$ miles 62. $1\frac{2}{3}$ cups sugar; $2\frac{1}{8}$ cups flour 63. $206\frac{1}{8}$ miles 64. 15 lengths 65. $9\frac{5}{8}$ liters

66. $227\frac{1}{2}$ minutes or 3 hours and $47\frac{1}{2}$ min. 67. $\frac{35}{8}$ or $4\frac{3}{8}$ cups; $7\frac{5}{8}$ cups 68. $1\frac{1}{16}$ inch 69. $\$242$ 70. 25 miles 71. $\frac{3}{7}$

72. $\frac{68}{75}$ 73. $1\frac{5}{12}$ 74. $\frac{24}{77}$ 75. $\frac{64}{343}$ 76. $\frac{15}{4}$ or $3\frac{3}{4}$ 77. 99 78. 48

How Am I Doing?   Chapter 2 Test

1. $\frac{3}{5}$ (obj. 2.1.1) 2. $\frac{311}{388}$ (obj. 2.1.3) 3. $\frac{3}{7}$ (obj. 2.2.2) 4. $\frac{3}{14}$ (obj. 2.2.2)

5. $\frac{9}{2}$ (obj. 2.2.2) 6. $\frac{34}{5}$ (obj. 2.3.1) 7. $10\frac{5}{14}$ (obj. 2.3.2) 8. 12 (obj. 2.4.2) 9. $\frac{14}{45}$ (obj. 2.4.1) 10. 14 (obj. 2.4.3)

11. $\frac{77}{40}$ or $1\frac{37}{40}$ (obj. 2.5.1) 12. $\frac{39}{62}$ (obj. 2.5.1) 13. $\frac{90}{13}$ or $6\frac{12}{13}$ (obj. 2.5.3) 14. $\frac{12}{7}$ or $1\frac{5}{7}$ (obj. 2.5.3) 15. 36 (obj. 2.6.2)

16. 48 (obj. 2.6.2) 17. 24 (obj. 2.6.2) 18. $\frac{30}{72}$ (obj. 2.6.3) 19. $\frac{13}{36}$ (obj. 2.7.2) 20. $\frac{11}{20}$ (obj. 2.7.2) 21. $\frac{25}{28}$ (obj. 2.7.2)

22. $14\frac{6}{35}$ (obj. 2.8.1) 23. $4\frac{13}{14}$ (obj. 2.8.2) 24. $\frac{1}{48}$ (obj. 2.8.3) 25. $\frac{7}{6}$ or $1\frac{1}{6}$ (obj. 2.8.3) 26. 154 square feet (obj. 2.9.1)

27. 8 packages (obj. 2.9.1) 28. $\frac{7}{10}$ mile (obj. 2.9.1) 29. $14\frac{1}{24}$ miles (obj. 2.9.1) 30. 40 oranges (obj. 2.9.1)

31. 77 candles; $\frac{25}{16}$ or $1\frac{9}{16}$ pounds (obj. 2.9.1)

Chapter 3

3.1 Exercises

1. A decimal fraction is a fraction whose denominator is a power of 10. $\frac{23}{100}$ and $\frac{563}{1000}$ are decimal fractions.

3. hundred-thousandths 5. fifty-seven hundredths 7. three and eight tenths 9. seven and thirteen thousandths 11. twenty-eight and thirty-seven ten-thousandths 13. one hundred twenty-four and $\frac{20}{100}$ dollars 15. one thousand, two hundred thirty-six and $\frac{8}{100}$ dollars

17. eighteen thousand, forty-five and $\frac{19}{100}$ dollars 19. 0.7 21. 0.96 23. 0.481 25. 0.006114 27. 0.7 29. 0.76 31. 0.01

33. 0.053 35. 0.2403 37. 10.9 39. 84.13 41. 3.529 43. 235.0104 45. $\frac{1}{50}$ 47. $3\frac{3}{5}$ 49. $7\frac{41}{100}$ 51. $12\frac{5}{8}$ 53. $7\frac{123}{2000}$

55. $8\frac{27}{2500}$ 57. $235\frac{627}{5000}$ 59. $\frac{1}{80}$ 61. (a) $\frac{263}{1000}$ (b) $\frac{121}{500}$ 63. $\frac{1}{250,000}$ 65. 56,800 66. 8,069,000 67. $\frac{9}{20}$ 68. $\frac{7}{20}$

Quick Quiz 3.1 *See Examples noted with Ex.*

1. five and three hundred sixty-seven thousandths (Ex. 1) 2. 0.0523 (Ex. 3) 3. $12\frac{29}{50}$ (Ex. 5) 4. See Student Solutions Manual

3.2 Exercises

1. > 3. = 5. < 7. > 9. < 11. > 13. < 15. > 17. = 19. > 21. 12.6, 12.65, 12.8
23. 0.007, 0.0071, 0.05 25. 8.31, 8.39, 8.4, 8.41 27. 26.003, 26.033, 26.034, 26.04 29. 18.006, 18.060, 18.065, 18.066, 18.606 31. 6.9
33. 29.0 35. 578.1 37. 2176.8 39. 26.03 41. 37.00 43. 156.17 45. 2786.71 47. 7.816 49. 0.0595 51. 12.01578
53. 136 55. $\$788$ 57. $\$15,021$ 59. $\$96.34$ 61. $\$5783.72$ 63. 0.636; 0.545 65. 365.24 67. 0.0059, 0.006, 0.0519, $\frac{6}{100}$, 0.0601, 0.0612, 0.062, $\frac{6}{10}$, 0.61 69. You should consider only one digit to the right of the decimal place that you wish to round to. 86.23498 is closer to 86.23 than to 86.24. 71. $12\frac{1}{8}$ 72. $10\frac{9}{20}$ 73. $\$32,800$ 74. (a) 0.07 (b) 0.145

Quick Quiz 3.2 *See Examples noted with Ex.*

1. 4.056, 4.559, 4.56, 4.6 (Ex. 3) 2. 27.18 (Ex. 4) 3. 155.525 (Ex. 5) 4. See Student Solutions Manual

3.3 Exercises

1. 76.8 3. 593.9 5. 296.2 7. 12.76 9. 36.7287 11. 67.42 13. 235.78 15. 1112.16 17. 21.04 ft
19. 8.6 pounds 21. $\$78.12$ 23. 47,054.9 25. $\$1411.97$ 27. 3.5 29. 25.93 31. 49.78 33. 508.313 35. 135.43 37. 4.6465
39. 6.737 41. 1189.07 43. 1.4635 45. 176.581 47. 41.59 49. 5.2363 51. 73.225 53. 7.5152 pounds 55. $\$36,947.16$

57. $45.30 **59.** 11.64 centimeters **61.** 2.95 liters **63.** 0.0061 milligram; yes **65.** $6.2 billion; $6,200,000,000 **67.** $162.1 billion; $162,100,000,000 **69.** $8.40; yes; $8.34; very close: the estimate was off by 6¢. **71.** $x = 8.4$ **73.** $x = 43.7$ **75.** $x = 2.109$ **77.** 20,288 **78.** 25 **79.** 400 **80.** 1000

Quick Quiz 3.3 *See Examples noted with Ex.* **1.** 72.981 (Ex. 1) **2.** 2.1817 (Ex. 4) **3.** 55.675 (Ex. 5) **4.** See Student Solutions Manual

3.4 Exercises **1.** Each factor has two decimal places. You add the number of decimal places to get 4 decimal places. Multiply 67×8 to get 536. Place the decimal point 4 places to the left to obtain the result, 0.0536 **3.** When you multiply a number by 100, move the decimal point two places to the right. The answer is 0.78. **5.** 0.12 **7.** 0.06 **9.** 0.00288 **11.** 54.24 **13.** 0.000516 **15.** 0.6582 **17.** 2738.4 **19.** 0.017304 **21.** 768.1517 **23.** 8460 **25.** 53.926 **27.** 6.5237 **29.** $9324 **31.** $420 **33.** 297.6 square feet **35.** $664.20 **37.** 514.8 miles **39.** 28.6 **41.** 5212.5 **43.** 22,615 **45.** 56,098.2 **47.** 1,756,144 **49.** 816,320 **51.** 593.2 centimeters **53.** $2710 **55.** $618.00 **57.** $62,279.00 **59.** To multiply by numbers such as 0.1, 0.01, 0.001, and 0.0001, count the number of decimal places in this first number. Then, in the other number, move the decimal point to the left from its present position the same number of decimal places as were in the first number. **61.** 201 **62.** 451 R 100 **63.** 500 **64.** $\frac{54}{5}$ or $10\frac{4}{5}$ **65.** 16.1 million or 16,100,000 **66.** 62.5 million or 62,500,000 **67.** 60.9 million or 60,900,000 **68.** 3.2 million or 3,200,000

Quick Quiz 3.4 *See Examples noted with Ex.* **1.** 0.0304 (Ex. 1) **2.** 3.2768 (Ex. 2) **3.** 51,620 (Ex. 7) **4.** See Student Solutions Manual

How Am I Doing? Sections 3.1–3.4 **1.** thirty-one and nine hundred three thousandths (obj. 3.1.1) **2.** 0.0567 (obj. 3.1.2) **3.** $4\frac{9}{100}$ (obj. 3.1.3) **4.** $\frac{19}{40}$ (obj. 3.1.3) **5.** 1.59, 1.6, 1.601, 1.61 (obj. 3.2.2) **6.** 123.5 (obj. 3.2.3) **7.** 8.0654 (obj. 3.2.3) **8.** 17.99 (obj. 3.2.3) **9.** 19.45 (obj. 3.3.1) **10.** 27.191 (obj. 3.3.1) **11.** 10.59 (obj. 3.3.2) **12.** 7.671 (obj. 3.3.2) **13.** 0.3501 (obj. 3.4.1) **14.** 4780.5 (obj. 3.4.2) **15.** 37.96 (obj. 3.4.2) **16.** 7.85 (obj. 3.4.1) **17.** 6.874 (obj. 3.4.1) **18.** 0.00000312 (obj. 3.4.1)

3.5 Exercises **1.** 2.1 **3.** 17.83 **5.** 10.52 **7.** 136.5 **9.** 5.412 **11.** 53 **13.** 18 **15.** 130 **17.** 5.3 **19.** 1.2 **21.** 49.3 **23.** 94.21 **25.** 13.56 **27.** 0.21 **29.** 0.081 **31.** 91.264 **33.** 123 **35.** 213 **37.** $82.73 **39.** approximately 27.3 miles per gallon **41.** 24 bouquets **43.** 182 guests **45.** 23 snowboards. The error was in putting one less snowboard in the box than was required. **47.** $n = 32.2$ **49.** $n = 975$ **51.** $n = 44$ **53.** 41 **54.** $3\frac{7}{40}$ **55.** $\frac{15}{16}$ **56.** $\frac{91}{12}$ or $7\frac{7}{12}$ **57.** 15 **58.** $32.5 billion or $32,500,000,000 **59.** $1.4 billion or $1,400,000,000 **60.** about 3.6 times **61.** about 14.6 times

Quick Quiz 3.5 *See Examples noted with Ex.* **1.** 0.658 (Ex. 4) **2.** 3.258 (Ex. 5) **3.** 6.58 (Ex. 2) **4.** See Student Solutions Manual

3.6 Exercises **1.** same quantity **3.** The digits 8942 repeat. **5.** 0.25 **7.** 0.8 **9.** 0.125 **11.** 0.35 **13.** 0.62 **15.** 2.25 **17.** 2.875 **19.** 5.1875 **21.** $0.\overline{6}$ **23.** $0.\overline{45}$ **25.** $3.58\overline{3}$ **27.** $4.\overline{2}$ **29.** 0.308 **31.** 0.905 **33.** 0.146 **35.** 2.036 **37.** 0.404 **39.** 0.944 **41.** 3.143 **43.** 3.474 **45.** < **47.** > **49.** 0.3125 inch **51.** 0.19 inch **53.** yes; it is 0.025 inch too wide. **55.** 2.3 **57.** 3.36 **59.** 0 **61.** 21.414 **63.** 0.0072 **65.** 0.325 **67.** 28.6 **69.** 20.836 **71.** 0.586930 **73. (a)** 0.16 **(b)** $0.144\overline{949}$ **(c)** (b) is a repeating decimal and (a) is a nonrepeating decimal **75.** $5\frac{3}{4}$ feet deep **76.** $32\frac{3}{10}$ feet

Quick Quiz 3.6 *See Examples noted with Ex.* **1.** 3.5625 (Ex. 1) **2.** 0.29 (Ex. 4) **3.** 6.1 (Ex. 6) **4.** See Student Solutions Manual

3.7 Exercises **1.** 700,000,000 **3.** 30,000 **5.** 8000 **7.** 15 **9.** $20,000 **11.** 2982 kroner **13.** 2748.27 square feet **15.** 96 molds **17.** 11.59 meters **19.** 24 servings **21.** $1263.09 **23.** $510 **25.** 3.303 million or 3,303,000 square kilometers **27.** $17,319; $5819 **29.** yes; by 0.149 milligram per liter **31.** 137 minutes **33.** 22.9 quadrillion Btu **35.** approximately 62.1 quadrillion Btu; 62,100,000,000,000,000 Btu **37.** $\frac{47}{6}$ **38.** $7\frac{2}{5}$ **39.** $\frac{1}{10}$ **40.** $\frac{19}{21}$

Quick Quiz 3.7 *See Examples noted with Ex.* **1.** 1.02 inches (Ex. 2) **2.** 23.2 miles per gallon (Ex. 3) **3.** $9918; $1918 (Ex. 2) **4.** See Student Solutions Manual

Use Math to Save Money **1.** SHELL: $4.55; ARCO: $4.88 **2.** SHELL: $13.65; ARCO: $13.74 **3.** SHELL: $18.20; ARCO: $18.17 **4.** SHELL: $45.50; ARCO: $44.75 **5.** 3.75 gallons **6.** SHELL **7.** ARCO **8.** Answers will vary. **9.** Answers will vary. **10.** Answers will vary.

You Try It **1.** three hundred thirty-two and one hundred ninety-four thousandths **2.** 0.054 **3.** $\frac{211}{250}$ **4.** 5.7, 5.713, 5.73, 5.735 **5. (a)** 1.35 **(b)** 9.0776 **6. (a)** 33.807 **(b)** 130.014 **7. (a)** 0.285 **(b)** 3.43 **8. (a)** 793 **(b)** 0.15 **(c)** 312.5 **(d)** 4119 **(e)** 237,500 **9. (a)** 3.8 **(b)** 1900 **10. (a)** 0.44 **(b)** 0.643 **11.** 4.7

Chapter 3 Review Problems **1.** thirteen and six hundred seventy-two thousandths **2.** eighty-four hundred-thousandths **3.** 0.7 **4.** 0.81 **5.** 1.523 **6.** 0.0079 **7.** $\frac{17}{100}$ **8.** $\frac{9}{250}$ **9.** $34\frac{6}{25}$ **10.** $1\frac{1}{4000}$ **11.** = **12.** > **13.** < **14.** 0.901, 0.918, 0.98, 0.981 **15.** 5.2, 5.26, 5.59, 5.6, 5.62 **16.** 2.3, 2.302, 2.36, 2.362 **17.** 0.6 **18.** 19.21 **19.** 9.8522 **20.** $156 **21.** 77.6 **22.** 152.81 **23.** 14.582 **24.** 113.872 **25.** 0.003136 **26.** 887.81 **27.** 2398.02 **28.** 0.613 **29.** 123,540 **30.** $8.73 **31.** 36.8 **32.** 232.9 **33.** 574.4 **34.** 0.059 **35.** $0.91\overline{6}$ **36.** 0.85 **37.** $1.8\overline{3}$ **38.** 0.786 **39.** 0.345 **40.** 3.391 **41.** 19.546 **42.** 3.538

43. 23.13   **44.** 439.19   **45.** 64.3   **46.** 4.459   **47.** 0.904   **48.** 20.004   **49.** 1.25   **50.** 112 people   **51.** 24.8 miles per gallon
52. no; by 0.0005 milligram per liter   **53.** 15.75 inches   **54. (a)** 55.8 feet   **(b)** 175.68 square feet   **55.** 1396.75 square feet   **56.** 6.1 miles
57. 259.9 feet   **58.** $12,750.00; $12,255.00; they should change to the new loan   **59.** $230.00   **60.** $31.67   **61.** $53.60   **62.** $50.80

How Am I Doing?   Chapter 3 Test   1. twelve and forty-three thousandths (obj. 3.1.1)   **2.** 0.3977 (obj. 3.1.2)   **3.** $7\frac{3}{20}$ (obj. 3.1.3)
4. $\frac{261}{1000}$ (obj. 3.1.3)   **5.** 2.19, 2.9, 2.907, 2.91 (obj. 3.2.2)   **6.** 78.66 (obj. 3.2.3)   **7.** 0.0342 (obj. 3.2.3)   **8.** 99.698 (obj. 3.3.1)
9. 37.53 (obj. 3.3.1)   **10.** 0.0979 (obj. 3.3.2)   **11.** 71.155 (obj. 3.3.2)   **12.** 0.5817 (obj. 3.4.1)   **13.** 2189 (obj. 3.4.2)   **14.** 0.1285 (obj. 3.5.2)
15. 47 (obj. 3.5.2)   **16.** $1.\overline{2}$ (obj. 3.6.1)   **17.** 0.875 (obj. 3.6.1)   **18.** 1.487 (obj. 3.6.2)   **19.** 6.1952 (obj. 3.6.2)   **20.** $26.95 (obj. 3.7.2)
21. 18.8 miles per gallon (obj. 3.7.2)   **22.** 3.43 centimeters (obj. 3.7.2)   **23.** $390.55 (obj. 3.7.2)

Cumulative Test for Chapters 1–3   1. thirty-eight million, fifty-six thousand, nine hundred fifty-four (obj. 1.1.3)   **2.** 479,587 (obj. 1.2.4)
3. 54,480 (obj. 1.3.3)   **4.** 39,463 (obj. 1.4.4)   **5.** 48,000 (obj. 1.4.3)   **6.** 316 (obj. 1.5.3)   **7.** 16 (obj. 1.6.2)   **8.** 237,000 (obj. 1.7.1)
9. 24,000,000,000 (obj. 1.7.2)   **10.** $\frac{2}{5}$ (obj. 2.2.2)   **11.** $\frac{35}{8}$ (obj. 2.3.1)   **12.** $\frac{22}{3}$ or $7\frac{1}{3}$ (obj. 2.4.3)   **13.** $8\frac{7}{24}$ (obj. 2.8.1)   **14.** $\frac{9}{35}$ (obj. 2.7.2)
15. $\frac{17}{20}$ (obj. 2.8.3)   **16.** 16 (obj. 2.5.3)   **17.** $\frac{33}{10}$ or $3\frac{3}{10}$ (obj. 2.5.3)   **18.** 0.039 (obj. 3.1.2)   **19.** 2.01, 2.1, 2.11, 2.12, 20.1 (obj. 3.2.2)
20. 26.080 (obj. 3.2.3)   **21.** 21.946 (obj. 3.3.1)   **22.** 13.118 (obj. 3.3.2)   **23.** 1.435 (obj. 3.4.1)   **24.** 182.3 (obj. 3.4.2)   **25.** 1.058 (obj. 3.5.2)
26. 0.8125 (obj. 3.6.1)   **27. (a)** 110.25 square feet   **(b)** 42 feet (obj. 3.7.2)   **28.** 60 months (obj. 3.7.2)

Chapter 4   **4.1 Exercises   1.** ratio   **3.** 5 to 8   **5.** $\frac{1}{3}$ **7.** $\frac{7}{6}$ **9.** $\frac{2}{3}$ **11.** $\frac{11}{6}$ **13.** $\frac{5}{6}$ **15.** $\frac{2}{3}$ **17.** $\frac{8}{5}$ **19.** $\frac{2}{3}$

21. $\frac{3}{2}$ **23.** $\frac{15}{19}$ **25.** $\frac{13}{1}$ **27.** $\frac{10}{17}$ **29.** $\frac{165}{285} = \frac{11}{19}$ **31.** $\frac{35}{165} = \frac{7}{33}$ **33.** $\frac{205}{1225} = \frac{41}{245}$ **35.** $\frac{450}{205} = \frac{90}{41}$ **37.** $\frac{1}{16}$ **39.** $\frac{\$7}{2 \text{ pairs of socks}}$
41. $\frac{\$85}{6 \text{ bushes}}$ **43.** $\frac{\$19}{2 \text{ CDs}}$ **45.** $\frac{410 \text{ revolutions}}{1 \text{ mile}}$ or 410 rev/mi   **47.** $\frac{\$27,500}{1 \text{ employee}}$ or $27,500/employee   **49.** $15/hour   **51.** 28 mi/gal
53. 70 people/sq mi   **55.** 70 books/library   **57.** 66 mi/hr   **59.** 19 patients/doctor   **61.** 3 geraniums/pot   **63.** $22/share
65. $4.50 profit per puppet   **67. (a)** $0.08/oz small box; $0.07/oz large box   **(b)** 1¢ per ounce   **(c)** The consumer saves $0.48.
69. (a) 13 moose/acre   **(b)** 12 moose/acre   **(c)** North Slope   **71. (a)** $40.95   **(b)** $52.80   **(c)** $11.85   **73.** increased by Mach 0.2   **75.** $2\frac{5}{8}$
76. 5   **77.** $\frac{5}{24}$ **78.** $1\frac{1}{48}$ **79.** $12.25/sq yard   **80.** $24,150; $16,800

Quick Quiz 4.1 *See Examples noted with Ex.*   **1.** $\frac{3}{5}$ (Ex. 1)   **2.** $\frac{340 \text{ square feet}}{11 \text{ pounds}}$ (Ex. 3)   **3.** 27.18 trees/acre (Ex. 4)   **4.** See Student
Solutions Manual

4.2 Exercises   1. equal   **3.** $\frac{6}{8} = \frac{3}{4}$ **5.** $\frac{20}{36} = \frac{5}{9}$ **7.** $\frac{220}{11} = \frac{400}{20}$ **9.** $\frac{4\frac{1}{3}}{13} = \frac{5\frac{2}{3}}{17}$ **11.** $\frac{6.5}{14} = \frac{13}{28}$ **13.** $\frac{3 \text{ inches}}{40 \text{ miles}} = \frac{27 \text{ inches}}{360 \text{ miles}}$
15. $\frac{\$40}{12 \text{ cars}} = \frac{\$60}{18 \text{ cars}}$ **17.** $\frac{3 \text{ hours}}{\$525} = \frac{7 \text{ hours}}{\$1225}$ **19.** $\frac{3 \text{ teaching assistants}}{40 \text{ children}} = \frac{21 \text{ teaching assistants}}{280 \text{ children}}$ **21.** $\frac{4800 \text{ people}}{3 \text{ restaurants}} = \frac{11,200 \text{ people}}{7 \text{ restaurants}}$
23. It is a proportion.   **25.** It is not a proportion.   **27.** It is not a proportion.   **29.** It is a proportion.   **31.** It is a proportion.
33. It is not a proportion.   **35.** It is a proportion.   **37.** It is not a proportion.   **39.** It is a proportion.   **41.** It is a proportion.
43. It is a proportion.   **45.** It is not a proportion.   **47.** no   **49. (a)** no   **(b)** The van traveled at a faster rate.   **51.** yes   **53. (a)** yes   **(b)** yes
(c) the equality test for fractions   **54.** 23.1405   **55.** 17.9968   **56.** 402.408   **57.** 25.8   **58.** $12\frac{3}{8}$ miles

Quick Quiz 4.2 *See Examples noted with Ex.*   **1.** $\frac{8}{18} = \frac{28}{63}$ (Ex. 1)   **2.** $\frac{13}{32} = \frac{3\frac{1}{4}}{8}$ (Ex. 1)   **3.** It is not a proportion. (Ex. 5)
4. See Student Solutions Manual

How Am I Doing?   Sections 4.1–4.2   1. $\frac{13}{18}$ (obj. 4.1.1)   **2.** $\frac{1}{5}$ (obj. 4.1.1)   **3.** $\frac{9}{2}$ (obj. 4.1.1)   **4.** $\frac{9}{11}$ (obj. 4.1.1)   **5. (a)** $\frac{2}{11}$
(b) $\frac{16}{165}$ (obj. 4.1.1)   **6.** $\frac{3 \text{ flight attendants}}{100 \text{ passengers}}$ (obj. 4.1.2)   **7.** $\frac{31 \text{ gallons}}{42 \text{ square feet}}$ (obj. 4.1.2)   **8.** 16.25 miles per hour (obj. 4.1.2)
9. $35 per MP3 player (obj. 4.1.1)   **10.** 160 cookies per pound of cookie dough (obj. 4.1.2)   **11.** $\frac{13}{40} = \frac{39}{120}$ (obj. 4.2.1)
12. $\frac{116}{148} = \frac{29}{37}$ (obj. 4.2.1)   **13.** $\frac{33 \text{ nautical miles}}{2 \text{ hours}} = \frac{49.5 \text{ nautical miles}}{3 \text{ hours}}$ (obj. 4.2.1)   **14.** $\frac{3000 \text{ shoes}}{\$370} = \frac{7500 \text{ shoes}}{\$925}$ (obj. 4.2.1)
15. It is a proportion. (obj. 4.2.2)   **16.** It is not a proportion. (obj. 4.2.2)   **17.** It is not a proportion. (obj. 4.2.2)
18. It is a proportion. (obj. 4.2.2)   **19.** It is a proportion. (obj. 4.2.2)   **20.** It is a proportion. (obj. 4.2.2)

4.3 Exercises   1. Divide each side of the equation by the number a. Calculate $\frac{b}{a}$. The value of n is $\frac{b}{a}$.   **3.** $n = 9$ **5.** $n = 5.6$ **7.** $n = 20$
9. $n = 8$ **11.** $n = 49\frac{1}{2}$ **13.** $n = 15$ **15.** $n = 16$ **17.** $n = 7.5$ **19.** $n = 5$ **21.** $n = 75$ **23.** $n = 22.5$ **25.** $n = 192$

27. $n = 31.5$ **29.** $n = 18$ **31.** $n = 8$ **33.** $n = 162$ **35.** $n \approx 30.9$ **37.** $n \approx 30.1$ **39.** $n \approx 5.5$ **41.** $n = 48$

43. $n = 1.25$ **45.** $n \approx 3.03$ **47.** $n = 3.75$ **49.** $n = 87.36$ **51.** $n = 80$ **53.** $n = 4\frac{7}{8}$ **55.** 11 inches   **57.** $n = 3\frac{5}{8}$

59. $n = 10\frac{8}{9}$ **60.** 76   **61.** 47   **62.** five hundred sixty-three thousandths   **63.** 0.0034   **64.** $1560   **65.** 56 games

Quick Quiz 4.3 *See Examples noted with Ex.* **1.** $n = 1.8$ (Ex. 5)   **2.** $n = 2$ (Ex. 6)   **3.** $n \approx 11.3$ (Ex. 8)
4. See Student Solutions Manual

4.4 Exercises **1.** He should continue with people on the top of the fraction. That would be 60 people he observed on Saturday night.

He does not know the number of dogs, so this would be n. The proportion would be: $\dfrac{12 \text{ people}}{5 \text{ dogs}} = \dfrac{60 \text{ people}}{n \text{ dogs}}$.   **3.** 161 cars   **5.** 3 cups

7. $7\frac{1}{2}$ kilometers   **9.** 5520 Indian rupees   **11.** 197.6 feet   **13.** 217 miles   **15.** $12\frac{3}{4}$ cups   **17.** 102 free throws   **19.** 18.75 gallons

21. 40 hawks   **23.** $3570   **25.** 270 chips   **27.** 1 cup of water and $\frac{3}{8}$ cup of milk   **29.** $3\frac{1}{3}$ cups of water and 1 cup of milk

31. Albert Pujols, approximately $306,964 for each home run; Ryan Howard, approximately $333,333 for each home run
33. Kobe Bryant, approximately $37,333 for each two-point shot; Tim Duncan, approximately $39,684 for each two-point shot   **35.** $17.25

36. $35   **37.** 56.1   **38.** 2.7490   **39.** (a) $\dfrac{19}{20}$ of a square foot   (b) 1425 square feet

Quick Quiz 4.4 *See Examples noted with Ex.* **1.** 240 pounds (Ex. 1)   **2.** 29.09 miles (Ex. 2)   **3.** 44 free throws (Ex. 4)
4. See Student Solutions Manual

Use Math to Save Money **1.** The Gold plan   **2.** The Silver plan   **3.** The Silver plan   **4.** The Bronze plan
5. $30   **6.** $60   **7.** $90   **8.** The Silver plan   **9.** $2319

You Try It **1.** (a) $\frac{1}{3}$ (b) $\frac{4}{5}$ (c) $\frac{7}{32}$ **2.** $\dfrac{3 \text{ teachers}}{80 \text{ students}}$ **3.** (a) 63 mi/hr (b) 2.5 lb/sq ft   **4.** $\dfrac{15}{60} = \dfrac{13}{52}$ **5.** (a) It is not a proportion.
(b) It is a proportion.   **6.** $n = 7$ **7.** $36

Chapter 4 Review Problems **1.** $\dfrac{11}{5}$ **2.** $\dfrac{4}{5}$ **3.** $\dfrac{10}{19}$ **4.** $\dfrac{28}{51}$ **5.** $\dfrac{14}{25}$ **6.** $\dfrac{2}{5}$ **7.** $\dfrac{1}{8}$ **8.** $\dfrac{7}{32}$ **9.** $\dfrac{\$25}{2 \text{ people}}$
10. $\dfrac{4 \text{ revolutions}}{11 \text{ minutes}}$ **11.** $\dfrac{5 \text{ heartbeats}}{4 \text{ seconds}}$ **12.** $17/share   **13.** $112/credit-hour   **14.** $13.50/square yard   **15.** (a) $0.74 (b) $0.58

(c) $0.16   **16.** $\dfrac{12}{48} = \dfrac{7}{28}$ **17.** $\dfrac{1\frac{1}{2}}{5} = \dfrac{4}{13\frac{1}{3}}$ **18.** $\dfrac{3 \text{ buses}}{138 \text{ passengers}} = \dfrac{5 \text{ buses}}{230 \text{ passengers}}$ **19.** $\dfrac{15 \text{ pounds}}{\$4.50} = \dfrac{27 \text{ pounds}}{\$8.10}$

20. It is not a proportion.   **21.** It is a proportion.   **22.** It is not a proportion.   **23.** It is a proportion.   **24.** It is not a proportion.
25. $n = 18$ **26.** $n = 7\frac{3}{5}$ or 7.6   **27.** $n = 22.1$ or $22\frac{1}{10}$ **28.** $n = 33$ **29.** $n = 7$ **30.** $n = 24$ **31.** $n = 5\frac{3}{5}$ or 5.6
32. $n \approx 5.0$ **33.** $n = 19$ **34.** $n \approx 5.9$ **35.** $n = 12$ **36.** $n = 550$ **37.** 15 gallons   **38.** 1691 employees   **39.** 2016 francs
40. 903.61 Swedish kronor   **41.** 600 miles   **42.** 120 feet   **43.** (a) 7.65 gallons (b) $32.13   **44.** 5.71 centimeters tall   **45.** 7.5 grams
46. 1680 students   **47.** Technically it is 8.4 gallons, but in real life 9 gallons of paint will be needed.   **48.** Technically it is 833.33 liters,
but in real life 834 liters will be needed.   **49.** approximately 113.93 feet   **50.** 86 goals   **51.** 13,680 people   **52.** 195 trips

How Am I Doing? Chapter 4 Test **1.** $\dfrac{9}{26}$ (obj. 4.1.1)   **2.** $\dfrac{14}{37}$ (obj. 4.1.1)   **3.** $\dfrac{98 \text{ miles}}{3 \text{ gallons}}$ (obj. 4.1.2)   **4.** $\dfrac{140 \text{ square feet}}{3 \text{ pounds}}$ (obj. 4.1.2)
5. 3.8 tons/day (obj. 4.1.2)   **6.** $8.28/hour (obj. 4.1.2)   **7.** 245.45 feet/pole (obj. 4.1.2)   **8.** $85.21/share (obj. 4.1.2)

9. $\dfrac{17}{29} = \dfrac{51}{87}$ (obj. 4.2.1)   **10.** $\dfrac{2\frac{1}{2}}{10} = \dfrac{6}{24}$ (obj. 4.2.1)   **11.** $\dfrac{490 \text{ miles}}{21 \text{ gallons}} = \dfrac{280 \text{ miles}}{12 \text{ gallons}}$ (obj. 4.2.1)   **12.** $\dfrac{3 \text{ hours}}{180 \text{ miles}} = \dfrac{5 \text{ hours}}{300 \text{ miles}}$ (obj. 4.2.1)
13. It is not a proportion. (obj. 4.2.2)   **14.** It is a proportion. (obj. 4.2.2)   **15.** It is a proportion. (obj. 4.2.2)   **16.** It is not a proportion. (obj. 4.2.2)
17. $n = 16$ (obj. 4.3.2)   **18.** $n = 22.5$ (obj. 4.3.2)   **19.** $n = 19$ (obj. 4.3.2)   **20.** $n = 29.4$ (obj. 4.3.2)   **21.** $n = 120$ (obj. 4.3.2)
22. $n = 70.4$ (obj. 4.3.2)   **23.** $n = 120$ (obj. 4.3.2)   **24.** $n = 52$ (obj. 4.3.2)   **25.** 6 eggs (obj. 4.4.1)   **26.** 80.95 pounds (obj. 4.4.1)
27. 19 miles (obj. 4.4.1)   **28.** $360 (obj. 4.4.1)   **29.** 136.6 miles (obj. 4.4.1)   **30.** 696.67 kilometers (obj. 4.4.1)   **31.** 88 free throws (obj. 4.4.1)
32. 32 hits (obj. 4.4.1)

Chapter 5 **5.1 Exercises** **1.** hundred   **3.** two; left; drop   **5.** 59%   **7.** 4%   **9.** 80%   **11.** 245%   **13.** 12.5%   **15.** $4\frac{1}{3}$%
17. 13%   **19.** 9%   **21.** 0.51   **23.** 0.07   **25.** 0.2   **27.** 0.436   **29.** 0.0003   **31.** 0.0072   **33.** 0.0125   **35.** 2.75   **37.** 74%
39. 50%   **41.** 8%   **43.** 56.3%   **45.** 0.2%   **47.** 0.57%   **49.** 135%   **51.** 516%   **53.** 27%   **55.** 20%   **57.** 94%   **59.** 231%
61. 10%   **63.** 8.9%   **65.** 0.62   **67.** 1.38   **69.** 0.003   **71.** 0.75   **73.** 57%   **75.** 1.15   **77.** 0.006   **79.** 0.19; 0.0032

81. 36% = 36 percent = 36 "per one hundred" = $36 \times \dfrac{1}{100} = \dfrac{36}{100} = 0.36$. The rule is using the fact that 36% means 36 per one hundred.

83. (a) 15.62 (b) $\dfrac{1562}{100}$ (c) $\dfrac{781}{50}$ **85.** $\dfrac{14}{25}$ **86.** $\dfrac{39}{50}$ **87.** 0.6875   **88.** 0.875   **89.** 5336 vases

Quick Quiz 5.1 *See Examples noted with Ex.* **1.** 0.7% (Ex. 6) **2.** 4.5% (Ex. 3) **3.** 0.0125 (Ex. 5) **4.** See Student Solutions Manual

5.2 Exercises **1.** Write the number in front of the percent symbol as the numerator of a fraction. Write the number 100 as the denominator of the fraction. Reduce the fraction if possible. **3.** $\frac{3}{50}$ **5.** $\frac{33}{100}$ **7.** $\frac{11}{20}$ **9.** $\frac{3}{4}$ **11.** $\frac{1}{5}$ **13.** $\frac{19}{200}$ **15.** $\frac{9}{40}$ **17.** $\frac{81}{125}$ **19.** $\frac{57}{80}$

21. $1\frac{17}{25}$ **23.** $3\frac{2}{5}$ **25.** 12 **27.** $\frac{29}{800}$ **29.** $\frac{1}{8}$ **31.** $\frac{11}{125}$ **33.** $\frac{101}{1000}$ **35.** $\frac{6}{125}$ **37.** 75% **39.** 70% **41.** 35% **43.** 72%

45. 27.5% **47.** 360% **49.** 250% **51.** 412.5% **53.** 33.33% **55.** 41.67% **57.** 425% **59.** 52% **61.** 2.5% **63.** 5.95%

65. $37\frac{1}{2}$% **67.** $7\frac{1}{2}$% **69.** $26\frac{2}{3}$% **71.** $22\frac{2}{9}$%

	Fraction	Decimal	Percent
73.	$\frac{11}{12}$	0.9167	91.67%
75.	$\frac{14}{25}$	0.56	56%
77.	$\frac{1}{200}$	0.005	0.5%
79.	$\frac{5}{9}$	0.5556	55.56%
81.	$\frac{1}{32}$	0.0313	$3\frac{1}{8}$%

83. $\frac{463}{1600}$ **85.** 15.375% **87.** $n = 5.625$ **88.** $n = 4$ **89.** 549,165 documents **90.** 4500 square feet

Quick Quiz 5.2 *See Examples noted with Ex.* **1.** $\frac{9}{20}$ (Ex. 1) **2.** $\frac{19}{250}$ (Ex. 4) **3.** 92% (Ex. 7) **4.** See Student Solutions Manual

5.3A Exercises **1.** What is 20% of $300? **3.** 20 baskets out of 25 shots is what percent? **5.** This is "a percent problem when we do not know the base."

Translated into an equation:
$$108 = 18\% \times n$$
$$108 = 0.18n$$
$$\frac{108}{0.18} = \frac{0.18n}{0.18}$$
$$600 = n$$

7. $n = 5\% \times 90$ **9.** $30\% \times n = 5$ **11.** $17 = n \times 85$ **13.** 28 **15.** 56 **17.** $51 **19.** 1300 **21.** 1300 **23.** $150 **25.** 84%
27. 11% **29.** 65% **31.** 31 **33.** 85 **35.** 12% **37.** 3.28 **39.** 64% **41.** 75 **43.** 0.8% **45.** 18.9 **47.** 80% **49.** 61.63%
51. 663 students **53.** 40 years **55.** $57.60 **57.** 2.448 **58.** 4.1492 **59.** 2834 **60.** 2.36

Quick Quiz 5.3A *See Examples noted with Ex.* **1.** 127.68 (Ex. 6) **2.** 9000 (Ex. 9) **3.** 17% (Ex. 12) **4.** See Student Solutions Manual

5.3B Exercises

	p	b	a
1.	75	660	495
3.	22	60	a
5.	49	b	2450
7.	p	50	30

9. 28 **11.** 84 **13.** 56 **15.** 80 **17.** 80 **19.** 600,000 **21.** 20 **23.** 20 **25.** 22 **27.** 40 **29.** 16.4%
31. 3.64 **33.** 25% **35.** 170 **37.** $960 **39.** 15% **41.** 18 gallons **43.** $1890 **45.** 18% **47.** 16.4%
49. $\frac{76}{45}$ or $1\frac{31}{45}$ **50.** $\frac{1}{26}$ **51.** $\frac{21}{5}$ or $4\frac{1}{5}$ **52.** $\frac{28}{15}$ or $1\frac{13}{15}$

Quick Quiz 5.3B *See Examples noted with Ex.* **1.** 15.3 (Ex. 4) **2.** 120 (Ex. 5) **3.** 22% (Ex. 7) **4.** See Student Solutions Manual

How Am I Doing? Sections 5.1–5.3 **1.** 17% (obj. 5.1.3) **2.** 38.7% (obj. 5.1.3) **3.** 795% (obj. 5.1.3) **4.** 1225% (obj. 5.1.3)
5. 0.6% (obj. 5.1.3) **6.** 0.04% (obj. 5.1.3) **7.** 17% (obj. 5.1.1) **8.** 89% (obj. 5.1.1) **9.** 13.4% (obj. 5.1.1) **10.** 19.8% (obj. 5.1.1)
11. $6\frac{1}{2}$% (obj. 5.1.1) **12.** $3\frac{5}{8}$% (obj. 5.1.1) **13.** 80% (obj. 5.2.2) **14.** 50% (obj. 5.2.2) **15.** 260% (obj. 5.2.2) **16.** 106.25% (obj. 5.2.2)
17. 71.43% (obj. 5.2.2) **18.** 28.57% (obj. 5.2.2) **19.** 75% (obj. 5.2.2) **20.** 25% (obj. 5.2.2) **21.** 440% (obj. 5.2.2) **22.** 275% (obj. 5.2.2)
23. 0.33% (obj. 5.2.2) **24.** 0.25% (obj. 5.2.2) **25.** $\frac{11}{50}$ (obj. 5.2.1) **26.** $\frac{53}{100}$ (obj. 5.2.1) **27.** $1\frac{1}{2}$ (obj. 5.2.1) **28.** $1\frac{3}{5}$ (obj. 5.2.1)
29. $\frac{19}{300}$ (obj. 5.2.1) **30.** $\frac{1}{32}$ (obj. 5.2.1) **31.** $\frac{41}{80}$ (obj. 5.2.1) **32.** $\frac{7}{16}$ (obj. 5.2.1) **33.** 42 (obj. 5.3.2) **34.** 21 (obj. 5.3.2)
35. 94.44% (obj. 5.3.2) **36.** 80% (obj. 5.3.2) **37.** 3000 (obj. 5.3.2) **38.** 885 (obj. 5.3.2)

5.4 Exercises **1.** 180,000 pencils **3.** $45 **5.** 20.57% **7.** $3.90 **9.** $550 **11.** 30% **13.** $9,600,000 **15.** 2.48%
17. 41,175,600 people **19.** $761.90 **21.** $150,000 **23.** 2200 pounds **25.** (a) $12,210,000 for personnel, food and decorations
(b) $20,790,000 for security, facility rental, and all other expenses **27.** $123.50 **29.** (a) $1320 (b) $7480 **31.** 1,698,000
32. 2,452,400 **33.** 1.63 **34.** 0.800 **35.** 0.0556 **36.** 0.0792

Quick Quiz 5.4 *See Examples noted with Ex.* **1.** (a) $166.88 (b) $429.12 (Ex. 5) **2.** 64.4% (Ex. 3) **3.** 15,000 people (Ex. 1)
4. See Student Solutions Manual

5.5 Exercises **1.** $3400 **3.** $4140 **5.** 20% **7.** 15.6% **9.** $140 **11.** $7.50 **13.** $1040 **15.** 0.6% **17.** $1,600,000
19. $39.75 **21.** 39 boxes **23.** 43.57% **25.** 80% **27.** (a) $85.10 (b) $3785.10 **29.** (a) $6.96 (b) $122.96 **31.** $10,830

33. (a) $27,920 (b) $321,080 **35.** 96.9% **37.** $941.01 **39.** 2 **40.** 120 **41.** $\frac{13}{18}$ **42.** 3.64

Quick Quiz 5.5 *See Examples noted with Ex.* **1.** $26,000 (Ex. 1) **2.** 71.875% (Ex. 2) **3.** $299 (Ex. 4) **4.** See Student Solutions Manual

Use Math to Save Money **1.** $14,970.16 **2.** $25,094.08 **3.** $25,970.16 **4.** $876.08 **5.** He will save $281.22 each month in car payments. **6.** To get the best overall price, Louvy should buy the car since he will save $876.08 on the total price. To get a lower monthly payment, Louvy should lease the car since he will save $281.22 each month in car payments. **7.** Answers will vary. **8.** Answers will vary. **9.** Answers will vary.

You Try It **1.** **(a)** 78% **(b)** 125% **(c)** 9% **(d)** $3\frac{1}{4}$% **(e)** 18.5% **2.** **(a)** 0.86 **(b)** 0.01 **(c)** 0.018 **(d)** 3.35 **(e)** 0.0754 **3.** **(a)** 12%
(b) 235% **(c)** 0.71% **(d)** 650% **(e)** 100.5% **4.** **(a)** $\frac{1}{5}$ **(b)** $\frac{9}{20}$ **(c)** $2\frac{3}{5}$ **(d)** $\frac{3}{200}$ **(e)** $\frac{73}{1250}$ **(f)** $\frac{29}{300}$ **5.** **(a)** $3\frac{3}{5}$ **(b)** $1\frac{17}{20}$ **(c)** $2\frac{2}{25}$
6. **(a)** 68% **(b)** 32% **(c)** 0.8% **(d)** 160% **(e)** 325% **7.** **(a)** $n = 36$ **(b)** $n = 750$ **(c)** $n \approx 35.56$% **8.** **(a)** $a = 57.6$ **(b)** $b = 31.25$
(c) $p = 40$% **9.** **(a)** $33.75 **(b)** $191.25 **10.** $1240 **11.** approximately 14.8% **12.** $1440

Chapter 5 Review Problems **1.** 62% **2.** 43% **3.** 37.2% **4.** 220% **5.** 252% **6.** 103.6% **7.** 0.6% **8.** 62.5%
9. $4\frac{1}{12}$% **10.** 317% **11.** 0.32 **12.** 0.1575 **13.** 2.36 **14.** 0.32125 **15.** 76% **16.** 55% **17.** 22.5% **18.** 58.33% **19.** 93.33%
20. 225% **21.** 375% **22.** 277.78% **23.** 250% **24.** 0.63% **25.** $\frac{18}{25}$ **26.** $1\frac{3}{4}$ **27.** $\frac{41}{250}$ **28.** $\frac{5}{16}$ **29.** $\frac{1}{1250}$ **30.** $\frac{1}{2500}$

	Fraction	Decimal	Percent
31.	$\frac{3}{5}$	0.6	60%
32.	$\frac{7}{10}$	0.7	70%
33.	$\frac{3}{8}$	0.375	37.5%
34.	$\frac{9}{16}$	0.5625	56.25%
35.	$\frac{1}{125}$	0.008	0.8%
36.	$\frac{9}{20}$	0.45	45%

37. 17 **38.** 23 **39.** 97.2 **40.** 60 **41.** 40 **42.** 160 **43.** 38.46% **44.** 38.89%
45. 20% **46.** 49 students **47.** 96 trucks **48.** $11,200 **49.** $80,200 **50.** 60%
51. 7.5% **52.** $183.50 **53.** $756 **54.** $5010 **55.** $1200 **56.** **(a)** $362.50 **(b)** $1087.50
57. 2% **58.** 32% **59.** **(a)** $3360 **(b)** $20,640 **60.** **(a)** $330 **(b)** $1320 **61.** **(a)** $60
(b) $720

How Am I Doing? Chapter 5 Test **1.** 57% (obj. 5.1.3) **2.** 1% (obj. 5.1.3)
3. 0.8% (obj. 5.1.3) **4.** 1280% (obj. 5.1.3) **5.** 356% (obj. 5.1.3) **6.** 71% (obj. 5.1.1)
7. 1.8% (obj. 5.1.1) **8.** $3\frac{1}{7}$% (obj. 5.1.1) **9.** 47.5% (obj. 5.2.2) **10.** 75% (obj. 5.2.2)
11. 300% (obj. 5.2.2) **12.** 175% (obj. 5.2.2) **13.** 8.25% (obj. 5.2.3) **14.** 302.4% (obj. 5.2.3)
15. $1\frac{13}{25}$ (obj. 5.2.3) **16.** $\frac{31}{400}$ (obj. 5.2.3) **17.** 20 (obj. 5.3.2) **18.** 130 (obj. 5.3.2)
19. 55.56% (obj. 5.3.2) **20.** 200 (obj. 5.3.2) **21.** 5000 (obj. 5.3.2) **22.** 46% (obj. 5.3.2)

23. 699.6 (obj. 5.3.2) **24.** 20% (obj. 5.3.2) **25.** $6092 (obj. 5.5.1) **26.** **(a)** $150.81 **(b)** $306.19 (obj. 5.4.3) **27.** 89.29% (obj. 5.4.1)
28. 23.24% (obj. 5.5.2) **29.** 12,000 registered voters (obj. 5.4.1) **30.** **(a)** $240 **(b)** $960 (obj. 5.5.3)

Chapter 6 **6.1 Exercises** **1.** We know that each mile is 5280 feet. Each foot is 12 inches. So we know that one mile is
$5280 \times 12 = 63,360$ inches. The unit fraction we want is $\frac{63{,}360 \text{ inches}}{1 \text{ mile}}$. So we multiply 23 miles $\times \frac{63{,}360 \text{ inches}}{1 \text{ mile}}$. The mile unit divides out.
We obtain 1,457,280 inches. Thus 23 miles = 1,457,280 inches. **3.** 1760 **5.** 2000 **7.** 4 **9.** 2 **11.** 7 **13.** 9 **15.** 108 **17.** 2
19. 12,320 **21.** 48 **23.** 12 **25.** 128 **27.** 68 **29.** 11 **31.** 16 **33.** 0.5 **35.** 6.25 **37.** 30 **39.** 36 **41.** 5.5 **43.** 138,336 feet
45. 6.79 miles **47.** $9.75 **49.** **(a)** 142 inches **(b)** $85.20 **51.** 28,800 cups **53.** $\approx$ 12,000 yards **55.** $\approx$ 6 miles **57.** $10,800
58. 85% **59.** 161 miles **60.** 104 students

Quick Quiz 6.1 *See Examples noted with Ex.* **1.** 7000 pounds (Ex. 4) **2.** 13.5 feet (Ex. 3) **3.** 1.5 pounds (Ex. 4) **4.** See Student
Solutions Manual

To Think About **1.** 18,000 **2.** 26,000 **3.** 0.000017 **4.** 0.000038 **5.** 1,200,000,000 **6.** 528,000,000 **7.** 78,900 **8.** 24,900,000,000

6.2 Exercises **1.** hecto- **3.** deci- **5.** kilo- **7.** 460 **9.** 2610 **11.** 12.5 **13.** 0.0732 **15.** 200,000 **17.** 0.078 **19.** 3.5; 0.035
21. 4500; 450,000 **23.** b **25.** c **27.** a **29.** a **31.** b **33.** 39 **35.** 8000 **37.** 0.482 **39.** 3255 m **41.** 183.2 cm **43.** 63.5 cm
45. 2.5464 cm or 25.464 mm **47.** 3.23 m **49.** 939.86 m **51.** 0.964 **53.** false **55.** true **57.** true **59.** false **61.** **(a)** 507,200 cm
(b) 5.072 km **63.** 0.00000000254 **65.** 106,000 m **67.** 0.415 megameter **69.** 260.4 mi **70.** 5000 **71.** 1.77 **72.** $528 **73.** $2160

Quick Quiz 6.2 *See Examples noted with Ex.* **1.** 4590 cm (Ex. 5) **2.** 0.283 mm (Ex. 5) **3.** 5.16 km (Ex. 6) **4.** See Student
Solutions Manual

6.3 Exercises **1.** 1 kL **3.** 1 mg **5.** 1 g **7.** 9000 **9.** 12,000 **11.** 0.0189 **13.** 0.752 **15.** 5,652,000 **17.** 82 **19.** 0.024418
21. 74,000 **23.** 0.216 **25.** 0.035 **27.** 6.328 **29.** 2920 **31.** 2400 **33.** 0.007; 0.000007 **35.** 0.084; 0.000084 **37.** 33; 33,000
39. 2580; 2,580,000 **41.** b **43.** a **45.** 83 L + 0.822 L + 30.1 L = 113.922 L or 113,922 mL **47.** 20 g + 0.052 g + 1500 g = 1520.052 g
or 1,520,052 mg or 1.520052 kg **49.** true **51.** false **53.** false **55.** true **57.** $71.92 **59.** $340,000 **61.** 2,500,000,000 metric tons
63. 17,200,000,000,000 kg **65.** about 17,300,000,000 metric tons **67.** 20% **68.** 57.5 **69.** $4536 **70.** $716.80

Quick Quiz 6.3 *See Examples noted with Ex.* **1.** 0.671 kg (Ex. 5) **2.** 8520 mL (Ex. 1) **3.** 0.04562 g (Ex. 5) **4.** See Student
Solutions Manual

How Am I Doing? Sections 6.1–6.3 **1.** 16 (obj. 6.1.2) **2.** 6 (obj. 6.1.2) **3.** 5280 (obj. 6.1.2) **4.** 9000 (obj. 6.1.2)
5. 1320 (obj. 6.1.2) **6.** 6 (obj. 6.1.2) **7.** $15.30 (obj. 6.1.2) **8.** 6750 (obj. 6.2.2) **9.** 7390 (obj. 6.2.2) **10.** 340 (obj. 6.2.2)
11. 0.027 (obj. 6.2.2) **12.** 529.6 (obj. 6.2.2) **13.** 0.482 (obj. 6.2.2) **14.** 2376 m (obj. 6.2.2) **15.** 91.7 m (obj. 6.2.2) **16.** 1.34 m or 134 cm
(obj. 6.2.2) **17.** 5660 (obj. 6.3.1) **18.** 0.535 (obj. 6.3.2) **19.** 0.0563 (obj. 6.3.2) **20.** 4800 (obj. 6.3.1) **21.** 0.568 (obj. 6.3.2)
22. 8900 (obj. 6.3.1) **23.** $116.25 (obj. 6.3.2) **24.** $227.50 (obj. 6.3.2) **25.** $7.20 (obj. 6.3.1) **26.** $19,200 (obj. 6.3.2)

6.4 Exercises
1. A meter is approximately the same length as a yard. A meter is slightly longer. **3.** An inch is approximately twice the length of a centimeter. **5.** 2.44 m **7.** 22.86 cm **9.** 34.88 yd **11.** 28.15 m **13.** 132.02 km **15.** 10.08 yd **17.** 6.90 in. **19.** 656 ft **21.** 3.1 mi **23.** 181.92 L **25.** 21.76 L **27.** 5.02 gal **29.** 4.77 qt **31.** 180.4 lb **33.** 59.02 kg **35.** 737.1 g **37.** 334.4 lb **39.** 5.58 oz **41.** 1066.8 cm **43.** 34.1 mi/hr **45.** 273 mi/hr **47.** 0.51 in. **49.** 185°F **51.** 53.6°F **53.** 60°C **55.** 35°C **57.** yes **59.** 18.85 liters **61.** 1188 lb **63.** 8.92 ft **65.** 66.2°F at 4 A.M.; 113°F after 7 A.M. **67.** 59,861 miles **69.** 180.6448 sq cm **71.** $896 for the American carpet; $802 for the German carpet; the German carpet is $94 cheaper. **73.** 169 **74.** 114 **75.** It is not a proportion. **76.** It is a proportion.

Quick Quiz 6.4 *See Examples noted with Ex.* **1.** 141.75 g (Ex. 2) **2.** 14.88 mi (Ex. 2) **3.** 6.36 qt (Ex. 2) **4.** See Student Solutions Manual

6.5 Exercises
1. 3 ft **3.** 53 yd **5.** $29.40 **7.** 8.1 m **9.** 880 yd is 4.32 m longer than 880 m. **11.** $3.37/gal; gasoline is more expensive in Mexico. **13.** 77°F; 9°F **15.** The difference is 6°F. The temperature reading of 180°C is hotter. **17.** (a) about 105 km/hr (b) Probably not. We cannot be sure, but we have no evidence to indicate that they broke the speed limit. **19.** 15 gallons **21.** $180 **23.** 2.08 oz **25.** (a) 4.34 quarts (b) $33.70 **27.** (a) $7.63 (b) about 132 mi/gal **29.** yes; 240,000 gal/hr is equivalent to $533\frac{1}{3}$ pt/sec **31.** 15.5 mi **32.** 41.25 yd

Quick Quiz 6.5 *See Examples noted with Ex.* **1.** The prediction was 7.6°F cooler than the actual temperature. (Ex. 2) **2.** 9.5 in. (Ex. 1) **3.** 43.5 min (Ex. 2) **4.** See Student Solutions Manual

Use Math to Save Money
1. $545.75 **2.** $578.06 **3.** He did not deposit enough money to cover the checks he wrote for May. But the $300.50 he already had in the bank will help to cover his expenses for May. **4.** $268.19 **5.** Eventually Terry will be in debt. **6.** Answers will vary. **7.** Answers will vary.

You Try It
1. (a) 135 ft (b) 5.5 lb **2.** (a) 1,500,000 m (b) 250 mm (c) 12.5 L **3.** (a) 5.941 m (b) 12.88 km **4.** (a) 164 ft (b) 74.4 mi/hr **5.** 104°F **6.** 29.44°C

Chapter 6 Review Problems
1. 11 **2.** 8800 **3.** 10.5 **4.** 13,200 **5.** 168 **6.** 14,000 **7.** 0.5 **8.** 56 **9.** 60 **10.** 15.5 **11.** 560 **12.** 176.3 **13.** 1325 **14.** 10 **15.** 9200 **16.** 24 **17.** 7.93 m **18.** 35.63 m **19.** 17,000 **20.** 8 **21.** 196,000 **22.** 0.095 **23.** 3.5 **24.** 15,100 **25.** 765 **26.** 423 **27.** 256 **28.** 92.4 **29.** 2.75 **30.** 72.45 **31.** 5.52 **32.** 9.08 **33.** 45.7 **34.** 49.6 **35.** 53.6° **36.** 89.6° **37.** 105° **38.** 0° **39.** 3.43 **40.** 25.54 **41.** (a) 200 m (b) 0.2 km **42.** (a) 17 ft (b) 204 in. **43.** 270 sq ft; 30 sq yd **44.** $2.88 **45.** yes; 112.7 km/hr **46.** 25°F. Too hot **47.** 380 cm **48.** 39.38 mi/hr **49.** about $3.98 **50.** approximately 516.4 square feet

How Am I Doing? Chapter 6 Test
1. 3200 (obj. 6.1.2) **2.** 228 (obj. 6.1.2) **3.** 84 (obj. 6.1.2) **4.** 7 (obj. 6.1.2) **5.** 30 (obj. 6.1.2) **6.** 0.75 (obj. 6.1.2) **7.** 0.5 (obj. 6.1.2) **8.** 16.5 (obj. 6.1.2) **9.** 9200 (obj. 6.2.2) **10.** 0.0988 (obj. 6.2.2) **11.** 4.6 (obj. 6.2.2) **12.** 1270 (obj. 6.2.2) **13.** 9.36 (obj. 6.2.2) **14.** 0.046 (obj. 6.3.1) **15.** 0.0289 (obj. 6.3.2) **16.** 0.983 (obj. 6.3.2) **17.** 920 (obj. 6.3.1) **18.** 9420 (obj. 6.3.2) **19.** 67.62 (obj. 6.4.1) **20.** 1.63 (obj. 6.4.1) **21.** 3.55 (obj. 6.4.1) **22.** 18.6 (obj. 6.4.1) **23.** 16.06 (obj. 6.4.1) **24.** 85.05 (obj. 6.4.1) **25.** 56.85 (obj. 6.4.1) **26.** 3.18 (obj. 6.4.1) **27.** (a) 20 m (b) 21.8 yd (obj. 6.5.1) **28.** (a) 15°F (obj. 6.5.1) (b) yes (obj. 6.4.2) **29.** 82.5 gal/hr (obj. 6.5.1) **30.** (a) 300 km (b) 14 mi (obj. 6.5.1) **31.** $5\frac{1}{4}$ lb (obj. 6.5.1) **32.** 104°F (obj. 6.4.2) (obj. 6.5.1)

Cumulative Test for Chapters 1–6
1. 21,570 [1.2.4] **2.** 10,747 [1.3.3] **3.** 2,272,704 [1.4.4] **4.** 2667 [1.5.2] **5.** 21 [1.6.2] **6.** 100 [1.7.2] **7.** $\frac{2}{5}$ [2.2.2] **8.** $8\frac{1}{2}$ [2.3.2] **9.** $\frac{35}{4}$ or $8\frac{3}{4}$ [2.4.3] **10.** $\frac{3}{2}$ or $1\frac{1}{2}$ [2.5.1] **11.** $10\frac{1}{6}$ [2.8.1] **12.** $\frac{13}{24}$ [2.8.3] **13.** $\frac{12}{25}$ [3.1.3] **14.** 1.864 [3.2.3] **15.** 361.07 [3.3.1] **16.** 0.35 [3.5.2] **17.** 0.65 [3.6.1] **18.** $7.35 [3.7.2] **19.** 32 points/game [4.1.1] **20.** Yes, it is a proportion. [4.2.2] **21.** $n = 6$ [4.3.2] **22.** 209.23 g [4.4.1] **23.** $\frac{17}{20}$ [5.2.1] **24.** 7.5% [5.2.2] **25.** 120 [5.3A.2] **26.** 200% [5.3.2] **27.** 20,000 [5.3.2] **28.** $1037 [5.4.3] **29.** 9.5 [6.1.2] **30.** 3700 [6.3.1] **31.** 0.05 [6.2.2] **32.** 672 [6.1.2] **33.** 11.88 m [6.4.1] **34.** 59°F; the difference is 44°F; the 15°C temperature is higher [6.4.2]

Chapter 7
7.1 Exercises
1. An acute angle is an angle whose measure is between 0° and 90°. **3.** Complementary angles are two angles whose measures have a sum of 90°. **5.** When two lines intersect, two angles that are opposite each other are called vertical angles. **7.** A transversal is a line that intersects two or more other lines at different points. **9.** ∠ABD, ∠CBE **11.** ∠ABD and ∠CBE; ∠DBC and ∠ABE **13.** There are no complementary angles. **15.** 90° **17.** 25° **19.** 110° **21.** 155° **23.** 59° **25.** 53° **27.** 34° **29.** 35° **31.** 25° **33.** ∠b = 102°; ∠c = ∠a = 78° **35.** ∠b = 38°; ∠a = ∠c = 142° **37.** ∠a = ∠c = 48°; ∠b = 132° **39.** ∠e = ∠d = ∠a = 123°; ∠b = ∠c = ∠f = ∠g = 57° **41.** 4° **43.** 53° north of east **45.** 778.5 km; 482.67 mi **46.** 36.0 mi **47.** 25.2 mi **48.** 12.5%

Quick Quiz 7.1 *See Examples noted with Ex.* **1.** 124° (Ex. 3) **2.** 56° (Ex. 2) **3.** 56° (Ex. 4) **4.** See Student Solutions Manual

7.2 Exercises
1. perpendicular; equal **3.** multiply **5.** 15 mi **7.** 23.6 ft **9.** 17.2 in. **11.** 1.92 mm **13.** 17.12 km **15.** 14.4 ft or 172.8 in. **17.** 0.272 mm **19.** 14 cm **21.** 35 cm **23.** 180 cm **25.** 6.25 ft² **27.** 12 mi² **29.** 117 yd² or 1053 ft² **31.** (a) 294 m² (b) 78 m **33.** $33,000 **35.** (a) 49 ft² (b) 28 ft **37.** (a) 1 × 7, 2 × 6, 3 × 5, 4 × 4; there are four possible shapes. (b) 7 ft², 12 ft², 15 ft², 16 ft² (c) Square garden measuring 4 ft on a side. **39.** $598.22 **41.** 223.3 **42.** 7.18 **43.** 21,842.8 **44.** approximately 1.5759

Quick Quiz 7.2 *See Examples noted with Ex.*   **1.** 7.6 cm (Ex. 1)   **2.** 121 mi^2 (Ex. 5)   **3.** $1056 (Ex. 3)   **4.** See Student Solutions Manual

7.3 Exercises   1. adding   **3.** perpendicular   **5.** 40.2 m   **7.** 49.6 in.   **9.** 354.64 m^2 **11.** 602 yd^2
13. $P = 48$ m; $A = 72$ m^2 **15.** $P = 9.6$ ft; $A = 3.6$ ft^2 **17.** 82 m   **19.** 55 ft + 135 ft + 80.5 ft + 75.5 ft = 346 ft   **21.** 118.8 yd^2
23. 76,850 m^2 **25. (a)** 718 m^2 **(b)** rectangle   **(c)** trapezoid   **27. (a)** 357 ft^2 **(b)** parallelogram   **(c)** trapezoid   **29.** $80,960
31. 10   **32.** 5   **33.** 144   **34.** 8200

Quick Quiz 7.3 *See Examples noted with Ex.*   **1.** 50 yd (Ex. 1)   **2.** 288 m^2 (Ex. 5)   **3.** 12 cm^2 (Ex. 2)   **4.** See Student Solutions Manual

7.4 Exercises   1. right   **3.** Add the measures of the two known angles and subtract that value from 180°.   **5.** You could conclude that
the lengths of all three sides of the triangle are equal.   **7.** true   **9.** true   **11.** false   **13.** false   **15.** 70°   **17.** 82.9°   **19.** 118 m

21. 116.75 in.   **23.** 10 mi   **25.** 56.25 in.^{2} **27.** 83.125 cm^2 **29.** $7\frac{7}{12}$ yd^2 **31.** 126.5 cm^2 **33.** 188 yd^2 **35.** 1740 ft^2 **37.** $21,060

39. 6.25%   **41.** $n = 12$ **42.** $n = 42$ **43.** 155 tons; 152.5 mi   **44.** 96 magazines

Quick Quiz 7.4 *See Examples noted with Ex.*   **1.** 81.4 m (Ex. 2)   **2.** 102 in.^{2} (Ex. 3)   **3.** 50.9° (Ex. 1)   **4.** See Student Solutions Manual

7.5 Exercises   1. $\sqrt{25} = 5$ because $(5)(5) = 25$ **3.** whole   **5.** Use a square root table or a calculator.   **7.** 3   **9.** 8   **11.** 12
13. 0   **15.** 13   **17.** 10   **19.** 10   **21.** 10   **23.** 3   **25.** 8   **27.** 22   **29. (a)** yes   **(b)** 16   **31.** 4.243   **33.** 8.718   **35.** 14.142
37. ≈ 5.831 m   **39.** ≈ 11.662 m   **41.** 10.472   **43.** 7.071   **45.** 104.7 ft   **47.** 127.3 ft   **49.** 39.299   **51.** 4800 in.^{2} **52.** 12.2 mi
53. 18.6 mi   **54.** about 6.7 in.

Quick Quiz 7.5 *See Examples noted with Ex.*   **1.** 8 (Ex. 1)   **2.** 18 (Ex. 2)   **3.** 14 ft (Ex. 5)   **4.** See Student Solutions Manual

How Am I Doing?   Sections 7.1–7.5   1. 18° (obj. 7.1.1)   **2.** 117° (obj. 7.1.1)   **3.** $\angle b = 136°$; $\angle a = \angle c = 44°$ (obj. 7.1.1)
4. 18 m (obj. 7.2.1)   **5.** 14 m (obj. 7.2.1)   **6.** 23.04 cm^2 (obj. 7.2.3)   **7.** 22.62 yd^2 (obj. 7.2.3)   **8.** 25.6 yd (obj. 7.3.1)   **9.** 79 ft (obj. 7.3.2)
10. 351 in.^{2} (obj. 7.3.1)   **11.** 171 in.^{2} (obj. 7.3.2)   **12.** 97 m^2 (obj. 7.3.2)   **13.** 103° (obj. 7.4.1)   **14.** 20 in. (obj. 7.4.2)
15. 105 m^2 (obj. 7.4.2)   **16. (a)** 592 ft^2 **(b)** 114 ft (obj. 7.4.2)   **17.** 8 (obj. 7.5.1)   **18.** 19 (obj. 7.5.1)   **19.** 13 (obj. 7.5.1)
20. 16 (obj. 7.5.1)   **21.** 6.782 (obj. 7.5.2)

7.6 Exercises   1. Square the length of each leg and add those two results. Then take the square root of the remaining number.   **3.** 15 yd
5. 15.199 ft   **7.** 11.402 m   **9.** 14.142 m   **11.** 10.247 ft   **13.** 9.798 yd   **15.** 15 m   **17.** 11.619 ft   **19.** 13 ft   **21.** 9.8 cm
23. 11.1 yd   **25.** 6.9 in.; 4 in.   **27.** 8.5 m   **29.** 25.5 cm   **31.** 7.1 in.   **33.** 0.47 mi   **35.** 14.866 cm   **37.** 341 m^2 **38.** 297.25 ft^2
39. 441 in.^{2} **40.** 4224 yd^2

Quick Quiz 7.6 *See Examples noted with Ex.*   **1.** ≈ 11.18 ft (Ex. 2)   **2.** 10 cm (Ex. 3)   **3.** ≈ 8.54 mi (Ex. 5)   **4.** See Student
Solutions Manual

7.7 Exercises   1. circumference   **3.** radius   **5.** Multiply the radius by 2 and then use $C = \pi d$.   **7.** 58 in.   **9.** 17 mm   **11.** 22.5 yd
13. 16.09 ft   **15.** 100.48 cm   **17.** 116.18 in.   **19.** 41.87 ft   **21.** 78.5 yd^2 **23.** ≈ 226.87 in.^{2} **25.** 803.84 cm^2 **27.** 452.16 ft^2
29. 6358.5 mi^2 **31.** 163.28 m^2 **33.** 30.96 m^2 **35.** 189.25 m^2 **37.** $1211.20   **39.** 9.42 ft   **41.** 141.3 ft   **43.** 1345.22 revolutions
45. (a) 25.12 ft   **(b)** 50.24 ft^2 **47.** 125,600 mi^2 **49. (a)** $1.50 per slice; ≈ 25.12 in.^{2} **(b)** $\approx$ $1.33 per slice; ≈ 18.84 in.^{2}
(c) For 12-in.: $0.07 per in.^{2}; for 16-in.: $0.06 per in.^{2}; 16-in.   **51.** 30   **52.** 0.3   **53.** 70   **54.** 3000

Quick Quiz 7.7 *See Examples noted with Ex.*   **1.** 28.26 in. (Ex. 1)   **2.** 379.94 m^2 (Ex. 3)   **3.** 3.79 cm^2 (Ex. 5)   **4.** See Student
Solutions Manual

7.8 Exercises   1. (a) sphere   **(b)** $V = \dfrac{4\pi r^3}{3}$ **3. (a)** cylinder   **(b)** $V = \pi r^2 h$ **5. (a)** cone   **(b)** $V = \dfrac{\pi r^2 h}{3}$ **7.** 540 mm^3

9. 226.1 m^3 **11.** 6459.0 m^3 **13.** 3052.1 yd^3 **15.** 210 ft^3 **17.** 0.216 cm^3 **19.** 65.94 yd^3 **21.** 718.0 m^3 **23.** 937.8 cm^3
25. 641.1 ft^3 **27.** 163.3 m^3 **29.** 373.3 m^3 **31.** 12 bags   **33.** 1004.8 in.^{3} **35.** 381,251,976,256,667 mi^3 **37.** 2928 in.^{3}
39. 7.1 in.^{3}; 162 cups   **41.** 413.8 cm^3 **43.** 263,900 yd^3 **45.** $9\frac{7}{12}$ **46.** $6\frac{3}{8}$ **47.** $\frac{135}{16}$ or $8\frac{7}{16}$ **48.** $\frac{25}{14}$ or $1\frac{11}{14}$ **49.** $\frac{23}{64}$ **50.** $\frac{115}{18}$ or $6\frac{7}{18}$

Quick Quiz 7.8 *See Examples noted with Ex.*   **1.** 267.95 cm^3 (Ex. 3)   **2.** 112 yd^3 (Ex. 5)   **3.** 367.38 m^3 (Ex. 2)   **4.** See Student
Solutions Manual

7.9 Exercises   1. size; shape   **3.** sides   **5.** $n = 8$ m   **7.** $n \approx 2.6$ ft   **9.** $n = 3.4$ yd   **11.** a corresponds to f, b corresponds to e,
c corresponds to d **13.** 3.4 m   **15.** 26 in.   **17.** 2.2 ft   **19.** 36 ft   **21.** 44 ft   **23.** 8.3 ft   **25.** 12 cm   **27.** 12   **28.** 32

29. 1   **30.** $\frac{1}{5}$

Quick Quiz 7.9 *See Examples noted with Ex.*   **1.** 85.71 ft (Ex. 3)   **2.** 3.21 m (Ex. 1)   **3.** 21 ft (Ex. 4)   **4.** See Student Solutions Manual

7.10 Exercises
1. (a) 50 mi/hr **(b)** ≈54.3 mi/hr **(c)** through Woodville and Palermo **3.** 80.5 min or 1 hr 20.5 min **5.** 4006 ft^2; 12 gal **7.** $510 **9.** $795.15 **11. (a)** 40,820 km **(b)** 20,410 km/hr **13.** ≈50,240 in.^{3} **15.** 128 **16.** 308 **17.** 0.25 **18.** 4.87

Quick Quiz 7.10
See Examples noted with Ex. **1.** 8962.5 yd^2 (Ex. 1) **2.** 638 ft^2 (Ex. 2) **3.** $700 (Ex. 2) **4.** See Student Solutions Manual

Use Math to Save Money
1. 750, 468.75, and 250 gallons respectively **2.** $2700.00, $1687.50, and $900.00 respectively **3.** $1012.50 **4.** About six and a half years **5.** $5062.50; $10,125.00 **6.** $1800 **7.** About one year and three months **8.** $9000; $18,000

You Try It
1. 34 ft **2.** 34 in. **3.** 30 m^2 **4.** 81 ft^2 **5.** 43 in. **6.** 78 cm^2 **7.** 64 ft^2 **8.** 79° **9.** 32.5 cm^2 **10. (a)** 1 **(b)** 3 **(c)** 12 **(d)** 15 **11. (a)** 4.243 **(b)** 11.832 **(c)** 6.245 **12.** 13.038 in. **13.** 16 ft **14.** 8.5 mi **15.** 7.5 m **16.** 14.14 in. **17. (a)** 14 ft **(b)** 11 in. **18.** approximately 18.8 m **19.** approximately 254.3 ft^2 **20.** 280 ft^3 **21.** approximately 339.1 in.^{3} **22.** approximately 523.3 ft^3 **23.** approximately 58.6 m^3 **24.** 120 ft^3 **25.** 12 ft **26.** 63 m

Chapter 7 Review Problems
1. 14° **2.** 104° **3.** $\angle b = 146°$, $\angle a = \angle c = 34°$ **4.** $\angle t = \angle x = \angle y = 65°$, $\angle s = \angle u = \angle w = \angle z = 115°$ **5.** 23.6 m **6.** 50.8 yd **7.** 16.5 cm^2 **8.** 51.8 in.^{2} **9.** 38 ft **10.** 58 ft **11.** 68 m^2 **12.** 63.5 m^2 **13.** 105 m **14.** 62 mi **15.** 3500 ft^2 **16.** 360 yd^2 **17.** 422 cm^2 **18.** 357 m^2 **19.** 60 ft **20.** 46.5 ft **21.** 107° **22.** 55° **23.** 52.3 m^2 **24.** 59.4 m^2 **25.** 450 m^2 **26.** 87 m^2 **27.** 9 **28.** 8 **29.** 11 **30.** 16 **31.** 18 **32.** 7.874 **33.** 12.845 **34.** 13.416 **35.** 5 km **36.** 5 yd **37.** 8.72 cm **38.** 9.22 m **39.** 6.4 cm **40.** 18.1 ft **41.** 6.3 ft **42.** 3.6 ft **43.** 106 cm **44.** 63 cm **45.** 62.8 m **46.** 56.5 in. **47.** 254.3 m^2 **48.** 58.1 ft^2 **49.** 201.0 m^2 **50.** 318.5 ft^2 **51.** 107.4 ft^2 **52.** 80.1 m^2 **53.** 1263.6 ft^3 **54.** 381.5 in.^{3} **55.** 21.2 ft^3 **56.** 245 m^3 **57.** 9074.6 yd^3 **58.** 30 m **59.** 3.3 m **60.** 348 cm **61.** 175 ft **62.** 147 yd^2 **63.** $V \approx 2034.7$ in.^{3}; $W = 32,555.2$ g **64.** $736 **65.** 381,510 m^3 **66.** ≈8.8 ft^3 **67.** ≈66 gal **68.** 1662.5 ft^2 **69. (a)** 50 km; 100 km/hr **(b)** 56 km; 70 km/hr **(c)** through Ipswich **70. (a)** ≈21,873.2 ft^3 **(b)** ≈17,498.6 bushels **71.** 3,429,708,000 ft^3 **72.** 17.1 ft **73.** 1116 lb; 130 gal **74.** 1728 in.^{3} **75.** 942 ft **76.** ≈106.5 yd **77.** 3 spools **78.** 13.7 ft

How Am I Doing? Chapter 7 Test
1. $\angle b = 52°$; $\angle c = 128°$; $\angle e = 128°$ (obj. 7.1.1) **2.** 40 yd (obj. 7.2.1) **3.** 25.2 ft (obj. 7.2.1) **4.** 20 m (obj. 7.3.1) **5.** 80 m (obj. 7.3.2) **6.** 137 m (obj. 7.4.2) **7.** 180 yd^2 (obj. 7.2.1) **8.** 104.0 m^2 (obj. 7.2.1) **9.** 78 m^2 (obj. 7.3.1) **10.** 144 m^2 (obj. 7.3.2) **11.** 12 cm^2 (obj. 7.4.2) **12.** 12 (obj. 7.5.1) **13.** 13 (obj. 7.5.1) **14.** 27° (obj. 7.1.1) **15.** 73° (obj. 7.1.1) **16.** 84° (obj. 7.4.1) **17.** 7.348 (obj. 7.5.2) **18.** 11.619 (obj. 7.5.2) **19.** 8.602 (obj. 7.6.2) **20.** 10 (obj. 7.6.2) **21.** 5.83 cm (obj. 7.6.3) **22.** 9 ft (obj. 7.6.3) **23.** 56.52 ft (obj. 7.7.1) **24.** 113.04 ft^2 (obj. 7.7.1) **25.** 107.4 in.^{2} (obj. 7.7.2) **26.** 144.3 in.^{2} (obj. 7.7.2) **27.** 700 m^3 (obj. 7.8.1) **28.** 803.8 m^3 (obj. 7.8.4) **29.** 113.0 m^3 (obj. 7.8.3) **30.** 508.7 ft^3 (obj. 7.8.2) **31.** 56 m^3 (obj. 7.8.5) **32.** 46.8 m (obj. 7.9.1) **33.** 42 ft (obj. 7.9.1) **34.** 6456 yd^2 (obj. 7.10.1) **35.** $2582.40 (obj. 7.10.1)

Chapter 8
8.1 Exercises
1. Multiply 25% × 4000, which is 0.25 × 4000 = 1000 students **3.** Divide the circle into quarters by drawing two perpendicular lines. Shade in one quarter of the circle. Label this with the title "within five miles = 1000." **5.** rent **7.** $200 **9.** $800 **11.** $\frac{13}{4}$ **13.** $\frac{10}{27}$ **15.** 85 years or older **17.** 47 million or 47,000,000 people **19.** 282 million or 282,000,000 people **21.** $\frac{33}{23}$ **23.** $\frac{89}{336}$ **25.** 19% **27.** reasonable prices and great food **29.** 343 people **31.** 13,776,000 vehicles **33.** 60.1% **35.** 48.3% **37.** 126,000 vehicles **39.** 120 ft^2 **40.** 204 in.^{2} **41.** 16 gal **42.** about 3 g

Quick Quiz 8.1
See Examples noted with Ex. **1.** 51% (Ex. 8) **2.** 382,500 vehicles (Ex. 8) **3.** 816,000 vehicles (Ex. 8) **4.** See Student Solutions Manual

8.2 Exercises
1. 16 million or 16,000,000 **3.** 24 million or 24,000,000 **5.** 1980–1990 and 2000–2010 **7.** $6000 **9.** $6400 **11.** 2003–04 **13.** $200 **15.** $37,200 **17.** about 17% **19.** $1.3 million or $1,300,000 **21.** 2003 to 2005 **23.** $0.5 million or $500,000 **25.** 2.5 in. **27.** October, November, and December **29.** 1.5 in. **31.** 115 **32.** $\frac{2}{9}$ **33.** 1,078,032 degrees **34.** 6960 mi

Quick Quiz 8.2
See Examples noted with Ex. **1.** 800 condominiums (Ex. 8) **2.** 800 homes (Ex. 8) **3.** 2000 (Ex. 9) **4.** See Student Solutions Manual

How Am I Doing? Sections 8.1–8.2
1. 14% (obj. 8.1.2) **2.** Great Smoky Mountain National Park (obj. 8.1.2) **3.** 24% (obj. 8.1.2) **4.** about 4,320,000 visitors (obj. 8.1.2) **5.** about 7,020,000 visitors (obj. 8.1.2) **6.** 450 housing starts (obj. 8.2.2) **7.** 550 housing starts (obj. 8.2.2) **8.** during the fourth quarter of 2009 (obj. 8.2.2) **9.** during the third quarter of 2010 (obj. 8.2.2) **10.** 250 more housing starts (obj. 8.2.2) **11.** 150 fewer housing starts (obj. 8.2.2) **12.** Aug. and Dec. (obj. 8.2.4) **13.** Dec. (obj. 8.2.4) **14.** Nov. (obj. 8.2.4) **15. (a)** 20,000 sets **(b)** 35,000 sets (obj. 8.2.4)

8.3 Exercises
1. The horizontal label for each item in a bar graph is usually a single number or a word title. For a histogram it is a class interval. The vertical bars have a space between them in a bar graph. For a histogram, the vertical bars join each other. **3.** A class frequency is the number of times a data value occurs in a particular class interval. **5.** 120 cities **7.** 10 cities **9.** 40 cities **11.** 170 cities **13.** 8000 books **15.** books costing $5.00–$7.99 **17.** 28,000 books **19.** 52,000 books **21.** 28.6%

	Tally	Frequency
23.	|||	3
25.	⊞|	6
27.	|||	3
29.	||	2

31.

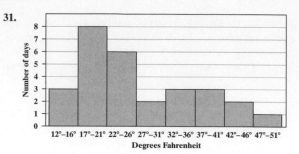

33. 17 days **34.** $m = 138.32$
35. $n = 7$ **36.** 615 mi **37.** 13.0 in.

Quick Quiz 8.3 *See Examples noted with Ex.* **1.** 700 people (Ex. 1) **2.** 2000 people (Ex. 4) **3.** 500 more people (Ex. 4) **4.** See Student Solutions Manual

8.4 Exercises 1. The median of a set of numbers when they are arranged in order from smallest to largest is that value that has the same number of values above it as below it. The mean of a set of values is the sum of the values divided by the number of values. The mean is most likely to be not typical of the values you would expect if there are many extremely low values or many extremely high values. The median is more likely to be typical of the value you would expect. **3.** 32 customers **5.** 5.0 in. **7.** 0.375 **9.** 155,800 **11.** 23.7 mi/gal **13.** 195 **15.** 12.0 **17.** $37,025 **19.** 21 **21.** $207 **23.** 2.2 **25.** $254,129 million or $254,129,000,000 **27.** $69,161.88 **29.** 4850 **31.** mean ≈ $104.74; median = $80.95 **33.** 60 **35.** 121 and 150 **37.** $269 **39.** mean ≈ 81.6 yr; median = 81.7 yr; mode = 80.9 yr **41. (a)** $2157 **(b)** $1615 **(c)** There is no mode. **(d)** The median because the mean is affected by the high amount, $6300. **43. (a)** 8.4 phone calls **(b)** 7 phone calls **(c)** 3 phone calls **(d)** The median. On three nights she gets more calls than 7. On 3 nights she gets fewer calls than 7. On one night she got 7 calls. The mean is distorted a little because of the very large number of calls on Sunday night. The mode is artificially low because she gets so few calls on Monday and Wednesday and it just happened to be the same number, 3. **45.** 19.3 in.² **46.** 20,096 gal/hr **47.** $330 **48.** about $0.04 per in.³

Quick Quiz 8.4 *See Examples noted with Ex.* **1.** 11.5 times (Ex. 2) **2.** 13.2 times (Ex. 1) **3.** 8 times (Ex. 4) **4.** See Student Solutions Manual

Use Math to Save Money 1. December $295.00, November $355.00 **2.** $1775.00 **3.** 4° lower = 8% savings, $1775.00 × 0.08 = $142 **4.** $355 = 20% of $1775; 20% = 10° lower, 72 − 10 = 62 degrees **5.** Answers will vary. **6.** Answers will vary.

You Try It 1. (a) 12% **(b)** 96 people **2. (a)** 5000 sets **(b)** 5000 sets **3. (a)** 6000 visitors **(b)** July and August **(c)** between July 2009 and August 2009 **4. (a)** 16 students **(b)** 48 students **5.** 29.5 **6. (a)** 19 **(b)** 55 **7. (a)** 8 **(b)** 9 and 12

Chapter 8 Review Problems 1. 13 computers **2.** 68 computers **3.** $\frac{13}{21}$ **4.** $\frac{43}{32}$ **5.** ≈17.9% **6.** ≈22.9% **7.** 48% **8.** 77% **9.** art **10.** art and education **11.** 1200 students **12.** 880 students **13.** $121 billion or $121,000,000,000 **14.** $552 billion or $552,000,000,000 **15.** $122 billion or $122,000,000,000 **16.** $83 billion or $83,000,000,000 **17.** 2000 to 2005 **18.** $\frac{32}{69}$ **19.** 12.1 billion or 12,100,000,000 bushels **20.** 3.2 billion or 3,200,000,000 bushels **21.** between 2008 and 2010 **22.** 16.7 billion or 16,700,000,000 bushels **23.** 2010 **24.** 2002 **25.** 11.32 billion or 11,320,000,000 bushels **26.** 15.9 billion or 15,900,000,000 bushels **27.** 350 students **28.** 500 students **29.** 200 students **30.** 2005 and 2006 **31.** about 408 students **32.** 37.5% **33.** 45,000 cones **34.** 30,000 cones **35.** 10,000 cones **36.** 30,000 cones **37.** 135,000 cones **38.** Since August was cold and rainy, significantly fewer people wanted ice cream during August. **39.** 3500 degrees **40.** 800 more degrees **41.** 2000 and 2002 **42.** 2004, 2006, and 2008 **43.** 2002 to 2004 **44.** 2002 to 2004 **45.** Between 2002 and 2004, the number of master's degrees in mathematics and statistics increased by 900. **46.** 6100 degrees **47.** 65 pairs **48.** 10 pairs **49.** 145 pairs **50.** 90% **51.** 45 pairs **52.** $\frac{5}{36}$

	Number of Defective Televisions (Class Intervals)	Tally	Frequency
53.	0–3	⊞ ⊞	10
54.	4–7	⊞ |||	8
55.	8–11	|||	3
56.	12–15	⊞	5
57.	16–19	||	2

58.

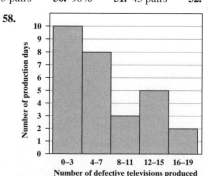

59. 18 times **60.** 90°
61. $89.67 **62.** 20,200 people **63.** 1353 employees
64. $36,000 **65.** $141,500

66. median = 30.5 years; mode = 28 years **67.** median = 18.5 deliveries; mode = 15 deliveries **68.** The median is better because the mean is skewed by the one low score, 31. **69.** The median is better because the mean is skewed by the one high data item, 39. **70. (a)** 5 hr **(b)** 4 hr **(c)** 2 hr **(d)** The median is the most representative. On three days she uses the computer more than 4 hours and on three days she uses the computer less than 4 hours. One day she used it exactly 4 hours. The mean is distorted a little because of the very large number of hours on Friday. The mode is artificially low because she happened to use the computer only two hours on Sunday and Tuesday. All other days it was more than this.

How Am I Doing?   Chapter 8 Test 1. 37% (obj. 8.1.2)   **2.** 21% (obj. 8.1.2)   **3.** 12% (obj. 8.1.2)   **4.** 90,000 automobiles (obj. 8.1.2)
5. 81,000 automobiles (obj. 8.1.2)   **6.** $12,000 (obj. 8.2.2)   **7.** $3000 (obj. 8.2.2)   **8.** $10,000 (obj. 8.2.2)   **9.** $3000 (obj. 8.2.2)
10. $11,000 (obj. 8.2.2)   **11.** $16,000 (obj. 8.2.2)   **12.** 20 yr (obj. 8.2.4)   **13.** 26 yr (obj. 8.2.4)   **14.** 12 yr (obj. 8.2.4)   **15.** age 35 (obj. 8.2.4)
16. age 65 (obj. 8.2.4)   **17.** 60,000 televisions (obj. 8.3.1)   **18.** 25,000 televisions (obj. 8.3.1)   **19.** 20,000 televisions (obj. 8.3.1)
20. 60,000 televisions (obj. 8.3.1)   **21.** 14.5 (obj. 8.4.1)   **22.** 14.5 (obj. 8.4.2)   **23.** 10 (obj. 8.4.3)   **24.** mean or median (obj. 8.4.3)

Chapter 9   9.1 Exercises 1. First, find the absolute value of each number. Then add those two absolute values. Use the common

sign in the answer.   **3.** $<$ **5.** $>$ **7.** $>$ **9.** $<$ **11.** 7   **13.** 16   **15.** -17 **17.** -7 **19.** 16.5   **21.** $\frac{17}{35}$ **23.** -3 **25.** 9

27. -5 **29.** 22   **31.** -3.25 **33.** $-\frac{2}{3}$ **35.** $\frac{5}{9}$ **37.** -22 **39.** -0.72 **41.** -363 **43.** 4   **45.** $-5\frac{9}{20}$ **47.** 6.24   **49.** $-\frac{11}{2}$ or $-5\frac{1}{2}$

51. -14.7 **53.** 0   **55.** -6 **57.** $-\frac{53}{75}$ **59.** $-$94,000$ **61.** $-$7500$ **63.** $8800   **65.** $-18°F$ **67.** $-1°F$ **69.** -0.82 **71.** -1 yd

73. $37.00   **75.** 904.3 ft^3 **76.** 210 m^3 **77.** 60   **78.** $40

Quick Quiz 9.1 *See Examples noted with Ex.* **1.** 6 (Ex. 7)   **2.** -1.5 (Ex. 5)   **3.** $-3\frac{1}{3}$ (Ex. 5)   **4.** See Student Solutions Manual

9.2 Exercises 1. -6 **3.** -5 **5.** -6 **7.** -17 **9.** 9   **11.** 85   **13.** -18 **15.** 3   **17.** -60 **19.** 556   **21.** -6.7 **23.** -8.7

25. -8.6 **27.** 32.94   **29.** 1   **31.** $-\frac{7}{6}$ or $-1\frac{1}{6}$ **33.** $1\frac{8}{15}$ **35.** $-\frac{31}{63}$ **37.** 15   **39.** 0   **41.** 46   **43.** -22 **45.** -2 **47.** -5.4

49. 14,776 ft   **51.** 42°F   **53.** $-13°F$ **55.** $-$16,000$ **57.** $+$24,900$ **59.** $13\frac{1}{2}$ or $13.50   **61.** 5   **62.** 6

Quick Quiz 9.2 *See Examples noted with Ex.* **1.** $\frac{15}{14}$ or $1\frac{1}{14}$ (Ex. 3)   **2.** -5.5 (Ex. 3)   **3.** -82 (Ex. 2)   **4.** See Student Solutions Manual

9.3 Exercises 1. To multiply two numbers with the same sign, multiply the absolute values. The sign of the result is positive.   **3.** 36   **5.** 60

7. -160 **9.** -66 **11.** -1.5 **13.** 28.125   **15.** $-\frac{6}{35}$ **17.** 3   **19.** -8 **21.** -8 **23.** 6   **25.** 5   **27.** $\frac{3}{4}$ **29.** $\frac{8}{7}$ or $1\frac{1}{7}$

31. -8.38 **33.** -5.7 **35.** 30   **37.** -45 **39.** 48   **41.** -5 **43.** 10   **45.** -2.8 **47.** -0.02 **49.** $\frac{3}{7}$ **51.** -8 **53.** 150

55. 84   **57.** 16   **59.** $-18,000$ **61.** 0   **63.** $-\frac{5}{12}$ **65.** He gained $74.   **67.** $-6.25°F$ **69.** 70 ft   **71.** $+33$ **73.** $+8$ **75.** $+8$
77. 0; at par   **79.** 90 in.^{2} **80.** 264 m^2

Quick Quiz 9.3 *See Examples noted with Ex.* **1.** 45 (Ex. 4)   **2.** -72 (Ex. 6)   **3.** 39 (Ex. 5)   **4.** See Student Solutions Manual

How Am I Doing?   Sections 9.1–9.3 1. -19 (obj. 9.1.1)   **2.** -4 (obj. 9.1.2)   **3.** 4.5 (obj. 9.1.2)   **4.** 0 (obj. 9.1.3)
5. $\frac{2}{9}$ (obj. 9.1.2)   **6.** $-\frac{7}{6}$ or $-1\frac{1}{6}$ (obj. 9.1.1)   **7.** -7 (obj. 9.1.1)   **8.** 1.7 (obj. 9.1.2)   **9.** -8 (obj. 9.2.1)   **10.** -41 (obj. 9.2.1)
11. $\frac{14}{17}$ (obj. 9.2.1)   **12.** -12 (obj. 9.2.1)   **13.** -16.3 (obj. 9.2.1)   **14.** -5.6 (obj. 9.2.1)   **15.** 42 (obj. 9.2.1)   **16.** $\frac{19}{15}$ or $1\frac{14}{15}$ (obj. 9.2.1)
17. 24 (obj. 9.3.1)   **18.** 4 (obj. 9.3.1)   **19.** -8 (obj. 9.3.1)   **20.** -10 (obj. 9.3.2)   **21.** -24 (obj. 9.3.1)   **22.** $\frac{8}{11}$ (obj. 9.3.1)
23. -64 (obj. 9.3.2)   **24.** -10 (obj. 9.3.1)   **25.** 24 (obj. 9.2.1)   **26.** 21 (obj. 9.3.1)   **27.** -1.5 (obj. 9.3.1)   **28.** -3.6 (obj. 9.1.3)
29. -0.6 (obj. 9.2.1)   **30.** $-\frac{11}{15}$ (obj. 9.1.1)   **31.** $\frac{1}{5}$ (obj. 9.3.1)   **32.** -11 (obj. 9.3.1)   **33.** $-3.5°F$ (obj. 9.3.2)

9.4 Exercises 1. 6   **3.** -8 **5.** 8   **7.** -12 **9.** -19 **11.** 3.2   **13.** -5 **15.** -6 **17.** -27 **19.** -102 **21.** -18
23. 1   **25.** 4   **27.** $-\frac{1}{5}$ **29.** 1   **31.** 0   **33.** 4   **35.** 123   **37.** -7 **39.** $-\frac{13}{50}$ **41.** 6.84   **43.** $-22°F$ **45.** $-1.7°F$ **47.** 20.5°F
49. 3.84 km   **50.** 36,800 mg

Quick Quiz 9.4 *See Examples noted with Ex.* **1.** -74 (Ex. 4)   **2.** -7.9 (Ex. 2)   **3.** $\frac{4}{9}$ (Ex. 3)   **4.** See Student Solutions Manual

9.5 Exercises 1. Our number system is structured according to base 10. By making scientific notation also in base 10, the calculations are easier to
perform.   **3.** The first part is a number greater than or equal to 1 but smaller than 10. It has at least one nonzero digit. The second part is 10 raised to
some integer power.   **5.** 1.2×10^2 **7.** 1.9×10^3 **9.** 2.63×10^4 **11.** 2.88×10^5 **13.** 1×10^4 **15.** 1.2×10^7 **17.** 9.31×10^{-2}
19. 2.79×10^{-3} **21.** 8.2×10^{-1} **23.** 5.4×10^{-4} **25.** 5.31×10^{-6} **27.** 8×10^{-6} **29.** 53,600   **31.** 5334   **33.** 4,600,000,000,000
35. 0.062   **37.** 0.00899   **39.** 900,000,000,000   **41.** 0.00000003862   **43.** 3.5689×10^4 **45.** 0.00033   **47.** 2.78×10^{-3} **49.** 1,880,000
51. 5.878×10^{12} mi   **53.** 9.2×10^{-14} L   **55.** 0.000000001 sec   **57.** 0.000075 cm   **59.** 14,000,000,000 t   **61.** 9.01×10^7 dollars
63. 1.068×10^{22} t   **65.** 3.624×10^8 ft   **67.** 6.0×10^6 mi^2 **69.** 1.176×10^{14} mi   **71.** 2.625   **72.** 0.258   **73.** $176   **74.** 589 ft

Quick Quiz 9.5 *See Examples noted with Ex.* **1.** 3.45×10^{-4} (Ex. 2)   **2.** 5.683×10^5 (Ex. 1)   **3.** 0.00000834 (Ex. 5)   **4.** See Student Solutions Manual

Use Math to Save Money 1. $\frac{8}{3}$ or $2\frac{2}{3}$ C   **2.** 56 oz   **3.** 3.5 lb   **4.** $9.31   **5.** $6.23   **6.** Lucy saves approximately 67%.
7. $162 \times 52 = $8424; $8424 \times 67\% = $5644.08   **8.** $1664

You Try It 1. (a) 10   **(b)** 18   **(c)** 2.5   **2. (a)** 17   **(b)** -17 **(c)** -6.7 **(d)** $-\frac{5}{8}$ **3. (a)** 1   **(b)** -17 **(c)** -4.8 **(d)** $-\frac{1}{3}$

4. (a) 9.4   **(b)** -13 **(c)** -18 **(d)** -0.7 **(e)** $\frac{1}{5}$ **5. (a)** -45 **(b)** -12 **(c)** -21 **(d)** -5 **6. (a)** 3   **(b)** 56   **(c)** 2

(d) 2   **7.** -26 **8.** $-\frac{1}{6}$ **9. (a)** 3.124×10^3 **(b)** 5.88×10^{-4} **(c)** 1.8×10^8 **(d)** 9×10^{-6} **10. (a)** 400,000   **(b)** 0.004
(c) 2,526,000   **(d)** 0.0813

Chapter 9 Review Problems 1. -15 **2.** -14 **3.** -8.8 **4.** 2.6   **5.** $-\frac{8}{15}$ **6.** $-\frac{8}{5}$ **7.** 6   **8.** 0   **9.** -11 **10.** -28

11. 20.5   **12.** -17.2 **13.** -12.3 **14.** $-\frac{1}{15}$ **15.** 13   **16.** -3 **17.** -9 **18.** $\frac{2}{35}$ **19.** -7.8 **20.** 3   **21.** 6   **22.** -5

23. 6   **24.** $-\frac{9}{2}$ or $-4\frac{1}{2}$ **25.** $\frac{3}{7}$ **26.** 30   **27.** -36 **28.** 0   **29.** -11 **30.** -5 **31.** -8 **32.** -0.5 **33.** -1.2 **34.** 2

35. $-\frac{3}{4}$ **36.** $\frac{1}{10}$ **37.** -32 **38.** 21   **39.** $-\frac{7}{45}$ **40.** 0.04   **41.** 63.72   **42.** 4.16×10^3 **43.** 3.7×10^6 **44.** 2×10^5 **45.** 7×10^{-3}
46. 2.18×10^{-5} **47.** 7.63×10^{-6} **48.** 18,900   **49.** 3760   **50.** 314,000   **51.** 0.0752   **52.** 0.00661   **53.** 0.0000009   **54.** 8.18×10^7
55. 9.98×10^{10} **56.** 1.44×10^{14} **57.** 6.8×10^{25} **58.** 4.9104×10^{11} ft   **59.** 2.604×10^{13} mi   **60.** 1.67×10^{-24} g; 9.1×10^{-28} g
61. 250,000,000 m   **62.** 384,400,000 m   **63.** total loss 6 yd   **64.** 2676 ft   **65.** $2   **66.** $-9.6°$F   **67.** 2 points above par

How Am I Doing?   Chapter 9 Test 1. -11 (obj. 9.1.2)   **2.** -43 (obj. 9.1.1)   **3.** 3.9 (obj. 9.1.2)   **4.** -6 (obj. 9.1.3)

5. $-3\frac{1}{2}$ (obj. 9.1.2)   **6.** $-\frac{7}{8}$ (obj. 9.1.1)   **7.** -38 (obj. 9.2.1)   **8.** 5 (obj. 9.2.1)   **9.** $\frac{17}{15}$ or $1\frac{2}{15}$ (obj. 9.2.1)   **10.** -43 (obj. 9.2.1)

11. 4 (obj. 9.2.1)   **12.** -11.3 (obj. 9.2.1)   **13.** $\frac{11}{12}$ (obj. 9.2.1)   **14.** 0 (obj. 9.2.1)   **15.** 120 (obj. 9.3.1)   **16.** -36 (obj. 9.3.1)

17. 10 (obj. 9.3.1)   **18.** -18 (obj. 9.3.2)   **19.** 3 (obj. 9.3.1)   **20.** $-\frac{7}{10}$ (obj. 9.3.1)   **21.** 18 (obj. 9.3.2)   **22.** -32 (obj. 9.3.1)

23. 17 (obj. 9.4.1)   **24.** 0.5 (obj. 9.4.1)   **25.** -8 (obj. 9.4.1)   **26.** -46 (obj. 9.4.1)   **27.** -7.2 (obj. 9.4.1)   **28.** -20 (obj. 9.4.1)

29. $-\frac{1}{7}$ (obj. 9.4.1)   **30.** $-\frac{1}{6}$ (obj. 9.4.1)   **31.** 8.054×10^4 (obj. 9.5.1)   **32.** 7×10^{-6} (obj. 9.5.1)   **33.** 0.0000936 (obj. 9.5.2)

34. 72,000 (obj. 9.5.2)   **35.** $-6.2°$F (obj. 9.2.3)   **36.** 2.72×10^{-4} m (obj. 9.5.3)   **37.** 186.9°F (obj. 9.2.2)

Chapter 10   10.1 Exercises
1. A variable is a symbol, usually a letter of the alphabet, that stands for a number.
3. All the exponents for like terms must be the same. The exponent for x must be the same. The exponent for y must be the same. In this case, x is raised to the second power in the first term but y is raised to the second power in the second term.   **5.** G, x, y **7.** p, a, b **9.** $r = 3m + 5n$

11. $H = 2a - 3b$ **13.** $10x$ **15.** $-x$ **17.** $\frac{1}{3}x$ **19.** $7x + 4$ **21.** $-1.1x + 6.4$ **23.** $37x + 9y - 11$ **25.** $\left(2\frac{1}{3}\right)x - 50$ or $\frac{7}{3}x - 50$

27. $-3a + 6b - 4c$ **29.** $-\frac{1}{4}x + \frac{8}{21}y$ **31.** $-3x - 6$ **33.** $3.1m - 11.1n + 1.2$ **35. (a)** $12x + 1$ **(b)** It is doubled to obtain $24x + 2$.

37. $n = 12$ **38.** $n = 8$ **39.** $n = 4.8$ **40.** $n = 8.1$ **41.** 8   **42.** 40

Quick Quiz 10.1 *See Examples noted with Ex.* **1.** $3a - 14b$ (Ex. 7)   **2.** $-\frac{5}{12}x - \frac{17}{25}y$ (Ex. 7)   **3.** $-18x + 15y - 47$ (Ex. 6)
4. See Student Solutions Manual

10.2 Exercises 1. variable   **3.** $3x$ and x; $2y$ and $-3y$ **5.** $27x - 18$ **7.** $-2x - 2y$ **9.** $14.4x - 30y$ **11.** $30x - 70y$

13. $48a - 40b$ **15.** $24y + 21z$ **17.** $4p + 36q - 40$ **19.** $\frac{3}{5}x + 2y - \frac{3}{4}$ **21.** $30a + 48b - 67.5$ **23.** $32a + 48b - 36c - 20$

25. $-2.6x + 17y + 10z - 24$ **27.** $x - \frac{3}{2}y + 2z - \frac{1}{4}$ **29.** $-3s + 10t + 21$ **31.** $P = 2l + 2w$ **33.** $A = \frac{hB + hb}{2}$ **35.** $27x - 39$

37. $-8a + 50b - 16$ **39.** $8.1x + 8.1y$ **41.** $-9a - 19b + 12c$ **43.** $A = a(b + c) = ab + ac$ **45.** 23 ft   **46.** 63.75 in.^{2}

Quick Quiz 10.2 *See Examples noted with Ex.* **1.** $\frac{5}{2}x - \frac{7}{4}y$ (Ex. 1)   **2.** $-7x + 10.5y - 3.5z + 14$ (Ex. 4)   **3.** $-16x + 59y$ (Ex. 6)
4. See Student Solutions Manual

10.3 Exercises 1. equation   **3.** opposite   **5.** $y = 32$ **7.** $x = 9$ **9.** $x = -18$ **11.** $x = -25$ **13.** $x = 19$ **15.** $9.8 = x$

17. $y = -11.6$ **19.** $x = 13.2$ **21.** $\frac{2}{5} = x$ **23.** $x = 1$ **25.** $x = -\frac{3}{2}$ or $-1\frac{1}{2}$ **27.** $y = \frac{17}{8}$ or $2\frac{1}{8}$ **29.** $x = 14$ **31.** $x = -13$

33. $x = 2$ **35.** $x = -9$ **37.** $y = \frac{13}{2}$ or $6\frac{1}{2}$ **39.** $-8 = z$ **41.** $y = 1.2$ **43.** $x = 6$ **45.** $x = 12$ **47.** $-7 = x$

49. To solve the equation $3x = 12$, divide both sides of the equation by 3 so that x stands alone on one side of the equation.   **51.** $3x + 3y + 3$ **52.** $-x + 21y + 3$ **53.** \$82   **54.** 86

Quick Quiz 10.3 *See Examples noted with Ex.* **1.** $x = -2.2$ (Ex. 2)   **2.** $x = 35$ (Ex. 3)   **3.** $x = -10$ (Ex. 3)   **4.** See Student Solutions Manual

10.4 Exercises **1.** A sample answer is: To maintain the balance, whatever you do to one side of the scale, you need to do the exact same thing to the other side of the scale.   **3.** $\dfrac{4}{3}$ **5.** $x = 9$ **7.** $y = -4$ **9.** $x = -\dfrac{16}{9}$ or $-1\dfrac{7}{9}$ **11.** $x = 12$ **13.** $16 = m$ **15.** $x = 10$

17. $7 = t$ **19.** $x = -13.5$ **21.** $x = 8$ **23.** $y = 10$ **25.** $n = \dfrac{5}{4}$ or $1\dfrac{1}{4}$ **27.** $x = -\dfrac{18}{5}$ or $-3\dfrac{3}{5}$ **29.** $x = -\dfrac{9}{2}$ or $-4\dfrac{1}{2}$ **31.** $z = 6$

33. $6 = x$ **35.** $x = -9$ **37.** $x = 0.03$ **39.** $-20 = x$ **41.** $4x - 7y + 6$ **42.** $-11a + 20b - 27c$ **43.** 59.4%   **44.** 81.1%; 18.9%

Quick Quiz 10.4 *See Examples noted with Ex.* **1.** $x = -\dfrac{15}{4}$ or $-3\dfrac{3}{4}$ (Ex. 2)   **2.** $7 = x$ (Ex. 3)   **3.** $x = -6$ (Ex. 4)   **4.** See Student Solutions Manual

10.5 Exercises **1.** You want to obtain the x-term all by itself on one side of the equation. So you want to remove the -6 from the left side of the equation. Therefore you would add the opposite of -6. This means you would add 6 to each side.   **3.** no   **5.** yes   **7.** $x = 3$

9. $x = -\dfrac{1}{2}$ **11.** $x = \dfrac{5}{6}$ **13.** $x = 2$ **15.** $x = 0.3$ **17.** $x = 33$ **19.** $x = 2$ **21.** $x = 1$ **23.** $z = -4$ **25.** $x = 11$

27. $x = -1$ **29.** $y = -6$ **31.** $y = -\dfrac{1}{3}$ **33.** $y = 5$ **35.** $x = 16$ **37.** $y = 0$ **39.** $x = -17$ **41.** $x = -\dfrac{11}{4}$ or $-2\dfrac{3}{4}$

43. $x = 10$ **45.** 407,513.4 cm^3 **46.** 23.4 in.^{2}

Quick Quiz 10.5 *See Examples noted with Ex.* **1.** $x = 4$ (Ex. 4)   **2.** $x = -4$ (Ex. 5)   **3.** $x = -6$ (Ex. 6)   **4.** See Student Solutions Manual

How Am I Doing? Sections 10.1–10.5 **1.** $-17x$ (obj. 10.1.2)   **2.** y (obj. 10.1.2)   **3.** $-3a + 2b$ (obj. 10.1.2)
4. $-6x + 11y - 10$ (obj. 10.1.2)   **5.** $16x - 2y - 6$ (obj. 10.1.2)   **6.** $4a - 12b + 3c$ (obj. 10.1.2)   **7.** $42x - 18y$ (obj. 10.2.1)
8. $-2a + b - 12$ (obj. 10.2.1)   **9.** $-3a - 6b + 12c + 10$ (obj. 10.2.1)   **10.** $-6x + 14y$ (obj. 10.2.2)   **11.** $-18x - 8y$ (obj. 10.2.1)
12. $-21x + 9y$ (obj. 10.2.1)   **13.** $x = 37$ (obj. 10.3.1)   **14.** $x = 3.5$ (obj. 10.3.1)   **15.** $y = -\dfrac{11}{10}$ or $-1\dfrac{1}{10}$ (obj. 10.3.1)
16. $8 = x$ (obj. 10.3.1)   **17.** $y = 8$ (obj. 10.4.1)   **18.** $y = 10$ (obj. 10.4.1)   **19.** $x = \dfrac{3}{2}$ or $1\dfrac{1}{2}$ (obj. 10.4.2)   **20.** $-12 = x$ (obj. 10.4.1)
21. $m = \dfrac{16}{3}$ or $5\dfrac{1}{3}$ (obj. 10.5.1)   **22.** $a = 1$ (obj. 10.5.2)   **23.** $x = 3$ (obj. 10.5.3)   **24.** $x = \dfrac{9}{4}$ or $2\dfrac{1}{4}$ (obj. 10.5.3)   **25.** $x = 8$ (obj. 10.5.3)
26. $x = \dfrac{25}{6}$ or $4\dfrac{1}{6}$ (obj. 10.5.3)   **27.** $y = 7$ (obj. 10.5.2)   **28.** $x = 4$ (obj. 10.5.2)

10.6 Exercises **1.** $h = 34 + r$ **3.** $b = n - 107$ **5.** $n = a + 14$ **7.** $l = 2w + 7$ **9.** $l = 3w - 2$ **11.** $m = 3t + 10$
13. $t + l = 32$ **15.** $ht = 500$ **17.** $p =$ airfare to Phoenix; $p + 135 =$ airfare to San Diego   **19.** $b =$ number of degrees in angle B; $b - 46 =$ number of degrees in angle A **21.** $w =$ height of Willis Tower; $w + 1267 =$ height of Burj Khalifa   **23.** $a =$ number of books Aaron read; $2a =$ number of books Nina read; $a + 5 =$ number of books Molly read   **25.** $h =$ height; $h + 5 =$ length; $3h =$ width   **27.** $x =$ 1st angle; $2x =$ 2nd angle; $x - 14 =$ 3rd angle   **29.** 8   **30.** -1 **31.** $x = -1$ **32.** $y = -14$ **33.** Kyle Korver (3-Point %) 0.536; Mike Miller (3-Point Attempts) 171; Daniel Gibson (3-Point Shots Made) 71

Quick Quiz 10.6 *See Examples noted with Ex.* **1.** $t = c - 12$ (Ex. 2)   **2.** $w =$ width; $2w + 3 =$ length (Ex. 4)
3. $c =$ number of compact cars; $0.5c$ or $\dfrac{1}{2}c$ or $\dfrac{c}{2} =$ number of SUVs; $c + 35 =$ number of trucks (Ex. 7)   **4.** See Student Solutions Manual

10.7 Exercises **1.** $x =$ length of shorter piece; $x + 5.5 =$ length of longer piece; 5.25 ft; 10.75 ft   **3.** $x =$ number of points scored by France, $x - 22 =$ number of points scored by Japan; France scored 51 points, Japan scored 29 points   **5.** $x =$ the number of cars in November; $x + 84 =$ the number of cars in May; $x - 43 =$ the number of cars in July; 119 cars in November; 203 cars in May; 76 cars in July
7. $x =$ length of shorter piece; $x + 4.7 =$ length of the longer piece; the shorter piece is 3.65 feet long; the longer piece is 8.35 feet long
9. $x =$ width; $2x - 4 =$ length; width is 14 in.; length is 24 in.   **11.** $x =$ length of the first side; $x + 20 =$ length of the second side; $x - 4 =$ length of the third side; 61 mm; 81 mm; 57 mm   **13.** $x =$ length of the first side; $2x =$ length of the second side; $x + 12 =$ length of the third side; 8 cm; 16 cm; 20 cm   **15.** $x =$ number of degrees in angle A; $3x =$ number of degrees in angle B; $x + 40 =$ number of degrees in angle C; angle A measures 28°; angle B measures 84°; angle C measures 68°   **17.** $x =$ total sales; \$76,000   **19.** $x =$ yearly rent; \$6000   **21.** $x =$ length of the adult section; $x + 6.2 =$ length of the child section; adult section is 12.9 ft; child section is 19.1 ft   **23.** 2211 heart transplants; 6320 liver transplants; 379 pancreas transplants   **25.** 2125 thousand or 2,125,000   **26.** 60%   **27.** 500   **28.** $x = 8$ **29.** 80 oz

Quick Quiz 10.7 *See Examples noted with Ex.* **1.** Barbara earns \$281; Melinda earns \$156 (Ex. 1)   **2.** 1440 students work full time; 2880 students work part time; 1680 students do not work (Ex. 2)   **3.** width is 27 yd; length is 61 yd (Ex. 3)   **4.** See Student Solutions Manual

Use Math to Save Money **1.** (\$2500 × 0.05) × 12 = \$1500   **2.** \$4500   **3.** \$3450   **4.** About 55 months, or 4 years and 7 months   **5.** About 28 months, or 2 years and 4 months   **6.** (\$2500 + (5800/12)) × 0.05 = \$149.17 a month   **7.** (\$2500 + (\$5800/12)) × 0.20 = \$596.67 a month   **8.** Answers will vary   **9.** Answers will vary   **10.** Answers will vary

You Try It **1.** (a) $12a$ (b) $3a - 2b$ (c) $19x - 7y - 6$ **2.** (a) $3a - 18b$ (b) $-8x + 12y + 4$ (c) $40a - 8b$ **3.** $-13x + 12y$
4. $x = 4.8$ **5.** $x = -6$ **6.** $x = -\dfrac{5}{6}$ **7.** $x = 3$ **8.** $a = f - 32$ **9.** $x =$ the number of degrees in the second angle;

$x + 12 =$ the number of degrees in the first angle; $x - 18 =$ the number of degrees in the third angle   **10.** 57 students on first floor; 114 students on second floor; 89 students on third floor

Chapter 10 Review Problems
1. $-13a + 3$ **2.** $\frac{5}{6}x + \frac{8}{9}$ **3.** $-2x - 7y$ **4.** $11x - 5y$ **5.** $-x - 12y + 6$
6. $13a - 12b - 4$ **7.** $-15x - 3y$ **8.** $-8x - 12y$ **9.** $2x - 6y + 8$ **10.** $30a - 40b + 25$ **11.** $-9a + 2b + 12$
12. $6x + 15y - 27.5$ **13.** $-2x + 14y$ **14.** $7x - 8y$ **15.** $-8a - 2b - 24$ **16.** $-7a + 8b + 15$ **17.** $x = 12$ **18.** $x = 11.7$
19. $x = 4$ **20.** $x = 7.4$ **21.** $x = -12.1$ **22.** $x = \frac{13}{4}$ or $3\frac{1}{4}$ **23.** $y = -\frac{3}{4}$ **24.** $x = 5$ **25.** $9 = y$ **26.** $x = -\frac{5}{2}$ or $-2\frac{1}{2}$
27. $y = -5$ **28.** $x = 6$ **29.** $y = 9$ **30.** $x = 8$ **31.** $x = \frac{5}{4}$ or $1\frac{1}{4}$ **32.** $x = 6$ **33.** $x = 3$ **34.** $x = -8$ **35.** $x = 3$
36. $x = -10$ **37.** $x = -2$ **38.** $x = 5$ **39.** $y = 0$ **40.** $w = c + 3000$ **41.** $e = 12 + a$ **42.** $A = 3B$
43. $l = 2w - 3$ **44.** $r =$ Roberto's salary; $r + 2050 =$ Michael's salary   **45.** $x =$ length of first side; $2x =$ length of second side
46. $d =$ the number of days Dennis worked; $2d + 12 =$ the number of days Carmen worked   **47.** $n =$ number of nonfiction books;
$n + 225 =$ number of fiction books   **48.** $x =$ length of shorter piece; $x + 6.5 =$ length of longer piece; 26.75 ft; 33.25 ft
49. $x =$ the experienced employee's salary; $x - 28 =$ the new employee's salary; \$192; \$220   **50.** $x =$ number of customers in February;
$2x =$ number of customers in March; $x + 3000 =$ number of customers in April; 10,550 in Feb.; 21,000 in Mar.; 13,550 in Apr.   **51.** $x =$ miles on
Friday; $2x =$ miles on Saturday; $x + 30 =$ miles on Sunday; 160 mi on Fri.; 320 mi on Sat.; 190 mi on Sun.   **52.** $x =$ width; $2x - 3 =$ length;
width $= 13$ in.; length $= 23$ in.   **53.** $z =$ measure of angle Z; $2z =$ measure of angle Y; $z - 12 =$ measure of angle X; $X = 36°$, $Y = 96°$,
$Z = 48°$ **54.** $x =$ length; $x - 67 =$ width; width is 53 yd; length is 120 yd   **55.** $x =$ miles on Saturday; $x + 106 =$ miles on Sunday;
352 mi on Sat.; 458 mi on Sun.   **56.** $x =$ number of applications in the first week; $x + 156 =$ number of applications in the second week;
$x - 142 =$ number of applications in the third week; 262 the first week; 418 the second week; 120 the third week   **57.** $x =$ total sales; \$19,375

How Am I Doing? Chapter 10 Test
1. $-6a$ (obj. 10.1.2)   **2.** $\frac{2}{15}x + \frac{9}{8}y$ (obj. 10.1.2)   **3.** $\frac{5}{8}a - \frac{2}{3}b$ (obj. 10.1.2)
4. $a - 8b$ (obj. 10.1.2)   **5.** $7x - 7z$ (obj. 10.1.2)   **6.** $-4x - 2y + 5$ (obj. 10.1.2)   **7.** $60x - 25y$ (obj. 10.2.1)   **8.** $2x - \frac{10}{3}y$ (obj. 10.2.1)
9. $-4.5a + 3b - 1.5c + 12$ (obj. 10.2.1)   **10.** $-11a + 14b$ (obj. 10.2.2)   **11.** $x = -8$ (obj. 10.5.1)   **12.** $x = -6.35$ (obj. 10.3.1)
13. $x = 15$ (obj. 10.5.2)   **14.** $x = -\frac{1}{2}$ (obj. 10.5.2)   **15.** $x = -5$ (obj. 10.5.2)   **16.** $x = -\frac{7}{10}$ (obj. 10.4.2)   **17.** $s = f + 15$ (obj. 10.6.1)
18. $n = s - 15,000$ (obj. 10.6.1)   **19.** $\frac{1}{2}s =$ measure of the first angle; $s =$ measure of the second angle; $2s =$ measure of the third angle
(obj. 10.6.2)   **20.** $w =$ width; $2w - 5 =$ length (obj. 10.6.2)   **21.** 87 acres on the Prentice farm; 261 acres on the Smithfield farm (obj. 10.7.1)
22. Marcia earns \$24,000; Sam earns \$22,500 (obj. 10.7.3)   **23.** 41 students in the morning class, 65 students in the afternoon class, 77 students in
the evening class (obj. 10.7.1)   **24.** width is 25 feet; length is 34 feet (obj. 10.7.2)

Practice Final Examination
1. eighty-two thousand, three hundred sixty-seven   **2.** 30,333   **3.** 173   **4.** 34,103   **5.** 4212
6. 217,745   **7.** 158   **8.** 606   **9.** 116   **10.** 32 mi/gal   **11.** $\frac{7}{15}$ **12.** $\frac{42}{11}$ **13.** $\frac{33}{20}$ or $1\frac{13}{20}$ **14.** $\frac{89}{15}$ or $5\frac{14}{15}$ **15.** $\frac{31}{14}$ or $2\frac{3}{14}$
16. 4   **17.** $\frac{14}{5}$ or $2\frac{4}{5}$ **18.** $\frac{22}{13}$ or $1\frac{9}{13}$ **19.** $6\frac{17}{20}$ mi   **20.** 5 packages   **21.** 0.719   **22.** $\frac{43}{50}$ **23.** $>$ **24.** 506.38   **25.** 21.77
26. 0.757   **27.** 0.492   **28.** 3.69   **29.** 0.8125   **30.** 0.7056   **31.** $\frac{1400 \text{ students}}{43 \text{ faculty}}$ **32.** no   **33.** $n \approx 9.4$ **34.** $n \approx 7.7$
35. $n = 15$ **36.** $n = 9$ **37.** \$3333.33   **38.** 9.75 in.   **39.** \$626.09   **40.** 1.6 lb   **41.** 0.63%   **42.** 21.25%   **43.** 1.64   **44.** 17.33%
45. 302.4   **46.** 250   **47.** 4284   **48.** \$17,296   **49.** 4500 students   **50.** 34.3%   **51.** 4.25 gal   **52.** 6500 lb   **53.** 192 in.   **54.** 5600 m
55. 0.0698 kg   **56.** 0.00248 L   **57.** 19.32 km   **58.** 0.0962 m   **59.** 15,840 ft   **60.** 1.36728 cm   **61.** 14.4 m   **62.** 206 cm
63. 5.4 ft^2 **64.** 75 m^2 **65.** 113.04 m^2 **66.** 56.52 m   **67.** 167.47 cm^3 **68.** 205.2 ft^3 **69.** 32.5 m^2 **70.** $n = 32.5$ **71.** \$8 million
72. \$1 million   **73.** 50°F   **74.** from 2000 to 2010   **75.** 600 students   **76.** 1400 students   **77.** mean ≈ 15.83; median $= 16.5$
78. 16   **79.** 11.091   **80.** 15 ft   **81.** -13 **82.** $\frac{1}{8}$ **83.** -3 **84.** -17 **85.** 24   **86.** $-\frac{8}{3}$ or $-2\frac{2}{3}$ **87.** 4   **88.** 27   **89.** 8
90. $\frac{1}{2}$ or 0.5   **91.** $-3x - 7y$ **92.** $-7 - 4a - 17b$ **93.** $-2x + 6y + 10$ **94.** $-11x - 9y - 4$ **95.** $x = 2$ **96.** $x = -2$
97. $x = -\frac{1}{2}$ or -0.5 **98.** $x = -\frac{2}{5}$ or -0.4 **99.** 122 students are taking history; 110 students are taking math   **100.** length is 37 m; width is 16 m

Appendix A.1 Balancing a Checking Account
Exercises **1.** \$555.12   **3.** \$2912.65
5. \$153.44; Yes, Justin can pay his car insurance.   **7.** The account balances.   **9.** \$949.88   **11.** Jeremy's account balances.

Appendix A.2 Determining the Best Deal When Purchasing a Vehicle
Exercises **1.** \$1295.94   **3.** \$379.98   **5.** \$4245   **7. (a)** \$1244.95; \$497.98 **(b)** \$27,741.93   **9. (a)** \$3135.93; \$895.98 **(b)** \$50,930.91
11. (a) \$7499.85 **(b)** \$42,499.15   **13.** Dealership 2   **15. (a)** Dealership 3; \$25,678.93 **(b)** Dealership 1; \$28,839 **(c)** The least expensive
purchase price does not guarantee the least expensive total cost. Many factors need to be considered to determine the best deal.

Basic College Mathematics Glossary

Absolute value of a number (9.1) The absolute value of a number is the distance between that number and zero on the number line. When we find the absolute value of a number, we use the $|\ |$ notation. To illustrate, $|-4| = 4, |6| = 6, |-20 - 3| = |-23| = 23, |0| = 0$.

Addends (1.2) When two or more numbers are added, the numbers being added are called addends. In the problem $3 + 4 = 7$, the numbers 3 and 4 are both addends.

Adjacent angles (7.1) Two angles that share a common side and a common vertex.

Algebraic expression (10.6) An algebraic expression consists of variables, numerals, and operation signs.

Altitude of a triangle (7.4) The height of a triangle.

Amount of a percent equation (5.3A) The product we obtain when we multiply a percent times a number. In the equation $75 = 50\% \times 150$, the amount is 75.

Angle (7.1) An angle is made up of two rays that start at a common endpoint.

Area (7.1) The measure of the surface inside a geometric figure. Area is measured in square units, such as square feet.

Associative property of addition (1.2) The property that tells us that when three numbers are added, it does not matter which two numbers are added first. An example of the associative property is $5 + (1 + 2) = (5 + 1) + 2$. Whether we add $1 + 2$ first and then add 5 to that, or add $5 + 1$ first and then add that result to 2, we will obtain the same result.

Associative property of multiplication (1.4) The property that tells us that when we multiply three numbers, it does not matter which two numbers we group together first to multiply; the result will be the same. An example of the associative property of multiplication follows: $2 \times (5 \times 3) = (2 \times 5) \times 3$.

Base (1.6) The number that is to be repeatedly multiplied in exponent form. When we write $16 = 2^4$, the number 2 is the base.

Base of a percent equation (5.3A) The quantity we take a percent of. In the equation $8 = 20\% \times 400$, the base is 400.

Billion (1.1) The number 1,000,000,000.

Borrowing (1.3) The renaming of a number in order to facilitate subtraction. When we subtract $42 - 28$, we rename 42 as 3 tens plus 12. This represents 3 tens and 12 ones. This renaming is called borrowing.

Box (7.8) A three-dimensional object whose every side is a rectangle. Another name for a box is a *rectangular solid*.

Building fraction property (2.6) For whole numbers a, b, and c, where neither b nor c equals zero,

$$\frac{a}{b} = \frac{a}{b} \times 1 = \frac{a}{b} \times \frac{c}{c} = \frac{a \times c}{b \times c}.$$

Building up a fraction (2.6) To make one fraction into an equivalent fraction by making the denominator and numerator larger numbers. For example, the fraction $\frac{3}{4}$ can be built up to the fraction $\frac{30}{40}$.

Caret (3.5) A symbol $\wedge$ used to indicate the new location of a decimal point when performing division of decimal fractions.

Celsius temperature (6.4) A temperature scale in which water boils at 100 degrees ($100°C$) and freezes at 0 degrees ($0°C$). To convert Celsius temperature to Fahrenheit, we use the helpful formula $F = 1.8 \times C + 32$.

Center of a circle (7.7) The point in the middle of a circle from which all points on the circle are an equal distance.

Centimeter (6.2) A unit of length commonly used in the metric system to measure small distances. 1 centimeter = 0.01 meter.

Circle (7.7) A two-dimensional figure for which all points are at an equal distance from a given point.

Circumference of a circle (7.7) The distance around the rim of a circle.

Commission (5.5) The amount of money a salesperson is paid that is a percentage of the value of the sales made by that salesperson. The commission is obtained by multiplying the commission rate times the value of the sales. If a salesman sells $120,000 of insurance and his commission rate is 0.5%, then his commission is $0.5\% \times 120,000 = \$600.00$.

Common denominator (2.7) Two fractions have a common denominator if the same number appears in the denominator of each fraction. $\frac{3}{7}$ and $\frac{1}{7}$ have a common denominator of 7.

Commutative property of addition (1.2) The property that tells us that the order in which two numbers are added does not change the sum. An example of the commutative property of addition is $3 + 6 = 6 + 3$.

Commutative property of multiplication (1.4) The property that tells us that the order in which two numbers are multiplied does not change the value of the answer. An example of the commutative property of multiplication is $7 \times 3 = 3 \times 7$.

Composite number (2.2) A composite number is a whole number greater than 1 that can be divided by whole numbers other than itself. The number 6 is a composite number since it can be divided exactly by 2 and 3 (as well as by 1 and 6).

Cone (7.8) A three-dimensional object shaped like an ice cream cone or the sharpened end of a pencil.

Cross-multiplying (4.3) If you have a proportion such as $\dfrac{n}{5} = \dfrac{12}{15}$, then to cross-multiply, you form products to obtain $n \times 15 = 5 \times 12$.

Cubic centimeter (6.3) A metric measurement of volume equal to 1 milliliter.

Cup (6.1) One of the smallest units of volume in the American system. 2 cups = 1 pint.

Cylinder (7.8) A three-dimensional object shaped like a tin can.

Debit (1.2) A debit in banking is the removing of money from an account. If you had a savings account and took $300 out of it on Wednesday, we would say that you had a debit of $300 from your account. Often a bank will add a service charge to your account and use the word *debit* to mean that it has removed money from your account to cover the charge.

Decimal fraction (3.1) A fraction whose denominator is a power of 10.

Decimal places (3.4) The number of digits to the right of the decimal point in a decimal fraction. The number 1.234 has three decimal places, while the number 0.129845 has six decimal places. A whole number such as 42 is considered to have zero decimal places.

Decimal point (3.1) The period that is used when writing a decimal fraction. In the number 5.346, the period between the 5 and the 3 is the decimal point. It separates the whole number from the fractional part that is less than 1.

Decimal system (1.1) Our number system is called the decimal system or base 10 system because the value of numbers written in our system is based on tens and ones.

Decimeter (6.2) A unit of length not commonly used in the metric system. 1 decimeter = 0.1 meter.

Degree (7.1) A unit used to measure an angle. A degree is $\frac{1}{360}$ of a complete revolution. An angle of 32 degrees is written as 32°.

Dekameter (6.2) A unit of length not commonly used in the metric system. 1 dekameter = 10 meters.

Denominator (2.1) The number on the bottom of a fraction. In the fraction $\frac{2}{9}$ the denominator is 9.

Deposit (1.2) A deposit in banking is the placing of money in an account. If you had a checking account and on Tuesday you placed $124 into that account, we would say that you made a deposit of $124.

Diameter of a circle (7.7) A line segment across the circle that passes through the center of the circle. The diameter of a circle is equal to twice the radius of the circle.

Difference (1.3) The result of performing a subtraction. In the problem $9 - 2 = 7$ the number 7 is the difference.

Digits (1.1) The symbols 0, 1, 2, 3, 4, 5, 6, 7, 8, and 9 are called digits.

Discount (5.4) The amount of reduction in a price. The discount is a product of the discount rate times the list price. If the list price of a television is $430.00 and it has a discount rate of 35%, then the amount of discount is 35% × $430.00 = $150.50. The price would be reduced by $150.50.

Distributive property of multiplication over addition (1.4) The property illustrated by the following: $5 \times (4 + 3) = (5 \times 4) + (5 \times 3)$. In general, for any numbers a, b, and c, it is true that $a(b + c) = a \times b + a \times c$.

Dividend (1.5) The number that is being divided by another. In the problem $14 \div 7 = 2$, the number 14 is the dividend.

Divisor (1.5) The number that you divide into another number. In the problem $30 \div 5 = 6$, the number 5 is the divisor.

Earned run average (4.4) A ratio formed by finding the number of runs a pitcher would give up in a nine-inning game. If a pitcher has an earned run average of 2, it means that, on the average, he gives up two runs for every nine innings he pitches.

Equal fractions (2.2) Fractions that represent the same number. The fractions $\frac{3}{4}$ and $\frac{6}{8}$ are equal fractions.

Equality test of fractions (2.2) Two fractions $\dfrac{a}{b}$ and $\dfrac{c}{d}$ are equal if the product $a \times d = b \times c$. In this case, a, b, c, and d are whole numbers and b and $d \neq 0$.

Equations (10.3) Mathematical statements with variables that say that two expressions are equal, such as $x + 3 = -8$ and $2s + 5s = 34 - 4s$.

Equilateral triangle (7.4) A triangle with three equal sides.

Equivalent equations (10.3) Equations that have the same solution.

Equivalent fractions (2.2) Two fractions that are equal.

Expanded notation for a number (1.1) A number is written in expanded notation if it is written as a sum of hundreds, tens, ones, etc. The expanded notation for 763 is $700 + 60 + 3$.

Exponent (1.6) The number that indicates the number of times a factor occurs. When we write $8 = 2^3$, the number 3 is the exponent.

Factors (1.4) Each of the numbers that are multiplied. In the problem $8 \times 9 = 72$, the numbers 8 and 9 are factors.

Fahrenheit temperature (6.4) A temperature scale in which water boils at 212 degrees (212°F) and freezes at 32 degrees (32°F). To convert Fahrenheit temperature to Celsius, we use the formula $C = \dfrac{5 \times F - 160}{9}$.

Foot (6.1) American system unit of length. 3 feet = 1 yard. 12 inches = 1 foot.

Fundamental theorem of arithmetic (2.2)　Every composite number has a unique product of prime numbers.

Gallon (6.1)　A unit of volume in the American system. 4 quarts = 1 gallon.

Gigameter (6.2)　A metric unit of length equal to 1,000,000,000 meters.

Gram (6.3)　The basic unit of weight in the metric system. A gram is defined as the weight of the water in a box that is 1 centimeter on each side. 1 gram = 1000 milligrams. 1 gram = 0.001 kilogram.

Hectometer (6.2)　A unit of length not commonly used in the metric system. 1 hectometer = 100 meters.

Height (7.3)　The distance between two parallel sides in a four-sided figure such as a parallelogram or a trapezoid.

Height of a cone (7.8)　The distance from the vertex of a cone to the base of the cone.

Height of a pyramid (7.8)　The distance from the point on a pyramid to the base of the pyramid.

Height of a triangle (7.4)　The distance of a line drawn from a vertex perpendicular to the other side, or an extension of the other side, of the triangle. This is sometimes called the *altitude of a triangle*.

Hexagon (7.3)　A six-sided figure.

Hypotenuse (7.6)　The side opposite the right angle in a right triangle. The hypotenuse is always the longest side of a right triangle.

Improper fraction (2.3)　A fraction in which the numerator is greater than or equal to the denominator. The fractions $\frac{34}{29}$, $\frac{8}{7}$, and $\frac{6}{6}$ are all improper fractions.

Inch (6.1)　The smallest unit of length in the American system. 12 inches = 1 foot.

Inequality symbol (3.2)　The symbol that is used to indicate whether a number is greater than another number or less than another number. Since 5 is greater than 3, we would write this with a "greater than" symbol as follows: 5 > 3. The statement "7 is less than 12" would be written as follows: 7 < 12.

Interest (5.4)　The money that is paid for the use of money. If you deposit money in a bank, the bank uses that money and pays you interest. If you borrow money, you pay the bank interest for the use of that money. Simple interest is determined by the formula $I = P \times R \times T$. Compound interest is usually determined by a table, a calculator, or a computer.

Invert a fraction (2.5)　To invert a fraction is to interchange the numerator and the denominator. If we invert $\frac{5}{9}$, we obtain the fraction $\frac{9}{5}$. To invert a fraction is sometimes referred to as *to take the reciprocal of a fraction*.

Irreducible (2.2)　A fraction that cannot be reduced (simplified) is called irreducible.

Isosceles triangle (7.4)　A triangle with two sides equal.

Kilogram (6.3)　The most commonly used metric unit of weight. 1 kilogram = 1000 grams.

Kiloliter (6.3)　The metric unit of volume normally used to measure large volumes. 1 kiloliter = 1000 liters.

Kilometer (6.2)　The unit of length commonly used in the metric system to measure large distances. 1 kilometer = 1000 meters.

Least common denominator (LCD) (2.6)　The least common denominator (LCD) of two or more fractions is the smallest number that can be divided without remainder by each fraction's denominator. The LCD of $\frac{1}{3}$ and $\frac{1}{4}$ is 12. The LCD of $\frac{5}{6}$ and $\frac{4}{15}$ is 30.

Legs of a right triangle (7.6)　The two shortest sides of a right triangle.

Length of a rectangle (7.2)　Each of the longer sides of a rectangle.

Like terms (10.1)　Like terms have identical variables with identical exponents. $-5x$ and $3x$ are like terms. $-7xyz$ and $-12xyz$ are like terms.

Line segment (7.3)　A portion of a straight line that has a beginning and an end.

Liter (6.3)　The standard metric measurement of volume. 1 liter = 1000 milliliters. 1 liter = 0.001 kiloliter.

Mean (8.4)　The mean of a set of values is the sum of the values divided by the number of values. The mean of the numbers 10, 11, 14, and 15 is 12.5. In everyday language, when people use the word *average,* they are usually referring to the mean.

Median (8.4)　If a set of numbers is arranged in order from smallest to largest, the median is that value that has the same number of values above it as below it. The median of the numbers 3, 7, and 8 is 7. If the list contains an even number of items, we obtain the median by finding the mean of the two middle numbers. The median of the numbers 5, 6, 10, and 11 is 8.

Megameter (6.2)　A metric unit of length equal to 1,000,000 meters.

Meter (6.2)　The basic unit of length in the metric system. 1 meter = 1000 millimeters. 1 meter = 0.001 kilometer.

Metric ton (6.3)　A metric unit of measurement for very heavy weights. 1 metric ton = 1,000,000 grams.

Microgram (6.3)　A metric unit of weight equal to 0.000001 gram.

Micrometer (6.2)　A metric unit of length equal to 0.000001 meter.

Mile (6.1)　Largest unit of length in the American system. 5280 feet = 1 mile. 1760 yards = 1 mile.

Milligram (6.3)　A metric unit of weight used for very, very small objects. 1 milligram = 0.001 gram.

Milliliter (6.3)　The metric unit of volume normally used to measure small volumes. 1 milliliter = 0.001 liter.

Millimeter (6.2)　A unit of length commonly used in the metric system to measure very small distances. 1 millimeter = 0.001 meter.

Million (1.1)   The number 1,000,000.

Minuend (1.3)   The number being subtracted from in a subtraction problem. In the problem $8 - 5 = 3$, the number 8 is the minuend.

Mixed number (2.3)   A number created by the sum of a whole number greater than 1 and a proper fraction. The numbers $4\frac{5}{6}$ and $1\frac{1}{8}$ are both mixed numbers. Mixed numbers are sometimes referred to as *mixed fractions*.

Mode (8.4)   The mode of a set of data is the number or numbers that occur most often.

Multiplicand (1.4)   The first factor in a multiplication problem. In the problem $7 \times 2 = 14$, the number 7 is the multiplicand.

Multiplier (1.4)   The second factor in a multiplication problem. In the problem $6 \times 3 = 18$, the number 3 is the multiplier.

Nanogram (6.3)   A unit of weight equal to 0.000000001 gram.

Nanometer (6.2)   A metric unit of length equal to 0.000000001 meter.

Negative numbers (9.1)   All of the numbers to the left of zero on the number line. The numbers $-1.5, -16, -200.5, -4500$ are all negative numbers. All negative numbers are written with a negative sign in front of the digits.

Number line (1.7)   A line on which numbers are placed in order from smallest to largest.

Numerator (2.1)   The number on the top of a fraction. In the fraction $\frac{3}{7}$ the numerator is 3.

Numerical coefficients (10.1)   The numbers in front of the variables in one or more terms. If we look at $-3xy + 12w$, we find that the numerical coefficient of the xy term is -3 while the numerical coefficient of the w term is 12.

Octagon (7.3)   An eight-sided figure.

Odometer (1.8)   A device on an automobile that displays how many miles the car has been driven since it was first put into operation.

Opposite of a number (9.2)   The opposite of a number is a number that has the same absolute value but the opposite sign. The opposite of -5 is 5. The opposite of 7 is -7.

Order of operations (1.6)   An agreed-upon procedure to do a problem with several arithmetic operations in the proper order.

Ounce (6.1)   Smallest unit of weight in the American system. 16 ounces = 1 pound.

Overtime (2.9)   The pay earned by a person if he or she works more than a certain number of hours per week. In most jobs that pay by the hour, a person will earn $1\frac{1}{2}$ times as much per hour for every hour beyond 40 hours worked in one workweek. For example, Carlos earns $6.00 per hour for the first 40 hours in a week and overtime for each additional hour. He would earn $9.00 per hour for all hours he worked in that week beyond 40 hours.

Parallel lines (7.3)   Two straight lines that are always the same distance apart.

Parallelogram (7.3)   A four-sided figure with both pairs of opposite sides parallel.

Parentheses (1.4)   One of several symbols used in mathematics to indicate multiplication. For example, $(3)(5)$ means 3 multiplied by 5. Parentheses are also used as a grouping symbol.

Percent (5.1)   The word *percent* means per one hundred. For example, 14 percent means $\frac{14}{100}$.

Percent of decrease (5.5)   The percent that something decreases is determined by dividing the amount of decrease by the original amount. If a tape deck sold for $300 and its price was decreased by $60, the percent of decrease would be $\frac{60}{300} = 0.20 = 20\%$.

Percent of increase (5.5)   The percent that something increases is determined by dividing the amount of increase by the original amount. If the population of a town was 5000 people and the population increased by 500 people, the percent of increase would be $\frac{500}{5000} = 0.10 = 10\%$.

Percent proportion (5.3B)   The percent proportion is the equation $\frac{a}{b} = \frac{p}{100}$ where a is the amount, b is the base, and p is the percent number.

Percent symbol (5.1)   A symbol that is used to indicate percent. To indicate 23 percent, we write 23%.

Perfect square (7.4)   When a whole number is multiplied by itself, the number that is obtained is a perfect square. The numbers 1, 4, 9, 16, 25, 36, 49, 64, 81, and 100 are all perfect squares.

Perimeter (7.2)   The distance around a figure.

Perpendicular lines (7.1)   Lines that meet at an angle of 90 degrees.

Pi (7.7)   Pi is an irrational number that we obtain if we divide the circumference of a circle by the diameter of a circle. It is represented by the symbol π. Accurate to eleven decimal places, the value of pi is given by 3.14159265359. For most work in this textbook, the value of 3.14 is used to approximate the value of pi.

Picogram (6.3)   A unit of weight equal to 0.000000000001 gram.

Pint (6.1)   Unit of volume in the American system. 2 pints = 1 quart.

Placeholder (1.1)   The use of a digit to indicate a place. Zero is a placeholder in our number system. It holds a position and shows that there is no other digit in that place.

Place-value system (1.1)   Our number system is called a place-value system because the placement of the digits tells the value of the number. If we use the digits 5 and 4 to write the number 54, the result is different than if we placed them in opposite order and wrote 45.

Positive numbers (9.1)   All of the numbers to the right of zero on the number line. The numbers 5, 6.2, 124.186, 5000 are all positive numbers. A positive number such as +5 is usually written without the positive sign.

Pound (6.1)   Basic unit of weight in the American system. 2000 pounds = 1 ton. 16 ounces = 1 pound.

Power of 10 (1.4)   Whole numbers that begin with 1 and end in one or more zeros are called powers of 10. The numbers 10, 100, 1000, etc., are all powers of 10.

Prime factors (2.2)   Factors that are prime numbers. If we write 15 as a product of prime factors, we have $15 = 5 \times 3$.

Prime number (2.2)   A prime number is a whole number greater than 1 that can only be divided by 1 and itself. The first fifteen prime numbers are 2, 3, 5, 7, 11, 13, 17, 19, 23, 29, 31, 37, 41, 43, and 47. The list of prime numbers goes on forever.

Principal (5.4)   The amount of money deposited or borrowed on which interest is computed. In the simple interest formula $I = P \times R \times T$, the P stands for the principal. (The other letters are I = interest, R = interest rate, and T = amount of time.)

Product (1.4)   The answer in a multiplication problem. In the problem $3 \times 4 = 12$ the number 12 is the product.

Proper fraction (2.3)   A fraction in which the numerator is less than the denominator. The fractions $\frac{3}{4}$ and $\frac{15}{16}$ are proper fractions.

Proportion (4.2)   A statement that two ratios or two rates are equal. The statement $\frac{3}{4} = \frac{15}{20}$ is a proportion. The statement $\frac{5}{7} = \frac{7}{9}$ is false, and is therefore not a proportion.

Pyramid (7.8)   A three-dimensional object made up of a geometric figure for a base and triangular sides that meet at a point. Some pyramids are shaped like the great pyramids of Egypt.

Pythagorean Theorem (7.6)   A statement that for any right triangle the square of the hypotenuse equals the sum of the squares of the two legs of the triangle.

Quadrilateral (7.3)   A four-sided geometric figure.

Quadrillion (1.1)   The number 1,000,000,000,000,000.

Quart (6.1)   Unit of volume in the American system. 4 quarts = 1 gallon.

Quotient (1.5)   The answer after performing a division problem. In the problem $60 \div 6 = 10$ the number 10 is the quotient.

Radius of a circle (7.7)   A line segment from the center of a circle to any point on the circle. The radius of a circle is equal to one-half the diameter of the circle.

Rate (4.1)   A rate compares two quantities that have different units. Examples of rates are $5.00 an hour and 13 pounds for every 2 inches. In fraction form, these two rates would be written as $\dfrac{\$5.00}{1 \text{ hour}}$ and $\dfrac{13 \text{ pounds}}{2 \text{ inches}}$.

Ratio (4.1)   A ratio is a comparison of two quantities that have the same units. To compare 2 to 3, we can express the ratio in three ways: the ratio of 2 to 3; 2 : 3; or the fraction $\frac{2}{3}$.

Ratio in simplest form (4.1)   A ratio is in simplest form when the two numbers do not have a common factor.

Ray (7.1)   A ray is a part of a line that has only one endpoint and goes on forever in one direction.

Rectangle (7.2)   A four-sided figure that has four right angles.

Reduced fraction (2.2)   A fraction for which the numerator and denominator have no common factor other than 1. The fraction $\frac{5}{7}$ is a reduced fraction. The fraction $\frac{15}{21}$ is not a reduced fraction because both numerator and denominator have a common factor of 3.

Regular hexagon (7.3)   A six-sided figure with all sides equal.

Regular octagon (7.3)   An eight-sided figure with all sides equal.

Remainder (1.5)   When two numbers do not divide exactly, a part is left over. This part is called the remainder. For example, $13 \div 2 = 6$ with 1 left over; the 1 is the remainder.

Repeating decimals (3.6)   Decimals that have a digit or a group of digits that repeat. The decimals 0.33333333333 … and 1.234234234234 … are repeating decimals. The pattern of repeating continues forever. Repeating decimals can be written in a form with a bar over the repeating digit(s). Thus the preceding decimals could be written as $0.\overline{3}$ and $1.\overline{234}$.

Right angle (7.1) and (7.4)   An angle that measures 90 degrees.

Right triangle (7.4)   A triangle with one 90-degree angle.

Rounding (1.7)   The process of writing a number in an approximate form for convenience. The number 9756 rounded to the nearest hundred is 9800.

Sales tax (5.4)   The amount of tax on a purchase. The sales tax for any item is a product of the sales tax rate times the purchase price. If an item is purchased for $12.00 and the sales tax rate is 5%, the sales tax is $5\% \times 12.00 = \$0.60$.

Scientific notation (9.5)   A positive number is written in scientific notation if it is in the form $a \times 10^n$ where a is a number greater than or equal to 1, but less than 10, and n is an integer. If we write 5678 in scientific notation, we have 5.678×10^3. If we write 0.00825 in scientific notation, we have 8.25×10^{-3}.

Semicircle (7.7)   One-half of a circle. The semicircle usually includes the diameter of a circle connected to one-half the circumference of the circle.

Sides of an angle (7.1)   The two line segments that meet to form an angle.

Signed numbers (9.1)   All of the numbers on a number line. Numbers like −33, 2, 5, −4.2, 18.678, −8.432 are all signed numbers. A negative number always has a negative sign in front of the digits. A positive number such as +3 is usually written without the positive sign in front of it.

Similar triangles (7.9)   Two triangles that have the same shape but are not necessarily the same size. The corresponding angles of similar triangles are equal. The corresponding sides of similar triangles have the same ratio.

Simple interest (5.4)   The interest determined by the formula $I = P \times R \times T$ where I = the interest obtained, P = the principal or the amount borrowed or invested, R = the interest rate (usually on an annual basis), and T = the number of time periods (usually years).

Solution of an equation (10.3)   A number is a solution of an equation if replacing the variable by the number makes the equation always true. The solution of $x - 5 = -20$ is the number −15.

Sphere (7.8)   A three-dimensional object shaped like a perfectly round ball.

Square (7.2)   A rectangle with all four sides equal.

Square root (7.5)   The square root of a number is one of only two identical factors of that number. The square root of 9 is 3. The square root of 121 is 11.

Square root sign (7.5)   The symbol $\sqrt{}$. When we want to find the square root of 25, we write $\sqrt{25}$. The answer is 5.

Standard notation for a number (1.1)   A number written in ordinary terms. For example, 70 + 2 in standard notation is 72.

Subtrahend (1.3)   The number being subtracted. In the problem $7 - 1 = 6$, the number 1 is the subtrahend.

Sum (1.2)   The result of an addition of two or more numbers. In the problem $7 + 3 + 5 = 15$, the number 15 is the sum.

Term (10.1)   A number, a variable, or a product of a number and one or more variables. $5x$, $2ab$, $-43cdef$ are three examples of terms, separated in an expression by a + sign or a − sign.

Terminating decimals (3.6)   Every fraction can be written as a decimal. If the division process of dividing denominator into numerator ends with a remainder of zero, the decimal is a terminating decimal. Decimals such as 1.28, 0.007856, and 5.123 are terminating decimals.

Trapezoid (7.3)   A four-sided figure with at least two parallel sides.

Triangle (7.4)   A three-sided figure.

Trillion (1.1)   The number 1,000,000,000,000.

Unit fraction (6.1)   A fraction used to change one unit to another. For example, to change 180 inches to feet, we multiply by the unit fraction $\dfrac{1 \text{ foot}}{12 \text{ inches}}$. Thus we have

$$180 \text{ inches} \times \frac{1 \text{ foot}}{12 \text{ inches}} = 15 \text{ feet}.$$

Variable (10.1)   A letter that is used to represent a number.

Vertex of a cone (7.8)   The sharp point of a cone.

Vertex of an angle (7.1)   The point at which two line segments meet to form an angle.

Volume (7.8)   The measure of the space inside a three-dimensional object. Volume is measured in cubic units such as cubic feet.

Whole numbers (1.1)   The whole numbers are the set of numbers 0, 1, 2, 3, 4, 5, 6, 7, 8, 9, 10, 11, 12, ... The set goes on forever. There is no largest whole number.

Width of a rectangle (7.2)   Each of the shorter sides of a rectangle.

Word names for whole numbers (1.1)   The notation for a number in which each digit is expressed by a word. To write 389 with a word name, we would write three hundred eighty-nine.

Zero (1.1)   The smallest whole number. It is normally written 0.

Index

Photo Credits

CHAPTER 1 CO Tim McCaig/iStockphoto **p. 79** Valerijs Kostreckis/Alamy **p. 84** Frank and Helena/Alamy **p. 92** Zen Shui/SuperStock **p. 94** Lev Olkha/Alamy **p. 104** Courtesy of authors

CHAPTER 2 CO Sculpies/iStockphoto **p. 128** Andystjohn/Dreamstime **p. 176** Rick Szczechowski/iStockphoto **p. 179** Thomasowen/Dreamstime **p. 183** Stephen Coburn/Shutterstock **p. 192** Courtesy of authors

CHAPTER 3 CO SNEHIT/Shutterstock **p. 202** Petographer/Alamy **p. 245** Jim West/The Image Works **p. 246** Stockbyte/Getty Images **p. 247** Dmitri Ogleznev/Shutterstock **p. 248** Roger Fletcher/Alamy **p. 251** ARENA/Shutterstock **p. 259** Courtesy of authors

CHAPTER 4 CO Leila Cutler/Alamy **p. 273** Digital Visionz/Thinkstock **p. 288** Technotr/iStockphoto **p. 290** Photos.com **p. 295** Webphotographeer/iStockphoto **p. 302** Courtesy of authors

CHAPTER 5 CO Chris Schmidt/iStockphoto **p. 325** Rebelml/Dreamstime **p. 332** Exactostock/SuperStock **p. 339** Natal'ya Markova/iStockphoto **p. 353** Eugene Choi/iStockphoto **p. 361** Courtesy of authors

CHAPTER 6 CO Toddmedia/iStockphoto **p. 366** Pixtal Images/Photolibrary New York **p. 385** Goldnelk/Dreamstime **p. 390** Tom McHugh/Photo Researchers, Inc. **p. 405** Creatas/Thinkstock **p. 411** Courtesy of authors

CHAPTER 7 CO Jonathan Larsen/iStockphoto **p. 417** Wally Stemberger/Shutterstock **p. 439** Rubberball/iStockphoto **p. 468** Comstock **p. 478** Photos.com **p. 491** Jens Stolt/Shutterstock **p. 497** Getty Images/Thinkstock **p. 510** Courtesy of authors

CHAPTER 8 CO Doug James/Shutterstock **p. 539** Shutterstock **p. 545** Tetra Images/Superstock **p. 557** Courtesy of authors

CHAPTER 9 CO Shannon Long/iStockphoto **p. 568** Andresr/Dreamstime **p. 581** Frenchmen77/Dreamstime **p. 591** NASA **p. 599** Andy Ward/Photolibrary New York **p. 606** Courtesy of authors

CHAPTER 10 CO U.S. Agency for International Development (USAID) **p. 648** Jeff Greenberg/Alamy **p. 652** Leigh Schindler/iStockphoto **p. 659** Courtesy of authors

METRIC SYSTEM MEASUREMENTS

Length

1 kilometer	(km)	=	1000 meters
1 hectometer	(hm)	=	100 meters
1 dekameter	(dam)	=	10 meters
1 meter	(m)	=	1 meter
1 decimeter	(dm)	=	0.1 meter
1 centimeter	(cm)	=	0.01 meter
1 millimeter	(mm)	=	0.001 meter

Weight

1 metric ton	(t)	=	1,000,000 grams
1 kilogram	(kg)	=	1000 grams
1 hectogram	(hg)	=	100 grams
1 dekagram	(dag)	=	10 grams
1 gram	(g)	=	1 gram
1 decigram	(dg)	=	0.1 gram
1 centigram	(cg)	=	0.01 gram
1 milligram	(mg)	=	0.001 gram

Volume

1 kiloliter	(kL)	=	1000 liters
1 hectoliter	(hL)	=	100 liters
1 dekaliter	(daL)	=	10 liters
1 liter	(L)	=	1 liter
1 deciliter	(dL)	=	0.1 liter
1 centiliter	(cL)	=	0.01 liter
1 milliliter	(mL)	=	0.001 liter

Temperature: Celsius Scale

$100°C$ = Boiling point of water

$-273.15°C$ = Absolute zero: coldest possible temperature

$0°C$ = Freezing point of water

$37°C$ = Normal human body temperature

AMERICAN SYSTEM MEASUREMENTS

Length

1 mile	(mi)	=	1760 yards (yd)
1 mile	(mi)	=	5280 feet (ft)
1 yard	(yd)	=	3 feet (ft)
1 foot	(ft)	=	12 inches (in.)

Volume

1 gallon	(gal)	=	4 quarts (qt)
1 quart	(qt)	=	2 pints (pt)
1 pint	(pt)	=	2 cups (c)

Weight

1 ton	(T)	=	2000 pounds (lb)
1 pound	(lb)	=	16 ounces (oz)

APPROXIMATE EQUIVALENT MEASURES FOR CONVERSION OF UNITS

	American to Metric	Metric to American
Units of Length	1 mile = 1.61 kilometers 1 yard = 0.914 meter 1 foot = 0.305 meter 1 inch = 2.54 centimeters	1 kilometer = 0.62 mile 1 meter = 3.28 feet 1 meter = 1.09 yards 1 centimeter = 0.394 inch
Units of Volume	1 gallon = 3.79 liters 1 quart = 0.946 liter	1 liter = 0.264 gallon 1 liter = 1.06 quarts
Units of Weight	1 pound = 0.454 kilogram 1 ounce = 28.35 grams	1 kilogram = 2.2 pounds 1 gram = 0.0353 ounce